WATER-SOLUBLE POLYMERS

WATER-SOLUBLE POLYMERS

Developments Since 1978

by

Yale L. Meltzer

NOYES DATA CORPORATION
Park Ridge, New Jersey, U.S.A.
1981

Library of Congress Catalog Card Number: 80-26174
ISBN: 0-8155-0834-4
Printed in the United States

Published in the United States of America by
Noyes Data Corporation
Noyes Building, Park Ridge, New Jersey 07656

Library of Congress Cataloging in Publication Data

Meltzer, Yale L
Water soluble polymers.

(Chemical technology review ; no. 181)
Includes indexes.
1. Water-soluble polymers--Patents. I. Title.
II. Series.
QD382.W3M43 1981 668.9'027 80-26174
ISBN 0-8155-0834-4

FOREWORD

The detailed, descriptive information in this book is based on U.S. patents, issued beginning with January 1979, that deal with water-soluble polymers. This title contains new developments since our previous title *Water-Soluble Polymers—Recent Developments.*

This book is a data-based publication providing information retrieved and made available from the U.S. patent literature. It thus serves a double purpose in that it supplies detailed technical information and can be used as a guide to the patent literature in this field. By indicating all the information that is significant, and eliminating legal jargon and juristic phraseology, this book presents an advanced commercially oriented review of recent developments in the field of water-soluble polymers.

The U.S. patent literature is the largest and most comprehensive collection of technical information in the world. There is more practical, commercial, timely process information assembled here than is available from any other source. The technical information obtained from a patent is extremely reliable and comprehensive; sufficient information must be included to avoid rejection for "insufficient disclosure." These patents include practically all of those issued on the subject in the United States during the period under review; there has been no bias in the selection of patents for inclusion.

The patent literature covers a substantial amount of information not available in the journal literature. The patent literature is a prime source of basic commercially useful information. This information is overlooked by those who rely primarily on the periodical journal literature. It is realized that there is a lag between a patent application on a new process development and the granting of a patent, but it is felt that this may roughly parallel or even anticipate the lag in putting that development into commercial practice.

Many of these patents are being utilized commercially. Whether used or not, they offer opportunities for technological transfer. Also, a major purpose of this book is to describe the number of technical possibilities available, which may open up profitable areas of research and development. The information contained in this book will allow you to establish a sound background before launching into research in this field.

Advanced composition and production methods developed by Noyes Data are employed to bring these durably bound books to you in a minimum of time. Special techniques are used to close the gap between "manuscript" and "completed book." Industrial technology is progressing so rapidly that time-honored, conventional typesetting, binding and shipping methods are no longer suitable. We have bypassed the delays in the conventional book publishing cycle and provide the user with an effective and convenient means of reviewing up-to-date information in depth.

The table of contents is organized in such a way as to serve as a subject index. Other indexes by company, inventor and patent number help in providing easy access to the information contained in this book.

16 Reasons Why the U.S. Patent Office Literature Is Important to You

1. The U.S. patent literature is the largest and most comprehensive collection of technical information in the world. There is more practical commercial process information assembled here than is available from any other source. Most important technological advances are described in the patent literature.
2. The technical information obtained from the patent literature is extremely comprehensive; sufficient information must be included to avoid rejection for "insufficient disclosure."
3. The patent literature is a prime source of basic commercially utilizable information. This information is overlooked by those who rely primarily on the periodical journal literature.
4. An important feature of the patent literature is that it can serve to avoid duplication of research and development.
5. Patents, unlike periodical literature, are bound by definition to contain new information, data and ideas.
6. It can serve as a source of new ideas in a different but related field, and may be outside the patent protection offered the original invention.
7. Since claims are narrowly defined, much valuable information is included that may be outside the legal protection afforded by the claims.
8. Patents discuss the difficulties associated with previous research, development or production techniques, and offer a specific method of overcoming problems. This gives clues to current process information that has not been published in periodicals or books.
9. Can aid in process design by providing a selection of alternate techniques. A powerful research and engineering tool.
10. Obtain licenses–many U.S. chemical patents have not been developed commercially.
11. Patents provide an excellent starting point for the next investigator.
12. Frequently, innovations derived from research are first disclosed in the patent literature, prior to coverage in the periodical literature.
13. Patents offer a most valuable method of keeping abreast of latest technologies, serving an individual's own "current awareness" program.
14. Identifying potential new competitors.
15. It is a creative source of ideas for those with imagination.
16. Scrutiny of the patent literature has important profit-making potential.

CONTENTS AND SUBJECT INDEX

INTRODUCTION

DEVELOPMENTS IN THE MARKET

In view of the rising problems of pollution, use of water-soluble polymers has taken on new importance. Water-soluble polymers provide systems for employing water instead of organic-based solvents, for example, in coatings. Water-based systems make it possible to reduce air and water pollution.

In addition, over the past 15 years, synthetic and semisynthetic water-soluble polymers have been found effective in reducing BOD (biological oxygen demand or biochemical oxygen demand). In contrast, some traditional starch products were found to cause stream pollution in certain industrial operations, e.g., textile mills.

Countries have moved in the direction of increasingly stringent environmental controls. In the United States, the Environmental Protection Agency (EPA) is illustrative of the trend with its history of environmental control based on laws passed by Congress. Water-soluble polymers, such as the semisynthetics (e.g., CMC and starch derivatives) and synthetics provide technically viable means for meeting some of the governmental antipollution regulations.

PRODUCT CHANGES

The synthetic and semisynthetic water-soluble polymers have been developing many new markets over the past few years and have had the advantage over the natural gums and traditional starches in being tailor-made for applications. Traditional starches, modified starches and starch derivatives, however, have regained some lost markets as a result of rising costs, by benefiting from their lower prices as compared to many synthetic and semisynthetic polymers.

PRODUCT CLASSIFICATIONS

Water-soluble polymers are also referred to in the trade as water-soluble resins, hydrogels, hydrocolloids and gums. More specific product classifications are as follows:

(1) synthetic water-soluble polymers, e.g., acrylamide polymers, acrylic acid polymers, ethylene oxide polymers, methacrylic acid polymers, polyethyleneimine (PEI) polymers, polyvinyl alcohol (PVA) polymers and polyvinylpyrrolidone (PVP) polymers

(2) semisynthetic water-soluble polymers, e.g., cellulose ethers, modified starches, starch derivatives and natural gum derivatives

(3) natural water-soluble polymers, e.g., natural gums and the traditional starches (e.g., arrowroot starch, corn starch, potato starch, rice starch, sorghum starch, tapioca starch and wheat starch).

NATURAL GUMS vs THE SYNTHETIC AND SEMISYNTHETIC WATER-SOLUBLE POLYMERS

Countries producing natural gums have been seeking to expand their markets and to regain some of the markets lost to the synthetic and semisynthetic water-soluble polymers. The countries producing natural gums are mainly Third World countries, especially those of Africa and Asia.

Natural gum production is, in general, more labor-intensive than the production of synthetic and semisynthetic water-soluble polymers. Although natural gum derivatives have been developed, which can be tailor-made for end-uses, natural gums are agricultural products and historically the demand for agricultural products grows more slowly than that for industrial products, which are more capital-intensive, such as exists with the synthetic water-soluble polymers and some semisynthetic water-soluble polymers.

In addition, agricultural products such as the natural gums, suffer from the uncertainties of the weather and other unpredictable factors. For example, gum arabic, one of the main natural gums, had a sharp decline in production as a result of a severe drought which ravaged the Sudan from 1968 to 1974 (the Sudan normally has approximately 80% of the world's production of gum arabic). The problems of gum arabic output were further compounded by political instability in the Sudan during 1968 to 1974.

As a result of the drought, prices for gum arabic rose sharply. Since 1974, however, the market has improved, with gum arabic prices stabilizing, especially in 1980. By the summer of 1980, gum arabic prices stabilized in the range of \$0.72 to \$1.37 per pound, depending on the type. There is also more political stability in the Sudan now than existed in 1974, which does aid gum arabic production.

Guar gum has had growing shortages in the summer of 1980 as a result of earlier floods and droughts in Asian countries. As a result, the price of guar gum doubled in about one year from the summer of 1979 to the summer of 1980, rising to \$1.10 per pound.

Meanwhile, gum tragacanth rose sharply in price in the summer of 1980, with the No. 1 grade as high as $38 per pound in August, 1980. Prices range broadly depending upon the grade. In August, 1980, the range was from $8.50 to $38 per pound, depending on the grade. Gum tragacanth is tied directly to the political and economic instability in Iran (Iran is the world's main producer of gum tragacanth.) (Please see the Market Survey for Guar Gum, Gum Arabic, Gum Tragacanth and Locust Bean Gum.)

MARKET GROWTH AND FUTURE PROSPECTS

Long-term prospects for sodium carboxymethylcellulose (CMC) look particularly promising because of expected expanded use of purified grades of CMC in oil well drilling muds, as well as expanded use of CMC in foods and pharmaceuticals. Some decline, however, in the long-term can be expected in the CMC detergent market. (Please see the Market Survey for Sodium Carboxymethylcellulose.)

PVA should expand its market as a result of increased use in adhesives and paper coatings. In the long-term, however, PVA consumption by the textile industry can be expected to decline. (Please see the Market Survey for Polyvinyl Alcohol.)

Acrylamide polymers, acrylic acid and methacrylic acid polymers, ethylene oxide polymers, derivatives of natural gums, PEG, PEI and starch derivatives can be expected to develop many specialized applications, although the natural gums can be expected to continue to suffer from uncertainties of production. (Please see the Market Survey for the respective products.)

In the near-term, traditional starch products, as well as modified starches and starch derivatives should benefit from lower prices, as compared to CMC, PVA and other more expensive water-soluble polymers. Nevertheless, CMC and PVA are the water-soluble polymers that should particularly benefit from market expansion in the medium-term (3 to 5 years) because of the depth and breadth of their markets.

Many technological innovations can be expected in all the water-soluble polymers in the next few years. The need for water-soluble polymers is growing. Countries producing natural gums are now working very hard to make their output more organized and systematic. They are seeking to compete with the synthetic and semisynthetic water-soluble polymers by providing more detailed technical information than in the past.

Advances in knowledge are constantly being made and new products are being brought to market. There are constant problems of investment of time, energy, effort and money. There are also opportunity costs. Particularly promising areas for the future appear to be inorganic water-soluble polymers and xanthan gum, a biopolymer based on the microorganism *Xanthomonas campestris.*

Yale L. Meltzer
November, 1980

MARKET SURVEY

ACRYLAMIDE POLYMERS

Product: Water-soluble anionic, cationic and neutral synthetic polymers.

Main Method of Manufacture: Polymerization of aqueous solutions of acrylamide.

Major Markets: Adhesives, cement, explosives, mine operations, oil drilling operations, paper processing, water pollution control.

Major Producers and/or Suppliers

Producer and/or Supplier	Country
American Cyanamid Co., Inc.	USA
BASF, AG	West Germany
Bayer, AG	West Germany
Betz Laboratories, Inc.	USA
Cassella Farbwerke Mainkur, AG	West Germany
Dow Chemical Co.	USA
General Mills, Inc. (General Mills Chemicals, Inc.)	USA
Halliburton Co.	USA
Hercules, Inc.	USA
Hodag Chemical Corp.	USA
Konan Chemical Co., Ltd.	Japan
Kyoritsu Yuki	Japan
Merck & Co., Inc. (Calgon Corp.)	USA
Nalco Chemical Co.	USA
Sanyo Chemical Industries, Ltd.	Japan
Stein Hall & Co., Inc.	USA

ACRYLIC ACID AND METHACRYLIC ACID POLYMERS

Product: Water-soluble polymers of acrylic and methacrylic acids.

Main Method of Manufacture: Polymerization of aqueous solutions of the monomers.

Major Markets: Adhesives, cosmetics, film, medical applications, metal manufacture, oil-well drilling muds, paints, paper manufacture, pharmaceuticals, rubber latexes, textile sizes.

Major US Producers and/or Suppliers

American Cyanamid Co.
Borden, Inc.
Diamond Shamrock Corp.
B.F. Goodrich
National Starch & Chemical Corp.
Polymer Research Corporation of America
Polysciences, Inc.
PVO International, Inc.
Rohm and Haas Co.
Union Carbide

ETHYLENE OXIDE POLYMERS

Product: Water-soluble high molecular weight polymers (molecular weights of 100,000 and over). (See Polyethylene Glycol.)

Main Method of Manufacture: Catalytic polymerization of ethylene oxide.

Major Markets: Adhesives, ceramics, coatings, detergents, films, foods, oil recovery operations, paints, printing.

Major Producers and/or Suppliers

Producer and/or Supplier	Country
BASF, AG	West Germany
Dow Chemical Co.	USA
Hoechst, AG	West Germany
Jefferson Chemical Co., Inc.	USA
Meisei Chemical Works, Ltd.	Japan
Olin Corp.	USA
Pechiney-Ugine-Kuhlmann	France
Petrochim	Belgium
Seitetsu Kagaku Co., Ltd.	Japan
Union Carbide	USA

GUAR GUM (NATURAL GUM)

Product: A water-soluble carbohydrate polymer of galactomannan obtained from *Cyamopsis tetragonoloba* (plant family: *Leguminosae*).

Main Method of Manufacture: Grinding the endosperms of the seeds of *Cyamopsis tetragonoloba.*

Major Markets: Cream cheese, explosives, ice cream, mining, oil-well drilling muds, paper, pharmaceuticals, textiles.

Major Producers of Guar Gum (by Country)

India
Pakistan
United States

Major US Producers and/or Suppliers of Guar Gum (by Company)

Celanese Corp. (Celanese Polymer Specialties Co.)
Colony Import & Export Corp.
CPC International Inc. (Penick Corp.)
General Mills, Inc. (General Mills Chemicals, Inc.)
Hercules, Inc.
Meer Corporation
Stein Hall & Co., Inc.
Tragacanth Importing Corporation

GUM ARABIC (NATURAL GUM)

Product: A heteropolysaccharide, which is mainly the calcium salt of arabic acid, obtained from the bark of various species of the *Acacia* tree (plant family: *Leguminosae*). Hydrolysis yields L-arabinose, D-rhamnose, D-galactose and D-glucuronic acid. Gum arabic is known by several names, e.g., gum hashab, Kordofan gum, Senegal gum and Sudan gum.

Main Method of Manufacture: Gum arabic is gathered at random from *Acacia* trees growing wild in various parts of Africa or systematically on plantations primarily in the Sudan. The gum is obtained by slitting the bark of the trees. Gum arabic is obtained primarily from the *Acacia senegal* tree, which grows along the southern border of the Sahara Desert.

Major Markets: Adhesives, beverages, cosmetics, foods, inks, lithography, paints, pharmaceuticals, textiles.

Major Producers of Gum Arabic (by Country)

Chad
Egypt
Ethiopia
Mali
Mauritania
Niger
Nigeria
Senegal
Somalia
Sudan*
Tanzania

*approximately 80% of world's supply.

Major US Suppliers of Gum Arabic (by Company)

Colony Import & Export Corp.
CPC International Inc. (Penick Corp.)
Meer Corporation

(continued)

Norda, Inc.
Stauffer Chemical Co.
Tragacanth Importing Corporation

GUM TRAGACANTH (NATURAL GUM)

Product: An exudate obtained from the shrubs of the *Astragalus* species (plant family: *Leguminosae),* especially from *Astragalus gummifer.* It is a mixture of acidic and neutral polysaccharides. Hydrolysis of the acidic component yields fucose, galactose, galacturonic acid and xylose, while hydrolysis of the neutral component yields L-arabinose and D-galactose.

Main Method of Manufacture: The gum is collected mainly from wild bushes in an unsystematic manner in desert and partially desert-like areas.

Major Markets: Beverages, candy, cheese, cosmetics, crayons, dietetic foods, hand lotions, ice cream, inks, medical applications, salad dressings, sherbet.

Major Producers of Gum Tragacanth (by Country)

Afghanistan
Greece
Iran (world's main producer)
Iraq
Israel
Pakistan
Soviet Union
Syria
Turkey

Major US Suppliers of Gum Tragacanth (by Company)

Colony Import & Export Corp.
CPC International Inc. (Penick Corp.)
Meer Corporation
Tragacanth Importing Corporation

LOCUST BEAN GUM (NATURAL GUM)

Product: A galactomannan polymer, obtained from the endosperm of the carob seed of the Mediterranean locust tree *Ceratonia siliqua* (plant family: *Leguminosae*). The gum is also referred to as carob gum, carob seed gum, gum gatto, gum hevo and tragosol. It is a carbohydrate, which upon hydrolysis yields mannose and galactose.

Main Method of Manufacture: Crushing and grinding of the carob seeds.

Major Markets: Antidiarrheic, cakes, cheese, cosmetics, ice cream, mining, oil-well drilling muds, paper processing, pharmaceuticals, salad dressings, textiles.

Major Producers of Locust Bean Gum (by Country)

Crete
Cyprus
Greece
North African Countries

(continued)

Portugal
Sardinia
Sicily
Spain

Major US Suppliers of Locust Bean Gum (by Company)

Colony Import & Export Corp.
CPC International Inc. (Penick Corp.)
General Mills, Inc. (General Mills Chemicals, Inc.)
Meer Corporation
Tragacanth Importing Corporation

METHYLCELLULOSE AND OTHER ALKYLCELLULOSES (CELLULOSE ETHERS)

Product: Water-soluble alkylcellulose ether semisynthetic polymers. The following are examples:

Ethylcellulose (EC)
Ethylhydroxyethylcellulose (EHEC)
Hydroxybutylmethylcellulose (HBMC)
Hydroxyethylcellulose (HEC)
Hydroxyethylmethylcellulose (HEMC)
Hydroxypropylcellulose (HPC)
Methylcellulose (MC)

Main Method of Manufacture: Cellulose (usually obtained from wood pulp or cotton linters) is reacted with an aqueous solution of sodium hydroxide and then an etherifying agent (e.g., cellulose when reacted with an aqueous solution of sodium hydroxide and then methyl chloride yields methylcellulose).

Major Markets: Adhesives, agriculture, ceramics, coatings, cosmetics, films, foods, hair preparations, inks, oil-well drilling muds, paints, paper processing, pharmaceuticals, shampoos, textiles.

Major Producers and/or Suppliers

Producer and/or Supplier	Country
British Celanese Ltd.	United Kingdom
Dow Chemical Co. (EC, HBMC, HPMC, MC)	USA
Henkel & Cie, GmbH	West Germany
Hercules, Inc. (EC, EHEC, HEC, HPC)	USA
Kalle, AG (subsidiary of Hoechst, AG)	West Germany
Nippon Soda	Japan
Shin-Etsu	Japan
Union Carbide (HEC)	USA
Wolff, AG	West Germany

POLYETHYLENE GLYCOL (PEG)

Product: Water-soluble synthetic polymers of ethylene oxide, which are differentiated from water-soluble ethylene oxide polymers on the basis of molecular weight. Polyethylene glycols usually have molecular weights up to 20,000 but are classified as polyethylene glycols up to 100,000. Ethylene oxide polymers have molecular weights of 100,000 and over.

Main Method of Manufacture: Ethylene oxide reacts with water to form PEG upon continuous addition. Ethylene glycol (a dihydric alcohol) reacts with ethylene oxide to yield diethylene glycol, subsequent addition of ethylene oxide yields triethylene glycol. Continuous addition of ethylene oxide yields higher and higher condensation products until PEG is obtained.

Major Markets: Adhesives, coatings, cosmetics, detergents, foods, inks, medical applications, oil recovery, paints, paper, textile sizes.

Major Producers and/or Suppliers

Producer and/or Supplier	Country
BASF Wyandotte Corp.	USA
Dow Chemical Co.	USA
Hoechst, AG	West Germany
Hüls, AG	West Germany
Imperial Chemical Industries	United Kingdom
Jefferson Chemical Co., Inc.	USA
Montedison	Italy
Olin Corp.	USA
Pechiney-Ugine-Kuhlmann	France
Sanyo Chemical Industries, Ltd.	Japan
Union Carbide	USA

POLYETHYLENEIMINE (PEI)

Product: A water-soluble alkyleneimine synthetic polymer that is highly hygroscopic.

Main Method of Manufacture: Addition polymerization of the monomer. Ethyleneimine polymerizes in the presence of acids to polyethyleneimine.

Major Markets: Adhesives, ion-exchange resins, paper manufacture, photography, textiles, water pollution control.

Major U.S. Producers and/or Suppliers

Arsynco Co., Inc.
Chemirad Corp.
Cordova Chemical Co.
M.W. Hardy & Co., Inc.

POLYVINYL ALCOHOL (PVA)

Product: Water-soluble linear polyhydroxy synthetic polymer.

Main Method of Manufacture: Hydrolysis of polyvinyl acetate.

Major Markets: Adhesives, binders, ceramics, cosmetics, films, packaging, paper sizes, steel manufacture, textile sizes.

Major Producers and/or Suppliers

Producer and/or Supplier	Country
Air Products & Chemicals, Inc.	USA
Borden, Inc.	USA
Denki Kagaku Kogyo KK	Japan
Du Pont	USA
Hoechst, AG	West Germany
Kuraray Co.	Japan
Monsanto	USA
Nippon Synthetic Chemical Industry	Japan

US Production of PVA

Producer	Capacity (1976)	Capacity (1979)
Air Products	40	55
Du Pont	80	100
Monsanto	16	18
Total	136	173

Source: *Chemical Marketing Reporter* issues of July 1, 1976 and April 1, 1979 (Data presented with permission of *Chemical Marketing Reporter,* Schnell Publishing Company, Inc.)

In the table above, capacity figures indicate the millions of pounds annually estimated for polyvinyl alcohol not used in the manufacture of polyvinyl butyral (PVB). Du Pont's capacity for PVA including that used for PVB is reported at 125 million pounds. Monsanto's overall capacity is estimated to be 45 million pounds. Du Pont produces only fully hydrolyzed grades of PVA and Monsanto produces only partially hydrolyzed grades; hence the two firms do not compete directly. Air Products produces both types. Capacities are somewhat variable as different grades of PVA can require different production times.

Price Data: The PVA market has often been plagued by price weakness. In the United States, for the period 1957 to August 8, 1980, the high price was $0.915/lb (fully hydrolyzed, medium viscosity, bags, truckloads, delivered); a low price of $0.32/lb, delivered, was reported.

Percentage Breakdown by Application of PVA in the US

	1963	1967	1969	1970	1973	1976	1979
Adhesives	50	42	25	25	25	25	25
Emulsifiers	NA	15	NA	NA	12	12	12
Film	16	6	NA	NA	NA	NA	NA
Paper coatings	15	10	15	15	13	13	13
Textile sizes	12	21	43	43	40	40	40
Other	7	6	17	17	10	10	10
Total	100	100	100	100	100	100	100

Note: NA = not available

Source: *Chemical Marketing Reporter* issues of July 29, 1963; February 27, 1967; November 10, 1969; April 1, 1970; January 29, 1973; July 1, 1976; April 1, 1979 (Data presented with permission of *Chemical Marketing Reporter,* Schnell Publishing Company, Inc.)

Market Growth and Forecasts: For the period 1962 to 1972, overall US PVA consumption grew at a rate of 8.8% per year. For the period 1968 to 1978, growth was 11.4% per year. Growth through 1983 is expected to be 6% to 7% per year. Growth of the PVA market is closely tied to developments in the textile and paper industries. As a result of decreases in those industries, demand for PVA sharply declined. In the next few years, the greatest growth in PVA is expected in adhesives and paper coatings. In the long-term, PVA consumption by the textile industry is expected to decline, as a result of new recycling processes. The Japanese share of the US PVA market is estimated to be 7% to 8%, which is not considered a serious threat to US domestic producers. The following table gives US total PVA demand with a forecast for the future.

Total US PVA Demand (with Forecast)

Year	Quantity (in pounds)
1972	80,000,000
1975	90,000,000
1978	133,000,000
1983 (Forecast)	182,000,000

Source: *Chemical Marketing Reporter* issues of July 1, 1976 and April 1, 1979 (Data presented with permission of *Chemical Marketing Reporter,* Schnell Publishing Company, Inc.)

POLYVINYLPYRROLIDONE (PVP)

Product: A water-soluble polyvinylamide.

Main Method of Manufacture: Polymerization of N-vinyl-2-pyrrolidone.

Major Markets: Adhesives, beverages, cosmetics, detergents, lithography, medical applications, paper manufacture, pharmaceuticals, photography, textiles.

Major Producers and/or Suppliers

Producer and/or Supplier	Country
Aldrich Chemical Co., Inc.	USA
BASF, AG	West Germany
BASF Wyandotte	USA
Dan River, Inc.	USA
GAF Corporation	USA

SODIUM CARBOXYMETHYLCELLULOSE (CMC) (CELLULOSE ETHER)

Product: Water-soluble anionic cellulose ether semisynthetic polymer.

Main Method of Manufacture: Cellulose (usually obtained from wood pulp or cotton linters) is reacted with an aqueous solution of sodium hydroxide and then sodium monochloroacetate.

Major Markets: Detergents, foods, oil-well drilling muds, paints, paper size, pharmaceuticals, pollution control, textiles.

Major Producers and/or Suppliers

Producer and/or Supplier	Country
Adachi Paste Co.	Japan
AKZO, NV	Netherlands
British Celanese, Ltd.	United Kingdom
Carbose	USA
Cellulose Products of India, Ltd.	India
Chemical Development of Canada	Canada
Color Prager, SA	Spain
Daicel, Ltd.	Japan
Dai-ichi Kogyo Seiyaku Co., Ltd.	Japan
Henkel, GmbH	West Germany
Hercules, Inc.	USA
Imperial Chemical Industries	United Kingdom
Industries of India, Ltd. Gujchem	India
Kalle, AG (Subsidiary of Hoechst, AG)	West Germany
Kyoto Gosei KK	Japan
Novacel	France
Phrix-Werke, AG	West Germany
Procter & Gamble (Buckeye Cellulose Corp.)	USA
Sanyo Kokusaku Pulp	Japan
Shikoku Chemicals Corp.	Japan
Snia Viscosa	Italy
Svenska Cellulosa	Sweden
Uddeholms, AB	Sweden

US Production of CMC (1980)

Producer	Capacity
Carbose	8
Hercules, Inc.	77
MAK Chemical	2.5
Procter & Gamble (Buckeye Cellulose Corp.)	7
Total	94.5

Source: *Chemical Marketing Reporter* (August 11, 1980) (Data presented with permission of *Chemical Marketing Reporter,* Schnell Publishing Company, Inc.)

In the above table, capacity figures indicate millions of pounds per year of sodium carboxymethylcellulose, 100% basis. Most producers make CMC of 65% to 70% assay. Hercules also produces purified grades. Tillett Chemical produces specialty grades of CMC at Pineville, NC.

Price Data: In the United States, for the period 1952 to August 11, 1980, the high price was $0.82/lb, detergent grade, 65% basis, truckloads, bags, freight allowed; a low price of $0.37/lb, same basis, was reported.

Percentage Breakdown by Application of CMC in the U.S.

	1963	1965	1969	1971	1974	1978	1980
Detergents	38	38	32	28	30	30	30
Foods	12	14	20	20	23	25	NA
Foods and pharmaceuticals	–	–	–	–	–	–	27
Oil well drilling muds	15.5	13	12	10	10	12	15
Paper size	9	8	10	5	3	3	3
Paints	2.5	3	NA	NA	NA	NA	NA
Pharmaceuticals	7	8	NA	NA	NA	NA	NA
Textiles	11	11	18	24	24	22	20
Miscellaneous*	5	5	8	13	10	8	5
Total	100	100	100	100	100	100	100

Note: NA = not available.

*Includes adhesives, ceramics, latex paints and welding rod coatings.

Source: *Chemical Marketing Reporter* issues of January 21, 1963; July 1, 1965; January 1, 1969; November 1, 1971; October 21, 1974; January 1, 1978; August 11, 1980. (Data presented with permission of *Chemical Marketing Reporter,* Schnell Publishing Company, Inc.)

Market Growth and Forecasts: For the period 1963 to 1973, growth of the overall US CMC market (in terms of consumption) was 3.5% per year. For the period 1969 to 1979, growth was 2.3% per year. Growth through 1984 is expected to be 2.5% per year. Use of CMC in detergents could decline as liquid consumer products replace heavy duty powders. CMC price increases are forcing some users to switch to the cheaper starch products. Use of purified grades of CMC in oil-well drilling muds is expected to grow at the rate of about 10% per year and the food and pharmaceutical markets are growing. Use of CMC as a tobacco substitute, which looked promising at one time, no longer looks promising after a poor reception in test markets in Europe. Carbose has recently established itself as a CMC producer in the US, which could lead to overcapacity.

In addition, US producers are concerned about possibly being underpriced by imports from Europe and Japan. The following table gives US total CMC demand with a forecast for the future.

Total US CMC Demand (with Forecast)

Year	Quantity (in pounds)
1976	70,000,000
1977	72,400,000
1979	78,000,000
1984 (Forecast)	88,000,000

Source: *Chemical Marketing Reporter* issues of January 1, 1978 and August 11, 1980 (Data presented with permission of *Chemical Marketing Reporter,* Schnell Publishing Company, Inc.)

STARCH

Product: A naturally occurring carbohydrate. It is a glucose polymer with an α-glucoside structure. Starches differ greatly in composition, with considerable variation in amylose and amylopectin components. Both components yield glucose upon hydrolysis. Amylose, however, is a linear polymer which consists of glucose linked through the 1,4 positions and its configuration is classified as an "α-1,4-linkage," while amylopectin is a highly branched molecule through the C-6 hydroxyl and its configuration is classified as an "α-1,6-linkage." Types of starches are: arrowroot, corn starch, potato starch, rice starch, sorghum, tapioca and wheat starch.

Main Method of Manufacture: Various methods of separation are used to obtain starch from its natural state. Slurry processes in various solvents are used and the starch is then washed and dried.

Major Markets: Adhesives, food, paper, textiles.

Major US Producers and/or Suppliers of Starch or Starch Products

American Maize Products Co.
Anheuser-Busch, Inc.
Clinton Corn Processing Company (a division of Standard Brands, Inc.)
CPC International, Inc.
General Mills Chemicals, Inc.
Grain Processing Corp.
Hubinger Co.
National Starch and Chemical Corp.
Pennwalt Corp.
Pfanstiehl Laboratories, Inc.
A.E. Staley Manufacturing Co.
Western Polymer Corp.

ACRYLAMIDE POLYMERS

PROCESSING

Gel Polymers of N-Methylol-Acrylamide

Highlights of the Technological Achievement: A process for preparing plates of a new gel polymer, the plates being adapted to the electrophoresis separation of the serum or plasma lipoproteins under a stepped gradient. The gel polymers are prepared by radical polymerization of N-methylol-acrylamide and of a bifunctional allylic or acrylic compound causing cross-linking to yield a tridimensional configuration polymer. The polymerization may optionally be catalyzed by peroxides and/or ultraviolet radiation. Anionic polysaccharides containing only COOH groups may be added to assist migration of lipoproteins in the gel.

Reference: E. Boschetti; U.S. Patent 4,189,370; February 19, 1980; assigned to Societe Sebia, France.

Background: It is known that the separation of human lipoproteins is carried out upon natural supports such as agar, agarose, or gelled starch, or synthetic supports such as cellulose acetate or gelled acrylamide.

In the case of supports made of agar, agarose and cellulose acetate, the electrophoretic separation shows four characteristic fractions named according to their increasing mobility: beta-lipoproteins, pre-beta-lipoproteins, alpha-lipoproteins and pre-alpha-lipoproteins.

When, in pathologic conditions there are, in the serum, micelles of fatty compounds, the chylomicrons, the electrophoresis normally shows a more or less noticeable trace at the beginning of the migration, the trace being revealed through liposoluble coloring reactants.

In the case of the classical supports of acrylamide and starch gel, the difficulty lies in the migration of the beta- and pre-beta-lipoproteins. In fact, due to the very small size of the pores of the gel, such molecules cannot move. Attempts have been made to make possible the migration of the lipoproteins in a discontinuous gel of polyacrylamide, by decreasing to the extreme the polymer concentration. Such gels which are too soft to be handled are cast inside a small glass tube in which the electrophoretic migration is carried out. Despite rather good results, the interpretation remains difficult in view of the fact that it is impossible to obtain an accurate quantitative measure of each colored fraction by a densitometric method.

Process Conditions: The plates of this process obviate those drawbacks. The chylomicrons are totally stopped at the starting tank, the pre-beta-lipoproteins are stopped at the discontinuous gradient, and the beta-, alpha- and pre-alpha-lipoproteins are fractionated through their difference of mobility in the more concentrated area of the gel-polymer. The densitometric reading of such a picture no longer suffers from the drawback of the cylinders and finally it is possible to air-dry the plate after use to obtain then a transparent and flexible film easy to stock.

The plate is made of a transparent polyester support. Upon one face is stratified the gel-polymer. The face of the plastic support receiving the gel is pretreated by the usual procedures making it wettable. The gel-polymer is divided into two areas of different concentrations, the first taking about ⅔ of the available area. The separation line of two gels of the same kind but of different concentrations is rectilinear.

The gel-polymer is obtained by radical polymerization of N-methylol-acrylamide and of a bifunctional allylic or acrylic compound causing the cross-linking, i.e., the tridimensional configuration of the polymer. The polymerization can be catalyzed by peroxides and/or under ultraviolet radiation.

The process of this formulation comprises: (a) preparing two hot aqueous or buffered solutions of strongly endosmotic polysaccharide each contining the requested elements for the polymer; (b) casting each solution at a well-fixed moment in a plastic mold containing the polyester film; (c) cooling and withdrawing the gel from the mold, then washing it in an isotonic solution; (d) recovering the buffered gel, of given thickness and shape, fixed upon the polyester film. Such product, a lipofilm, can be kept at 4°C in any closed container, such as a plastic box of a polyethylene or aluminum sealed bag.

Catalytic Systems: The monomer concentrations are 2.5 and 7.5% of cross-linking ratio for the first layer and 2.0 and 10 to 17% of cross-linking ratio for the second layer. The polysaccharide concentrations can vary between 0.1 and 1%.

As catalytic systems, there can be used for example: N,N,N',N'-tetramethylethylenediamine-ammonium persulfate, and N,N,N',N'-tetramethylethylenediamine-riboflavine-hydrogen peroxide-ultraviolet.

The bifunctional allylic or acrylic compounds used as cross-linking agents can vary according to the requested solubility of the gel. Such a characteristic is determined by the process selected for the quantitative evaluation of the lipoprotein fractions: (1) an insoluble gel is requested for a densitometric study, and (2) a gel soluble in proper chemical solvents is requested for a study through elution.

As examples it can be cited: N,N'-methylenebisacrylamide leads to an insoluble gel; diallyltartramide leads to a gel soluble in 2% periodic acid; and ethylenediacrylate leads to gels soluble in basic media.

Product: According to a further feature of this process the plates of gel-polymers can also be treated to be dehydrated and rehydratable through simple chemical processes. Actually it has been found that if the gel is washed in a solution prepared from a mixture of a simple sugar, a diol or a triol and a carboxylic polysaccharide, previously treated by a reducing agent, such a washed gel can be air-dried, then regain its original characteristic through rehydration in distilled water.

A simple sugar can be, for example, sorbitol used at a 10% or less concentration. A carboxylic polysaccharide can be, for example, purified sodium or ammonium alginates or soluble sodium carboxymethylcellulose. Such products are characterized by a complete solubility in aqueous medium and by a viscosity less than 30 cp (1.2% aqueous solution at 20°C). Their concentration in the washing solution is at most 0.5%. The diol and the triol can be respectively ethylene glycol and glycerol used at a 1% concentration. The reducing agents can be inorganic salts such as sodium or potassium borohydrides, or aluminum-lithium hydride, used at a 0.02% concentration.

The reduction should be carried out when the compounds are well dissolved in water, and it starts as soon as the reducing agent is added, as shown by more or less of a release of native hydrogen.

Producing Microcapsules by Complex Coacervation

Highlights of the Technological Achievement: Aqueous polymer solutions containing water and 5 to 30 wt % of a polymeric mixture of: Copolymer (A), 65 to 90 mol % of whose structural units, based on polymer, consist of statistically distributed residues of acrylamide and 10 to 35 mol % of residues of maleic acid or maleic acid anhydride, and which has an intrinsic viscosity η of 0.05 to 1 dl/g; and Copolymer (B) whose structural units consist of statistically distributed polymerized residues of acrylamide, acrylic acid and maleic acid, at least some of the maleic acid residues being present in salt form, which has an intrinsic viscosity η of from 0.05 to 1.5 dl/g and which contains from 65 to 90 mol %, based on polymer, of residues of acrylamide and residues of acrylic acid together and from 10 to 35 mol% of residues of maleic acid, at least some of which are present in salt form.

The ratio of weight of Copolymer (A) to Copolymer (B) is from 1:2-20:1.

Reference: B. Bomer, H. Schnoring and R. Schmidt; U.S. Patent 4,181,639; January 1, 1980; assigned to Bayer AG, Germany.

Background: The copolymerization of acrylamide and maleic acid anhydride is best carried out by precipitation polymerization in organic solvents which are inert to anhydride groups. Acetone and acetic acid esters of alcohols containing from 1 to 3 carbon atoms are particularly favorable solvents by virtue of their high dissolving power on the monomers and not least by virtue of their ready volatility.

The polymerization reaction may be initiated with the usual radical initiators, such as aliphatic azo compounds, or organic percompounds, such as peroxides or peresters. The polymerization temperature is determined by the decomposition properties of the initiator and, where azoisobutyrodinitrile is used in batch operation, is in the range of from 50° to 80°C. In this process, the copolymers are obtained in the form of finely divided powders and may readily be filtered off and freed from residual monomers by washing.

Process Conditions: The copolymerization of acrylamide and maleic acid to form Copolymers (A) may be carried out both by precipitation polymerization in organic solvents and by solution polymerization in water.

In the precipitation copolymerization of acrylamide and maleic acid, lower aliphatic alcohols containing from 1 to 4 carbon atoms may also be successfully used in addition to acetone and acetic acid esters of alcohols containing from 1 to 3 carbon atoms. The polymerization reaction itself and the isolation of the polymer powders are carried out in the same way as for acrylamide/maleic acid anhydride copolymers.

After drying or while they are still moist with solvent, the precipitation copolymers may be dissolved in water to form 5 to 30 wt % solutions, the anhydride groups of the acrylamide/maleic acid anhydride copolymers being converted into maleic acid groups. Where the polymers used are still moist with solvent, the organic solvent is subsequently separated off by decantation and/or distillation.

The radical initiators are used in quantities of from 0.1 to 10 wt % and preferably in quantities of from 0.3 to 3.0 wt %, based on monomer. In the case of precipitation polymerization, the polymer yields generally amount to between 75 and 95%. The Copolymers (A) have intrinsic viscosities η of from 0.05 to 1.0 dl/g and preferably from 0.08 to 0.4 dl/g, as measured in a 0.9% aqueous sodium chloride solution.

The solutions of the Copolymers (B) may be obtained by heating the aqueous solutions of the Copolymers (A) over a relatively long period to temperatures of from 60° to 150°C, optionally under pressure, in an autoclave. Hydrolysis under reflux conditions at normal pressure is particularly preferred. Partial hydrolysis of acrylamide units occurs during heating. The ammonia formed neutralizes some of the maleic acid units and, in doing so, increases the pH value of the solution.

As a result of hydrolysis, the acrylamide/maleic acid copolymer is converted into a copolymer which consists of acrylamide, acrylic acid and maleic acid units and in which at least some of the maleic acid units are present in salt form.

Acrylamide (% by wt/mol %)	Maleic Acid Anhydride (% by wt/mol %)	Maleic Acid (% by wt/mol %)	η* (dl/g)	pH**	Hydrolysis to pH	Mixture*** (g/g)	pH†	Encapsulation Behavior
80/84.7	20/15.3	–	0.24	3.2	4.5	40/80	3.8	good
77/84.5	–	23/15.5	0.20	3.3	4.6	30/90	3.8	good
83/88.9	–	17/11.1	0.28	3.7	4.5	20/120	3.9	average
89/93.0	–	11/7.0	0.39	3.7	4.5	20/120	3.9	poor
95/96.3	5/3.7	–	0.60	3.9	4.9	20/120	4.0	poor

*Intrinsic viscosity, as measured in 0.9% aqueous sodium chloride solution.
**Fresh solution.
***Mixture of hydrolysis solution/fresh solution.
†Of mixture.

Inhibiting Unwanted Polymerization

Highlights of the Technological Achievement: Use of copper ion in acrylamide solution during concentration to inhibit polymerization.

Reference: K. Matsuda, M.S. Butensky, K.H. Tsu and R.J. Munch; U.S. Patent 4,163,755; August 7, 1979; assigned to American Cyanamid Company.

Background: This process relates to concentrating aqueous acrylamide solutions by evaporation of water from solution. An object of this process is to provide a process for concentrating acrylamide solutions without causing unwanted polymerization of acrylamide in the solution during the concentration process.

Cupric ion is a known polymerization inhibitor for use in aqueous acrylamide solutions but it seems impractical at first to incorporate soluble salts of copper in the solution because they were found to precipitate as basic salts and cause serious corrosion of stainless steel surfaces in heated lines and vessels containing the acrylamide solutions. It has been discovered, however, that when aqueous acrylamide solutions containing cupric ion are maintained at pH value of pH 5 or less the precipitation of the copper salt and, therefore, the corrosion of stainless steel surfaces are effectively inhibited.

Process Conditions: A 30% acrylamide aqueous solution is concentrated to make 50% acrylamide aqueous solution in laboratory apparatus. The recirculating solution to the evaporator is continuously adjusted to pH 5 with sulfuric acid. Cupric sulfate is added to maintain concentration of 50 ppm Cu^{+2}/AMD in the evaporator. The feed is led into a vacuum flash evaporator in which the pressure has been reduced to maintain a boiling point of acrylamide aqueous solution at 80°C inside the vessel (about 300 mm Hg).

Liquid residue from the evaporation is removed to atmospheric pressure and contacted with air to restore oxygen to saturation in the solution. Rate of recycle of liquid residue to the evaporator, which is preheated to 85°C before reentry is adjusted to make a constant 50% acrylamide residue at the selected fresh feed rate, which is balanced with the designed water vaporization rate and the product withdrawal rate. In runs where sulfuric acid was not added (solution pH about 6 to 7) it was found that stainless steel surfaces contacting the process solution were corroded by pitting, particularly at heat exchange surfaces where cupric solids were precipitated. Adjustment to pH 5 in the liquid residue was found to eliminate this corrosion problem.

Because of the presence of cupric ion it is easier to avoid polymerization in the evaporator at higher operating temperatures so it becomes more practical to operate the flash evaporator with higher boiling temperature at less vacuum, if desired. It is quite possible even to operate the flash evaporator at atmospheric pressure with superheated feed, although this is not usually the most preferred mode of operation. This process can be used at any evaporation pressure from 70 mm Hg absolute up to atmospheric pressure, or even higher if desired.

Under the operating conditions contemplated, the mol ratio of air to water vapor in the vapor phase inside the flash evaporation chamber is much below 0.1, usually in the magnitude about 0.005 and generally in the range of 0.05 to 0.001. The boil-up rate is generally in the range of ⅓ to 4 pbw water per

100 parts of liquid circulated. The boil-up rate is controlled by the recycle feed entry temperature, usually in the range from 5° to 20° higher than the boiling point in the evaporator. Calculated residence time is preferably about 20 min but may range up to as much as 250 min. Cupric ion should be kept to the minimum practical concentration, usually in the range from 5 to 200 ppm.

Stable Partially Hydrolyzed Water-Soluble Acrylamide Polymers

Highlights of the Technological Achievement: A stable, water-in-oil emulsion of a partially hydrolyzed carboxamide polymer is prepared by contacting a water-in-oil emulsion of an inherently water-dispersible polymer containing carboxamide groups, e.g., polyacrylamide, with a water-in-oil emulsion of a hydrolyzing agent such as sodium hydroxide. By emulsifying the hydrolyzing agent prior to hydrolysis, the hydrolysis reaction can be carried out over a wider range of reaction conditions such as temperature and the like without destroying the water-in-oil emulsion of the polymer.

Reference: H. Volk and J.C. Lamphere; U.S. Patent 4,151,140; April 24, 1979; assigned to The Dow Chemical Company.

Background: The use of partially hydrolyzed polyacrylamides and other carboxamide polymers (often called anionic polyacrylamides) in such applications as enhanced recovery of petroleum, flocculation of finely divided solids from aqueous suspension and the like is well known. In such applications it is often desirable to convert from 1 to 65 mol % of the carboxamide groups of the polymer to carboxy groups.

For the production of high molecular weight, partially hydrolyzed polyacrylamides and other carboxamide polymers, it has been suggested that the carboxamide polymer in aqueous solution be hydrolyzed to the desired extent by contacting the aqueous solution with alkali such as sodium hydroxide. However, the preparation of a homogeneous hydrolyzed carboxamide product by such method is difficult because of the high viscosities of aqueous solutions of such polymers. Such high viscosities hinder and often prevent uniform mixing of the alkali with the aqueous solution of the polymer. Thus, in such a method of operation, it is generally found that some of the polymer is hydrolyzed to a greater extent than desired while other portions of the polymer undergo little or no hydrolysis.

Process Conditions: The water-in-oil emulsions of the carboxamide polymers are suitably any in which the disperse phase is the aqueous phase containing the carboxamide polymer and the continuous phase is a water-immiscible, inert liquid medium.

It is generally desirable to minimize the proportion of oil phase and thereby maximize the concentration of the carboxamide polymer in the emulsion. For similar reasons, it is desirable that the concentration of the carboxamide polymer in the disperse phase be as concentrated as possible, i.e., at least 35 wt % of polymer based on the weight of the aqueous phase.

The concentration of the polymer in the aqueous phase is less than that which appreciably destabilizes the water-in-oil emulsion. Preferably, based on total volume of the water-in-oil emulsion, the disperse phase constitutes from 20 to 80 volume %

and the continuous oil phase constitutes from 80 to 20 volume %. Preferably, the concentrated aqueous phase containing the carboxamide polymer contains from 35 to 65 wt % of the polymer.

The carboxamide polymer is suitably any polymer of ethylenically unsaturated monomer or monomers wherein at least 50 mol % of the polymerized monomer units (so-called mers) have pendant carboxamide groups.

The carboxamide polymer is further characterized as being at least inherently water dispersible and preferably water soluble and insoluble in the water-immiscible, inert liquid of the continuous oil phase. Most preferably, such polymers are generally linear in structure. For the purposes of this process, the carboxamide group is represented by the formula:

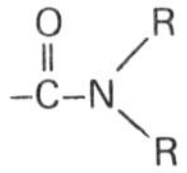

wherein each R is individually hydrogen; alkyl; aminoalkyl, particularly dialkylaminomethyl; hydroxyalkyl; or $-R'-N(R'')_3X^-$ wherein R' is alkylene, preferably methylene; each R'' is individually hydrogen, alkyl or hydroxyalkyl; and X^- is a neutralizing anion such as chloride, bromide, methylsulfate anion, hydroxide and the like. In preferred carboxamide polymers, R is hydrogen or alkyl, most preferably hydrogen.

Preferred carboxamide polymers are the polymers of acrylamide and methacrylamide, with the homopolymers of acrylamide and copolymers thereof containing at least 85 mol % of acrylamide being especially preferred.

Mixing the Particles of Hydrous Gel

Highlights of the Technological Achievement: Acrylamide polymers or copolymers are partially hydrolyzed by mixing the particles of hydrous gel of the polymer in an apparatus which has a mixing mechanism comprising blades having a blank space, and the face of the apparatus where the polymer is contacted is made of or coated with synthetic resin.

Reference: A. Tago, H. Kudomi and Y. Nakanishi; U.S. Patent 4,146,690; March 27, 1979; assigned to Mitsubishi Chemical Industries, Ltd., and Kyoritsu Yuki Co., Ltd., both of Japan.

Background: Partially hydrolyzed acrylamide polymers are generally used as flocculent and paper-strengthening agents. The partially hydrolyzed acrylamide polymers are produced usually by treating an acrylamide polymer with an alkaline material such as, for example, caustic soda to hydrolyze the amide groups. Acrylamide polymers are usually produced by polymerizing the monomers in a 10 to 35 wt % aqueous solution, but as the acrylamide polymers thus obtained are in the form of a hard hydrous rubber-like gel, no uniform contact of such gel with an alkaline material is obtained when mixing them by a common method, and hence it is hardly possible with such method to accomplish uniform hydrolysis of a desired degree.

As a means for hydrolyzing such hydrous polymer gel, there has been proposed a method of kneading the gel and an alkaline material by using a rotary screw extruder. This method is capable of accomplishing hydrolysis steadily and uniformly, with no risk of causing local hydrolysis, but it has a serious drawback in that there could occur shear degradation of molecular weight or quality deterioration of the polymer owing to the strong shearing force that develops in the presence of an alkaline material, resulting in the reduced commercial value of the products.

The process applies to the study aimed at overcoming such problems and it has been found that it is possible to accomplish hydrolysis of a desired degree, without causing any deterioration of the polymer quality or local hydrolysis, if the gel used for the hydrolysis is previously pulverized into particles and then mixed with an alkaline material by using a specific apparatus, and that the hydrolysis can be advanced favorably by subjecting the mixture to drying under heating immediately following the abovedescribed operation.

Process Conditions: The process resides in a method of producing the partially hydrolyzed acrylamide polymers characterized in that the finely divided particles of acrylamide polymers with water content of 65 to 90 wt % are mixed with an alkaline material under agitation in an apparatus equipped with a mixing mechanism comprising blades or flights having blank space such as ribbon blades (or flights), the apparatus being made of or coated with a synthetic resin at its surface contacted with the particles.

Figure 3.1: Ribbon Blade for Partially Hydrolyzed Acrylamide Polymers

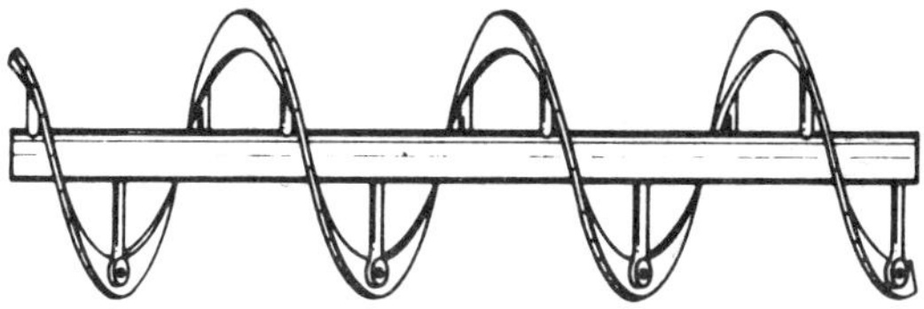

Source: U.S. Patent 4,146,690

Catalytic Hydration of Acrylonitrile to Form Acrylamide Aqueous Solutions

Highlights of the Technological Achievement: A process for preparing an aqueous solution of acrylamide, which comprises: (a) catalytically hydrating acrylonitrile in the presence of a metal-containing catalyst to form a mixed aqueous solution containing 15 to 35 wt % of acrylamide and 0.5 to 20 wt % of acrylonitrile; and (b) concentrating the mixed aqueous solution to a high concentration aqueous solution of acrylamide having an acrylonitrile concentration of 0.1 wt % or less using a centrifugal film evaporator in which a stream of mixed aqueous solution and a stream of evaporated vapor flow countercurrently to each other.

Reference: Y. Yamaguchi and K. Kohno; U.S. Patent 4,188,339; February 12, 1980; assigned to Nitto Chemical Industry Co., Ltd., Japan.

Background: Acrylamide is useful as a starting material for the production of polymers. For use as such a starting material, acrylamide must have the lowest possible acrylonitrile content, and the concentration of residual acrylonitrile should normally be less than 0.1 wt %.

For heat evaporation, a stirred tank-type evaporation still, a forced circulation-type evaporation still or a horizontal cocurrent centrifugal evaporator has previously been used. When such a heat evaporator is used, it is difficult to reduce the concentration of acrylonitrile remaining in the solution to 0.1% or less. It has been necessary, therefore, to reduce the acrylonitrile concentration by rectification.

A packed tower or a plate tower is used in the rectification section, and during rectification, the solution adheres to the inner wall of the rectifying tower, the inner wall of the packing, the undersurface of the trays, etc. Thus, these portions are not refreshed, and a polymerization of the adhering acrylamide may occur. The acrylamide polymer is water-soluble, and the inclusion of even a small amount of this polymer in the acrylamide aqueous solution decreases the quality. Furthermore, in order to prevent the polymerization of acrylamide during heat concentration, the concentrating operation must be performed at lower pressures and temperatures. Since a pressure drop occurs at the rectification section, the pressure at the condenser should be further reduced, making it difficult to condense the acrylonitrile.

Process Conditions: The conversion of acrylonitrile increases if the residence time in the reaction apparatus is prolonged. However, since with longer residence times, the amount of by-product ammonium acrylate increases, the conversion of acrylonitrile is preferably 20 to 95%, and up to 98% at the highest. The mixed solution obtained under these preferred conditions contains 15 to 35% acrylamide and 0.5 to 20% of acrylonitrile.

Examples of the metallic component of the metal-containing catalyst used in the hydration of the acrylonitrile are metals such as Cu, Ag, Zn, Cr, Mn, Fe, Co, Ni, Ru, Rh, Pd and Pt, and the oxides and the salts of these metals. Copper-containing catalysts, especially metallic copper, are preferred.

The hydration reaction is carried out at room temperature (20° to 30°C) to 300°C, especially 60° to 150°C. The residence time is adjusted such that the conversion of acrylonitrile becomes 20 to 98%. The solution so obtained is a mixed aqueous solution containing 15 to 35% of acrylamide and 0.5 to 20% acrylonitrile and is substantially free from impurities ascribable to side-reactions.

The mixed aqueous solution, either as such or after having been concentrated by flash evaporation is introduced into a countercurrent centrifugal film evaporator. The term countercurrent centrifugal film evaporator as used herein generically denotes a centrifugal film evaporator of the type in which a stream of a mixed aqueous solution and a stream of an evaporated vapor thereof flow countercurrently to each other. Where flash evaporation is used the mixed aqueous solution is introduced into the flash evaporator at a temperature of 60° to 300°C and removed as a more concentrated solution at a temperature of 35° to 70°C. The pressure generally employed in the flash evaporator is 50 to 760 mm Hg.

Figure 3.2: Centrifugal Film Type of Evaporators

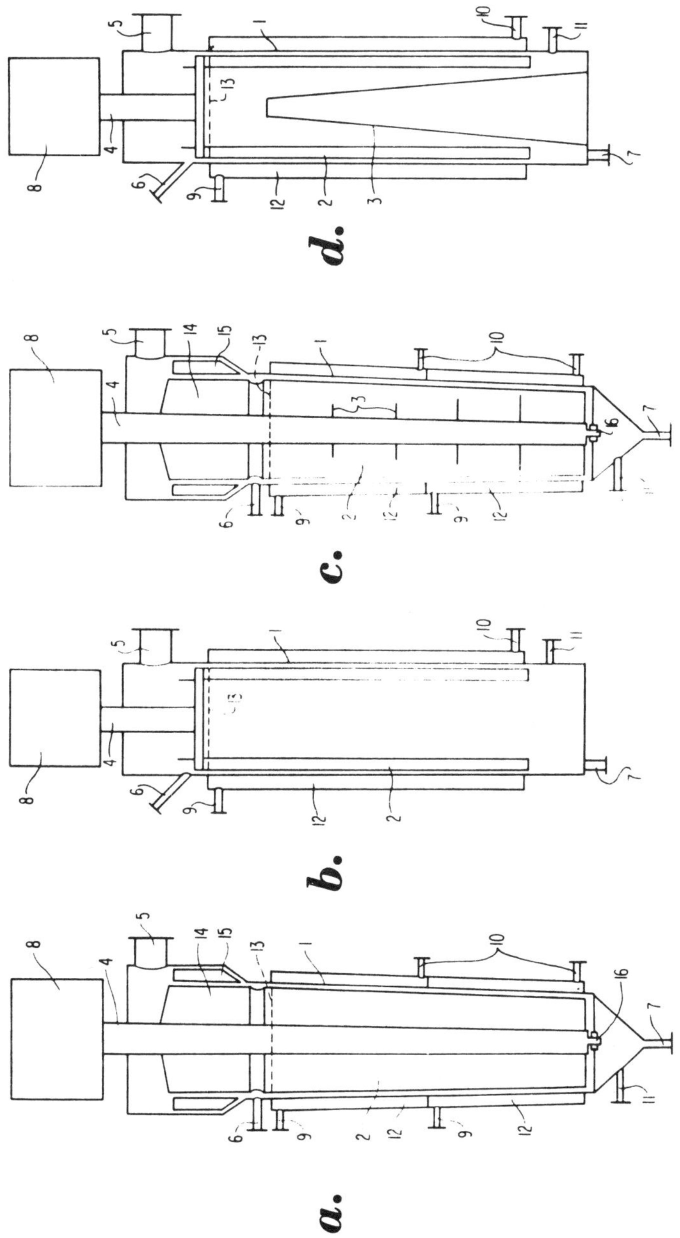

Source: U.S. Patent 4,188,339

Drawings: Examples of a centrifugal film type of evaporator are shown in Figure 3.2. The evaporator which contains restrictive members as shown in Figures 3.2c and 3.2d is more effective to achieve the object of this process as compared with the evaporator as shown in Figures 3.2a and 3.2b (which does not contain any restrictive members).

Preparation with Ammonium Type Monomers

Highlights of the Technological Achievement: In the production of a water-soluble, cationic high polymer of (A) at least one of acrylamide and methacrylamide; and (B) at least one of ammonium type monomers a process is described which comprises initiating the polymerization of the monomeric Component (A) with or without a portion of the monomeric Component (B) in an aqueous medium containing an organic solvent selected from the group consisting of acetone, acetonitrile, t-butanol, tetrahydrofuran and dioxane until the polymerization proceeds to a certain extent and then continuing polymerization while adding the remaining portion of the monomeric Component (B) under the occasional supplementation of a water-miscible organic solvent so as to keep an appropriate viscosity of the reaction system to produce the high polymer in a high yield with ease.

Reference: S. Kawakami, T. Shibata, S.-I. Isaoka and T. Shintani; U.S. Patent 4,138,446; February 6, 1979; assigned to Sumitomo Chemical Company, Limited, Japan.

Background: As dehydration-promoting agents for treatment of surplus sludge and digested sludge which are produced by activated sludge treatments of various wastewaters exhausted from excretion-treating works, drain-treating works and factories, there have hitherto been employed some inorganic materials such as ferrous sulfate, ferric chloride, quicklime and slaked lime. However, these inorganic materials must be used in large amounts so that the amount of sludge becomes increased to decrease the burning-up effect and to require large-sized apparatus.

Recently, the treatment of such sludge has become an important problem in the field of wastewater treatment, and there has been raised a demand for a dehydration-promoting agent which can be employed in a small amount and afford an excellent dehydrating effect without decrease of the burning-up effect.

For this purpose, it has been proposed to use, as a dehydration-promoting agent, organic high polymer coagulants, for example, cationic high polymer coagulants such as cationic modified polyacrylamides, cationic polycondensation products and polyamines, but sufficient effect has not yet been attained therewith.

Process Conditions: Polymerization of (A) at least one of acrylamide and methacrylamide and (B) at least one of ammonium type monomers of either one of the formulas:

$$(1)\quad CH_2{=}\overset{\displaystyle R_1}{\overset{|}{C}}{-}COY(CH_2)_nN^+\!\begin{matrix} \diagup R_2 \\ -R_3\cdot X^- \\ \diagdown R_4 \end{matrix} \qquad \text{and} \qquad (2)\quad CH_2{=}\overset{\displaystyle R_1}{\overset{|}{C}}{-}COY(CH_2)_nN^+\!\begin{matrix} \diagup R_2 \\ -R_3 \\ \diagdown R_5^- \end{matrix}$$

wherein R_1 is a hydrogen atom or a methyl group, R_2 and R_3 are each a lower alkyl group, R_4 ia a hydrogen atom, a lower alkyl group, an aryl group, a hydroxy(lower)alkyl group, a benzyl group or a group of the formula:

$$-CH_2COO(CH_2)_mCH_3$$

(m being an integer of 0 or 1), R_5^- is a group of either one of the formulas: $-(CH_2)_2COO^-$ and $-(CH_2)_3SO_3^-$, X is a halogen atom or an acid residue, Y is –O– or –NH– and n is an integer of 1 to 4, which comprises initiating the polymerization of the monomeric Component (A) with or without a portion of the monomeric Component (B) in an aqueous medium containing an organic solvent selected from the group consisting of acetone, acetonitrile, t-butanol, tetrahydrofuran and dioxane at a concentration of 15 to 70 wt % until the polymerization proceeds to a certain extent and then continuing the polymerization while adding the remaining portion of the monomeric Component (B) under the occasional supplementation of a water-miscible organic solvent thereto so as to keep an appropriate viscosity of the reaction system.

The monomer concentration at the beginning of the polymerization is desired to be from 10 to 50 wt % with respect to the reaction system. The reaction temperature may be usually not higher than about 55°C, preferably not lower than 5°C. With a lower monomer concentration and/or a lower reaction temperature, the reaction rate becomes extremely small, and a longer time is required for obtaining a viscosity of the reaction system sufficient for the addition of the ammonium type monomer(s), thus causing an extremely long duration of time for completion of the polymerization, which is a great disadvantage from the economical viewpoint.

The incorporation of a water-soluble, protective colloid into the reaction system at the beginning of the polymerization is usually preferred for preventing the sticking of the produced polymer on the wall of the reaction apparatus or the inner surface of the pipes. As the water-soluble protective colloid, there may be exemplified polyvinyl alcohol, polyvinyl pyrrolidone, carboxymethylcellulose, and hydroxyethylcellulose.

Reducing the Amount of Water-Insolubles

Highlights of the Technological Achievement: A method for reducing the amount of water-insolubles and free monomer in an aqueous acrylamide polymer gel whereby the gel is heated at from 80° to 150°C for at least 30 min while the water content of the gel is maintained substantially constant.

Reference: M.W.C. Coville; U.S. Patent 4,132,844; January 2, 1979; assigned to American Cyanamid Company.

Background: High molecular weight polymers of acrylamide have found use over the years as flocculants for settling industrial slurries and removing suspended matter from municipal or process water. To be as effective as possible, these polymers should be water-soluble and have a high molecular weight.

One major problem which has been plaguing the industry for many years is the difficulty with which recovery of these high molecular weight polymers in a dry state is achieved. They are normally prepared as very viscous, aqueous gels which must be dried before transportation of the polymers is economical.

Drying, however, usually tends to deleteriously alter the properties of the polymers to the extent that water-insolubles form therein and the molecular weight tends to decrease. The reduction in molecular weight is particularly bothersome because the effectiveness of these polymers depends proportionally upon the molecular weight.

Process Conditions: The acrylamide polymer gels, having been produced by any well-known polymerization procedure such as suspension, emulsion, precipitation procedures, etc. are treated by subjecting them to a temperature from 80° to 150°C for at least 30 min. It has generally been found that the higher the temperature of the heat treatment, the shorter the contact time so that at 150°C a contact time of 30 min would be sufficient while at 80°C a contact time of 4 hr would generally be used.

The heating step must be conducted under such conditions that substantially no drying whatsoever of the polymer gel occurs. If any substantial amount of drying is effected, it has been found that the undesirable production of large amounts of insoluble material results. Therefore, it is necessary to simultaneously maintain the water content of the gel undergoing heating at substantially the same concentration as that of the gel being charged.

Any effective means of maintaining the water concentration of the gel substantially uniform may be used. Satisfactory procedures which have proven useful include placing the gel in a saturated steam atmosphere such as by sealing the gel in a water impermeable plastic material and placing it in water or sealing the gel in a water impermeable plastic material and placing it in a circulating air oven or placing the gel in an air impermeable atmosphere such as a noncirculating air oven to which has been added a highly humid atmosphere.

As can be seen, many different techniques may be utilized to accomplish this desired result, the examples above constituting the most economical techniques. Other techniques could also be employed without departing from the scope of the process, the crux of the process residing in the simultaneous maintenance of the polymer gel water contained at a substantially constant level and the heating of the gel under the conditions specified above.

Aqueous Gels of Acrylamide Copolymers

Highlights of the Technological Achievement: Water-insoluble, tridimensional crosslinked random copolymers, which contain in copolymerized form the following.

(a) 25 to 98 wt % of N-[tris(hydroxymethyl)methyl] acrylamide or N-[tris(hydroxymethyl)methyl] methacrylamide or a mixture of these two compounds;

(b) 2 to 50 wt % of one or more monomers having several polymerizable ethylenic double bonds and one or more hydroxy groups, and free from NH_2 or COOH groups; and

(c) 0 to 50 wt % of one or more monomers having a polymerizable ethylenic double bond and one or more amino or carboxy groups. These copolymers may be used in the form of aqueous gels, as supports in techniques of separation such as gel permeation chromatography and in techniques of immobilization of natural substances.

Reference: E. Brown, E. Boschetti, M.M. Corgier and J. Touet; U.S. Patent 4,192,784; March 11, 1980; assigned to Mar-Pha Societe d'Etudes et d'Exploitation de Marques, France.

Background: Gel permeation chromatography is a current method of fractionation and separation of molecules according to their size. Molecules of a size greater than the largest pores of the gel are displaced by the mobile liquid phase, although molecules of a smaller size penetrate more or less deeply into the pores of the gel, which constitutes the stationary phase, and are then more or less retained. If then a liquid phase, in which a sample containing molecules of different sizes has been injected, is made to circulate on a column filled with gel, the speed of migration of these molecules towards the bottom of the column will vary from one molecule to another and the molecules will then be separated.

The principal hydrogels used at present as supports in the techniques described above are on the one hand gels of natural polysaccharides possibly crosslinked (agarose gels, dextran gels–Sephadex) and, on the other hand, gels of synthetic or semisynthetic polymers such as, for example, the gels of polyacrylamide or the polyacrylamide-agarose mixed gels. These supports are not free of defects. The gels of polyacrylamide or polyacrylamide-agarose are chemically unstable in a basic medium owing to hydrolysis of the amide groups into carboxylic acid groups, which limits their field of use.

Certain gels of dextran have moderate mechanical properties, and hence it is impossible to make columns thereof having a relatively high and constant rate of flow. Finally, the agarose gels have a poor resistance to heat and to agents for weakening the hydrogen linkages (urea, guanidine, etc.), and further, like all gels of natural origin, are very sensitive to attack by bacteria or certain enzymes.

Process Conditions: The monomers (b), which are cross-linking agents, correspond preferably to one of the general formulas (1) or (2) below:

(1)

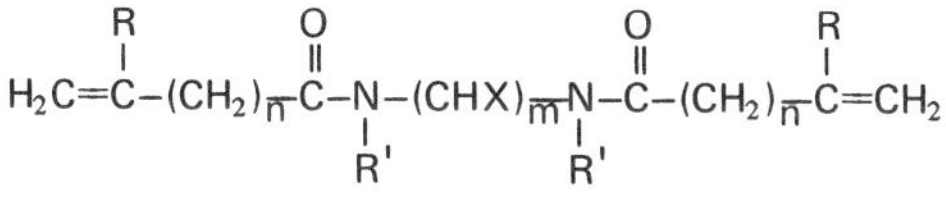

(2)

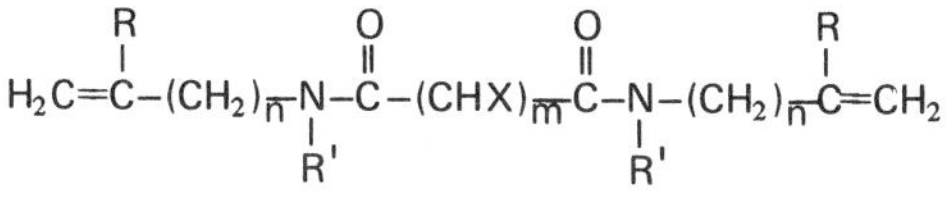

in which R is a hydrogen atom or a methyl group, R' is a hydrogen atom or a hydroxymethyl group; X is a hydrogen atom or an OH group, n and m are whole numbers from 0 to 6; with the restriction that X and R' cannot be simultaneously hydrogen atoms.

Examples of monomers (b) include especially N,N'-diallyltartardiamide, N,N'-methylene-bis-hydroxymethylacrylamide and glyoxal-bis-acrylamide (also called N,N'-dihydroxyethylene-bis-acrylamide).

Examples of compounds (c) are, among others, N-[(N-acryloyl)-glycyl] glycine, N-[(N-methacryloyl)-glycyl] glycine, and N-acryloyl-ϵ-aminocaproic acid, N-acryloyl glycine and acrylamido glycolic acid.

Although the process relates to the whole of the copolymers defined above, it is more particularly related to those of these copolymers which contain, in copolymerized form, 50 to 98 wt % of monomers (a), the remaining 50 to 2% being formed by one or more monomers (b), and most particularly to those which contain 50 to 98 wt % of N-[tris-(hydroxymethyl)-methyl] acrylamide, the complement to 100% being formed by one or more monomers (b).

The crosslinked hydrophilic copolymers according to this process can be prepared by free radical polymerization of the various monomers in aqueous solution. The polymerization is effected at a temperature from 0° to 100°C.

APPLICATIONS

Fracturing Subterranean Formations

Highlights of the Technological Achievement: A method and composition for fracturing a subterranean formation which comprises: (a) contacting a subterranean formation with an aqueous liquid comprising a propping agent and a composition containing water, acrylamide:methacrylate copolymer crosslinked with chromium(III) ion, in presence of carbonate ion, oxalate ion, and optionally, persulfate ion, (b) applying pressure to the liquid, (c) maintaining the pressure to fracture the formation, and (d) the gel breaking within 24 hr of gel formation.

Reference: H.S. Golinkin; U.S. Patent 4,137,182; January 30, 1979; assigned to Standard Oil Company (Indiana).

Background: The use of polymer thickened fluids and gels is widespread, particularly in the oil industry where such fluids are used for plugging or fracturing operations in oil wells and in flooding for secondary recovery. In many cases, the thickened polymer needs to continue to serve in a thickened state. In others, after the thickened polymer has served its intended purpose, it is often desirable to remove it from the well or formation being treated. To this end, it is necessary to be able to decrease the viscosity of the fluid to a more pumpable state.

Desirable properties of a hydraulic fracturing fluid include high viscosity, low fluid loss, low friction loss during pumping into the well, stability under the conditions of temperature, use and contaminants found, such as in high temperature deep wells, and ease of removal from the fracture and well after the operation is complete.

Process Conditions: The amount of copolymer used is sufficient to increase viscosity of the aqueous fluid as a gel to a certain initial amount. For purposes of this process, the term acrylamide:methacrylate copolymer is defined as a high molecular weight water-soluble polymeric salt of ammonia, sodium or potassium, of a molecular weight range of greater than 1,000,000 consisting essentially of 50 to 90 wt % acrylamide and from 10 to 50 wt % methacrylate as the sodium salt. The most preferred copolymer consists of 65 wt % acrylamide and 35 wt % sodium methacrylate. The defined acrylamide/sodium methacrylate copolymers are further characterized as having a molecular weight sufficient to provide a brine viscosity of at least 1.8 cp (measured in Cannon viscometer) when dissolved in a concentration of 500 ppm in an aqueous brine containing from 3,650 to 3,750 ppm sodium chloride and from 365 to 375 ppm calcium chloride made with deionized water.

The copolymers can be produced by any known method of conducting polymerization reactions provided substantially no cross-linking or formation of water-insoluble polymer occurs. Solution, suspension or emulsion techniques can be used. The physical form of the copolymer is not critical but it is used in accordance with conventional practices known in the art.

A preferred water-soluble trivalent chromium salt is chrome alum, chromium potassium sulfate, $KCr(SO_4)_2 \cdot 12H_2O$.

Carbonate ion can be incorporated into the aqueous gel as an aqueous solution of any carbonate compound in which the carbonate ion appears as $CO_3^=$ or HCO_3^-, and in which the cation of the carbonate compound does not exhibit a detrimental effect upon the aqueous gel.

Examples 1 through 5: The following examples show the result of increasing the amount of Na_2CO_3 beyond a maximum amount which in turn slowed the gelation rate. The amount of sodium carbonate is critical to the gelation rate as is evidenced in Examples 1 through 5. It is unknown whether the critical composition lies in the overall concentration of the carbonate ion or in the ratio of the carbonate ion to the chromium ion or polymer content.

A 1.0 N NaCl brine solution, 297 g, containing 0.4 wt % copolymer of the final total gel weight was treated with amounts of Na_2CO_3 ranging from 0.0033 to 0.17 wt % of total gel weight. Then 0.64 g of chrome alum in 3.2 ml of a 20% chrome alum solution, 0.21 wt % of total gel weight, were added. Viscosities were determined with a Fann viscometer. The results are shown in the following table.

	Example No.				
	1	2	3	4	5
Copolymer					
Weight, g	1.2	1.2	1.2	1.2	1.2
Percent gel weight	0.4	0.4	0.4	0.4	0.4
Na_2CO_3					
Weight, g	0.01	0.03	0.10	0.30	0.50
Percent gel weight	0.0033	0.010	0.033	0.10	0.17
pH before chrome alum	6.0-6.4	6.8	8.5-9.0	–	10.5
Chrome alum					
Weight, g	0.64	0.64	0.64	0.64	0.64
Percent gel weight	0.21	0.21	0.21	0.21	0.21
Gel viscosity, poises					
1 minute	0.38	0.42	5.9	1.3	0.59
2 minutes	0.36	3.8	8.04	8.04	0.58
3 minutes	0.42	8.04	–	–	0.55
4 minutes	3.8	–	–	–	0.52
5 minutes	4.4	–	–	–	0.52

(continued)

	Example No.				
	1	2	3	4	5
6 minutes	4.0	–	–	–	0.51
7 minutes	5.4	–	–	–	0.50
8 minutes	5.5	–	–	–	0.50
9 minutes	5.0	–	–	–	0.50
10 minutes	4.8	–	–	–	0.48

Firm Gels Within a Short Time at Ambient Temperature

Highlights of the Technological Achievement: Aqueous solutions of amide polymers, such as polyacrylamide when treated with small amounts of a water-soluble polyaldehyde and a hypohalite salt at an alkaline pH, react to form firm gels within a short time at ambient temperature. Such gels, which are stable under alkaline as well as acidic conditions, are usefully employed to plug porous subterranean formations, for grouting of leaking soil pipes or wells, and to otherwise render porous structures impermeable to the passage of liquids such as water.

Reference: R.J. Pilny and T.W. Regulski; U.S. Patent 4,199,625; April 22, 1980; assigned to The Dow Chemical Company.

Background: In view of the deficiencies of prior art methods for treating porous structures with polymeric materials to render them impermeable to passage of liquids such as water, it would be highly desirable to provide an improved method for so treating porous structures whereby the polymer to be employed is readily pumped into or otherwise incorporated in the porous structure to be rendered impermeable without polluting the porous structure and its surrounding environment with toxic substances.

Process Conditions: Gelation in accordance with this process can be obtained with amide polymers having such low molecular weights as to be characterized by a viscosity of only about 50 cp for an aqueous 40 wt % solution of the polymer. On the other hand, gelation can be obtained with amide polymers of very high molecular weight characterized, for example, by a viscosity of 60 cp for an aqueous 0.2 wt % solution.

In practice, to provide the firmness, abrasion resistance and structural stability desired in gels employed for plugging porous structures, for example, in and around sewer pipes, it is preferred to produce a gel containing from 5 to 10% or more of amide polymer solids. Thus, for ease of handling and placing in the porous structure, it is preferred to employ amide polymers characterized by viscosities of from 100 to 15,000 cp for an aqueous 20 wt % solution thereof.

To form the desired gel capable of rendering the porous structure impermeable, the amide polymer preferably contains from 50 to 100 mol % of one or more ethylenically unsaturated carboxamide monomers.

Raw Materials: In the gelable aqueous reaction mixture, the concentration of the amide polymer varies depending upon the molecular weight of the polymer and the firmness of the gel desired. If the amide polymer has a relatively low molecular weight, the concentration of amide polymer is advantageously from 0.25 to 30, preferably from 3 to 20, most preferably from 5 to 15 wt % based on the reaction mixture.

For the purposes of this process, an amide polymer has a relatively low molecular weight if a 40 wt % aqueous solution of the polymer exhibits a viscosity from 50 to 15,000 cp as determined by a Brookfield LVT viscometer (#2 spindle, 15 rpm, 25°C). It is understood, however, that somewhat lower concentrations than the aforementioned can be employed with higher molecular weight amide polymers.

The concentration of polyaldehyde in the gelable reaction mixture is at least an amount sufficient to cause the reaction mixture to form a water-insoluble, three dimensional gel within 5 min after the mixture is subjected to gelation conditions, up to the saturation concentration of the polyaldehyde in the reaction mixture. Advantageously, however, the gelable reaction mixture contains enough polyaldehyde to provide from 1 to 300 mmol of aldehyde moiety per mol of carboxamide moiety in the amide polymer.

When the polyaldehyde is glyoxal, the concentrations of glyoxal sufficient to provide the aforementioned mol ratios are within the range from 0.004 to 1.2 wt % based on the gelable reaction mixture. For sewer grouting and similar subterranean plugging applications, it is preferred to employ from 15 to 60 mmol of polyaldehyde per mol of carboxamide moiety in the amide polymer.

The concentration of the hypohalite salt in the gelable reaction mixture is an amount sufficient to provide the resulting gel with alkaline stability. A gel possesses the requisite alkaline stability if it does not dissolve in an aqueous medium having a pH of 8 in a period of at least a week after the gel is placed in the aqueous medium. Gels having preferred alkaline stability do not dissolve in an aqueous medium having a pH of 8 to 10 for a period of six months, those gels which are insoluble in 5 N NaOH for a week being most preferred. The concentration of hypohalite salt is in the range from 2 to 600 mmol of the hypohalite anion per mol of carboxamide moiety in the amide polymer.

In the following table alkaline stability is defined as follows. One volume of the gel is placed in 1.5 volumes of 5 N NaOH for 17 days and then observed and rated as follows: G-gel absorbs up to ~90% of NaOH but remains as a firm, insoluble gel; VG-gel absorbs up to ~75% NaOH but remains as a firm, insoluble gel; E-gel absorbs less than ~50% NaOH and remains as a firm, insoluble gel.

	. . . . Glyoxal.		. .Sodium Hypochlorite. .			
Sample No.	40% Solution (ml)	Glyoxal/ Amide (mmol/mol)	5.25% Active Solution (g)	Hypochlorite/ Amide (mmol/mol)	Gel Time (sec)	Alkaline Stability
1	0.16	9.9	6.4	40.5	20-22	G
2	0.16	9.9	9.6	60.7	22-25	VG
3	0.16	9.9	12.8	81.0	20-25	E
4	0.24	14.9	6.4	40.5	17-19	G
5	0.24	14.9	9.6	60.7	15-18	VG
6	0.24	14.9	12.8	81.0	18-20	E
7	0.32	19.8	6.4	40.5	18-20	G
8	0.32	19.8	9.6	60.7	15-19	VG
9	0.32	19.8	12.8	81.0	18-20	E

Aqueous Can Coatings of Improved Impermeability

Highlights of the Technological Achievement: There is disclosed a metal container containing a beverage or food having its internal surface coated with an adherent sanitary liner, the coating being a cured layer of a water-based coating composition consisting essentially of from 5 to 60 wt % of an amine-neutralized or partially neutralized interpolymer formed in the presence of a vinyl polymerization catalyst and in the absence of mercaptan, the interpolymer being comprised of 10 to 30% of acrylamide or methacrylamide in units of the structure:

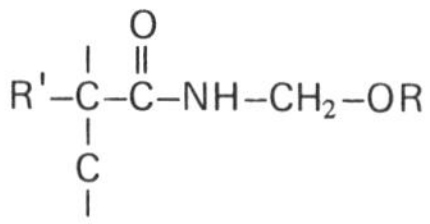

where R' is methyl or hydrogen and R is hydrogen or lower alkyl, from 5 to 15% of units formed by vinyl polymerization of monoethylenically unsaturated carboxylic acid; from 10 to 30% of units derived by vinyl polymerization of monovinyl aromatic monomer; from 10 to 30% of units formed by vinyl polymerization of flexibilizing monomer selected from lower alkyl acrylates such as ethyl acrylate; and from 20 to 45% of nitrile, such as acrylonitrile; wherein the interpolymer is solubilized by neutralizing acid groups of the interpolymer with a sufficient amount of monomeric amine to produce at least 0.200 meq of salt per gram of resin solids; and from 40 to 95 wt % of a liquid medium consisting of a mixture of water-soluble or water-miscible organic solvents and water, wherein at least 60 wt % of the mixture is water.

These cured coatings are characterized by improved impermeability to lubricants and flavor components and also better resist water at elevated temperature so as to also be useful as an exterior coating.

Reference: G.L. Brown and A.T. Spencer; U.S. Patent 4,195,006; March 25, 1980; assigned to Mobil Oil Corporation.

Background: An impermeable film is needed for the following two reasons. First, foods and beverages contain components which provide flavor, and it is desired that these and similar components not be extracted by contact with the film. Second, the metal surface which is coated frequently contains lubricants, and it is desired to prevent these lubricants from passing through the film to contaminate the food or beverage and spoiling the flavor. It is the capacity of agents to move into and through the film which must be measured.

Since beverages are primarily contemplated, especially beer, it is convenient to measure the impermeability of the film by its capacity to pick up and remove the n-heptyl ester of p-hydroxybenzoic acid which is an oily material commonly used to inhibit microbial growth in packaged beer. This agent which is commercially available as Staypro WS-7 is frequently extracted from beer when an off taste is encountered and it is concluded that coatings which extract Staypro will also extract flavor-producing constituents and allow oily lubricants to migrate.

Process Conditions: The development of water-based coating compositions for use as internal sanitary liners for metal containers has received considerable attention, and one system for providing such coating compositions is set forth in U.S. Patent 3,991,216.

The styrene-ethyl acrylate-isobutoxymethyl acrylamide-methacrylic acid interpolymer described in U.S. Patent 3,991,216 has a Staypro rating of 3.7. In this process a Staypro rating of at least 6 is desired, preferably at least 9.

In accordance with this process, the monomer mixtures contemplated by U.S. Patent 3,991,216 are modified by the addition of 20 to 45% of a monoethylenic nitrile selected from acrylonitrile and methacrylonitrile. Acrylonitrile is preferred. Comparative data which shows the new result is set forth in Table 1, the wedge bend rating being of significance included in the table, where minimum acceptable rating is 6 on a 1 to 10 scale, and 10 is perfect.

Table 1

Monomer (wt %)					Staypro	Wedge Bend
AN	STY	EA	IBMAM	MAA	Rating	Rating
0	35	35	20	10	3.2	5
10	30	30	20	10	5.2	7
20	25	25	20	10	8.6	6
30	20	20	20	10	10.9	5
40	15	15	20	10	12.0	4
0	45	25	20	10	2.7	6.5
30.5	20.3	20.3	20.3	8.5*	10.2	6.5
30	20	23**	20	7.0*	4.1	5.0

*Acrylic acid was used.
**n-Butyl methacrylate was used.

In the above table, AN is acrylonitrile, STY is styrene, EA is ethyl acrylate, IBMAM is isobutoxymethyl acrylamide and MAA is methacrylic acid. With the data of Table 1 in mind, this process uses an interpolymer of monomer components as tabulated in Table 2.

Table 2

Monoethylenic Monomer Type	Illustration	Proportions (%)	Preferred Proportions (%)
Nitrile	Acrylonitrile*, methacrylonitrile	20-45	25-40
Vinyl aromatic	Styrene*, p-methyl styrene, vinyl toluene	10-30	15-25
Flexibilizing C_{1-4} acrylate	Methyl, ethyl*, propyl or butyl acrylate	10-30	15-25
N-methylol amide	N-methylol acrylamide, acrylamide or ether thereof such as the propyl, butyl, or isobutyl* ether	10-30	15-25
Carboxy-functional	Acrylic acid*, methacrylic acid, crotonic acid, fumaric acid, itaconic acid	5-15	7-12

*Materials which are preferred.

Dye-Containing Polymer Composition

Highlights of the Technological Achievement: The preparation of dye-containing compositions, in which the dye is copolymerized with an acrylic based monomer, is disclosed. A hydrophilic copolymer is formed in which the dye is present in a nondiffusible form. The dye-acrylic polymer composition can be produced in the form of microbeads by emulsion polymerization. In addition, light scattering particles may be added in the form of a material such as polystyrene microspheres.

Reference: J.I. Peterson; U.S. Patent 4,194,877; March 25, 1980; assigned to United States of America.

Background: Previous methods for the bonding of dyes with other molecules include such known procedures as the use of a Mannich-type reaction, i.e., formaldehyde condensation of the dye with an amine group, as in preparation of the complexones.

In U. S. Patent 4,017,476, a polymer of a very specific type is formed in the usual way from a monomer, but the polymer forming reaction, i.e., formation of molecular chains, is not ended in the usual way, but rather the polymer is left with reactive sites, i.e., reactive ends of the chains, which will react with a subsequently added dye molecule to combine the dye chemically with the polymer. This prior art method is distinguished from this process in which the dye is included in the monomer solution so that the dye molecule is essentially another monomer which becomes a part of the chain during polymerization.

Thus, in effect, this prior art patent covers methods of treating a preformed polymer in such a manner that dyes combine with it. Such a prior art method is not a copolymerization with the dye as is the method of this process.

Raw Materials: The acrylic based monomer which is employed in this process may be, for example, acrylamide, methyl methacrylate or hydroxymethyl methacrylate (Hydron). Other suitable acrylic based monomers may be used which, like those monomers specifically mentioned, will copolymerize with the dye to provide a hydrophilic polymer.

The copolymerization of the dye with the acrylic based monomer may be carried out under the conditions which are generally employed in the polymerization of such acrylic based monomers alone, with the feature of this process residing in the simple inclusion of a soluble dye of the right type in the polymerization solution with the acrylic based monomer. Thus, for example, in the case where the acrylic based monomer is acrylamide, the procedure employed is the common method of using a persulfate and an amine (tetramethylethylenediamine).

Process Conditions: These dye-containing copolymer compositions can be produced in the form of microbeads by the use of known procedures of emulsion (water-in-oil) polymerization. As an additional aspect of the process, light scattering materials such as, for example, polystyrene microspheres or other similar material may be added to the dye-containing copolymer. Such light scattering materials are particularly desirable when the copolymer composition is to be employed in a fiber optic pH probe.

Example: *Method of Making Dye Polyacrylamide Microspheres* – Dissolve 10 mg of phenol red in 2.0 ml of 6 M acrylamide (containing 0.06M N,N'-methylene bisacrylamide). Dissolve about 50 mg ammonium persulfate in this solution; add 0.5 g of emulsifier mixture (Span 80:Tween 80, 85:15 pbw). Then add 10 ml of toluene and shake well in a test tube. Bubble nitrogen through the solution for 5 min, add 4 drops of tetramethylethylenediamine (TMED), shake well, continue slow bubbling of nitrogen for 15 min. The heat of reaction should be evident after about 1 min following addition of TMED. Wash the resulting microspheres in a centrifuge as follows:

(1) Three times with ethanol:water (1:1 ratio), with sodium hydroxide solution added as needed to make the dye basic (purple color);

(2) Once with acetic acid solution added to make the dye acidic (red color); and

(3) Twice with ethanol and filter by suction until dry.

Elastic Tubes from Hydrogels

Highlights of the Technological Achievement: There is disclosed a method of producing hydrogel tubes comprising partially hydrolyzing an acrylonitrile-containing polymer in an inorganic acidic solvent solution at temperatures of −10° to 50°C, extruding the hydrolyzed polymer solution through a circular nozzle into an aqueous coagulating bath, the same or similar coagulating bath simultaneously being fed through a central tube or pin of the nozzle.

Preferably, the surface of the tube is made very slippery in wet condition, by treating it with chemical agents causing either a further hydrolysis or other reaction leading to carboxylic, sulfonic or sulfuric groups, or an esterification with polyols such as glycerol, glycols or polyvinyl alcohol, preferably with simultaneous sulfonation or sulfuration.

Reference: O. Wichterle and A. Stoy; U.S. Patent 4,183,884; January 15, 1980; assigned to Ceskoslovenska akademie ved, Czechoslovakia.

Background: Previously, elastic tubes used in surgery as, for example, sounding tubes, catheters, cannulae and the like have been made from various hydrophobic and impervious materials which materials have physiologically disadvantageous properties; exemplary of the same is a product made from rubber or polyvinyl chloride which has been plasticized.

They possess sufficient strength and flexibility, on the one hand, however, when introduced into the body over a long period of application they cause painful and slow healing injuries, hemorrhages and irritation, all of which are hardly bearable by patients, particularly infants. Inflammations often occur, causing reduction of the diameter of the tubular organ after healing, to such a degree that further surgery is often necessary afterwards. It has been suggested to remove the shortcomings by coating rubber tubes such as catheters with a thin layer of a sparingly crosslinked hydrogel produced from glycol methacrylate. Thereby, the introduction thereof into the body is reputably made easier and it is supposedly possible to impregnate the hydrogel layer with drugs but only in small amounts due to the physical dimensions of the layer (i.e., thickness).

Further shortcomings of such gel coatings are comparatively low strength and poor adhesion to the tube body, making for necessary intermediate layers. These and other shortcomings are reduced by this process and furthermore, the amount of absorbed drugs is increased and a difficulty encountered at the introduction of the tubes, caused by excessive friction in contact with mucous membranes is also reduced.

Process Conditions: A method of producing hydrogel tubes comprising partially hydrolyzing an acrylonitrile-containing polymer in an inorganic acidic solvent solution at temperatures of -10° to 50°C, preferably from 0° to 25°C, extruding the hydrolyzed polymer through a circular nozzle into an aqueous coagulating bath, the same or similar coagulating bath simultaneously being fed into the lumen of the formed tube through a central tube or pin of the nozzle.

The hydrolyzed viscous polyacrylonitrile solution is extruded. The hydrolysis is carried out until the polyacrylonitrile solution coagulates in aqueous liquids to a rubbery hydrogel, absorbing from 30 to 75 wt % water at swelling equilibrium. The time needed, therefore, is determined by many variables, such as, temperature, pH value, solution employed, concentration of the latter and the like. The extrusion may be carried out through a circular nozzle into a coagulating bath.

The extrusion, furthermore, may be carried out either directly or after coagulating the copolymer and redissolving it in organic solvent such as wet dimethyl sulfoxide, wet dimethyl formamide or wet cyclic ethylene carbonate, and then passing it through a circular nozzle into the coagulating bath, the same or similar coagulating bath simultaneously being fed through a central tube or pin of the nozzle.

Raw Materials: Treatment with a hot mixture of concentrated sulfuric acid with a lower alkyl polyhydric alcohol, i.e., either glycerol or ethylene, diethylene, or triethylene glycols or polyvinyl alcohol is preferred. Hydrogels containing from 30 to 75 wt % of water may be easily treated with the reagent, notwithstanding the initial thought that water will suppress, by its diluting effect, the hydrolytic and esterifying power of sulfuric acid, thereby thus making the building of glycerol or glycol esters impossible.

Moreover, sulfuric acid alone dissolves the same hydrogel, even if diluted to 70 to 90% with water. Glycerol alone removes water from the hydrogel rapidly so that the latter hardens, particularly at increased temperatures. It previously was not known apparently that the conflicting forces would mutually compensate for each other so that a thin, smooth and slippery layer would be formed on the water-swollen hydrogel by a treatment lasting only several minutes, such that the layer is inseparably connected with the tube and, moreover, the stability of the hydrogel against deterioration is increased by crosslinking.

The thin layer not only does not hinder the diffusion of drugs from the hydrogel itself, on the contrary, it increases the diffusion rate and simultaneously enhances the sorption capacity for drugs having comparatively large molecules which would penetrate only with some difficulty or not at all into the untreated hydrogel. Moreover, the surface treatment raises the concentration of negatively charged groups, whereby the possibility of blood coagulation is lowered. This is of importance particularly in the application for blood-vessel cannulae.

The above treatment with sulfuric acid contemplates using the same at a concentration of preferably 90 to 97%, at a temperature of preferably 20° to 100°C. The ratio of polyhydric alcohol to acid is preferably 4:1.

The time period for the treatment in the acid-alcohol mixture to take place can vary from 1 sec to 60 min, advantageously 20 sec to 30 min.

Washable Nonsplatter Ink

Highlights of the Technological Achievement: A washable nonsplatter ink for use in oscillographic type pen recorders is disclosed. The ink is characterized by being stable at room temperature for extended periods of time, thus not requiring refrigeration for the maintenance of stability. The ink contains a high molecular weight polyacrylamide rather than polyethylene oxide as a pituity agent to stop splatter of the ink at fast writing speeds. The polyacrylamide allows the use of washable acid dyes. The inclusion of an unsaturated alcohol, such as butyne diol acts to retard the oxidation of the pituity, thus stabilizing the pituity against aging degradation.

Reference: W.C. Corwin; U.S. Patent 4,163,738; August 7, 1979; assigned to Beckman Instruments, Inc.

Background: While the inclusion of high molecular weight polyethylene oxide into ink has bestowed the desired benefits of nonsplattering and nonskipping desired, it is not without attendant drawbacks. Under normal circumstances, other additives to the ink, including the dye thereof, have a tendency to oxidize the polyethylene oxide thus destroying its long chain molecules and attendant filament forming qualities. This is particularly noticeable with the sulfone groups of an acid dye.

The known formulations of nonsplatter ink incorporating polyethylene oxide require dyes with amine, $-NH_2$ or dimethyl amine, $-N(CH_3)_2$, groups or basic dyes such as basic violet 2, basic violet 3, or basic green 4, in order not to oxidize the polyethylene oxide. The basic dyes do not exhibit the tendency to rapidly oxidize the polyethylene oxides as do the acid dyes.

On the other hand, the basic dyes stain the users hands and the instrument badly while the acid dyes are water soluble and stain very little. Thus, to the user, inks incorporating acid dyes are preferred for cleanliness. Additionally, it is often desirable to have a single dye component black ink which is available only as an acid dye.

With the use of a pituity agent that is easily oxidizable, a washable ink that is stable at room temperature for more than approximately two months is not possible. The ink can be refrigerated to stop the deterioration process, but such refrigeration facilities are often not available at the sites wherein the recorders are desired to be incorporated.

Process Conditions: The object of this process has been met by the ink of this method wherein a polyacrylamide pituity agent has been added in place of polyethylene oxide. In a preferred embodiment, polyacrylimide is used as a pituity agent along with an acid dye, a bacterial agent incorporated in a water based solvent, and 2-butyne-1,4-diol is used as an antioxidant.

A typical batch of the preferred embodiment of the ink of this process is prepared according to the following table.

Component	Quantity	Function
Water	50 ml	Solvent (100 ml total)
Ethylene glycol methyl ether	4 ml	
Ethylene glycol	45 ml	
Glycerin	1 ml	
Polyacrylamide	10 mg	Pituity agent
Formaldehyde-producing preservative	20 mg	Antibacterial agent
2-Butyne-1,4-diol	20 mg	Antioxidant
Acid Black 2	7 g	Dye

In the ink according to the foregoing composition, the Acid Black 2 (or alternately, nigrosine) dye is soluble to 0.12 g of dye per milliliter of solvent in water, 0.07 g/ml in ethylene glycol methyl ether, and 0.035 g/ml in ethylene glycol. Water alone as a solvent has too high a surface tension and evaporates too quickly to make a good ink solvent.

Ethylene glycol methyl ether and ethylene glycol are used to reduce the surface tension. Glycerin, ethylene glycol, Polyglycol 200 or a solvent with a very low vapor pressure and properties as a humectant is needed to prevent rapid clogging at the pen tip or in the polyvinyl chloride tube from the pen to ink bottle.

The polyacrylamide, preferably Separan AP273 by Dow Chemical Company, is used as a pituity agent to stop splatter of the ink at fast writing speeds. A formaldehyde producing preservative, preferably Dowacil 75 by Dow Chemical Company, is used to keep bacteria from consuming the pituity agent.

An antioxidant 2-butyne-1,4-diol is also added to retard oxidation of the pituity agent. The result is an ink with a pituity of 2±1 second, a viscosity of 10±5 cp as measured with a Brookfield LVF viscosimeter operating at 60 rpm with a No. 1 spindle, and a pH equal to 10±1.

ACRYLIC ACID AND METHACRYLIC ACID POLYMERS

PROCESSING

Preparing a Stable Dispersion in an Aqueous Medium

Highlights of the Technological Achievement: A method of preparing a stable dispersion in an aqueous medium of particles of polymer in which stabilization of the particles against flocculation is effected by a steric mechanism.

Reference: A.J. Backhouse; U.S. Patent 4,209,435; June 24, 1980.

Background: It is a matter of practical interest to be able to prepare aqueous polymer dispersions by a route which is equally applicable to water-insoluble and water-soluble monomers and which enables charge-stabilization of the particles of the disperse phase to be supplanted either completely or substantially by a stabilization mechanism not requiring the presence of electrically charged species.

Process Conditions: A method of preparing a stable dispersion in an aqueous medium of particles of polymer is provided, the method comprising the steps of:

(1) forming a stable dispersion of particles of a first polymer in a nonaqueous liquid in the presence of a steric stabilizer for the particles, the first polymer being insoluble both in the nonaqueous liquid and in water;

(2) polymerizing in the same nonaqueous liquid, in the presence of the particles of the first polymer and of a steric stabilizer, one or more monomers capable of giving rise to a second polymer which is soluble in the chosen aqueous medium;

(3) transferring the resulting polymer particles from the non-aqueous medium to the aqueous medium.

In those cases where the monomers being polymerized in step (1) are capable of forming a crosslinked polymer, it does not necessarily follow that the whole of the first polymer so produced is crosslinked; since the crosslinking monomer will normally be present in a minor proportion of the total monomers, reaction statistics will result in some noncrosslinked polymer being produced. The mode of stabilization of the particles of the first polymer in the nonaqueous liquid is, however, the same whether a particle consists entirely of crosslinked polymer or partly of crosslinked polymer and partly of noncrosslinked polymer.

In addition to the requirement that the polymer of the dispersion formed in step (1) should be insoluble in the nonaqueous liquid, it is also necessary that this polymer should be insoluble in water. This further condition may be satisfied in either of two ways. Firstly, the chemical composition of the polymer may be such that it is inherently insoluble in water, which is the case with the polymers derived from the acrylic and methacrylic ester monomers.

Secondly and alternatively, the polymer may be one which is inherently soluble in water but is rendered insoluble by being crosslinked. For example, a copolymer of one of the ester monomers just referred to with a substantial proportion of a hydrophilic monomer, such as a monoester of acrylic acid or methacrylic acid with a polyglycol (e.g., polyethylene glycol), or vinyl pyrrolidone, may be actually soluble in water, but by also including a minor proportion of a polyfunctional monomer such as ethylene glycol dimethacrylate the copolymer may be caused to be lightly crosslinked and it will then be insoluble in water, as indeed in nonaqueous liquids also.

The steric stabilizer employed in step (2) of the process will normally be the same as that employed in step (1), and this second step is most straightforwardly carried out simply by continuing the dispersion polymerization process with the appropriate monomer or mixture of monomers immediately after completion of the polymerization of the monomer or monomers of step (1).

A principal advantage of the process is that the monomers which are polymerized to form the water-insoluble first polymer, of which the disperse particles chiefly consist, may include water-soluble monomers such as monoesters of acrylic acid or methacrylic acid with a polyglycol, or vinyl pyrrolidone, without incurring difficulties which are usually encountered when employing such monomers in an emulsion polymerization procedure. Thus it is possible according to the process to obtain aqueous dispersions which are valuable as additives to other water-based film-forming systems by reason of the fact that the polymer particles are swollen by the aqueous medium.

This imparts rheological properties to the dispersions, such as high degree of thixotropy, which make them useful, for example, in the control of flow of aqueous paints to be applied by spray or by brush. The method of the process enables the degree of swelling of the particles by the aqueous medium and the molecular weight of the second polymer formed in step (2) to be controlled so as to secure the optimum effect in any given system.

Acrylic Acid Containing Little or No Low Boiling Aldehydes

Highlights of the Technological Achievement: A process for purifying crude acrylic acid characterized in that high-purity purified acrylic acid containing little or no low boiling aldehydes is obtained by rectifying a crude acrylic acid consisting essentially of acrylic acid and containing low boiling substances.

Reference: T. Ohrui, M. Kato, M. Ayano, T. Kawaguchi and T. Abe; U.S. Patent 4,199,410; April 22, 1980; assigned to Sumitomo Chemical Company, Limited, Japan.

Background: Hitherto, acrylic acid has been produced by the Reppe process or hydrolysis of acrylonitrile. Recently, however, production of acrylic acid by direct oxidation of propylene was developed, and this process is mainly employed. Since, however, various kinds of low boiling or high boiling aldehydes are produced as by-products in the course of the oxidation of propylene, the crude acrylic acid produced by this process contains, even though purified, various kinds of aldehydes of considerably high concentration as compared with the purified acrylic acid resulting from the aforementioned two conventional processes, particularly hydrolysis of acrylonitrile.

Consequently, the purified acrylic acid resulting from oxidation of propylene is generally inferior to the purified acrylic acid prepared by the hydrolysis of acrylonitrile in polymerization performances, and the latter purified acrylic acid has been mainly used as a material for high polymerization.

These low boiling aldehydes may be considered as easily separable by the conventional rectification processes, since they have a sufficiently large relative volatility as compared with acrylic acid. The rectification carried out by the process, however, showed that removing the aldehydes to less than 100 ppm by rectification alone was very difficult. The content of the low boiling aldehydes in acrylic acid for high polymerization is generally less than 10 ppm, preferably less than 5 ppm. Consequently, the process does not effect the production of purified acrylic acid for high polymerization by mere rectification of the crude acrylic acid resulting from the direct oxidation of propylene.

Figure 4.1 shows one example for carrying out the purification of crude acrylic acid according to the process.

Figure 4.1: Purification of Crude Acrylic Acid

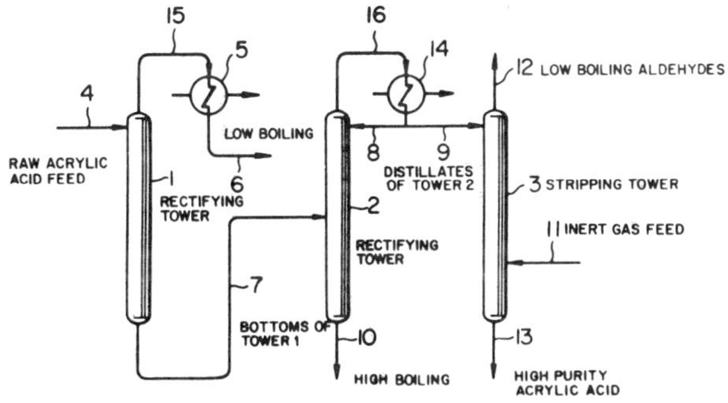

Source: U.S. Patent 4,199,410

In the drawing, **1** is a rectifying tower for separation of low boiling substances, **2** is a rectifying tower for separation of high boiling substances, **3** is a stripping tower for removal of low boiling aldehydes, **5** and **14** are condensers, and **4, 6, 7, 8, 9, 10, 11, 12, 13, 15** and **16** are liquid or gas lines.

Raw Materials: The crude acrylic acid is one which is obtained by the direct oxidation of propylene, and it contains low boiling compounds (e.g., acetaldehyde, acrolein, water, acetic acid) and/or high boiling compounds (e.g., maleic acid, phenol, benzaldehyde, acrylic acid dimer) as impurities. The content of these impurities is not constant since it varies with the condition of oxidation of propylene. But, the process is not affected by the variation of the content of the impurities.

Process Conditions: Separation of the low boiling substances and/or high boiling substances by rectification is satisfactorily carried out by separating firstly the former substances and then the latter ones. The separation may be carried out by means of any form of rectifying tower and in any form of operation. For example, the rectifying tower may be one in which the substances can be withdrawn as a side cut, or composite towers comprising a plural number of rectifying towers, and the operation may be carried out in a continuous form or batchwise form.

The operating pressure in the rectifying tower is less than 400 mm Hg, preferably less than 100 mm Hg, and in this case combined use of a suitable polymerization inhibitor is of course necessary since the acrylic acid to be treated is a substance having a property to violently polymerize by heat.

Both the low boiling fraction and high boiling residue produced as by-products in the distillation contain a considerable amount of acrylic acid, and therefore, they can be used as a material for production of acrylic acid esters such as methyl acrylate.

The acrylic acid fraction obtained by distillation is subjected to stripping with an inert gas such as nitrogen or air for removal of the low boiling aldehydes. The stripping is carried out at preferably 20° to 50°C.

The amount of the inert gas used for stripping the low boiling aldehydes is properly determined from the economical viewpoint of the stripping tower: always, as the amount increases, the tower height necessary to obtain a required rate of stripping becomes small, but at the same time, the tower diameter becomes large. Too large amounts of the inert gas are, however, disadvantageous, since the amount of acrylic acid accompanying the inert gas, i.e., the loss of the acid, becomes large. Accordingly, it is desirable that the molar ratio of liquid to gas in the stripping tower is selected within the range of 2 to 20.

Polyelectrolyte Hydrogels

Highlights of the Technological Achievement: A predetermined mass comprising an acrylate salt and acrylamide is exposed to a controlled intensity and dose of ionizing radiation effecting simultaneous crosslinking and polymerization thereof to form a polyelectrolyte hydrogel. The resulting hydrogel is an insoluble hydrophilic copolymer which can contain or, when dried, absorb large quantities of aqueous fluids.

Reference: J.A. Ward; U.S. Patent 4,192,727; March 11, 1980; assigned to Union Carbide Corporation.

Background: This process is based upon the discovery that hydrophilic copolymers which are water-insoluble and possess superior absorption properties can be prepared in a single step by subjecting a predetermined mass comprising an aqueous solution of a mixture of monomers to a controlled dose and controlled intensity of ionizing radiation. This mixture of monomers comprises acrylamide and at least one acrylate salt.

The method of forming an insoluble polyelectrolyte polymer of this process comprises: (a) forming a predetermined mass comprised of a mixture of an acrylate salt and acrylamide in a ratio of acrylate salt to acrylamide between 0.15:1 and 20:1, the predetermined mass having a pH in solution between 6 and 12; and (b) exposing the predetermined mass to a dose rate up to 10 Mrad per second or at a dose rate varying in intensity between a first level and a second level for a time period sufficient to form a water-insoluble hydrogel which is substantially free of unreacted monomer.

Raw Materials: The polyelectrolyte polymer comprises a crosslinked random copolymer comprised of an acrylate salt and acrylamide wherein the crosslink is a direct covalent bond joining carbon atoms of separate polymer chains and there are no chemical groups such as crosslinking agents between such separate polymer chains. The copolymer comprises acrylate units having the formula:

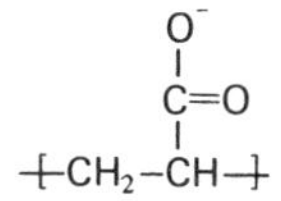

and acrylamide units having the formula:

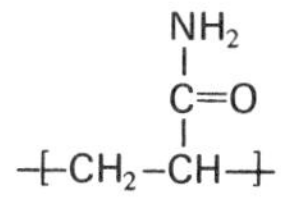

The acrylate units and acrylamide units are present in the copolymer in a random fashion. The ratio of acrylate units to acrylamide units in the copolymer should be up to about 20:1. Preferred ratios depend upon the end application of the hydrogels. Where firmer gels are desired, ratios of acrylate units to acrylamide units should be most preferably between 0.4 and 0.6:1. Where higher capacities for fluids are desired, ratios of acrylate units to acrylamide units can be as low as 1:1 and as high as 20:1. Preferably, they are as low as 5:1 and as high as 11:1, and most preferably are about 10:1.

The crosslinked random copolymer comprised of an acrylate salt and acrylamide of this process differs structurally from a copolymer of acrylate salt and acrylamide produced by chemical crosslinking. For example, the crosslinks of the copolymer of this process can be illustrated by the following unit:

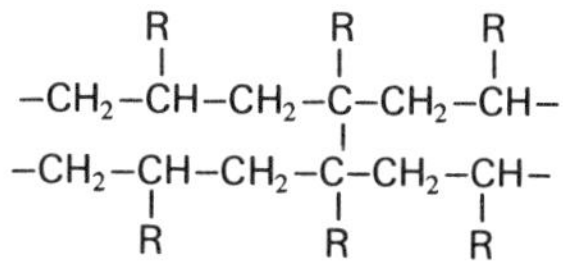

wherein R represents an amide or carboxyl group.

Thickening Agents: It has been found that high molecular weight thickening agents such as polyacrylamide, polyethylene oxide and hydroxyethylcellulose provide excellent viscosity building characteristics for the aqueous solutions made according to the methods of this process. Lower molecular weight versions of these three thickening agents may also be employed, but in higher concentrations to achieve comparable viscosities for the thickened solution. Moreover, Gelgard particles, XD-1300 (Gelgard is a crosslinked, partially hydrolyzed polyacrylamide, 40% hydrolyzed and sized finer than 100 mesh) and ground up fine particles of the polyelectrolyte polymer made according to the methods of this process (finer than 60 mesh) also provide excellent viscosity building characteristics for the aqueous solutions made according to the methods of this process.

In general, high molecular weight versions of water-soluble polymers that are capable of increasing the viscosity of an aqueous solution to above about 5,000 cp would be suitable as thickening agents for the methods of this process. Illustrative of water-soluble polymers which are believed to be suitable thickening agents are the following: sodium carboxymethylcellulose (cellulose gum); partially hydrolyzed polyacrylamide; polyacrylic acid; polyvinyl alcohol; naturally occurring water-soluble polymers such as guar gum and starch; polyvinylpyrrolidone; xanthan gum; polyethyleneimine; polyethylene sulfonate; polystyrene sulfonate; hydroxypropylcellulose; and methylcellulose.

Effectiveness of Gels as Diaper Absorbents: One of the possible uses of these polyelectrolyte hydrogels is as absorbent in diapers. In many cases, the ultimate equilibrium capacity of a gel is not a good measure of its performance in diapers.

The table below shows that the more highly crosslinked gels absorb as much or more saline solution in these tests as do gels having higher capacities. The saline solution used was 0.3 N NaoH and capacity equaled weight of fluid absorbed divided by weight of gel.

Absorption of Acrylic Gels* in Simulated Diaper Tests

Dose Rate (Mrad/sec)	.... Capacity in Diapers.		
	1 Mrad	2 Mrad	3 Mrad
12	ND	ND	7.2
1.2	9.7	16.7	19.5
0.3	19.4	19.2	19.9
0.12	20.1	23.0	23.5
0.03	24.6	21.9	20.4

Note: ND is not determined.

*32.3% sodium acrylate, 8.7% acrylamide by weight at pH 10.7.

Viscosity of Thickened Monomer Solutions as a Function of Concentration of Polyacrylamide or Hydroxyethylcellulose: 6 ℓ (6,600 g) of a stock solution consisting of 19 weight percent potassium acrylate and 35 weight percent acrylamide (hereinafter referred to as "the 19-35 stock solution") were made. To six separate 880 g portions of this solution, high molecular weight polyacrylamide (American Cyanamid P-250) was added in amounts sufficient to provide six solutions whose polyacrylamide concentrations were 1, 2, 3, 4, 5, and 6 weight percent respectively.

Similarly, 3 additional liters of the 19-35 stock solution were prepared. To four separate 500 g portions of this solution, high molecular weight hydroxyethylcellulose (Cellosize QP, 100 M) was added in amounts sufficient to provide four solutions whose hydroxyethylcellulose (hereinafter referred to as HEC) concentrations were 1, 2, 3 and 4 weight percent, respectively.

Viscosity of the ten thickened solutions was measured with a Brookfield viscometer. The results are summarized in the table below. It is noted that HEC thickened these monomer solutions considerably better than it thickens water. This additional viscosity building characteristic of HEC was unexpected. Also, smaller concentrations of HEC (compared with polyacrylamide) provided thickened solutions with larger viscosities.

The table summarizes the results of the viscosity tests hereinbelow.

Viscosity of Monomer Solutions* Thickened with Polyacrylamide

Concentration (wt %)	.........Viscosity (cp).........	
	Polyacrylamide	HEC
1	175	60,600
2	1,800	824,000
3	30,000	2,550,000
4	77,000	5,720,000
5	356,000	ND
6	610,000	ND

Note: ND is not determined.

*19-35 stock solution.

Water-Soluble Acrylic Polymers by Photopolymerization

Highlights of the Technological Achievement: Water-soluble acrylic polymers are prepared by a sequential photopolymerization process. A monomer solution is deposited on a support which is subsequently irradiated to effect polymerization. Photopolymerization promoters are incorporated in the monomer solution to facilitate polymerization.

Reference: J. Boutin and J. Neel; U.S. Patent 4,178,221; December 11, 1979; assigned to Rhone-Poulenc Industries, France.

Background: There remains a need in the art for an economically advantageous means of preparing high molecular weight water-soluble acrylic polymers lacking insoluble fractions and unpolymerized monomer. This need is especially acute in the field of water purification and treatment where regulations exist which limit the quantity of residual monomer which may be present in the polymer system. For example, the FDA has established 0.05% as the maximum residual monomer concentration.

Process Conditions: The process comprises:

(a) continuously depositing on a support a thin layer of an aqueous solution comprising acrylic monomer or monomers and preferably 0.01 to 0.05%, by weight relative to the monomer or monomers of a photopolymerization promotor, the solution containing preferably less than 0.5 mg of dissolved oxygen per liter of solution;

(b) subjecting the thin liquid layer on the support to radiation of wavelengths preferably between 330 and 400 mμ, in a gaseous atmosphere having an oxygen concentration preferably less than 0.5%, for preferably 3 to 7 minutes, wherein the average power of the active radiation is between 20 and 300 W/m^2, while cooling the support to remove the heat of reaction;

(c) maintaining the cooled support in an atmosphere substantially free of oxygen as in step (b) while subjecting it to radiation of wavelengths between 300 and 450 mμ for preferably 3 to 10 minutes, wherein the average power of the active radiation is greater than 300 but less than 2,000 W/m^2, and

(d) detaching the resultant, partially solidified thin layer from the support and subjecting it to radiation of wavelengths preferably between 330 and 400 mμ for preferably 40 to 90 minutes, in an atmosphere containing preferably cool air, wherein the average of the active radiation is between 20 and 500 W/m^2.

The abovedescribed process produces the acrylic polymer or copolymer directly in the form of a relatively thin layer or film.

Thermosetting Acrylic Copolymer Compositions

Highlights of the Technological Achievement: Thermosetting acrylic copolymer compositions useful in coating applications are disclosed. The acrylic polymer has a weight average molecular weight of 15,000 to 100,000 and contains from 2 to 20 percent by weight of an alkyl ester of an unsaturated carboxylic acid containing from 14 to 20 carbon atoms in the alkyl group, for example, stearyl methacrylate. The compositions are particularly useful as water-based exterior coatings for automobiles and trucks.

Reference: S.K. Das and C.M. Kania; U.S. Patent 4,177,178; December 4, 1979; assigned to PPG Industries, Inc.

Background: Coating compositions based on acrylic polymers are well known in the art. The coatings are hard, curable and have excellent humidity resistance making them particularly desirable as exterior or topcoats for the automotive (including truck) market. Both thermoplastic and thermosetting acrylic polymers are employed for this use. However, for the automotive market, the acrylic polymers are organic solvent-based and present environmental problems with regard to solvent emissions. These problems can be overcome by using water-based acrylic polymers; but, until this process, water-based acrylics have not been considered acceptable for automotive topcoat applications. It had been thought that water-based acrylic polymers had poor rheological properties and had insufficient gloss and humidity resistance for automotive topcoat use.

It has been found that certain water-based acrylic polymers of a specific molecular weight range and prepared with a specific class of higher alkyl esters of alpha, beta-ethylenically unsaturated carboxylic acids can be used in aqueous-based thermosetting compositions for automotive topcoats. The resultant coating compositions have good rheological properties and sufficient gloss and humidity resistance for automotive usage. The thermosetting coating compositions of the process can be applied as aqueous-based compositions at high solids contents with good metallic pigment pattern control, gloss and distinctness of image (DOI).

Process Conditions: The process provides for an improved thermosetting composition comprising an acrylic polymer containing active hydrogens selected from the class consisting of hydroxyl, primary amine, secondary amine, amido, thiol and carboxylic acid including mixed groups and a curing agent stable in the presence of the acrylic polymer at room temperature (i.e., 20° to 25°C) but reactive with the active hydrogens of the acrylic polymer at an elevated temperature to form a cured product. The improvement of the process resides in using as the acrylic polymer one which has a weight average molecular weight of 15,000 to 100,000, preferably 15,000 to 60,000, and which is prepared from copolymerizing with one or more different acrylic monomers 2 to 20 percent by weight of an alkyl ester of a copolymerizable alpha,beta-ethylenically unsaturated carboxylic acid containing from 14 to 20 carbon atoms in the alkyl group, preferably stearyl methacrylate.

Aqueous Emulsions of Acrylate Copolymers in the Presence of Degraded Starch

Highlights of the Technological Achievement: Nonblocking coatings are disclosed, which coatings comprise an acrylate ester emulsion polymer prepared by polymerization in the presence of a degraded starch which, in a 10% solution in 1 N NaOH, has a viscosity not greater than 15 mPa-s at a velocity gradient $D = 532.1\ s^{-1}$.

Reference: W. Elser, K. Huebner, H. Ottofrickenstein and H. Boessler; U.S. Patent 4,171,407; October 16, 1979; assigned to Rohm GmbH, Germany.

Background: The blocking point, that is, the temperature at which the coatings become tacky, can be raised by using monomers such as styrene, acrylonitrile, or methyl methacrylate in preparing the emulsion polymer to impart hardness to the polymer. However, this measure simultaneously increases the minimum film-forming temperature (MFT), so that the dispersions do not dry to a clear coherent film at room temperature.

Thus, the problem is posed of increasing the temperature differences between the blocking point and the minimum film-forming temperature. The blocking point is to be understood as the highest temperature at which two coating layers, each 30 μ thick and applied to paper, can be laid one upon the other for 60 minutes under a load of 50 g/cm^2 and can then subsequently again be separated without damage to the surface. The MFT is that limiting temperature above which a synthetic resin dispersion forms a crack-free film on drying under specified conditions.

The possibilities known for increasing the temperature separation between the blocking point and the MFT are not satisfactory. Crosslinking of the synthetic resin subsequent to film formation can only be carried out if the film on the coated substrate can be heated to temperatures above 100°C, which is often not the case. Further, tackiness of the film is not completely overcome by crosslinking. Volatile plasticizers have also already been combined, prior to film formation, with dispersions of synthetic resins having a sufficiently high blocking point in order to decrease the MFT into the room-temperature region. This process cannot be used if the film is to be nonblocking already shortly after drying or if the coated objects are to be used in interior spaces in which the air would be contaminated by the evaporating plasticizer.

Process Conditions: A method has been found for the preparation of nonblocking coatings in which an aqueous agent is applied to the substrate to be coated, which agent contains a protective colloid and, as a coating material or bonding agent, an emulsion polymer having a minimum film-forming temperature not above 50°C and which comprises at least 80 percent by weight of alkyl esters of acrylic acid, alone or in combination with alkyl methacrylates and/or styrene, of which at least 30 percent by weight are acrylate esters having 1 to 8 carbon atoms in the alkyl portion thereof.

The balance of up to 69.5 percent by weight of this component, to the extent it is not all an acrylate ester or esters having 1 to 8 carbon atoms in the alcohol portion, comprises acrylate esters of longer-chain alcohols, alkyl methacrylates, and/or styrene. The emulsion polymer further comprises from 0.5 to 5 percent by weight of an unsaturated polymerizable carboxylic acid, which may be present in the form of its salt.

Raw Materials: To produce a large effect, for example, increases in viscosity, thixotropy, structural viscosity, and the like, as a rule the highest possible molecular weight protective colloids must be used. The protective colloids in these cases are added after the conclusion of polymerization. Common protective colloids are polyvinyl alcohol, polyvinylpyrrolidone, polyacrylic acid salts, or cellulose ethers, for example.

Natural products have also already been used as protective colloids, for example, casein, alginates, dextrins, or starch which has been made water-soluble by enzymatic or hydrolytic treatment. Paint dispersions comprising vinyl acetate have also been prepared in the presence of amylopectin. Amylopectin is a very high molecular weight, strongly-branched component of natural starch. Synthetic resin dispersions have also been prepared from vinyl esters in the presence of water-soluble starch derivatives. Branched water-soluble starch, such as amylopectin, is used together with dextrins.

The heart of the process involves the recognition that certain specific starch degradation products have the property of clearly increasing the blocking point of films of acrylate dispersions without simultaneously increasing the MFT. However, this effect only occurs if the starch derivatives in question are present before or during the emulsion polymerization. If they are first added after the polymerization process is concluded, an increase in the blocking point does not occur. The effect is, further, associated with certain viscosity properties in the starch derivatives. Namely, their viscosity must not exceed 15 mPa-s measured in the form of a 10% solution in 1 N sodium hydroxide at a velocity gradient $D = 532.1\ s^{-1}$.

Preparing an Aqueous Dispersion from an Acrylic-Derived Polymer

Highlights of the Technological Achievement: A process for producing a resin dispersion characterized in that an ethylenically unsaturated monomer capable of forming a polymer insoluble in an aqueous medium is polymerized in a solution formed by dissolving in the aqueous medium a water-soluble polymer derived from:

(1) 10 to 90% by weight of a hydroxyalkyl (meth)acrylate expressed by the following general formula:

$$CH_2{=}\overset{\displaystyle R}{\overset{|}{C}}{-}COO{-}C_xH_{2x}{-}OH$$

wherein R is a hydrogen atom or a methyl group, and x is an integer of 2, 3, or 4;

(2) 10 to 90% by weight of a polyalkylene glycol mono(meth)acrylate expressed by the following general formula:

$$CH_2{=}\overset{\displaystyle R}{\overset{|}{C}}{-}COO{-}(C_mH_{2m}O)_n{-}OH$$

wherein R is a hydrogen atom or a methyl group, m is an integer of 2 or 3, and n is an integer of 2 to 20, and

(3) 0 to 30% by weight of another ethylenically unsaturated monomer.

Reference: K. Mori, H. Tanaka, H. Ozawa and Y. Murakami; U.S. Patent 4,170,582; October 9, 1979; assigned to Dainippon Ink and Chemicals, Inc., Japan.

Background: It is well known that the dispersion polymerization of an ethylenically unsaturated monomer capable of forming a water-insoluble polymer using an ordinary emulsifier or a protective colloid agent as a dispersant provides a resin dispersion in which a polymer is dispersed in water. Though the resulting composition has gained acceptance for application for a paint, it has poor stability and in addition, its cured coating has low water resistance on account of the nature of the dispersant used for the polymerization.

To cope with this problem, a process has been proposed which uses a water-soluble polymer obtained by polymerizing a hydroxyalkyl (meth)acrylate as a dispersant (Japanese Patent Publication No. 41457/71). According to this process, however, it is not possible to increase nonvolatile components of the resulting resin dispersion up to about 50% by weight because the polymer used as the dispersant has low water-solubility, resulting thereby in the formation of a coagulated precipitate during the reaction. Hence, the process provides only such a resin dispersion having at most about 25% by weight of the nonvolatile components. In addition, this resin dispersion of a low solid content has inferior stability and can not be optionally diluted with water.

Raw Materials: The water-soluble polymer to be used in the process is soluble in an aqueous medium used for the contemplated resin dispersion of the process and functions as a dispersant and at the same time, as a film-forming component in combination with the resulting polymer. The number average molecular weight of the water-soluble polymer (by gel permeation chromatography) generally ranges preferably from 2,000 to 20,000. As to water-solubility of this polymer, a 50 weight percent solution of the polymer in the aqueous medium has generally a water tolerance (at 25°C) of preferably 1,000% or more.

Process Conditions: The water-soluble polymer is obtained by polymerizing these monomers (1), (2) and (3) in accordance with the known solution polymerization method. A water-soluble organic solvent or an aqueous medium may be used as a solvent for the polymerization. The polymerization is preferably performed at a temperature of 200°C or below, preferably for 2 to 15 hours. Examples of polymerization initiators are oil-soluble or water-soluble initiators such as azobisisobutyronitrile, benzoyl peroxide, di-tert-butyl peroxide, cumene hydroperoxide, hydrogen peroxide and ammonium persulfate.

The suitable amount of the polymerization initiator is from 0.1 to 10% by weight of the resulting polymer by adding to the polymerization system during the polymerization such a compound as n-octyl mercaptan, n-dodecyl mercaptan, thioglycollic acid, 2-ethylhexyl thioglycolate, allyl alcohol, methyl methacrylate dimer, or α-methylstyrene dimer.

Photopolymerization of an Alkaline Metal Salt of Acrylic Acid

Highlights of the Technological Achievement: Highly water absorbent films and fibers are prepared by photopolymerizing an alkaline metal salt of acrylic acid, at least one alkyl acrylate or methacrylate wherein the alkyl group contains 10 to 30 carbon atoms, at least one alkyl acrylate or methacrylate wherein the alkyl group contains 1 to 4 carbon atoms and a photoinitiator. The resulting interpolymers are particularly useful in the preparation of highly absorbent personal hygiene and care products.

Reference: P.J. George; U.S. Patent 4,167,464; September 11, 1979; assigned to The B.F. Goodrich Company.

Background: U.S. Patent 4,062,817 discloses polymers of unsaturated copolymerizable carboxylic acids, at least one alkyl acrylate or methacrylate wherein the alkyl group has 10 to 30 carbon atoms and another alkyl acrylate or methacrylate wherein the alkyl group has 1 to 8 carbons. This composition alleviated many of the deficiencies of the earlier compositions. Further improvements in the hydrophilic properties were obtained by compositions disclosed in U.S. Patent 4,066,583. This patent discloses a composition comprising: (1) a copolymer of the type disclosed in U.S. Patent 4,062,817, except that after copolymerization 30 to 90% of the carboxylic groups were neutralized with an alkali metal or ammonia and (2) an aliphatic glycol, a plasticizer which is important in facilitating extrusion of the polymer.

Process Conditions: This process is directed to a photopolymerized interpolymer which has outstanding absorption and retention properties of water and ionic solutions such as urine or blood. The interpolymer is prepared from a monomer mixture comprising:

(a) 50 to 90 weight percent of acrylic acid, most preferably 80 to 100% of the carboxylic groups having been neutralized with an alkali metal hydroxide or ammonia base prior to polymerization,

(b) 2 to 25 weight percent of a higher acrylic ester monomer of the formula:

$$(1) \qquad CH_2{=}\overset{\overset{\displaystyle R'}{|}}{C}{-}\overset{\overset{\displaystyle O}{\|}}{C}{-}OR$$

wherein R' is hydrogen, methyl or ethyl and R is an alkyl group of 10 to 30 carbon atoms,

(c) 5 to 30 weight percent of a lower acrylic ester monomer of the formula:

$$CH_2{=}\overset{\overset{R'}{|}}{C}{-}\overset{\overset{O}{\|}}{C}{-}OR'' \qquad (2)$$

wherein R'' is a lower alkyl group having 1 to 8 carbon atoms, 0 to 50 percent of the lower acrylic ester being replaced by acrylic or methacrylic nitrile or amide, and

(d) 0.01 to 5 weight percent of a photoinitiator.

The higher acrylic ester monomers are those which have a long chain aliphatic group and may be represented by the formula (1) wherein R is an alkyl group having from 10 to 30 carbon atoms, preferably 10 to 18 carbon atoms and R' is hydrogen, methyl or ethyl group.

The lower acrylic esters can be represented by the formula (1) wherein R is an alkyl, alkoxy, haloalkyl, cyanoalkyl, hydroxyalkyl and like groups having from 1 to 8 carbon atoms and R' is hydrogen, methyl or ethyl group. Preferably R' is hydrogen or methyl and R is alkyl, most preferably methyl, and the lower ester is present in the amount of from 5 to 20 and most preferably 7 to 17 weight percent.

Fiber Preparation: The aqueous monomer dispersion is thickened to the desired degree with a nonreactive thickening agent such as a cellulose derivative as, for example, hydroxypropylcellulose, high molecular weight polyvinylpyrrolidone and the like; natural gums such as guar gum, locust bean gum, gum tragacanth; agar, naturally occuring hydrocolloids such as alginates and the like. Fibers are then spun from a spinneret in a regular manner and immediately exposed to UV irradiation.

Examples 1 through 10: To further illustrate the process the following examples are presented in table form below. Copolymers and films were prepared from the ingredients listed.

	Example No.					
	1	2	3	4	5	6
Sodium hydroxide, g	43.6	43.6	46.1	43.3	34.6	32.5
Neutralization, %	95	95	100	100	75	75
Water, g	152.0	152.0	160.0	159.0	149.0	148.0
Acrylic acid, g	83.0	78.0	83.0	78.0	83.0	78.0
Lauryl methacrylate, g	7.0	7.0	7.0	7.0	7.0	7.0
Methyl methacrylate, g	10.0	15.0	10.0	15.0	10.0	15.0
Aerosol A 102*, g	1.24	1.24	1.24	1.24	1.19	1.18
Irgacure 651**, g	0.62	0.62	0.62	0.62	0.60	0.59

*Disodium ethoxylated alcohol half ester of sulfosuccinic acid, a dispersant.
**2,2-Dimethoxy-2-phenylacetophenone, a photoinitiator.

	Example No. 7	8	9	10	11	12
Sodium hydroxide, g	43.6	43.6	–	–	43.6	46.1
Ammonium hydroxide, g	–	–	67.6	63.5	–	–
Neutralization, %	95	95	100	100	95	100
Water, g	152.0	152.0	27.2	29.2	152.0	160.0
Acrylic acid, g	83.0	83.0	83.0	78.0	83.0	83.0
Lauryl methacrylate, g	7.0	–	–	–	7.0	*
Stearyl methacrylate, g	–	7.0	7.0	7.0	–	–
Methyl methacrylate, g	10.0	10.0	10.0	15.0	**	***
Aerosol A102, g	1.24	1.24	1.24	1.24	1.24	1.24
Irgacure 651, g	–	–	0.62	0.62	0.62	0.62
Photoinitiator, g	1.0†	0.83††	–	–	–	–

*Tridecyl acrylate (7.0 g).
**2-Hydroxyethylmethacrylate (10.0 g).
***n-Butyl acrylate (10.0 g).
†N,N'-bis(diethylamino)benzophenone (0.5 g) and benzophenone (0.5 g).
††Benzoin ethyl ether.

Forming Stable Thixotropic Gels

Highlights of the Technological Achievement: A stable, compatible thixotropic gel of the type gelled from a mixture including a solvent, a cosolvent and a gelling agent. The gelling agent comprises a copolymer of ethylene and at least one comonomer selected from the group consisting of acrylic acid, methacrylic acid, ethacrylic acid, esterified acrylic acid, vinyl acetate, vinyl formate, vinyl propionate and mixtures thereof. The copolymer has a number average molecular weight between about 500 and about 10,000.

Reference: I.B. Chang; U.S. Patent 4,200,561; April 29, 1980; assigned to Allied Chemical Corporation.

Background: Gels are used for a variety of products having cosmetic, medicinal, veterinary, paint, printing and other applications. In many such applications, thixotropic gels are preferred, having the property of being semisolid colloids (gels) when undisturbed, but flowing readily when shaken, stirred, sheared or otherwise subjected to stress. Such gels typically are prepared by mixing a solvent, cosolvent, gelling agent (also called gellant) and optionally additional oils and waxes, active ingredients, preservatives, pigments or other coloring matter and other additives under conditions in which the mixture is either a liquid solution or a suspension.

In the production of such gels, particularly in cosmetic gels such as lipsticks and deodorant and antiperspirant sticks, the gelling agent is frequently carnauba wax, candelilla wax, beeswax or other organically derived waxes which contain a mixture of paraffins, fatty esters, fatty alcohols and sometimes fatty acids. Such gelling agents are often inconsistent from one batch to the next, requiring extra care in formulating cosmetic products. Such waxes also have relatively low melting points such that gels having such waxes become liquid and flow when warmed to 50°C or above. Often, gellants such as paraffin, insect and microcrystalline waxes suffer from the same defects as do the synthetic organophilic bentonite gellants, such as those which are conventionally used by the cosmetic industry.

Thus a need exists for a gelling agent of consistent quality which can form stable, compatible thixotropic gels with common solvents, cosolvents and other additives of the type now used in thixotropic gels.

Process Conditions: The process includes a stable, compatible thixotropic gel having in the oil phase between 5 and 60 wt % gellant, between 5 and 50 wt % of a solvent incompatible at room temperature with the gellant and between 20 and 90 wt % of a cosolvent compatible with both the gellant and the solvent, at least 25 wt % of the gellant being a copolymer having between 40 and 96 mol % ethylene monomer and between 4 and 60 mol % of comonomer selected from the group consisting of acrylic acid, methacrylic acid, ethacrylic acid, esterified acrylic acid, vinyl acetate, vinyl formate, vinyl propionate and mixtures thereof. The copolymer has a number average molecular weight between 500 and 10,000, the copolymer being present in a proportion great enough to maintain the gel composition stable from 10° to 50°C.

Example: The three ingredients shown in Figure 4.2 were heated and mixed in a glass beaker until a homogeneous solution resulted. The mixture was then cooled and a sample of the cooled product was then placed between microscope slides and held up to a light. A clear film was an indication of compatibility. Excessive cloudiness or graininess was taken as an indication that the mixture was incompatible.

Figure 4.2: Stable Thixotropic Gel

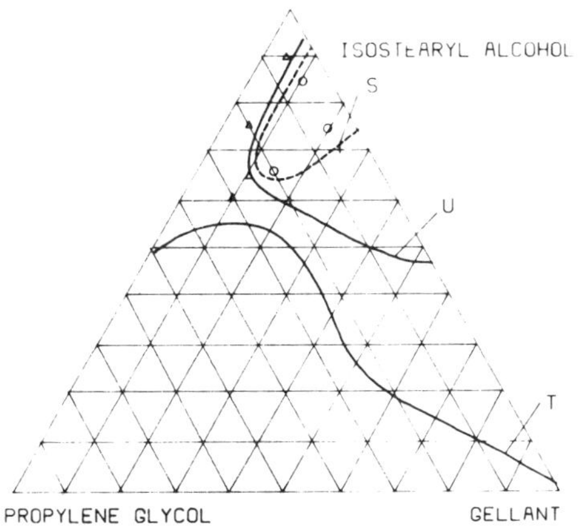

Source: U.S. Patent 4,200,561

Several points were selected on the ternary diagram designed to delineate the line between compatible and incompatible mixtures. Another group of points was selected to delineate the line between "interesting" gels of the type that are stable and spread easily and other gels which are either too thick or too thin. This determination was somewhat subjective, but corresponds in general to the property of not separating when held at 50°C for one or more days.

Following the procedure described above, 9 gels were prepared with propylene glycol as solvent, isostearyl alcohol as cosolvent and, as gellant, an ethylene acrylic acid copolymer having an acid number of 40, a softening point of 108°C, a Brookfield viscosity at 140°C of about 500 cp, and a number average molecular weight about 3,000. The region bounded by line **S** was determined to contain the "interesting" gels and the region below line **T** was determined to contain the incompatible mixtures. Line **U** has been drawn to further include gels marginally either too thin or too thick.

APPLICATIONS

Water or Alcohol-Soluble Printing Ink Composition

Highlights of the Technological Achievement: A water-soluble or alcohol-soluble printing ink composition comprising:

(A) 10 to 50% by weight of a resin, the resin being prepared by radical-polymerizing a mixture of (a) an unsaturated aromatic hydrocarbon, (b) a monoacrylate or monomethacrylate of a polyhydric alcohol and (c) an α,β-unsaturated carboxylic acid in the presence of an organic solvent, the resin having an acid value of 5 to 150 and a hydroxyl value of 30 to 120;

(B) 30 to 80% by weight of a solvent; and

(C) 1 to 40% by weight of a pigment or dye.

Reference: H. Sato; U.S. Patent 4,177,076; December 4, 1979; assigned to Nippon Oil Company, Ltd., Japan.

Background: Efforts have been made to develop water and/or alcohol-soluble resins, since they may be used for preparing printing inks such as water-soluble gravure inks or flexo inks or coating compositions such as water or alcohol-soluble paints, which are advantageous in preventing pollution problems for the reason that they cause lesser air pollution as compared to the inks and paints containing hydrocarbon solvents such as toluene.

The conventionally known water-soluble resins include, for example, maleic acid resins modified with rosin having high acid values and shellac, the waterproof and antifriction properties of which are, however, not satisfactory. Besides they are disadvantageous in that the costs and supply thereof are unstable since they are made from natural resources. Synthetic resins such as copolymers of styrene and acrylic acid have been developed. However, they have disadvantages in that the stability, antifriction property and waterproof property thereof are unsatisfactory when used for inks because of their high acid values.

While maleic acid resins having high acid values, nitrocellulose and polyamide resins have been known as the conventional alcohol-soluble resins, they are not sufficiently satisfactory because of the disadvantages in that their pigment dispersion properties are poor so that they are not conveniently used for high speed printing.

Process Conditions: A process for preparing a resin is provided which comprises radical-polymerizing a mixture of (a) an unsaturated aromatic hydrocarbon, (b) a monoacrylate or monomethacrylate of a polyhydric alcohol and (c) an α,β-unsaturated carboxylic acid in the presence of an organic solvent to prepare a water and/or alcohol-soluble resin having an acid value of 5 to 150 and a hydroxyl value of 30 to 120. There is also provided a coating composition containing the above resin.

Water-Based Coatings from Acrylic Copolymers

Highlights of the Technological Achievement: An aqueous copolymer coating system, adapted for forming abrasion-resistant coatings for application to wood, plastic, metal, paper and other substrates using either printing or coating processes; the copolymer employed in the coating system includes the following:

(a) a four component (monomer) copolymer comprising (1) an alkyl methacrylate, (2) an alkyl acrylate, (3) an unsaturated nitrile, (4) unsaturated organic acid, this copolymer adapted for use in an aqueous system;

(b) at least one coalescing agent for the ether-alcohol type;

(c) at least one surfactant agent which may be either of the substituted polyether nonionic type or one of a group of selected anionic agents; and

(d) one or more additives as may be necessary to impart color, control foam, and prevent mildew and/or fungus attack.

Reference: J.M. Keeler; U.S. Patent 4,172,064; October 23, 1979; assigned to Claremont Polychemical Corporation.

Background: The use of water-based coatings by industry is becoming increasingly desirable, since applications in which water-based coatings are employed minimize the emission of hydrocarbon or solvent materials during drying and other processes. This has numerous advantages, both in minimizing the economic penalty of lost materials and also in ecological (air and water pollution) advantages. In many jurisdictions, the emission of volatile organic products is limited by law, so that users of such solvents are faced with the cost of installation of recovery equipment or of incinerator devices. The use of solvents based on hydrocarbons may become an economic luxury that cannot be afforded in the future. In addition, the potential fire hazard of flammable solvent systems is well known to the industry.

Previously, if a hard tough polymer film based upon an acrylic ester polymer was desired, the normal procedure was to proceed by way of the thermosetting acrylic route. In this approach, a reactive monomer system was copolymerized to form an acrylic copolymer, as a latex. This material was then mixed with a water-soluble curing agent, so that after the initial film drying process, a second-stage baking step at elevated temperature would be conducted to give the ultimate physical properties to the system. This two-stage process was time-consuming, expensive as to both energy and equipment, and could not be adapted for a substrate which could not endure the thermal cure. For this reason, the materials of the process, which give good physical properties without requiring an extended thermal cure represent an advance in the state of the art.

Process Conditions: It has been discovered that superior film-forming compositions which are tough and durable films at ambient temperatures can be formulated using certain selected mixtures containing at least one component from each of the following three groups:

(1) acrylate-methacrylate copolymers which are present as aqueous dispersions;

(2) coalescing agents to promote film formation during drying; and

(3) surfactants to assist the wetting of substrate and printing or coating devices which are used in the water-based coating systems.

The compositions of the process are compositions of matter which are capable of forming coating films having superior properties when coated on a substrate. The specific areas of the process are believed to include at least the three categories below:

(1) the copolymers, formed by free radical copolymerization of several monomeric materials in combination, the copolymers so formed having unique properties in their performance. Among these unique properties are film-forming capability at temperatures less than 80°C which makes them useful under drying conditions practical with water-based systems; satisfactory self-bonding film-forming properties such that the resulting films have abrasion-resistance to physical abrasion both when they are dry and when exposed to either organic solvents or commonly used aqueous detergent systems; good adhesion properties for substrates onto which they would ordinarily be printed or coated;

(2) the dispersions, in aqueous suspension, of the polymer backbone system so prepared; and

(3) the formulated aqueous coating compositions formed by the dispersion of the polymer, in combination with coalescing agents, cosolvents, surfactants and auxiliary materials, including pigments that are capable of forming abrasion-resistant and detergent-resistant coatings on nonporous substrates at ambient temperatures, i.e., no high temperature thermal curing required.

Dyeing and Printing a Textile Substrate

Highlights of the Technological Achievement: Disclosed is a process for dyeing or printing a textile substrate employing an aqueous paste comprising dyestuff and a synthetic thickener, which process comprises the step of applying an electrolyte to the substrate prior to or after application of the aqueous paste, the paste being free from added electrolyte.

Reference: P. Koltai and G. Robert; U.S. Patent 4,181,498; January 1, 1980; assigned to Sandoz Ltd., Switzerland.

Background: The process is suitable for the dyeing or printing of a wide variety of textile substrates; it has particular application in the dyeing and printing of pile and tufted fabrics and floor coverings, particularly pile and tufted carpets.

The aqueous paste comprising dyestuff and a synthetic thickener, hereinafter called "the dye paste," may be applied to the substrate in conventional manner, e.g., over the whole surface thereof or over discrete areas thereof to obtain a desired pattern, e.g., using screen, rotary film, rotary relief, spray (e.g., "militron" technique) and drop (e.g., "Tak" technique) methods. Indeed, a combination of methods may be employed, for example, by applying a first color dye paste over the entire surface, followed shortly thereafter by one or more different colored further dye pastes over selected, optionally overlapping areas of the substrate, to obtain multicolor effects.

Process Conditions: The synthetic thickening agents used in the process are those which are sensitive to, and lose their thickening properties on contact with, electrolytes. Indeed, however, all the synthetic thickening agents currently available to and used by the dryer and printer are of this type. They are generally mixtures of long-chain polymers which contain neutralized acid groups, such as those described as carboxylated acrylate polymers, carboxylated vinyl polymers and ethylene maleic anhydride polymers.

The exact make-up of the dye paste is not critical. The preferred pastes, however, have a viscosity of 10,000 to 30,000 cp, when applied to the substrate, as measured on a Haake Viskotester VT-02, spindle diameter 24 mm, spindle height 53 mm, at 62.5 rpm at room temperature.

Modifier for Carpet Shampoos

Highlights of the Technological Achievement: As a modifier for carpet shampoos and the like, an aqueous composition containing a polymer component comprising an aqueous dispersion of an acrylic polymer, polyvalent metal ions, ammonia or a volatile amine to solubilize the copolymer, and optionally $CO_3^{=}$, HCO_3^- or an anion of an amino acid.

Reference: W.A. Kirn and D.R. Gehman; U.S. Patent 4,203,859; May 20, 1980; assigned to Rohm and Haas Company.

Background: It has been discovered that utilizing a particular polymer composition having critical ratios of certain monomers, critical molecular weights, a critical maximum pK_a, a critical ratio of polyvalent metal ions to carboxyl groups, and the like, gives a much improved shampoo modifier, and an improved shampoo. For example, polymers with substantial proportions of isobutyl acrylate in place of butyl acrylate, ethyl acrylate in place of butyl acrylate, methyl methacrylate in place of butyl acrylate or styrene, or butyl methacrylate in place of butyl acrylate, give relatively poor or only fair soil retardancy, or do not provide stable solutions in the presence of large quantities of a detergent such as sodium lauryl sulfate, at a representative polymer:detergent ratio of 1:1 by weight.

Process Conditions: The improved shampoo modifier of the process is an aqueous composition containing:

(1) a polymer component comprising an aqueous dispersion of a low molecular weight acrylic addition copolymer consisting essentially of polymerized units of (a) butyl acrylate, (b) styrene, (c) methyl methacrylate, and (d) an acid monomer selected from methacrylic acid, acrylic acid, itaconic acid and

any mixture of two or more thereof, in the ratios by weight of 20-60:0-25:0-15:40-60, the polymer having a number average molecular weight of between 2,500 and 100,000;

(2) polyvalent metal ions;

(3) ammonia or a volatile amine; and

(4) optionally, an anion in the form of $CO_3^{=}$, HCO_3^{-} or the anion of an amino acid.

The pK_a of the polymer is <6.7 and it contains no more than 1 part styrene to 1 part butyl acrylate by weight. The pH of the composition is 7.5 to 11, there being at least about 0.8 equivalents of polyvalent metal ion per carboxyl group in the polymer. In addition to being used in an amount to solubilize the polymer, the amount of the ammonia or volatile amine will also be selected to solubilize the polyvalent metal or polyvalent metal compound which supplies the metal ions, if the metal compound is insoluble or only marginally soluble.

Quick Drying Aqueous Resin Compositions Crosslinkable by Oxidation

Highlights of the Technological Achievement: An aqueous resin composition crosslinkable by oxidation and prepared by dissolving a radical-polymerizable polyfunctional monomer in an aqueous solution of at least one of water-soluble resins having a drying oil fatty acid group or semidrying oil fatty acid group and neutralized maleates of synthetic petroleum resins having an unsaturated bond of the vinyl or nonvinyl type, and polymerizing the monomer in the solution.

Reference: Y. Nakayama and T. Aihara; U.S. Patent 4,199,487; April 22, 1980; assigned to Kansai Paint Co., Ltd., Japan.

Background: This process relates to quick drying aqueous resin compositions which are crosslinkable by oxidation. Water-soluble resins which are crosslinkable by oxidation heretofore known and used include maleinized drying oils, maleinized semidrying oils, alkyd resins, modified with drying or semidrying oil fatty acids, maleinized alkyd resins, polybutadiene, epoxy resins modified with drying or semidrying oil fatty acids and acrylic or vinyl resins modified with drying or semidrying oil fatty acids. Coating compositions prepared from these resins have the drawbacks that the compositions, when applied, form coatings of poor initial hardness and are prone to result in excessive surface drying (i.e., the phenomenon in which the interior of the coating remains in the state of semiliquid owing to delayed curing).

Most preferably such drawbacks can be overcome by the use of resins of increased molecular weight, but the resins of the type described invariably involve limitations in the degree to which they are polymerizable because of high viscous property and gelling tendency. Accordingly, even when polymerized to the highest possible degree, the resins still remain to be improved to fully overcome the foregoing drawbacks. Moreover, resins of increased molecular weight are not readily soluble in water. Although it has been attempted to eliminate the drawbacks by styrenating resins of the oxidation crosslinkable type in the case of those composed of drying oils or alkyd resins, the same difficulties as with the polymerized resins are encountered with the resulting resins.

Process Conditions: Examples of useful synthetic petroleum resins having an unsaturated bond of the vinyl or nonvinyl type are polymers of diene monomers having 4 to 8 carbon atoms, such as butadiene, pentadiene, cyclopentadiene, hexadiene, heptadiene, octadiene, etc., and copolymers of at least 10 wt % of these dienes and unsaturated monomers such as olefins having 2 to 8 carbon atoms, styrene, α-methylstyrene, vinyltoluene, acrylonitrile, methacrylonitrile, chloroprene, and esters of alcohols having 1 to 18 carbon atoms and acrylic acid or methacrylic acid.

According to this process, at least one of the water-soluble resins having a drying or semidrying oil fatty acid group and neutralized maleates of the synthetic petroleum resins having a vinyl or nonvinyl type unsaturated bond is used as dissolved in water or in a mixture of water and an organic solvent miscible with water. Useful organic solvents are a wide variety of those miscible with water. The aqueous solution should have a concentration of preferably 15 to 70% by weight.

The radical-polymerizable polyfunctional monomers useful in this process must be highly amenable to radical polymerization. Examples of useful monomers are esters of polyhydric alcohols having 2 to 6 functional groups and acrylic acid or methacrylic acid; esters of glycidyl acrylate or glycidyl methacrylate and acrylic acid or methacrylic acid; esters of polycarboxylic acids having 2 to 4 functional groups and glycidyl acrylate, glycidyl methacrylate, hydroxyethyl acrylate, hydroxyethyl methacrylate, hydroxypropyl acrylate or hydroxypropyl methacrylate; esters of polyepoxy compounds having 2 to 4 functional groups and hydroxyethyl acrylate, hydroxyethyl methacrylate, hydroxypropyl acrylate, hydroxypropyl methacrylate, acrylic acid or methacrylic acid; divinylbenzene; adducts of polyisocyanates having 2 to 4 functional groups and hydroxyethyl acrylate, hydroxyethyl methacrylate, hydroxypropyl acrylate, hydroxypropyl methacrylate, acrylic acid or methacrylic acid; etc.

Also usable are adducts of epoxy polymers of relatively low molecular weight (up to 2,000, preferably up to 500, in number average molecular weight) or compounds containing a hydroxyl, epoxy or carboxyl group, and acrylic acid, methacrylic acid, glycidyl acrylate, glycidyl methacrylate, hydroxyethyl methacrylate, hydroxypropyl acrylate or hydroxypropyl methacrylate. Further usable are adducts prepared from the hydroxyl-containing monomers mentioned above and methylolated melamine, urea or ethyl silicate by ether exchange reaction.

Preferable among these radical-polymerizable polyfunctional monomers are the esters of polyhydric alcohols having 2 to 6 functional groups and acrylic acid or methacrylic acid; esters of glycidyl acrylate or glycidyl methacrylate and acrylic acid or methacrylic acid; esters of polycarboxylic acid having 2 to 4 functional groups and glycidyl acrylate, glycidyl methacrylate, hydroxyethyl acrylate, hydroxyethyl methacrylate, hydroxypropyl acrylate or hydroxypropyl methacrylate; esters of polyepoxy compounds having 2 to 4 functional groups and acrylic acid or methacrylic acid; and divinylbenzene.

Acrylic Hair-Setting Resins Having High Resistance to Moisture and Rapid Removability from the Hair with Shampoo

Highlights of the Technological Achievement: Acrylic polymer solutions having superior properties when used as hair-setting or holding resins are disclosed. Specific polymers disclosed are derived from (1) alkyl acrylates with 2 to 8 carbon alkyl groups, (2) methyl methacrylate, (3) hydroxyethyl methacrylate, and (4) methacrylic acid in proportions yielding a film forming resin readily soluble in shampoo, with a Tg between 40°C and 80°C, a molecular weight between 25,000 and 250,000 and not more than 50 ppm of unreacted alkyl acrylate monomer.

Reference: D.R. Gehman and W.A. Kirn; U.S. Patent 4,196,190; April 1, 1980; assigned to Rohm and Haas Company.

Background: Cosmetic hair spray resins are polymers which when suitably formulated are applied to the hair to impart curl retention and "hold" to the coiffure. The first synthetic resin, polyvinylpyrrolidone, was introduced in 1953 and is still employed in hair sprays. Its performance has been shown to be marginal, at best, in that it offers only fair curl retention and at conditions of high relative humidity is extremely tacky.

Copolymers have been developed of polyvinylpyrrolidone/vinyl acetate and polyvinyl ether/maleic anhydride which offer improved curl retention but still impart a sticky feel to the hair at high relative humidity. Since hair sprays should be readily removable by shampoo washing, this problem is a direct result of the compromise of adequate shampoo removability with water and humidity resistance.

Acrylic solution copolymers have also been marketed for use as hair sprays. However, such products have never been commercially successful, in part because of the offensive monomer odor which accompanies these products.

It has been discovered that an optimum hydrophobe/hydrophile balance through proper acrylic monomer selection will produce a hair spray resin which imparts improved curl retention at high relative humidity, superior moisture resistance, excellent shampoo removability and good aesthetic properties to the hair. Such a balance of performance properties is obtainable through the judicious incorporation of hydrophilic and hydrophobic acrylic monomers in the same backbone.

A polymer of this preferred composition, when incorporated into an aerosol or pump type hair spray formulation has been shown to offer an improved performance profile over current competitive products. Further, by reduction of the unreacted alkyl acrylate monomer in the final copolymer, a composition which avoids the offensive odors of the prior art acrylate hair resins has been produced.

Process Conditions: In general, the alkyl acrylate component of the copolymer plasticizes or reduces the brittleness of the copolymer film, the methyl methacrylate provides hardness to give the film its nontacky holding properties and the functional monomers (hydroxyethyl methacrylate) (HEMA) and methacrylic acid (MAA) produce shampoo removability. The preferred alkyl acrylate is butyl acrylate with shorter chain acrylates, e.g., ethyl acrylate being preferred over long-chain acrylates, e.g., 2-ethylhexyl acrylate. A combination of the functional monomers, such as HEMA/MAA in a proportion of 10%/18% by weight of the copolymer, is much preferred over either alone. Increasing the MAA level above 18% further improves shampoo removability but begins to adversely affect curl retention performance. A HEMA level of approximately 10% is the preferred level as will be seen hereinafter from the data relating to free film and curl retention studies. Higher levels of HEMA produce adverse effects on free films at high relative humidity as well as lower curl retention.

Water-Dilutable Acrylic Resin Having Functional Carboxylic Groups

Highlights of the Technological Achievement: There is provided a coating composition comprising polyvinyl chloride homopolymer or copolymer dispersed in an aqueous solution of solubilized carboxyl functional acrylic resin, a triazine-formaldehyde crosslinking resin, and a metal carboxylate stabilizer such as iron neodecanoate.

Reference: H.J. Salmon; U.S. Patent 4,192,787; March 11, 1980; assigned to Mobil Oil Corporation.

Background: In U.S. Patent 4,116,905 there is disclosed an aqueous solution of water-dilutable resin, solubilized with ammonia or an amine, into which is dispersed PVC. This dispersion can be admixed with an aminoplast, applied to a substrate, and heat cured. It has been found, however, that as a result of heat curing at elevated temperatures (375° to 750°F) the film becomes badly discolored and film properties are relatively poor.

It is the discovery of this process that such heat discoloration can be prevented by incorporating into the PVC dispersion formulation a small amount of an iron carboxylate.

Figure 4.3 shows the relationship between the Hunter scale a and b values for heat cured coatings obtained from PVC formulations containing different metal carboxylates, A-K, E being iron neodecanoate,and the relative color degradation resulting from each as measured on the Hunter Color Difference Meter.

Figure 4.3: Relationship Between Hunter Scale Values and Color Degradation for PVC Formulations Containing Metal Carboxylates

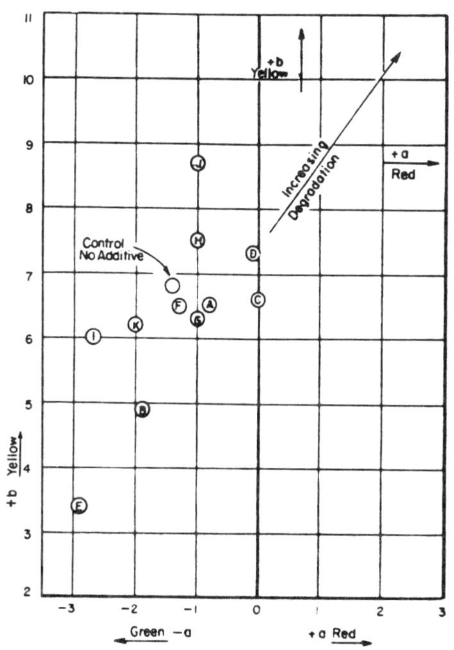

Source: U.S. Patent 4,192,787

Process Conditions: In preparing the coating formulations of this process, generally there is prepared a solution of water-dilutable acrylic resin in an aqueous solution. To this solution there is added polyvinyl chloride in powder form, as well as other components, so that it is dispersed in the aqueous solution.

The water-dilutable acrylic resin can be any acrylic resin that has carboxyl functionality. Such resins include copolymers of acrylic acid or methacrylic acid with acrylate esters or methacrylate esters and can also contain small amounts of styrene monomer. Polyacrylic acid and polymethacrylic acid are also contemplated. In general, the copolymers will contain at least 10 to 20 wt % acrylic acid or methacrylic acid. A typical copolymer can contain methyl methacrylate/ethyl acrylate/acrylic acid in the proportions: 0-80/1-85/10-20 in parts by weight. The carboxyl functional acrylic resins contemplated herein are well known in the art and are generally available commercially.

The water-dilutable acrylic resin is solubilized with ammonia, NH_4OH, or amines such as lower alkyl and/or alkylol amines. The weight ratio of PVC to water-dilutable acrylic resin will be preferably between 10:1 and 1:1. The total solids content in the finished coating (PVC + water-dilutable resin) will be between 15 and about 65 weight percent.

The coating formulation is prepared by adding the PVC powder and other components to the solution of water-dilutable acrylic resin with stirring. Alternatively the PVC powder may be combined with the water-dilutable acrylic resin or resins and the other components prior to neutralization and water addition. The resultant formulation is a suspension of PVC particles in the solution.

Water-Soluble Salts of Low Molecular Weight Polyacrylic Acids for Preparing Portland Cement

Highlights of the Technological Achievement: Water-soluble salts of low molecular weight polyacrylic acids function synergistically with inorganic salts to obtain superior reduction of water demand of dispersed inorganic solids. The disclosed process is particularly adapted for use in the wet-grinding process for preparing portland cement.

Reference: D.C. MacWilliams and J.R. Wirt; U.S. Patent 4,175,975; Nov. 27, 1979; assigned to The Dow Chemical Company.

Background: It is known that polyelectrolytes such as polyacrylic acid salts, copolymers of acrylic acid and acrylamide, hydrolyzed polyacrylonitrile and the like which are flocculants at higher molecular weights show different properties and act as dispersants at lower molecular weights. Such polymers having molecular weights from a few thousand up to about 50,000, for example, have been recommended for use in various dispersant applications.

Many substances with dispersant activity are available and have been tried in order to decrease the water demand of suspended inorganic solids in various high solids water suspensions or slurries for various applications. Most of these, particularly in the wet process for making portland cement, have proven relatively ineffective or undesirable for one reason or another. Complex phosphates are undesirable because they tend to hydrolyze at the warm temperatures developed during grinding and in storage of the slurry and because of the adverse effect of residual phosphate on the properties of the final portland cement product. Lignosulfonates, alone and in combination with inorganic salt dispersants, have been tried for this use, but these require high addition levels for only marginal improvement. They also lose their activity rapidly during storage of the slurry. Many materials are also too expensive to be economically utilized in such operations.

It has been found, however, that the water demand of suspended inorganic solids necessary to make a pumpable slurry in the wet grinding process for making cement can be significantly reduced by use of water demand reducing agent combination as described below.

Process Conditions: The polyacrylic acid salt which can be employed in this process can be any such polymer salt having an average polymer molecular weight in the range of from 2,000 to 50,000. An especially preferred molecular weight is in the range of from about 5,000 to about 10,000. The preferred polymer molecules are further characterized in that about one-eighth to one-half of them have terminal sulfonate groups.

In another embodiment, polymers having a viscosity of preferably from 75 to 110 cp at 30% by weight solids and a molecular weight range from 5,000 to 10,000 are preferred. The salt of the acid may be that of alkali metal or ammonium salts, such as sodium or potassium. However, the sodium and ammonium salts are preferred over other alkali metal salts.

Polymers of acrylic acid which are useful in this process are readily prepared from the monomer by the action of heat, light, and/or catalysts. Catalysts which are particularly effective for this polymerization are the organic peroxides.

Certain preferred polyacrylates of the process are most advantageously prepared by polymerizing acrylic acid in aqueous solution at 50° to 170°C in the presence of a redox polymerization catalyst system. The acrylic acid and a peroxy catalyst are separately and continuously dispersed into the aqueous medium at rates such that an effective and substantially constant concentration of the catalyst system is maintained in contact with the acrylic acid throughout the polymerization.

A sulfite reducing agent, the preferred other component of the redox catalyst system, can be combined with the acrylic acid and the two added as a single aqueous solution, but preferably, the reducing agent is added separately and continuously as a third stream. Suitable sulfite reducing agents include sodium metabisulfite, sodium sulfite or bisulfite, sodium formaldehyde sulfoxylate, sodium formaldehyde hydrosulfite.

Grouting Composition Containing Cupric Acrylate

Highlights of the Technological Achievement: An aqueous solution of (a) acrylic acid; (b) cupric acrylate; (c) methylenebisacrylamide; and (d) a water-soluble polymerization catalyst which also is a reducing agent; with the ratio of (a) to (b) being between 10:0.1 to 10:1.5, which solution forms a water resistant gel.

Reference: K.G. Phillips; U.S. Patent 4,173,555; November 6, 1979; assigned to Nalco Chemical Company.

Background: Commercially, grouting is accomplished using methylenebisacrylamide with acrylamide. Soils treated with solutions of these two monomers which are subsequently polymerized in contact with the soil produce dense, friable masses of soil which can be chipped or broken. When sodium acrylate is substituted for acrylamide in making these gels, the resultant gels tend to become less rigid as they are exposed to moisture, thus rendering their usability limited in treating soils.

Process Conditions: The concentration of the above ingredients in water used to produce the gels of the process may vary although approximately 10% acrylic acid, 0.1 up to 1.5% cupric acrylate, and 1% methylenebisacrylamide give optimum results. Slightly lower or greater dosages may be used. The soils treated with these solutions and gelled are resilient and have a different character than the methylenebisacrylamide treated soils.

As indicated, one of the most interesting and unique features of the process is that when cupric acrylate is used, it is necessary to utilize a water-soluble polymerization catalyst which is also a reducing agent. Such catalysts are ammonium, sodium or potassium bisulfate.

Examples: To illustrate the process, the following are presented by way of examples. In the table set forth below, the penetration values were obtained using a standard grease penetrometer. Any value below 250 is considered acceptable.

It should be particularly noted that very little water is picked up by the gels based on percent by weight increase after standing in water for prolonged periods of time. This is surprising since most acrylate-methylenebisacrylamide gels pick up substantial quantities of water under the test conditions used in the table.

Acrylic Acid (%)	Acrylate Salt (%)	Methylene-bisacrylamide (%)	Catalyst	Water (%)	Set Time (sec)	Penetration	Water Pickup (% increase). 24 hr	120 hr
10	0.1 Cu^{2+}	1.0	R*	88.9	240	205	7.2	23
10	0.2 Cu^{2+}	1.0	R	88.8	300	225	18.1	49
10	0.3 Cu^{2+}	1.0	R	88.7	600	–	–	–
10	0.4 Cu^{2+}	1.0	R	88.6	900	–	–	–

*R = 0.5% $NaHSO_3$ was used as the catalyst and reducing agent in all samples.

Red Mud Treatment

Highlights of the Technological Achievement: A method of fluidizing concentrated aqueous suspensions of red mud which comprises uniformly admixing with such suspensions at least 0.1 pound per metric ton of an acrylic acid polymer which has a molecular weight within the range of 5,000 to 30,000.

Reference: J. Kane; U.S. Patent 4,169,824; October 2, 1979; assigned to Nalco Chemical Company.

Background: The almost universally used process for the manufacture of alumina is the Bayer process. In its broadest aspects, this method is carried out almost exclusively in aqueous solution, and is achieved by reaction of bauxite and a strong base such as caustic soda or lime in steam heated autoclaves whereby the alumina is transformed into a soluble aluminate form. In this step, a considerable amount of insoluble impurities result or are released from the bauxite, which must be separated from the desired alumina constituent. These residues, commonly known as red muds, include iron oxides, sodium alumino-silicate, titanium oxide and other materials. Generally, these muds appear as very fine particles which are difficult to separate out. Yet, the red muds which usually constitute about 10 to 55% by weight of the ore must be rapidly and cleanly separated from the solubilized alumina liquor. It is important that the excess caustic also be recovered and reused in subsequent runs.

The separation of red mud from alumina and caustic solution is generally effected by first treating the red mud suspension or slurry with chemicals to promote flocculation of solids contained therein. The initial separation in the primary settlers is followed by a wash step. Specifically, the primary settler solid, normally a 15 to 30% solid slurry, is washed with oncoming liquor from the washing cycle, i.e., countercurrent displacement wash circuit or sand filter, whichever is being utilized. The resultant underflow generally having from 15 to 40% solids is subsequently washed a plurality of times. The wash waters containing suspended red mud and dissolved alumina and caustic are then also treated with chemicals to promote settling.

After the final wash step, the mud is normally concentrated via vacuum filtration or other means to a plastic mass which must be transported and acted upon by chemical processing equipment such as conveyors, mixers, raking devices and the like. The concentration of the red mud contained in these solid-like suspensions can range between 50 and 65%. The handling of such unworkable masses has always been a problem to the aluminum manufacturing industry. If it were possible to readily decrease the viscosity of these materials so they could be acted upon by means of pumps and other fluid-handling devices, a great improvement would be made in this art.

Process Conditions: It has been found that concentrated aqueous slurries of red mud can be fluidized by uniformly admixing with such suspensions at least 0.1 to 1.6 pounds per dry metric ton of red mud of an acrylic acid polymer which has a molecular weight within the range of 5,000 to 30,000. Preferably, from 0.10 to 1.0 pound of polymer on a dry basis will be used to treat each metric ton of dry red mud. The abovedescribed polymeric materials are generally used in the form of an aqueous solution. When used, as for example, in a 20% by weight aqueous solution, from 0.5 to 7.0 and preferably 0.5 to 1.5 pounds of the 20% aqueous solution of the acrylic acid polymer will be utilized for each dry metric ton of red mud.

As indicated above, the molecular weight of these polymers can vary between 5,000 and 30,000. A preferred molecular weight range of the polymer is between 8,000 and 20,000.

Raw Materials: The acrylic acid polymers used in the practice of the process may be either homopolymers or copolymers of acrylic acid. When used as copolymers, they should contain at least 50% by weight of acrylic acid. These acrylic acid polymers are preferably used in the form of their alkali metal or ammonium salts.

Preferred copolymers are copolymers of acrylic acid which have been polymerized with from 5 to 40% by weight of methyl methacrylate. Other comonomers that may be used in preparing the acrylic acid copolymers are acrylamide, maleic anhydride and other monomers used in such an amount and of such a type so as not to render the polymers water-insoluble.

Oligomeric Aqueous Finishes

Highlights of the Technological Achievement: This process discloses the composition and use of salts of low molecular weight polymers of methacrylic and acrylic esters formulated with wetting, emulsifying, plasticizing or coalescing agents as aqueous finish compositions for hard surfaces. The polymers have molecular weights between 700 and 5,000 and acid numbers between 50 and 275. The finishes produced by the compositions are water clear and are self-dispersible. Thus, on reapplication, the previous film is dissolved, the embedded soil removed and a fresh film left.

Reference: S.N. Lewis, R.A. Haggard and D.R. Gehman; U.S. Patent 4,168,255; September 18, 1979; assigned to Rohm and Haas Company.

Background: The hard surface cleaner-polish of this process differs from conventional all-purpose cleaners in leaving a film on the cleaned surface. This film restores the luster of the surface and seals it, making the substrate more resistant to new soil. The film, formed on drying is soluble and dispersible in fresh cleaner-polish, so on recleaning, the old film dissolves, releasing accumulated soil. Use of the cleaner-polish thus results in less scrubbing effort in cleaning as well as reducing the wear and damage to the substrate by facilitating the removal of oily and greasy soils, particulate dirt, water-soluble matter and other types of household soils.

Other polymers of relatively low molecular weight and aqueous alkali solubility, such as rosin acid adducts, styrene/acrylic acid, and styrene/maleic anhydride resins, are found to impart good leveling and flow properties to polishes containing higher molecular weight polymeric vehicles but are unsatisfactory when used solely as the major polymeric constituent in aqueous soluble floor polishes. Their defects include: unacceptably low water resistance, unsatisfactory recoatability, poor wear properties, and poor initial color and color stability. Although the addition of low levels of polyvalent metal ions (e.g., Zn) may be used to improve these properties, it does so at the expense of gloss, so polishes based on those other polymers remain inferior to polishes of the polymers disclosed herein.

Process Conditions: The aqueous polishing compositions of the process can generally be defined in terms of the following proportions of the main constituents:

(A) 25 to 100 parts by weight of an addition polymer comprising mers derivable from an α,β-monoethylenically unsaturated acid and at least 25 mol % of mers having the structure

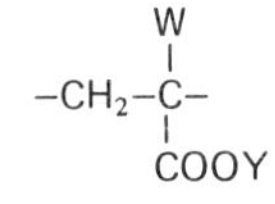

wherein W is hydrogen or methyl, Y is alkyl or substituted alkyl; the polymer having an acid number from about 50 to about 275, a weight average molecular weight from about 700 to about 5,000, and being at least partly neutralized;

(B) 0 to 25 parts by weight of a wax;

(C) 0 to 50 parts by weight of an alkali-soluble resin;

(D) 0.1 to 90% by weight of the sum of (A), (B) and (C) of one or more wetting, emulsifying, plasticizing or coalescing agents;

(E) 0 to 5% by weight of (A) of a polyvalent metal compound; and

(F) water to make the total solids 1 to 45% by weight.

In a preferred embodiment, the aqueous finish composition comprises: (A) an at least partly neutralized polymer having mer units of methacrylic acid and at least one ester of methacrylic acid, having an average carboxyl content of at least about 2, an average chain length, $\bar{n}$, of from about 6 to about 35 mer units per molecule, and (B) one or more agents selected from the group of wetting, emulsifying, plasticizing and coalescing agents in the amount of 0.1 to 90% of (A) by weight.

This polymer is an alkoxide-polymerized polymer with a narrow molecular weight distribution having at least about 85% by weight of the molecules with chain lengths of about $\sqrt{2\bar{n}}$ to about $2\bar{n}$ mers.

Salts of the bis-, tris-, tetra- and higher carboxyl containing methacrylate polymers must fall within specific molecular weight ranges and acid number ranges to afford improved finishes. The acid number is the number of milligrams of potassium hydroxide required to neutralize the acidic constituents in one gram of the acid form of the polymer. The acid number range for the polymer is given in the table below as 50 to 80 for the industrial polish and 80 to 150 for the household polish, corresponding to 50 to 150 for the overall range. This acid number range is encompassed by about 6% to about 23%, by weight, of acrylic, methacrylic or itaconic acids or a mixture thereof; these being the preferred acid monomers of this process.

Autoredispersible Polishes

Property	Household Floor Polish	Industrial Floor Polish
Detergent sensitivity	Sensitive	Resistant
Acid Number		
Broad range	80-150	50-80
Preferred range	100-135	55-70
Average MW		
Broad range	1,000-5,000	1,500-5,000
Preferred range	1,300-2,500	2,000-4,000
Average carboxyls per molecule		
Broad range	2.0-4.5	2.0-4.0
Preferred range	3.5-4.2	2.5-3.5

Imparting High Gloss to Water-Based Acrylic Paint Systems

Highlights of the Technological Achievement: An aqueous slurry of from 60 to 80% titanium dioxide is prepared from titanium dioxide of uniform and small particle size. At least 80% by weight of pigment is incorporated into an aqueous system containing suitable dispersant which slurry is then reduced to the final slurry concentration with water.

Reference: R.L. DeColibus; U.S. Patent 4,177,081; December 4, 1979; assigned to E.I. Du Pont de Nemours and Company.

Background: In recent years, substantial research has been undertaken to develop titanium dioxide slurries that were suitable for use in paper and water-based paint manufacture. The economics and ease of use of such slurries has been a substantial incentive for the development of titanium dioxide slurries.

Hall and Looby disclose in U.S. Patent 3,702,773 that a stabilized titanium dioxide slurry of from 60 to 82% by weight of titanium dioxide for use in paper manufacturing could be prepared. Roberts and Rowland disclose in U.S. Patent 3,758,322 that a process for producing slurries of from 60 to 80% titanium dioxide content can be prepared from a pigment slurry containing 20% solids. Daubenspeck and Jett disclose in U.S. Patent 3,847,640 that a pigment slurry containing 60 to 65% titanium dioxide is adapted to be used in latex flat paint formulations.

This process provides another step forward in the development of titanium dioxide slurries. This process provides a slurry of very high solids content that can be used to prepare high gloss water-based acrylic paint formulations.

Process Conditions: In an aqueous slurry of titanium dioxide which when dried from a water-based acrylic coating composition forms a high-gloss film, the aqueous slurry contains from 60 to 80% titanium dioxide, from 0.2 to 1% based upon the weight of titanium dioxide of a compound of the formula:

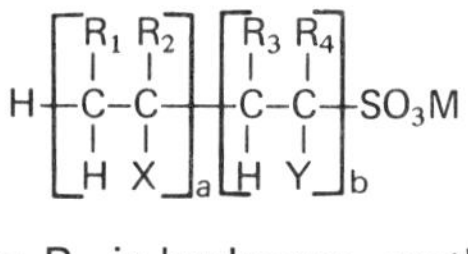

where M is alkali metal; R_1 to R_4 is hydrogen, methyl or ethyl; X is cyano or $-COOR_5$; where R_5 is H, or lower alkyl of 1 to 4 carbon atoms; Y is $-COOH$ or $-COONa$; a and b are 10 to 100; and from 0.1 to 0.6% based on the weight of titanium dioxide of 2-amino-2-methyl-1-propanol.

This process also relates to a process for producing a high solids content titanium dioxide aqueous slurry which comprises preparing a solution of 0.2 to 1% of a compound of the above formula and 0.1 to 0.6% of 2-amino-2-methyl-1-propanol, based on the weight of the titanium dioxide in about 10 to 25% of water, adding 75 to 90% by weight of titanium dioxide, grinding with a disc mill until uniform and adding sufficient water to reduce the titanium dioxide content to form 60 to 80% by weight of the finished slurry.

Aqueous Dispersion Paint Having a High Wet Adhesion

Highlights of the Technological Achievement: Emulsifier-containing plastics dispersions which serve as a binder for dispersion paints, should contain polymers forming a film at room temperature and having a high wet adhesion. Suitable polymers are prepared by polymerization of olefinically unsaturated monomers initiated by free radical-forming agents, especially in the presence of an emulsifier. As emulsifier there is used a salt of an acyl ester of the hydroxymethanephosphonic acid. The emulsifier may remain in the dispersion ready for use without involving disadvantages for the wet adhesion.

Reference: E. Nölken; U.S. Patent 4,169,086; September 25, 1979; assigned to Hoechst AG, Germany.

Background: Plastics dispersions, which contain polymers forming a film at room temperature and which are used as latices in paints, papers, textile articles and adhesives, in most cases contain the emulsifier which has been used during their preparation. A further step, for example, dialysis, would be necessary to remove this emulsifier; for a number of applications, however, this is by no means required. However, when using plastics dispersions containing an emulsifier in paints, which are applied on nonabsorbent surfaces, for example, lacquer coatings, the wet adhesion obtained is unsatisfactory so that the coating of paint detaches from the substrate.

Process Conditions: Plastics dispersions containing polymers which form a film at room temperature and which have a high wet adhesion, consist of the following components:

(a) polymers or copolymers of vinyl esters, acrylic and methacrylic acid esters, styrene, vinyl halides, vinyl ethers, diolefins, α-olefins or nitrile groups containing monomers, alone or in admixture with one another; and

(b) optionally buffer substances, alkalies and water-soluble polymers

and as emulsifiers, salts of acyl esters of hydroxymethanephosphonic acid of the formula:

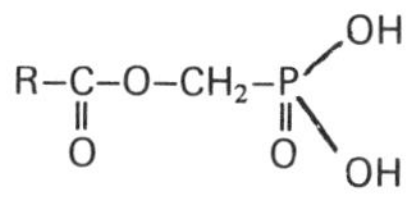

in which R is a hydrocarbon radical having from 5 to 23 carbon atoms, in an amount of from 0.1 to 5% by weight, calculated on the total quantity of the monomers used.

The polymerization is carried out at a temperature preferably of from 40° to 75°C. Low- or high-molecular weight water-soluble compounds which do not detrimentally affect the wet adhesion may be added to the latex to improve its resistance to storage, shearing and freezing and for improving its compatibility with pigments. There may be mentioned, by way of example, ammonium laurate, hydroxyethylcellulose, polyvinyl alcohol, poly-α-oxyacrylic acid, the ammonium salts of copolymers of styrene, ethylene or vinyl methyl ether with maleic acid and anhydride. These compounds may be added to the liquor prior to polymerization, or during polymerization or be added to the latex upon completion of the polymerization.

The plastics dispersions prepared in this manner are suitably used as binders in dispersion paints which adhere after wetting. The pigments are thereby frequently used in the form of pastes, for example, titanium dioxide dispersed in water. The pigment pastes may contain thickeners, for example, cellulose derivatives, such as methyl, ethyl, hydroxyethyl, or carboxymethylcellulose, polyvinylpyrrolidone, polyvinyl alcohol, polyacrylic acid, or polymethacrylic acid.

Aqueous Heat-Curable Particulate Coating Composition of Elastomeric Emulsion Copolymer and Thermosetting Copolymer

Highlights of the Technological Achievement: Protective and decorative coatings for metal are prepared using aqueous polymer compositions consisting essentially of a thermosetting vinyl or acrylic polymer which has been polymerized in the presence of a preformed, crosslinked, elastomeric copolymer.

Reference: H.H. Leiner, L.E. Koepke, D.R. Gelarden and J.L. Gardon; U.S. Patent 4,174,336; November 13, 1979; assigned to M & T Chemicals Inc.

Background: Thermosetting addition type polymers, including acrylic and vinyl polymers, are useful materials for coating metal substrates. The polymer chains contain reactive groups such as hydroxyl groups which are reacted with an external latent crosslinking agent, such as the N-methylol melamines, N-alkoxy melamines, N-methylol ureas or N-alkoxy ureas that is included in the coating formulation.

One objective of this process is to provide a coating composition wherein the film-forming component is a copolymer containing a latent, heat-activated crosslinking agent. The copolymers do not exhibit the undesirable characteristics of prior art materials. Coatings incorporating the copolymers exhibit superior solvent and impact resistance.

Process Conditions: This process provides improved, heat-curable coating compositions consisting essentially of an aqueous medium containing dispersed polymer particles which, in turn, consist essentially of an elastomeric emulsion copolymer A and a thermosetting copolymer B, wherein:

(1) copolymer B is prepared in the aqueous medium and in the presence of preformed copolymer A;

(2) the rate of addition of the constituent monomers of copolymer B is substantially equal to the reaction rate of the monomers to form copolymer B;

(3) copolymer B constitutes from 10 to 70% of the combined weight of copolymers A and B, exhibits a weight average molecular weight of from 5,000 to 50,000 and a glass transition temperature from 40° to 75°C;

(4) the repeating units of copolymer B are derived from:

  (a) at least one monoethylenically unsaturated compound containing from 2 to 20 carbon atoms;

  (b) from 2 to 20%, based on the weight of copolymer B, of at least one monoester of a diol containing from 2 to 4 carbon atoms and acrylic or methacrylic acid, the monoester being copolymerizable with the monoethylenically unsaturated compound, and

(c) from 5 to 40%, based on the weight of copolymer B, of at least one latent curing agent selected from the group consisting of N-alkoxymethyl acrylamides and N-alkoxymethyl methacrylamides wherein the alkoxy group contains from 1 to 8 carbon atoms;

(5) copolymer A exhibits a glass transition temperature of from -15° to +25°C;

(6) the repeating units of copolymer A are derived from:

(a) at least two monoethylenically unsaturated compounds containing from 2 to 20 carbon atoms, and

(b) from about 0.5 to 5.0%, based on the weight of copolymer A, of a polyfunctional compound selected from the group consisting of diesters and triesters derived from acrylic or methacrylic acid and a diol containing from 2 to 8 carbon atoms or a triol containing from 3 to 8 carbon atoms, divinyl benzene and allyl methacrylate.

Improved Latex Binder for Nonwoven Webs

Highlights of the Technological Achievement: An improved latex binder for nonwoven webs comprising an acrylate- and/or vinyl acetate-based copolymer.

Reference: R.J. Caimi and W.F. Schlauch; U.S. Patent 4,176,108; November 27, 1979; assigned to National Starch and Chemical Corporation.

Background: Nonwoven fiber materials, generally prepared by randomly depositing individual fibers to form a web, impregnating the web with a binder to hold the individual fibers together, and then heating the web at a set temperature, are known to possess distinct advantages over conventional woven materials. Among the advantageous properties characterizing the nonwoven fiber materials are absence of ravelling, smoother surfaces, increased softness, improved hand, greater absorbency, higher loft, and the like.

The binders used to impregnate nonwoven webs are typically aqueous dispersions of copolymer latices which form rubbery or elastic films upon drying. A wide variety of copolymers are known in the art to be used in latex binders. Typical of these copolymers are butadiene copolymers such as butadiene-styrene-acrylonitrile, alkyl acrylate copolymers, rubbery chloroprene copolymers, vinyl and vinylidene ester copolymers, and the like. These copolymers may contain (i.e., be prepared with) any of a wide variety of polymerizable and/or crosslinkable comonomers such as, for example, substituted alkyl acrylate esters such as 2-hydroxyethyl acrylate; vinyl halides; vinyl esters; ethylenically unsaturated amides such as acrylamide and N-methylolacrylamide; and the like. Typical copolymers of this type employed in latex binders for nonwoven fiber material are described in, for example, U.S. Patents 3,985,929; 3,231,533; and 2,984,588.

Process Conditions: The copolymers employed in the latex binders of this process are copolymers in latex form of which at least 86% by weight, and preferably

95% by weight, is an acrylate or methacrylate ester or vinyl acetate, or mixtures thereof. Copolymers of the type disclosed herein are well known in the art.

Suitable acrylate or methacrylate monomers for the copolymers herein are esters of (meth)acrylic acid of the general formula:

$$\mathrm{CH_2{=}\overset{\overset{R'}{|}}{C}{-}\overset{\overset{O}{\|}}{C}{-}OR}$$

wherein R is an alkyl, alkenyl or aryl group and R' is hydrogen or a methyl group. Representative of such esters are methyl acrylate, ethyl acrylate, allyl acrylate, butyl acrylate, hexyl acrylate, 2-ethylhexyl acrylate, methyl methacrylate, and the like. Mixtures of two or more of these esters may also be present in the copolymer, if desired.

The copolymer used in the latex binder of this process must contain from 0.1 to 10% by weight of the mers derived from 2-hydroxyethyl acrylate (or equivalent monomers) in order to obtain a binder with the required coagulation properties. It has been found that omission of this monomer from the copolymer results in a latex binder which will coagulate in an uncontrollable manner only at very high temperatures, if at all. However, if the copolymer contains more than about 10% of 2-hydroxyethyl acrylate or its equivalent, the binder prepared therefrom will be too viscous and will coagulate at temperatures of less than 40°C. Thus, the specific amount of 2-hydroxyethyl acrylate to be present in the copolymer will depend on the coagulation temperature and latex viscosity desired. Use of 2-hydroxyethyl acrylate monomer is preferred over use of the equivalent acrylate monomers referred to above.

When mixed with 1 to 40 wt % based on latex solids, of a water-soluble electrolyte salt, the latex binder is stable at room temperature and may be stored indefinitely at that temperature until ready to be used. The binder should not, however, be stored at temperatures of over 30°C. Upon heating to a temperature within the range of 40° to 85°C, the latex binder will coagulate, with the particular coagulation temperature depending on the copolymer used, the amount of latex solids, the type and relative amounts of surfactants, the type and amount of salt used, etc.

Sizing Agent for Use in Making Paper

Highlights of the Technological Achievement: This process relates to a neutral sizing agent of the emulsion type to be added to paper pulp for facilitating the disaggregation and recovery of wastepaper in making paper including paper board.

Reference: S. Yoshioka and K. Okada; U.S. Patent 4,200,562; April 29, 1980; assigned to Seiko Kagaku Kogyo Co., Ltd., Japan.

Background: It is a well-known fact that, as a loading material for making paper, calcium carbonate has many advantageous features not only in its properties, but also from a viewpoint of economy. In the meantime, since the rosin sizing agent which is commonly used in paper making must be adapted for use under acid conditions, a paper making process with the use of calcium carbonate as a loading material cannot use the rosin sizing agent.

The sizing agent for paper making according to the process is obtained by dissolving in an aqueous medium a water-soluble cationic copolymer compound having principal units of a monomer comprising a predominant amount of a compound given by the formula:

(1) $$[H_2C{=}C(R){-}\overset{\overset{\large O}{\|}}{C}{-}O{-}A{-}\underset{\underset{\large CH_3}{|}}{\overset{\overset{\large CH_3}{|}}{N^+}}{-}CH_2{-}CH(OH){-}CH_2X]\,Y^-$$

and a minor amount of a compound given by the formula:

(2) $$[H_2C{=}C(R){-}\overset{\overset{\large O}{\|}}{C}{-}O{-}A{-}\underset{\underset{\large CH_3}{|}}{\overset{\overset{\large CH_3}{|}}{N^+}}{-}CH_2{-}\underbrace{CH{-}CH_2}_{O}]\,Y^-$$

wherein R is H or CH_3; X is a halogen atom; A is a (C_{2-6}) alkylene group, and Y is an anion such as halogen ion (Cl^-, Br^- or I^-) or an anion of an acid salt such as phosphate, acid phosphate, sulfate, bisulfate, methyl sulfate, carboxylate, sulfonate, sulfamate, acetate, formate, oxalate, acrylate, or α-methacryloxyacetate or, more preferably, an anion of an acid having an ionization constant of 5.0 or less, namely, an anion of acid which is dissociated at a hydrogen ion and a monomer given by the formula:

(3)

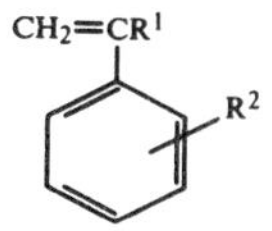

wherein R^1 and R^2 is H or CH_3; dispersing therein a polymerizable hydrophobic monomer such as styrene or 2-ethylhexyl methacrylate, and then subjecting the resultant dispersion to a radical emulsion polymerization.

Also, the sizing agent according to the process is obtained by dispersing a polymerizable hydrophobic monomer in an aqueous medium containing as a principal emulsifying agent, a water-soluble cationic copolymer of the monomers given by the formulas (1) and (2) with the monomer of the formula (3) and as an auxiliary emulsifying dispersing agent, a compound given by the formula:

(4) $$R^3{-}O{-}(CH_2{-}CH_2{-}O)_nSO_3M$$

wherein R^3 is a higher aliphatic or cycloaliphatic hydrocarbon residue; n is an integer equal to or larger than 2; and M is a monovalent metallic element or NH_4. The resultant dispersion is then subjected to a radical emulsion polymerization.

As disclosed in U.S. Patent 3,678,098, the monomer given by the formula (1) can be produced by causing a hydrogen acid salt of a basic ester given by the formula:

$$H_2C{=}C(R){-}C(O){-}O{-}A{-}N(CH_3)_2{\cdot}HY^-$$

wherein R is H or CH_3; A may be a (C_{2-6}) alkylene group interposed between adjacent oxygen and nitrogen atoms or a polyoxyethylene group represented by the following formula:

$$(CH_2CH_2O)_xCH_2CH_2$$

wherein x is 1 to 11 or more; and Y is a halogen or an anion of acid to react with epihalohydrin under acid conditions at room temperature to 80°C, the basic ester being produced by copolymerizing an aromatic vinyl hydrocarbon such as styrene, a methacrylate, such as dimethylamino methacrylate and a nitrile compound such as azobisisobutyronitrile. In this case, the total amount of reactants is not converted into the monomer of the formula (1), but some units of formula (2) are produced along therewith. The relative amounts of the monomers (1) and (2) will vary depending on the reaction but, in typical cases, they will be present in roughly equivalent amounts, and it is known that as the pH decreases the proportion of the monomer (2) decreases, while it increases at neutral or in the alkaline side.

The water-soluble cationic copolymer compound constituting the principal component of the emulsifying dispersing agent can be readily obtained by copolymerizing one or more monomers given by the aforementioned formulas (1) or (2) with the monomer given by formula (3). Alternatively, it can be produced by copolymerizing the monomer of formula (3) with a monomer given by the formula:

$$(5) \qquad [H_2C{=}C(R){-}\overset{\overset{\large O}{\|}}{C}{-}O{-}A]{-}\underset{\underset{\large CH_3}{|}}{\overset{\overset{\large CH_3}{|}}{N}}{\cdot}HY$$

wherein R is hydrogen or CH_3; A is a (C_{2-6}) alkylene group; and Y is an anion similar to that of formulas (1) and (2). The resultant copolymer is then caused to react with epihalohydrin at 80° to 85°C under acid conditions and, more preferably, at pH 6 or less.

In producing the aforementioned copolymer, the principal component of the emulsifying dispersing agent according to the process may comprise a water-soluble cationic copolymer compound which is obtained by copolymerizing predominant amounts of those monomers given by formulas (1), (2), (3) and (5) with a small amount of other copolymerizable monomers such as acrylic ester, methacrylic ester and aromatic vinyl hydrocarbon added thereto as required.

Sizing Staple Fiber Yarns

Highlights of the Technological Achievement: A process for sizing staple fiber yarns by applying an aqueous sizing liquor which contains an alkaline earth metal salt of a copolymer containing from 90 to 30% by weight of acrylic acid and/or methacrylic acid and from 10 to 70% by weight of acrylonitrile and/or methacrylonitrile, which copolymer has a viscosity of from 50 to 1,000 cp in 15% strength aqueous solution at 85°C, and drying the treated yarns. The sized films obtained are moisture-insensitive and have good heat stability.

Reference: W. Schenk, J. Stedefeder, H. Leitner and H.-U. Schenck; U.S. Patent 4,167,485; September 11, 1979; assigned to BASF AG, Germany.

Background: It is conventional practice in the textile industry to treat cotton yarns or yarns of other cellulosic staple fibers with hot liquors containing water-soluble, natural or synthetic products, before the yarns are run on a loom. This yarn pretreatment, referred to as sizing, is carried out to increase the mechanical strength of the spun yarns so that they are better able to withstand the high stresses during weaving than would be possible in the raw and untreated state. It is an object of the process to provide a water-soluble polymer based on acrylic or methacrylic acid which can be used as a size and which, unlike conventional sizes, produces moisture-insensitive films on the sized staple fiber yarns.

Components: Preferably, the compounds of the process contain from 35 to 85% by weight of acrylic acid as component (a) and from 15 to 65% by weight of acrylonitrile as component (b). They may or may not contain up to 20% by weight of acrylamide and/or methacrylamide. The viscosity of the alkaline earth metal polymer salts of the process in 15% strength aqueous solution at 85°C is from 50 to 1,000 cp, preferably from 200 to 600 cp. The viscosities were in each case measured by means of a Brookfield RVT viscometer at 100 rpm.

Prevention of $CaCO_3$ Scale in Cooling Water

Highlights of the Technological Achievement: When certain low molecular weight polyacrylic acid salts are combined with an organic phosphate ester, an improved composition is afforded for preventing scale in industrial cooling systems.

Reference: G.T. Farley; U.S. Patent 4,168,230; September 18, 1979; assigned to Nalco Chemical Company.

Background: It is known that certain phosphate esters are capable of being added to hardness-containing industrial cooling waters to prevent scale formation on various heat transfer surfaces. These esters somehow prevent hardness-forming salts, e.g., calcium, magnesium, barium, iron and the like from precipitating from these waters even though their normal saturation solubilities have been exceeded.

By interfering with these hardness-forming ions, it is possible to operate industrial cooling systems such as cooling towers with inordinately high amounts of hardness and alkalinity present without scale formation occurring. This process is predicated on the discovery that by adding a small amount of a particular polymeric material in combination with the phosphate esters, it is possible to substantially increase the amount of supersaturation of potential scalants that may be carried in industrial cooling waters without deposition on heat transfer surfaces occurring.

Process Conditions: The process provides a composition for preventing scale in industrial cooling water systems which comprises: (1) a water-soluble polyacrylic acid having a molecular weight between 3,000 and 12,000 and (2) a phosphate ester of a polyol of the formula:

$$(HO-)_x R[-O(R_1O-)_z CH_2CH_2OH]_y$$

wherein R is a saturated, hydrocarbon radical having 2 to 6 carbon atoms, R_1 is a member selected from the group consisting of:

$$-CH_2CH_2- \text{ and } \left(\begin{array}{c}-CH_2CH- \\ \quad\;\; | \\ \quad\;\; CH_3\end{array}\right)$$

wherein x is a number average in the range of 0 to 5, inclusive, y is a number average in the range of 1 to 6, inclusive, and the sum of x plus y equals 3 to 6, and z is a number average in the range of 0 to 3, inclusive, with the weight ratio of (1):(2) being within the range of 1:5-5:1.

Raw Materials: The water-soluble polyacrylic acid is always used in the form of its alkali metal ammonium or amine salts since the acid per se is relatively insoluble. These polymers have an average molecular weight based upon intrinsic viscosity within the range of 3,000 to 12,000 with a preferred molecular weight range being 3,000 to 9,000. These materials are readily prepared by solution polymerization of acrylic acid or they may be prepared by the hydrolysis of low molecular weight polyacrylonitriles.

The phosphate esters are described in U.S. Patent 3,462,365. These esters comprise phosphate esters of a phosphated composition derived by the reaction of phosphorus pentoxide with a polyol composition of the formula:

$$(HO-)_x R[-O(R_1O-)_z CH_2CH_2OH]_y$$

wherein R is a saturated, hydrocarbon radical having 2 to 6 carbon atoms, R_1 is a member selected from the group consisting of:

$$-CH_2CH_2- \text{ and } -\underset{\displaystyle CH_3}{\underset{|}{C}}HCH_2-$$

wherein x is a number average in the range of 0 to 5 inclusive, y is a number average in the range of 1 to 6, inclusive, preferably 3 to 5, and the sum of x plus y equals 3 to 6, and z is a number average in the range of 0 to 3, inclusive.

Preferred embodiments include those in which the polyol is derived by oxyethylation of glycerol with 1.5 to 2.5 mols of ethylene oxide per mol of glycerol; those in which the polyol is derived by the oxyethylation of mannitol with 2 to 20 mols of ethylene oxide per mol of mannitol; those in which the polyols are derived by the oxyethylation of sorbitol with 2 to 20 mols of ethylene oxide per mol of sorbitol; those in which the polyol is derived by the oxyethylation of trimethylolpropane with 1.5 to 2.5 mols of ethylene oxide per mol of trimethylolpropane. Examples of these materials prior to reaction with either P_2O_5 or polyphosphoric acid are set forth in Table 1 of U.S. Patent 3,462,365.

ALKYDS

PROCESSING

Water-Soluble Alkyd Resins with Reduced Yellowing and Improved Weatherability

Highlights of the Technological Achievement: Preparing an alkyd resin by forming a polyester having OH groups, an acid value of 5 or less, an OH value of 40-200 and an average MW of 600-2,500; reacting the OH groups with an aromatic polycarboxylic acid anhydride to adjust the acid value to 25-80 and form a carboxyl group-containing alkyl resin; and neutralizing the resin with a base.

Reference: M. Kubo, M. Nakanishi and S. Watanabe; U.S. Patent 4,200,560; April 29, 1980; assigned to Daicel Ltd., Japan.

Background: Water-soluble alkyd resins are obtained usually by treating oil-soluble resins to render them water-soluble. Oil- and fat-modified, water-soluble alkyd resins contain a fatty acid ester at a side chain and are obtained by reacting dicarboxylic acids such as phthalic anhydride, polyhydric alcohols such as glycerol and fatty acids or fatty acid glycerides. Baked coatings of paints obtained by using these oil- and fat-modified alkyd resins as raw materials have excellent properties, but have the defect that they yellow markedly and have poor weatherability.

One known water-soluble alkyd resin prepared by using a 1,2-epoxyalkane is a reaction product of 1,2-octene oxide, phthalic anhydride, and dipentaerythritol (e.g., as described in British Patent 1,298,773). This alkyd resin is prepared by the ring-opening esterification of phthalic anhydride with dipentaerythritol and then the ring-opening esterification reaction of 1,2-octene oxide with the resulting product. Baked coatings from this resin are hard and brittle and this resin lacks compatibility with other resins. Hence, this resin is not entirely satisfactory as a resin for paints.

It has been found that the defects of the prior art oil- and fat-modified resins of marked yellowing and poor weatherability can be eliminated by using a 1,2-epoxyalkane or a straight-chain internal epoxyalkane as one component of the alcohol as a raw material for the alkyd resin instead of the unsaturated fatty acid which is a cause of these defects. This finding has led to the accomplishment of the method.

Process Conditions: A process for preparing a water-soluble alkyd resin, which comprises (1) reacting

(A) At least one of (a) a 1,2-epoxyalkane containing 6 to 28 carbon atoms; (b) a straight-chain internal epoxyalkane containing 6 to 20 carbon atoms; (c) a mixture of one or more of the 1,2-epoxyalkane (a) and/or the internal epoxyalkane (b); or (d) a mixture of one or more of the 1,2-epoxyalkane (a) and/or the internal epoxyalkane (b) and an unsaturated fatty acid containing 12 to 18 carbon atoms in a molar ratio of about 4:1 to 2:1 of the 1,2-epoxyalkane (a) and/or the internal epoxyalkane (b) to the unsaturated fatty acid; with

(B) At least one of (a') an aromatic carboxyl compound selected from the group consisting essentially of phthalic anhydride, isophthalic acid, terephthalic acid, an ester of isophthalic acid and an ester of terephthalic acid; (b') a mixture of two or more of the aromatic carboxyl compounds (a'); or (c') a mixture of one or more of the aromatic carboxylic compounds (a') and an aliphatic dicarboxylic acid having 3 or more carbon atoms or an anhydride thereof; and

(C) A polyhydric alcohol containing 2 to 10 carbon atoms to form a polyester containing hydroxyl groups and having an acid value of about 5 or less, a hydroxyl value of 40 to 200 and an average molecular weight of 600 to 2,500;

(2) Reacting the hydroxyl groups of the polyester with an aromatic polycarboxylic acid anhydride to adjust the acid value of the polyester to 25 to 80 and form a carboxyl group-containing alkyd resin; and

(3) Neutralizing the carboxyl group-containing alkyd resin with a base.

Examples of 1,2-epoxyalkanes containing 6 to 28 carbon atoms which can be used in the process include 1,2-epoxyhexane; 1,2-epoxyheptane; 1,2-epoxyoctane; 1,2-epoxynonane; 1,2-epoxydecane; 1,2-epoxyundecane; 1,2-epoxydodecane; 1,2-epoxytetradecane; 1,2-epoxyhexadecane; and 1,2-epoxyoctadecane. These 1,2-epoxyalkanes can be represented by the general formula:

(1)

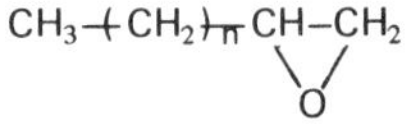

where n is an integer of from 3 to 25. Within this general formula (1) 1,2-epoxyalkanes with any number of carbon atoms can be used. However, to obtain alkyd resins with good water-solubility 1,2-epoxyalkanes with 12 to 18, especially 12 to 14, carbon atoms are preferred.

The straight-chain internal epoxyalkane which can be used in this process is an epoxyalkane in which 6 to 20 carbon atoms are bonded in a straight chain without branching and the oxirane oxygen is bonded to adjacent carbon atoms other than the terminal 1,2-positions (internal epoxyalkanes). Examples of such internal epoxyalkanes which can be used in the process include 2,3-epoxyhexane; 3,4-epoxyoctane; 3,4-epoxydecane; 2,3-epoxyundecane; 3,4-epoxyundecane;

5,6-epoxydodecane; 3,4-epoxytridecane; 5,6-epoxytetradecane; 4,5-epoxyoctadecane; and 8,9-epoxyeicosane. These internal epoxyalkanes have the general formula (2):

(2)

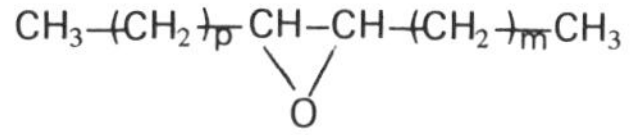

where p and m represent 0 or a positive integer and satisfy the relationship $2 \leqslant m + p \leqslant 16$. Within this general formula (2), straight-chain internal epoxyalkanes with any number of carbon atoms can be used. For the preparation of especially good alkyl resins, internal epoxyalkanes having 8 to 14 carbon atoms are preferred.

These internal epoxyalkanes can be obtained, for example, by converting the corresponding n-paraffins into straight-chain internal olefins by catalytic dehydrogenation, and epoxidizing the straight-chain internal olefins with organic peroxides such as peracetic acid or perbenzoic acid or other peroxides such as hydrogen peroxide.

First Step – The reaction for forming the alkyd resin comprises two steps. In the first step, while a mixture of the acid, i.e., the aromatic carboxyl compound and the aliphatic dicarboxylic acid, if present, and polyhydric alcohol, is being heated to 100° to 150°C with stirring, the 1,2-epoxyalkane or the straight-chain internal epoxyalkane is gradually added to the mixture and reacted. A solvent azeotroping with water such as xylene is then added in an amount of about 5 to 20% by weight, and while heating the mixture at 150° to 250°C with stirring, water is removed from the reaction system. In this way, the dehydroesterification is allowed to proceed to produce a hydroxyl group-containing polyester having an acid value of about 5 or less.

Second Step – The second step of the reaction comprises adding an acid anhydride such as phthalic anhydride or trimellitic anhydride to the hydroxyl-containing polyester obtained in the first step and heating the mixture at 100° to 150°C with stirring to obtain a carboxyl group-containing alkyd resin. When this alkyd resin has an acid value of at least about 25, the resin becomes soluble in water by the addition of a partially or completely water-miscible organic solvent and neutralization with a base to a pH of 7.0 to 9.0. Generally, alkyd resins having an acid value of 25 to 80 are suitable.

Water-Solubilization: The water-solubilization treatment of the alkyd resin obtained in the second step is carried out by dissolving the alkyd resin in at least one partially or completely water-miscible organic solvent. A base is then added to the extent of neutralization and the resulting neutralized alkyd resin solution is diluted with water. A suitable temperature for the neutralization ranges from 0° to 50°C. Of course, the viscosity of the solution is most dependent upon the molecular weight of the alkyd resin. The viscosity is also determined depending on the type and amount of the organic solvent used.

Aqueous Nitrocellulose/Alkyd Resin Dispersion

Highlights of the Technological Achievement: Aqueous dispersion of a binder for coating compositions comprising (a) 5 to 35% by weight of nitrocellulose; (b) 0 to 30% by weight, based on (a) of plasticizer; (c) 5 to 50% by weight of alkyd resin; (d) 30 to 80% by weight of water; (e) 0.5 to 10% by weight of emulsifier; and, optionally, (f) 4 to 40% by weight of a urea resin and/or melamine resin.

Reference: H.-J. Traenckner; J. Goossens, W. Kremer and H. Krönke; U.S. Patent 4,177,172; December 4, 1979; assigned to Bayer AG, Germany.

Background: Aqueous coating compositions, for example for lacquering furniture, have hitherto generally contained synthetic resins such as acrylic polymers or cellulose esters, as the film-forming agents.

Nitrocellulose, usually combined with alkyd resins, is used for lacquering wood or similar materials. This combination leads to coatings of high gloss, adequate hardness, good elasticity and excellent flow, if processed from organic solutions. All attempts to emulsify alkyd resins and nitrocellulose have led either to emulsions of inadequate stability or to emulsions with particles which are too coarse and which do not permit levelling to give glossy coatings.

Process Conditions: The emulsifier is chosen from a partially saponified polyvinyl acetate having a degree of hydrolysis of 70 to 97, preferably 80 to 92, in particular 86 to 89%; a polyvinylpyrrolidone; and a hydrolyzed vinyl acetate/vinylpyrrolidone copolymer, the particle size of the dispersed phase being from 0.1 to 1.5 μ.

The dispersions are stable and can give coatings of substantially faultless gloss.

By alkyd resins there are to be understood polycondensates produced by polycondensation of alcohols and carboxylic acids. In the alkyd resins which are used for the coating dispersions according to the process, the proportion of oil can be 5 to 50%, preferably 20 to 45%, by weight, calculated as triglyceride and relative to the alkyd resin.

The molecular weight, determined as number average of the alkyd resins, is 2,000 to 10,000. (Molecular weights below 5,000 are determined by vapor pressure osmosis in dioxane and acetone; if both values differ from each other, the lower value is considered to be the more accurate one; molecular weights above 5,000 are determined by membrane osmosis in acetone.)

The alkyd resins are prepared according to known methods by condensation. The starting materials are usually reacted at temperatures of 140° to 250°C in an inert gas atmosphere, e.g., nitrogen, with the splitting-off of water until the desired acid number has been reached. These alkyd resins can optionally be mixed with other components, for example, also with oil-free polyesters.

The aqueous dispersions according to the process can be produced by dissolving the alkyd resin described above, the nitrocellulose and the plasticizer in a polar solvent which is not soluble in water, and which can be removed by distillation from the emulsion to be formed, combining the solution with an aqueous solution of the emulsifier and thus forming an oil-in-water emulsion by means of high shearing forces. The organic solvent is then removed by distillation, on its own or in the form of an aqueous azeotrope.

Solvents which are particularly suitable are alkyl acetates and alkyl propionates. The aqueous dispersion according to the process which remains can be diluted with more water and/or a suitable coalescing agent can be added to it; the dispersions of the process are starting materials for the preparation of lacquers which can be produced by the addition of water, coalescing agents, dyestuffs, pigments and other auxiliaries.

Aqueous Dispersions of Alkyd and/or Amino Resins Containing Amine Oxides

Highlights of the Technological Achievement: Aqueous dispersions of crosslinkable alkyd and/or amino resins, which contain surface-active amine oxides of the formula:

$$\begin{array}{c} R_2 \\ | \\ R_1-N\rightarrow O \\ | \\ R_3 \end{array}$$

where R_1 is a hydrocarbon radical of 8 to 30 carbon atoms which, optionally, contains 1 to 3 ether groups and/or up to 4 hydroxyl groups and R_2 and R_3 are independently an aliphatic or cycloaliphatic hydrocarbon radical of 1 to 16 carbon atoms which, optionally, are interrupted by heteroatoms and/or together contain up to 8 hydroxyl groups, are useful as coating agents or varnish binding agents.

Reference: H. Eschwey, W.-D. Willmund and W. Gress; U.S. Patent 4,171,293; October 16, 1979; assigned to Henkel Kommanditgesellschaft auf Aktien, Germany.

Background: Alkyd resins having high acid numbers are known to be used for the production of varnish binding agents, which have good water-solubility in the form of their amine salts. The disadvantages of systems of this kind, apart from their unsatisfactory waterproofness, are to be seen, in the majority of cases, in their considerable content of volatile amines and, in some cases, of toxic auxiliary solvents. When the varnish layer is baked, these substances must be trapped by means of suitable absorption systems.

It is further known to modify the varnish binding materials by the incorporation of hydrophilic chains, particularly polyethylene glycol chains, in such a manner that self-emulsifying alkyd resins or melamine resins result. Furthermore, even without incorporation in the resin molecule, it is possible to use polyglycol ether derivatives, such as the adducts of ethylene oxide to fatty alcohols, as low molecular weight, nonionic emulsifiers. In both cases, however, the hydrophilic radical remains active in the baked varnish layer and, due to its hydrophilic nature, gives rise to a susceptibility to penetration by water. In addition, a marked reduction in film hardness also often occurs.

Process Conditions: The amine oxide compounds can advantageously contain hydrocarbon radicals which are, optionally, substituted by hydroxyl groups and/or contain a heteroatom or heteroatoms in the hydrocarbon chain. Amine oxide compounds of this type decompose at temperatures between 80° and 200°C into nonemulsifying fragments which, in certain cases, react with the binding agent by means of additional functional hydroxyl groups, whereby they can be chemically incorporated into the varnish layer.

As the coating or varnish binding agent of the process, crosslinkable alkyd resins can be used which cannot be converted by amine neutralization into a water-soluble or dilutable form and are thus such as are normally applied in organic solvents. These alkyd resins suitably have an acid number of preferably from 10 to 25. Their fatty acid content should be between 20 and 45 wt %, relative to the solid resin. As the crosslinkable aminoplast resin component of the process, those known preliminary resin condensates can be used which are also used in a conventional manner in systems containing a solvent, as, for example, the alkyl ethers of hexamethylolmelamine. Examples of such ethers of hexamethylolmelamine are hexamethyl ether and hexabutyl ether. Particularly favorable results are obtained when the alkyd resins and aminoplast resins are combined with each other.

In practice resin combinations of 45 to 85%, more particularly 60 to 85%, by weight of alkyd resin with 55 to 15%, more particularly 40 to 15%, by weight of aminoplast resin have proven quite effective. In general, the total amount of resin used is such that the aqueous dispersion of the process has a resin solids content of 30 to 70% by weight.

Water-Soluble Alkyds Modified with Multifunctional Monomers

Highlights of the Technological Achievement: Improved water-reducible alkyd resins are made by incorporating α,β-ethylenically unsaturated acid esters of polyols into the final resin product.

Reference: E.J. Kuzma; U.S. Patent 4,145,319; March 20, 1979; assigned to Celanese Corporation.

Background: It has been found that incorporation of certain multifunctional polyol derivatives into the water-reducible alkyd resins improves the properties of the resins, particularly in film-forming compositions, e.g., paints, as evidenced by rapid drying in short time periods and significant improvement in hardness, gloss and drying times of enamels prepared therewith. It has also been discovered that the addition of the multifunctional polyol derivatives to the reaction can be critical in that, when added to the initial alkyd resin reaction mixture, it may lead to gelling of the reaction mixture, or increasing the viscosity of the alkyd resin which adds considerable difficulty in the second stage reaction to impart water-solubility.

It has been found that addition of the multifunctional polyol just prior to, if not simultaneous with, the trimellitic acid addition substantially obviates the aforementioned difficulties and permits the reaction to proceed to desired acid numbers (as a measure of free carboxyl groups) and desired high, but workable, viscosities of the final product.

Process Conditions: The preferred multifunctional polyol derivatives are acrylic acid or methacrylic acid esters of the selected polyol in which at least 2 hydroxyl groups are esterified.

The multifunctional polyol derivatives can be used at any level to attain the desired result. As little as 0.5% by weight, based on the weight of the reaction mixture, will provide some beneficial results. Generally, from 1 to 10% can be used for most compositions, although larger amounts can be used.

For convenience, the initial alkyd resin is prepared in batchwise fashion by condensing the selected polyol with the monocarboxylic acid and polycarboxylic acid in a single reactor. The mixture of starting materials is heated at elevated temperature until the acid number is preferably less than 10. Usually, reaction temperatures between 180° and 250°C for time periods of from 6 to 12 hours are sufficient.

Since water is evolved, the course of the reaction can be monitored by trapping the water evolved. As can be expected, the reaction temperature and time required for any specific mixture will vary, depending upon the starting materials. Of course, when low reaction temperature is used, the time of reaction will be longer than with higher reaction temperature.

After the initial alkyd resin formation is complete, trimellitic acid, preferably in the form of the anhydride, and the multifunctional polyol derivative are added to the mixture and the mixture heated to form the final product.

The heating is continued until the acid number of the resin is about 50 to 60 and preferably until the viscosity is at least at a value of T on the Gardner scale. The reaction is conveniently monitored by measuring the water formed during the course of the reaction.

When alkyd resin formation starts, water is given off and continues until the reaction is complete. Thus, the reaction mixture is heated to a temperature where water evolution commences and the reaction is followed by merely collecting the water in a suitable trap.

Heating is continued until the acid number (solids) reaches the desired value. For most such reactions, temperatures between 180° and 220°C are adequate to provide substantially complete reaction in time periods of from about one to several hours.

Examples 1 Through 6: Alkyd resins are prepared in accordance with the following procedures with results tabulated in the table. They were formulated into white gloss paint formulations and tested, the results also being tabulated in the table.

Procedure A – Soya fatty acid, trimethylolpropane, isophthalic acid and trimethylolpropane triacrylate are charged into a three-neck flask, heated under nitrogen sparge and reacted to an acid number of less than 10. The contents of the flask are cooled to 180°C and trimellitic anhydride is added. The contents are then heated to 190° to 200°C and reacted to an acid number of 50 to 60. The resin is cooled to 100°C and diluted to 50% solids with an 80/20 water/monobutyl ether of ethylene glycol solution [Glycocel EB (Celanese Corporation)]. The solution is then neutralized to a pH of 7 to 8.5 with NH_4OH.

Procedure B – Soya fatty acid, trimethylolpropane and isophthalic acid are charged into a three-neck flask, heated under nitrogen sparge and reacted to an acid number of less than 10. The contents of the flask are cooled to 180°C and trimethylolpropane triacrylate and trimellitic anhydride are added. The contents are then heated to 190° to 200°C are reacted to an acid number of 50 to 60. The resin is cooled to 100°C and diluted to 50% solids with an 80/20 water/Glycocel EB solution. The solution is then neutralized to a pH of 7 to 8.5 with NH_4OH.

	Sample Number 1	2	3	4	5	6
Components, wt %						
Soya fatty acids	357	357	357	357	357	357
Trimethylolpropane	335	335	335	335	–	334.99
1,6-Hexanediol	–	–	–	–	–	–
Pentaerythritol (mono)	–	–	–	–	127.51	–
Ethylene glycol	–	–	–	–	116,26	–
Isophthalic acid	299	299	299	299	298.80	149.40
Adipic acid	–	–	–	–	–	131.40
Phthalic anhydride	–	–	–	–	–	–
Trimellitic anhydride	99	99	99	99	98.99	98.99
Trimethylolpropane triacrylate	55	22	22	55	19.97	53.58
Procedure used	A	B	A	B	B	B
Viscosity at 50% in 80/20 water/Glycocel EB	Gelled	Z-2	Z-6+	Z-4	Z-1	X-¾
Physical properties						
Set to touch, minutes	7	8	7	8	7	8
Tack-free cure before testing, minutes	43	39	34	45	*	*
Tukon hardness						
1 Day	2.7	2.8	3.0	2.1	1.7	1.4
3 Days	5.3	5.6	8.0	3.4	2.9	2.1
5 Days	6.6	7.2	9.6	5.2	2.8	2.1
7 Days	9.1	9.1	11.4	7.5	3.1	2.2
Crosshatch adhesion (pass), %	100	100	100	100	100	100
Conical mandrel	Passed	Passed	Passed	Passed	Passed	Passed
Reverse impact (pass/fail)	0/4	0/4	0/4	0/4	144/148	156/160
Gloss, 60%	94.8	95.2	96.8	93.8	86.4	85.0

*Overnight.

Acrylate-Modified Alkyd Resin

Highlights of the Technological Achievement: An acrylate-modified alkyd resin obtained by the process consisting of the steps (a) forming an alkyd resin by reacting a conjugated fatty acid and an aromatic dicarboxylic acid or anhydride and dimethylolpropionic acid with an excess of polyol; and (b) reacting the alkyd so-formed with an alkyl-substituted or unsubstituted acrylic acid or ester thereof or mixture thereof.

Reference: R.R. Harris and W.J. Pollack; U.S. Patent 4,133,786; January 9, 1979; assigned to International Minerals & Chemical Corp.

Background: Vegetable oil acids suitable for use in the practice of this process include aliphatic conjugated, unsaturated fatty acids or mixtures thereof, generally equivalent to linoleic acid in unsaturation. Preferred acids include, but are not limited to, tall oil fatty acids and linseed oil fatty acids. Both acids are well-known to those skilled in the art.

Tall oil fatty acids are particularly preferred; they are derived from tall oil, which is a by-product of the paper industry. Also, blends of dehydrated castor and soybean oil or a blend of tung oil fatty acids and soybean fatty acids having equivalent conjugated unsaturation can be used.

The amount of vegetable oil acid used is generally from 18 to 35% by weight, preferably 18 to 25%, based on total ingredients; 19 to 21% is particularly preferred. Although tall oil fatty acids are preferred for the practice of this process, linseed fatty acids or linoleic acid can also be used.

Process Conditions: The alkyd portion of the resin is prepared by charging the above ingredients, along with 1 to 2% of reflux solvent (e.g., an aromatic hydrocarbon; e.g., toluene, xylene or ethyl benzene; or a ketone such as methylpropyl ketone, methyl isobutyl ketone or methyl-n-butyl ketone, preferably xylene) to a reaction vessel and after mixing well the ingredients are heated to about 210° to 215°C while removing water of reaction until an acid value of 38±2 is reached. At this point, there are still unreacted hydroxy groups present.

The reaction mixture is allowed to cool to 138°C and 5% of additional ketone or aromatic hydrocarbon, preferably ethyl benzene, is added. The reaction mixture is well mixed and then a mixture of an aliphatic olefinic acid, preferably methacrylic acid, and substituted or unsubstituted acrylic ester, preferably methyl methacrylate, is added over a period of about 5 hours, while holding the temperature at 138°C. Methacrylic acid is used in an amount of about 2 to 4%, preferably 2.25 to 3.5% and 2.5 to 3.0% is more preferred.

When acrylic acid or other substituted acrylic acid is used, the amount is selected on an equimolar basis to the methacrylic acid. Methyl methacrylate is used in an amount of 15 to 25%, preferably 19 to 21%. Other acrylic esters such as methyl acrylate, ethyl acrylate or ethyl methacrylate can be substituted for methyl methacrylate, when desired.

Preferably, a polymerization catalyst is used in the mixture. Any catalyst known in the art can be used, the catalyst not being regarded as a part of the process. Di-tert-butyl peroxide is a preferred catalyst and is used in an amount of about 0.7 to 0.9%. After all the acrylic monomers have been added, the mixture is heated for about another 2 hours at 138°C; the temperature is then raised to 155°C for 1 hours for complete conversion.

When the reaction is complete, the acid value is 45 to 50 and the Gardner viscosity is Z_5 to Z_7. The mixture is then cooled to 138°C and monoalkyl ether of ethylene glycol, e.g., the ethyl or butyl monoether (17 to 18%) and n- or isobutyl alcohol (5 to 6%) is added. When cool, the mixture is ready for use.

Product: The acrylic-modified alkyd resin of this method has particular utility as a fast, air-dry, water-soluble vehicle in high gloss enamels intended for topcoats. It is especially useful in white and light-colored enamels intended for application on implements.

Thermoplastic Powder Coating Systems

Highlights of the Technological Achievement: A thermoplastic powder coating composition is provided, comprising particles of polymeric material comprising a mixture of polymer A and polymer B, polymer A and polymer B being selected from the group consisting of polymer of an alkyd, polyester and polyester urethane. One of either polymer A or polymer B contains an average of about 0.9 to 1.1 reactive groups per molecule and the other of polymer A or polymer B contains an average of 1 to 10 reactive groups per molecule. The reactive groups on polymer A are coreactive in a condensation reaction with the reactive groups on polymer B; and polymer A and polymer B are compatible. Any alkyl resin used in baking type coating compositions may be used.

Reference: E.B. FitzGerald and W.C. Meyer; U.S. Patent 4,200,566; April 29, 1980; assigned to E.I. DuPont de Nemours and Company.

Background: There has been substantial research to develop powder coating paint systems. One reason for this is to eliminate the need for organic liquid carrier for the film-forming polymer which, when vaporized, may cause pollution problems. With a powder coating system, particles of polymer are applied to a substrate such as by spraying. Often, the particles are caused to adhere to the substrate by suitable means, such as by imparting electrostatic charge to the particles of polymer. Afterwards, the particles of polymer are caused to coalesce by suitable means such as by baking.

One of the problems that has been encountered in existing powder coating systems is that if the particles of polymer are of low enough molecular weight, they will flow and coalesce when baked. The resulting baked paint film may not have acceptable toughness. One solution of this problem has been through the use of thermosetting powder coating systems. Such systems, however, once baked, are no longer suitable in applications requiring subsequent flow.

Process Conditions: Generally, polymer A and polymer B should have a low enough glass transition temperature to allow the particles comprised of them to melt, flow out and coalesce at the prevalent baking conditions. Often, polymer A and polymer B will have a glass transition point of 20° to 115°C.

Generally, polymer A and polymer B should have a molecular weight such that when polymer A and polymer B react with each other, the thermoplastic condensation polymer so-produced will have a weight average molecular weight of 40,000 to 200,000. Often polymer A and polymer B will have a weight average molecular weight of 5,000 to 180,000.

One of either polymer A or polymer B should contain an average of at least one reactive group per molecule and the other of either polymer A or polymer B should be substantially monofunctional, which, for the purposes of this process, is defined to be a polymer that contains an average of 0.5 to 1.3 reactive groups per molecule.

Under these most preferable conditions, the product of polymer A with polymer B can be substantially uncrosslinked. This means that the final coating is substantially thermoplastic and is capable of subsequent reflow in certain applications.

Preferably, the number of reactive groups contained in either of polymer A or polymer B that contains at least about one reactive group per molecule should be about 1 to 20, preferably about 1 to 10, and most preferably about 2 to 5. There should be present a sufficient number of equivalents of the substantially monofunctional polymer to theoretically react with at least about 50% of the reactive groups of the other polymer. The following are examples of pairs of coreactive groups, the preferred being carboxyl with epoxy:

Carboxyl with epoxy
Carboxyl with isocyanate
Carboxyl with hydroxyl
Carboxyl with aziridinyl
Acid anhydride with hydroxyl
Acid chloride with hydroxyl
Acid chloride with amine
Isocyanate with hydroxyl
Amine with epoxy

APPLICATIONS

Alkyd Resin Primer

Highlights of the Technological Achievement: An aqueous primer composition useful for priming substrates such as steel, polyester reinforced with fiber glass, acrylonitrile/butadiene/styrene plastics that are used in automobile and truck bodies and can be dried at ambient temperatures.

Reference: M.H. Verma; U.S. Patent 4,180,610; December 25, 1979; assigned to E.I. DuPont de Nemours and Company.

Background: It has become increasingly important for paint manufacturers to provide coating compositions that are non-air-polluting and that can be dried at ambient temperatures to save energy. Along with the above characteristics, the automotive and truck body repair industry have requirements that primers have excellent adhesion to all types of painted and unpainted substrates, provide good resistance to corrosion, can be easily sanded to give a smooth surface and provide a surface to which conventional refinish paints will adhere.

Conventional alkyd resin primers are well known, but none have a combination of properties that meet the above requirements, whereas the aqueous primer composition of this process meets these requirements.

Process Conditions: An aqueous primer composition useful for priming substrates contains 5 to 50% by weight of film-forming binder and 50 to 95% by weight of an aqueous medium, wherein the binder consists essentially of an alkyd resin which is the esterification product of drying oil fatty acids, an aromatic dicarboxylic acid, an anhydride of an aromatic tricarboxylic acid and a polyhydric alcohol; and 0.05 to 1.0% by weight, based on the weight of the composition, of an anionic polyelectrolyte dispersant. The composition contains a sufficient amount of ammonia or an amine to provide the composition with a pH of 6.5 to 8.5.

A more detailed description of the process is as follows. The aqueous primer composition contains about 5 to 50% by weight of film-forming binder. Generally, 20 to 40% by weight of binder is used. The remainder of the composition is an aqueous medium that contains water and dispersants. Optionally, the aqueous medium contains water-soluble or water-miscible solvents and dispersants.

Usually the primer contains dispersed pigments in a pigment-to-binder weight ratio of about 300:100 to 10:100. Typical pigments that are used are talc, barytes, zinc molybdate, calcium molybdate, basic zinc/calcium molybdate, zinc oxide, calcium phosphosilicate and mixtures thereof.

The alkyd resin used in the primer composition is the esterification product of drying oil fatty acid, an aromatic dicarboxylic acid, an anhydride of an aromatic tricarboxylic acid and a polyhydric alcohol. To prepare the alkyd resin, the above constituents, along with an esterification catalyst, are charged into a reaction vessel. Either a conventional fusion or solution process using conventional equipment can be used to prepare the alkyd resin. Generally, reaction temperatures of 200° to 275°C for 1 to 5 hours are used to prepare the resin.

The resulting alkyd resin has an acid number of 20 to 100.

One particularly preferred alkyd resin that forms a high quality composition is the esterification product of soya oil fatty acids/phthalic acid/trimellitic anhydride/trimethylolpropane that has an acid number of about 20 to 100.

The alkyd resin is dispersed in water by mixing the resin in an amine or ammonia and then with water. A sufficient amount of amine or ammonia is used to provide the resulting composition with a pH of about 6.5 to 8.5. Generally, the pH of the primer composition is about 7 to 8. Triethylamine and N,N-dimethylethanolamine are preferred amines, since these form good quality products.

Generally, 0.05 to 1.0% by weight, based on the weight of the composition, of a dispersant is added. Typically useful dispersants are anionic polyelectrolytes such as Polywet ND-1 and Polywet ND-2. Preferred anionic polyelectrolyte dispersants are polymers of acrylonitrile, a polyacrylate such as a metal polyacrylate and are terminated with sodium sulfonate. One particularly useful dispersant of this type has the formula:

$$\text{-(}CH_2-CHCN\text{)}_n\text{-(}CH_2-\underset{\displaystyle COOH}{\underset{|}{CH}}\text{)}_m-SO_3M$$

where n and m are sufficiently large to provide a molecular weight of about 500 to 2,500 and M is an alkali metal ion such as sodium. Preparation of these dispersants are described in U.S. Patent 3,646,099.

Example: A water-based pigment dispersion is prepared as follows. Portion 1 contains 1,730 pbw alkyd resin solution (80% solids alkyd resin of soya oil fatty acids/phthalic acid/trimethylolpropane/trimellitic anhydride having an acid number of 50 to 60 in a propylene glycol monopropyl ether solvent); 35 pbw ethylene glycol monobutyl ether; 8 pbw anionic polyelectrolyte dispersant (Polywet ND-1, which is believed to be a polymer of acrylonitrile/polyacrylate terminated with sodium sulfonate, prepared according to U.S. Patent 3,646,099); and 54 pbw loss of dry inhibitor solution metallic soap of calcium and zinc in mineral spirits (Nyact).

Portion 2 contains 2,284 pbw deionized water; and 92 pbw triethylamine. Portion 3 contains 814 pbw talc pigment; 3,475 pbw barytes pigment; 927 pbw basic zinc/calcium molybdate pigment [Molywhite 212 (Sherwin Williams)]; and 581 pbw iron oxide.

Portion 1 is charged into a mixing vessel and thoroughly mixed. Portion 2 is added with mixing and mixing is continued for 15 minutes. The pH of the resulting mixture is adjusted to 7.0 to 7.2 by the addition of triethylamine. Portion 3 is then added with mixing and mixing is continued until a smooth mixture is obtained. This mixture is charged into a conventional ball mill and ground for 48,000 cycles. The resulting mixture is a pigment dispersion having a weight solids of about 72%, a pH of about 7.5 to 7.8; and a pigment to binder ratio of 408.96:100, where the binder is the alkyd resin in the dispersion.

A primer is prepared as follows. Portion A contains 1,205 pbw alkyd resin solution (described above); 319 pbw ethylene glycol monobutyl ether; 26 pbw cobalt naphthenate solution (54% solids in mineral spirits and contains 6% cobalt); 8 pbw drier stabilizer solution (35% solution of 1,10-phenanthroline and 2-ethylhexoic acid in n-butanol); and 8 pbw methyl ethyl ketone.

Portion B contains 1,291 pbw deionized water; 90 pbw triethylamine; and 7,053 pbw pigment dispersion (described above).

Portion A is charged into a mixing vessel and thoroughly mixed and then Portion B is added and thoroughly mixed to form a primer. The pH of the primer is adjusted to 7.7 by the addition of triethylamine.

The resulting primer has a weight solid of 60% and a pigment-to-binder ratio of 206:100, where the binder is the alkyd resin in the primer. Sufficient deionized water is mixed with the primer to reduce the viscosity of the primer to 30 seconds, determined by a No. 2 Zahn cup at 25°C.

Stoving Lacquers Containing Reactive Diluents

Highlights of the Technological Achievement: An alkyd resin stoving lacquer with decreased loss of reactive diluent yields outstandingly hard coatings if a reactive diluent mixture comprising a lactam and a hydroxyl compound are used.

Reference: R. Dhein, P. Höhlein, R. Küchenmeister and W. Beer; U.S. Patent 4,171,294; October 16, 1979; assigned to Bayer AG, Germany.

Background: There are two basic difficulties in using lacquers containing reactive diluents–firstly, very hard coatings have not yet been successfully prepared using alkyd resins or oil-free polyesters with a relatively low molecular weight; and, secondly, it is not possible to avoid loss of the reactive diluents, in particular, in the case of nonaqueous systems during stoving. It has been found that the difficulties mentioned can be avoided when a mixture of a lactam and a hydroxy compound is used as the reactive diluent.

Process Conditions: The process relates to stoving lacquers consisting of (a) 40 to 94%, preferably 50 to 80%, by weight of alkyd resins containing carboxyl and hydroxyl groups (with an oil content of up to 40% by weight) or oil-free polyesters with an acid number of 1 to 100 and average molecular weights between 400 and 10,000 (it being possible for the OH number to be between 30 and 400); (b) 5 to 40%, preferably 10 to 30%, by weight of an aminoplast resin; (c) 1 to 20% by weight of reactive diluent; (d) 0 to 35% by weight of organic solvents; and (e) optionally, pigments and/or further auxiliaries and additives, characterized in that a mixture of a lactam and a hydroxy compound with at least two hydroxyl groups per molecule is used as the reactive diluent and the molar ratio lactam:hydroxy compound should be n:1 and n can assume values from $^1/_{20}$ to 10, preferably $^1/_5$ to 5. The percentage data relate to the sum of components (a), (b) and (c).

In general, in this reaction the mixtures of raw materials are reacted at temperatures from 140° to 250°C in an inert gas atmosphere, for example N_2, with the splitting-off of water until the desired acid number is achieved.

All alcohols which are known as starting components for the polyesters and the preparation of the alkyd resin are suitable as the hydroxy compound component of the reactive diluent. For the purpose of crosslinking reactions and thus for improving the surface hardness, it is advisable to use hydroxy compounds with at least three hydroxyl groups. This is particularly advantageous when relatively low molecular alkyd resins or polyesters are used.

Compounds of the formula:

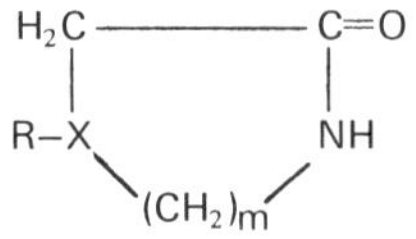

where X denotes nitrogen or CH; R denotes hydrogen, an aliphatic radical with 1 to 30 carbon atoms, an araliphatic radical with 3 to 12 carbon atoms or a pyridine radical which is optionally substituted by C_{1-4} alkyl radicals; and m denotes an integer from 1 to 9, are suitable as the lactam component of the reactive diluent.

Water-Based Coating Composition Containing Polyvinyl Chloride

Highlights of the Technological Achievement: A water-based coating composition is provided that is suitable for food packaging, such as in coating metal containers. The composition comprises a water-dilutable resin, solubilized with ammonia or an amine, dissolved in water or mixtures of water and an alcohol or an ether alcohol into which is dispersed polyvinyl chloride.

Reference: K.G. Davis; U.S. Patent 4,165,304; August 21, 1979; assigned to Mobil Oil Corporation.

Background: Vinyl resins have been used extensively in the packaging industry because of their desirable properties of hardness, flexibility, product resistance, water-resistance, and freedom from taste. Such resins are applied as solutions in organic solvents, which have to be removed, e.g., by evaporation into the air. Because, however, of air pollution regulations and standards, it has become desirable to use water as at least the major portion of the volatile vehicle.

Most vinyl polymers, particularly polyvinyl chloride, do not have functional groups on the polymer that can react with a solubilizing agent such as ammonia or an amine to convert the resin into a water-soluble material. It has been proposed to apply PVC as a hydrosol, i.e., PVC dispersed in water by use of wetting agents (surfactants). Films from such hydrosols have poor water-resistance, probably because the wetting agent remains in the dried film and causes poor performance.

It is the discovery of this process that PVC can be dispersed in solution of a water-dilutable resin other than vinyl chloride-maleic anhydride solubilized with ammonia, NH_4OH or an amine, the solvent being water or a mixture of water and an alcohol or ether alcohol. When applied to a substrate and dried, there is obtained a film suitable for packaging applications.

Process Conditions: In preparing the coating formulas of this process, there is first prepared a solution of water-dilutable resin in an aqueous vehicle. To this solution there is added polyvinyl chloride in powder form so that it is dispersed in the aqueous solution.

The water-dilutable resin will be any resin or mixture of resins having a functional group, such as carboxyl or hydroxyl, that will react with a solubilizing agent, such as ammonia or an amine. A wide variety of such resins have been found utilizable and most are well-known in the art and commercially available.

Typical types of water-dilutable resins include:

> Acrylic resins, both thermoplastic and thermosetting, which resins include copolymers of acrylic acid or methacrylic acid with acrylate and methacrylate resins, polyacrylic acid, polymethacrylic acid and polyglycidyl methacrylate;
>
> Alkyd resins, both oil-modified and oil-free, which alkyds are prepared from polyesterifying dicarboxylic acids such as phthalic anhydride, maleic anhydride, succinic anhydride, adipic acid and sebacic acid, with polyalcohols such as glycerol, glycols, pentaerythritol and sorbitol. These are oil-modified with fatty substances such as oleic acid;
>
> Epoxy resins having an epoxy equivalent weight of 175 to 200; and
>
> Blends of alkyd resins and phenolic resins, which include blends of the aforementioned alkyd resins with phenolic resins such as phenol-formaldehyde resins.

The coating formulation is prepared by adding the PVC powder to the solution of water-dilutable resin with stirring. Alternatively, the PVC powder may be combined with the water-dilutable resin or resins prior to neutralization and water addition. The resultant formulation is a suspension of PVC particles in the solution.

Example: In a mixing vessel were placed 20.97 pbw of an alkyd resin prepared from tall oil fatty acids, isophthalic acid, dimethylolpropane and trimellitic anhydride (75 wt % solids in butoxyethanol, acid number about 10), 1.89 pbw of n-butyl alcohol, and 0.29 pbw of butoxyethanol (butyl Cellosolve). This mixture was stirred at room temperature until uniform.

While continuing agitation, there were added 1.54 pbw dimethylethanolamine until uniform. 23.85 pbw of deionized water were then slowly added to obtain an aqueous solution of the alkyd resin.

Agitation was increased to a high rate and 39.27 pbw of powdered polyvinyl chloride were added and stirring continued until a fineness of grind of greater than 4¾ is obtained as measured on a North Standard gauge. Care was taken that the temperature did not exceed 140°F. Thereafter, another 12.2 pbw of deionized water were slowly added under normal (slow) agitation and the suspension was stirred to afford the final product. Solids content was 55%; viscosity was 90 to 115 sec (No. 4 Ford cup).

The coating formulations can be applied to substrates, such as paper, metal foil, and stock metals. Suitable metals include aluminum, steel and tinplated steel. The coating can be applied by spray-coat, direct or reverse roll coat, flow coat, or gravure coat. After application, the coating is baked for between about 5 seconds and 10 minutes at between about 375° and 750°F.

The above dispersion coating formulation was applied to aluminum test panels and to tinplated steel test panels 9 to 10 mils thick. Application was by roll coat at film weights of 4 to 10 mg/in^2 and the coated panels were baked 10 minutes at 400°F. The baked panels were tested for adhesion. Pertinent results are set forth in the table.

Adhesion Test – A baked coated area on the test panel is crosshatched with individual score lines about 1/16" apart. Scotch tape is then firmly applied to the crosshatched area and removed with a quick snap. The amount of coating remaining on the substrate is observed visually and rated on a scale of zero to ten, with ten being perfect, i.e., no coating was pulled off with the tape.

Pasteurization is carried out by immersing the coated panel in water at 155°F for 30 minutes. The panel is then wiped dry with an absorbent towel and the adhesion test is carried out as described above. Blush, i.e., clouding of the film, is rated on a zero to ten scale, with ten representing no blush. In a more severe test, pasteurization can be carried out at 170°F for 45 minutes.

	Adhesion Rating
Aluminum	
Dry Adhesion	10
Pasteurized	
Adhesion	10
Blush	9
Tinplate	
Dry Adhesion	10
Pasteurized	
Adhesion	10
Blush	9

Water-Dispersible Epoxy Modified Alkyd Resins

Highlights of the Technological Achievement: A water-dispersible epoxy modified alkyd resin is produced by reacting an oil (or oil fatty acid), a polyhydric alcohol, a monobasic acid, a polybasic acid (or its anhydride), a polyoxyalkylene glycol and an epoxy compound.

Reference: R. Ukita, A. Isobe, T. Hirayama and S. Tanaka; U.S. Patent 4,154,709; assigned to Hitachi Chemical Company, Ltd., Japan.

Background: Attention has been given to aqueous coatings which can reduce fire risk and air pollution due to organic solvents in conventional organic solvent type coatings and can prevent the waste of resources due to the use of water as a diluent. However, there are many problems to be solved in using the aqueous coatings, such as drying of coating films being slow, water-resistance of coating films being poor, corrosion-resistance of coating films being poor, and the like.

Process Conditions: The water-dispersible epoxy modified resin can be produced by reacting:

(a) 0 to 50 pbw of at least one oil or oil fatty acid;

(b) 10 to 50 pbw of at least one polyhydric alcohol having 2 to 6 hydroxyl groups in the molecule, except for the component (e), mentioned below;

(c) 0 to 25 pbw of at least one monobasic acid having 6 to 18 carbon atoms in the molecule;

(d) 15 to 70 pbw of at least one polybasic acid having 4 to 10 carbon atoms in the molecule or its anhydride;

(e) 5 to 25 pbw of at least one polyoxyalkylene glycol having a molecular weight of 600 to 20,000; and

(f) An epoxy compound, where the ratio of the hydroxyl groups to the carboxyl groups in the components (a) through (e) is in the range of 0.625 to 1.8, and the amount of the component (f) is 3 to 50 pbw/100 pbw of the total weight of the components (a) through (e).

The water-dispersible epoxy modified alkyd resin can be produced by one-step reaction, two-step reaction or three-step reaction. A preferable example for producing the water-dispersible epoxy modified alkyd resin in the case of using an oil as the component (a) is to react in a first step the component (a), two-thirds to one-third of the component (b) and the component (c) under nitrogen at 170° to 270°C with stirring so as to produce ester exchange while removing water until a suitable acid number is obtained and then to cool the reaction mixture, and subsequently, to produce condensation in a second step by adding the residue of the component (b), the component (d), the component (e) and the component (f) to the reaction mixture and heating at 170° to 270°C under nitrogen with stirring until a suitable acid number and a viscosity are obtained and then to cool the reaction product.

Other methods selected from the following reaction methods can also be used for the production of the water-dispersible epoxy modified alkyd resin. In the case of using at least one oil selected from the group consisting of drying oils, semidrying oils and nondrying oils as the component (a), the component (e) can be added in at least one step among the first step and the second step as mentioned above or in a third step newly added after the second step, and the component (f) can be added in at least one step among the second step and the newly added third step, in order to carry out the reaction. Adding the component (f) in the third step is one preferable method. It is also preferable to use the component (c) in an amount of 5 to 25 pbw.

Raw Materials: The components (a) through (e) should be compounded so that the ratio of the hydroxyl groups:carboxyl groups is in the range of 0.625–1.8:1, preferably 1.0–1.6:1. If the ratio is outside the abovementioned range, only the alkyd resin having a low molecular weight is obtained and this is unfavorable.

As the component (f), there can preferably be used epoxy resins having two or more epoxy groups per molecule. Preferable examples of the epoxy resins are epoxy resins obtained by reacting aromatic polyols or polyhydroxy compounds such as bisphenol with epihalohydrins, e.g., bisphenol-type epoxy resins such as a reaction product of 2,2-bis(4-hydroxyphenyl)propane with an epihalohydrin.

Commercially available bisphenol-type resins are Epikote 828, Epikote 834, Epikote 1001, Epikote 1004, Epikote 1007, Epikote 1009 (Shell Chemical Co.); DER 660, DER 661 J, DER 662, DER 664 J, DER 667 J, DER 668, DER 669 J (The Dow Chemical Co); EPS 485, EPX 43 (Asahi Denka Kogyo Co.);

EPOMIK 300, EPOMIK 301, EPIMOK 304, EPIMOK 307, EPIMOK 309 (Mitsui Kanebo Co.) and the like.

Other alicyclic-type epoxy resins such as Yunocks 201, Yunocks 206, Yunocks 221, Yunocks 289 (Union Carbide Co.); polyethylene glycol-type resins such as Epikote 812 (Shell Chemical Co.); Epolite 40E, Epolite 200E, Epolite 400E (Kyoei-sha Co.); epoxidized polybutadienes such as BF-1000 (Adeka Ahgas Kagaku Co.); and the like can also be used. It is preferable to use epoxy compounds having softening points of 60°C or more.

BUTADIENES

PROCESSING

Butadiene and α,β-Ethylenically Unsaturated Dicarboxylic Acid Compound

Highlights of the Technological Achievement: There is disclosed a water-soluble coating composition having excellent properties including corrosion resistance, adhesion and impact resistance. The coating composition comprises an adduct of a butadiene lower polymer or copolymer and α,β-ethylenically unsaturated dicarboxylic acid compound resulting from the special addition reaction, a hydrophilic solvent and a neutralizer.

Reference: A. Kaiya, Y. Otsuki and H. Horii; U.S. Patent 4,197,226; April 8, 1980; assigned to Nippon Oil Co., Ltd., Japan. Other References: U.S. Patent 4,145,501; U.S. Patent 4,175,068; U.S. Patent 4,196,112 and U.S. Patent 4,196,113.

Background: Heretofore, as raw materials for a water-soluble coating composition there have been used natural oils. However, use of natural oils for the above purposes has been regarded to be disadvantageous because their supply is precarious and they cause pollution in respect to smell and smoke. Therefore, instead of natural oils a butadiene low polymer has been given much consideration as a film-forming material for a water-soluble coating composition since the coating film formed from a butadiene low polymer or copolymer has very excellent film properties in respect of chemical resistance, solvent resistance, water resistance, corrosion resistance and levelling property as compared with those prepared from the conventional film-forming materials such as natural oils.

In order to prepare a film-forming material for a water-soluble coating composition, it is necessary to introduce hydrophilic groups such as carboxyl, hydroxyl, ether, amino and sulfonic acid groups into the molecular structure of the material.

With regard to the introduction of the carboxyl group into the butadiene lower polymer or copolymer, it is well-known that an adduct can be prepared by heating it with α,β-ethylenically unsaturated dicarboxylic acid compound. The addition reaction between butadiene lower polymer or copolymer and α,β-ethylenically unsaturated dicarboxylic acid is carried out through a radical or thermal reaction. Both reactions are accompanied by crosslinking reactions among molecules of starting polymer so that gelation of the adduct occurs, which is disadvantageous.

In order to eliminate the above-noted difficulties there are known many methods for preparing such an adduct by the addition of certain compounds such as hydroquinone, 2,6-di-tert-butyl-4-methylphenol, urethane compounds or metallic salts to the thermal reaction system. According to these methods the amount of addition of α,β-ethylenically unsaturated dicarboxylic acid compounds is very small, the adduct is greatly colored, the viscosity of the adduct is too high to be treated and the molecular weight of the usable starting polymer is limited to a very low range so that the adduct produced by the above conventional methods has insufficient properties as a film-forming material for a water-soluble coating composition.

Process Conditions: The concept of this process resides in a water-soluble coating composition comprising an adduct (1) of a butadiene lower polymer or lower copolymer (A) and α,β-ethylenically unsaturated dicarboxylic acid compound (B) prepared by the reaction between the (A) and (B) in the presence of one or more compounds (C) selected from p-phenylenediamine derivatives, catechol derivatives, pyrogallol derivatives, nitrosamines, quinoline derivatives and naphthol derivatives; a hydrophilic solvent (2); and a neutralizer (3).

Raw Materials: The butadiene lower polymers used in the process are butadiene polymers containing a large proportion of 1,2-double bonds or a large proportion of 1,4-double bonds or large proportions of both 1,2- and 1,4-double bonds. The butadiene polymers which are prepared by polymerizing butadiene alone or with other monomers in the presence of alkali metal or organic alkali metal compound as a catalyst, are typical ones used in the process.

In order to regulate the molecular weight, to reduce the gel content and to form a light-colored polymer, the polymerization is typically a living polymerization which is carried out in a tetrahydrofuran medium or chain transfer polymerization in which ethers such as dioxane and alcohols such as isopropyl alcohol are added and aromatic hydrocarbons such as toluene and xylene are used as the chain transfer agent or the solvent.

A lower polymer in which most of the double bonds in the butadiene units are 1,2-double bonds can be prepared by the above methods, and this lower polymer can be used in the process. A butadiene lower polymer in which most of the double bonds in the butadiene units are 1,4-double bonds and which can be used in the process can be prepared by polymerizing butadiene alone or with other monomers in the presence of a catalyst comprising a compound of a metal of Group VIII of the periodic table and an alkyl aluminum.

A lower copolymer as referred to in the process means copolymers of butadiene with conjugated diolefins other than butadiene such as isoprene, 2,3-dimethylbutadiene and piperylene or with vinyl-substituted aromatic compounds such as α-methylstyrene, vinyltoluene and divinylbenzene as the comonomers. Butadiene lower copolymers containing less than 50% by weight of the comonomers may be preferably used.

Further, in the process, modified products of the butadiene lower polymer or copolymer which are prepared by heating and partially oxidizing the butadiene lower polymer or copolymer with blowing off air in the presence of carboxylic acid metallic salts such as cobalt naphthenate or manganese octenoate, or prepared by thermal treatment of the butadiene lower polymer or copolymer in the presence of an organic peroxide, may also be used.

The butadiene lower polymer or copolymer in the process is liquid or semisolid at room temperature, and the number average molecular weight thereof is in the range of 200 to 10,000.

The α,β-ethylenically unsaturated dicarboxylic acid compounds in the process may be represented by the following general formula:

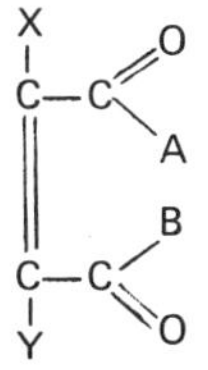

in which X and Y are hydrogen atoms or alkyl groups, and may be the same or different; and A and B are hydroxyl groups, alkoxyl groups or an –O– bond formed by linking A and B together. The α,β-ethylenically unsaturated dicarboxylic acid compounds include anhydrides of maleic acid such as maleic anhydride, citraconic anhydride and 1,2-diethylmaleic anhydride, and esters of maleic acid such as monomethylmaleic acid, dimethylmaleic acid and diethylmaleic acid, and those having 12 or less carbon atoms in each molecule may be advantageously used.

Addition Reaction: In the addition reaction of a butadiene lower polymer or copolymer, and α,β-ethylenically unsaturated dicarboxylic acid compound of the process, one or more compounds selected from (1) p-phenylenediamine derivatives, (2) catechol derivatives, (3) pyrogallol derivatives, (4) quinoline derivatives, (5) N-nitrosamines and (6) naphthol derivatives are used.

The abovementioned p-phenylenediamine derivatives are represented by the following general formula:

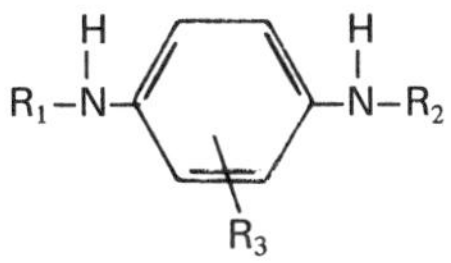

in which R_1 and R_2 are the same or different saturated hydrocarbon groups each having 1 to 20 carbon atoms, or monocyclic and polycyclic aromatic hydrocarbon groups, and R_3 is a hydrogen atom or a saturated hydrocarbon group having 1 to 20 carbon atoms.

Further, the catechol derivatives in the process are represented by the following general formula:

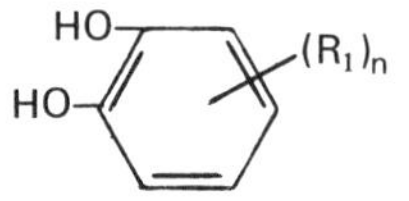

in which R_1 is a saturated hydrocarbon group or aromatic hydrocarbon group having 1 to 10 carbon atoms, and n is an integer which is 0, 1, 2, 3 or 4. As the catechol derivatives, there are, for example, alkyl catechols such as catechol itself, p-methylcatechol, p-ethylcatechol, p-propylcatechol, p-isopropylcatechol, p-n-butylcatechol, p-sec-butylcatechol, p-tert-amylcatechol and 1,2-dihydroxy-3-tert-butyl-5-methylbenzene; and aromatic group-substituted catechols such as p-phenylcatechol and p-(p-methylphenyl)catechol.

Furthermore, the pyrogallol derivatives in the process are represented by the following general formula:

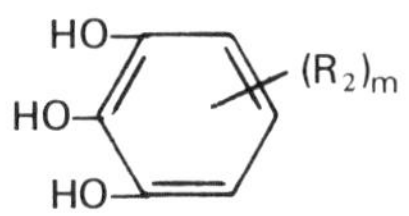

in which R_2 is a saturated hydrocarbon group or aromatic hydrocarbon group having 1 to 10 carbon atoms, and m is an integer which is 0, 1, 2, or 3. As the pyrogallol derivatives, there are, for example, pyrogallol and 1,2,3-trihydroxy-5-methylbenzene.

Furthermore, the N-nitrosamines in the process are represented by the following general formula:

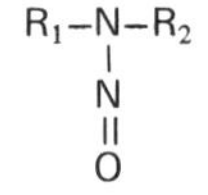

in which R_1 and R_2 are the same or different alkyl, cycloalkyl, aryl or naphthyl groups, each having 1 to 20 carbon atoms and each having or not having amino and alkoxy groups as substituent groups. As the N-nitrosamines, there are, for example, N-nitroso dialkylamines such as N-nitrosodimethylamine, N-nitrosodiethylamine, N-nitroso-di-n-propylamine.

Still further, the quinoline derivatives in the process are represented by the following general formula:

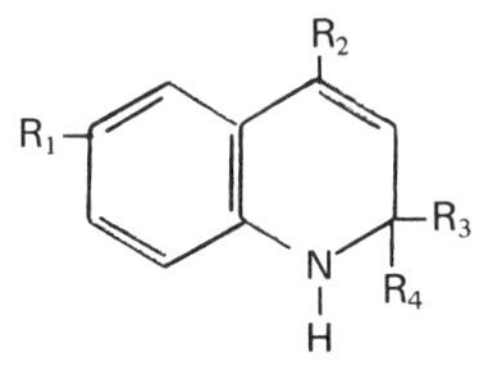

in which R_1 is a hydrogen atom or an alkoxy group having 1 to 5 carbon atoms, and R_2, R_3 and R_4 are the same or different alkyl groups each having 1 to 5 carbon atoms. As the quinoline derivatives, there are, for example, 1,2-dihydro-2,2,4-trimethylquinoline, 1,2-dihydro-2,2,4-triethylquinoline, 1,2-dihydro-2,2,4-

tri-n-propylquinoline, 1,2-dihydro-2,2,4-tri-n-butylquinoline, 1,2-dihydro-2,2-dimethyl-4-ethylquinoline, 1,2-dihydro-2,2,4-trimethyl-6-methoxyquinoline, 1,2-dihydro-2,2,4-trimethyl-6-propoxyquinoline and 1,2-dihydro-2,2,4-trimethyl-6-n-butoxyquinoline.

Still further, the naphthol derivatives in the process are represented by the following general formula:

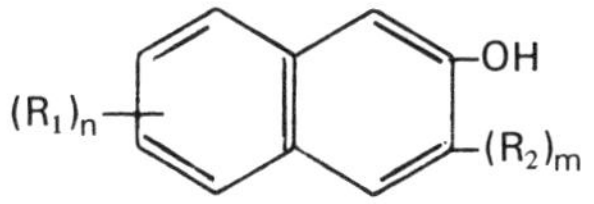

in which R_1 and R_2 are hydrogen atoms, alkyl groups each having 1 to 20 carbon atoms, nitro groups, hydroxyl groups or amino groups; n is an integer which is 1, 2 or 3; m is an interger which is 1, 2, 3 or 4; and one of $(R_1)_n$ and $(R_2)_m$ is one or more of nitro, hydroxyl or amino groups. As the above naphthol derivatives, there are, for example, dihydroxynaphthalenes such as 1,2-dihydroxynapththalene, 1,3-dihydroxynaphthalene, 1,4-dihydroxynaphthalene.

Product: The water-soluble coating composition according to this process serves as an electrodeposition paint or an immersion deposition paint and forms a tough film by being heated at 100° to 250°C for 2 to 90 min.

It is excellent in stability in water and the film of the coating composition coated on both a nontreated steel plate and a chemically treated steel plate is excellent with respect to corrosion resistance, throwing power, impact resistance, adhesion, leveling property and hardness.

Example: 1,000 pbw of a butadiene lower polymer or copolymer having a number average molecular weight of 1,000 and a viscosity of 42 poises (25°C), 333 pbw of xylene, 163 pbw of maleic anhydride and 2 pbw of N-phenyl-N'-isopropyl-p-phenylenediamine were fed into a 2-ℓ stainless steel autoclave provided with a magnetic stirrer. The air in the reaction system was sufficiently displaced by dried nitrogen gas. The reaction mixture was stirred at 190°C for 8 hr to cause the addition reaction between the polybutadiene and maleic anhydride. After the reaction, the autoclave was cooled to room temperature.

Then it was treated in a reduced pressure of 1 mm Hg at 150°C to distill off the solvent, namely, xylene and a small amount of remaining maleic anhydride, whereby an adduct of maleic anhydride having an acid value of 80.0 and a viscosity of 9,500 poises (25°C) was obtained. 500 pbw of the adduct thus obtained were fed to a 1-ℓ separable flask having a reflux condenser. After the adduct was heated to 70°C 40 pbw of water and 10 pbw of triethylamine were added thereto.

The mixture was heated at 80°C for 2 hr to open the rings of the anhydride groups. After 80 pbw of ethylene glycol monoethyl ether was fed to the separable flask, it was cooled to room temperature. Then the reaction mixture was neutralized with a deionized water solution of triethylamine in such a manner that the pH of the obtained solution will be 8.2 when it contains 10 pbw of solids. After that, the solution was diluted with deionized water so as to contain 10 pbw of solids. Thus a clear electrodeposition solution was prepared.

Steel test pieces (0.8 mm x 70 mm x 150 mm) treated with zinc phosphate were immersed in the thus-prepared clear electrodeposition solution at 25°C and electric voltages of 80, 100, 120, 150, 200, 250 and 300 volts, respectively, were applied for 3 min under stirring and with 10 cm of a distance between electrodes. The test pieces were thus electrodeposition-coated and taken out from the solution.

Then the test pieces were washed with water to remove the superfluous electrodeposition solution thereon and were air-dried in a room having a temperature of 25°C and a relative humidity of 75%. After air-drying, these were put in an oven circulating hot air of 160°C for 30 min to harden the films coated thereon. Measurements of the thicknesses of the obtained film showed the following results: (voltage/film thickness) 80 V/10 μ, 100 V/12 μ, 120 V/13 μ, 150 V/15 μ, 200 V/20 μ, 250 V/23 μ, 300 V/25 μ. All of the films showed excellent film surfaces and there were no craters or other defects thereon.

The coated film of 20 μ thickness was subjected to various performance tests and exhibited a 4H pencil hardness, more than 240 hr of corrosion resistance, an Erichsen test value of 6.5, a DuPont impact test value of 40 cm and a crosscut test value of 100/100.

APPLICATIONS

Producing a Water-Dispersible Film from a Butadiene Polymer

Highlights of the Technological Achievement: A process for producing a film-forming material, which comprises graft-polymerizing a monomer having a radical polymerizable ethylenically unsaturated group in the presence of a radical initiator with a butadiene polymer of a low degree of polymerization.

Reference: Y. Otsuki, H. Horii and Y. Araki; U.S. Patent 4,137,282; January 30, 1979; assigned to Nippon Oil Company, Ltd., Japan.

Background: Water-base paints are classified into aqueous solution and emulsion types. Since these paints use water as a thinner, they have the advantage of being free from toxicity and flammability and cheaper than organic solvent-base paints.

Conventional aqueous solution-type water-base paints containing resins of high molecular weights are useless for practical purposes because of their high viscosity. Accordingly, only relatively low-molecular-weight resins can be used, and thus coated films prepared from such aqueous solution-type water-base paints containing low-molecular-weight resins have low hardness, poor physical properties, and poor chemical properties such as water resistance and chemical resistance.

Process Conditions: A monomer having a radical polymerizable ethylenically unsaturated group is graft-polymerized in the presence of a radical initiator with a butadiene polymer of a low degree of polymerization having a number average molecular weight of 300 to 10,000, a butadiene copolymer of a low degree of polymerization containing not more than 50% by weight of a copolymer component, or a mixture of these, in the form of an aqueous solution or aqueous dispersion, the butadiene polymer or copolymer in aqueous solution or aqueous dispersion containing 0.05 to 0.5 mol, per 100 g of the polymer, of a basic group of the general formula:

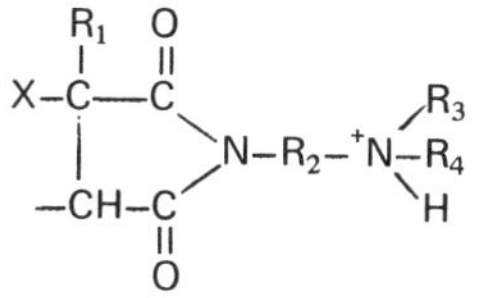

where R_1 represents a hydrogen atom, a halogen atom or an organic group containing 1 to 8 carbon atoms; R_2 and R_3 are identical or different and represent an organic group containing 1 to 20 carbon atoms; R_4 represents a hydrogen atom or an organic group containing 1 to 20 carbon atoms; and X represents a hydrogen atom or a bond, and when X is a bond, the carbon atom attached to R_1 and the adjacent carbon atom to which hydrogen is attached may form a part of the main chain.

The basic group-containing butadiene low polymer or copolymer may contain not more than 0.2 mol/100 g of the polymer, of a group expressed by the formula:

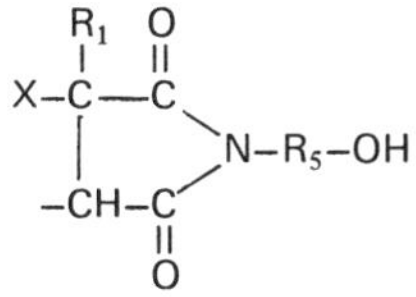

wherein R_1 and X are as defined above and R_5 represents an organic group containing 1 to 20 carbon atoms.

The basic group-containing low polymer or copolymer of butadiene used in this process can be prepared by reacting (A) an organic polymer (dicarboxylic acid-modified butadiene polymer or copolymer) composed of (a) a low polymer of butadiene having a molecular weight of 300 to 10,000, a low butadiene copolymer containing not more than 50% by weight of a copolymer component, or a mixture of these and (b) an acid group bonded to the low butadiene polymer and/or copolymer through a carbon-carbon bond, the acid group being expressed by the formula:

(1) 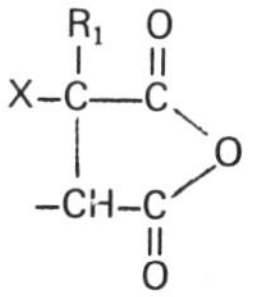 or (2) 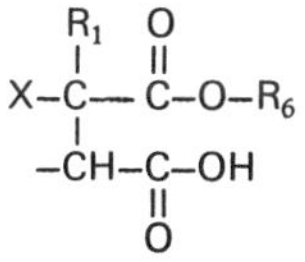

wherein R_1 and X are as defined above, and R_6 represents a hydrogen atom or an organic group containing 1 to 10 carbon atoms, and contained in an amount of 0.05 to 0.7 mol/100 g of the resin, with (B) a diamine compound of the general formula:

(3) 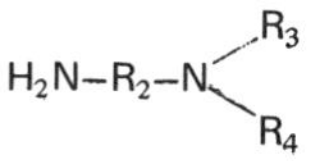

wherein R_2 and R_3 are identical or different and represent an organic group containing 1 to 20 carbon atoms, and R_4 represents a hydrogen atom or an organic group containing 1 to 20 carbon atoms, to form a resinous material containing the basic group and unsaturated group, and then neutralizing the basic group with an acid.

The above reaction may be carried out in the presence of an aqueous medium or an organic liquid medium.

In the production of the resinous material, an alcohol amine compound (C) of the following general formula:

(4) $$N_2H{-}R_5{-}OH$$

wherein R_5 represents an organic group containing 1 to 20 carbon atoms, may be used together with the diamine compound of formula (3).

Butadiene Polymer: The dicarboxylic acid-modified butadiene polymer or copolymer used in the process can be prepared by introducing an acid group with one of the following formulas

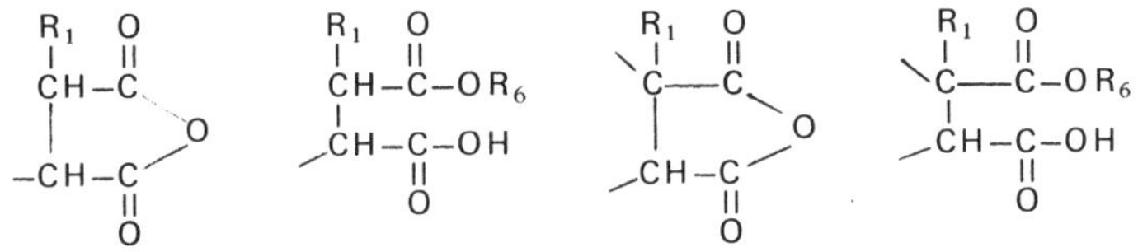

wherein R_1 represents a hydrogen atom or an organic residue containing 1 to 3 carbon atoms, and R_6 represents a hydrogen atom or an organic residue containing 1 to 20 carbon atoms, into a low polymer or copolymer of butadiene.

Imidization Reaction: The imidization reaction can be carried out in the presence or absence of a solvent. When the polymer (A) containing the acid group and the double bond to be imidized has a low viscosity, it is preferred not to use a solvent. The solvent which is optionally used is a solvent miscible with the polymer (A), such as a hydrocarbon solvent (e.g., benzene, toluene, cyclohexane, or xylene), a hydrophilic alcohol solvent (e.g., butyl Cellosolve), and an ether-type solvent (e.g., diglyme). It is preferred to use a hydrophilic solvent such as butyl Cellosolve, and to water-solubilize the resin directly.

This reaction affords a resin having an amido group and a secondary or a tertiary amino group with a side chain of the following formula:

(5)

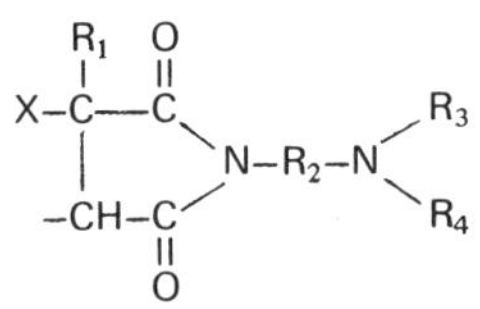

and/or a side chain of the formula

(6)

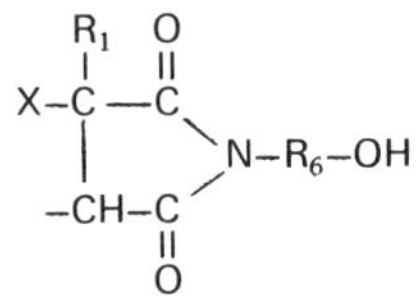

In the above R_2 and R_6 are as defined above.

By the above imidization reaction, the acid group of the component (A) can be completely consumed to leave no acid group in the polymer after the imidization reaction. A part of the acid group may be left in the form of:

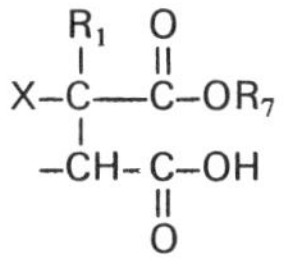

where R_1 represents a hydrogen atom or a methyl group, R_7 represents a hydrogen atom or an organic group with 1 to 8 carbon atoms, and X represents a hydrogen atom or a bond, and when X is a bond, the carbon atom bonded to R_1 and the adjacent carbon atom to which hydrogen is attached may form a part of the main chain. The amount of the acid group left is preferably not more than 0.2 mol/100 g of the resin.

Water-solubilization of the resulting resin is performed by neutralizing it with 0.2 to 2.0 mols, preferably 0.5 to 1.0 mol/mol of the group of formula (5) of a mineral acid such as hydrochloric acid or sulfuric acid, or a water-soluble organic acid such as formic acid, acetic acid or propionic acid.

This water-solubilization gives a group of the formula:

(7)

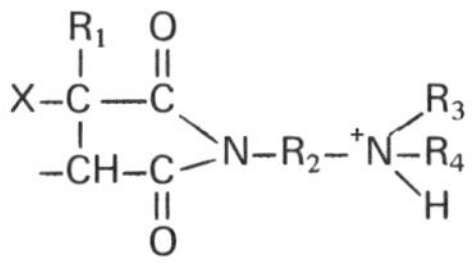

In water-solubilizing the resin in accordance with this process, it is preferred to use 100 to 10 g/100 g of the resin, of an organic solvent being water-soluble and capable of dissolving the resin, such as ethyl Cellosolve.

Examples of the monomer are compounds of the formula:

$$CH_2{=}\underset{\displaystyle R_1'}{\underset{|}{C}}{-}COOR_2'$$

wherein R_1' represents a hydrogen atom or a methyl group, and R_2' represents an organic group containing 1 to 20 carbon atoms, such as methyl (meth)acrylate, butyl (meth)acrylate, methoxybutyl (meth)acrylate, diethylaminoethyl (meth)acrylate, ethylene glycol di(meth)acrylate, and styrene or styrene derivatives of the general formula:

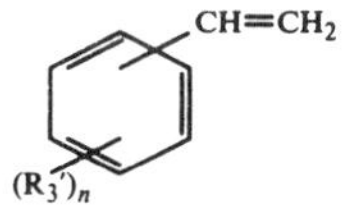

wherein n is an integer of 1 to 4, R_3' represents a hydrogen atom, an alkyl group containing 1 to 10 carbon atoms, a vinyl group, or a halogen atom, such as α-methylstyrene, vinyltoluene or divinylbenzene. Two or more of these monomers may be used as a mixture.

CARBOXYLIC PRODUCTS

PROCESSING

Phosphorus-Containing Carboxylic Polymers

Highlights of the Technological Achievement: The compositions are formed as the reaction product, preferably in an organic solvent, of phosphorous acid or PCl_3 and a water-soluble carboxyl polymer. A precursor of phosphorous acid, which precursor is capable of generating phosphorous acid in aqueous solution, may be used, and preferably, the water-soluble carboxyl polymer has a relatively low molecular weight, preferably from 400 to 6,000.

Reference: W.F. Masler III and D.C. Spaulding; U.S. Patent 4,207,405; June 10, 1980; assigned to The B.F. Goodrich Company.

Background: Polyphosphates are used to inhibit and prevent scale and deposit formation in aqueous systems, mainly because they are effective in substoichiometric or threshold quantities, and they are relatively inexpensive. Their proclivity to hydrolyze somewhat unpredictably has initiated their substitution by phosphonic acids and other polyelectrolytes which are stable in aqueous solutions. It is also known that substoichiometric quantities of polyacrylic acid and polymethacrylic acid, hereinafter together referred to as poly(meth)acrylic acid, inhibit scale formation in aqueous systems.

It is also known that mixtures of carboxylic polymers having a molecular weight of from 1,000 to 15,000,000 and aminopoly(methylenephosphonic)acid (British Patent 1,208,827), and, mixtures of lower carboxylic acids such as maleic acid and other organophosphonic acids (British Patent 1,258,067) are useful to inhibit the deposition of scale. These compositions require relatively small quantities of the water-soluble polycarboxylic acid or a derivative thereof.

Similarly, U.S. Patent 3,784,469 teaches that cyclohexane-1,2,3,4,5-pentacarboxylic acid is a scale and deposit suppression agent, particularly in the presence of certain organophosphonic acids. Improved compositions for these uses are desired.

Process Conditions: The materials of this process find utility in water treatment, detergent formulations, agricultural applications and complexation of trace elements, agricultural micronutrients, soil stabilizers, selective flocculants, stabilization of peroxides against metal ion catalyzed decomposition, electroplating baths, soaps and cosmetics, and the like.

The low molecular weight organic water-soluble carboxyl polymer is reacted with phosphorous acid or a precursor capable of generating phosphorous acid in aqueous solution, provided the proportions of the reactants lie within a range of molar ratios of titratable carboxyl groups to phosphorous acid from 10:1 to 1:1. If phosphorous trichloride, or other precursor phosphorous compound is used, its concentration is calculated as phosphorous acid.

When used in a water system, the proportion by weight of the reaction product may be as high as 1 gram on a solids basis, per liter of water and even higher concentrations may be used. By solids basis is meant the reaction product essentially free of organic solvent, or the hydrolyzed reaction product essentially free of water, recognizing that the solids may be in a semisolid or gumlike state.

More preferably the composition of this process is formed by the reaction of from 1 to 10 mols of titratable carboxyl groups in the water-soluble carboxyl polymers, per mol of phosphorous acid or precursor thereof present in a polar organic solvent.

Typically, the polymer has a molecular weight of from 400 to 6,000, and has repeating units conforming to the structure (1)

(1) 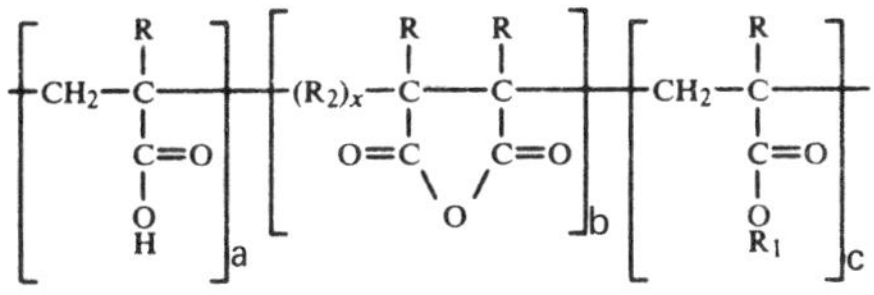

where R represents H, CH_3 or $CH_2{-}COOH$, R_1 represents alkyl having from 1 to 18 carbon atoms, R_2 represents alkylene having from 2 to 4 carbon atoms, and vinyl having the structure

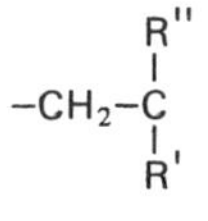

where R' represents O–R''' and NH–C(O)R''', R'' may be H or R_1, and R''' may be H or R_1, x may be from 0 to 5, a and b each represent from 0 to 100% by weight based on the combined weight of a + b + c, the sum of the numerical values of a + b + c is always 100, and the monomeric groups are present in a heterogenous relative order, i.e., the order in which the monomeric groups appear is immaterial.

Most preferred are water-soluble carboxyl polymers having repeating units conforming to the structure (2) on the following page:

(2) 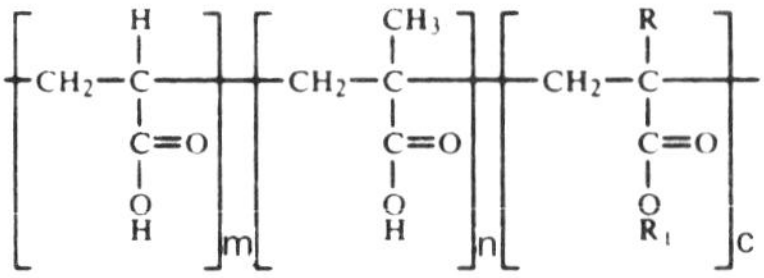

where R represents H, CH_3 and CH_2–COOH, R_1 represents alkyl having from 1 to 18 carbon atoms, c is from 0 to 30% by weight, m is from 0 to 100% by weight, and correspondingly, n is from 100 to 0% by weight, and the sum of m + n + c is always 100.

The polymers may be prepared from a batch charge of the monomers in an excess of a lower aliphatic monohydric alcohol or water at a temperature of from 20° to 200°C in the presence of from 0.25 to 20 parts by weight of a free radical polymerization initiator such as acetyl benzoyl peroxide.

Reaction Product: Preparation of the reaction product may leave a residue of phosphorus compound(s) not bound to the polymer. It is desirable for better product performance in water treatment that the residual phosphorus compounds be reacted with a lower carboxylic acid, acid chloride or acid anhydride or mixtures thereof wherein the carboxylic acid has from 2 to 12 carbon atoms.

Useful materials include acetic acid, acetic anhydride, acetyl chloride, and the like in amounts to react with the residual phosphorus compounds. ^{31}P NMR data indicate a reaction product which contains pendant carboxyl as well as some hydroxymethine diphosphonic acid groups probably in this structure:

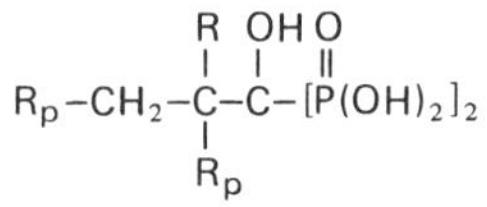

where R represents H or CH_3 and R_p represents the remaining polymer.

It is preferred that the reaction be carried out under anhydrous conditions, and either H_3PO_3, or PCl_3, or both are normally used. Under certain circumstances it may be desirable to use PCl_3 as a precursor of H_3PO_3, and enough water is added to react with the PCl_3 forming H_3PO_3 so that there is no free water present. Thus, although some water may be added to a reaction in which PCl_3 is a reactant, the reaction with a relatively low molecular weight carboxyl polymer occurs under essentially anhydrous conditions. PBr_3 also may be used.

The reaction products may contain from 1% bound phosphorus up to 20% by weight. A useful range is from 2 to 15% by weight.

Example 1: Polyacrylic acid (MW 1,700) and PCl_3 in a molar ratio of 2.0 are reacted in sulfolane to prepare a solvent-free reaction product as follows. A 1 liter flask is charged with 53 g (0.6 mol) of polyacrylic acid (MW 1,700), 130 g sulfolane and 3.3 g water. Gradually, over a period of 2 minutes, 26 ml PCl_3 (0.3 mol) are stirred into the mixture to form a homogeneous reaction mixture. This reaction mixture is heated to 100°C for 3 hours and allowed to cool to below 60°C.

The cooled mixture is poured, while stirring, into a 1 liter flask containing 500 ml $CHCl_3$. A precipitate is formed which is filtered and washed with 200 ml $CHCl_3$. Residual $CHCl_3$ is removed at room temperature under vacuum of 29" Hg. A light yellow powder weighing 70 g is isolated. Sulfolane is recovered for reuse from the chloroform solution.

Example 2: Polymaleic anhydride (MW 700) and H_3PO_3 in a molar ratio of carboxyl groups to H_3PO_3 of 2.98 are reacted in sulfolane to prepare a solvent-free reaction product as follows. A 500 ml flask is charged with 49 g (0.5 mol) of polymaleic anhydride (MW 700), 75 g sulfolane, and 27.5 g H_3PO_3 (0.335 mol). This mixture is heated at 140°C for 3 hours with stirring, which produces a homogeneous liquid reaction mass. The sulfolane is removed from the reaction mixture by vacuum distillation at 115°C and 0.5 mm Hg.

The semisolid product is homogenized with a Waring blender in 200 ml CH_2Cl_2. The slurry is filtered and washed with 200 ml CH_2Cl_2. The resulting paste is dried in a vacuum oven at room temperature and 29" Hg vacuum to give 42.3 g of a fine light tan powdery product. The sulfolane is recovered for reuse from the CH_2Cl_2 solution.

Example 3: Polyacrylic acid (MW 1,700) and PCl_3 in a molar ratio of carboxyl groups to a PCl_3 of 1.67 are reacted in sulfolane to prepare a solvent-free reaction product as follows. A 1 liter flask is charged with 87.3 g (1 mol) of polyacrylic acid (MW 1,700), 175 g sulfolane, and 6 g (0.33 mol) water. Gradually, over 10 minutes, 52 ml (0.6 mol) of PCl_3 are added. There is initial foaming and HCl evolution.

The homogeneous reaction mixture is heated at 115°C for 3 hours during which time it remains homogeneous. The reaction mixture is allowed to cool to less than 60°C and the product is precipitated with $CHCl_3$. The solid product is filtered and washed with $CHCl_3$ and is then dried at room temperature in a vacuum of 29" Hg. The sulfolane is recovered for reuse from the chloroform solution. The yield of light yellow dry powdery product is 124 g.

Test Data for Sequestering Power ($CaCO_3$) of Solvent-Free Reaction Products

Product (Ex. No.)	Components	Mol Ratio CO_2H/PCl_3	Reaction Solvent	Power*
1	Polyacrylic acid** + PCl_3	2	Sulfolane	750
2	Polymaleic anhydride*** + H_3PO_3	2.98	Sulfolane	700
3	Polyacrylic acid** + PCl_3	1.69	Sulfolane	750

*Power = mg $CaCO_3$ sequestered per gram of sample.
**MW 1,700 ***MW 700

Aqueous Metal Working Fluid Containing Carboxylic Acid Group Terminated Diesters of Polyoxyalkylene Diols

Highlights of the Technological Achievement: Aqueous metal working fluids exhibiting high lubricity and high stability are provided. These fluids comprise water and an acid terminated diester of a secondary diol or the amine or alkali metal salts of the ester.

Reference: W.E. Rieder; U.S. Patent 4,172,802; October 30, 1979; assigned to Cincinnati Milacron Inc.

Background: Oil-based and aqueous metal working fluids have long been known in the art and used in metal working processes. Such fluids are known to have lubricating and cooling functions which reduce friction and dissipate heat in metal working processes. This reduction of friction and dissipation of heat promotes long tool-life, increases production and allows the attainment of high quality finished metal products.

Aqueous metal working fluids, notwithstanding their effectiveness in metal working processes, have been known to have problems which reduce their effectiveness and/or useful life in metal working processes. Among such problems is the problem of hydrolytic instability. The art has continuously strived to overcome these problems, particularly hydrolytic instability, in view of the economic, safety and environmental attractiveness of the aqueous metal working fluids.

It is, therefore, an object of this process to provide an aqueous metal working fluid composition having improved hydrolytic stability.

Process Conditions: It has been discovered that many of the drawbacks of prior art metal working fluid compositions can be overcome and that highly stable, effective aqueous metal working fluid compositions are provided by this process which comprise (a) water and (b) a carboxylic acid group terminated diester of a polyoxyalkylene diol having two terminal secondary alcohol groups or the organic amine salt or alkali metal salt of the diester.

Further, in accordance with this process, there are provided metal working compositions comprising (a) water and (b) a carboxylic acid group terminated diester of a polyoxyalkylene diol having two terminal secondary alcohol groups, the diester having at least two terminal carboxylic acid groups per molecule, or the organic amine salt or alkali metal salt of the diester.

Additionally, in accordance with this process, there are provided metal working compositions comprising (a) water and (b) a carboxylic acid group terminated diester of a polyoxyalkylene diol having two terminal secondary alcohol groups per molecule, the diester having from two to four terminal carboxylic acid groups per molecule, or the organic amine salt or alkali metal salt of the diester.

Still further in accordance with this process, there are provided metal working compositions comprising (a) water and (b) a carboxylic acid group terminated diester of a polyoxyalkylene diol having two terminal secondary alcohol groups per molecule, the diester having two terminal carboxylic acid groups per molecule, or the organic amine salt or alkali metal salt of the diester. Mixtures of different carboxylic acid group terminated diesters of polyoxyalkylene diols having two terminal alcohol groups or the organic amine or alkali metal salts thereof may be used in accordance with this process.

Further, there may be used in accordance with this process, combinations of carboxylic acid group terminated diesters of polyoxyalkylene diols having two secondary alcohol groups and organic amine or alkali metal salts of carboxylic acid group terminated diesters of polyoxyalkylene diols having two terminal secondary alcohol groups.

Raw Materials: The carboxylic acid group terminated diesters and salts thereof of the above disclosed metal working compositions of this process may be described by the following general formula

$$\begin{array}{ccccc} {}_{n}(HOOC)-\underset{|}{R'}-COO-R-OOC-\underset{|}{R''}-(COOH)_{m} \\ (COO^{-}X^{+})_{a} \qquad\qquad (COO^{-}X^{+})_{b} \end{array}$$

where R is the divalent radical residue of a secondary alcohol terminated polyoxyalkylene homopolymer or copolymer diol absent the terminal hydroxyl groups, R' and R'' are the same or different and are selected from aliphatic, aromatic, cycloaliphatic, aryl aliphatic, alkyl aromatic, heteroaliphatic having S or O-heteroatoms, heteroaromatic having O or N heteroatoms, halogen substituted aliphatic or halogen substituted aromatic radicals having a free valence of a + n + 1 and b + m + 1 respectively, n is 0 to 3, m is 0 to 3, X is organic amine cation or alkali metal ion, a is 0 to 3, b is 0 to 3, a + n is 0 to 3, b + m is 0 to 3 and a + b + m + n is 1 to 6.

In a particular aspect of this process, the above general formula and the definitions pertaining thereto may be substituted for the described carboxylic acid group terminated diesters of the above metal working compositions of this process. It is, of course, to be recognized that the carboxylic acid group terminated diester in accordance with the above general formula may, in keeping with the previous disclosure herein, be used in the metal working compositions of this process in the form of its organic amine salt or its alkali metal salt, preferably the alkanol amine salt or the sodium or potassium salts.

APPLICATIONS

Dental Cement Containing Polycarboxylic Acid, Chelating Agent and Glass Cement Powder

Highlights of the Technological Achievement: A polycarboxylate cement pack comprising a water-soluble polycarboxylic acid having a relative viscosity of from 1.05 to 2.00, a chelating agent and a cement powder which will react with the polycarboxylic acid in the presence of the chelating agent and water to give a plastic mass which rapidly hardens to form a polycarboxylate cement.

Reference: A.D. Wilson and S. Crisp; U.S. Patent 4,209,434; June 24, 1980; assigned to National Research Development Corporation, England.

Background: The materials known as dental cements have many applications in dentistry including use as filling materials for restoring teeth and for cementing inlays and crowns into place in the tooth, providing a base and/or lining in a tooth cavity, providing a temporary fixing for the bonds of orthodontic appliances to the teeth and sealing root canals after endodontic treatment.

The traditional phosphate dental cements, i.e., zinc phosphate dental cement and dental silicate cement, have been to some extent displaced for many applications by the polycarboxylate dental cements, in which the cement-forming liquid is an aqueous solution of a polycarboxylic acid.

The polycarboxylate cements have improved acid and stain resistance over conventional dental cements and have the additional advantage that they do not irritate pulpal tissues. However, it has been found in practice that polycarboxylate cements occasionally have a slower hardening rate than conventional materials and an improvement in this respect would be desirable.

Process Conditions: Polycarboxylate cement packs in accordance with this process preferably comprise a polycarboxylic acid in the form of an aqueous solution containing from 20 to 65% by weight of the polycarboxylic acid. The polycarboxylate cement pack may be a two-part pack in which the weight ratio of powder to liquid in the two parts is preferably from 0.5:1 to 5:1 so that when the entire contents of the pack are mixed together a rapidly hardening cement is obtained.

In another embodiment the pack may contain the powder and the liquid in separate capsules, the total amount of powder in the pack and the total amount of liquid in the pack being in the appropriate ratio. In a further embodiment, both components may be encapsulated in the same capsule, in the desired ratio, provided that steps are taken to prevent premature reaction.

In a still further embodiment the pack may be a one-part pack containing an intimately blended mixture of the cement powder, solid water-soluble polycarboxylic acid and chelating agent, the ratio of powder to polycarboxylic acid being from 1:1 to 10:1, the mixture requiring the addition of water to produce the cement. The mixed powder and the water may be contained in the same capsules provided steps are taken to prevent premature reaction, for example, by dividing the capsule. This latter procedure generally requires some form of mechanical mixing.

In the abovementioned embodiments, the cement powder is from 15 to 85% by weight, the polycarboxylic acid is from 3 to 50% by weight, and the water is from 5 to 70% by weight of the total composition.

It is found that when the components are mixed together a plastic mass is obtained which sets rapidly in the mouth (1.5 to 10 minutes following completion of preparation).

Polycarboxylate Cements

Highlights of the Technological Achievement: A polycarboxylate cement pack comprises as one component a water-soluble polycarboxylic acid or a precursor thereof, and as another component an aluminosilicate glass, the ratio of acidic to basic oxides in the glass being selected such that the glass will react with a polycarboxylic acid in the presence of water to form a polycarboxylate cement.

Reference: S. Crisp and A.D. Wilson; U.S. Patent 4,143,018; March 6, 1979; assigned to National Research Development Corporation, England.

Background: Polycarboxylate cements are formed by the reaction of a polycarboxylic acid and an ion-leachable inorganic compound. The polycarboxylate cements were originally developed primarily for dental and surgical applications, in which comparatively small quantities of material are used.

However, the relatively high fluorine content previously required in these glasses made them rather difficult and expensive to manufacture on a large scale, and thus it has not proved possible to develop polycarboxylate cements for large volume applications. It has been found that the fluorine content of the ion-leachable aluminosilicate glass may be reduced to a low value, and in some cases even eliminated, by appropriate control of the basicity of the glass.

Process Conditions: This process provides a method for the preparation of a polycarboxylate cement which comprises mixing a water-soluble polycarboxylic acid or a precursor thereof with an aluminosilicate glass in the presence of water, the ratio of acidic to basic oxides in the glass having been selected such that the glass will react with the polycarboxylic acid to form a polycarboxylate cement. Preferably the ratio by weight of acidic to basic oxides in the glass is from 0.1 to 3.0 and most preferably from 0.2 to 2.5.

The aluminosilicate glasses of the process may be prepared by fusing mixtures of the components in the appropriate proportions at temperatures above 900°C and preferably 1050° to 1550°C. The mixture is preferably fused for 1 to 4 hours.

The preferred polycarboxylic acids are those prepared by the homopolymerization and copolymerization of unsaturated aliphatic carboxylic acids, for example, acrylic acid, itaconic acid, mesaconic acid, citraconic acid and aconitic acid, and copolymerization of these acids with other unsaturated aliphatic monomers, for example, acrylamide and acrylonitrile.

The polycarboxylic acid or precursor thereof is preferably linear although branched polymers may also be used, and preferably has an average molecular weight of from 1,000 to 250,000 and most preferably from 5,000 to 100,000. The average molecular weight is that measured by an absolute method such as light scattering or ultracentrifuge sedimentation.

Product: Cement packs in accordance with this process preferably comprise the polycarboxylic acid in the form of an aqueous solution containing from 20 to 65% by weight of the polycarboxylic acid. The cement pack may be a two-part pack in which the weight ratio of aluminosilicate glass to liquid in the two parts is from 0.5:1 to 5:1 and preferably from 1.5:1 to 4.5:1, so that when the entire contents of the pack are mixed together a rapidly hardening cement is obtained.

The properties of the cements are set out in the table below. All the set cements prepared are found to be hydrolytically stable.

	Cements		
	1	2	3
Setting time, minutes*	2.25	3.75	4.0
Compressive strength, N/mm^2**	74	61	56
Powder/liquid ratio, g/ml	2.5	2.75	3.0

*Using a 1 lb Gilmore needle at 37°C.
**Developed after 24 hr at 37°C.

Ionized Rosin-Based Resinous Mixtures as Tackifiers for Latex

Highlights of the Technological Achievement: A resinous polyelectrolyte, especially suitable for use as a tackifier in latex adhesives, comprises an ionizable resinous mixture which is ionized with a volatile ionizing agent, the ionizable resinous mixture being the reaction product of a rosin, a carboxylic acid, and a C_1 to C_{18} monoalcohol, the ionizable resinous mixture having a softening point not substantially above 125°C and an acid number of between 30 and 150, the reaction product comprising the carbocyclic nucleus of rosin having at least one carboxylic acid group and at least one carboxylic acid ester group of the monoalcohol.

Reference: J.M. Evans and W.W. Spangler, Jr.; U.S. Patent 4,183,834; Jan. 15, 1980; assigned to Sylvachem Corporation.

Background: Latex adhesives typically contain tackifiers (e.g., rosins, hydrocarbons, etc.) which are emulsified in the adhesive formulation with the aid of a surfactant (e.g., potassium oleate or the like) and a stabilizing additive (e.g., casein or the like). However, such surfactants and stabilizers contribute to poor adhesive properties and to increased water sensitivity of the latex adhesive formulation.

The instant ionizable resinous mixture can be stably dispersed in the latex adhesive formulation and is quite compatible with the latex, is converted to a hydrophobic resinous material upon volatilization of the volatile ionizing agent therefrom, provides excellent adhesion, tack, sheer, and peel to the adhesive formulation, and is substantially insensitive to water.

Process Conditions: The ionizable resinous mixture can be anionic or cationic for especially good compatibility with the latex in the latex adhesive formulation, be the latex anionic, cationic or nonionic. The essential ingredients for the ionizable mixture are rosin, an unsaturated carboxylic acid (hereinafter referred to as grafting acid), and a monoalcohol.

An unsaturated carboxylic acid compound is grafted or reacted onto the rosin. Preferably, the carboxylic acid is α,β-ethylenically unsaturated for grafting onto the rosin by reaction with the unsaturation contained in the resin acid found in the rosin. Additionally, one may conduct a conventional Diels-Alder reaction for grafting an unsaturated carboxylic acid onto the rosin. The precise nature of the graft of the ionizable compound onto the rosin need not be precisely characterized as it is necessary only that the grafting acid be chemically united with the rosin for purposes of this process.

Representative acids include, for example, fumaric acid and its half ester, maleic acid (including maleic anhydride) and its half ester, acrylic acid, methacrylic acid and related alkyl and aryl acrylic acids, itaconic acid (and anhydride), and oligomers and copolymers of acrylics and vinyls with ethylenically unsaturated acids (e.g., styrene/acrylic acid copolymers, etc.).

Paper Sizing Agents

Highlights of the Technological Achievement: A paper sizing composition which comprises a mixture of

(A) An alkaline aqueous dispersion of acid-group containing aliphatic hydrocarbon resins prepared by polymerizing a C_4 cut or a C_5 cut or mixtures thereof, obtained during thermal cracking of a hydrocarbon stream and modifying the product by addition thereto of an unsaturated polycarboxylic acid anhydride, wherein the resins have an acid number between 10 and 130 and are present in the dispersion with a solids content of 10 to 70%; and

(B) An alkaline solution of a copolymer of an α,β-unsaturated polycarboxylic acid and a C_5-α-olefin, wherein the solution has a solids content of 5 to 50% by weight; and

wherein (A) and (B) are employed in a ratio of 99:1 to 1:99 relative to the solids content.

Reference: F. Schülde and V. Kulisch; U.S. Patent 4,207,223; June 10, 1980; assigned to Chemische Werke Huls AG, Germany.

Background: It is known that resin sizings, based upon colophonium resin can be used, if one modifies the colophonium resin by reacting it with an α,β-unsaturated carboxylic acid or its anhydride. This reinforced colophonium can be extended by the addition of neutral hydrocarbon resins. This resin mixture is recommended, in combination with a volatile base in the form of aqueous dispersions, which require the addition of a protective colloid, such as casein, as a paper sizing agent.

Process Conditions: Hydrocarbon resins, suitable as component (A) in the sense of this process are those that are accessible through polymerization (in the presence of Friedel-Crafts catalysts) of a C_4- and/or C_5-cut that has been distilled off during thermal cracking of a hydrocarbon stream, possibly with the addition of a C_4-raffinate, that consists of isobutene, n-butene-1, cis,trans-n-butene-2, n-butane, isobutane and possibly traces of butadiene. This may have been obtained by the isolation of butadiene from a hydrocarbon-C_4-cut.

These hydrocarbon resins are modified after polymerization by reaction (addition) with unsaturated dicarboxylic acid units, such as maleic acid anhydride, so that they show the following characteristics:

	Resins Based on C_4-Cuts	Resins Based on C_5-Cuts
Acid number	10-100	10-130
Softening point (Ring and Ball)	70-130	90-140

The synthetic resins that are, possibly, used in component (A), may be, for example, neutral aliphatic hydrocarbon-natural resins or indene-cumarin resins, and as natural resins, for example, colophonium- or root-resins may be used.

For component (B) in the sense of the process, copolymers from α,β-unsaturated dicarboxylic acid units, such as, e.g., maleic acid anhydride, fumaric acid, with α-olefins with up to 5 carbon atoms or vinyl monomers, such as styrene and its homologues; vinyl esters; vinylpyridine; vinyl amides; vinyl halides; vinyl alcohols; allyl alcohols; vinyl alkyl ethers; (meth)acrylic acid, esters, amides; nitriles; indene; cyclopentadiene and dicyclopentadiene can be employed.

In this, the dicarboxylic acid units can be copolymerized with other monomers in the ratios of between 1:5 to 5:1 and the copolymers are used in the form of their reaction products in an aqueous or water-free medium with basic components like alkali hydroxides, alkali carbonates, NH_3, primary or secondary aliphatic amines or diamines.

The preparation of the sizing agent according to this process takes place in a simple manner. Initially, the acid hydrocarbon resin and the, possibly, also used, natural resin are melted at 130°C and mixed in the melt. After cooling, the solidified resin mixture is crushed. In a suitable vessel, water and alkali are prepared, the crushed resin mixture is added, and the mixture is heated under stirring. The dispersion temperature is at 90° to 100°C. The addition of usual dispersants is not necessary.

After obtaining a homogeneous dispersion, stirring continues until the material is cool. Subsequently, the aqueous solution of the copolymer-conversion product is added and uniformly stirred. The sizing agent thus obtained is of good fineness and storage stability without requiring the addition of a protective colloid.

The paper sizing agents according to this process can be combined with the anionic starches that are usual in paper sizing and can be processed on the sizing presses that are usual in paper production.

Example: *Sizing Agent 1* – A hydrocarbon resin that was polymerized in the presence of Friedel-Crafts catalysts, consisting of 75% of a hydrocarbon-C_4-cut and 25% of a butadiene-free C_4-raffinate, was modified by the addition of maleic acid anhydride in such a manner that it showed the following properties: acid number, 57 mg KOH/g; saponification number, 90 mg KOH/g; softening point, 72°C (ring and ball); and Gardner color number, 13.

42.6 parts by weight of this hydrocarbon resin were dispersed with 6.4 parts by weight of a 20% sodium hydroxide solution and 51.0 parts by weight of water. Added to this dispersion were 4.7 parts by weight of a 25% aqueous solution of a maleic acid anhydride copolymer of the formula

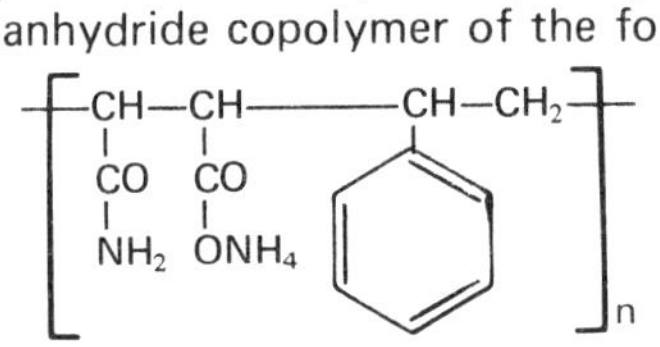

and the sizing agent, thus obtained, was thinned to a solids content of 20% by weight with water.

Sizing Agent 2 – 37.7 parts by weight of the hydrocarbon resin used in sizing agent 1 were melted with 4.9 parts by weight of a Portuguese balsam resin of type WW with an acid number of from 165 to 171 and a ring and ball softening point of 71° to 75°C. After solidification, it was crushed and this resin mixture was formulated analogously to sizing agent 1, into a 20% sizing agent.

Sizing Agent 3 (Comparison) – A resin, produced analogously to the hydrocarbon resin that was used in sizing agent 1, but without addition of maleic acid anhydride, with the following properties: acid number, <1 mg KOH/g; saponification number, <1 mg KOH/g; softening point, <94°C (ring and ball); and Gardner color number, 4, was dispersed corresponding to the sizing agent 2 and adjusted to a solids content of 20%.

Sizing agents 1 and 3 were combined with an anion-active cornstarch that is commercially available as Amisol 5591, into a 10% sizing float. The solids ratio of sizing agent to starch amounted to 1.5:8.5. These floats were applied on a laboratory press onto an unsized 80 g paper. Drying of the paper in the sizing press took place at 140° to 160°C within 5 seconds. The sizing values obtained are contained in the table below.

	. . . Sizing Agent . . .			
	1	2	3	Without
Water uptake (DIN 53 132)				
seconds	120	120	*	<1
g/m^2	23	22	*	89
Degree of sizing, sec (DIN 53 145)	290	415	*	0
Suction height, mm (DIN 53 106)	3	2	*	44
Smoothness (Beck), sec (DIN 53 107)	8	7	*	19

*Incompatible with the starch, precipitates.

It can be seen that sizing agents 1 and 2, produced in the sense of this process, provide the paper with excellent water resistance, while the comparative sizing agent 3 was totally unable to be formulated into a sizing float with the starch.

Aqueous Can Coatings of Improved Impermeability

Highlights of the Technological Achievement: There is disclosed a metal container containing a beverage or food having its integral surface coated with an adherent sanitary liner, the coating being a cured layer of a water-based coating composition consisting essentially of from 5 to 60% by weight of an amine-neutralized or partially neutralized interpolymer formed in the presence of a vinyl polymerization catalyst and in the absence of mercaptan, the interpolymer being comprised of from 10 to 30% of acrylamide or methacrylamide in units of the structure

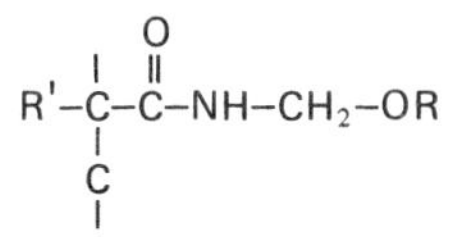

where R' is methyl or hydrogen and R is hydrogen or lower alkyl, from 5 to 15% of units formed by vinyl polymerization of monoethylenically unsaturated carboxylic acid; from 10 to 50% of units derived by vinyl polymerization of isobutyl methacrylate; and from 20 to 45% of nitrile, such as acrylonitrile, wherein the interpolymer is solubilized by neutralizing acid groups of the interpolymer with a sufficient amount of monomeric amine to produce at least 0.200 meq of salt per gram of resin solids; and from 40 to 95% by weight of a liquid medium consisting of a mixture of water-soluble or water-miscible organic solvents and water, wherein at least 60% by weight of the mixture is water.

These cured coatings are characterized by improved impermeability to lubricants and flavor components and also better resist water at elevated temperature so as to also be useful as an exterior coating.

Reference: G.L. Brown and A.T. Spencer; U.S. Patent 4,195,005; March 25, 1980; assigned to Mobil Oil Corporation.

Background: The development of water-based coating compositions for use as internal sanitary liners for metal containers has received considerable attention, and one system for providing such coating compositions is set forth in U.S. Patent 3,991,216. However, typical systems in accordance with the teachings of this patent lack desired film impermeability which is a characteristic which is particularly difficult to obtain when the cured coatings are adequately flexible. This process is directed to overcoming the inadequacies which characterize these typical systems.

Process Conditions: In accordance with this process, the monomer mixtures contemplated by U.S. Patent 3,991,216 are modified by the addition of from 20 to 45% of a monoethylenic nitrile selected from acrylonitrile and methacrylonitrile. Acrylonitrile is preferred. Comparative data which show the result are set forth in Table 1. The Staypro rating is based on a scale of 0 to 12, 12 denoting total impermeability, and 0 denoting total permeability. The wedge bend rating is based on a scale of 1 to 10 where 10 is perfect and 1 is the lowest limit of acceptability.

Table 1

Run No.	Components, %. AN	STY	EA	NBMA	IBMA	IBMAM	MAA (AA*)	Staypro Rating	Wedge Bend Rating
1	0	45	25	0	0	20	10	3.7	6.5
2	35	0	0	0	41.5	15	8.5*	6.1	8.3
3	40	0	35	0	0	20	5.0*	4.8	6.3
4	30	20	0	23	0	20	7.0*	4.1	5.0

*AA is preferred.

In the above table, AN = acrylonitrile, STY = styrene, EA = ethyl acrylate, NBMA = n-butyl methacrylate, IBMA = isobutyl methacrylate, IBMAM = isobutoxymethyl acrylamide, MAA = methacrylic acid, and AA = acrylic acid.

Runs 1 and 2 show a contrast between a typical system in accordance with the teachings of U.S. Patent 3,991,216 (see run 1), and a typical system in accordance with the teachings of this process (run 2).

Further preferred interpolymers are presented in Table 2.

Table 2

Run No.	AN	IBMAM	IBMA	MAA
1	35	15	40	10
2	35	17.5	37.5	10
3	37.5	15	37.5	10

The headings in Table 2 are explained in the text following Table 1. These three interpolymers are similar to run 2 of Table 1, but proportions are balanced and methacrylic acid is selected in order to maximize Staypro rating while maintaining the wedge bend rating at a minimum acceptable rating in the range of 6 to 7.

With the data of Tables 1 and 2 in mind, this process uses an interpolymer of monomer components as tabulated in Table 3.

Table 3

Run No.	Monoethylenic Monomer Type	Illustration	Proportions (%)	Preferred Proportions (%)
1	Nitrile	Acrylonitrile* methacrylonitrile	20–45	25-40
2	Isobutyl methacrylate	–	10–50	20–45
3	N-methylol amide	N-methylol acrylamide, acrylamide or ether thereof, e.g., propylbutyl or isobutyl* ether	10–30	12–25
4	Carboxy-functional	Acrylic acid, methacrylic acid*, crotonic acid, fumaric acid, itaconic acid	5-15	7-12
5	Other nonreactive	Ethyl acrylate, methyl methacrylate, styrene	0–10	0

*Indicates materials preferred. The most preferred interpolymer involves the use of from 30 to 40% acrylonitrile and from 35 to 45% of isobutyl methacrylate.

Water-Miscible Carboxylic Acid with Inorganic Fillers and Silanes

Highlights of the Technological Achievement: What is disclosed is a method of treating particulate, inorganic fillers with certain silanes in a certain manner such that when the fillers are used in organic resins, the viscosity of the normally hard to handle resins is significantly reduced so that the resin is easier to handle.

Reference: W.T. Collins and M.D. Fey; U.S. Patent 4,162,245; July 24, 1979; assigned to Dow Corning Corporation.

Background: Fabricating and processing techniques such as molding, extruding, sheet forming, casting and embedding require that the resins used in these processes be of a kind that can be easily handled. Thus, many methods have developed which attempt to improve the handling characteristics of the resins.

One such technique is to use a much lower molecular weight resin, or a monomer with a concomitant low viscosity, in conjunction with a high molecular weight resin to give a dilution effect and thus easier handling. This approach suffers from the standpoint that the products prepared from such blends often have unacceptable physical properties and there is a tendency for the low molecular weight materials to bleed from the formed products.

A second manner of reducing the viscosity of organic resins is to reduce the amount of filler actually used in the resin. It is obvious, however, that the use of lesser amounts of filler will lead to reduced benefit of the mineral. What is needed are organic resins that have enhanced handling properties but which retain good mechanical properties.

Process Conditions: Any organic resin that can contain adequate filler to give good physical properties while having a lower viscosity for better handling is very desirable.

Such resins, and a method for their preparation, can be found herein wherein the method consists of (A) acidifying water to a pH of 3 to 5 using a water-miscible organic carboxylic acid; (B) adding thereto, with constant agitation, a silane selected from a group consisting of

$$ClCH_2CH_2CH_2Si(OCH_3)_3$$

$$F_3CCH_2CH_2Si(OCH_3)_3$$

$$C_{18}H_{37}Si[OCH_2CH_2N(CH_2CH_3)_2 \cdot HCl]_3$$

$$CH_2{=}C(CH_3)C(=O)OCH_2CH_2CH_2Si(OCH_3)_3$$

and thereafter stirring the components (A) and (B) until the solution clears; (C) adding to the solution a particulate, inorganic solid and thoroughly mixing the combination of (A), (B) and (C) at a high speed for a time sufficient to treat the particulate, inorganic solid with the hydrolysis product of component (B); (D) removing volatile materials from the combination to obtain a particulate, inorganic hydrate; (E) mixing the product (D) with an organic resin capable of being heat cured and thereafter (F) removing entrapped air from (E) to obtain an organic resin having a reduced viscosity over that of an unmodified organic resin.

The method consists of steps (A) through (F) and more specifically, the method consists of prehydrolyzing the silanes, contacting the hydrolysis products with the particulate, inorganic material under conditions in which the particulate, inorganic material is chemically bound with the hydrolysis product and the so-treated particulate, inorganic material is then added to an organic resin whereby the viscosity of the resin is reduced rather than increased.

Step (A) requires that the water of hydrolysis be acidified to a pH of 3 to 5 using a water-miscible carboxylic acid. The water required for hydrolysis of the silane is first calculated and then sufficient acid is added to the water to give a pH of 3 to 5 which stabilizes the silanol that is formed by the hydrolysis.

Particulate, inorganic solids contemplated within the scope of this process include alumina trihydrate, i.e., $Al_2O_3 \cdot 3H_2O$, zinc borate, silica, magnesium oxide, talc, clay, calcium carbonate, novaculite, mica and silicates in general. Generally, from 50 to 75% by weight of treated solid is used based on the weight of the solid and the resin.

A preferred embodiment of this process is the treatment of alumina trihydrate with chloropropyl trimethoxysilane using acetic acid and water as the hydrolysis medium wherein 1% by weight of the silane is used. The treated filler is then used to fill polyester resins to reduce the viscosity thereof.

Hydrophilic Polyolefin Fibers for Use in Papermaking

Highlights of the Technological Achievement: Hydrophilic polyolefin fibers may be prepared, for example, by discharging a heated and pressurized solution of an anionic polyolefin composition containing carboxylic functionality into a zone of reduced temperature and pressure, and then modifying the fibers so produced by treatment with an aqueous admixture of selected cationic and anionic water-soluble, nitrogen-containing polymers. Blends of the hydrophilic fibers with wood pulp provide paper products having improved physical properties.

Reference: T.W. Rave; U.S. Patent 4,156,628; May 29, 1979; assigned to Hercules Incorporated.

Background: This process relates to a method for the preparation of hydrophilic polyolefin fibers which are readily dispersible in water and which can be blended with wood pulp fibers to provide a pulp which can be made into high quality paper using conventional papermaking techniques.

More particularly, this process relates to the formation of polyolefin-based fibers containing carboxylic functionality and treatment of these fibers with blends of certain water-soluble, nitrogen-containing polymers, one of which is cationic and the other of which is anionic. The process also relates to specific blends of the cationic and anionic polymers.

Process Conditions: As an example of this process, polypropylene and an ethylene-acrylic acid copolymer are dispersed in a solvent such as methylene chloride, and the dispersion is heated in a closed system to a temperature of 190°C to dissolve the polymer components in the solvent. Under these conditions, the pressure generated by the methylene chloride vapors is 600 psi. After introducing nitrogen to increase the vapor pressure of the system to a pressure of 1,000 psi, the resulting solution is vented to the atmosphere through an orifice, resulting in evaporation of the methylene chloride solvent and formation of the fiber product.

The fiber product is then suspended in an aqueous medium formed by blending a dilute aqueous solution of, for example, epichlorohydrin-modified poly(diethylenetriamine-adipic acid) with a dilute aqueous solution of, for example, glyoxal-modified poly(acrylamide-coacrylic acid), and the components of the resulting suspension are brought into intimate contact with each other by refining in a disc refiner. The treated fibers may then be isolated and stored in wet cake form, or the suspension containing the fibers may be used directly in a papermaking process.

Water-Dispersible Dye/Resin Compositions

Highlights of the Technological Achievement: Disclosed are water-dispersible dye/resin compositions in solid, semisolid or viscous liquid form, useful, for example, in the preparation of water-based heat transfer printing inks and aqueous dye dispersions for conventional dyeing. These compositions are solutions (this term as used herein encompasses true solution, colloidal dispersion and suspension) of disperse dyes in, for example, copolyesters of 5-sodiosulfoisophthalic acid optionally blended with certain aliphatic or cycloaliphatic dicarboxylic acids.

Exemplary are solutions of dyes in the copolyesters of isophthalic acid, 5 to 20 mol percent based on mols of glycol of 5-sodiosulfoisophthalic acid, and diethylene glycol. Aqueous dye dispersions can be made with these dye resin compositions without grinding of the dye beyond rough pulverization of the dye cake.

Reference: S.E. Blackwell and K.R. Barton; U.S. Patent 4,148,779; April 10, 1979; assigned to Eastman Kodak Company.

Background: In the manufacture of disperse dyes for use in heat transfer printing inks or for use in conventional dyeing, the customary practice is to grind or ball mill the disperse dye with dispersing aids for extended periods to reduce its primary particle to less than 25 μ and insure at least a reasonably stable dye suspension in the aqueous bath. This procedure is costly, time consuming and inconvenient.

Process Conditions: This process may be defined as water-dispersible compositions comprising:

(A) From 5 to 60% by weight of disperse dye; and

(B) From 95 to 40% by weight of linear, water-dispersible copolyester; or

(C) From 95 to 40% by weight of a blend of the copolyester of (B) with up to 40% by weight based on total weight of the blend of a saturated aliphatic or cycloaliphatic dicarboxylic acid having 6 to 12 carbon atoms, the copolyester being comprised of

(D) At least one aliphatic, cycloaliphatic or aromatic dicarboxylic acid component;

(E) A diol component of from 2 to 28 carbons, at least 20 mol percent of which is a poly(ethylene glycol) having the formula $H{+}OCH_2CH_2{+}_nOH$ wherein n is an integer of from 2 to 14; and

(F) At least one difunctional dicarboxylic acid containing a $-SO_3M$ group attached to an aromatic nucleus, wherein M is hydrogen, Na^+, Li^+ or K^+, and constituting at least 8 to 45 mol percent of the total mols of components (D) and (F).

Example 1: *Preparation of Water-Dispersible Resin* – 60 parts of a 90/10 copolyester of isophthalic acid/5-sodiosulfoisophthalic acid, and diethylene glycol (inherent viscosity 0.39) are mixed with 40 parts of adipic acid and 10 parts of stearic acid at 165°C under an atmosphere of dry nitrogen until all solid materials have melted and the blend is homogeneous. The blend is then extruded onto granules of solid carbon dioxide and ground to 3 mm particle size in a Wiley mill. The inherent viscosity of the blend corrected for the low molecular weight diluents, adipic and stearic acids, is 0.38.

Example 2: 2 parts of the resin described in Example 1 are melted at 165°C under a dry nitrogen atmosphere. 1 part of C.I. Disperse Red 60 dye containing no dispersing agent or other diluent is slowly added with stirring and allowed to mix until all the dye is fully dissolved.

The molten solution is poured out onto a Teflon base and allowed to cool, forming a solid sheet which can be granulated readily. 2 g of this solid solution of dye are dispersed in 120 ml of water at 80°C. A stable dispersion is formed in 20 minutes and 2.8 g of high viscosity hydroxyethylcellulose is dissolved in the mixture to form a thick paste ink suitable for screen printing. When the ink has cooled to room temperature, screen prints are made onto paper sheets and allowed to dry. Heat transfer prints are made by contacting this paper with 100% polyester fabric under 2 to 6 psi pressure for 30 seconds at 210°C. Sharp, clear prints with good dye transfer are obtained. The prints are fast to light, washing, and rubbing.

Two-Pack Type Aqueous Paint

Highlights of the Technological Achievement: An improvement in a process for coating a surface with an aqueous paint composition containing as the primary film-forming components: a synthetic polycarboxylic acid resin as the main binder component; an amino resin as a heat curing crosslinking agent for the polycarboxylic acid resin and a neutralizing agent for neutralizing the acid groups of the polycarboxylic acid resin, is provided.

The improvement is applying the aqueous paint composition to the surface as two compositions (A) comprising a synthetic polycarboxylic acid resin having a neutralization degree of 30 to 70% and an amino resin; and (B) comprising a synthetic polycarboxylic acid resin having a neutralization degree 30 to 80% higher than that of the polycarboxylic acid resin of (A) and an amino resin, or (C) a neutralizing agent comprising ammonia or an organic amine.

Reference: K. Toyoshima, M. Koyanagi and T. Suzuki; U.S. Patent 4,132,686; January 2, 1979; assigned to Shinto Paint Co., Ltd., Japan.

Background: Generally, for paint compositions, there are used organic solvents and such organic solvent type paints are not desirable from the viewpoint of air pollution, material economy and safety of work. Therefore, there is a tendency toward transition from organic solvent type paints to aqueous type paints. Such aqueous paints are generally of the one-pack type and are applied to objects to be painted, directly as such or after dilution with water when they are thick compositions.

In conventional one-pack aqueous paint compositions, there is used as the main binder component a natural or synthetic resin. Polycarboxylic acid resins have become to be used in many cases. These polycarboxylic acid resins are completely neutralized to an excess of the equivalent for the carboxyl groups in the resin with an alkaline substance (usually ammonia or an amine) in order to dissolve the resin completely in water and to prepare a stable solution.

However, by using such an alkaline substance, the viscosity of the resulting paint composition increases and this makes the painting operation difficult. Further, it becomes necessary to use a relatively large amount of a diluent such as water to lower the viscosity, and this causes sag of the paint and unevenness of the sprayed surface, making the painting operation extremely difficult.

The object of this process, therefore, is to provide a two-pack aqueous paint composition which overcomes the abovementioned disadvantages of conventional aqueous paint compositions.

Process Conditions: This process provides a two-pack aqueous paint composition which comprises (1) an aqueous paint (Paint A) composed mainly of a synthetic polycarboxylic acid resin having a neutralization degree of not higher than

70% and (2) an aqueous paint (Paint B) composed mainly of a synthetic polycarboxylic acid resin having a neutralization degree higher than that of Paint A, or a neutralizing agent comprising ammonia and/or an amine (Neutralizing Agent C). When the Paint A and Paint B, or Paint A and Neutralizing Agent C are mixed into contact upon application, the viscosity of the coated film is increased.

The synthetic polycarboxylic acid resins having a neutralization degree of not higher than 70% to be used in Paint A include (a) polymers or copolymers obtained by polymerizing or copolymerizing one or more of carboxyl group-containing monomers such as acrylic acid, methacrylic acid, itaconic acid, fumaric acid, synthetic resin acid, vinyl acetate or one or more monomers such as acrylic acid esters, methacrylic acid esters, styrene, etc., copolymerizable with the carboxyl group-containing monomers, the polymers or copolymers having a number average molecular weight of above 2,000 and an acid value of 10 to 150, and neutralized with a neutralizing agent for not higher than 70% of the acid groups present in the molecule; and (b) oil-free alkyd resins or modified alkyd resins synthesized by using an acid such as maleic acid anhydride, phthalic acid anhydride, adipic acid, tetrahydrophthalic acid anhydride, trimellitic acid, etc., and a a polyhydric alcohol such as ethylene glycol, propylene glycol, trimethylolpropane, sorbitol, glycerin, pentaerythritol, etc., and if necessary, a fatty acid, fat or oil, petroleum resin or epoxy resin, the alkyd resins or modified alkyd resins having a number average molecular weight of above 2,000 and an acid value of 10 to 150, and neutralized with a neutralizing agent for not higher than 70% of the acid groups.

Examples 1 through 3: The order of preparing paints is to mix a synthetic polycarboxylic acid resin with a pigment first, if any pigment should be contained in the paint, and finally another resin, neutralizing agent and additive are added to regulate the viscosity.

Paints A and B have the compositions shown in the following table.

	...... Example No..		
	1	2	3
Paint A composition			
Pigment (Al powder)	1.9	1.9	1.9
Synthetic polycarboxylic acid (acrylic resin)			
70% NV	48.21	60.0	–
80% NV	–	–	42.19
Amino resin (hexamethoxymethylmelamine)	11.25	11.25	11.25
Neutralizing agent (dimethylethanolamine)	1.88	0.40	1.69
Diluent (water)	33.76	26.45	42.97
Paint A properties			
Neutralization degree, %	50	30	30
Nonvolatile matter, %	46.9	55.2	46.9
Viscosity, poises/20°C	3.6	4.1	2.8
Synthetic polycarboxylic acid average MW	15,000	15,000	2,500
Acid value of synthetic polycarboxylic acid	70	20	105
Paint B composition			
Synthetic polycarboxylic acid (acrylic resin)			
70% NV	48.21	30.0	–
50% NV	–	–	21
Amino resin (hexamethoxymethylmelamine)	11.25	11.25	11.25
Neutralizing agent (dimethylethanolamine)	5.26	0.6	4.2
Diluent (water)	35.28	58.15	63.55

(continued)

	Example No. 1	2	3
Paint B properties			
Neutralization degree, %	140	90	150
Nonvolatile matter, %	45.0	32.3	28.1
Viscosity, poises/20°C	46.6	50.3	49.7
Synthetic polycarboxylic acid average MW	5,000	3,500	2,500
Acid value of synthetic polycarboxylic acid	70	20	105

The following solutions were prepared as Neutralizing Agent C: dimethylethanolamine, 100%; monoisopropanolamine, 100%; dimethylethanolamine, 40% solution in butyl glycol; and dimethylethanolamine, 45% aqueous solution.

In the painting operation, Paint A was applied on a zinc phosphate-treated iron plate by means of a spray gun such that the thickness of the film became 40 μ. Subsequently, within 60 seconds, Paint B or Neutralizing Agent C was sprayed thereon by means of a spray gun such that it formed a film having a thickness of 3 or 1 μ, respectively.

The painting conditions were as follows: spray booth temperature, 25°C; and spray booth humidity, 40% RH, 75% RH and 90% RH. The baking conditions were as follows: setting–in the atmosphere at 60°C for 7 minutes; baking–the atmosphere was raised to 160°C in 10 minutes; and curing of the film–160°C for 20 minutes.

Detergent Compositions Comprising Polyacetal Carboxylates

Highlights of the Technological Achievement: Ether and acetal carboxylates useful as detergent builders to replace sodium tripolyphosphate (STP) in detergent formulations are known in the art. Certain polymeric polycarboxylates are also known but all of these products suffer from either having poor builder properties or they are not readily biodegradable.

According to this process, polyacetal carboxylates are provided which are stable under laundry use conditions, but depolymerize in acid media, making the fragments more readily biodegradable in waste streams.

Reference: M.M. Crutchfield, V.D. Papanu and C.B. Warren; U.S. Patent 4,146,495; March 27, 1979; assigned to Monsanto Company.

Background: It can be seen that there is a need for a new class of materials with builder properties equivalent to STP, which does not contain phosphorus, which is water-soluble, and which achieve environmental acceptability by being readily biodegradable. According to this process, there is provided a new class of polyacetal carboxylates which are equal to, if not superior to, STP and which will depolymerize rapidly in a nonalkaline medium to form low molecular weight components which are readily biodegradable.

Process Conditions: These and other advantages are achieved by a detergent composition comprising at least 5% by weight of a surfactant selected from the group consisting of anionic, nonionic, zwitterionic, ampholytic and amphoteric surfactants; and at least 1% by weight of a stabilized, water-soluble polymer comprising polyacetal carboxylate segments having the following structure;

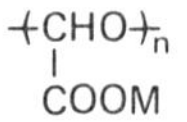

where M is selected from the group consisting of alkali metal, ammonium, tetralkyl ammonium and alkanol amine groups having from 1 to 4 carbon atoms in the alkyl chain; and n averages at least 4.

Any number of esters of glyoxylic acid can be used to prepare the polyacetal carboxylates of this process. Such esters can be made by the reaction of an alcohol containing from 1 to 4 carbon atoms with glyoxylic acid hydrate under conditions known to those skilled in the art. Thereafter, the ester hemiacetal can be converted to the corresponding aldehyde ester by any number of techniques known to those skilled in the art, such as the reaction of the ester hemiacetal with phosphorus pentoxide. The product of the above reaction is then polymerized by techniques known to those skilled in the art using an initiator in accordance with the following general equation:

The resulting polyacetal carboxylate ester is then reacted at its termini with a reagent which produces a chemically stable end group to stabilize the polyacetal carboxylate against rapid depolymerization in alkaline solution.

Particularly suitable end groups include alkyl groups and cyclic alkyl groups containing oxygen, such as oxyalkyl groups like methoxy, ethoxy and the like; carboxylic acids such as $-CH_2COOM$,

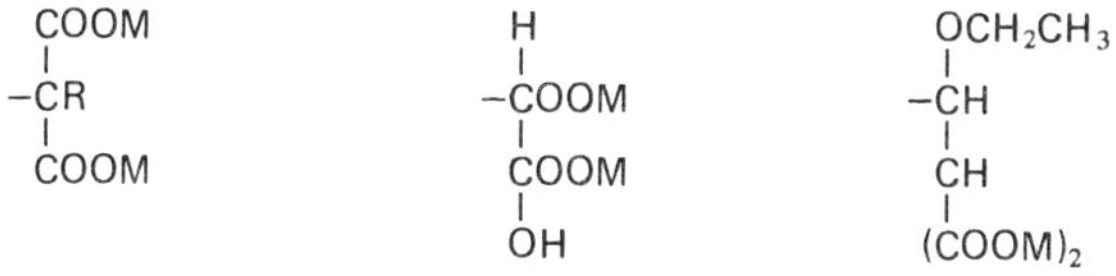

and the like; aldehydes, ethers and other oxygen-containing alkyl groups such as $-OCHCH_3OC_2H_5$, $-(OCH_2CH_2)_{1-4}OH$, $-(CH_2CH_2O)_{1-4}H$,

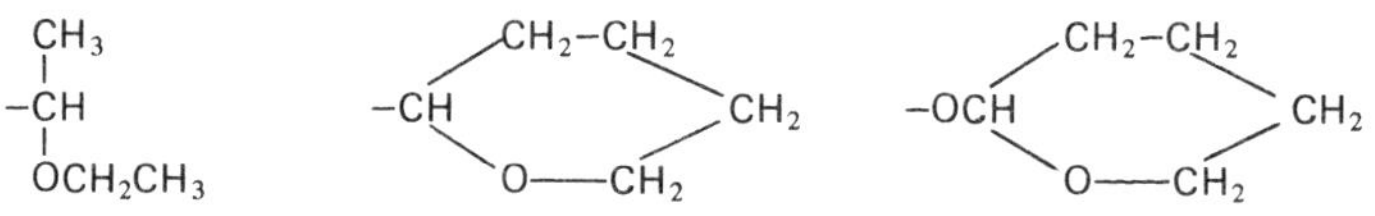

and the like. In the above examples of suitable end groups, M is alkali metal, ammonium, alkanol amine, alkyl groups having 1 to 4 carbon atoms, tetralkyl ammonium groups and alkanol amine groups having from 1 to 4 carbon atoms in the alkyl chain, and R is hydrogen or alkyl group of 1 to 8 carbon atoms.

It can be seen that in one embodiment of this process there is provided a detergent composition comprising at least 5% by weight of a surfactant selected from the group consisting of anionic, nonionic, zwitterionic, ampholytic and amphoteric surfactants; and at least 1% by weight of a water-soluble polyacetal carboxylate having the structure:

$$R_1\!\left[\begin{array}{c}\mathrm{CHO}\\ |\\ \mathrm{COOM}\end{array}\right]_n\!R_2$$

where M is selected from the group consisting of alkali metal, ammonium, tetralkyl ammonium groups and alkanol amine groups having from 1 to 4 carbon atoms in the alkyl chain; n averages at least 4; and R_1 and R_2 are individually any chemically stable group which stabilize the polyacetal carboxylate against rapid depolymerization in alkaline solution.

Product: It can be seen that in one embodiment of this process there is provided a detergent composition comprising at least 5% by weight of a surfactant selected from the group consisting of anionic, nonionic, zwitterionic, ampholytic and amphoteric surfactants; and at least 1% by weight of a stabilized water-soluble polymer comprising polyacetal carboxylate segments having the general formula:

$$\left[Y\left(\begin{array}{c}\mathrm{CHO}\\ |\\ \mathrm{COOM}\end{array}\right)_p\right]_q$$

where Y is at least one chain extending agent, preferably alkyl or oxyalkyl having 1 to 4 carbon atoms, p averages at least 4, q is at least 1 and M is selected from the group consisting of alkali metal, ammonium, tetralkyl ammonium groups and alkanol amine groups having from 1 to 4 carbon atoms in the alkyl chain. Furthermore, the polyacetal carboxylates having a chain extending agent can be stabilized against rapid depolymerization in alkaline solution by the same techniques used above using suitable reagents or polymers as described above.

Aqueous Emulsion Thermosetting Coating Compositions

Highlights of the Technological Achievement: Aqueous emulsion thermosetting coating compositions possessing superior flexibility and room temperature coalescence are provided by combining certain aqueous emulsion copolymers including hydroxy and carboxy functionality, aminoplast resin and water-dispersible high boiling polyhydric adduct of an aliphatic polyhydric alcohol containing from 2 to 4 hydroxyl groups with a monoepoxide or a lactone.

Reference: D.J. Berenschot, D.F. Anders and F.D. Hawker; U.S. Patent 4,137,205; January 30, 1979; assigned to DeSoto, Inc.

Background: Aqueous coating compositions of the type defined intended to be thermoset on baking are subject to numerous disadvantages. Most importantly, the cured films at desired film hardness tend to be unduly brittle and the gloss is lower than desired. Also, and particularly when using aqueous emulsion copolymer particles which do not adequately coalesce on air drying and which do not sufficiently reflow on baking, it is hard to obtain smooth and uniform cured films. This process is concerned with countering the above noted as well as other difficulties associated with thermosetting aqueous emulsion coatings.

Process Conditions: Certain aqueous emulsion copolymers containing carboxyl and hydroxyl functionalities are combined in aqueous medium with an aminoplast resin and a water-dispersible high boiling polyhydric adduct of an aliphatic polyhydric alcohol containing from 2 to 4 hydroxyl groups and preferably having from 2 to 6 carbon atoms with a C_2 to C_4 monoepoxide or a lactone containing 3 to 6 carbon atoms.

More particularly, the aqueous emulsion copolymer consists essentially of from 0.75 to 25% of nonnitrogenous hydroxy functional monoethylenically unsaturated monomer, from 0.1 to 15% of monoethylenically unsaturated carboxylic acid, and the balance of the copolymer consists essentially of monoethylenically unsaturated monomer free of functional groups other than the single ethylenic group.

It is found in these emulsion systems that the adduct, which is a polyether or polyester, improves room temperature coalescence as well as reflow on baking, desired hardness is maintained, and flexibility is enhanced. Also, the gloss is increased and the adduct becomes an integral part of the cured film, thus minimizing the need for volatile components in the aqueous coating.

In this description, all parts or proportions are by weight, unless otherwise stated.

Referring first to the components of the copolymer, from 0.75 to 25%, preferably from 2 to 15%, of the copolymer is required to be a nonnitrogenous hydroxy functional monoethylenically unsaturated monomer. The hydroxy functional monomer is included in the copolymer by having it present in the monomer mixture which is copolymerized. Below 2%, the film hardness falls off slightly, but this can be tolerated since the other properties are so advantageous.

These hydroxy functional monomers must be monoethylenically unsaturated, though the monomer may contain one or several hydroxy groups. The N-methylol group includes the OH moiety, but compounds containing this group do not react in the same way as the hydroxy group. To positively exclude the N-methylol group, the hydroxy monomers of this process are nonitrogenous. The hydroxy group may be more precisely, alcoholic. The hydroxy functional monomers which are used in this process are preferably illustrated by hydroxy alkyl esters of monoethylenically unsaturated monocarboxylic acids, such as hydroxy ethyl acrylate. A small proportion of ethylenically unsaturated hydroxy functional polyester having an hydroxyl number of from 100 to 350 may be included with the monomers.

The copolymer should also include at least 0.1%, up to 15%, but more preferably from 0.3 to 10%, and most preferably from 0.5 to 8%, based on the weight of the copolymer, of monoethylenically unsaturated carboxylic acid, such as acrylic acid, methacrylic acid, crotonic acid, fumaric acid, monobutyl maleate, and the like. The presence of a small amount of acid helps to stabilize the aqueous dispersions in the presence of ammonia or an amine. Larger amounts, above 2%, provide an additional basis for the ultimate cure with aminoplast resin.

Particularly preferred products are dihydric water-soluble polyether adducts having a molecular weight of from 400 to 5,000. In addition to the polyether adducts, polyester adducts of the same molecular weight range can also be used, these being provided by the adduction of a lactone, such as ϵ-caprolactone, with the same aliphatic polyhydric alcohols noted hereinbefore.

The lactone-based polyesters are in commerce and are usually based on ϵ-caprolactone which has the formula:

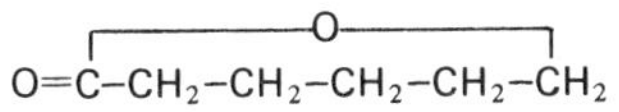

These lactones are reacted with polyhydric alcohols in known fashion to form adducts having the formula:

$$R\text{-}(O\text{-}C\text{-}CH_2\text{-}CH_2\text{-}CH_2\text{-}CH_2\text{-}CH_2\text{-}OH)_n$$

where R is the residue of the aliphatic polyhydric alcohol, preferably propylene glycol or other glycol. Thus, n is preferably 2, though it may range from 2 to 4.

CELLULOSE ETHERS AND OTHER CELLULOSE PRODUCTS

PROCESSING

Use of Sodium Carboxymethylcellulose to Form Aqueous Dispersions of Materials That Are Difficult to Disperse

Highlights of the Technological Achievement: Oleaginous and other materials that are difficult to form into aqueous dispersions are rendered easily dispersible in cold water by partially encapsulating them around particles comprising from 70 to 99 parts of disintegrated β-1,4-glucan and from about 1 to 30 parts of a water-soluble polymer intimately associated therewith and especially where the particles comprise from 85 to 95 parts of disintegrated β-1,4-glucan and from 5 to 15 parts of sodium carboxymethylcellulose.

Reference: E.J. McGinley; U.S. Patent 4,199,368; April 22, 1980; assigned to FMC Corporation.

Background: There are numerous instances in manufacturing, notably in food processing but also in operations entirely unrelated to food, where it is desirable to obtain a uniform stabilized dispersion of a powdered or granular form of an oleaginous or other difficult-to-disperse material in a cold aqueous system. For example, in making bread it would be desirable to have a dry, free-flowing mono- and diglyceride powder which could be added directly to bread dough, the bread dough being a cold, aqueous system. Mono- and diglycerides are widely used in making bread but normally a predispersed hydrate form is necessary to achieve dispersion in the bread dough cold aqueous system.

Process Conditions: The basic unit of the process is a particle comprising by weight from about 70 to 99 parts of disintegrated β-1,4-glucan and from about 1 to 30 parts of a water-soluble polymer intimately associated therewith. In the preferred form of the process the basic unit comprises by weight from 85 to 95 parts of a disintegrated β-1,4-glucan intimately associated with from 5 to 15 parts of sodium carboxymethylcellulose, especially sodium carboxymethylcellulose having a degree of substitution of 0.75 ± 0.15. The nature of the preferred particulate material and a method of preparation thereof is described in U.S. Patent

3,539,365. These basic units or particles are water-insoluble but are known to be readily dispersible in water to form a thixotropic gel. Thus, this process is not directed toward forming a dispersion of the water-insoluble β-1,4-glucan-containing particles, but involves using these particles as a means of attaining a dispersion of other water-insoluble materials as well as difficult-to-disperse water-soluble materials. When these basic particles are put in water, irrespective of temperature, they swell or expand rapidly and break into much smaller particles, many of which are less than one micron in diameter.

Raw Materials: The basic particles comprising β-1,4-glucan, having intimately associated therewith carboxymethylcellulose or other water-soluble polymer, e.g., xanthan gum, sodium alginate, carrageenan, tragacanth gum, pregelatinized waxy maize starch, are coated with or encapsulated by the material it is desired to disperse. The coating or encapsulating material should normally be in liquid form so if it is, for example, a hard wax material, it may be necessary to apply heat and melt it before mixing it with the basic β-1,4-glucan-containing particles. If the material to be dispersed is polyethylene glycol 4000, which is water-soluble, it is liquefied by heating or dissolved in ethanol, ethanol not being effective to cause rapid expansion of the basic β-1,4-glucan-containing particles. The basic particles are added to the material to be dispersed in sufficient quantity so as to absorb and adsorb the liquid material and form, upon stirring, a free-flowing powder or granular material.

When the material prepared as abovedescribed is put into water, the water penetrates the minute openings in the partially encapsulating skin material and contacts the basic or core particles comprising β-1,4-glucan and the water-soluble polymer, whereupon the basic particles expand almost instantly to a considerable extent.

There is a substantial amount of prior art where one form of β-1,4-glucan is employed as a carrier or adsorbent to produce a powdered form of certain materials which may be reconstituted to approximately their original nonpowdered form. This material is variously referred to as microcrystalline cellulose, level-off DP cellulose and cellulose crystallite aggregates and is not suitable for carrying out this process because the particles of such material do not swell or expand when put in water and therefore do not produce fragmentation of the material adsorbed thereon.

Increasing the Viscosity of Hydroxyalkylcellulose with Benzoquinone

Highlights of the Technological Achievement: Viscosity of hydroxyalkylcellulose solutions is increased by treatment with benzoquinone under controlled conditions of pH.

Reference: T.M. Podlas; U.S. Patent 4,183,765; January 15, 1980; assigned to Hercules Incorporated.

Background: The hydroxyalkyl ethers of cellulose are used as viscosity-building ingredients in aqueous solution in a wide range of applications. These ethers are available in low, medium, high and even ultrahigh molecular weight grades. (These are conventionally referred to as low, medium, high, and ultrahigh viscosity materials, referring to the viscosity of their aqueous solutions.) There are applications, however, for the hydroxyalkylcellulose ethers where even the ultrahigh viscosity grades are not high enough to accomplish the end desired. In some other applications, it is desirable to use a material of a low viscosity which is pumpable but which can develop very high viscosity or even form a gel after it has been transferred to the location where it is to accomplish the purpose intended for it. A method of increasing the viscosity or crosslinking a hydroxyalkylcellulose ether is much to be desired in such cases.

However, practical methods have not been known to effect viscosity enhancement of hydroxyalkyl ethers of cellulose in aqueous solution. U.S. Patent 3,378,070 details crosslinking of hydroxyethylcellulose with certain metal ions, but is concerned with the formation of tough, rubbery gels. U.S. Patent 4,040,484 and U.S. Patent 4,068,720 describe treating cellulose ethers with polyvalent, reducible metal ions and reducing agents capable of reducing these ions in situ to a valence state at which they can cause crosslinking to take place. These latter patents purport to be applicable to all cellulose ethers but are demonstrated only for the ionic species.

Process Conditions: It has been found that water-soluble hydroxyalkyl ethers of cellulose can readily be increased in viscosity in aqueous solution with p-benzoquinone to form high viscosity solutions or even gels. In accordance with this process, a hydroxyalkyl ether of cellulose is increased in viscosity or crosslinked by a method which comprises contacting a solution containing at least 0.075% by weight of such ether with 0.4 to 75% by weight, based on the weight of the ether, of benzoquinone, the solution being one having a pH greater than 6.4 after addition of the benzoquinone and having sufficient buffering capacity to retain a pH greater than 6.4 for at least 24 hr after benzoquinone is added thereto.

Cellulose Ethers: The method of the process is applicable to water-soluble cellulose ethers generally, such as, for example, methylcellulose, carboxyalkylcellulose and the hydroxyalkylcelluloses. It is particularly useful with the hydroxyalkylcelluloses which have a relatively high concentration of unsubstituted hydroxyl groups. Thus hydroxyethylcellulose, hydroxypropylcellulose, methylhydroxypropylcellulose, carboxymethylhydroxyethylcellulose, dihydroxypropylcellulose and ethylhydroxyethylcellulose, by way of example, are readily modified by the method of the process.

Benzoquinone: The benzoquinone can be brought into contact with the cellulose ether after the ether is dissolved in the water. Either particulate benzoquinone can be added to the solution or a prepared solution can be added. Alternatively the cellulose ether can be added to a solution of the benzoquinone. For greatest ease of handling, it is also desirable on some occasions that the benzoquinone and hydroxyalkylcellulose be dry blended in the appropriate proportions prior to incorporation of either component to the water.

pH: In increasing the viscosity with benzoquinone according to this process, the pH of the solution is the most critical factor. Normally, the pH of an aqueous solution of a commercial cellulose ether will be at least slightly basic regardless of the pH of the water used as the solvent. In the presence of the benzoquinone, the pH immediately begins to decrease toward the acidic range.

At acid pH levels, the benzoquinone may cause a decrease in the viscosity of the cellulose ether solution. If the pH decrease is excessive, this effect is greater than the viscosity increasing effect and no measurable viscosity increase takes place.

By observation, it has been established that the pH decrease is excessive if the pH of the cellulose ether solution is less than 6.4 after 24 hr in the presence of the benzoquinone.

Other Factors: It is preferred to use water which initially is of basic pH or near neutral to carry out the process. Normal tap water or groundwater which is neutral or slightly basic is usually sufficiently buffered that its pH will remain within the operative range. If it does not, it can be buffered artificially via addition of an electrolyte, preferably an alkali metal salt such as sodium carbonate or sodium bicarbonate. Likewise, if the original pH of the water is not within the basic range, the water can be made basic by addition of caustic, such as sodium or potassium hydroxide or by the addition of a basic electrolyte such as potassium or sodium carbonate. Such an acidic water will then be usable in the process if there is sufficient buffering to assure that the previously defined pH condition is met.

The concentration of benzoquinone required to effect the viscosity increase or crosslinking varies to some extent with the cellulose ether being treated, but more so with the concentration of the ether in solution.

With respect to type of cellulose ether being modified, the difference in required concentration apparently arises from the differences in available hydroxyl groups remaining on the ether. Thus, those ethers such as carboxymethylcellulose, methylcellulose and methylhydroxypropylcellulose which have a nonhydroxylic substituent are more difficult and need a higher concentration of benzoquinone to modify them.

It has also been observed that the rate of viscosity increase can be retarded by the presence of formaldehyde. Thus, a means is provided of further control over viscosity increase or gelling rate. The required formaldehyde concentration which will give noticeable retardation is about 1 to 6 parts of formaldehyde per part of benzoquinone.

The process is illustrated in the following example. Parts and percentages are by weight unless otherwise indicated. Viscosities were measured using a Brookfield viscometer at 60 rpm, No. 2 spindle.

Example 1: A solution was prepared of 0.25% high molecular weight hydroxyethylcellulose (HEC) having MS of 2.5 (i.e., 2.5 mols of hydroxyethyl substitution per cellulosic anhydroglucose unit) in well water having a pH of 8.3. Upon complete dissolution of the HEC, this solution had a viscosity of 31 cp.

To 100-ml aliquots of the HEC solution was added, respectively, 5 ml and 10 ml of a 1% solution of technical grade benzoquinone in water. These were allowed to sit until gelling took place. Periodic viscosity and pH readings showed the following:

Benzoquinone	pH			Viscosity (cp)				
Content (ml)	Immediately	1 Day	7 Days	1 Day	3 Days	4 Days	5 Days	6 Days
10	8.1	7.0	6.4	44	55	Gel	–	–
5	8.0	7.2	6.8	32	36	270	Gel	–

The above experiments were repeated using distilled water of pH 6.9. After 24 hr, the pH had dropped to 5.0 and 5.4, respectively. After 17 days, no evidence of crosslinking was visible, the viscosity having decreased to the 6 to 7 cp range for both quinone addition levels. Substantially the same pH drop resulted and no viscosity increase occurred when the pH of the distilled water was increased to 8.3 with sodium hydroxide prior to preparing the HEC solution, indicating the lack of buffering of the distilled water.

Hydrophilic Film from a Water-Soluble Cellulose Ether

Highlights of the Technological Achievement: This process relates to a hydrophilic shaped article of a water-insoluble polymer selected from the group consisting of regenerated cellulose, cellulose ether, cellulose ester and polyalkylene, and capable of forming fibers and film, the article containing amorphous particles of an originally water-soluble cellulose ether prepared from natural cellulose and rendered by chemical modification at least partly water-insoluble while remaining water-absorbent.

Reference: A. Holst, M. Kostrzewa and H. Lask; U.S. Patent 4,194,024; March 18, 1980; assigned to Hoechst AG, Germany.

Background: In U.S. Patents 3,259,537 and 3,275,580, there is described an article of a mixture of an organic polymer and cellulose crystallite aggregates or derivatives. Those crystallite aggregates have an average level-off DP value in the range of from about 15 to 375 anhydroglucose units and the specific value will be dependent upon the source material, which in all instances has an actual DP greater than its average level-off DP and was modified by the controlled acid hydrolysis of cellulose to reflect a destruction of the original amorphous structure of the cellulosic source material. The derivatives of cellulose crystallite aggregates possess physical characteristics and properties similar to those of cellulose crystallite aggregates.

Process Conditions: The particles are generally finely pulverulent to granular. Their particle size is in the range of 0.01 to 2 mm and depends upon the end use of the structure. The particles are composed of cellulose ethers, such as for example carboxymethylcellulose, hydroxyethylcellulose or methylhydroxyethylcellulose, which are modified. Examples of possible modifying agents for crosslinking are: dimethylolmethylenebisacrylamide; methylenebisacrylamide; trichloropyrimidine, and tetrachloropyrimidine; cyanuric chloride; epichlorohydrin; dichloroacetic acid, diepoxides or their precursors di-α-halogenohydrins. Such materials are disclosed in U.S. Patents 3,589,364, and 3,936,441.

A parameter which also can be varied within wide limits is the degree of modification, for instance, of crosslinking of the cellulose ether of which the particles are composed. When using minor amounts of, for instance, crosslinking agents, for example, from 3 to 10% by weight of epichlorohydrin, relative to sodium carboxymethylcellulose (Na CMC), when producing Na CMC fibers, highly swellable fibers are obtained. At higher degrees of crosslinking, the swellability of the cellulose ether greatly decreases and its water-insoluble content increases. If, for example, about 33% by weight of these highly swellable Na CMC fibers (about 20-fold water absorption) are incorporated into a regenerated cellulose film, the swelling factor of the film rises from about 150% (pure regenerated cellulose) to about 350%.

In order to make modified water-insoluble ionic cellulose ethers more readily accessible for ion exchange processes it is sometimes advisable to locate them on the surface of a structure, for example, of a film. A regenerated cellulose film of which the surface is covered with the particles is produced, for example, by sprinkling finely ground crosslinked Na CMC homogeneously through a sieve onto an as yet not regenerated web of viscose and then regenerating the viscose web in the usual manner to form regenerated cellulose. This gives a hydrated cellulose film coated on one side with crosslinked Na CMC.

Example: 1 g of finely fibrous water-insoluble crosslinked Na CMC, which was crosslinked with epichlorohydrin, contained about 25% by weight of soluble constituents and had a water retention capacity of 20 times its weight and a theoretical exchange capacity of about 3.5 meq/g, was homogeneously incorporated, in a 3-roll mill, into 100 g of spinnable viscose containing about 10% by weight of cellulose.

A layer of 0.2 to 0.4 mm thickness of the modified viscose was cast on a glass plate and regenerated in the usual manner. A transparent regenerated cellulose film was obtained, which as a result of the embedded water-insoluble crosslinked Na CMC fibers had a total theoretical exchange capacity of about 0.32 meq/g. The water retention capacity of the film was about 250% by weight. A cellulose hydrate film manufactured under the same conditions had a water retention capacity of 125% by weight.

Continuous Process for Cellulose Saccharification

Highlights of the Technological Achievement: A continuous process for effecting the acid hydrolysis of cellulosic waste materials, at high solids and in an economically desirable manner. The process may be controlled to produce, as primary products, glucose, furfural and/or xylose. By appropriate control of water and temperature, through the use of direct steam injection, high solids feedstocks may quickly be converted, in good yields, to valuable products, thus making the process very attractive from a practical commercial standpoint.

Reference: J.A. Church, D. Wooldridge, R.L. Burroughs, A.A. Strzepek and W.J. Thompson; U.S. Patent 4,201,596; May 6, 1980; assigned to American Can Company.

Background: Traditionally, used paper and agricultural by-products such as sawdust, wood waste, corncobs, straw, sugar cane bagasse and rice hulls and the like, have been regarded essentially as waste materials, and have been disposed of through incineration or by other, similarly unproductive means. It is well-known that the cellulosic constitutents of such materials can be hydrolyzed to produce more valuable products; however, such operations are in limited use, due largely to the relatively low return-on-investment which they are capable of generating.

The capital expenditures required to design and construct the facilities necessary for carrying out such recovery operations tend to be very significant, thus demanding that relatively high conversion rates be attainable, in order to justify the expense involved. Moreover, the ready availability of the same or similar reaction products from alternative sources and at relatively low cost renders it that much more difficult to justify the adoption and use of any new scheme.

The known methods of hydrolyzing cellulosic materials are not, by-and-large, considered to be commercially feasible, for a number of reasons. In certain instances, the processes are not readily adapted to continuous operation; they typically require excessively long reaction times, and the reactions are normally carried out at low solids concentrations, all of which seriously limit productivity. Moreover, the known methods do not afford such control as would enable utilization for the production of end products having the highest value and greatest salability.

Process Conditions: A continuous process for the saccharification of cellulosic materials utilizes, as the feedstock, an aqueous mixture of such a material containing from 15 to 45 weight percent of solids. The mixture is substantially continuously introduced into one end of a tubular reactor, while the other end of the reactor is constricted so as to develop a back pressure therein. A strong mineral acid is admixed with the cellulosic material in a concentration appropriate to catalyze the hydrolysis thereof, and steam is substantially continuously injected into the mixture at a location downstream from the "one" end of the reactor.

The steam produces a reaction mass containing 10 to 25 weight percent solids, and is injected in an amount sufficient to maintain the temperature of the reaction mass at 160° to 240°C to thereby effect hydrolysis of the material in the presence of the acid; solids content of the reaction mass will be 50 to 80% of that of the aqueous feed mixture. Thereafter, the reaction mass is substantially continuously passed into and through a reaction zone of the reactor, located downstream from the steam injection location.

The length of the reaction zone and the rate of feed mixture introduction are so selected and controlled as to afford an average residence time of about 1 to 10 min for each increment of the reaction mass in the reaction zone, with the back pressure therewithin being 90 to 400 psi gauge. Finally, the reaction mass is substantially continuously discharged from the reactor, and the products of saccharification are recovered therefrom.

In one embodiment of the process, the steam is injected in an amount sufficient to maintain the reaction mass at a temperature of at least 190°C so as to optimize the production of glucose and furfural.

Normally in the discharging step of the process, the reaction mass will be subjected to an abrupt pressure reduction (with or without auxiliary cooling), thereby cooling the mass and terminating the reactions therein, whereby a fraction of the hydrolyzate vaporizes and may be separately recovered. In the event that the hydrolysis is effected at 190°C or above, the vaporized fraction will comprise furfural as the desired product, although a major amount of steam and some acetic acid, will also be present. Most desirably, the hydrolyzate will be neutralized to a pH of about 4 to 7, and generally calcium carbonate will be utilized for that purpose.

Drawing: Figure 8.1 is a schematic illustration of a reactor system suitable for use in carrying out the process.

Figure 8.1: Continuous Process for Cellulose Saccharification

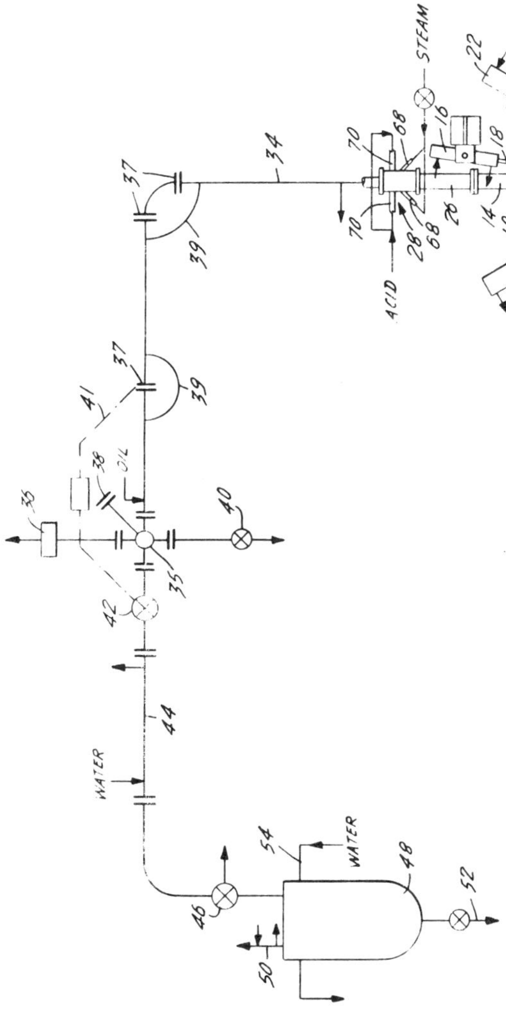

Source: U.S. Patent 4,201,596

Preparing Cellulose Derivatives Having Functional Moieties

Highlights of the Technological Achievement: Production of cellulose derivatives substituted by groups having functional moieties by reaction of a hydroxyl group(s) of the cellulose with an at least bifunctional reagent. The reagent is reacted with cellulose dissolved in dimethyl sulfoxide containing polyhydroxymethylene.

Reference: H.L. Hülsmann and G. Renckhoff; U.S. Patent 4,137,399; Jan. 30, 1979; assigned to Dynamit Nobel AG, Germany.

Background: Cellulose contains for each glucose structural unit of its macromolecule three free but poorly reactive hydroxyl groups, a primary one on carbon atom 6 and two secondary ones on carbon atoms 2 and 3. If their reactivity is compared with that of low-molecular alkanols, the first difference is found to be that the primary hydroxyl groups are less reactive than the secondary. Then, the secondary hydroxyl group on carbon atom 3 is thus less reactive for sterical reasons than the one on carbon atom 2.

Lastly, the rest of the oxygen atoms of cellulose are in the form of an intramolecular bridge between carbon atoms 1 and 5 or of ether bridges to additional glucose units, and do not participate in reactions.

The reactions at the hydroxyl groups of cellulose are thus generally much more difficult than those of alkanols.

In order to improve the reactivity of cellulose, the hydrogen bridges must be substantially broken up and also the amorphous areas must be expanded. Attempts have been made to bring this about by swelling the cellulose, but not many swelling agents for cellulose have become known. In practice, probably only water can be used, since the other agents are too expensive or too difficult to use. This not only narrows the temperature range that can be chosen, but also makes it impossible to perform reactions with reagents sensitive to water. In attempts to alter swollen cellulose chemically through its alkali salts, its great sensitivity to oxidizing influences is a great drawback.

In any case, the reaction takes place on solid structures of cellulose, which in some cases may be loosened up by swelling, i.e., it takes place in a heterogeneous phase. The disadvantages of such a method of reaction are generally well known.

It would be far more advantageous to introduce groups more capable of forming bonds into the cellulose in the dissolved form instead of the merely swollen form. The two solvents used on a large technical scale for cellulose, namely aqueous ammoniacal solutions of copper(II) tetrammine hydroxide and aqueous, alkaline solutions of alkali salts of thiocarboxylic acids, are unsuitable, not merely on account of their water content, as set forth above, but also because, in many cases, a chemical reaction takes place between these solvents and the reagents provided for the functionalization of the cellulose.

Process Conditions: The subject of the process is a method of preparing cellulose derivatives with chemically bound groups rendered capable of forming bonds, which is characterized in that the reactions are performed with bifunctional or polyfunctional reagents, preferably with different reactivity of the groups, on cellulose dissolved in dimethyl sulfoxide containing polyhydroxymethylene.

The solvent dimethyl sulfoxide containing polyhydroxymethylene decisively promotes the reactivity of cellulose. The solvent apparently exposes preferentially the primary hydroxyl groups of the cellulose.

Cellulose Derivatives: Cellulose derivatives having chemically bonded groups capacitated for the formation of bonds in the meaning of the process are those cellulose derivatives in which one or more of the free hydroxyl groups in the cellulose are bonded to the functional group of a substituent, this substituent bearing one or more additional combining functional groups. In general, the functionalized glucose units of the cellulose derivatives prepared in accordance with the process are accordingly formulated as $Gl(-O-R)_{1-3}$ wherein Gl represents a glucose moiety of cellulose and R, the moiety of the reagent used which is bonded to the oxygen of the hydroxy group; wherein R has the structure

$$-(A_{0-1}-\underset{\underset{A'_{0-2}}{|}}{R'}-B_n)$$

wherein B represents functional groups of the reagent which have not entered into reaction with cellulose and which remain preferentially unaltered in the reaction or, in some cases, have been transformed into another functional group; n represents the basically unlimited number of groups B, which amounts preferably to 1, in some cases to 2, or in special cases as much as 4; A represents a functional group of the reagent which is bound to cellulose and whose index in the case of the ether bond is 0; A' represents a sometimes-present group equal to or different from A, which is capable of the reaction with cellulose, but for reasons of the spatial arrangement does not enter into reaction or, in some cases, reacts with another hydroxyl group of cellulose, and R' represents the moiety of the reagent which contains carbon, or carbon and silicon.

Reagents: The reagents of the method are accordingly selected such that at least two functional groups are present, namely, one to three groups A, of which at least one becomes bonded to cellulose in the performance of the process, and at least one second group B identical with A or preferably different therefrom, which under the reaction conditions of Group A is preserved as the same or a different functional group and therefore is contained in the functionalized celluloses as a group or groups rendered capable of combination. For this reason, reagents having varying functional group reactivity are greatly preferred, especially those whose group B remains unaltered under the reaction conditions necessary for group A.

Suitable A groups are, for example, halogen alkyl moieties or carboxylic acid chloride groups which react with cellulose with the formation of an ether bond or ester group.

Since the reaction of the dissolved cellulose can be performed under mild conditions, the reaction in an open vessel surmounted by a reflux condenser is greatly preferred. The temperatures then will be between room temperature and the boiling point of the solvent, preferably between 40° and 186°C. However, it is possible to apply slight overpressures up to, say, 50 bars, for the purpose of reacting the less reactive reagents, or of reacting additionally the abovementioned less reactive hydroxyl groups of cellulose.

After the reaction is completed, the cellulose derivatives are precipitated from the reaction mixtures.

Products: The products are especially usable for the preparation of immobilized bioactive proteins, e.g., enzymes or other soluble biological active pharmaceuticals, which form a solid body together with the product, the bioactivity of this immobilizate being greater respective the volume and better standardizable and more resistant against deactivation than the enzyme itself.

Example: 10 g of powdered linters cellulose "MN 200," (Macherey, Nagel & Co., Duren) were dissolved in a 500-ml flask equipped with stirrer, thermometer and electrical heater, in a mixture of 190 g of dimethyl sulfoxide and 10 g of paraformaldehyde, at 50°C. At this temperature, after the addition of 0.1 g of finely powdered sodium hydroxide, a solution of 10 g of p-chloromethylbenzoic acid methyl ester in 30 g of dimethyl sulfoxide was added in portions over a period of 30 min.

Then the mixture was stirred for 4 hr at 50°C. The reaction mixture, cooled to room temperature, was then poured into ethanol while the latter was being stirred vigorously. The solid was suction filtered, suspended repeatedly in ethanol until all the soluble substance had been removed, and dried. The yield was 10.6 g. The product was free of chlorine and acid.

According to analysis, there was a content of 8.2 weight percent of carbomethoxy groups in the functionalized cellulose. Accordingly, 28.4% of the free hydroxyl groups of the cellulose had been etherified by the entry of the group:

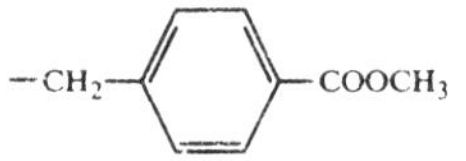

Process for Making Solutions of Cellulosic Compounds

Highlights of the Technological Achievement: A solution of cellulose or cellulose derivative suitable for spinning, extruding or other shaping process is prepared by steeping unaged cellulose or a cellulosic derivative a plurality of times, consecutively, subsequently shredding and then aging prior to xanthation or the like. The concentration of alkali metal hydroxide of the solution used for resteeping is lower than that in the first steeping step.

Reference: N.E. Franks; U.S. Patent 4,136,255; January 23, 1979; assigned to Akzona Incorporated.

Background: It is an object of this method to provide a process for making solutions of cellulosic materials which is advantageous over the modified process. Another object of the method is to provide a process for making viscose which can be practiced with steeping in the conventional presses or with slurry steeping. Still another object of the method is to provide a process for making viscose in improved yields in shorter xanthation and viscose ripening times.

Process Conditions: By postponing aging of the alkali cellulose until after two or more steepings in aqueous caustic solution, the loss in yield of cellulose is significantly reduced and the amount of carbon disulfide required in preparing viscose can be less than that required in a conventional process in which the aged product is xanthated with the carbon disulfide.

Further, the use of this consecutive procedure involves fewer handling steps of the alkali cellulose and a smoother mechanical flow in the process. Moreover, the process is adaptable for using the press-type steeping apparatus as well as a stirred vessel for slurry steeping. The combination of the two types of steeping in the abovementioned order is preferred because of improved equilibration of NaOH at the lower concentration, i.e., more uniform dispersion and better diffusion of alkali into the cellulose.

Although described in terms of cellulose xanthate, similar advantages can be realized for other cellulose derivatives which can be made in like manner. It is intended to include alkali solutions of cellulose as well as cellulose ether derivatives such as those mentioned hereinbelow and fibrous alkyl or hydroxyalkylcellulose products are intended to be included within the terms regenerated cellulosic material or cellulosic material. These would include methylcellulose, ethylcellulose, hydroxyethylcellulose, hydroxypropylcellulose, cyanoethylcellulose, or combinations of these ethers like methylhydroxypropylcellulose.

Cellulose ethers are useful, in addition to making shaped articles, as viscosity control agents for various coatings applications, e.g., in paints, gels, lacquers, and in binders, emulsifiers, etc. The presence of lesser amounts of alkali metal hydroxides as provided in the process is advantageous since these reaction mixtures need to be neutralized before recovery of the desired cellulose ether.

The concentration of the solutions used for steeping the cellulosic compound may vary, but is preferably about 18 to 22% NaOH based on the weight of the solution with the remainder being water, in the first steeping step and lower NaOH concentration in the additional steeping steps. For example, the last steeping solution should contain from about 10 to 12% by weight NaOH and 90 to 88% water. A more preferred range of NaOH based on the weight of the solution for the first steeping step is from 18 to 19%.

After the first steeping step, the wet sheet is pressed to reduce the amount of liquid to obtain an intermediate product containing from about 32.5% cellulose, about 14.7 to 15.7% NaOH and the remainder water. After the last steeping step, and usually only two steeping steps will be needed, the wet sheet of cellulosic material is pressed to reduce the amount of liquid to provide a product containing from about 32 to 37% by weight cellulose, about 10 to 13.5% by weight NaOH and the remainder water.

This product is then shredded to form a crumb and aged preferably at ambient temperature or greater for from about 20 to 45 hr, reacted with carbon disulfide, and dissolved in dilute aqueous sodium hydroxide. The resulting solution of the cellulose derivative should have low gel/fiber ratio to help assure easy filtration in the subsequent processing steps.

APPLICATIONS

Print Paste Formulations with Hydroxyalkylcarboxyalkylcellulose

Highlights of the Technological Achievement: A print paste formulation, useful in the printing of carpets and fabrics, comprising hydroxyalkylcarboxyalkylcellulose as the thickening agent, a dye and water as the carrier.

Reference: W.C. Kuryla and R.H. Raines; U.S. Patent 4,192,647; March 11, 1980; assigned to Union Carbide Corporation.

Background: In the printing of natural or synthetic textile fabrics, the use of flat-bed and rotary screen printing has become quite prevalent.

Screen printing procedures have imposed stringent performance requirements on the print paste formulations employed. These must be readily applicable to the fabric through the design screen to produce a clearly defined pattern and the resultant printing or design must be highly reproducible, preferably with a minimum of shutdowns for maintenance of screens and squeegees.

The major constituents of the print pastes are a dye and water. Additionally, a chemical thickener is usually present to impart the desired viscosity to the print paste. Hydroxyalkylated cellulose, such as hydroxyethylcellulose, is a commonly employed thickening agent in many applications, particularly in water-based compositions since the hydroxyalkyl group is hydrophilic. However, a problem has occurred in attempts to use hydroxyalkylcellulose as a thickener in certain print paste formulations in that it has caused the formation of a water-insoluble sticky residue in the print paste. The formation of the residue resulted in clogging of the screens necessitating interruption of production operations in order to remove the residue from screens and squeegees.

The abovedescribed residue problem has been noted with a number of print pastes, but it is particularly severe when the print paste contains one of the so-called "metallized" or "premetallized" dyes and especially when dyeing is carried out under conditions of low pH (i.e., on the order of about 2.0 to 5.0). In the carpet dyeing industry metallized dyes are frequently used and dyeing is frequently carried out under low pH conditions in order to obtain a high level of colorfastness; thus the residue formation problem is frequently encountered.

A further problem encountered with hydroxyalkylated cellulose is that it is sometimes difficult to dissolve in print pastes containing metallized dyes, requiring surfactants to be used to achieve a homogeneous composition.

Process Conditions: In some instances the reaction of the 1,2-alkylene oxide with the hydroxyl groups of the cellulose results in a chaining-out reaction that produces substituents of the form

$$-\!\!\left(\underset{\substack{|\\ R}}{CH}-CH_2-O\right)_{\!x}\!H$$

wherein R is hydrogen or alkyl and wherein $x > 1$. It is understood that such substituents, while they may not strictly be termed hydroxyalkyl groups (being more properly called hydroxyl-terminated polyoxyalkylene groups), can be present in hydroxyalkylcarboxyalkylcellulose and as used herein, the term "hydroxyalkyl" includes substituents having the above formula. Where such groups are present, each alkylene moiety thereof has from 2 to 5 carbon atoms.

The molar substitution (also referred to hereinafter as MS) of the hydroxyalkylcarboxyalkylcellulose is defined as the average number of oxyalkylene groups per anhydroglucose unit which is present in the hydroxyalkyl groups. The degree of substitution (also referred to hereinafter as DS) is defined as the average number of carboxyalkyl groups per anhydroglucose unit. Suitable methods for determining MS and DS values are described in ASTM D 2364 and ASTM D 1439, respectively.

From a standpoint of inhibiting residue formation, providing water solubility, and providing a print paste formulation with a viscosity useful in commercial printing operations, it is usually desired to employ hydroxyalkylcarboxyalkylcellulose wherein the MS value is at least 0.8 and the DS value is at least 0.3.

The hydroxyalkylcarboxyalkylcellulose is present in the improved print paste formulation of this process in an amount sufficient to impart the desired viscosity to the print paste formulation, which will vary somewhat depending on the substrate to be printed, the mesh size of the printing screens, the MS and DS values, etc. In commercial print paste formulations the hydroxyalkylcarboxyalkylcellulose is at a concentration of from 0.05 to 25 weight percent, preferably from about 0.1 to 5 weight percent, based on the total weight of the print paste formulation.

The improved print paste formulations of this process have Brookfield viscosities at 25°C of from 100 cp to 20,000 cp. The particular viscosity required will depend on the structure of the substrate that is to be dyed and the design to be imparted to it.

These formulations can be used to dye shag carpets with short or full pile, plush or level-loop carpets or conventional fabrics.

Substantially Water-Insoluble Carboxymethylcellulose Absorbents

Highlights of the Technological Achievement: A process for the preparation of substantially water-insoluble, particulate sodium carboxymethylcellulose, suitable for use as an absorbent in tampons and other catamenial devices, or the like. The process involves treating solid, water-soluble sodium carboxymethylcellulose having a degree of substitution of at least 0.4 with hydrogen chloride gas and heating the carboxymethylcellulose, either after the hydrogen chloride treatment or concurrently therewith, to produce a crosslinked, insoluble, partially acid form carboxymethylcellulose suitable for absorbent applications.

Reference: H.L. Marder, N.D. Field and M. Shinohara; U.S. Patent 4,200,737; April 29, 1980; assigned to International Playtex, Inc.

Background: For purposes of the process the proportion of carboxyl groups in the acid form, as a percentage of the carboxylate moieties in the sodium carboxymethylcellulose salt, is termed the degree of acidification (DA) of the particular CMC salt referenced.

One procedure which has been proposed for preparing water-insoluble CMC in the partially acid form involves slurrying water-soluble CMC, e.g., sodium CMC, with hydrochloric acid in a water-miscible, organic solvent, e.g., isopropanol, and thereafter insolubilizing the isolated material merely by heat-catalyzed crosslinking. Commercial application of such a procedure necessitates the application of solvent recovery techniques to facilitate economic production, and poses the risk of CMC losses in residues from distillation operations which may, for example, be utilized in such techniques.

Moreover, the use of solvent recovery operations and the disposal of residues formed therein may additionally pose environmental pollution problems.

A principal object of the process is to provide an improved process for the preparation of water-insoluble, acid-form CMC materials which may be readily utilized as absorbents for catamenial devices and other absorbent dressings.

Process Conditions: In accordance herewith, a process is provided for the preparation of substantially water-insoluble, particulate CMC suitable for application as an absorbent material. The process involves the direct treatment of a solid water-soluble particulate sodium CMC having a DS value of at least about 0.4 with hydrogen chloride gas, and heating the CMC to convert the same to the desired partially acid-form, crosslinked and substantially insolubilized material.

By thus proceeding, water-soluble sodium CMC having DS values of as low as about 0.4, and generally from about 0.5 to 1.2 may be readily insolubilized. Further, by reacting the CMC in the specific manner indicated, materials are produced having DA values of up to about 80%. Suitable regulation of the DS values of the materials reacted, and the DA values of the partially acid form materials produced, facilitates the formation of insoluble CMC which exhibit swell ratios of from as little as 5 to as much as 50, and which demonstrate percent extractables (soluble contents) of less than 40%.

This process can be carried out employing lower heat treat temperatures and reaction times than are necessary, for example, to insolubilize CMC in the salt form to produce absorbents of comparable characteristics. The combined HCl gas/heat treatment technique appears to effect the desired insolubilization of the initially soluble CMC salt more efficiently, with less concomitant problems, and more rapidly (under the same reaction conditions) than required by prior art processes.

When the acidification step and heat treatment are conducted sequentially the hydrogen chloride gas is most conveniently reacted with the sodium CMC at initial temperatures less than about 100°C. Employing such conditions, the exothermic acid-forming reaction increases the temperature of the acidified material sometimes to as much as 100° to 125°C.

Heat crosslinking may similarly be carried out over a broad temperature range varying from ambient temperatures to as much as 200°C, preferably, from 105° to 200°C. The reaction periods will vary inversely with temperature, and may be further dependent upon the desired reaction rate and the quantity of material to be reacted.

Acidification of CMC and CO_2 Gas/Heat Treatment

Highlights of the Technological Achievement: The process involves heating solid, water-soluble sodium carboxymethylcellulose, having a degree of substitution of at least 0.4, with carbon dioxide gas to substantially insolubilize the carboxymethylcellulose and convert it to a form suitable for absorbent applications.

Reference: M. Shinohara and N.D. Field; U.S. Patent 4,200,736; April 29, 1980; assigned to International Playtex, Inc.

Background: It is a principal object of the process to provide an improved process for the preparation of water-insoluble CMC materials in the salt form, the products of which may be readily utilized as absorbents for catamenial devices and other absorbent dressings.

Process Conditions: The combined CO_2 gas/heat treatment technique of the process thus appears to effect the desired insolubilization of the initially soluble CMC salt treated more efficiently, with less concomitant problems, and more rapidly (under the same reaction conditions) than required by prior art processes.

It is possible that insolubilization of the CMC occurs by a crosslinking reaction which may be characterized, at least in part, as follows:

$$\text{cellulose–OH} + \text{cellulose–COONa} \rightleftharpoons \text{cellulose–O–}\overset{\displaystyle O}{\overset{\|}{C}}\text{–cellulose} + \text{NaOH}$$

The CO_2 gas added in the practice of the process may react with the sodium hydroxide thus produced, forming sodium bicarbonate, and shifting the equilibrium of the crosslinking reaction toward completion.

It is also possible that the CO_2 contact step of the process has a purely catalytic effect on the insolubilization reaction. Thus, in marked contrast to prior art procedures in which CO_2 has been added as a precipitant to CMC-containing dispersions (U.S. Patent 2,447,757) or in which it may be generated from a neutralizing or forming agent added at other stages in the processing of CMC materials (U.S. Patent 2,711,407 and U.S. Patent 3,122,479), the heterogeneous solid-vapor contact effected during the combined CO_2/heat treatment of the process may have a catalytic effect where insolubilization occurs in the polymer.

Heating of the carboxymethylcellulose with simultaneous contact with carbon dioxide gas is carried out, in accordance with the process, at temperatures preferably within the range of from 120° to 190°C, and for reaction times varying inversely with the temperature. In contrast, it has been found that the same catalytic effect is not achieved when, for example, the soluble CMC is contacted with other gases under elevated temperature conditions, or when the CMC is contacted with CO_2 under ambient conditions prior to effecting heat crosslinking.

The soluble CMC may be contacted with carbon dioxide either by passing the gas over or through the particulate CMC material, or by maintaining a stagnant charge or blanket of the CO_2 gas over the solid material. In either event, the CO_2 may comprise from 20 to 100% by volume of the ambient atmosphere in the reaction zone, as may be desired.

The following examples illustrate the enhanced insolubilization of dry, particulate sodium CMC (Hercules' 7HCF, having a DS of 0.7 and a water content of 7.1%) by heating in CO_2.

Table 1: Comparative Data for Heat Treatment with CO_2 and in Air to Produce Insolubilized CMC

	...... Reaction Conditions			Swell	
	Atmosphere	Temperature (°C)	Time (hr)	Ratio (ml/g)	Extractables (%)
Example 1	CO_2	130	2⅓	36	–
Example 2*	CO_2	145	2	25	17

(continued)

Table 1: (continued)

	...... Reaction Conditions.			Swell	
	Atmosphere	Temperature (°C)	Time (hr)	Ratio (ml/g)	Extractables (%)
Example 3	CO_2	160	1 1/12	26	–
Example 4	CO_2	175	½	27	–
Example 5	CO_2	190	1/6	27	–
Control A-1**	Air	145	5	40	30
Control A-2	Air	145	6	30	–
Control B***	Air	160	2	29	21
Control C	Air	175	⅔	31	–
Control D	Air	190	1/10	33	–

*The flow rate of CO_2 was 1 ℓ/min instead of 4 ℓ/min, and its swell ratio was determined at 21 hr.
**The flow rate of air was 1 ℓ/min instead of 4 ℓ/min.
***Total reaction time in an air circulating oven, and its swell ratio was determined at 19.5 hr.

The conditions required to convert soluble CMC (7HCF) to insolubilized CMC having SR of 12 to 16 ml/g are illustrated in these examples.

Table 2: Comparative Data for Heat Treatment with CO_2 and in Air to Produce CMCs Having SRs of About 13

	...... Reaction Conditions.			Swell	
	Atmosphere	Temperature (°C)	Time (hr)	Ratio (ml/g)	Extractables (%)
Example 6*	CO_2	145	5	14	10
Example 7	CO_2	175	2½	12	11
Example 8	CO_2	190	⅔	12	14
Control E	Air	160	7	16	13
Control F	Air	175	3½	13	14
Control G	Air	190	1**	13	–

*The CO_2 flow rate was 1.0 to 1.4 ℓ/min rather than 4 ℓ/min, and the swell ratio was determined at 21 hr.
**Total residence time in an air circulating oven.

Low Irritant Conditioning Shampoo Composition

Highlights of the Technological Achievement: Shampoo compositions for human hair having low eye irritancy and conditioning properties are prepared. These compositions are comprised of water having dispersed therein a complex consisting essentially of a water-soluble cationic cellulose ether derivative and an amphoteric surfactant.

Reference: E.N. Khalil and A.N. Syed; U.S. Patent 4,205,063; May 27, 1980; assigned to Johnson Products Co., Inc.

Background: A problem found generally with shampoos alone is that they tend to irritate the eye when inadvertently dripped or splashed therein. Sodium chloride (salt) is frequently used as a shampoo viscosity builder to help avoid the problem of shampoo dripping into the user's eyes. However, when salt is used in sufficient quantities to build the shampoo viscosity and thereby lessen the tendency for splashing or dripping, the stability of the system on aging often suffers and salt itself can also be an eye irritant.

Process Conditions: The process relates to detangling hair shampoo compositions having low eye irritancy and a pH of 6.0 to 7.0. These compositions are comprised of water having dispersed therein a complex consisting essentially of a water-soluble cationic cellulose ether derivative and an amphoteric surfactant. The two components of the complex are present in the composition at 0.25 to 2.5 and 5 to 12 weight percent, respectively.

This complex is dispersed in the shampoo composition by a mixture of nonionic surfactants consisting essentially of 1 to 10 weight percent C_{12-18} fatty acid N,N-diethanolamide and 0.5 to 7 weight percent polyoxyethylene sorbitan fatty acid ester, the ester being selected from the group consisting of C_{12-18} fatty acid monoesters and triesters. All of the above weight percentages are based upon the total weight of the shampoo composition.

The water-soluble cationic cellulose ether derivative is a compound having a chain of anhydroglucose units with substituent groups pendant therefrom and spaced along the chain. These compounds also contain a plurality of quaternary nitrogen-containing groups having from zero to 3 mols of quaternary nitrogen-containing groups per anhydroglucose unit. The average molecular weights of these cationic cellulose ether derivatives is about 10,000 to about 4,000,000.

Cellulose Ether: These cationic cellulose ether derivatives are Polymer JR (Union Carbide Corporation). The available materials include Polymer JR-125, Polymer JR-400 and Polymer JR-30M. Of these three polymers, Polymer JR-30M, having an average molecular weight of about 600,000 is preferred as it provides better conditioning properties than do the other derivatives.

The water-soluble cationic cellulose ether derivative may be present in the composition at 0.25 to 2.5 weight percent of the total composition. Preferably it is present at 1.5 weight percent of the total composition. If less than 0.25 weight percent of the total composition of the cationic cellulose ether derivative is used, there is little complex formation and little conditioning effect observed. When greater than 2.5 weight percent of the total is present, the composition becomes difficult to rinse, gives the hair an oily feel and leads to build-up on the hair.

Surfactant: The second essential component of the hair conditioning complex is comprised of an amphoteric surfactant of the formula:

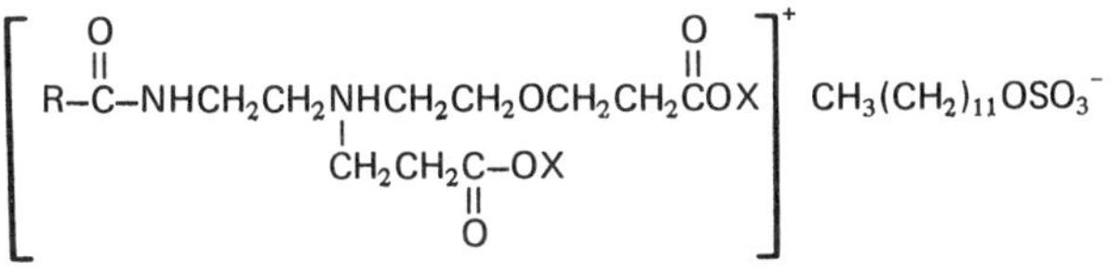

where

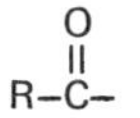

is selected from the group consisting of C_{12-18} fatty acids. While it may be selected from single fatty acids in the group C_{12-18} including unsaturated fatty acids, or groups thereof, preferably it includes those fatty acids derived from coconut oil. X is selected from the group consisting of alkali metal cations. While X may be an alkali metal cation, such as lithium, sodium or potassium, preferably X is the sodium cation. The preferred amphoteric surfactant is Miranol 2MCA-ESF.

Hydrophilic Articles of a Water-Insoluble Polymer from a Water-Soluble Cellulose Ether

Highlights of the Technological Achievement: Hydrophilic films, fibers and spongy articles of a water-insoluble polymer selected from the group consisting of regenerated cellulose, cellulose ether, cellulose ester and polyalkylene, and capable of forming fibers and films, are produced by distributing in a mass of the polymer up to 50% of pulverulent or granular amorphous particles of an originally water-soluble cellulose ether prepared from natural cellulose and rendered by chemical modification at least partly water-insoluble while remaining water-absorbent. The resulting mixture is then processed by conventional methods into the desired article.

Reference: A. Holst, M. Kostrzewa and H. Lask; U.S. Patent 4,200,558; April 29, 1980; assigned to Hoechst AG, Germany.

Process Conditions: The structures or articles constructed according to the process may be shaped by any process suitable for the shaping of the base polymer. A pulverulent or granular material of modified cellulose ether may, for example, be added to the polymer mass before shaping and uniformly distributed therein. Thus, for example, the customary precipitation processes are used if filaments, films or sponges of regenerated cellulose are being manufactured, or the casting process is used if films of cellulose acetate are being manufactured, while melt extrusion is used principally if structures of polyethylene or other polyalkylenes or other extrusible polymers are being manufactured.

Articles which have the particles as a surface covering are suitably manufactured by sprinkling the supporting article, composed of the polymer, with the particles, if the surface being sprinkled is either provided with an adhesive or is in a state in which the surface itself acts as an adhesive.

The particles are generally finely pulverulent to granular. Their particle size is in the range of 0.01 to 2 mm and depends upon the end use of the structure. The particles are composed of cellulose ethers such as, for example, carboxymethylcellulose, hydroxyethylcellulose or methylhydroxyethylcellulose, which are modified. Examples of possible modifying agents for crosslinking are: dimethylolmethylenebisacrylamide; methylenebisacrylamide; trichloropyrimidine, and tetrachloropyrimidine; cyanuric chloride; epichlorohydrin; dichloroacetic acid, diepoxides or their precursors di-α-halogenohydrins.

A parameter which also can be varied within wide limits is the degree of modification, for instance, of crosslinking of the cellulose ether of which the particles are composed. When using minor amounts of, for instance, crosslinking agents, for example, from 3 to 10% by weight of epichlorohydrin, relative to Na CMC, when producing Na CMC fibers, highly swellable fibers are obtained. At higher degrees of crosslinking, the swellability of the cellulose ether greatly decreases and its water-insoluble content increases. If, for example, about 33% by weight of these highly swellable Na CMC fibers (about 20-fold water absorption) are incorporated into a regenerated cellulose film, the swelling factor of the film rises from about 150% (pure regenerated cellulose) to 350%.

Example: 1 g of finely fibrous water-insoluble crosslinked Na CMC, which was crosslinked with epichlorohydrin, contained about 25% by weight of soluble constituents and had a water retention capacity of 20 times its weight and a theoretical exchange capacity of about 3.5 meq/g, was homogeneously incorporated, in a 3-roll mill, into 100 g of spinnable viscose containing about 10% by weight of cellulose.

A layer of 0.2 to 0.4 mm thickness of the modified viscose was cast on a glass plate and regenerated in the usual manner. A transparent regenerated cellulose film was obtained, which as a result of the embedded water-insoluble crosslinked Na CMC fibers had a total theoretical exchange capacity of about 0.32 meq/g. The water retention capacity of the film was about 250% by weight. A cellulose hydrate film manufactured under the same conditions has a water retention capacity of 125% by weight.

A Hydrophilic Shaped Article from a Cellulose Ether

Highlights of the Technological Achievement: This process relates to a hydrophilic shaped article of a water-insoluble polymer selected from the group consisting of regenerated cellulose, cellulose ether, cellulose ester and polyalkylene, and capable of forming fibers and films, the article containing amorphous particles of an originally water-soluble cellulose ether prepared from natural cellulose and rendered by chemical modification at least partly water-insoluble while remaining water-absorbent.

Reference: A. Holst, M. Kostrzewa and H. Lask; U.S. Patent 4,186,238; Jan. 29, 1980; assigned to Hoechst AG, Germany.

Background: The process provides articles, which have been rendered hydrophilic, of water-insoluble polymers which can be manufactured from polymeric material other than regenerated cellulose material and can be manufactured with a great range of hydrophilic properties. The process provides an article or shaped structure of a polymer capable of forming fibers and films and being water-insoluble, which contains, distributed within its mass or coated on its surface, particles of cellulose ethers that are inherently water-soluble but have been rendered at least partly, preferably more than 50% by weight, water-insoluble by chemical modification, but are still capable of absorbing water.

Process Conditions: The amount of added modified cellulose ether particles can be varied within wide limits depending upon what degree or what type of hydrophilic character, for example, of swelling capacity or ion exchange capacity, is desired. However, in general, the maximum amount of additive should be limited such that the mechanical strength of the film or other structure is not significantly reduced.

For example, for Na CMC which is slightly modified and thus swells very greatly in water, this maximum is at 50% by weight of the total in a base of regenerated cellulose. In the case of hydroxyethylcellulose (HEC) of low substitution and cellulose acetate a maximum is reached at 30 to 40% by weight of the total. The lower concentration limit depends upon the nature and on the action which the particles of modified cellulose ethers are to bring about.

The following table summarizes the essential data of examples performed relative to the manufacture of films coated on one side with crosslinked cellulose ethers.

Crosslinked Cellulose Ethers, Sprinkled onto the Surface of Cellulose-Based Films

Ex. No.	Polymeric Carrier Material	Cross-linked Cellulose Ether	Crosslinking Agent	Theoretical Ion Exchange Capacity (meq/g of film)	Film Water-Retention Capacity (% by wt)
1	Hydrated cellulose	Na CMC	Epichlorohydrin	1.1	440
2	Cellulose acetate	Na CMC	Epichlorohydrin	0.9	33
3	Hydroxyethyl-cellulose	Na CMC	Epichlorohydrin	1.0	130
4	Hydrated cellulose	Na CMC	Cyanuric chloride	1.2	280

Example 5: To show the difference between modified cellulose ethers made from natural cellulose and those made from cellulose crystallite aggregates in any case, 100 g of cellulose crystallite aggregates in 300 g of 87% isopropanol were alkalized with 46 g of aqueous NaOH solution (50%) by kneading them for 45 min at 20°C. Then, the mixture was etherified by adding 55 g of sodium monochloroacetate and kneading it for another hour at 70°C.

Then it was crosslinked with 50 g of aqueous dimethylolmethylenebisacrylamide solution (30%) while it was kneaded for 1 hr at 50°C. Then it was neutralized with acetic acid, washed salt-free with 80% methanol, and dried. The following table shows two of the physical/chemical properties of the product in each case:

Source Material	Water-Retaining Capacity (g water/100 g of modified cellulose ether)	Water-Soluble Portion (% by wt)
Natural cellulose	2,895	21
Cellulose crystallite aggregates	unmeasurably low	78

Absorbent Product Including Grafted Insolubilized Cellulose Ether

Highlights of the Technological Achievement: An insoluble etherified cellulose graft copolymer is provided comprising etherified cellulose which is soluble in water in the absence of grafting and is chosen from the group consisting of the alkali metal salts of carboxyalkylcellulose, sulfoalkylcellulose and phosphonoalkylcellulose. The etherified cellulose has grafted onto its cellulose backbone side chains of polymer moieties in sufficient quantities to render the grafted etherified cellulose insoluble. The products of this process are used alone or mixed with other absorbent materials such as unmodified cellulose, in the manufacture of absorbent napkins, tampons, sponges and the like.

Reference: P.K. Chatterjee and R.F. Schwenker, Jr.; U.S. Patent 4,200,557; April 29, 1980; assigned to Personal Products Company.

Background: Cellulose fiber and regenerated cellulose are raw materials for many commercial absorbent products including, for example, such products for absorbing body fluids as catamenial napkins and tampons, diapers and surgical dressings. While, in the main, unmodified cellulose has proven useful in such products, in an effort to improve product quality and economy, the art has searched for improved materials. It was discovered, for example, that alkali metal salts of cellulose ethers such as sodium carboxymethylcellulose exhibited increased absorption and retention properties for body fluids, these desirable properties increasing with the degree of ether substitution (DS).

Process Conditions: The particular etherified cellulose may be selected from the group consisting of alkali metal salts of carboxyalkylcellulose, phosphonoalkylcellulose, sulfoalkylcellulose and a sufficient quantity of such etherifying groups are present per anhydroglucose unit in the cellulose chain to render such salts water-soluble if the grafted polymer groups were not present. Such a degree of ether substitution of at least 0.35 ether groups per anhydroglucose unit is generally sufficient. Such alkali metal salts are, for example, the Li, Na, K, or Cs salts.

The salts of the cellulose ethers are rendered insoluble by grafting, to the cellulose backbone, side chains of homopolymer or copolymer moieties which side chains may be made up of hydrophilic, hydrophobic or both hydrophilic and hydrophobic polymeric moieties and should be present in sufficient quantities to render the salts water-insoluble. The quantity of side chains which must be grafted to the backbone will vary in accordance with the type of substituted group and the DS of the ungrafted cellulose ether salt.

For example, in the case of alkali metal carboxymethylcellulose at a DS of 0.4, the side chains should constitute approximately at least 10% by weight of the grafted cellulose ether salt. At a DS of 1.2, the side chains should be at least about 25% by weight of the grafted cellulose ether salt. The side chains could constitute as much as 90% or more.

Phosphonoalkylcellulose may be made by employing a phosphonoalkylating reagent solution comprising a basic aqueous solution of a compound of the structure:

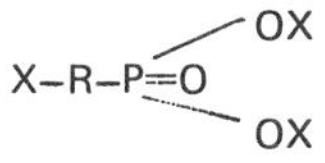

wherein X is a halogen and R is an alkylene radical. A preferred reagent is chloromethylphosphonic dichloride.

Sulfoalkylcellulose may be made in a similar manner employing sulfoalkylating reagent solution comprising a basic aqueous solution of a compound of the structure:

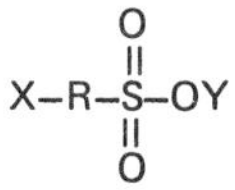

wherein X is a halogen, R is an alkylene radical and Y is chosen from the group consisting of a halogen, hydrogen, or alkali metal atom. A preferred reagent is chloroethylsulfonic acid.

The resulting salt of cellulose ether should have sufficient etherifying groups to render it water-soluble in the absence of other treatment. Generally, for an alkali metal salt of carboxymethylcellulose ether, a DS of at least about 0.35 will accomplish this result.

The pH range used for the reaction depends on the particular initiator used. One could use anywhere from a highly acidic pH (0.8 to 2.3) to a highly basic pH (12 to 14), depending on the choice of initiator. For the preferred ceric ion initiator, the pH should be acidic, i.e., less than seven and preferably 0.8 to 2.3.

The grafting reaction may be carried out at temperatures ranging from room temperature (i.e., 20° to 30°C) to the normal boiling point of the lowest boiling component of the mixture. If the reaction is carried out under greater than atmospheric pressure, the temperature could then be raised above the normal boiling point of the lowest boiling component of the mixture. The reaction mixture may also be cooled below room temperature, if desired.

The following table illustrates this relationship using sodium carboxymethylcellulose of varying DS as exemplary cellulose ethers.

Minimum Polymer Grafting to Insolubilize Sodium Carboxymethylcellulose

Degree of substitution of CMC	0.3	0.4	0.5	0.6	0.7	0.8	0.9	1.0	1.1	1.2	1.3
Minimum polymer add-on, %	0	9	12	13	14	15	17	19	21	23	27

Example: A series of samples of sodium carboxymethylcellulose powder having a DS of 0.4 to 1.2 (Hercules, Inc.) were used in this example. 20 g of each sample is converted to the acid form by treating with 250 ml of methanol-nitric acid solution (100 ml HNO_3 in 1 ℓ MeOH) for 24 hr at 25°C. The resulting acidified material is washed thoroughly with distilled water and then placed in a reactor with 1,000 ml of distilled water. Oxygen is removed from the system by purging with nitrogen for ½ hr. 10 ml of a ceric ammonium nitrate solution (0.1 M Ce(IV) in 1 N nitric acid) was then added and, after 5 min, a predetermined quantity of acrylonitrile (varying from 5 to 30 ml) is added. The reaction is allowed to proceed for 2 hr, at 25°C, under a nitrogen atmosphere.

The resulting grafted cellulose ethers are transferred to a Buchner funnel and washed thoroughly with water and acetone. The washed ethers are then converted to the alkali metal salt by being treated with a 5% solution by weight of potassium hydroxide in methanol for 24 hr at 25°C. The product is washed with methanol and dried at 105°C.

The resulting series of hydrophobic polymer-grafted alkali metal cellulose ethers are tested to determine the weight percent of grafted polymer and the water and salt solution absorption and retention properties. The polymer concentration of these samples is determined by nitrogen analysis wherein the nitrogen content of the grafted ethers is obtained by the Kjeldahl method, using an ammonia electrode, this method being described in *Official Methods and Analysis of the Association of Official Analytical Chemists,* Washington, D.C., 12th Edition, edited by W. Horowitz, 1975.

Aqueous liquid absorbency is determined by the "XOW Test," wherein an approximately ½-g sample of the test material is accurately weighed and stirred into a beaker containing 100 ml of the aqueous test fluid. After 20 min, the mixture is filtered through a piece of nylon tricot fabric and allowed to drain for 5 min. The filtrate is collected and measured in the graduated cylinder. The XOW, the absorption capacity of the material expressed in units of grams of absorbed liquid per gram of test material, is calculated as follows:

$$XOW = \frac{100 - \text{filtrate (ml)}}{\text{weight of test material (dry)}}$$

Aqueous fluid retention capacity is determined by using a porous plate testing apparatus, as described in detail in *Textile Res. J.*, 37, pp 356 to 366, 1967. Briefly, this test involves placing the test material in what is essentially a Buchner funnel having a porous bottom plate and holding the sample in place by applying thereon a standard weight to maintain a standardized confining pressure. The porous plate is placed in contact with a reservoir of liquid and the sample is allowed to absorb liquid through the porous plate until saturated.

By maintaining the sample at essentially the level of the reservoir, the liquid absorbed is subjected to essentially a zero hydraulic head with respect to the reservoir. To determine liquid retention, the saturated sample is elevated with respect to the liquid reservoir thereby imposing a hydraulic head upon the liquid absorbed, the head arbitrarily chosen as 35.5 cm of fluid. The apparatus is provided with means for directly measuring the volume of liquid retained under this hydraulic head. Retention values are reported as the volume retained per unit weight of the sample.

The results of these tests are tabulated in the table below using variously, aqueous salt solutions and water. As a control, ethers, as commercially obtained, are likewise tested and reported as "unmodified." As a further control, ether salts which are acidified following the abovedescribed procedure and then converted back to the salt form without undergoing the grafting reaction are tested and also reported in the table as, "treated controls." For comparative purposes, wood pulp fibers are also tested and reported as "cellulose fiber."

Absorbency of Hydrophobic Polymer (PAN)*
Grafted CMC (K-Salt) Powder

Carboxymethylcellulose Powder DS	Polymer Add-On (%)	Absorption Properties: 1% NaCl Solution Retention (cc/g) Porous Plate	Absorption Properties: 1.5% NaCl Solution XOW (g/g)	Absorption Properties: Water XOW (g/g)
0.4**	0	NA	NA	NA
0.4***	0	NA	NA	NA
0.4	3.9	NA	NA	NA
0.4	17.2	8.1	22	56
0.4	37.9	4.6	10	11
0.4	38.2	4.4	16	18
0.7**	0	NA	NA	NA
0.7***	0	NA	NA	NA
0.7	3.7	NA	NA	NA
0.7	13.4	6.5	22	60

(continued)

Carboxy-methyl-cellulose Powder DS	Polymer Add-On (%)	Absorption Properties: 1% NaCl Solution Retention (cc/g) Porous Plate	1.5% NaCl Solution XOW (g/g)	Water XOW (g/g)
0.7	19.5	5.6	18	34
0.7	41.9	3.4	10	11
0.9**	0	NA	NA	NA
0.9***	0	NA	NA	NA
0.9	12.7	NA	NA	NA
0.9	25.5	5.8	18	18
0.9	34.6	3.8	14	19
1.2**	0	NA	NA	NA
1.2***	0	NA	NA	NA
1.2	0.7	NA	NA	NA
1.2	15.0	NA	NA	NA
1.2	27.4	3.3	12	20
1.2	41.4	2.3	9	12
CF†	–	1.5	16	16

Note: NA means not applicable because the material was soluble or gelled in fluid.

*PAN is polyacrylonitrile.

**Unmodified.

***Treated control.

†CF is cellulose fiber (reference).

As can be seen from the results of the table, the grafting technique of this process has insolubilized the etherified cellulose and has generally endowed the resulting product with absorbency properties (XOW values) at least as good as those of wood pulp. The retention values have increased markedly.

Gelled Nail Polish Remover

Highlights of the Technological Achievement: A solvent-based nail polish remover comprising 85.0 to 99.5% acetone or ethyl acetate solvent and 0.5 to 15% gelling agent, preferably hydroxypropylcellulose acetate gelling agent. The gelled remover has a viscosity of 1,000 to 10,000 cp and possesses little odor due to the low rate of solvent evaporation when compared to acetone alone. The consistency of the gel allows it to be packaged in a collapsible tube to avoid breakage or spillage. The process includes the steps of adding the gelling agent to the acetone or ethyl acetate solvent after the addition of the other ingredients.

Reference: A. Minton, J.H. Baker and J. Teng; U.S. Patent 4,197,212; April 8, 1980; assigned to Anheuser-Busch, Incorporated.

Background: In the cosmetic field, it is desirable to have a nail polish remover that is easy to use and convenient, that is, one that can be easily spread on the nail and carried in a purse or suitcase without the danger of breaking or spilling, and which, when placed on the nail does not spread or run unduly. The nail polish remover should provide a proper foundation for nail polish without excessive drying of the nails while maintaining a glossy appearance. It is also desirable to reduce the unpleasant odor of nail polish removers and to avoid waste due to excessive application and spillage. Heretofore, no nail polish remover has been found which meets all of these requirements.

Process Conditions: A nail polish remover comprises 85.0 to 99.5% nail polish solvent and 0.5 to 15% gelling agent, the remover having a viscosity of 1,000 to 10,000 cp. The preferred solvent is acetone and the preferred gelling agent is hydroxypropylcellulose acetate.

Many gelling agents can be used with conventional nail polish removers and are categorized as follows: (A) cellulose derivatives such as hydroxypropylcellulose, cellulose acetate butyrate, and ethyl hydroxyethylcellulose, (B) synthetic derivatives such as polystyrene, and methyl vinyl ether and maleic anhydride copolymers, and (C) inorganic materials such as colloidal silica, and Bentone clay, a complex of bentonite and various quaternary compounds (NL Corporation). The preferred gelling agent is an acetate ester of hydroxypropylcellulose. The hydroxypropylcellulose acetate is particularly useful as a gelling agent for organic solvents such as acetone or ethyl acetate commonly used as solvents in nail polish removers.

The reactants include only acetic anhydride, pyridine and hydroxypropylcellulose. The reactants in some situations can be dissolved in the material to be gelled and the gel forms in situ.

The hydroxypropylcellulose ester to be used as the gelling agent in this process can have a degree of MS of 3 to 5. The MS determines the number of mols of

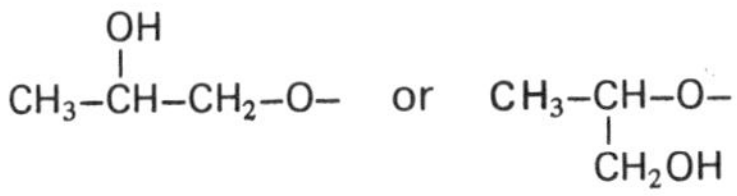

groups per anhydroglucose unit of the cellulose.

The esters of hydroxypropylcellulose can have a DS of 0 to 3. The theoretical maximum is a DS of 3. With the acetate, the degree of substitution for effective compounds is about 1.0 to 2.4.

There are three hydroxyl groups in each anhydroglucose unit in the cellulose molecule. DS is the average number of hydroxyl groups substituted in the cellulose per anhydroglucose unit. For the hydroxyalkyl derivatives of cellulose, the MS is generally greater than the DS. The reason for this is that the hydroxyalkyl group is capable of homopolymerization.

Example: The preferred embodiment of the process is formulated as follows:

	Percent by Weight
Acetone	98.0
Durlin Red Grind #50	0.1
Rona Non-Lead Pearl NLY-L-2X (70% in castor oil)	0.4
Gelling Agent*	1.5

*Hydroxypropylcellulose acetate with MS of 4.0 and DS of 1.3.

Contact Lenses Formed from Mixtures of Cellulose Esters and Vinyl/Acetate Copolymers

Highlights of the Technological Achievement: Contact lenses and scleral lenses formed of mixtures consisting essentially of from 99 to 70% by weight of a cellulose ester of one or more aliphatic carboxylic acids, and from 1 to 30% by weight of an ethylene/vinyl acetate copolymer containing from 30 to 98% by weight of incorporated vinyl acetate.

Reference: F. Wingler, T. Eicher, F. Muller, W. Fischer and R. Prinz; U.S. Patent 4,193,897; March 18, 1980; assigned to Bayer AG, Germany.

Background: Contact lenses are generally divided into two main types, namely hard lenses and soft, hydrophilic lenses. Lenses of the first type are normally produced from polymethyl methacrylate which may be cut and polished to give lenses having good optical properties. However, hard polymethyl methacrylate lenses are attended by the known disadvantage that they may irritate the cornea and the inside of the eyelid. If, therefore, a patient is to wear hard lenses, a considerable period of acclimatization is required during which the patient wears the lenses for gradually increasing periods up to a maximum of about eight hours' continuous wearing. On account of this disadvantage, hard lenses are worn by only a small number of patients.

Recent years have seen the introduction of soft contact lenses consisting of hydrogels of hydrophilic polymers. Soft contact lenses have largely been able to overcome the problem of irritation of the eyes, so that patients may now acclimatize themselves much more quickly to the wearing of contact lenses. Just like hard contact lenses, however, conventional soft lenses have another serious disadvantage which is well-known among experts. This disadvantage lies in the fact that neither type of lens is capable of supplying the cornea with enough oxygen to maintain its normal metabolic rate. It is known that the surface of the cornea has to receive a reasonable supply of oxygen for the cornea to remain transparent and healthy.

It would be advantageous to use polymeric materials which do not cause any irritation, are chemically inert and comfortable to wear, show high permeability to oxygen and do not take up any bacteria or fungi.

Process Conditions: It has been found that contact lens materials having excellent compatibility and high permeability to oxygen may be obtained by using polymer mixtures of cellulose esters of aliphatic carboxylic acids and ethylene-vinyl acetate copolymers as the light-refracting material. In addition, these mixtures have the advantage of being free from plasticizers and processing stabilizers, because the thermoplastic processing does not require such additives.

Suitable cellulose esters for producing the light-refracting materials according to the process are cellulose esters of aliphatic carboxylic acids containing from 1 to 5 carbon atoms, preferably cellulose acetate, cellulose acetopropionate and cellulose acetobutyrate.

Effervescent Product with a Water-Soluble Cellulosic Adhesive Base

Highlights of the Technological Achievement: An effervescent product useful for forming a holding adhesive, especially a wall-covering adhesive, is disclosed.

Reference: H.F. Zoumut; U.S. Patent 4,184,888; January 22, 1980; assigned to DAP Inc.

Background: Adhesive products for adhering types of wall coverings have classically been formulated from such natural sources as starch, dextrin, gum arabic, gum tragacanth and certain modified casein or other animal protein bases. Most are dispersed as powders in water and develop, after application, their strength by drying. In recent years such adhesive products have been formulated with synthetic polymers to provide enhanced water resistance.

Adhesive products suitable for use as wall-covering adhesives are designated as holding adhesives, i.e., adhesives that are intended primarily for merely attaching one adherent to another and holding it in place without requiring any significant resistance to external stressing.

Preparing a lump-free paste mixture often involves portionwise addition of the dry powder to the volume of water coupled with vigorous and continuous stirring.

Process Conditions: The effervescent composition may be packaged as a premeasured powdered mixture or alternatively, compressed into a tablet form. The effervescent product of the process is an article of manufacture which comprises by weight: 20 to 50% of a water-soluble polymeric adhesive base; and 45 to 80% of a compatible effervescent source consisting essentially of an acid component and a carbonating component whereby the acid component and carbonating component are present in amounts sufficient to afford a water solution of approximate pH of 3.5 to 9.5.

A preferred effervescent product in powdered form of this process comprises by weight: 30 to 50% of a water-soluble polymeric adhesive base; and 50 to 70% of a compatible effervescent source. The latter liberates carbon dioxide in the form of bubbles when placed in an aqueous medium, thereby producing the effervescence, and does not chemically react with the other components of the composition.

In a preferred effervescent product of this process the adhesive base is hydroxyethylcellulose, and the effervescent source consists essentially of citric acid and sodium carbonate.

The adhesive base of this composition is the component which is primarily responsible for the adhesive forces which hold the one adherent to the other. Suitable water-soluble polymeric adhesives are those such as the starches, e.g., corn, tapioca and sago, dextrins and the water-soluble celluloses, such as hydroxyethylcellulose, hydroxypropylcellulose, methylcellulose and ethyl hydroxyethylcellulose. A preferred water-soluble, cellulosic adhesive base for use in this process is hydroxyethylcellulose having a degree of substitution from 1.5 to 3.0. A suitable hydroxyethylcellulose source is Natrosol (Hercules Chemical Co.), and especially Natrosol 250.

Tablet Binder: To further facilitate tableting, a conventional tableting binder may optionally be employed. It should be soluble in the foregoing anhydrous solvent and inert to the other components of the mixture. Suitable binders include gum acacia, gelatin, gum tragacanth, guar gum, sodium carboxymethylcellulose, ethylcellulose, hydroxypropylmethylcellulose, starch glucose or a carboxypolymethylene polymer (carbomer). The tableting binder is preferably water-soluble. The carboxypolymethylene polymers, or carbomers (*CTFA Cosmetic*

Ingredient Dictionary 1973), are preferred tableting binder agents for use in this process. Those resins consist essentially of a colloidally water-soluble polymer of acrylic acid crosslinked with from 0.72 to 2% of a crosslinking agent selected from the class consisting of polyalkylsucrose and polyallylpentaerythritol. Suitable carboxypolymethylene polymers, or carbomers, are Carbopol resins (B.F. Goodrich Company). A preferred polymer is the carbomer 934, which is a water-soluble polymer of acrylic acid crosslinked with about 1% of a polyallyl ether of about 5.8 allyl groups for each molecule of sucrose and having a molecular weight of about 3,000,000.

Gelled Lacquer

Highlights of the Technological Achievement: An overprint lacquer comprising a solvent, a resin, a plasticizer, and a gelling agent. The gelling agent is hydroxypropylcellulose acetate. The coatings of this gel-containing lacquer show improved adhesive characteristics and improved durability when compared with conventional overprint lacquers that have higher nitrocellulose content. Also, the viscosity of the lacquer is more readily controllable due to the presence of the gelling agent.

Reference: J.M. Lucas and J. Teng; U.S. Patent 4,183,764; January 15, 1980; assigned to Anheuser-Busch, Incorporated.

Background: One of the current lacquer formulations used to over-coat aluminum sheeting contains the following components in the amounts listed:

	Weight Percent
Nitrocellulose resin	15.00
Di(n-butyl) phthalate	3.75
Toluene	20.31
Isopropyl acetate	60.94

Nitrocellulose resin acts as a coating agent. Other suitable coating agents include a copolymer of vinyl acetate and vinyl chloride, polyvinyl acetate, polyvinyl formal, polyvinylidene fluoride. Di(n-butyl) phthalate acts as a plasticizer for the resin. Other suitable plasticizers include dioctyl phthalate and di(n-hexyl) phthalate. The ratio, by weight, of di(n-butyl) phthalate to nitrocellulose resin (plasticizer to resin) is about 1 to 4. Toluene and isopropyl acetate function as solvents for the resin. Other suitable solvents include butyl acetate, xylene, and ethyl acetate.

Process Conditions: This process involves an overprint lacquer comprising a solvent or mixture of solvents, a resin, preferably nitrocellulose resin, a plasticizer for the resin, and a gelling agent. The gelling agent is hydroxypropylcellulose acetate. The adhesive properties of the coatings formed from the gelled lacquer can be seen to be improved over the properties associated with conventional lacquers, which have a higher nitrocellulose content. Moreover, the gelling agent permits more control of the viscosity of the lacquer.

The following components are mixed together in the amounts indicated:

	Parts by Weight
Toluene	32-34

(continued)

	Parts by Weight
Isopropyl acetate	39-43
95% ethyl alcohol	8-9
Di(n-butyl) phthalate	3-2
Hydroxypropylcellulose acetate*	0.05-0.25

*The degree of substitution of acetyl groups can range from 1.0 to 2.0; the degree of molecular substitution of hydroxypropyl groups can range from 3.5 to 4.5.

The mixture is stirred and heated for 15 to 20 min at 50° to 60°C in a reactor until the solution is complete and the liquid is smooth. The heating is discontinued and 18 to 12 pbw of nitrocellulose resin, R.S. ½ sec, containing about 30 weight percent of 95% ethyl alcohol, is added to the solution. The heating is renewed, and the mixture is stirred for an additional 20 to 30 min until the solution is complete and the lacquer is smooth and free from lumps. The lacquer is then allowed to cool.

The lacquer thus produced has a viscosity ranging from 20 to 25 sec (as measured by a No. 3 Zahn cup). This value is approximately 0 to 25% greater than the viscosity of a similarly prepared lacquer which does not contain hydroxypropylcellulose acetate.

Mixture of Hydroxypropylcellulose and Poly(Maleic Anhydride/Alkene-1) as a Hydrocolloid Gelling Agent

Highlights of the Technological Achievement: This process provides a fluid composition which comprises an aqueous medium which is thickened with a synergistic viscosity mixture of hydroxypropylcellulose and poly(maleic anhydride/alkene-1).

The thickened aqueous medium is applicable as a vehicle for explosive slurries, printing pastes, and paint formulations. A high-viscosity aqueous medium of this process is uniquely adapted for use as a hydraulic fluid in well-drilling operations and in secondary oil and gas recovery from subterranean formations.

Reference: R.N. DeMartino; U.S. Patent 4,172,055; October 23, 1979; assigned to Celanese Corporation.

Background: The disadvantage of employing natural gum as a gelling agent in hydraulic fracturing fluids is the lessening of fluid loss control caused by plugging of apertures which reduces the permeability of a fractured formation. In the case where a natural resin is employed as a gelling agent in combination with an enzyme breaker additive in a hydraulic fracturing fluid, the hydraulic gel system is often unsatisfactory because the hydrolysis residue of the natural gum is present in sufficient quantity to create a plugging effect and reduces the permeability of a fractured formation.

There remains a need for an efficient and economic hydraulic fluid medium which exhibits improved properties in well-treating applications.

Process Conditions: In the practice of the process, a synergistic effect can be observed when a mixture of hydroxypropylcellulose and maleic anhydride/alkene-1 copolymer is employed as a gelling agent in an aqueous medium in a quantity between 0.05 and 5 weight percent, based on the weight of water.

The relative ratio between the hydroxypropylcellulose and maleic anhydride/alkene-1 copolymer components in the gelling agent can vary over a broad range between 95:5 and 5:95 weight percent of hydroxypropylcellulose to maleic anhydride/alkene-1 copolymer.

In order to achieve an incremental synergistic viscosity increase of at least 50% over the calculated additive viscosities of the individual gelling agent components as measured in centipoises at 25°C, it is preferred to maintain the relative ratio of the hydroxypropylcellulose to maleic anhydride/alkene-1 copolymer in the gelling agent in the range between 70:30 and 30:70 weight percent. An incremental synergistic viscosity increase of at least 100% can be obtained when the gelling agent components are in the range between 60:40 and 40:60 weight percent.

In the practice of the process, the mixture of hydroxypropylcellulose and poly(maleic anhydride/alkene-1) can be dry-blended to produce a convenient form of the process gelling agent. The gelling agent can then be incorporated into an aqueous medium by the simple expediency of mixing the gelling agent in the medium with stirring. Hydration of the two hydrocolloid components of the gelling agent is usually complete within a period of several hours at room temperature.

In order to achieve solubilization of the poly(maleic anhydride/alkene-1) component of the thickener composition in an aqueous medium, it is advantageous to add an alkaline reagent to the aqueous medium to convert the carboxylic acid anhydride groups of the copolymer into the form of neutral salts. It is desirable that the ultimate pH of the thickened aqueous composition be in the range between 4 and 13, and preferably in the range between 6 and 11.

The adjustment of pH, and the hydration of the copolymer component, are facilitated by the addition of an appropriate quantity of alkali metal hydroxide such as sodium hydroxide. If the hydroxypropylcellulose hydrocolloid component of the gelling agent contains an alkaline residue as an impurity from the synthesis reaction of cellulose with propylene oxide in the presence of an alkaline catalyst to produce the hydroxypropylcellulose, then it is an economic advantage to permit the alkaline residue to aid and effect adjustment of pH of the thickened aqueous composition into the desired range.

Hydroxypropylcellulose Component: Hydroxypropylcellulose is commercially produced by reacting alkali cellulose with propylene oxide at elevated temperatures and pressures. An idealized structural model for hydroxypropylcellulose with MS of 3 can be represented by the chemical formula:

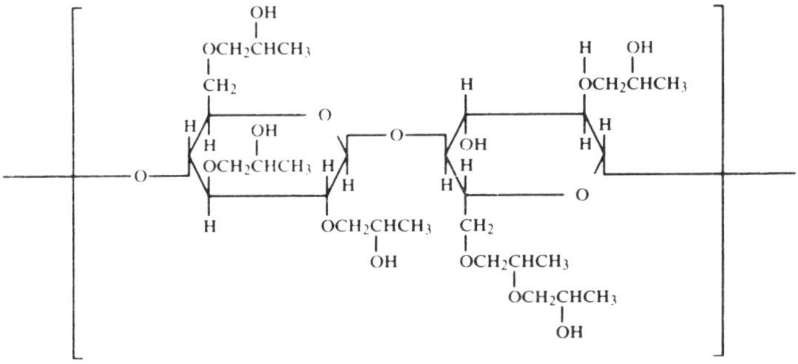

The term "molecular substitution" refers to the average number of mols of hydroxypropyl substituted in the cellulose per anhydroglucose unit. For the purposes of the process the average MS in the hydroxypropylcellulose hydrocolloid component preferably will vary in the range between about 2 and 5, and most preferably in the range between about 2.5 and 4.

The synergistic viscosity enhancement observed in the high-viscosity aqueous fluid medium of the process appears to increase as the molecular weight of each hydrocolloid component increases. Hence, in the practice of the process it is preferred that the molecular weight of the hydroxypropylcellulose component be at least 300,000, and most preferably be at least 800,000. Klucel H (Hercules) is a commercial hydroxypropylcellulose product which is characterized as having a nominal molecular weight of about 1,000,000.

Poly(Maleic Anhydride/Alkene-1) Component: The term "maleic anhydride" as employed herein is meant to include α,β-olefinically unsaturated dicarboxylic acid anhydride comonomers represented by the structural formula:

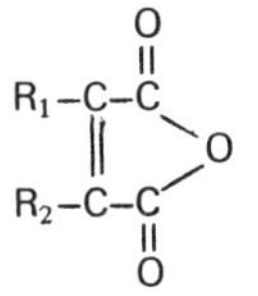

wherein R_1 and R_2 are independently selected from the group consisting of hydrogen, halogen, cyano, and aliphatic and aromatic substituents such as alkyl, aryl, alkaryl, aralkyl, cycloaliphatic, and the like, containing between one and about ten carbon atoms.

Compounds corresponding to the above formula include maleic anhydride, chloromaleic anhydride, 2,3-dichloromaleic anhydride.

Example: This example illustrates the large viscosity increase in an aqueous medium provided by a synergistic mixture of hydroxypropylcellulose and maleic anhydride/alkene-1 copolymer as a gelling agent.

A 1% aqueous solution of each of hydroxypropylcellulose (Klucel H) and maleic anhydride/isobutylene copolymer Isobam HH (Kuraray Co. Ltd.) is prepared. The viscosity of each solution is determined in centipoises with a Brookfield viscometer Model RVF, spindle No. 4 at 20 rpm.

In the preparation of the solutions, dissolution of the maleic anhydride/isobutylene copolymer is aided by adjusting the pH of the aqueous medium into the range of about 6 to 11 with sodium hydroxide.

Equal volumes of each solution are blended together and the viscosity of the blend solution is measured.

Initial Viscosities	
Hydroxypropylcellulose (1%)	2,000
Maleic anhydride/isobutylene copolymer (1%)	9,100

(continued)

Blend viscosity (1:1)	
Calculated	4,200
Observed	9,200

Under comparable experimental conditions, the following viscosity data are obtained for the indicated hydrocolloid mixtures:

	Viscosities	
1:1 of 1% Solutions	Calculated	Observed
HPC/Cyanamer P-250 (3%)	2,932	1,800
HPC/PEO (2%)	2,720	1,500
HPC/Kelzan XC	3,367	5,000
Isobam HH/PEO (2%)	5,562	7,000
Isobam HH/MC	5,479	9,000
Isobam HH/HEC	5,137	7,000
Isobam HH/CMC	4,266	4,300

In the above table, HPC is hydroxypropylcellulose, Klucel H; Cyanamer P-250 is a polyacrylamide; PEO is a polyethylene oxide, Polyox Coagulant; Kelzan XC is a xanthan gum; Isobam HH is a maleic anhydride/isobutylene copolymer from Kuraray Co. Ltd.; MC is a methylcellulose, Methocel J 75 MS; HEC is a hydroxyethylcellulose, Natrosol 250HH; and CMC is a carboxymethylcellulose, CMC-7H from Hercules. The calculated viscosity of a blend is determined by means of the following equation:

$$\text{Log } \eta S = \frac{X \log \eta^1 + (100 - X) \log \eta^2}{100}$$

where

ηS = viscosity of blend
X = weight percent of first component of blend
η^1 = viscosity of first component of blend
η^2 = viscosity of second component of blend

A Fluid Composition Containing Hydroxypropylcellulose and Poly(Maleic Anhydride/Alkyl Vinyl Ether)

Highlights of the Technological Achievement: In one embodiment this process provides a fluid composition which comprises an aqueous medium which is thickened with a synergistic viscosity mixture of hydroxypropylcellulose and poly(maleic anhydride/alkyl vinyl ether) as a hydrocolloid gelling agent.

An aqueous medium thickened with a gelling agent is applicable as a vehicle for explosive slurries, printing pastes, paint formulations, adhesives, and the like. A thixotropic aqueous medium of this process is uniquely adapted for use as a hydraulic fluid in well-drilling operations and in secondary oil and gas recovery from subterranean formations.

Reference: R.N. DeMartino; U.S. Patent 4,169,818; October 2, 1979; assigned to Celanese Corporation.

Background: Well-treating hydraulic fluids commonly employ a viscosity-increasing agent such as a water-soluble polymer to improve the flow characteristics and the suspending ability of the fluids. Carboxymethylcellulose, polyacrylamide, polyvinylpyrrolidone, guar, tragacanth, Irish moss and modified starch have been employed as water-soluble or water-dispersible additives to provide pumpable gels having solids suspending properties. U.S. Patent 3,483,121; 3,757,864; 3,779,914; 3,818,998; 3,900,069; 3,971,440; and references cited therein, describe well-treating hydraulic fluid compositions which contain polymeric friction-reducing gelling components.

It is an object of this process to provide a partially synthetic hydrocolloid thickened aqueous medium having improved viscosity and suspending properties for application as a well-treating hydraulic fluid composition.

Process Conditions: The process is based on the discovery that unlike most combinations of natural and synthetic hydrocolloid gums and resins employed for thickening aqueous solutions, a mixture of hydroxypropylcellulose and maleic anhydride/alkyl vinyl ether copolymer is capable of inducing an extraordinary viscosity enhancement effect, which usually is at least 100% greater than the calculated additive viscosity contributions of the respective hydrocolloid components.

It was found that the specific combination of hydroxypropylcellulose with maleic anhydride/alkyl vinyl ether copolymer unexpectedly provides a substantial synergistic effect on the viscosity of an aqueous medium, in contradistinction to various other combinations of similar types of synthetic and natural hydrocolloid mixtures which exhibit less viscosity synergism or no viscosity synergism.

The term "alkyl vinyl ether" as employed herein is meant to include comonomers represented by the formula, $CH_2{=}CH{-}O{-}R$, wherein R is preferably limited to alkyl groups such as methyl, ethyl, isobutyl, pentyl, octyl, decyl, and the like, which contain between one and about 10 carbon atoms.

The ratio of the comonomers in the maleic anhydride/alkyl vinyl ether copolymer hydrocolloid usually is in a 1:1 ratio, and in some cases the ratio will vary between 5:4 and 4:5. The molecular weight of the copolymer may vary over a wide range between several thousand and several million.

Illustrative of a preferred copolymer hydrocolloid is one corresponding to the structural formula:

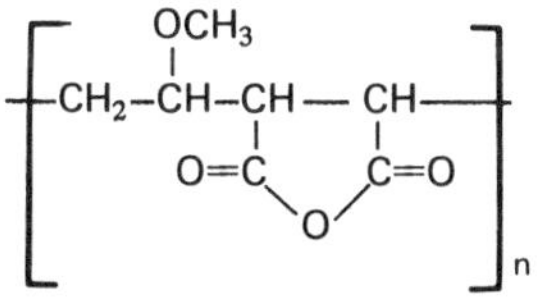

where n is an integer between 11 and 700. A particularly preferred poly(maleic anhydride/alkyl vinyl ether) hydrocolloid is a high molecular weight maleic anhydride/methyl vinyl ether copolymer having a specific viscosity of at least 2.5 as determined in a solution of 1 g of copolymer in 100 ml of methyl ethyl ketone (MEK) at 25°C.

Well-Treating Hydraulic Fluids: This process accomplishes an important objective by the provision of a hydraulic fluid composition of high viscosity which comprises an aqueous solution having incorporated therein (1) a hydrated mixture of hydroxypropylcellulose and poly(maleic anhydride/alkyl vinyl ether) hydrocolloids as a gelling agent, and (2) a breaker additive for subsequent reduction of solution viscosity. For the purpose of well-fracturing applications, the preferred hydraulic fluid compositions contain additionally a propping agent such as sand or comminuted walnut shells.

The quantity of gelling agent incorporated in a well-treating hydraulic composition can vary in the range between 0.05 and 5 weight percent based on the weight of the water component. A preferred range is between 0.1 and 2 weight percent of gelling agent, based on the weight of water.

Example: This example illustrates the large viscosity increase in an aqueous medium provided by a synergistic mixture of hydroxypropylcellulose and maleic anhydride/alkyl vinyl ether copolymer as a gelling agent.

A 3% aqueous solution of maleic anhydride/methyl vinyl ether copolymer [Gantrez AN-179 (GAF)] and a 1% aqueous solution of hydroxypropylcellulose [Klucel H (Hercules)] are prepared. Dissolution of the methyl vinyl ether copolymer is aided by adjusting the pH of the aqueous medium into the range of 6 to 11 with sodium hydroxide.

The viscosity of each solution is determined in centipoises with a Brookfield viscometer Model RVF, spindle No. 4 at 20 rpm.

Equal volumes of each solution are blended together and the viscosity of the blend solution is measured.

Initial Viscosities	
Hydroxypropylcellulose (1%)	1,800
Methyl vinyl ether copolymer (3%)	2,200
Blend Viscosity (1:1)	
Calculated	1,989
Observed	5,900

Under comparable experimental conditions, the following viscosity data are obtained for the indicated hydrocolloid mixtures:

	.... Viscosities	
1:1 of 1% Solutions	Calculated	Observed
HPC/Cyanamer P-250 (3%)	2,932	1,800
HPC/PEO (2%)	2,720	1,500
HPC/Kelzan XC	3,367	5,000
Isobam HH/PEO (2%)	5,562	7,000
Isobam HH/MC	5,479	9,000
Isobam HH/HEC	5,137	7,000
Isobam HH/CMC	4,266	4,300
Gantrez AN-179 (3%)/Cyanamer P-250	2,932	3,000
Gantrez AN-179 (3%)/CMG	3,382	1,200

In the above table, CMG is a carboxymethylguar; DS of 0.6 from Stein Hall. (See page 162 for meanings of other abbreviations.)

Controlled-Release Pharmaceutical Dosage Component Containing an Ethylcellulose Film

Highlights of the Technological Achievement: A controlled-release pharmaceutical dosage component comprising a microencapsulated bronchodilator agent of the β-adrenergic agonist type is prepared by organic phase separation. Pharmaceutical formulations containing these components exhibit prolonged bronchodilation with minimal cardiac stimulation.

Reference: P.M. John; U.S. Patent 4,153,677; May 8, 1979; assigned to Sterling Drug Inc.

Background: Bronchodilator agents of the β-adrenergic agonist type constitute a well-known class of pharmaceutical compounds (*AMA Drug Evaluation,* 3rd Edition, Publishing Sciences Group Inc., Littleton, Mass., 1977, pp 631-636 and W.C. Cutting, *Cutting's Handbook of Pharmacology,* 5th Edition, Meridan Corp., New York, 1972, pp 424-435). These compounds usually affect the adrenergic receptors of both the lungs (β-2) and the heart (β-1), thus producing bronchial relaxation (bronchodilation) accompanied by cardiac stimulation (tachycardia). Published reports show that much effort has been expended in attempting to separate the β-1 and β-2 effects of β-adrenergic agents with only partial success and accordingly, research directed at divorcing the cardiac effects from the bronchial effects continues.

Process Conditions: It has been found that bronchodilator agents of the β-adrenergic agonist type, when microencapsulated in polymeric film, produce bronchodilation of extended duration with an unexpected attenuation of the usually accompanying tachycardia. The process is applicable to any bronchodilator of the β-adrenergic agonist type which acts primarily on β-adrenergic receptors of the lungs (β-2) and to a lesser extent on β-adrenergic receptors of the heart (β-1). The process resides in a controlled-release pharmaceutical dosage component comprising a microcapsule of a film of hydrophobic polymeric material encapsulating a core comprising an agent of the β-adrenergic agonist type mixed with a pharmaceutically acceptable excipient.

Preferred agents have Formula 1 or 2

(1)

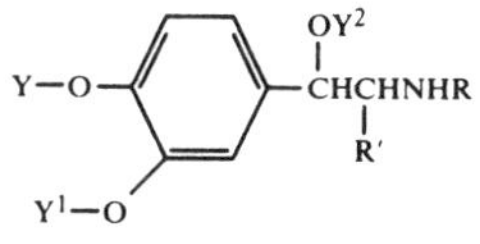

(2)

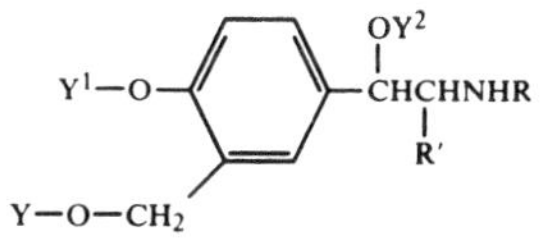

wherein in each of the above formulas:

R is hydrogen, alkyl having 1 to 4 carbon atoms, or cycloalkyl having 3 to 6 carbon atoms;

R' is hydrogen or alkyl having 1 to 3 carbon atoms;

Y is an acyl member which is alkanoyl having 1 to 22 carbon atoms, alkenoyl having one or two double bonds and having 4 to 22 carbon atoms, cycloalkyl $-C_nH_{2n}-CO-$ having a total of 4 to 10 carbon atoms of which 3 to 7 are ring carbon atoms in cycloalkyl and wherein n is zero, one or two, 1- or 2-adamantanecarbonyl, phenoxyacetyl, naphthalenecarbonyl, pyridinecarbonyl, or $Z-C_nH_{2n}-CO$ wherein n is zero, one or two and Z is phenyl or phenyl substituted by 1 to 3 members of the group consisting of alkyl having 1 to 4 carbon atoms, alkoxy having 1 to 4 carbon atoms, halo, trifluoromethyl, dialkylamino having 2 to 8 carbon atoms, and alkanoylamino having 1 to 6 carbon atoms; and

Y^1 and Y^2 are the same or different and are hydrogen or one of the acyl members defined by Y; or an acid-addition salt thereof.

The resulting microcapsules are generally spherical with a mean volume-surface diameter of about 700 to 900 μ and have a glossy surface free from gross irregularities. The ethylcellulose constitutes 1 to 5 percent by weight of the microcapsule and ranges in thickness from 5 to 11 μ.

The ethylcellulose content of the microcapsules depends on the ratio of core material to ethylcellulose employed in the microencapsulation process. For example, a ratio of core material to ethylcellulose of 30:1 afforded microcapsules having a film content of 3.2 percent by weight whereas reducing the ratio to 25:1 produced microcapsules with a film content of 4.0 percent by weight.

Fabric Conditioning Compositions Containing Methylcellulose Ether

Highlights of the Technological Achievement: Fabric conditioning compositions contain a cellulose ether having a molecular weight of from 3,000 to 10,000 and a methyl degree of substitution of from 1.8 to 2.7. The presence of the cellulose ether provides a soil release benefit to fabrics contacted with the disclosed compositions.

Reference: H.J. Pracht and M.E. Burns; U.S. Patent 4,136,038; January 23, 1979; assigned to The Procter & Gamble Company.

Background: Fabric conditioning compositions containing various soil release agents are known. For example, see U.S. Patent 3,920,561, and U.S. Patent 3,928,213. While certain of the prior art cellulose ethers have been able to efficiently and effectively deposit upon fabrics and thus provide a soil release benefit, there are consequent negatives involved. It has been found that deposition of certain of the prior art cellulose ethers on fabrics causes a negative in particulate soil removal. More specifically, when such treated fabrics are subsequently stained with a particulate soil, e.g., clay, and washed, the stain is more difficult to remove. It is believed the cellulose ether provides a bonding site for the particulate soil, thus causing its removal to be more difficult.

Process Conditions: A fabric conditioning composition having the ability to impart an oily soil release benefit to fabrics contacted therewith consists essentially of (a) from 0.05 to 10% of a methylcellulose ether having a DS methyl of from 1.8 to 2.7 and a molecular weight of from 3,000 to 10,000 and (b) the balance a carrier. (As used herein, all percentages and ratios are by weight unless otherwise indicated.) Such a composition can contain a builder, bleach, enzyme, water, etc. and is useful as a presoak laundry composition or as a rinse additive composition.

The cellulose ethers of this process have a low molecular weight and a high degree of methyl substitution. The cellulose ethers have the following structure

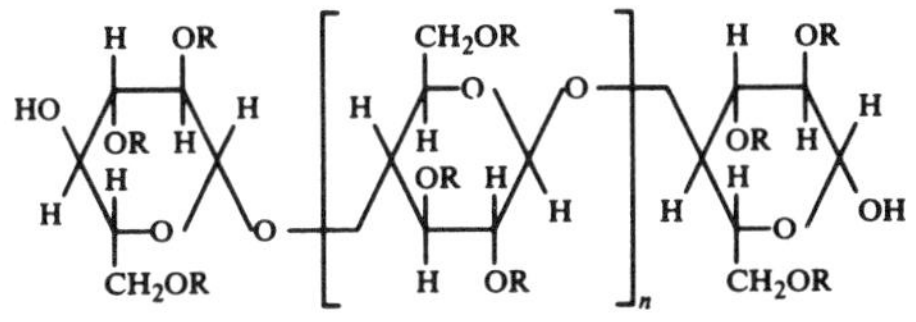

wherein R is hydrogen or methyl with the proviso that the average degree of methyl substitution on the anhydroglucose unit is from 1.8 to 2.7; n in the above formula ranges from 14 to 54. This corresponds to a molecular weight of from 3,000 to 10,000 as measured by known ultracentrifuge techniques. Preferably the molecular weight ranges from 3,500 to 8,000, more preferably from 4,000 to 5,500.

Natural cellulose materials such as wood pulp, cotton linters, etc. have a molecular weight ranging from 250,000 to 1,000,000 or greater. The cellulose ethers of this process are prepared from the natural cellulose materials. The cellulose material is methylated and then has its degree of polymerization reduced or it is initially reduced in degree of polymerization and then methylated.

Known processes for methylating cellulose are used. Such processes involve, for example, simply combining methyl chloride with the cellulose under alkaline conditions. Such a process results in a DS methyl of below 2 and most generally a DS methyl of about 1.5. Higher degrees of methyl substitution are obtained by exhaustive methylation of cellulose using a methyl halide, preferably methyl chloride and caustic, preferably sodium hydroxide in a pressure vessel in a manner well known in the art for preparing the lower degrees of methyl-substituted celluloses; the methylation procedure is simply repeated and continued until the desired degree of substitution is secured.

The progress of the methylation reaction is monitored by periodically sampling the reaction product and determining the degree of methylation. The cellulose ethers of this process have a DS methyl of from 1.8 to 2.7, preferably from 2.1 to 2.4.

EPOXY PRODUCTS

PROCESSING

Hardening Agent for Aqueous Epoxy Resin Compositions

Highlights of the Technological Achievement: A hardening agent for an aqueous epoxy resin composition comprising a reaction product of: (a) at least one polyepoxide compound; (b) at least one polyalkylene polyether polyol; and (c) at least one compound selected from the group consisting of aliphatic, cycloaliphatic and heterocyclic polyamine and wherein component (c) additionally may be reacted with (d); and (d) an unsaturated compound before or after the reaction with the adduct of components (a) and (b); and also a process for preparing the hardening agent and hardenable film forming coating compositions comprising as essential ingredients the hardening agent, at least one polyfunctional epoxy compound (a) and water.

Reference: W. Becker and H. Karasmann; U.S. Patent 4,197,389; April 8, 1980; assigned to Hoechst AG, Germany.

Background: Swiss Patent 487,955 describes hardenable aqueous dispersions, containing an epoxy resin hardener, which is liquid at room temperature, hydrogen being bonded in the hardener directly to nitrogen. The hardener may be obtained by addition of a nonaliphatic epoxide containing only one 1,2-epoxide group in a terminal position per molecule or of an aliphatic epoxide containing (on average) one or more 1,2-epoxide groups in a terminal position per molecule to a stoichiometric excess of a poly(aminoamide) condensation product obtained from an alkylene polyamide of formula:

$$H_2N\text{—(—alkylene—NH—)}_n\text{—H}$$

(in which the alkylene radical is an ethylene, propylene, butylene, pentylene or hexylene radical, and n represents an integer greater than zero) and a copolymer of an unsaturated conjugated fatty acid or an amide-forming acid derivative with an aromatic vinyl compound.

These aqueous dispersions also preferably contain a hardening accelerator. Suitable accelerators are compounds containing at least one tertiary amino nitrogen atom and a phenolic hydroxyl group. Furthermore, the dispersions contain orthophosphoric acid or a liquid aliphatic monocarboxylic acid, e.g., acetic acid. These epoxy resin hardener dispersions are also insufficiently stable and provide only relatively soft and mechanically sensitive protective coatings. Also, the presence of dissociating acids leads to instability in the dispersion, since the acid anion (A^-) may react with the epoxide group according to the following reaction and thereby consume epoxide groups.

$$\underset{\diagdown\,O\,\diagup}{RCH{-}CH_2} + A^- + H_2O \longrightarrow \underset{\displaystyle OH}{RCH}{-}CH_2A + OH^-$$

For the preparation of coatings, it is necessary to add to the mixture of epoxy resin prepolymer and hardening agent an organic solvent miscible with water, in order to reduce the viscosity of the composition and, therefore, make it more readily applied. Although the addition of organic solvents facilitates the application of the composition and produces films with good properties, they present a series of problems. Firstly, the solvents are relatively expensive and often toxic and/or inflammable. Secondly, pollution by the solvents during preparation, storage, transport and final use of the compositions, solvent vapors, solvent waste from any necessary purification plant and spillage of solvent are further problems which are often difficult and expensive to solve.

Due to these disadvantages, the use of known epoxy coating compositions is subject to a number of restrictions, especially when they are to be used in connection with foodstuffs. In addition, the cleaning of apparatus which has come into contact with such compositions presents considerable difficulties.

Process Conditions: The hardening agent comprises a reaction product which has been obtained from 20 to 80 wt % of an adduct derived from (a) 2 to 4 equivalents of a polyepoxide compound and (b) 0.2 to 1.5 equivalents of a polyalkylene polyether polyol, by reaction with 80 to 20 wt % of component (c) comprising an aliphatic, cycloaliphatic and/or heterocyclic polyamine or polyamine mixture, the quantity of amine used being in 2 to 10 times excess, relative to the reacting epoxide groups and hydrogen atoms bonded to the nitrogen.

Preferably, 50 to 70 wt % of an adduct consisting of (a) 2.3 to 3.0 equivalents of a polyepoxide compound and (b) 0.3 to 0.8 equivalents of a polyalkylene polyether polyol are reacted with 50 to 30 wt % of component (c) comprising an aliphatic, cycloaliphatic and/or heterocyclic polyamine or polyamine mixture, the quantity of amine used being in excess by 5 to 8 times relative to the reacting epoxide groups and amino hydrogen atoms.

In a further embodiment of this process, the polyamine component (c) has added to it (d) unsaturated compounds which may undergo addition reactions, whereby for each hydrogen atom bonded to the nitrogen, component (c) is further reacted before or after reaction with the adduct of components (a) and (b) with preferably 0.1 to 0.3 equivalents of an unsaturated compound (d), which may undergo addition reactions, of the formula:

$$\underset{\displaystyle R_1}{HC}{=}\underset{\displaystyle R_2}{C}{-}X$$

wherein R_1 represents a hydrogen atom or methyl or ethyl group, R_2 represents a hydrogen atom or methyl group, and X represents a cyano, carboxy, methoxycarbonyl, ethoxycarbonyl, propoxycarbonyl, butoxycarbonyl, carbamoyl, N-methylcarbamoyl or N,N-dimethylcarbamoyl group. From a further aspect, the process provides a process for the preparation of a hardening agent for aqueous epoxy resin compositions wherein in a first step, an adduct consisting of (a) a polyepoxide compound and (b) a polyalkylene polyether polyol is prepared, generally using a BF_3-amine complex as a catalyst, which adduct is subsequently reacted with (c) a polyamine which is optionally reacted with (d) an unsaturated compound which may undergo addition reactions of the following formula.

$$HC(R_1)=C(R_2)-X$$

An especially preferred epoxy resin for use in the process is a polyphenol glycidyl ether, e.g., a reaction product of epichlorohydrin and 2,2-bis-(4-hydroxyphenyl)-propane (that is, bisphenol A), this epoxy resin having the following theoretical structural formula:

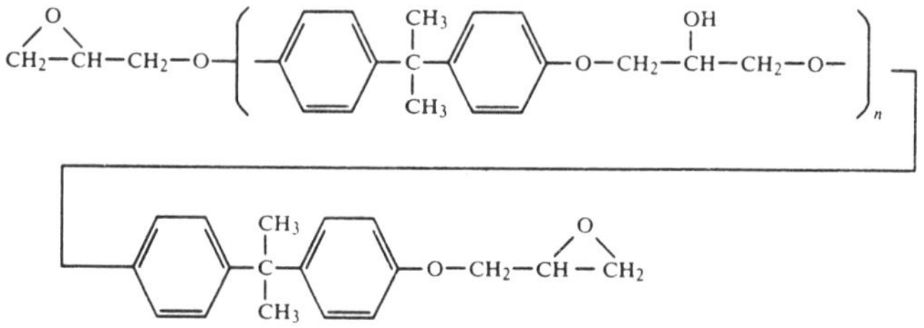

or the homologue thereof based on bisphenol F:

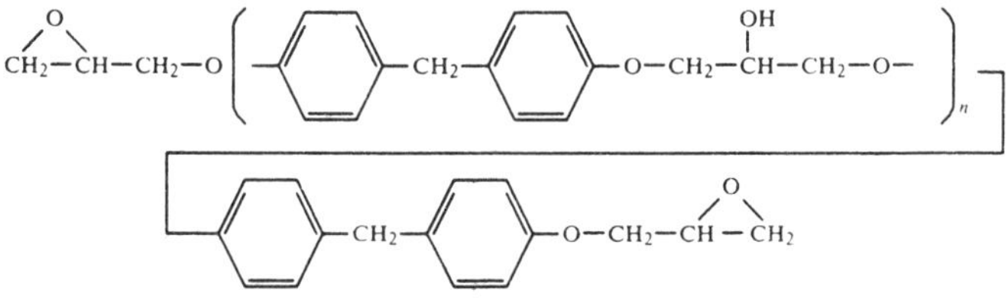

individually or in admixture with bisphenol A. In the above formulas, n represents 0 or an integer up to 10. Normally, n is not greater than 2 or 3 and preferably 1 or less. The epoxy resin has an epoxy equivalent of 160 to 500.

Examples of polyamines which may be used as component (c) are: aliphatic, saturated or unsaturated bifunctional amines, or polyoxypropylene polyamines and diamines of general formula $H_2N-CH_2-X-CH_2-NH_2$ (in which X represents phenylene or cyclohexylene).

For 1 g equivalent of hydroxyl group to be reacted with 1 g equivalent of an epoxide group, 100 to 1, preferably 15 to 2 mmol, of the BF_3-amine complex catalyst are conveniently used. The reaction of the hydroxyl groups with the epoxide groups can be carried out at a temperature in the range from 20° to 200°C. The reaction temperature is generally dependent on the particular BF_3-amine complex used.

For example, when BF_3-monoethylamine or BF_3-benzylamine are used, the reaction temperature is preferably about 130°C.

Epoxy Resin Composition Containing Glycidyl Ether-Metal Carboxylate Catalyst

Highlights of the Technological Achievement: This process relates to an epoxy resin composition suitable for impregnating coils of electrical apparatus such as motors, generators and the like. More particularly, this process relates to an epoxy resin composition having excellent workability and a long pot life, which composition is prepared by adding the reaction product of the glycidyl ether of a monoalcohol and a metal carboxylate, which works as a reactive diluent and accelerator, to an epoxy resin having at least two epoxy groups per molecule and containing a curing agent.

Reference: K. Shibayama, K. Okahashi and O. Hayashi; U.S. Patent 4,192,786; March 11, 1980; assigned to Mitsubishi Denki KK, Japan.

Background: Previously, a typical example of an epoxy resin composition suitable for impregnation was a bisphenol type epoxy resin containing an acid anhydride, e.g., methyltetrahydrophthalic anhydride, as a curing agent and a tertiary amine, imidazole or metal chelate compound as an accelerator. This conventional composition generally has a short pot life and a viscosity incompatible with workability, i.e., the viscosity is too high. In order to improve workability (to lower the viscosity), monoglycidyl ethers such as butyl glycidyl ether, allyl glycidyl ether and the like, or diglycidyl ethers such as ethylene glycol diglycidyl ether, propylene glycol diglycidyl ether and the like, can be used. However, there is the disadvantage that an improvement in pot life cannot be expected.

One of the easiest methods to overcome the abovementioned disadvantage is to add a curing agent and an accelerator to the epoxy resin at the time of using the resinous composition. However, this method is not employed to apply insulation to the coils of electrical machinery and apparatuses on an industrial scale since the operation is complicated and work efficiency is very low.

It is required that an epoxy resin composition having not only good electrical and mechanical properties but also excellent workability and a long pot life be provided.

Process Conditions: It has been found that the reaction product of the glycidyl ether of a monoalcohol and a metal carboxylate not only works as a reactive diluent and accelerator to an epoxy resin composition containing a curing agent but also prolongs the pot life and provides an appropriate viscosity to the composition. Besides, the reaction product provides excellent electrical and mechanical properties to the composition when it is cured.

Thus, the epoxy resin composition of this process is prepared by mixing an epoxy resin having not less than two epoxy groups per molecule, an acid anhydride curing agent and the reaction product of the glycidyl ether of a monoalcohol and a metal carboxylate.

Preferably, the epoxy resin composition of the process comprises a mixture of 40 to 90 pbw of the epoxy resin, 70 to 100 pbw of the acid anhydride curing agent and 60 to 10 pbw of the reaction product.

In the reaction of the glycidyl ether of a monoalcohol and a metal carboxylate, the latter is used preferably in an amount of 0.2 to 1.0 wt % of the former.

Examples of epoxy resin having at least two epoxy groups per molecule used as the first ingredient in the epoxy resin composition of this process preferably include bisphenol type epoxy resin as represented by the following formula:

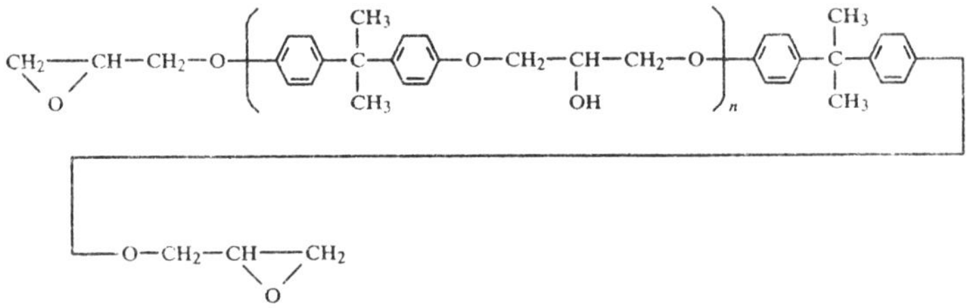

where n represents an integer of up to about 15. Examples of this epoxy resin are available as Epikote 828, and 826 (by Shell Chemical Corp.); DER 332, 330 and 331 (by Dow Chemical Co.); and Araldite GY 250, and 260 (by Ciba Co.).

In admixture with the bisphenol type epoxy resin, cycloaliphatic type epoxy resin as represented by the following structural formula may be used.

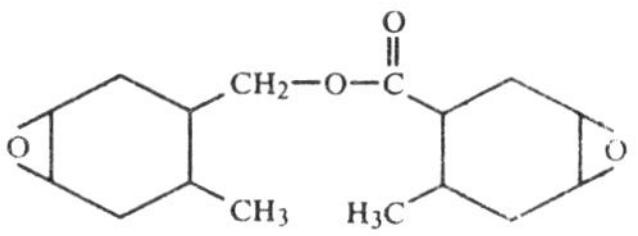

Examples of this epoxy resin include ERL-4201, and ERL-4221 (by Union Carbide Corp.).

The reaction product can be obtained, for example, by adding a metal carboxylate to the monoglycidyl ether and reacting the mixture at about 50°C for 2 to 4 hr in an atmosphere of nitrogen with stirring. In this reaction, the optimum reaction temperature ranges from 35° to 70°C, at which the effect of the catalyst clearly appears. The suitable amount of the metal carboxylate added is 0.2 to 1.0 wt % of the glycidyl ether of the monoalcohol. If the amount added is more than 1.0%, the pot life becomes short, while if it is less than 0.2%, gelation is unsuitably retarded. Thus, if the specific ingredient is previously reacted as mentioned above, the system becomes stabilized.

The epoxy resin composition of this process can be obtained by mixing the abovementioned epoxy resin, acid anhydride and reaction product in an appropriate ratio, the generally preferable mixing ratio in view of workability and other properties of the cured product being 40 to 90 pbw, 70 to 100 pbw and 60 to 10 pbw, respectively.

Weak Base Ion Exchange Resins of a Polyalkylenepolyamine and Polyepoxide

Highlights of the Technological Achievement: The cured reaction product of a polyalkylenepolyamine and a polyepoxide is useful as a weak base ion exchange resin. For example, the reaction of triethylenetetramine and diglycidyl ether of bisphenol A forms a weak base resin having an excellent total weak base capacity.

Reference: E.L. Ward; U.S. Patent 4,186,086; January 29, 1980; assigned to Dow Chemical Company.

Background: Weak base ion exchange resins, herein referred to as weak base resins, are generally solid materials which carry exchangeable anions. They possess the ability to exchange anions with a liquid without substantial structural alteration of the solid resin. For this reason, weak base resins are widely used in waste treatment, e.g., in the removal of undesired components from water and other liquids.

Generally, the most effective weak base resins are substantially insoluble in water but are swellable to a limited degree therein, e.g., 30 to 70%, due to their ability to absorb polar solvents.

However, it remains highly desirable to provide an improved method for removing anions from solution using the reaction product of a polyalkylenepolyamine and a polyepoxide which does not require frequent regeneration of the resin.

Process Conditions: This process is an improved method for removing anions from an anion-containing liquid solution of the type which comprises contacting the anion-containing liquid solution with a weak base ion exchange resin. The improvement comprises using a cured reaction product of a polyalkylenepolyamine and a polyepoxide having a molar ratio (expressed as the mols of polyepoxide to mols of polyalkylenepolyamine) from 0.1:1 to 1.5:1 as the weak base resin.

In a preferred embodiment, the cured reaction product is in the form of spheroidal beads. The weak base resins employed in the process exhibit excellent operating capacity and desirable water swellability properties.

The method of this process is particularly useful for removing a variety of anions from solution in operations such as deionizing water, separating strong and weak acids, and other purification processes. This method is also useful for removing undesired components from liquids in such fields as medicine, food processing and electroplating.

Polyepoxides suitably employed in the practice of this process are organic compounds which contain two or more oxiranyl groups that are reactive with the amino hydrogens of the polyalkylenepolyamine. Of particular interest in this process are diglycidyl phenyl ether, epoxidized butadiene, limonene dioxide, bisepoxydicyclopentyl ether of ethylene glycol, etc.; and the polyepoxides having two or more

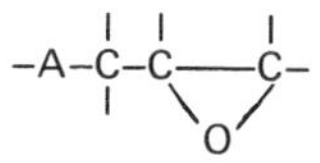

groups wherein each –A– is independently an electron donating substituent such as: >NQ–, –O–, –COO–, –S–, –SO–, $-SO_2-$ or $>NSO_2Q-$ or a covalent bond and Q is an alkyl, alicyclic or aromatic hydrocarbon.

Suspending Agents: Optionally, the reaction medium of aqueous solution of concentrated metal hydroxide also contains a suspending agent. Suspending agents suitably employed in the practice of this process are anionic or nonionic compounds which reduce agglomeration of the dispersed blend. Advantageously, the suspending agent is soluble or dispersible in the concentrated aqueous solution of the alkali metal hydroxide. Suitable suspending agents include hydroxyalkyl celluloses such as hydroxypropyl methylcellulose, hydroxyethyl cellulose and ethyl hydroxyethyl cellulose; carboxylated methylcellulose; carboxymethylated methylcellulose; modified starches such as white dextrin and canary dextrin; and xanthum gum.

Cationic Epoxide Resinous Composition

Highlights of the Technological Achievement: Polyepoxide resins are reacted with a molar excess of a primary amine to form an amine terminated resin, the amine groups of which are then reacted with a monoepoxide. The primary amine is a mixture of an aliphatic monoamine and an aliphatic diamine which contains one primary amine group and one tertiary amine group. The resinous reaction products can be salted with an acid and can be dissolved or dispersed in water. The aqueous dispersions or solutions can then be formulated into primer coatings for metal objects.

Reference: D.D. Hicks; U.S. Patent 4,182,831; January 8, 1980; assigned to Celanese Polymer Specialties Company.

Background: Cationic coating compositions generally are derived from resinous compositions containing a basic nitrogen atom which can be salted with an acid and then be dissolved or dispersed in water. Sufficient basic nitrogen atoms should be present so that dispersibility or solubility can be obtained with a minimum amount of acid. If the coating bath is very acidic, considerable corrosion problems with the metal tanks, piping systems and other equipment are encountered.

Process Conditions: The soluble and fusible resinous composition of this process is made from a polyepoxide resin, a mixture of primary amines and a monoepoxide. The polyepoxide resin is derived from a diol and an epihalohydrin and has a 1,2-epoxide equivalent weight of about 180 to 650.

The compositions of this process are the reaction products of polyepoxide resins and mono primary amines further reacted with a monoepoxide. These compositions can be represented by the formula $M(AE)_nAM$ wherein A represents a reacted primary amine, E represents a reacted polyepoxide resin, M represents a reacted monoepoxide, and n represents an integer of 2 to 5.

In the above formula, the AE linkages and the AM linkages which are formed by the reaction of an amine group and an epoxide group, can be represented by the skeletal formula:

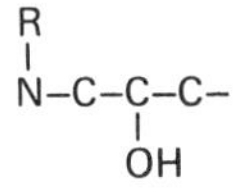

wherein R is a hydrocarbon group. The nitrogen atom is a tertiary amine group and can be salted with an acid. The OH group is available for cross-linking reactions.

The aliphatic diamines contain only one primary amine group and only one tertiary amine group and no other groups which are reactive with epoxide groups. These diamines can be represented by the formula:

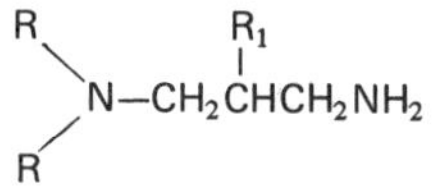

wherein each R is an alkyl group which contains 2 to 6 carbon atoms and R_1 is hydrogen or an alkyl group which contains 1 or 2 carbon atoms. The preferred diamine is 3-diethylaminopropylamine.

In preparing the compositions of this process, the reaction of the primary amine mixture and the polyepoxide resin is conducted with the amines in excess to prevent gelation of the reaction. The molar ratio of polyepoxide resin to primary amines will vary from about 5:6-2:3 and, preferably about 3:4. As calculated in this process, this molar ratio is actually the equivalent ratio, i.e., the ratio of 5:6-2:3 is based on 1,2-epoxide groups and primary amine hydrogen atoms.

After substantially all of the epoxide groups of the polyepoxide resin have reacted, the resulting resinous amine terminated polymer is then reacted with a monoepoxide. The amount of monoepoxide used is one mol of monoepoxide per each amine hydrogen which remains in the polymer, i.e., 2 mols of monoepoxide per mol of primary amine/polyepoxide reaction product. The molar ratio of polyepoxide resin to primary amine to monoepoxide is 5:6:2-2:3:2.

In order to minimize the epoxy-hydroxyl reaction and to prevent gel formation, the compositions of this process are made by adding the epoxide compounds to the amine mixtures at the reaction temperature of 50° to 150°C.

Electrodeposition of Aqueous Dispersions of Copolymers of Polyethylenically Unsaturated Epoxy-Amine Adducts

Highlights of the Technological Achievement: Nongelled, amine functional polymers dispersible in water with the aid of a solubilizing acid are provided by copolymerizing (a) an ethylenically unsaturated hydroxy functional amine adduct free of epoxy groups and containing from 1.4 to 2.0 ethylenically unsaturated amine groups per molecule, formed from a polyepoxide having a 1,2-epoxy equivalency of from 1.4 to 2.0 and at least a stoichiometric amount of a secondary amine with two unsaturated groups; and (b) copolymerizable monoethylenically unsaturated monomers, a portion of which is amine-functional. The water solutions are particularly useful to enable corrosion resistant coatings to be electrodeposited at the cathode.

Reference: K. Sekmakas and R. Shah; U.S. Patent 4,172,062; October 23, 1979; assigned to DeSoto, Inc.

Background: This process relates to copolymer of polyethylenically unsaturated epoxy-amine adducts, to stable aqueous dispersions containing the same at a pH close to neutral, and to the electrodeposition of such aqueous dispersions at the cathode of a unidirectional electrical system.

Process Conditions: The stoichiometry is between the amine group and the oxirane group, and the reaction is a simple addition reaction which forms a hydroxy group and a tertiary amine group, and all of the epoxy groups are consumed in the addition reaction which forms the addition reaction product (adduct).

The copolymer is stably dispersible in water at a pH in excess of 5.0, up to 7.5, and it can be electrodeposited at the cathode, preferably together with an aminoplast curing agent, and cured to provide cured coatings which have many desirable properties. The high corrosion resistance is particularly outstanding (salt spray and detergent resistance), and the high gloss and hardness combined with reasonable flexibility are also impressive.

Referring more particularly to the organic diepoxide which is used, these should have a molecular weight of at least 350. Since commercial epoxy resins are sometimes mixtures of diepoxides and monoepoxides, these can be used so long as the epoxy functionality is at least about 1.4. Polyepoxides having an epoxy functionality above 2.0 lead to gelation.

Raw Materials: The polyepoxides which are preferred are hydroxy functional polyepoxides having a linear aromatic backbone and a molecular weight in the range of preferably 700 to 4,000. These include a significant hydroxy value (equivalents per 100 g) of from 0.2 to 0.4 which participates in the final cure. These polyepoxides are more preferably diglycidyl ethers of bisphenols, such as bisphenol A, and have a 1,2-epoxy equivalency of 1.4 to 2.0.

From the standpoint of the commercially available materials, the Shell product Epon 1001 is preferred, this material having an average molecular weight of 1,000, a melting point of 67° to 74°C, an epoxy value of 0.20 equivalent per 100 g, and a hydroxyl value of 0.32 equivalent per 100 g. Other commercial epoxy resins of known characteristics which are also appropriate for use herein are Epon 834, Epon 864, Epon 1004 and Epon 1007.

The reaction between the secondary amine and the epoxy groups is itself conventional, the reactants being cooked together at moderate temperature (60° to 175°C).

Monoethylenically unsaturated tertiary amino amides are also useful in this process, and these have the formula:

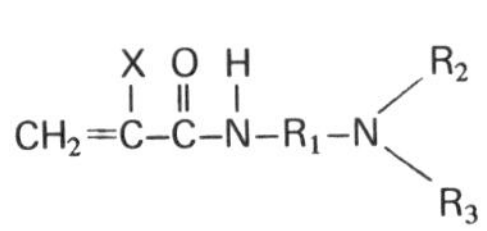

where X is hydrogen or methyl, R_1 is C_{1-4} alkylidene, and R_2 and R_3 are C_{1-4} alkyl, preferably both methyl.

These unsaturated amino amides are illustrated by dimethylaminopropyl methacrylamide which has the following formula.

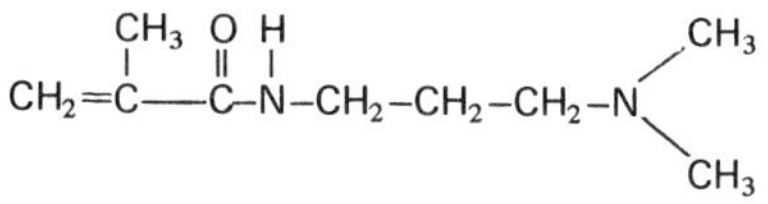

The corresponding acrylate is considered to be equivalent and the use of other alkylidene groups such as the methylidene, ethylidene, and butylidene groups, does not significantly alter the properties. While the dimethyl amino species is distinctly preferred, the corresponding diethyl, dipropyl, or dibutyl amino compounds are also useful.

Air Curable Resin Solution for Surface Coating

Highlights of the Technological Achievement: An air curable resin solution useful for surface coating and impregnation comprising: (1) the reaction product of (A) from 50 to 65 wt % based upon the total weight of (A) and (B) of an epoxy resin ester of a partially conjugated unsaturated fatty acid and (B) from 50 to 35 wt % based upon the total weight of (A) and (B) of a blend of reactive monomer possessing reactive double bonds, at least one of which must be an unsaturated mono-basic acid in the presence of (2) an alcohol ether of a glycol and subsequently reacted with (3) an amine and then (4) dispersed in water.

Component (A) is the reaction product of 2,2-bis(4-hydroxyphenyl)dimethyl methane and epichlorohydrin, and has a melting point from 130° to 230°F and an epoxide equivalent weight within the range from 400 to 1,100. The fatty acids used are straight chain monobasic acids of 18 carbon length having double bonds arranged in the chain in amount and position to give an iodine number of 125 to 185, an acid number of from 180 to 210 and a percentage of conjugation of the double bonds between 20 and 25%. The reaction is carried out to an acid number below 10. The monomer (B) portion consists of a mixture of 20 to 28% of unsaturated monobasic acids having a polymerizable double bond and 80 to 72% reactive monomers having a polymerizable double bond.

Reference: C.J. Meeske, E.H. Van der Tuin and M.J. Racey; U.S. Patent 4,166,054; August 28, 1979; assigned to Reichhold Chemicals, Inc.

Background: The replacement of organic solvents with water in surface coatings has become a prime interest of surface coating producers and applicators.

The major reasons for using water as the solvent are: (1) to reduce the danger of fire due to the extremely flammable solvent vapors released during application of coatings by spray, flow coat and dip processes, (2) to reduce the great expense of eliminating the vapors of organic solvents from reaching the atmosphere and causing air pollution, (3) to reduce the consumption of solvents based on petroleum of the decreasing petroleum available and the increasing costs of petroleum based products.

A partially esterified ester can be neutralized in a water/amine solution to become water-soluble. This composition will cure to a tough film if baked at 150° to 175°C. One drawback of this method is the great danger of gelation during the esterification step and the need for baking the film in order for it to cure.

In contradistinction, this process does not require baking and the reaction is easy to control in order to obtain a uniform polymer.

Process Conditions: This process relates to a water-soluble or dilutible amine salt of an epoxy ester copolymer which serves as a component for a surface coating which can be pigmented. The process comprises (A) an epoxy resin which is a reaction product of 2,2-bis(4-hydroxyphenyl)dimethyl methane and epichlorohydrin having an epoxide equivalent value of 450 to 525, a melting point of 65° to 75°C and a viscosity of E to J in a test solution at 40% nonvolatile in monobutylether of diethylene glycol; which is esterified with a drying oil fatty acid having an acid number of 190 to 200 and an iodine value between 125 and 185, preferably having a minimum of 20% of its double bonds in a conjugated configuration; the esterification is continued to an acid number below 10.0 and is then dissolved in a solvent which is an ether of a short chain alcohol and ethylene glycol to 60% nonvolatile which solution has a viscosity of K to N (Gardner-Holdt).

The 60% nonvolatile solution of epoxy ester (A) is reacted with (B) a mixture of unsaturated monomers, 25% of which must consist of an unsaturated monobasic acid.

The copolymer of (A) and (B) will be a 70% nonvolatile solution with a viscosity of Z_5-Z_7 and an acid number of 35 to 45 which, after addition of an amine compound to adjust the pH to a range of 6 to 9, will be dilutible with water and, when applied as a thin film, will cure at room temperature to a hard glossy water resistant film.

Water-Thinnable Mixtures of Base-Neutralized Products of Reaction of H_3PO_4 with Polyether Epoxides and Other Type Epoxides

Highlights of the Technological Achievement: The utility, as coatings, of water-thinnable, base-neutralized H_3PO_4/polyether epoxide reaction products, is improved by admixture therewith of the base-neutralized reaction products of phosphoric acid with various other kinds of epoxides. Of particular value are such mixtures in which the latter products are derived from methylol- or lower alkoxymethyl-substituted epoxides.

Reference: P.H. Martin; U.S. Patent 4,164,487; August 14, 1979; assigned to The Dow Chemical Company.

Process Conditions: Base-neutralized reaction products of phosphoric acid with various methylol- or lower alkoxymethyl-substituted epoxides have been found to be compatible with the water-thinned polyether epoxide (E^1) derived resin dispersions and to render them rapidly self-converting. It has also been found that the ability of the dispersions to wet various types of substrates can be manipulated to advantage by choice of other substituents in melamine resin curing agents. Other properties of the aqueous dispersions of the neutralized acid/E^1 products, and/or of the cured coatings derivable therefrom, can be modified to advantage by incorporation of (neutralized) reaction products of phosphoric acid with various other (E^2) type epoxides.

The composition of the process is a mixture of base-neutralized reaction products of H_3PO_4 with polyether epoxides and any of various other types of mono- or polyfunctional epoxides.

The preferred method of this process is to coreact the polyether (E^1) and other (E^2) epoxides with an orthophosphoric acid source material and to neutralize the resulting mixed product with a base, preferably a fugitive base. However, this process also comprises the method of preparation in which separately prepared E^1 and E^2 reaction products with phosphoric acid are combined.

The coatings are those formed on various substrates (preferably metallic) from aqueous dispersions of the preceding compositions.

The preferred method is more precisely definable as the process for preparing water-thinnable, base-salified reaction products of orthophosphoric acid and polyether epoxides which comprises: (A) reacting orthophosphoric acid with:

(1) A polyether epoxide resin E^1 consisting essentially of molecules, each of which is of the formula:

(a) 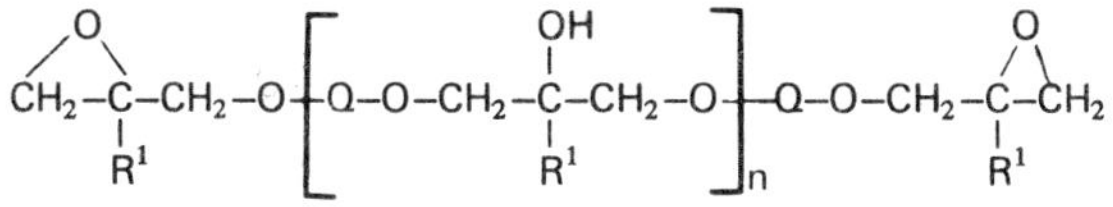

or the formula:

(b)

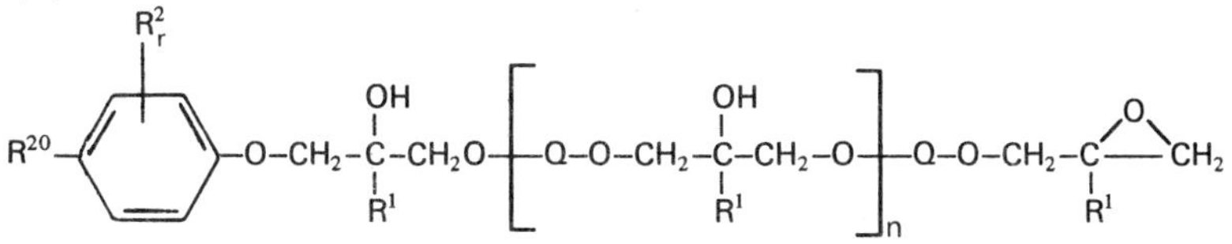

wherein Q, independently, in each occurence, is:

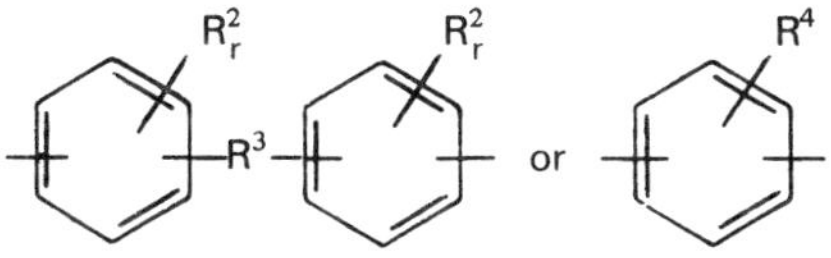

n is an integer of from 0 to 40, r is 0, 1 or 2 and, independently in each occurrence; R^1 is H, methyl or ethyl; R^2 is –Br, –Cl or a C_{1-4} alkyl or alkenyl group; R^3 is a C_{1-4} alkylene or alkenylene group, $>C(CF_3)_2$, –CO, $-SO_2-$, –S–, –O– or a valence bond; and R^4 is –Br, –Cl or a C_{1-4} alkyl or alkenyl group; R^{20} is H or alkyl of 1 to 12 carbons; and

(2) E^2, a vicinal epoxide, other than one of formula (a) or (b), which has an EEW (epoxide equivalent weight) within the range of from 90 to 2,000 and is convertible to a water dispersible material by reaction with orthophosphoric acid and neutralization with a base.

The reaction being carried out by contacting E^1 and E^2 with an orthophosphoric acid source material and from 0 to 25 molecular proportions of water per molecular proportion of H_3PO_4 provided by the source material until the fraction of the oxirane groups in E^1 and E^2 converted is at least sufficient to render the resulting mixed product water-thinnable when contacted with a base, the amount of orthophosphoric acid included as such in the source material, or obtainable therefrom by hydrolysis, being such as to provide at least 0.3 P–OH groups per oxirane group, and the mol ratio of E^1 to E^2 epoxides being from 0.1 to 100; and

(B) Contacting the resulting mixed reaction product with at least enough of a base to render it water-thinnable.

The mixed, neutralized, epoxide/acid reaction products of this process may be more precisely defined as a water-thinnable, resinous phosphate composition comprising:

(A) Resin molecules, each of which is derivable by conversion to 1,2-glycol- or to beta-hydroxy phosphomonoester groups of the oxirane groups in an E^1 epoxide represented by one of the preceding formulas (a) and (b);

(B) Other molecules, each of which is derivable by conversion to 1,2-glycol- or to beta-hydroxy phosphomonoester groups of the oxirane groups in E^2, a vicinal epoxide other than those of formulas (a) and (b), having an EEW within the range of 90 to 2,000, the mol ratio of the E^1-derivable molecules to the E^2-derivable molecules being within the range of 0.1 to 100, and the number ratio of glycol to monoester groups in each of the types of molecules being within the range of 0 to 18;

(C) From 0 to 85 pbw of orthophosphoric acid (H_3PO_4) per 100 pbw of the E^1- and E^2-derivable molecules; and

(D) One or more bases, in such amount that at least enough of the P–OH moieties in the E^1- and E^2-derivable molecules are salified thereby to render them dispersible together in water.

Curing Agent for Water-Based Epoxy Resins

Highlights of the Technological Achievement: This process relates to curing agents for epoxy resins. More particularly, the process relates to water-soluble or water-dispersible curing agents which are mixed with water-dispersed or water-emulsified epoxy resins to form coating compositions which are essentially free of volatile organic solvents or cosolvents and which have extended pot life at room temperature, but which cure under ambient conditions in less than 24 hr.

Reference: I.P. Thomassen; U.S. Patent 4,152,285; May 1, 1979.

Background: In addition to the fact that the prior art compositions have a short pot life, the gloss of coatings produced therefrom varies during the life of the mixed batch. The coatings resulting from coating compositions produced from the initial mixing of the prior art curing agents and the resin emulsion or dispersion have a high gloss, but as the work progresses during the application, the gloss of the coating decreases and the coatings produced from compositions near the end of their pot life have low gloss. This problem is especially noticeable when large areas are being coated and is further manifested when touch-up work is undertaken, wherein the touch-up paint will not match the original paint with respect to the gloss.

The curing agents of this process have greatly increased solubility or dispersibility in water, which allows the reaction between the new curing agent and the epoxy water emulsion (or dispersion) to proceed at a controlled rate.

Process Conditions: This process provides curing agents which are water-soluble or water-dispersible hydroxylated polyamines, which optionally may contain amide groups. The curing agents of this process are prepared by the reaction of aliphatic polyamine compounds with aliphatic or cycloaliphatic polyepoxy materials, and may be produced by a variety of different processes. Generally speaking, the curing agents are alkaline products having a pH of above 9.0, which are produced by treating the polyepoxy materials with a polyamine which may be an alkanol polyamine and may have amide groups therein.

The basic nature of the curing agents of this process permits the curing reaction with the epoxy resin to proceed smoothly at room temperature, thus making the use of elevated temperatures unnecessary. Polybasic acids may be used to increase the molecular weight of the curing agent through reactions with polyamine or the alkanol polyamine or by reacting with the hydroxylated polyamines, but care must be taken to avoid reducing the pH below 9.0, as is now fully explained below.

The curing agents of this process are prepared using sufficient polyamine to react with all of the epoxy groups present on the aliphatic or cycloaliphatic polyepoxy material. Further, sufficient polyamine or alkanol polyamine must be used to neutralize any polybasic acid which may be used and to maintain the pH above 9.0. Thus, the curing agents of this process are hydroxylated polyamines or hydroxylated amido polyamines, which are substantially free of unreacted carboxylic acid groups, are substantially free of unreacted epoxy groups, and are preferably free of ester groups.

The curing agents must have a pH above 9.0 and have amine values in the range of 150 to 800 and preferably in the range of 500 to 700, on a solids basis, although the curing agents (and their amine values) are usually reduced with water prior to use. The curing agents of this process generally have a molecular weight of 280 to 1,500, with the range of 500 to 700 being preferred. Preferably the curing agents of this process contain from 2 to 10 hydroxyl groups per molecule.

Coatings Based on Hydrolyzed Epoxy Resins

Highlights of the Technological Achievement: Poly-1,2-epoxy resins, such as diglycidyl ethers of bisphenols, are hydrolyzed at high solids content in organic solvents, such as hexoxyethanol. The hydrolyzed epoxy resins are used in coating formulations (aqueous emulsions or solutions) together with curing agent and, optionally, a catalyst for crosslinking.

Reference: A.C. Chen; U.S. Patent 4,145,324; March 20, 1979; assigned to Mobil Oil Corporation.

Background: The preparation of hydrolyzed epoxy resins has been described in U.S. Patent 3,405,093. The process, however, has been carried out at about 35% solids, whereas the process utilized here is carried out at 60 to 90% solids. Insofar as is known, it has not been proposed to utilize hydrolyzed poly-1,2-epoxy resins in water-based systems (emulsions) or in organic solution systems as coating formulations, particularly for metal containers.

Process Conditions: This process comprises a coating formulation of (a) an oil-in-water emulsion of a solution of a hydrolyzed diglycidyl ether of a bisphenol, a nonionic surfactant and a curing agent; or (b) a solution of a hydrolyzed diglycidyl ether of a bisphenol and a curing agent in alkoxyethanol or a mixture of alkoxyethanol with alcohols, ketones, carbitols, or their acetates, and optionally, liquid aromatic hydrocarbon solvents.

The 1,2-epoxy resin can be any polyglycidyl ether of polyhydric organic compounds. Particularly preferred are the glycidyl ethers of bisphenols, a class of compounds which are constituted by a pair of phenolic groups linked through an intervening aliphatic bridge. While any of the bisphenols can be used, the compound 2,2-bis(p-hydroxyphenyl)propane, commonly known as bisphenol A, is more widely available in commerce and is preferred. While polyglycidyl ethers can be used, diglycidyl ethers are preferred. The most preferred and commercially available poly-1,2-epoxy resins are diglycidyl ethers of bispehnol A.

Depending upon the degree of polymerization, the epoxy resins will have an epoxy equivalent weight of between 500 and 5,000. It is also within the contemplation of this process to use a mixture of two or more poly-1,2-epoxy resins in the emulsion coating formulations of this method, provided each has an epoxy equivalent weight falling within the range of 500 to 5,000. The solvents utilizable in the hydrolysis are alkoxyethanols that have the formula $ROCH_2CH_2OH$, wherein R is C_{5-7} alkyl, preferably 2-hexoxyethanol.

It has also been found that mixtures of alkoxyethanol with other solvents are utilizable. Such other solvents include C_{4-6} alcohols, C_{4-6} ketones, C_{5-10} Carbitols, and acetates of the alcohols, ethoxyethanol, and Carbitols. Such mixtures, however, should contain at least 50 wt % alkoxyethanol. A preferred combination is 2-hexoxyethanol and n-butanol.

The amount of solvent or solvent mixture to be used should be sufficient to provide a solids content of at least 60%. In general the amount of solvent or mixture of solvents used will be between 10 and 60 parts per 100 parts of epoxy resin.

The hydrolysis reaction is catalyzed by acid hydrolysis catalysts. Suitable acids include sulfuric acid, hydrochloric acid, phosphoric acid, alkylphenylpolyoxyethylene phosphate, benzenesulfonic acid, and p-toluenesulfonic acid. Generally, 0.1 pbw acid catalyst will be used per 100 pbw epoxy resin.

The hydrolysis is carried out at temperatures between 90° and 130°C using an amount of water in excess over the stoichiometric amount to effect substantially complete hydrolysis of the epoxy groups. In general, 5 to 10 pbw water per 100 pbw of epoxy resin are used. The time of hydrolysis will be between 4 and 24 hr.

Aqueous Coatings Based on Epoxy Ester Copolymers

Highlights of the Technological Achievement: Aqueous coatings particularly adapted for interior sanitary can use are provided in which the resin solids are constituted by a solution copolymer of a resinous polyepoxide esterified to be epoxy-free with from 0.3 to 1.5 mols of at least one monocarboxylic acid containing conjugated ethylenic unsaturation per mol of polyepoxide.

Reference: K. Sekmakas and R. Shah; U.S. Patent 4,145,323; March 20, 1979; assigned to DeSoto, Inc.

Background: In accordance with this process, a resinous polyepoxide, and preferably a bisphenol-based polyepoxide having a 1,2-epoxy equivalency of 1.4 to 2.0, is reacted (esterified) with at least one ethylenically unsaturated monocarboxylic acid to substantially consume the epoxy functionality and at least a portion of the monocarboxylic acids contains conjugated ethylenic unsaturation. In preferred practice, a second monocarboxylic acid containing nonterminal nonconjugated ethylenic unsaturation is used, but this is not essential.

Process Conditions: The resulting unsaturated epoxy ester contains conjugated unsaturated side chains, and this enables a small proportion of a monoethylenically unsaturated dicarboxylic acid containing nonterminal unsaturation to be incorporated by a Diels-Alder reaction while leaving some of the conjugated unsaturation unconsumed. The epoxy ester adduct is then copolymerized in organic solvent solution, the solvent being preferably water miscible, with monoethylenically unsaturated monomer containing terminal ethylenic unsaturation, especially with styrene, vinyl toluene or a mixture thereof.

This forms a normally solid nongelled organic solvent soluble copolymer with the conjugated unsaturation in the epoxy ester while leaving the nonterminal nonconjugated unsaturation supplied by the dicarboxylic acid reaction product or by the second monocarboxylic acid unconsumed in a copolymer containing carboxyl groups which permits dispersion in water by at least partial neutralization with a volatile base to form an aqueous coating solution in which aminoplast or phenoplast resin can be dispersed to provide a coating composition which thermosets at very low temperature. If desired, the aminoplast or phenoplast resin can be omitted and a room temperature oxidative cure will thermoset the coating.

The retained unsaturation has the capacity to combine with atmospheric oxygen so that a deposited coating cures both by reaction of methylol groups in the epoxy ester portion of the copolymer as well as by an oxidative cure which is speeded by the moderate heat used.

These low temperature curing aqueous coatings are remarkable because the aminoplast cure of hydroxy functional resins normally requires a curing temperature of 325° to 400°F to produce a good corrosion-resistant film. In contrast, this process consistently lowers the temperature which is needed, by at least 25°F, and the proportion of reactive monomer in the copolymer can be kept below 10%, and is preferably below 6%, based on the weight of the copolymer.

Raw Materials: The preferred starting materials are diepoxides or include diepoxides in admixture with monoepoxides. The epoxy esters of the process are preferably derivatives of the diepoxide component, and have the following formula:

$$R_1-\overset{\overset{\displaystyle O}{\|}}{C}-O-CH_2-\overset{\overset{\displaystyle OH}{|}}{CH}-Z-\overset{\overset{\displaystyle OH}{|}}{CH}-CH_2-O-\overset{\overset{\displaystyle O}{\|}}{C}-R_2$$

where R_1 is the residue of a fatty acid containing conjugated unsaturation, R_2 is the residue of a fatty acid containing nonterminal nonconjugated unsaturation, and Z is the residue of an organic diepoxide. The fatty acid would normally contain 8 to 22 carbon atoms, preferably 10 to 18 carbon atoms in the molecule.

The dicarboxylic acids are preferably selected from fumaric acid or maleic acid. The dicarboxylic acid is used in an amount of 3 to 10%, preferably 4 to 6%, based on the weight of the copolymer. As will be evident, selection of a dicarboxylic acid also minimizes the weight percent of acid needed for dispersibility in water with the aid of a base.

APPLICATIONS

Cationic Epoxide-Amine Reaction Products

Highlights of the Technological Achievement: Polyepoxide resins are reacted in approximately equivalent ratios with primary amines to form cationic resinous compositions. The primary amine is a mixture of an aliphatic monoamine and an aliphatic diamine which contains one primary amine group and one tertiary amine group. The resinous reaction products can be salted with an acid and can be dissolved or dispersed in water. The aqueous dispersions or solutions can then be formulated into primer coatings for metal objects.

Reference: D.D. Hicks; U.S. Patent 4,182,833; January 8, 1980; assigned to Celanese Polymer Specialties Company.

Background: There are certain disadvantages in anodic deposition processes. Anodic deposition on ferrous metals tends to discolor the electrodeposited film and phosphate conversion coatings, which are commonly applied to a metal surface before an organic coating composition is deposited thereon, tend to be stripped from the metal under anodic deposition conditions. In addition, it is a peculiarity of anodic electrodeposition coating methods that nascent oxygen is produced at the anode which can react with the resinous coating composition to produce bubbles or voids in the deposited coatings.

The use of cathodic electrodepositable compositions tends to alleviate the discoloration problems and to give improved resistance properties. Although nascent hydrogen develops at the cathode during the electrodeposition process, no metal ions pass into the coating solution or are present in the deposited film. Generally, the amount of nascent hydrogen produced at the cathode does not have the same deleterious effect on the properties of the deposited film as does the nascent oxygen produced during anodic deposition.

Cationic coating compositions generally are derived from resinous compositions containing a basic nitrogen atom which can be salted with an acid and then be dissolved or dispersed in water. Sufficient basic nitrogen atoms should be present so that dispersibility or solubility can be obtained with a minimum amount of acid. If the coating bath is very acidic, considerable corrosion problems with the metal tanks, piping systems and other equipment are encountered.

Process Conditions: The soluble and fusible resinous composition of this process is made from a polyepoxide resin and a mixture of primary amines. The polyepoxide resin is derived from a diol and an epihalohydrin and has a 1,2-epoxide equivalent weight of 150 to 300. One of the primary amines of the mixture is an aliphatic monoamine containing 8 to 18 carbon atoms, one primary amine group and no other groups which are reactive with epoxide groups under the conditions of the reaction.

The other primary amine is an aliphatic diamine which contains one primary amine group and one tertiary amine group and no other groups reactive with epoxide groups. The amines in the mixture are within the mol ratios of 2:1 aliphatic monoamine:aliphatic diamine to 1:3 aliphatic monoamine:aliphatic diamine. In the composition, the polyepoxide resin and the primary amine mixture are reacted in the mol ratio of 10:9-9:10.

The resinous compositions of this process contain tertiary amine groups within the polymeric chain as well as pendant tertiary amine groups. The compositions also contain aliphatic hydroxyl groups which can be present in the starting polyepoxide resin and which are formed when the primary amines react with the epoxide groups. Upon partial salting with an acid, the resins are readily dissolved or dispersed in water. Such aqueous compositions can be formulated into coating compositions which are particularly useful as primers for metals.

Raw Materials: The compositions of this process are the reaction products of polyepoxide resins and primary amines. The linkage formed from the reaction of the epoxide group and the amine group can be represented by the skeletal formula:

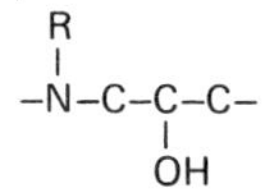

wherein R is a hydrocarbon group. The nitrogen atom is tertiary amine group and can be salted with an acid. The OH group is available for crosslinking reactions.

The aliphatic diamines contain only one primary amine group and only one tertiary amine group and no other groups which are reactive with epoxide groups. These diamines can be represented by the formula:

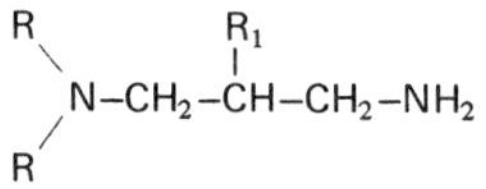

wherein each R is an alkyl group which contains 2 to 6 carbon atoms and R_1 is hydrogen or an alkyl group which contains 1 or 2 carbon atoms. The preferred diamine is 3-diethylaminopropylamine.

Water-Based Epoxy Resin Coating Composition

Highlights of the Technological Achievement: A water-based epoxy resin is disclosed. The epoxy resin is useful in protective coatings. It should have little or no utility in casting, encapsulating, etc. applications because of difficulty in removal of solvent, i.e., water. The cured epoxy resin is made by curing an epoxy resin of the polyhydric phenol, polyglycidyl ether type by combining the resin with a reactive water-compatible polyamide in an aqueous medium. The polyamide is prepared from fatty acids and an aminopropyl derivative of a polyoxyalkylenepolyamine.

Reference: H.G. Waddill; U.S. Patent 4,179,418; December 18, 1979; assigned to Texaco Development Corp.

Background: This process relates to the field of water-based epoxy resins. Prior art water-based epoxy resins generally utilize an emulsion and an organic solvent in their formulation. The presence of organic solvent provided a problem because of volatile solvents escaping into the atmosphere.

Process Conditions: This process is concerned with a water-based epoxy resin and the resulting cured resin. The resin may be made by curing an epoxy resin of the polyhydric phenol, polyglycidyl ether type by mixing with a reactive water compatible polyamide prepared from fatty acids (or derivatives of fatty acids) and an aminopropyl derivative of a polyoxyalkylenepolyamine in an aqueous medium.

The polyoxyalkylenepolyamines are those which have been modified to form an aminopropyl derivative of a polyoxyalkylenepolyamine. Compounds of the following general types have been found to be useful as the aminopropyl derivatives of polyoxyalkylenepolyamines in this process:

$$H_2NCH_2CH_2CH_2NHCH(CH_3)CH_2[OCH_2CH(CH_3)]_xNHCH_2CH_2CH_2NH_2$$

where x is about 2 to 6. These products are generally referred to as bis(propylamines) of polyoxypropylenepolyamines. The reaction of the carboxylic acids and the polyoxyalkylenepolyamine may be carried out under the following conditions, for example, the temperature is increased to 160° to 175°C and water begins to distill overhead. The temperature is maintained at 175° to 220°C until 90% of the water is collected and then is raised to maximum of 250° to 265°C during the final stages of reaction. The final temperature may be raised to 320°C if desired.

Aqueous Cold Setting Paint

Highlights of the Technological Achievement: In an aqueous cold setting paint, a specific copolymer and an epoxy resin being soluble in an alcoholic solvent are incorporated. The specific copolymer is obtained by a solution polymerization of: (a) a monomer having ethylenic unsaturated double bond and a nitrogen-containing base; (b) a monomer having ethylenic unsaturated double bond and a phosphoric acid group or a salt thereof or an ester thereof; (c) a monomer having ethylenic unsaturated double bond and carboxylic acid group or a salt thereof; and (d) a monomer having ethylenic unsaturated double bond which is copolymerizable with a component (a), (b) or (c) or further treatment of the resulting copolymer with a base.

Reference: K. Kurihara, S. Kasahara and T. Suzuki; U.S. Patent 4,177,179; assigned to Toyo Soda Manufacturing Co., Ltd., Japan.

Background: Aqueous compositions containing a drying oil, a semidrying oil and an unsaturated aliphatic acid or a derivative thereof or processes for preparing the compositions have been disclosed in Japanese Patents 32414, 1972; 4214, 1975; and 18503, 1975. However, the compositions obtained by these processes have disadvantages of inferior alkali resistance and weathering resistance.

In order to overcome these disadvantages, it has been found that aqueous paints prepared by admixing a specific acryl type copolymer with a water soluble epoxy resin have excellent solvent resistance, weathering resistance, chemical resistance, water resistance, adhesiveness, processibility and economical advantage.

Process Conditions: The monomers having ethylenic unsaturated double bond and a nitrogen-containing base include:

(1) aminoalkyl acrylates (monomers) having the formula:

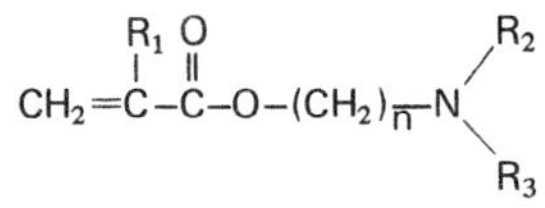

wherein R_1 represents hydrogen or a halogen atom or a C_{1-5} alkyl group; and R_2 represents hydrogen atom or a C_{1-6} alkyl group and R_3 represents hydrogen atom or a C_{1-6} alkyl group and n represents an integer of 1 to 6; for example, acrylates such as dimethylamino methyl acrylate, diethylamino methyl acrylate, dibutylamino methyl acrylate.

(2) the vinylpyridines such as 4-vinylpyridine, 2-vinylpyridine and 2-methyl-5-vinylpyridine.

The amount of the monomers is usually in the range of 0.01 to 20 wt %, preferably 0.5 to 8 wt % based on the total monomers.

The monomers having ethylenic unsaturated double bond and a phosphoric acid group, a salt thereof or an ester thereof include monomers having the formula:

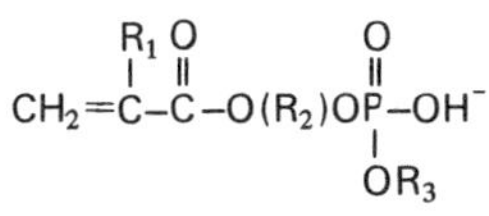

wherein R_1 represents hydrogen atom or a C_{1-5} alkyl group and R_2 represents a C_{2-10} straight chain or branched chain alkylene group or a halogen substituent of the alkylene group and R_3 represents hydrogen atom or a C_{1-10} alkyl group and salts such as ammonium, amine, sodium and potassium salts thereof and esters thereof, for example, primary phosphoric esters of acrylates and methacrylates having hydroxyl group such as mono(2-hydroxyethyl acrylate) acid phosphate.

The amount of the monomer is usually in the range of 0.01 to 20 wt %, preferably 0.5 to 8 wt % based on the total monomers.

The monomers having ethylenic unsaturated double bond and a carboxylic acid group or salts thereof include acrylic acid, methacrylic acid, maleic acid, and itaconic acid and salts thereof. The amount of the monomer is usually in a range of 0.01 to 20 wt %, preferably 0.5 to 8 wt % based on the total monomers. The monomers having ethylenic unsaturated double bond which is copolymerizable with the component (a), (b) or (c) include:

(1) nonfunctional monomers, for example, acrylates, methacrylates, aliphatic acid vinyl esters, aromatic monoolefins, vinyl ethers, and others such as acrylonitrile and methacrylonitrile; and

(2) functional monomers, for example, hydroxyalkyl acrylates, hydroxyalkyl methacrylates, hydroxy group containing unsaturated compounds, sulfur or nitrogen containing unsaturated compounds, and salts thereof.

The amount of the monomer is usually in a range of 0 to 20 wt %, preferably 0 to 8 wt % based on the total monomers.

Cathodic Electrocoating Process for Forming Resinous Coating Using an Aqueous Dispersion of Polyether Resinous Adduct and Acid-Functional Aminoplast

Highlights of the Technological Achievement: An acid-solubilized aqueous dispersion particularly suitable for use in the cathodic electrocoating art and stable up to pH of 9 comprises the product of reacting substantially all of the epoxide groups of a polyether diepoxide with 1.25 to 2 mols of a polyamine having 2 to 5 amine groups to form a precursor adduct and subsequently reacting this precursor adduct with 0.2 to 2 mols of a monoepoxide having a hydrolysis-resistant normal alkyl group of at least 4 carbon atoms to form a substantially epoxy-free polyamino polyhydroxy polyether resinous adduct. The coating cathodically electrodeposited from such dispersions exhibit superior corrosion and detergent resistance when cured at moderate temperatures.

Reference: V.W. Ting and J.M. Evans; U.S. Patent 4,159,233; June 26, 1979; assigned to SCM Corporation.

Background: Aminoplast resins are well known as curing agents for resins having active hydrogens. Both U.S. Patents 3,471,388 and 3,894,993 disclose cathodic electrocoating compositions which contain polyamine-aminated epoxy resins having both primary and secondary amine groups and also comprising etherified formaldehyde-aminotriazine adducts as curing agents, but neither suggests having carboxylic acid groups associated with the aminoplast resin for catalyzing cure.

U.S. Patent 3,519,627 discloses etherified benzoguanamine- or melamine-formaldehyde adducts having aliphatic carboxylic acid groups for use as curing agents with anionic resins in anodic electrocoating compositions, the carboxylic acid functionality of the aminoplast being described only as aiding in the codeposition of the aminoplast on the anode-substrate; there is no suggestion of any use with cationic resins or in particular in cathodic electrocoating.

Process Conditions: The product is a dispersion suitable for use in the cathodic electrocoating art and comprising:

(a) an aqueous medium having pH from 3 to 9 and containing an acid with a pK_a not greater than 5;

(b) a substantially epoxy-free, nongelled, resinous adduct of (1) a substantially epoxy-free precursor adduct of a polyether diepoxide having molecular weight from 300 to 10,000 and a polyamine having molecular weight from 60 to 600, and (2) a monoepoxide having molecular weight from 100 to 500 and having a normal alkyl group of at least about 4 carbon atoms pendant through only hydrolysis-resistant linkages from the epoxy group of the monoepoxide, the mol ratio of residues of the monoepoxide to residues of the diepoxide in the resinous adduct being from 0.2:1-2:1, the mol ratio of residues of the polyamine to residues of the diepoxide in the resinous adduct being from 1.25:1-2:1, and the polyamine having from 2 to 5 amine groups per molecule and at least sufficient amino hydrogens to react with substantially all of the epoxy groups of both the diepoxide and the monoepoxide; and

(c) from 0.05 to 0.4 weight parts, per weight part of the resinous adduct, of an aminoplast resin having at least about 1 gram-equivalent per 1,000 g of the aminoplast resin, of carboxylic acid groups with pK_a from 2 to 6 for catalyzing the heat-induced cure of the resinous adduct with the aminoplast resin.

The process further comprises a method for forming a cured resinous coating on a conductive substrate.

Use of Aqueous Resin Solutions as Binders and Impregnating Agents

Highlights of the Technological Achievement: An aqueous solution of a glycidyl compound containing a N-heterocyclic group and of an amine hardener is used for the modification, sealing or binding of substrates.

Reference: W. Huber-Nuesch; U.S. Patent 4,143,188; March 6, 1979; assigned to Ciba-Geigy Corporation.

Background: There has been no suggestion in prior art of the use of specific amines with specific epoxides, dissolved both together in water, and applied in this form for hardening. It has been found that an aqueous solution of water-soluble N-heterocyclic epoxide compounds and specific water-soluble amines, when used as a binder or impregnating agent, is able to impart to the substrates sealed or modified therewith excellent mechanical, chemical and in particular, water-resistant properties. It is unexpected that water-resistant addition products are obtained from two water-soluble constituents in water.

Process Conditions: The process relates to the use of an aqueous solution of epoxide resins, which are completely water-soluble at room temperature and which have at least one N-heterocyclic ring in the molecule, and amines containing at least three hydrogen atoms bound to nitrogen, the solubility of which epoxide resins in water is at least 70 wt % as binder and impregnating agent for the modifying, sealing or binding of substrates, especially as binder for loose aggregates and as sealing agent for porous materials, which aqueous solution is characterized in that as amines there are used those which satisfy the condition of Formula 1: $(a \cdot w/c) < 14$, wherein a represents the H-active amine equivalent, w represents the sum of the amino groups, the hydroxyl groups and the ether groups, with two ether groups counting as one group, and c represents the number of carbon atoms in the molecule, or the water-soluble salts thereof.

When the resin and hardener are dissolved in water, the clear solution clouds up within a few minutes to two hours, depending on concentration and temperature. The milky cloudiness (emulsion) then precipitates as a liquid phase from the aqueous phase and, unexpectedly, hardens or sets, in the presence of water, to form a compact, water-resistant, infusible substance.

The dissolving of the two reactants in water is performed as a rule at room temperature. The hardening in the aqueous solution is effected as a rule at room temperature, but it can be performed at temperatures from below 0° to 40°C, or at still higher temperatures. The provision of heat is not necessary. For example, a 50% solution of hydantoin resin can be hardened at -5°C, although the solution is frozen. After the water has melted, the major part of the reaction product is in the form not of a compact solid but of a hard network or hard sponge-like structure.

Applicable liquid epoxide resins are, e.g., the following: N-glycidyl compounds of hydantoin and also glycidyl ethers of addition products of alkylene oxides with hydantoins. Water-soluble hydantoin epoxides of this kind are described, for example, in the Swiss Patents 471811 and 523273, in the French Patents 2,163,160 and 2,080,885, as well as mixtures thereof in the Austrian Patent 298807.

The following water-soluble resins are preferably used: (a) N,N'-diglycidyl-5,5-dimethyl-hydantoin; (b) N,N'-diglycidyl-5-methyl-5-ethyl-hydantoin; (c) 1-glycidyl-3-(2-glycidyloxy-n-propyl)-5,5-dimethyl-hydantoin; (d) a mixture of about 70% of polyepoxide (a) and 30% of polyepoxide (c); and (e) 1,3-bis-(1-glycidyl-5,5-dimethyl-hydantoinyl-3)-propanol-2-glycidyl ether.

Applicable solid epoxide resin which is soluble at room temperature in water is triglycidyl ether of cyanuric acid.

Polyamines which fulfill the conditions of the Formula 1 are, for example, the following.

	Value of the Formula 1
1-Methyl-4-(1,1-dimethyl-aminomethyl)-cyclohexylamine	8.5
2-Aminomethyl-cyclopentylamine	9.5
2-Oxo-1,3-hexahydropyrimidine-di-neo-pentylamine	9.6
3,5,5-Trimethyl-3-(aminomethyl)-cyclo-hexylamine	8.5
3-Amino-1-1cyclohexylaminopropane	11.6

Particularly preferred are those polyamines which satisfy the Formula 2: $(a \cdot w)/c = 7$ to 12. There are preferably used the salts of such amines, especially those salts which are formed from 1 mol of amine and 0.5 to 1 mol of monobasic acid.

Air Drying Thermosetting Aqueous Epoxy-Acrylic Copolymer Coating Systems

Highlights of the Technological Achievement: Thermosetting aqueous epoxy-acrylic copolymer coating systems adapted to cure on air drying are provided using a first package constituted by acid-neutralized organic solvent soluble tertiary amine copolymer comprising at least 40% of nonreactive monoethylenic monomer, from 5 to 35% of monoethylenic hydroxy-functional monomer, and from 5 to 35% of monoethylenic tertiary amine monomer, the copolymer being in solution in a water-miscible organic solvent and a second component comprising a resinous polyepoxide in solution in at least partially water miscible organic solvent.

Reference: K. Sekmakas, R. Shah and J.F. DeRossi; U.S. Patent 4,133,790; January 9, 1979; assigned to DeSoto, Inc.

Background: In the aqueous coating compositions of U.S. Patent 3,719,629, the epoxy resin component is combined in an aqueous medium with an aminoethylated vinyl polymer having pendant aminoethyl groups, these groups having the formula: $-COOCH_2CH_2NH_2$, and the cure is between the oxirane (epoxy) group in the epoxy resin and the amino hydrogen atoms in the aminoethyl group stated.

In this process, the amino hydrogen atoms which are essential in the prior art are not present. Moreover, the production of the pendant aminoethyl groups in the prior art required the use of expensive and hazardous ethylene imine, and this is avoided in this process.

From the broader standpoint, and in addition to the aqueous epoxy system noted previously, other resin systems have been employed to provide air drying porch and floor paints. Thus, the systems of this process are intended to replace two existing aqueous systems, one based on an alkyd resin and another based on an aqueous copolymer latex, and also to replace an organic solvent-based epoxy system. The process exhibits significant superiority to all three as will be documented hereinafter.

Process Conditions: Referring more particularly to the tertiary amine copolymers, there are organic solvent-soluble copolymers comprising three copolymerized monoethylenically unsaturated components, namely: (1) nonreactive monomer, (2) hydroxy-functional monomer, and (3) tertiary amine monomer.

The solubilizing acid is used in an amount to provide a colloidal dispersion having a preferred pH in the range of 6.0 to 7.0, though a pH in the range of 3 to 8 is broadly useful.

The epoxy resins which are used in this process can be broadly referred to as resinous polyepoxides, and those which possess a 1,2-epoxy equivalency in the range of 1.4 to 2.0 are preferred. Polyepoxides which are diglycidyl ethers of bisphenols are particularly preferred, especially those having an average molecular weight in the range of 350 to 4,000.

The polyepoxide is desirably used in an amount providing approximate stoichiometry between epoxy groups in the polyepoxide and tertiary nitrogen atoms in the amine copolymer, ±50%, more preferably ±20%.

The advantages of this process will be apparent from the tabulated data presented below in which X indicates a positive feature, 0 indicates a negative feature, and 1 indicates neither.

		. . . . Competitive Materials in Commerce			
Property	This Process	Latex	Alkyd	Two-Component Epoxy	U.S. Patent 3,719,629 (Dow)
Water based	X	X	–	–	X
Solvent based	–	–	0	0	–
Price	X	X	X	0	0
One package	–	X	X	–	–
Two package	0	–	–	0	0
Ease of cleanup	X	X	0	0	X
Use on previous paints	X	X	X	0	X
Flammable	X	X	0	0	X
Use on interior	X	X	0	0	X
Hazardous	–	–	1	0	–
Use on porches	X	X	X	0	?
One coat hide	X	0	0	X	X
Alkali resistance	X	X	0	X	X
Abrasion resistance	X	1	1	X	X
Chemical resistance	X	1	0	X	X
Hardness	X	0	0	1	1
Good adhesion	X	X	X	X	X
Tire track adhesion	X	0	1	X	X
Tint retention	1	X	1	0	0
Chalking resistance	1	X	1	0	0
Cracking resistance	1	1	1	?	?

Aqueous Coating Agent for Stoving

Highlights of the Technological Achievement: The subject of the process is an aqueous coating agent for stoving, based on aqueous dispersions of epoxide resin mixtures, epoxide resin curing agents and optionally pigments and other customary additives.

Reference: W.-D. Görlitz and B. Broecker; U.S. Patent 4,132,692; January 2, 1979; assigned to Hoechst AG, Germany.

Background: In U.S. Patent 3,336,253 it is stated that the stability in storage of the epoxide/alkanolamine reaction products can be increased by removing all unreacted epoxide groups, using various compounds, such as, for example, dialkanolamines. This procedure is undesirable since, in the course thereof, all the reactive sites for subsequent crosslinking, which are necessary for the manufacture of a durable, solvent-free coating, are removed.

In addition, these products can only be manufactured if very large quantities of dialkanolamines are employed (for example, up to 28% of diethanolamine is used in U.S. Patent 3,336,253). The resulting coatings are so unstable towards aqueous media that they are completely unserviceable for many purposes. The products manufactured according to DT-OS 2,415,100 also contain stoichiometric quantities of alkanolamines, relative to epoxide resin. Although the quantities of amine are reduced to approximately 5% in DT-OS 2,426,996, the dispersions still contain considerable quantities of solvent.

The process mentioned is also not suitable for manufacturing solvent-free dispersions. Epoxide resin derivatives which are used in this process are, however, not described in the literature.

Example: 300 g of an epoxide resin having a softening point of 118°C by the Durrans method and an epoxide equivalent weight of 1,865 (EP 307, Hoechst AG) were melted at a temperature rising to approximately 130°C, and 75 g of an epoxide resin of improved elasticity and which had an epoxide equivalent weight of 470 (EP 151, Hoechst AG) were added. 60 g of toluene and 90 g of n-butanol were added to the melt at 130°C in the course of 1 hr, and the temperature fell gradually to 80°C. 18 g (0.046 mol/100 g of EP resin) of diethanolamine were then added in the course of 15 min.

The mixture was left to react further for 1 hr at 80°C. 180 g of phenolic resol (prepared from: 2,2-bis[4-hydroxyphenyl]-propane resol, etherified with n-butanol, charge ratio 1:4, condensation reaction carried out for 4 days at 25°C; catalyst: NaOH; neutralized with HCl; acidified with H_3PO_4; degree of etherification: 35 parts, viscosity at 25°C of a 60% strength solution in n-butanol equals 215), dissolved to form a 70% strength solution in n-butanol, were then added in the course of 15 min.

21.7 g of 90% strength lactic acid were now added at 70°C in the course of 1 hr. A solution is obtained, the epoxide resin component of which contains the unreacted epoxide resins and the reaction products of the epoxide resins used with diethanolamine and lactic acid, the reaction products being the product of the process.

The mixture was cooled to 40°C and 750 g of distilled water were stirred in in the course of 1 hr. Finally, the solvent mixture was removed by azeotropic distillation with water in vacuo at 40°C. The resulting dispersion was filtered through a 56 μ gauze. The characteristic values are: solids content 45.1%, viscosity at 25°C, 48 mPas (Brookfield, Spindel 2/60 U), and a pH value of 4.3.

Films approximately 30 μm thick were applied by means of a Handcoater to electrolytic sheet metal and were exposed to the air for 10 min. Then they were stoved for 10 min at 200°C. A glossy, transparent film of good elasticity and good resistance to sterilization, even in the presence of organic acids, was obtained.

ETHYLENE OXIDE POLYMERS AND RELATED PRODUCTS

PROCESSING

Improved Method for Ethylene Oxide Manufacture Using a Silver Catalyst

Highlights of the Technological Achievement: An improved method is disclosed for the vapor phase epoxidation of ethylene to ethylene oxide which includes intimately contacting an ethylene-containing feed stream with an epoxidizing amount of molecular oxygen epoxidizing agent in the presence of a catalytically effective amount of a supported, activated silver catalyst.

Reference: S.B. Cavitt; U.S. Patent 4,206,128; June 3, 1980; assigned to Texaco Development Corporation.

Background: The preparation of ethylene oxide by the oxidation of ethylene in the presence of suitable catalysts is well known. These known processes can be generally separated into two groups; the first utilizes air and the second utilizes molecular oxygen, e.g., from about 85 mol % to about 99 mol %. Silver catalysts are utilized with both groups.

It has been discovered that an extremely stable, physically durable, supported silver catalyst used in accordance with commercially known processes such as described in U.S. Patent 3,119,837, British Patent 1,314,613 and British Patent 1,132,095, results in a process which gives higher unit productivity and selectivity to the desired product at lower operating temperatures. The selectivity is particularly important in air processes which are not closed systems because some proportion of the unreacted ethylene is lost by venting excess gas.

Process Conditions: Ethylene is epoxidized to ethylene oxide substantially in the vapor phase by intimately contacting ethylene with an epoxidizing amount of molecular oxygen in the presence of a catalytically effective amount of an activated silver containing catalyst at temperatures of 200° to 300°C. Preferably, the ethylene is contacted with the molecular oxygen in the presence of an effective amount of at least one inhibitor which retards combustion of ethylene to carbon dioxide.

The activated silver-containing catalyst is prepared by impregnating certain porous, inorganic substrates with a silver carboxylate/amine complex impregnating solution and heating the impregnated substrate at temperatures of 50° to 300°C to evaporate volatiles, decompose the complex, and activate the catalyst.

The impregnating solution is formed by dissolving a silver carboxylate in a solubilizing amount of an amine-containing complexing agent selected from:

(a) Alicyclic diamines wherein at least one amino moiety is primary or secondary, but no more than one is primary;

(b) Polyamines containing at least three amino moieties wherein at least one is primary or secondary; or

(c) Amino ethers containing at least one ether linkage and at least one amino moiety which is primary or secondary.

Ethylene Oxide Production Using a Special Catalytic Structure

Highlights of the Technological Achievement: A method of producing a substrate having a particulate coating of high surface area to mass ratio, the method comprising in sequence the steps of:

(1) Disposing between a substrate and an intermediate body a mixture comprising (a) hard particles which are harder than the substrate and are harder than the intermediate body; and (b) soft particles which are softer than the hard particles and are softer than the substrate;

(2) Compressing the substrate and intermediate body with particles therebetween whereby the intermediate body pushes the hard particles into the substrate; and

(3) Removing the intermediate body from the particles leaving the hard particles embedded in the substrate and the soft particles adhering to the substrate.

The coated substrates produced by the process find utility as catalytic devices to accelerate or retard chemical reactions.

Reference: P. della Porta, A. Cantaluppi, B. Ferrario and P. Montalenti; U.S. Patent 4,145,355; March 20, 1979; assigned to SAES Getters SpA, Italy.

Background: Processes have been described for producing substrates having hard metallic particles embedded therein. (See U.S. Patent 3,856,709 of Dec. 24, 1974 and Italian Patent 746,551 issued Feb. 1, 1967.) However, previously it had been thought that such processes were limited to those wherein the particles were harder than the substrate. Although there are many advantages to such processes such as (1) the absence of a binder, (2) the ability to practice the process at ambient temperatures, (3) the use of rolls while avoiding wear of the rolls and (4) the production of a structure having a higher surface area to mass ratio, these processes require that the particles be harder than the substrate. See U.S. Patent 3,652,317.

It is therefore an object of the process to provide a method for producing a substrate having thereon a coating of particles which are softer than the substrate.

Drawings: Figure 10.1a is a schematic representation of an apparatus suitable for practicing the method of the process; and Figure 10.1b is a cross-sectional view with an enlargement of approximately 300 diameters of the structure produced by the process. Figure 10.1b is taken along line 2–2 of Figure 10.1a.

Figure 10.1: Ethylene Oxide Production Using a Special Catalytic Structure

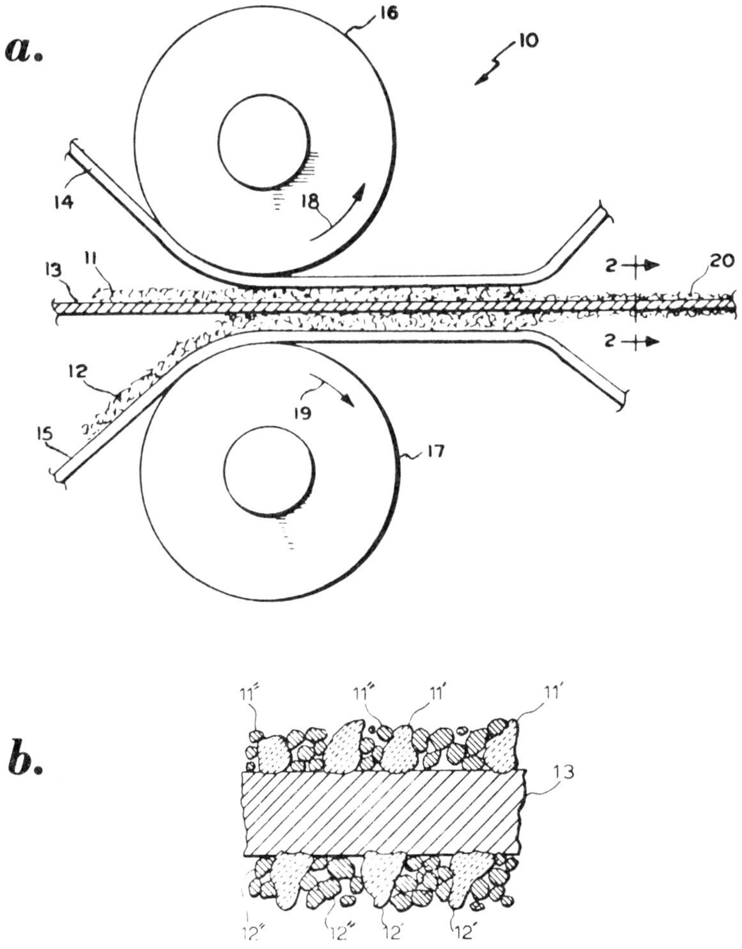

(a) Apparatus
(b) Catalytic Structure

Source: U.S. Patent 4,145,355

Process Conditions: Broad and preferred ranges of Vickers hardness for the intermediate body, the particles and the substrate are given in the following table:

Component	Vickers Hardness Broad Range (kg/mm^2)	Vickers Hardness Preferred Range (kg/mm^2)
Intermediate body	10-600	100-300
Hard particles	100-∞	200-800
Soft particles	0.1-200	1-100
Substrate	1-400	10-200

The soft particles can be those of any metal or material. In that preferred embodiment wherein the resultant structures are catalytic structures, the soft particles are those of a catalytic metal. Of course the catalytic metal is chosen with respect to the chemical reaction to be catalyzed depending upon chemical process considerations not part of this process. However, representative examples of suitable catalytic metals include among others, silver, gold, platinum, mixtures thereof and alloys thereof with one another and with other metals.

Polyisocyanate Crosslinked Alkylene Oxide-Linear Nonionic Polysaccharide Reaction Product

Highlights of the Technological Achievement: A method for the recovery of oil from an oil-containing subterranean formation by waterflooding employing as an injection medium an aqueous solution of a viscosifier comprising the water-soluble reaction product of an organic polyisocyanate and the addition product of an alkylene oxide and linear, nonionic polysaccharide, the addition product having a molecular weight of at least 100,000.

Reference: S. Stournas; U.S. Patent 4,174,309; November 13, 1979; assigned to Mobil Oil Corp.

Background: In the recovery of oil from oil-bearing reservoirs, it usually is possible to recover only a minor portion of the original oil in place by the so-called primary recovery methods which utilize only the natural forces present in the reservoir. Thus, a variety of supplemental recovery techniques have been employed in order to increase the recovery of oil from subterranean reservoirs. In the supplemental techniques, which are commonly referred to as secondary recovery operations although in fact they may be primary or tertiary in sequence of employment, fluid is introduced into the reservoir in order to displace the oil therein to a suitable production system through which the oil may be withdrawn to the surface of the earth.

One difficulty which often is encountered in secondary recovery operations is the relatively poor sweep efficiency of the injected displacing liquid. That is, the displacing liquid exhibits a tendency to channel through certain portions of the reservoir and to bypass other portions.

Even when the reservoir exhibits a relatively uniform permeability throughout, a situation referred to as instability fingering may develop in those instances where the viscosity of the injected displacing fluid is significantly less than the viscosity of the in situ reservoir oil.

Various techniques have been proposed in order to improve the sweep efficiency of the injected displacing fluid and thus avoid premature breakthrough. For example, it has been proposed in waterflooding operations to add thickening agents to at least a portion of the aqueous flooding medium in order to increase the viscosity. The viscosity of the flooding medium may be increased prior to its injection into the reservoir or alternatively the viscosity may be increased in situ in order to avoid a reduction in injectivity at the injection well. For example, in U.S. Patent 3,208,518 there is disclosed a waterflooding process wherein the viscosity of the flooding medium is increased in situ through the use of high molecular weight polymers, such as ionic polysaccharides produced by the fermentation of carbohydrates by bacteria of the genus *Xanthomonas*, under controlled pH conditions.

Process Conditions: In some cases, a total concentration of as small as 0.005% by weight will be satisfactory. Generally speaking, a total concentration ranging from 0.01 to 1% is employed.

In accordance with the process, the organic polyisocyanate is reacted with the addition product of an alkylene oxide and a linear, nonionic polysaccharide, the addition product having a molecular weight of at least 100,000 and preferably of at least 300,000. The alkylene oxides useful in the production of such addition products will generally include compounds having the formula

(1)

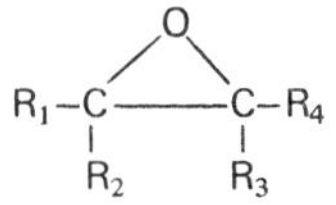

wherein R_1, R_2, R_3 and R_4 are the same or different and are selected from the class consisting of hydrogen and lower alkyl groups containing from 1 to 5 carbon atoms, such as methyl, ethyl, propyl, butyl and pentyl groups. Preferably, at least 1 of R_1, R_2, R_3 and R_4 is hydrogen and the total number of carbons comprising R_1, R_2, R_3 and R_4 attached to the oxirane ring does not exceed 5. Examples of suitable alkylene oxides useful in the practice of the process are ethylene oxide, propylene oxide, butylene oxide, pentylene oxide, 1-methyl-2-pentene oxide, trimethylene oxide and tetramethylene oxide.

The polysaccharides useful in the practice of the process are the linear, nonionic polysaccharides. Suitable polysaccharides include cellulose and starches, such as, for example, amylose. Generally speaking, the polysaccharide used will be formed of repeating structural units of the formula

(2)

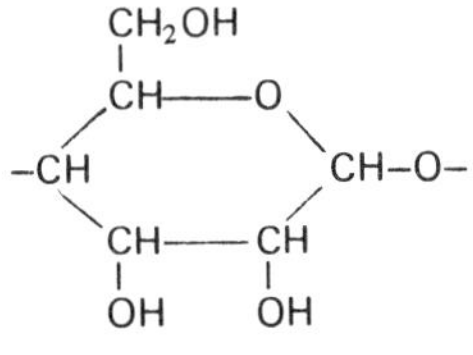

Generally, an amount of polyisocyanate is employed such that there is a minimum of an average of 1 crosslink derived from the polyisocyanate for every 2,000 repeating structural units of the type described in Formula (2) above, up to a maximum of an average of 1 crosslink for every 20 of such repeating structural units.

Stated otherwise, the molar amount of polyisocyanate that should be used to crosslink the alkylene oxide-polysaccharide addition product will range from a value of (0.005n/m) to a value of (0.5n/m) where n is the number of repeating units of Formula (2) above in the polysaccharide, and m is the average number of isocyanato groups per polyisocyanate molecule.

Examples of suitable polysaccharide ethers which are the product of the reaction of a linear nonionic polysaccharide and an alkylene oxide are hydroxyalkylcelluloses, such as hydroxyethylcellulose and hydroxypropylcellulose; alkylhydroxyalkylcelluloses, such as methylhydroxypropylcellulose, and mixed hydroxyalkylcellulose ethers such as hydroxyethyl, hydroxypropylcellulose ethers.

Solutions of the crosslinked compositions were prepared using both distilled water and brine as the solvents. The brine solution contained 10% by weight of solids. The composition of the brine solids, calculated as chlorides, was: 25,304 ppm Na^+, 9,272 ppm Ca^{++}, and 2,552 ppm Mg^{++}.

The viscosity measurements were made on the various solutions, including solutions of the untreated hydroxyethylcellulose to afford a comparison and demonstrate the improvement in viscosity that was realized using the composition of the process. The viscosity measurements were obtained at different shear rates with the cone and plate Brookfield viscometer, Model LTV, at temperatures of 25° and 65°C. The solutions in each case included 0.4% by weight of either the viscosifier of the process or of the untreated hydroxyethylcellulose. The results obtained are set forth in the table below. The term "HEC" signifies the uncrosslinked hydroxyethylcellulose. The term "X" signifies the polyisocyanate crosslinked hydroxyethylcellulose prepared in accordance with the process.

Viscosities of Cellulose Derivatives (0.4% Solutions)

Solvent	Shear Rate (cm^{-1})	Temperature (°C)	Viscosity (cp) HEC	Viscosity (cp) X
Brine	46	25	2	10.0
Brine	115	25	2	8.5
Brine	230	25	2	7.6
Distilled water	45	25	2	11.2
Distilled water	115	25	2	10.3
Brine	46	65	1	2.4
Brine	115	65	1	2.1
Brine	230	65	1	1.9

As is apparent from the data reported in the table above, the solutions of the composition (X) were characterized by significantly improved values of viscosity in comparison to the viscosities attained by the use of hydroxyethylcellulose alone.

APPLICATIONS

Surfactant for an Automatic Dishwasher

Highlights of the Technological Achievement: A surfactant which is the condensation product of a fatty alcohol containing from 17 to 19 carbon atoms with from 6 to 15 mols of ethylene oxide, the distribution of the ethylene oxide condensation products being such as to give more than about 50% by weight within ±2 ethylene oxide moieties of the average, preferably, no single condensation product (a single specific number of ethylene oxide moieties) constituting more than 50% by weight of the product; automatic dishwasher compositions containing the surfactant.

Reference: L.A. Halas; U.S. Patent 4,188,305; February 12, 1980; assigned to The Procter & Gamble Company.

Background: This process relates to certain specific surfactant compounds which are particularly suitable for use in automatic dishwashers. These surfactants are extremely low foaming thus requiring a minimal amount of suds suppressor. Similar products prepared by condensing 1 mol of octodecanol and 9 or 10 mols of ethylene oxide have been disclosed in U.S. Patent 3,630,923, (Dec. 28, 1971) and in U.S. Patent 3,544,473, (Dec. 1, 1970).

In this process it has been found that a very specific tight distribution of ethylene oxide homologs gives preferred performance in that less suds suppressor is required to give a desirable low sudsing automatic dishwasher composition.

Process Conditions: This process is based upon the discovery that a nonionic surfactant derived by condensing very specific fatty alcohols, preferably straight chain, with a specific amount of ethylene oxide to obtain, either directly or indirectly after further treatment, a condensation product having a narrow distribution of homologs having different ethylene oxide contents is a superior surfactant for use in automatic dishwashing detergents.

More specifically, this process relates to a low foaming nonionic surfactant prepared by condensing a C_{17}-C_{19} alcohol, preferably straight chain, preferably saturated, preferably having an essentially pure (>90%, preferably >95%) chain length, more preferably C_{18}, with from 6 to 15 mols of ethylene oxide per mol of fatty alcohol to give, either directly or indirectly after further treatment, a distribution of ethylene oxide homologs in which >50% of the product by weight is within ±2 ethylene oxide moieties from the average (preferably, no more than 50% of the surfactant comprises a single ethylene oxide homolog). A more preferred product contains <2% unethoxylated alcohol and >80% by weight of the product should contain <13% ethylene oxide moieties.

The process also relates to automatic dishwashing detergent compositions consisting essentially of: from 10% to 90%, preferably from 20% to 70% of a detergency builder; available chlorine at a level of from 0% to 3%, preferably from 1% to 2.5%; and from 2.5% to 30%, preferably above 5%, most preferably below 10% of the abovedescribed low foaming, biodegradable nonionic surfactant.

Spray-Dried Detergent Compositions

Highlights of the Technological Achievement: Detergent compositions containing organic surface active agents, water-insoluble aluminosilicate ion exchange materials, sodium silicate having an SiO_2:alkali metal oxide weight ratio of 1.4:1 to 1.9:1, and a hydratable salt of a water-soluble weak organic acid. The composition is produced by drying an aqueous slurry. The resulting composition is an effective laundry detergent in the form of crisp, free-flowing granules which are highly soluble when made and which maintain their solubility on aging.

Reference: R.A. Llenado; U.S. Patent 4,180,485; December 25, 1979; assigned to The Procter & Gamble Company.

Background: This process relates to granular detergent compositions which are capable of providing superior performance during conventional textile laundering and cleaning operations. The compositions contain as essential components a water-insoluble aluminosilicate ion exchange material, an organic surface active agent, an alkali metal oxide silicate solid with a SiO_2:alkali metal oxide weight ratio of from about 1.4:1 to 1.9:1 and a hydratable water-soluble salt of an organic carboxylic acid having at least one ionization constant of $<1.0 \times 10^{-3}$.

Process Conditions: The process is based on the discovery that cleaning compositions can contain aluminosilicate detergency builders and also relatively high levels of alkali metal oxide silicate solids, if the SiO_2:alkali metal oxide weight ratio is from 1.4:1 to 1.9:1 and the composition is prepared by drying a slurry containing, on a finished product basis, from 0.5% to 20% of a hydratable water-soluble salt of an organic acid characterized by having at least one ionization constant of $<1.0 \times 10^{-3}$. In particular, the compositions comprise a spray-dried detergent composition in the form of crisp, free-flowing granules comprising:

(a) From about 5% to about 40% of a water-soluble organic surface active agent selected from the group consisting of anionic, nonionic, zwitterionic, and ampholytic detergents;

(b) From about 10% to about 60% of a finely divided aluminosilicate ion exchange material selected from the group consisting of:

  (1) Crystalline aluminosilicate material of the formula: $Na_z[(AlO_2)_z \cdot (SiO_2)_y] \cdot xH_2O$, wherein z and y are at least 6, the molar ratio of z to y is from 10 to 264 and

  (2) Amorphous hydrated aluminosilicate material of the empirical formula: $M_z(zAlO_2 \cdot ySiO_2)$, wherein M is sodium, potassium, ammonium, or substituted ammonium, z is from about 0.5 to about 2, y is 1 and the material has a magnesium ion exchange capacity of at least about 50 mg equivalents of $CaCO_3$ hardness per gram of anhydrous aluminosilicate and

  (3) Mixtures thereof;

(c) From about 5% to about 30% of an alkali metal oxide silicate solid having a weight ratio of SiO_2:alkali metal oxide of from about 1.4:1 to about 1.9:1, and

(d) From about 0.5% to about 20% of a hydratable water-soluble salt of an organic carboxylic acid wherein the acid has at least one ionization constant below about 1×10^{-3}.

Other anionic detergent compounds herein include the sodium alkyl glyceryl ether sulfonates, especially those ethers of higher alcohols derived from tallow and coconut oil; sodium coconut oil fatty acid monoglyceride sulfonates and sulfates; and sodium or potassium salts of alkyl phenol ethylene oxide ether sulfate containing 1 to 10 units of ethylene oxide per molecule and wherein the alkyl groups contain about 8 to about 12 carbon atoms.

Water-soluble nonionic synthetic detergents are also useful as the detergent component of the instant composition. Such nonionic detergent materials can be broadly defined as compounds produced by the condensation of alkylene oxide groups (hydrophilic in nature) with an organic hydrophobic compound, which may be aliphatic or alkyl aromatic in nature. The length of the polyoxyalkylene group which is condensed with any particular hydrophobic group can be readily adjusted to yield a water-soluble compound having the desired degree of balance between hydrophilic and hydrophobic elements.

Free-Flowing Granular Detergent Compositions Containing Olefin Sulfonates

Highlights of the Technological Achievement: A granular detergent composition comprising α-olefin sulfonate and polyoxyethylene-polyoxypropylene alkyl ether at the weight ratio in the range of from 4:1 to 1:1 and further comprising an inorganic builder possesses satisfactory powder properties and is sufficiently easy to rinse out after being used for washing.

Reference: M. Nakamura, M. Yazaki and T. Ogoshi; U.S. Patent 4,186,114; January 29, 1980; assigned to The Lion Fat & Oil Co., Ltd, Japan.

Background: It is very important to provide satisfactory easiness in rinsing of heavy duty detergents because it shortens the washing cycle and also renders it possible to economize on electric power as well as rinsing water. As the conventional means for improving the easiness in rinsing with respect to granular heavy duty detergents, there are known the art of adding soap, the art of adding silicone and/or wax, and so forth. However, the art of adding soap is not an art which promises sufficient improvement of the easiness in rinsing, and the art of adding silicone and/or wax is defective in that the silicone and/or wax adheres to the washed clothes because of their water insolubility.

Process Conditions: In an attempt to improve the powder properties of granular detergents comprising nonionic surface active agent as active ingredient, without resorting to any special means such as hydration and/or employment of an anticaking agent, it has been found that the use of a mixture obtained by mixing a specific polyoxyethylene-polyoxypropylene alkyl ether with an α-olefin sulfonate at a specific ratio as active ingredient provides a granular detergent which is superior in powder properties and is satisfactory with respect to easiness in rinsing.

To be precise, a granular detergent composition according to the process comprises (a) an α-olefin sulfonate which has 12 to 18 carbon atoms, (b) a polyoxyethylene-polyoxypropylene alkyl ether which has alkyl radical having 9 to 15 carbon atoms and contains 6 to 22 ethylene oxide units and 2 to 8 propylene oxide units added thereto and (c) an inorganic builder, and the weight ratio of (a) to (b) therein is in the range of from 4:1 to 1:1.

As for the polyoxyethylene-polyoxypropylene alkyl ether used as the ingredient (b), it can be prepared by the known process of adding ethylene oxide units and propylene oxide units to straight chain or branched chain aliphatic alcohols having 9 to 15 carbon atoms by employing a basic or acid catalyst. As to the way of adding alkylene oxides, it will do either to apply the process of first effecting block addition of one of the alkylene oxides and next effecting block addition of the other thereof, or to apply the process of adding ethylene oxide and propylene oxide simultaneously to the reaction system, thereby effecting random addition.

Examples 1 through 10: A variety of granular detergents having the following compositions were prepared.

Component	Percent (%)
AOS*	15
Nonionic surface active agent**	5
Sodium silicate***	15
Sodium pyrophosphate	13
Water	8
Carboxymethylcellulose	0.6
Fluorescent whitening agent	0.4
Perfume	0.2
Sodium sulfate	balance

*C_{14-18} α-olefin sulfonate (85% C_{16-18}; 5% disulfonate)
**Polyoxyethylene-polyoxypropylene alkyl ether (C_{9-11} alkyl)
***SiO_2/Na_2O = 3.1

The thus obtained granular detergent compositions were subjected to evaluation of their powder properties as well as ease in rinsing. The evaluation of ease in rinsing was conducted by using a detergent concentration of 0.133%. The grading system is: A–no foam; B–trace of foam; C–substantial foam. The results are shown in the table below.

	Example No.									
	1	2	3	4	5	6	7	8	9	10
Average number of propylene oxide units	0	0	1	4	10	2	2	2	4	4
Average number of ethylene oxide units	8	15	15	15	15	5	6	10	20	24
Powder properties										
Angle of repose, °	80	50	50	40	55	60	50	45	40	40
Compression-caking, kg	≥5	3.0	3.0	2.0	4.5	2.0	2.0	2.0	2.0	2.0
Hydrocaking, %	90	50	50	30	50	30	35	30	30	30
Ease in rinsing	C	C	B	A	A	A	A	A	A	B

As is evident from the showing in the table above, in the case of a granular detergent composition in which the average number of propylene oxide units added and that of ethylene oxide units added are in the range specified in the process (Examples 4, 7 through 9), both the powder properties and the ease in rinsing thereof are satisfactory, while in the case of a granular detergent composition in which the average number of propylene oxide units added is 1 or less (Examples 1 through 3), it is inferior in ease in rinsing, and in the case of a granular detergent composition in which the average number of propylene oxide units added is 9 or more (Example 5), it is inferior in powder properties, particularly with respect to the compression-caking property.

Further, in the case of a granular detergent composition in which the average number of ethylene oxide units added is 5 or less (Example 6), it is particularly inferior with respect to the hygrocaking property among powder properties, and a granular detergent composition in which the average number of ethylene oxide units added is 23 or more (Example 10) is undesirable because it leaves fine foams at the time of rinsing.

Incorporating Polyethylene Oxide into Soap

Highlights of the Technological Achievement: High molecular weight polyethylene oxide is incorporated into soap by first forming a blend of polyethylene oxide and molten higher fatty acids, preferably containing water.

Reference: D.P. Joshi; U.S. Patent 4,169,066; September 25, 1979; assigned to Colgate-Palmolive Company.

Background: This process relates to the making of soap bars containing high molecular weight polyethylene oxide. Soap bars containing this polymer have been made commercially by adding powdered polyethylene oxide to soap chips in a soap amalgamator; such addition requires special precautions in order to avoid formation of specks in the final bar. Incorporating 2% powdered polyethylene oxide into a neat soap containing 30% water (kettle soap) resulted in a mixture which was lumpy and poorly dispersed and not pumpable.

Process Conditions: A superior blend is obtained by incorporating the polyethylene oxide, desirably in powder form, into a melt of all, or a portion of, the higher fatty acids which are to be used for superfatting the soap. Best results are obtained when this blend also contains a minor amount of water. As will be seen in the Example below, the presence of the water has a marked effect on the nature of the blend of fatty acid and high molecular weight polyethylene oxide, significantly decreasing its viscosity; the presence of the water also tends to opacify the blend which may indicate that a water-in-oil type of dispersion is formed.

The water-containing blend can be pumped readily; its viscosity is well below the upper viscosity limit (which is 7,000 or 8,000 cp) of materials that can be handled readily by ordinary reciprocating pumps. It can thus be easily pumped into admixture with the kettle soap. The polyethylene oxide-containing soap bars made by this procedure have outstanding properties; they lather well, have an especially pleasant feel during and after use, have good slough resistance and resistance to wet-cracking.

All proportions are by weight unless otherwise specified in the following example.

Example: 5 parts of commercial stearic acid (MP 54°C) and 5 parts of coco fatty acids (MP 25°C; iodine value 6) are melted together at a temperature of about 102°C. 1 part of powdered high molecular weight polyethylene oxide (Polyox WSR N-750, MP about 65°C) is added thereto with stirring. The viscosity of the mixture at 102°C is about 750 cp, at 88°C it is about 4,000 cp and at 54°C it is about 5,000 cp (as determined with a Brookfield viscosimeter).

1 more part of the same Polyox is then added with stirring. The viscosity of the mixture is about 18,000 cp at 94°C, about 19,000 cp at 88°C, about 33,500 cp at 66°C and about 42,500 cp at 54°C.

Then 0.5 part of lanolin is added with stirring. The resulting blend now has a viscosity of about 15,000 cp at 94°C and a similar viscosity at 70°C. Next 1 part of water is added. As a result, the formerly clear blend becomes cloudy or opaque. Its viscosity is about 5,000 cp at 77°C, about 4,000 cp at 71°C and about 3,250 cp at 50° to 63°C.

Thereafter an additional 1 part of water is added with stirring. The resulting blend has a viscosity of about 2,250 cp at 84°C, 1,500 cp at 81°C, 930 cp at 71°C, 470 cp at 67°C, 234 cp at 60°C and 114 cp at 52°C. Next 2.5 parts of a solution of sodium caseinate (containing 20% casein) is added. The resulting blend is white. Its viscosity is about 2,900 cp at 81°C, 2,520 cp at 71°C, 1,700 cp at 66°C and 1,800 cp at 60°C.

It will be seen that the addition of the water gives a blend whose viscosity at, say, 70°C is greatly decreased, (e.g., to well below half its value before the addition of the water). Also the blends containing water show a remarkable decrease in viscosity with decrease in temperature, (e.g., when temperature is reduced, from 70° or 80°C, by 10°C the viscosity drops by well over 10%).

The high molecular weight polyethylene oxide has an average molecular weight of at least 100,000. Examples of such compounds are those sold as Polyox. These polymers are nonionic materials, soluble in water and their molecular weights range from 100,000 to 5,000,000 or more. It is preferred to employ polymers having average molecular weights not above 600,000 such as 300,000 to 400,000.

For the material having an average molecular weight of about 300,000 a proportion in the neighborhood of 2% has given excellent results. This 300,000 molecular weight material (Polyox WSR N-750) has a viscosity at 25°C, for a 2% aqueous solution, of about 40 cp (Brookfield Spindle No. 1 at 10 rpm); for a 5% solution this viscosity is about 600 to 1,000 cp. Use of 2% of extremely high molecular weight polyethylene oxide, e.g., of 4,000,000 average molecular weight, in a bar causes the lather to be pituitous, which is less desirable. Soap typically has a pH in 1% aqueous solution of about 10, (e.g., 10.2), while the superfatted soaps generally have pHs somewhat below 10.

Shampoo

Highlights of the Technological Achievement: An improved nonirritating detergent composition is disclosed which has a high viscosity and good foaming properties. The composition is a mixture of (1) an amphoteric surfactant combined with an anionic surfactant and (2) a nonionic surfactant which is a 16 to 18 carbon atom fatty monoester of an aliphatic polyhydric alcohol reacted with 60 to 100 mols of ethylene oxide.

Reference: J.M. Walts; U.S. Patent 4,177,171; December 4, 1979; assigned to Johnson & Johnson.

Background: Nonirritating detergent compositions have been known and have been in use for some time. U.S. Patents 2,999,069 and 3,055,836 are representative of such prior art nonirritating detergent compositions. These compositions generally comprise an amphoteric surfactant combined with an anionic surfactant and a nonionic surfactant in admixture with other ingredients. The amphoteric-anionic surfactant consists essentially of:

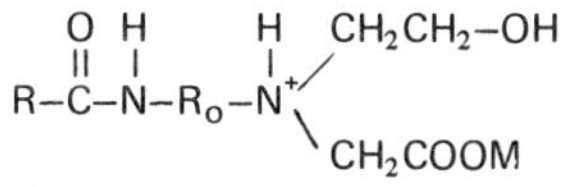

$$R_1-(OCH_2CH_2)_n-O-SO_3^-$$

where R is a 9 to 17 carbon alkyl radical, R_0 is an alkylene group of 2 to 4 carbon atoms, R_1 is a member of the group consisting of $C_{13}H_{27}$ and $C_{12}H_{25}$, n is 2 to 6, and M is a member of the group consisting of alkali metals, triethanolamine, mixtures of an alkali metal with hydrogen and mixtures of triethanolamine with hydrogen. The nonionic surfactant portion of the composition is usually a derivative of a 9 to 18 carbon atom fatty acid monoester of an aliphatic polyhydric alcohol reacted with 10 to 20 mols of ethylene oxide.

These prior art compositions, although they are nonirritating and have good foaming characteristics, have very low viscosities. Increasing the viscosity of the compositions by employing viscosity building additives or thickeners results in a deterioration of the foaming characteristics of the compositions.

Increased viscosity without foam degeneration is desirable in order to formulate detergent compositions, especially nonirritating shampoo compositions, that may be marketed as concentrates in tube type containers. The tubed type formulations have found wide acceptance with consumers because of the inherent ease and control of the application of such high viscosity formulations to the hair. However, no tube formulations of the nonirritating type shampoo currently exist in the marketplace.

Process Conditions: In formulating detergent compositions of the type discussed, by selecting a particular group of fatty acids as the fatty acid portion of the nonionic surfactant and reacting the resultant monoester with a relatively high mol ratio of an alkylene oxide, it is possible to obtain detergent formulations of high viscosity, that is, greater than 4,000 cp, 22° to 25°C, and which are also nonirritating and have excellent foaming properties.

The particular fatty acids which have been found to produce this result are those containing 16 to 18 carbon atoms. These include acids such as palmitic acid, stearic acid, isostearic acid and the like. These fatty acids are reacted with a polyhydric alcohol to form a hydrophobic ester and subsequently reacted with an alkylene oxide to produce the hydrophilic portion of the surfactant composition.

Generally, the polyoxyalkylene compounds are prepared by dehydration of known carbohydrates, preferably a 4 to 6 carbon atom carbohydrate, such as sorbitol to sorbitan in known fashion. The sorbitan is then esterified with a fatty acid to produce an ester. The latter product is then treated with alkylene oxide under known conditions with the resultant introduction of long oxyalkylene chains at the hydroxyl groups. The usual alkylene oxide used is ethylene oxide although propylene oxide or a mixture of both may be used. The total number of oxyalkylene units introduced is between 60 and 100.

The nonionic surfactant that is preferred is sorbitan monopalmitate having 80 mols of ethylene oxide. The amphoteric surfactant in combination with the anionic surfactant that is preferred in the composition is one of the formula shown above.

Examples 1 through 6: 6 detergent formulations were prepared from aqueous solutions of an amphoteric-anionic surfactant complex and nonionic surfactants. The amphoteric portion of the surfactant was:

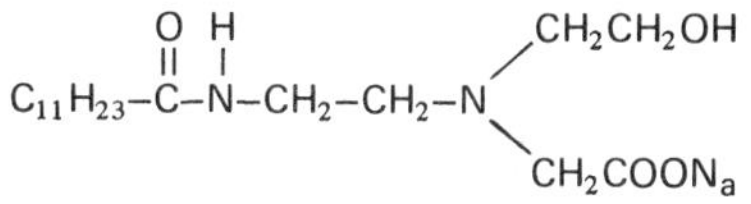

and was reacted with sodium tridecyloxydiethyleneoxyethyl sulfate.

The nonionic portion of the surfactant system was either the stearate or palmitate or laurate ester of sorbitan reacted with either 20 or 80 mols of ethylene oxide. The amounts of the amphoteric-anionic surfactant, shown as a 37% active aqueous solution, and the nonionic surfactant 100% active, are shown in grams in the table. The pH of each composition was adjusted to 7.0±0.1 by the addition of hydrochloric acid. Sufficient deionized water was added to each sample so that the total sample weighed 100 g. The viscosity of each sample was determined by the Brookfield method. The samples were also tested for initial foaming and the percentage decay of the foam after 5 minutes by the modified Ross-Miles method.

	Example No.					
	1	2	3	4	5	6
Components						
Amphoteric/anionic complex	30.1	30.1	30.1	30.1	30.1	30.1
Sorbitan monostearate*	12.6	–	–	–	–	–
Sorbitan monopalmitate*	–	12.6	–	–	–	–
Sorbitan monolaurate*	–	–	12.6	–	–	–
Sorbitan monostearate**	–	–	–	12.6	–	–
Sorbitan monopalmitate**	–	–	–	–	12.6	–
Sorbitan monolaurate**	–	–	–	–	–	12.6
Deionized H_2O, qs	100	100	100	100	100	100
Physical properties						
pH	7.02	7.01	7.01	7.02	7.02	7.00
Viscosity, cp	7.2	7.8	8.8	10,120	4,120	123
Ross Miles foam						
Initial, mm	85	82	99	102	107	119
5 minute decay, %	25	13	18	5.9	7.5	2.5

*Plus 20 mols of ethylene oxide.
**Plus 80 mols of ethylene oxide.

Examples 1 through 6 show the effect on viscosity and foaming characteristics by the increased polyoxyethylene content of the nonionic surfactant component of the composition. A comparison of Example 4 or 5 with Example 6 shows the effect of using the 16 to 18 carbon atom fatty acid on increasing viscosity without significant decreasing initial foam and foam decay.

Aqueous Dyestuff Dispersions

Highlights of the Technological Achievement: Aqueous dyestuff dispersions which contain water-insoluble dyestuffs and water-soluble oxyalkylation products of compounds which are obtainable by the addition of styrene or its derivatives onto phenols or naphthols, or derivatives of such oxyalkylation products, are suitable for pigmenting natural or synthetic materials.

Reference: V. Aign, H. Schulze, F. Bremer, R. Hörnle, K. Wolf, W. Berlenbach and P. Rasche; U.S. Patent 4,186,029; January 29, 1980; assigned to Bayer AG, Germany.

Background: Derivatives of styrene which may be mentioned as components or the addition products are, in particular, vinyltoluene, α-methylstyrene and divinylbenzene, and possible phenols are, in addition to phenol itself, above all cresols, xylenols and hydroxydiphenyls. The addition products can be prepared in a known manner, appropriately in the presence of catalysts, such as sulfuric acid, p-toluenesulfonic acid or zinc chloride. The addition products can be subjected to the oxyalkylation as single substances, but it is also possible to use mixtures, such as are in general initally obtained during the preparation.

Process Conditions: Preferred oxyalkylation products are those which are obtained by reacting condensation products of 1 to 3 mols of styrene or vinyltoluene and 1 mol of phenol with 12 to 100 mols of ethylene oxide.

Compounds of the formula

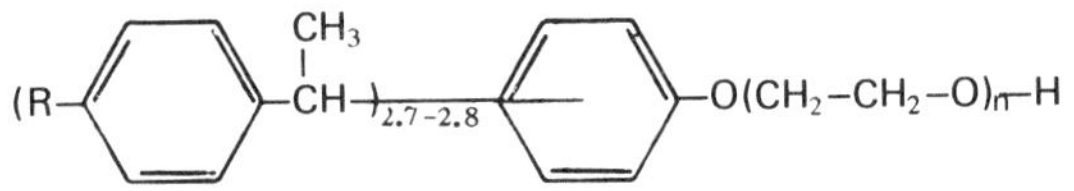

wherein R denotes hydrogen or methyl and n denotes a number from 25 to 30, are very particularly preferred.

The dyestuff dispersions according to the process can also contain other customary additives, such as agents which prevent drying out, for example, formamide, glycol or diglycol, anionic or nonionic surface-active agents, for example, ethanolammonium dodecylbenzenesulfonate or surface-active addition products of ethylene oxide, and preservatives, for example, sodium pentachlorophenolate. Examples of possible water-insoluble dyestuffs are pigments, disperse dyestuffs, transfer printing dyestuffs and water-insoluble optical brighteners.

These dyestuff dispersions according to the process can be universally employed and are suitable for the preparation of emulsion paints based on polyvinyl acetate, polyvinyl acetate copolymers, styrene-butadiene copolymers or acrylonitrile-butadiene copolymers, polyvinyl propionates, acrylic acid ester copolymers and methacrylic acid ester copolymers, water-soluble and water-dispersible alkyd resins and emulsions based on drying oils.

They are useful for the preparation of wallpaper paints based on cellulose derivatives, such as methylcellulose, hydroxymethylcellulose and carboxymethylcellulose, and for the preparation of printing inks which contain, as the binder material, predominantly saponified natural resins, such as shellac, or water-soluble synthetic resins, such as, for example, saponified polyvinyl acetate or copolymers which are based on acrylic esters and carry carboxyl groups or hydroxyl groups. However, these aqueous dyestuff dispersions are preferably used for pigmenting aqueous systems based on aqueous dispersion binders, such as are employed in textile pigment printing.

Pigment Dispersion for Use in Hydrophilic and Hydrophobic Media

Highlights of the Technological Achievement: Dispersions of inorganic and organic pigments containing an alkylene oxide adduct on long-chain aliphatic amines and an anionic surfactant having an aliphatic radical of 3 to 40 carbon atoms can easily be distributed in hydrophilic and hydrophobic media and yield colorations of high tinctorial strength and purity of shade.

Reference: E. Dietz, R. Gutbrod and M. Maikowski; U.S. Patent 4,156,616; May 29, 1979; assigned to Hoechst AG, Germany.

Background: This process relates to pigment dispersions suitable for pigmenting hydrophilic as well as hydrophobic media. In addition to the pigments, the dispersions contain alkylene oxide adducts on long chain, preferably primary, amines and anionic surfactants having at least 1 aliphatic radical with 3 to 40, preferably 6 to 22 carbon atoms. The anionic surfactants can be added in the form of salts or of acids. Especially suitable are monobasic anionic surfactants. Besides the aforesaid constituents, the pigment dispersions may further contain nonionic surfactants, preserving agents, water and/or substances preventing the dispersion from drying up.

Process Conditions: Suitable alkylene oxide addition products on long chain amines are preferably those having aliphatic chains, especially alkyl and alkenyl chains of from 6 to 30 carbon atoms, and ethylene oxide chains composed of 1 to 30 monomer units, for example, ethylene oxide adducts on oleyl amine, stearyl amine, tallow fat amine or coconut oil amine.

Preferred anionic surfactants contain $-COO^-$, $-OSO_3^-$ or $-SO_3^-$ as anion-forming groups and correspond especially to the formula

$$R-A_x-B_y-C_z-D$$

in which

R represents an aliphatic radical having from 3 to 40, preferably 6 to 22 carbon atoms, preferably an alkyl or alkenyl radical which may be substituted by hydroxy and/or carboxy groups;

A represents an aromatic radical, preferably a benzene radical, which may be substituted by 1 or 2 aliphatic radicals having from 1 to 20 carbon atoms, especially alkyl or alkenyl radicals;

B represents a bivalent group, preferably –O–, –CO–, $-SO_2-$, –NR'– or a combination of such groups, for example, –CO–O–, –O–CO–, –CO–NR'–, –NR'–CO–, $-SO_2-NR'-$ or $-NR'-SO_2-$, in which R' stands for hydrogen or a short chain alkyl radical preferably having from 1 to 4 carbon atoms;

C represents an alkylene group having from 1 to 6 carbon atoms or an alkylene oxide chain, preferably a propylene oxide and more preferably an ethylene oxide chain, of 1 to 30 monomer units;

D represents a preferably monobasic, anion-forming group, preferably $-COO^-$, $-SO_3^-$, or $-OSO_3^-$; and

x, y, and z are 0 to 1.

As anionic surfactants carrying aliphatic radicals preferably having from 6 to 22 carbon atoms there are mentioned, by way of examples, sulfuric acid semiesters of alkoxylated alkyl naphthols and especially of alkyl phenols and of fatty alcohols or alkoxylated fatty alcohols with alkylene oxide chains, preferably ethylene oxide chains, of 1 to 30 monomer units, as well as reaction products of fatty acids or fatty acid derivatives with lower hydroxy-alkane sulfonic acids, aminosulfonic acids and amino acids.

Product: With paints containing such flocculated coloring pigments the full coloring power of the pigments is not utilized. After having spread the paint on the area to be painted, the paint viscosity strongly increases by evaporation and absorption of the liquid in the ground. If the slightly dried and viscous coat of paint is then subjected to shearing forces, for example, by brushing over or rubbing with a finger, the flocculated pigment can be deagglomerated at least partially in these areas. As a result, the rubbed area has a more intense color than the adjacent area which has not been rubbed. It is quite obvious that a paint having such detrimental properties is quite useless. A surface coated therewith would have a spotty appearance when the paint is not uniform everywhere, apart from the fact that it is uneconomical not to use the full tinctorial strength of the pigment.

Jet Printing Ink Composition Containing an Ethylene Oxide Adduct of Acetylenic Diol

Highlights of the Technological Achievement: An ink composition for use in jet printing comprising an aqueous solution of a water-soluble dye and a humectant consisting of ethylene oxide adducts of at least 1 acetylenic diol in the absence of any glycol or glycol ether.

Reference: E.T. Bradley; U.S. Patent 4,184,881; January 22, 1980; assigned to M & T Chemicals, Inc.

Background: The principle underlying jet printing with ink is based on driving ink through a fine nozzle of metal or glass, so that uniform ink droplets are ejected from the nozzle tip. The ink droplets are electrostatically deflected with a deflecting electrode, arranged in front of the nozzle, so that symbols are printed with a preliminarily determined matrix.

In this printing process the ink in the nozzle is divided in fine ink droplets and the direction of the ink jet is controlled by deflecting plates or through nozzle movement. This printing process is highly dependent on the properties of the ink. The ink must be capable of forming stable ink droplets under pressure and under an electrical field, which have a high deflection sensitivity. Further, the ink must readily emerge from the nozzle. In addition, clogging of the nozzle by drying ink must be prevented.

Commercial nozzle jet inks are divided into aqueous and nonaqueous inks. Nonaqueous inks have the disadvantage of low stability ink droplets because of low surface tension. As a result, while in flight the ink droplets form a fog. In addition, such a nonaqueous ink has a high electrical resistance and a low deflecting sensitivity. At the same time, these ink droplets are particularly unstable under a high electrical charge. Consequently, it has been difficult to print clear and distinct symbols with this method.

Process Conditions: The concepts of the process reside in a new and improved ink composition suitable for use in jet printing comprising an aqueous solution of a water-soluble dye and a humectant consisting of ethylene oxide adducts of at least 1 acetylenic diol in the absence of any glycol or glycol ether. The preferred ethylene oxide adduct of at least 1 acetylenic diol is the ethylene oxide adduct of 2,4,7,9-tetramethyl-5-decyne-4,7-diol.

It has been found that printing inks formulated in accordance with this process have the desired viscosity for use in jet printing, and that the viscosity of the composition is subject to little variation with use as water is lost by evaporation during recirculation of the ink composition through the jet printer. In addition, the humectant of this process substantially prevents or minimizes tip drying of the printing ink in the orifice or nozzle during down time of the printer, such as when the printer is shut down overnight.

As the coloring material in the ink composition, use should be made of a water-soluble dye or mixture of dyes. The use of pigments or other forms of particulate matter is preferably avoided since such materials tend to clog the very small orifice of the jet printer.

Preferred dyes are the commercially available direct dyes which generally contain an ionizable inorganic salt, such as Glauber salts, or sodium chloride which has been added during manufacture in order to standardize the dye lot. The inorganic material thus renders the dye conductive whereby the ink jet is capable of being deflected in an electromagnetic field during jet printing. Such direct dyes are commercially available, as represented by the C.I. direct black 38 dyes [e.g., two that can be deflected in an electromagnetic field are Direct Black GW (Tenneco Color Division of Tenneco Chemicals) and Capamine Black ESA (Capital Color and Chemical Co.)].

The ink composition can be prepared by thoroughly mixing the components to insure uniform mixing and complete dissolution of the water-soluble dye, and then filtering the resulting composition to remove any particulate contamination.

Filtration of the composition after mixing the components is highly desirable in order to remove from the ink composition particulate matter, such as contamination or undissolved dye, which might otherwise serve to clog or obstruct the jet of a jet printer during use. It is frequently preferred to filter the composition of the process to remove particulate matter having a diameter $>2\ \mu$, and preferably $1\ \mu$, to insure that obstruction of the jet of the jet printer will be avoided. The following compositions are expressed in parts by weight:

Example 1:

	Percent (%)
Distilled water	61
Ethylene oxide adduct*	14
Direct Blue dye solution**	15
Direct Black***	10

*Adduct of 2,4,7,9-tetramethyl-5-decyne-4,7-diol (Surfynol 485–Air Products and Chemicals, Inc.)
**(C.I. 218, Pontamine Bond Blue B Liquid–E.I. DuPont de Nemours & Co.)
***(C.I. 38, Pontamine Black SP Liquid–E.I. DuPont de Nemours & Co.)

This ink has a viscosity of 3.5 cp. It prints satisfactorily, and has idled on and off in the prototype jet printer for 6 days without clogging. Printer startups, after long periods of idling, are immediate.

Example 2:

	Percent (%)
Distilled water	75
Ethylene oxide adduct*	5
Direct Blue dye solution**	10
Direct Black***	10

*Adduct of 2,4,7,9-tetramethyl-5-decyne-4,7-diol (Surfynol 485–Air Products and Chemicals, Inc.)
**(C.I. 218, Pontamine Bond Blue B Liquid–E.I. DuPont de Nemours & Co.)
***(C.I. 38, Pontamine Black SP Liquid–E.I. DuPont de Nemours & Co.)

This ink has a viscosity of 2.0 cp.

Aqueous Printing Inks with Improved Transfer Properties

Highlights of the Technological Achievement: The transfer properties of aqueous printing inks are improved by including in the ink formulation a complex of a polyethylene oxide, a urea, and an alkali metal salt.

Reference: T. Sulzberg; U.S. Patent 4,173,554; November 6, 1979; assigned to Sun Chemical Corporation.

Background: Among the criteria for an aqueous printing ink is its ability to be transferred from the inking roller to a wide variety of substrates, such as corrugated papers and cartons, plastic film, metal foils, newsprint, and so forth. The print should dry quickly; be strong, clear, sharp, resistant to rubbing and scuffing, and resistant to water; and possess high gloss.

It is known that the transfer properties of an aqueous printing ink can be improved by the addition of polyethylene oxide. In U.S. Patent 4,014,833, for example, ink compositions are claimed that contain 0.1 to 1.5 wt % of polyethylene oxide having an average molecular weight between 100,000 and 350,000.

When, however, it was attempted to improve the transfer onto corrugated paper stock of a flexographic printing ink based on, e.g., a styrene-acrylic ester-acrylic acid terpolymer latex by adding thereto a polyethylene oxide, the results were unsatisfactory for commercial use. With additions of 0.1 to 0.5 wt % of the polyethylene oxide, the ink transfer was poor, resulting in weak prints with low color intensity. Additions of 1.0 to 1.5 wt % of the polyethylene oxide greatly improved the transfer and the print strength was excellent; however, in a fast roll-out there was considerable flying of the ink at the edges of the print, giving it a speckled appearance (misting).

Process Conditions: It has been found that the addition to a flexographic ink of a modified polyethylene oxide resin system not only improves the ink transfer and thus the print strength, but also eliminates the objectionable flying and misting phenomena.

The polyethylene oxide resins suitable for the process are generally nonionic homopolymers having average molecular weights between 100,000 and 400,000. They are characterized by behaving like electron donors with a wide variety of electron acceptors, such as urea; thiourea; alkali metal salts, e.g., potassium chloride, potassium bromide, potassium oxalate, and potassium tartrate; polyacrylic acid; phenolic resins; cellulosics; and the like.

Complexes are formed when the polyethylene oxide is reacted with one or more of these electron acceptors. The complexes may be prepared in any convenient manner, such as in solution, in the vapor phase, or in high shear apparatus, e.g., a two-roll mill. The temperature, pressure, and other reaction conditions vary, depending upon the selected components and the end-use application.

While improved results are obtained with complexes of the polyethylene oxide with any one of the electron acceptors, it has been found that the incorporation in an aqueous printing ink formulation of a complex of a polyethylene oxide with a urea and an alkali metal salt results in a product that has excellent print strength and no flying or misting.

These complexes have the added advantage of permitting the use of decreased levels of the polyethylene oxide, excellent results being obtained with amounts within the range of about 0.8 to 2.0 parts or about 0.04 to 0.1% of a 5% solution in water of the polyethylene oxide, based on the total weight of the ink.

The amounts of the urea and the alkali metal salt are based on the amount of polyethylene oxide, the ratio of the amounts of 5% polyethylene oxide solution: alkali metal salt:urea being about 1:1-2:2-4. Thus the following parts are generally used:

	Polyethylene Oxide	Alkali Metal Salt	Urea
Minimum	0.8	0.8–1.6	1.6–3.2
Maximum	2.0	2.0–4.0	4.0–8.0

Coating Composition Comprising an Aqueous Emulsion of Polyethylene Containing a Water-Soluble Polyethylene Oxide

Highlights of the Technological Achievement: A coating composition for imparting improved scuff resistance to a substrate and a process for its use are disclosed. The coating composition comprises an aqueous emulsion of polyethylene containing from 0.5% to 15% of a water-soluble polyethylene oxide by weight of the polyethylene emulsion. The preferred substrate is a paper product such as paper, paperboard and linerboard.

Reference: J.C. Jones; U.S. Patent 4,169,823; October 2, 1979; assigned to Owens-Illinois, Inc.

Background: Substrates of various kinds are often subjected to abrasion and scuffing. As an instance, paper, paperboard and linerboard products are commercially used in the manufacture of containers such as setup boxes, folding cartons, fiber cans and tubes, corrugated and solid fiber boxes, drums, etc. Paperboard and like products have a tendency to form quills or peels when subjected to scuffing or rubbing, as in passing through a corrugator, on a conveyor system, or in freight cars when one box rubs against another. Also, during shipment, an article of manufacture within a container often strikes the paperboard such that its surface becomes marked. The coarse, relatively abrasive surface of the container inflects wear spots, scuffs, scratches, and other undesirable flaws on the surface of the article. Under such conditions, the shipped article has to be repaired or resurfaced prior to sale. This problem is particularly acute with items such as clothes, bowling balls, furniture, appliances and the like.

Process Conditions: It has been discovered that an initial application of a metal compound like a tin compound and the use of an organic solvent for such compound can be entirely eliminated if metal-free polyethylene oxide is added directly to the aqueous emulsion of polyethylene, and the resultant emulsion is applied to the substrate in a single step.

In one form, the coating composition comprises an aqueous emulsion of polyethylene containing from 0.5% to 15% of a water-soluble polyethylene oxide by weight of the polyethylene emulsion. The coating compositions may be applied to a substrate by several means, such as by roller application.

The coating compositions comprise an aqueous emulsion of polyethylene to which an effective amount of polyethylene oxide has been added. The emulsion preferred is an aqueous emulsion containing polyethylene commercially designated as AC Polyethylene, Product No. 629, (Semet-Solvay Division of Allied Chemical Corporation).

The polyethylene oxide which may be used is homopolymer, Polyox WSR N-80 (Union Carbide). Rheological measurements indicate that one resin used had a relatively narrow distribution of molecular weights and an average molecular weight of about 200,000. The viscosity range, at 25°C, for this product, using a Brookfield spindle No. 1, 50 rpm, was given as 81 to 105 cp for a 5% solution. Polyethylene oxide polymers are linear resins that may be characterized by a relatively large number of polar groups in the molecule. Such polyethylene polymers can be prepared by polymerizing ethylene oxide in contact with a catalyst, such as alkaline earth amide or hexammoniates, as is known in the polymer art.

For each test, the coating composition was applied to the paperboard at a rate of about 1.6 lb/1,000 sq ft on a dry basis. The table lists the results.

	Example No.	
	1	2
Coating	Control—silica	Polyethylene and polyethylene oxide
Coefficient of friction	0.65 at 33° incline	0.25 at 14° incline
Scuff dusting, cycles	6	42
Scuff failure, cycles	50	*

*Testing terminated after 500 cycles with no failure.

The coefficients of friction given were measured at the degree of plane indicated. Paperboard having the control coating began to dust early after only 6 cycles, whereas paperboard having the present coating composition did not begin to dust until much later at 42 cycles. It is usual for dusting to begin at about 10 cycles. Similarly, the control failed the scuff test at 50 cycles, while the present coating composition had not failed even at 500 cycles when the test was halted.

Perfluoroalkyl Substituted Anhydrides and Polyacids and Derivatives

Highlights of the Technological Achievement: This process describes new perfluoroalkyl substituted esters, diesters and polyesters which contain at least one cyclic 5-membered anhydride group or two carboxy groups; as well as half amides and half esters thereof; their synthesis and their use as surface-active reactants in polycondensate resin systems.

Reference: K.F. Mueller; U.S. Patent 4,153,590; May 8, 1979; assigned to Ciba-Geigy Corporation.

Background: Perfluoroalkyl (R_f) surfactants which are otherwise like their hydrocarbon analogues in that they are either anionic, cationic, nonionic or amphoteric, but which contain perfluoroalkyl groups reduce the surface tension of aqueous or organic liquids to extremely low levels, down to 15 dynes/cm, as compared to 25 to 30 dynes/cm obtainable with conventional surfactants. Such low surface tensions allow these liquids, which may be molten polymers or polymer solutions or emulsions, to wet substrates which are otherwise impossible to wet.

Therefore, properties which depend on good wetting are often substantially improved, such as adhesion and surface smoothness, and such coating deficiencies as crawling, "fisheyes," "orange peel," etc. are largely eliminated. Numerous R_f-surfactants have been described in U.S. Patents 2,915,554; 3,274,244; 3,621,059; 3,668,233; and German Patent 2,215,388.

R_f-surfactants of the prior art cited above are used for coating systems to help in wetting and to prevent crawling and other side effects of poor coatings. These R_f-surfactants are nonionic in nature because ionic compounds are poorly compatible with resins. All the R_f-surfactants described are nonreactive.

Process Conditions: This process pertains to perfluoroalkyl substituted esters, diesters and polyesters which contain at least one cyclic 5-membered anhydride group or two carboxy groups as well as half amides and half esters derived therefrom.

These compounds are useful as surface-active reactants in polycondensation resin systems. For incorporation into aqueous resin solutions, the compounds are reacted with water-solubilizing alcohols or amines such as polyethylene oxide, bisamino propyl ether of polyethylene oxide, or N,N-dialkylaminealcohols, or N,N-dialkylaminoalkyl diamines.

One example of a particularly preferred R_f dianhydride is the compound

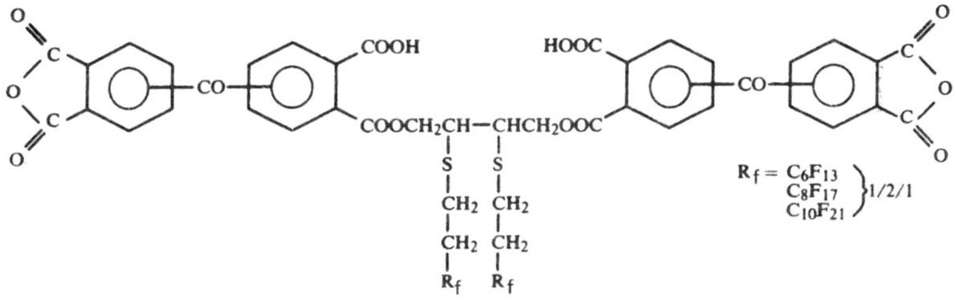

Although the anhydrides and dianhydrides of this process are insoluble in polyhydric resins, such as polyols or polyamines, they can be solubilized by ring opening reaction with these compounds. For instance, in order to incorporate the R_f-moieties into a polypropylene oxide diol, 5 g of a 33% solution of the new R_f-anhydride compound are heated with 5 to 20 g of the polyol until a clear product is obtained; the solvent can be evaporated and the remaining R_f-modified resin, which contains 1 to 10% fluorine, can be further diluted with unmodified resin to the desired fluorine level, usually between 0.001% and 0.5%.

Many reactive polymers or prepolymers can be treated and modified with this method: polyethylene oxide diols, polypropylene oxide diols, bis-2-aminopropyl ethers of polyalkylene oxides, poly-n-butylene oxide diols; polyester diols from dibasic acids and diols.

A particularly preferred derivative of the dianhydride illustrated above is prepared by reaction of 1 mol of the dianhydride with 2 mols of an N,N-substituted aminoalcohol or diamine such as N,N-dimethylethanolamine or N,N-dimethylpropane-1,3-diamine. One such derivative has the structure below:

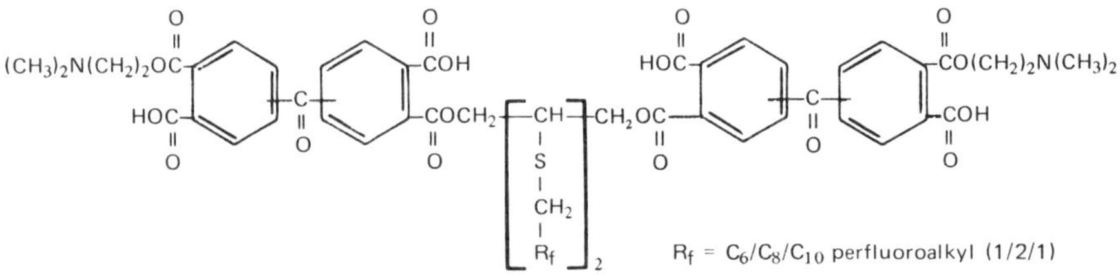

R_f = $C_6/C_8/C_{10}$ perfluoroalkyl (1/2/1)

This derivative of an $(R_f)_2$-dianhydride is especially designed for incorporation into aqueous and highly polar resin systems. It is soluble or dispersible in aqueous or polar organic resin solutions, as well as in resins themselves. However, its solubility sometimes is limited to low molecular weight and polar components of resin systems (amine-hardeners; alkyd resins; low molecular weight polyols; Methendic Anhydride). If it is used with water-based coatings, such as aqueous epoxies, it is preferably used in combination with an ionic R_f-surfactant designed to reduce the surface tension of water.

Restoring Agent for Soft Contact Lenses

Highlights of the Technological Achievement: Soft contact lenses, which have become deteriorated as a result of prolonged use, are treated with an aqueous solution of ethylene oxide/propylene oxide block copolymer and ethylenediaminetetraacetic acid or water-soluble salt thereof.

Reference: P. Cordrey and S.N. Ganju; U.S. Patent 4,152,283; May 1, 1979; assigned to Sauflon International, Inc.

Background: Soft contact lenses, that is, contact lenses formed from hydrogels having a very high water content, are particularly susceptible to deterioration after prolonged use. Soft contact lenses are commercially available products which are made of hydrophilic polymers as described in, for example, U.S. Patent Reissue 27,401, and U.S. Patents 3,822,089, 3,943,045 and 4,032,599.

This deterioration occurs in relation to both the optical and the physical properties of the hydrogel. Soft contact lenses which had deteriorated have been discarded because there was no treatment effective to restore them to their original condition.

Process Conditions: According to the process there is provided a composition for treating soft contact lenses, the composition comprising an ethylene oxide-propylene oxide block copolymer and ethylenediaminetetraacetic acid or water-soluble salt thereof, present in amounts effective to remove residues from the lenses and restore the physical and optical properties of the lenses substantially to their original new condition.

In this formulation the block copolymer preferably has two terminal hydroxyl groups. Ethylene oxide-propylene oxide block copolymers are known nonionic surfactants having the general formula $HO(CH_2CH_2O)_a(C_3H_6O)_b(CH_2CH_2O)_cH$, wherein b is at least 15 and a and c are such that the polyoxyethylene units are from 20 to 90% of the total weight of the block copolymer. The molecular weight of the block copolymers varies from about 1,100 to about 15,000.

Ethylenediaminetetraacetic acid and water-soluble salts thereof can be used in the composition according to the process. As the salts there can be mentioned the mono-, di-, tri-, and tetra-alkali metal salts, preferably sodium salts, and ethylenediaminetetraacetic acid disodium salt is especially preferred. It has also been found particularly advantageous to include sodium perborate and/or sodium chloride in the composition.

Example: A soft contact lens, which has been worn and which exhibited a characteristic ultraviolet absorbance due to its being coated with lacrymal secretions and residues of agents used for daily cleaning of the lenses, is treated by soaking it overnight in an aqueous solution containing 35% of ethylene oxide-propylene oxide block copolymer (Monolan E 80), 1% of ethylenediaminetetraacetic acid and 14% of sodium perborate. The lens is then boiled in physiological saline solution for 30 minutes. The cleansed lens does not exhibit the original characteristic ultraviolet absorbance when tested in an ultraviolet spectrophotometer.

FLUOROPOLYMERS

PROCESSING

Preparation of Hydrophilic Fluoropolymers from Sulfur- and Phosphorus-Containing Compounds

Highlights of the Technological Achievement: Certain fluoropolymers, when chemically modified by reacting them with sulfur- or phosphorus-containing compounds, become hydrophilic materials useful for making ion exchange membranes, especially diaphragms for electrolytic cells, particularly chlor-alkali cells, used in the production of chlorine, hydrogen and sodium hydroxide from brine.

Reference: J.C. Fang; U.S. Patent 4,189,369; February 19, 1980; assigned to E.I. DuPont de Nemours and Company. (Other references: U.S. Patents 4,164,463; 4,169,024 and 4,186,084.)

Background: In commercial chlor-alkali cells used for the production of chlorine, hydrogen and sodium hydroxide from brine, asbestos diaphragms are ordinarily used as anion exchange membranes between the anolyte and catholyte compartments. Diaphragms of this sort are generally satisfactory, but their electrical resistance is high and chlor-alkali cells which use such diaphragms therefore require more electric current for operation than is desirable.

Process Conditions: It has been found that by chemically modifying certain fluoropolymers with sulfur- or phosphorus-containing compounds, the fluoropolymers are made hydrophilic. This makes them especially suited for use in making diaphragms for electrolytic cells.

Such diaphragms have less electrical resistance than those made solely of asbestos. There is, accordingly, less power loss in a chlor-alkali cell which uses this diaphragm and the cell is therefore more efficient. It has been observed that the chlorine, hydrogen and caustic production of a cell using this diaphragm can generally be held at the same level as one using an asbestos diaphragm while operating at 10 to 15% lower voltage.

These fluoropolymers can contain up to 75 mol % of units derived from other ethylenically unsaturated monomers. These hydrophilic fluoropolymers contain nonterminal units represented by the structure:

(1) 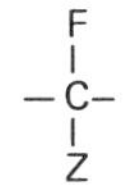

where Z is $-SO_2Br$, $-SCl$, $-SOCl$, $-SO_2Cl$, $-S_xM$, $-S_xBr$, S_xCl, $-S-C(S)-OR$, $-S-C(S)-SR$, $-S-P(SM)_2{=}S$, $-P(OR)_2{=}O$, $-P(OM)_2{=}O$, $-HP(OM){=}O$, $-PCl_2$ or $-POCl_2$ where M is hydrogen, sodium, potassium, lithium, calcium or magnesium; R is an alkyl radical of 1 to 12 carbon atoms or a cycloalkyl radical of 3 to 12 carbon atoms; and x is 1, 2, 3, 4 or 5.

Hydrophilic fluoropolymers containing nonterminal units of structure (1) where Z is

(2) 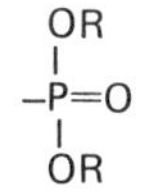

are preferred for use as diaphragm materials in electrolytic cells because of their low electrical resistance. The hydrophilic fluoropolymers can contain more than one type of unit of structure (1). The hydrophilic fluoropolymer will contain enough units of structure (1) to give it a sulfur or phosphorus content of 0.1 to 10% by weight, but not so many that more than about 1% by weight of the polymer will dissolve in water at 20°C.

Raw Materials: When these other ethylenically unsaturated monomer units are present in a fluoropolymer starting material which is to be modified and made into a diaphragm for an electrolytic cell, it is preferred that the units of structure (1) be present in the form:

(3) 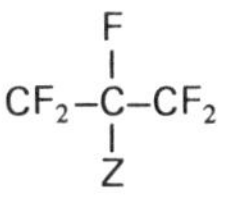

to avoid possible alpha, beta elimination of hydrogen fluoride under extreme conditions of heat and alkalinity found in some electrolytic cells.

The fluoropolymers preferred as starting materials because of the low electrical resistance of the hydrophilic fluoropolymers which result when they are used are polytetrafluoroethylene (PTFE), polychlorotrifluoroethylene (PCTFE), polybromotrifluoroethylene (PBTFE) and copolymers (in all monomer unit weight ratios) of (a) tetrafluoroethylene (TFE) and hexafluoropropylene (HFP), (b) TFE and chlorotrifluoroethylene (CTFE), (c) TFE and bromotrifluoroethylene (BTFE), (d) CTFE and BTFE, and (e) TFE and perfluoroalkyl vinyl ether (especially perfluoromethyl vinyl ether).

In the first method of preparation, the fluoropolymer starting material, in whatever form, is exposed to radiation while it is in contact with a suitable modifying compound, or the fluoropolymer and modifying compound are brought together in the presence of a peroxy compound such as benzoyl peroxide or tert-butyl peroxide.

Any compound which contains a sulfur or phosphorus atom which can react with the fluoropolymer starting material can serve as a modifying compound. Mixtures of modifying compounds can also be used.

The liquid containing the fluoropolymer starting material and modifying compound is then exposed to about 0.1 to 20 Mrad of electron radiation or to about 0.1 to 50 W/cm^2 of liquid surface of UV light at 20°C.

In the second preparative method, the fluoropolymer starting material and modifying compound are brought into contact with each other, under conditions suitable for reaction and kept together until the reaction is complete. This method can only be used with homopolymers and copolymers of CTFE and BTFE as starting materials. Mixtures of these compounds can also be used.

The fluoropolymer starting material, preferably in the form of a porous unsupported film or a porous reinforced structure, is placed in a 2 to 10% by weight solution of modifying compound in dimethylformamide or dimethylacetamide. As in the first preparative method, only enough of the solution is used to thoroughly wet the fluorocarbon polymer starting material. If desired, a small amount of cesium fluoride or potassium fluoride can be added to the solution to suppress side reactions.

The solution, with the fluoropolymer starting material in it, is then brought to a temperature within the range of about 20°C to about the boiling point of the solvent used, to bring about the reaction. The solution is held at that temperature for about 4 to 200 hours, the reaction at lower temperatures requiring more time for completion than one carried out at higher temperatures. When the initially colorless fluoropolymer starting material turns black, the reaction is complete.

In both methods of preparation, the modifying compounds used will provide radicals which will be attached directly to the fluoropolymer chain.

Process for Obtaining Colloidal Dispersion of Polymeric Tetrafluoroethylene

Highlights of the Technological Achievement: Presence of a selected carboxylic acid in the aqueous polymerization mixture during polymerization of tetrafluoroethylene to obtain a colloidal dispersion of polymeric tetrafluoroethylene has been found to reduce the amount of coagulum formed during the polymerization.

Reference: S.V. Gangal; U.S. Patent 4,186,121; January 29, 1980; assigned to E.I. DuPont de Nemours & Company.

Background: The homopolymerization of tetrafluoroethylene and the copolymerization of tetrafluoroethylene and ethylenically unsaturated comonomers in aqueous medium can be carried out by two distinctly different procedures. In one procedure, little or no dispersing agent is used and agitation is carried out vigorously in order to produce a precipitated resin, commonly referred to as granular resin.

In the other procedure, sufficient dispersing agent is used and agitation is mild in order to produce small colloidal size particles dispersed in the aqueous reaction medium. In this second procedure referred to as aqueous dispersion polymerization, precipitation, i.e., coagulation, of the resin particles is avoided.

The two procedures produce distinctly different products. The granular product can be molded in various forms, whereas the resin produced by the aqueous dispersion method cannot be molded, but is fabricated by dispersion coating or by converting to powder for paste extrusion with a lubricating medium. In contrast, the granular resin is incapable of being paste extruded or dispersion coated.

A disadvantage which occurs during aqueous dispersion polymerization is the formulation of coagulum, i.e., unwieldy large clumps of resin particles, which settle out of the polymerization medium and may clog the equipment as well as reduce the yield of suspended, colloidal size resin particles. This coagulum formation is particularly evident when ionic polymerization initiators are used.

Process Conditions: In this method, formation of coagulum during polymerization of tetrafluoroethylene in an aqueous medium to obtain a colloidal dispersion of a tetrafluoroethylene polymer is reduced by carrying out the polymerization process in the presence of an aliphatic, substantially nontelogenic carboxylic acid having a -log K of between 1.5 and 6.0.

Initiators used herein are ionic initiators. Examples of ionic polymerization initiators include inorganic persulfates such as ammonium persulfate or alkali metal persulfates such as potassium persulfate and the like.

The dispersing agent is an anionic, substantially nontelogenic dispersing agent. Commonly used dispersing agents are fluorinated carboxylates containing 7 to 20 carbon atoms, such as ammonium polyfluorocarboxylates. The amount of dispersing agent present is ordinarily between 800 and 4,000 ppm based on weight of water used in the aqueous dispersion.

Tetrafluoroethylene is either used alone, to produce the homopolymer (polytetrafluoroethylene), or is used with at least one other copolymerizable, ethylenically unsaturated comonomer. Ordinarily, the amount of comonomer used in the process will not be so great as to result in tetrafluoroethylene copolymers having more than 35% by weight of comonomer units. (If more than 35% is used, the copolymer tends to lose its nonelastomeric nature.)

Mixing is ordinarily carried out by mildly agitating the aqueous polymerization mixture. Agitation is controlled to aid in preventing premature coagulation of resin particles produced in the polymerization. Polymerization is ordinarily conducted until the solids level (i.e., polymer content) of the aqueous mixture is between 15 and 60% by weight of the mixture.

The aliphatic, substantially nontelogenic, carboxylic acid contains 1 to 6 carbon atoms and is one having a -log K of between about 1.5 and 6.0, preferably between 2.5 and 5.5. The term "K" represents the dissociation constant of the acids. In the case of acids with two or more carboxyl groups, K is the first dissociation constant. It has been found that when -log K is below 1.5, the acids tend to become too strong, while above 6.0, the acids tend to be too weak. The acid used should, preferably, be present in an amount between 10 and 150 times the amount of initiator present based on weight.

The dispersions are useful in making coated articles in which the dispersion is applied to a substrate by spraying, coating or dipping to obtain heat-resistant materials. The dispersions are also useful for conversion to fine powder for paste extrusion applications.

Preparation of Aqueous Dispersions of Tetrafluoroethylene Polymer Using Sodium Phosphate

Highlights of the Technological Achievement: The only practical way known to decrease the viscosity of an ionically thickened tetrafluoroethylene polymer dispersion has, heretofore, been by diluting it with water. This decreases the solids level, which can have an undesirable effect on subsequent processing.

It has been discovered that if the tetrafluoroethylene polymer dispersion contains a sodium phosphate in which the ratio of $Na_2O:P_2O_5$ is between 1:1 and 3:1, the viscosity of the dispersion can be decreased by adding selected inorganic electrolytes.

Reference: R.F. Richter; U.S. Patent 4,169,087; September 25, 1979; assigned to E.I. DuPont de Nemours and Co.

Background: Inorganic electrolytes have been used to thicken polytetrafluoroethylene aqueous dispersions, as evidenced by Ribbans in U.S. Patent 3,830,770. Thickening of aqueous dispersions, i.e., increasing the viscosity of the dispersions, is ordinarily carried out to aid in maintaining stability of the dispersions, especially during storage. In particular, thickening of the dispersions aids in retarding the settling out of various ingredients, such as inert filler materials.

Addition of an inorganic electrolyte to such a dispersion results in an increase in the viscosity of the dispersion and thus causes it to thicken. Heretofore, once a dispersion had been thickened, viscosity could only be lowered by adding water. Addition of water dilutes the dispersion and results in a change in the concentration of the ingredients. Since many applications involve the use of uniform batches of dispersion, it would be desirable to be able to lower the viscosity of a dispersion without changing the volume, i.e., concentration.

It has been discovered that if a selected sodium phosphate is used as the inorganic electrolyte to increase the viscosity of the dispersion, the viscosity can subsequently be lowered by adding certain other solid inorganic electrolytes in small amounts. Since no liquid is added to the thickened dispersion, the viscosity is lowered without substantially changing the volume of the dispersion.

Specifically, the thickened dispersion in which viscosity can be lowered in the manner described above is an aqueous dispersion consisting essentially of (a) water; (b) 45 to 75% by weight (based on weight of polymer, water and nonionic surfactant) of a tetrafluoroethylene polymer having a melt viscosity greater than 1×10^9 poises at 380°C; (c) 3 to 11% by weight (based on weight of polymer) of a nonionic surfactant; and (d) 0.2 to 90 g/kg of dispersion of a sodium phosphate having a mol ratio of $Na_2O:P_2O_5$ of between 1:1 and 3:1, inclusive.

Raw Materials: The tetrafluoroethylene polymer used is the non-melt-fabricable class of tetrafluoroethylene polymer, as opposed to the lower melting, melt-fabricable class of tetrafluoroethylene polymer. The difference between these classes of tetrafluoroethylene polymers can be determined by melt viscosity. Non-melt-fabricable polymers have high melt viscosities, e.g., 1 x 10^9 poises or more at 380°C, while melt-fabricable polymers have much lower melt viscosities, e.g., on the order of 1 x 10^4 to 1 x 10^6. The tetrafluoroethylene polymers used are prepared by the aqueous dispersion preparative method, rather than by the suspension polymerization method.

The nonionic surfactant is any nonionic surfactant which is soluble in water at room temperature (20° to 25°C) at the concentration desired. The surfactant can be composed of a single nonionic surfactant or a mixture of nonionic surfactants. Typically, the nonionic surfactants are prepared as reaction products of ethylene oxide, which imparts a hydrophilic moiety to the surfactant, with other compounds which impart hydrophobic moieties to the surfactant, such as propylene oxide, amines, saturated and unsaturated aliphatic alcohols and aliphatic acids, and alkyl phenols.

Example: *Thickening Effect of Sodium Phosphate of $Na_2O:P_2O_5$ Ratio of 1.15 to 1* – To 374.6 g of polytetrafluoroethylene homopolymer dispersion (polymer melt viscosity greater than 1 x 10^9 poises at 380°C) containing 70.9% solids and 4.0% (based on polymer) Triton X-100 (nonionic surfactant) were added 75.5 g demineralized water (used to adjust concentration without introducing impurities). After this dispersion was warmed to about 35°C, 3.41 g of Triton X-100 were added and the dispersion stirred 5 minutes to completely dissolve the surfactant. The dispersion was then cooled to room temperature (22°C). The dispersion contained 58.5% solids and 5.5% Triton X-100. Its viscosity was determined.

This dispersion was stirred using a propeller-type stirrer at a speed adequate to mix the ingredients without whipping air into the dispersion. To this stirred dispersion, 0.2 g of sodium phosphate having an $Na_2O:P_2O_5$ ratio of 1.15:1, buffered by sodium carbonate and sodium bicarbonate, was added and the stirring was continued for 5 minutes before the viscosity was measured. The mixture of sodium phosphate and buffer used was Calgon.

The addition of 0.2 g of the mixture was repeated nine times. After each addition, the viscosity was measured. The addition was then increased to 0.4 g and the procedure was repeated four times. Viscosity was measured after each addition. The viscosity of the original dispersion and the viscosity measured after each addition of phosphate is shown as follows:

. . . . Amount of Sodium Phosphate . . . . Present (g)	Viscosity (cp)
0	12
0.2	10
0.4	12
0.6	12
0.8	22
1.0	45
1.2	73
1.4	100
1.6	125
1.8	140
2.0	160

(continued)

. . . . Amount of Sodium Phosphate. . . .	
Present (g)	Viscosity (cp)
2.4	185
2.8	205
3.2	223
3.6	241

To a polytetrafluoroethylene homopolymer dispersion (454 g), as used above except that it contained 55.0% solids and 5.16% nonionic surfactant (Triton X-100), were added 8.0 g of sodium phosphate having a $Na_2O:P_2O_5$ ratio of 1.15:1, present in the form of Calgon. The ingredients were agitated. The viscosity rose to 155 cp. Two more additions of sodium phosphate ($Na_2O:P_2O_5$ ratio 1.15), 0.7 and 0.8 g respectively, resulted in viscosities of 177 and 205 cp respectively. Additionally, 20 g of sodium phosphate ($Na_2O:P_2O_5$ ratio 1.15) added to the thickened dispersion resulted in a viscosity of 1,600 cp. The total amount of sodium phosphate added was about 65 g/kg of dispersion.

Spheroidal Particulate Polytetrafluoroethylene Dispersion Resin

Highlights of the Technological Achievement: Particulate polytetrafluoroethylene dispersion resin, characterized by high molecular weight, relatively large particle size, uniform spherical particle shape and narrow particle size distribution, which is suitable for making stable concentrated aqueous polytetrafluoroethylene dispersions of concentrations higher than heretofore available, is made by polymerization of tetrafluoroethylene in aqueous dispersion in the conjoint presence of liquid octafluorocyclobutane and certain polyfluorinated carboxylic acid salt dispersing agents, using water-soluble peroxide initiators.

Reference: M.B. Mueller, P.P. Salatiello and D.L. Sawhney; U.S. Patent 4,145,502; March 20, 1979; assigned to Allied Chemical Corporation.

Background: Production of dispersion polytetrafluoroethylene resins of large, uniform particle size has, heretofore, been considered desirable because they were thought to have improved extrusion characteristics. While the resins of the process, which combine large average particle size and uniformity of particle shape with narrow particle size distribution, have good extrusion characteristics when paste-extruded after admixture with lubricants, such as deodorized kerosene, in an amount of about 20% by weight in the manner known to those skilled in the art, their extrusion characteristics do not appear to be substantially superior to those of known resins. The resins of the process, however, have dispersion characteristics which are substantially superior to those of known products.

Aqueous dispersions of the product of the process, as obtained from the polymerization process, can be concentrated by simple procedures to form stable concentrated aqueous dispersions containing more than about 75% by weight of polytetrafluoroethylene resin. Stable concentrated aqueous polytetrafluoroethylene dispersions of such high concentrations have, heretofore, not been available.

Process Conditions: The particulate polytetrafluoroethylene dispersion resin of this process is characterized by (a) having a surface area of at least 9 m^2/g; (b) predominantly comprising spheroidal particles having an average particle diameter of greater than 0.35 μ; (c) having a narrow particle size distribution corresponding to a distribution function of less than 0.4; and (d) having a molecular weight of at least 2.0 x 10^6 corresponding to a standard specific gravity of 2.23 or less. Preferred polytetrafluoroethylene of the process has an average particle diameter of greater than 0.4 μ, and especially those having, in addition, a particle size distribution corresponding to a distribution function of less than 0.25.

Standard specific gravity (SSG) is determined by the method set forth in ASTM D 1457, or equivalent. Molecular weight as used is the number average molecular weight and is determined from the standard specific gravity in accordance with the empirical equation:

$$\log \overline{Mn} = \frac{0.466 - \log SSG}{0.0187}$$

The particulate polytetrafluoroethylene resin of the process is further characterized by its capacity for forming stable concentrated aqueous dispersions containing at least 75% by weight of polytetrafluoroethylene when the latex, as obtained from the polymerization reaction which may contain as little as about 15% by weight of polytetrafluoroethylene, is concentrated by the following method. To the latex, as obtained from the polymerization, there is added a polyethylene glycol-p-octyl phenol ether (Triton X-100) having the formula:

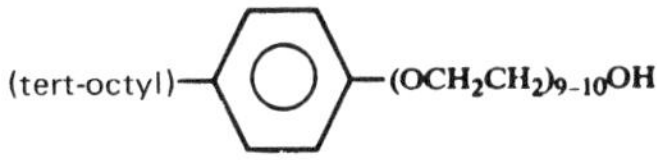

in the form of a 20% by weight aqueous solution, in an amount sufficient to incorporate in the latex 2% by weight of the polyethylene glycol-p-octyl phenol ether, based on the combined weight of the water. The pH of the latex is adjusted to 5 by addition of aqueous ammonia and a 1% aqueous solution of ammonium alginate is dispersed in the latex with gentle agitation in an amount sufficient to incorporate therein 0.1% ammonium alginate, based on the combined weight of the water.

The latex is allowed to stand for 16 hours at ambient temperature (22°C) during which time it will separate into a concentrated latex phase and a supernatant clear aqueous phase. Upon separation of the supernatant aqueous phase, the concentrated latex phase contains at least 75% by weight of polytetrafluoroethylene.

APPLICATIONS

Fire-Retardant Polyester-Polytetrafluoroethylene Compositions

Highlights of the Technological Achievement: A method for the production of thermoplastic polyester compositions comprising mixing a normally flammable thermoplastic polyester, preferably containing at least 80% by weight of repeating units of ethylene terephthalate, tetramethylene terephthalate or ethylene-1,2-diphenoxyethane-4,4'-dicarboxylate units, an effective amount of fire-retardant additives and an aqueous colloidal dispersion of PTFE, subjecting the mixture to conditions under which the polyester becomes molten and removing volatile materials from the melt.

Reference: C.P. Smith; U.S. Patent 4,191,678; March 4, 1980; assigned to Imperial Chemical Industries, Limited, Great Britain.

Background: British Patent 1,360,121 describes glass-filled, fire-retardant thermoplastic polyester compositions in which a polytetrafluoroethylene resin is used to render the compositions nondripping as assessed by the Underwriters Laboratories' Bulletin No. 94, using specimens ⅛" thick. This specification states that degradation occurs, resulting in a deterioration in physical properties of the composition, if the ingredients of the composition are not thoroughly dried before compounding.

It has been found that aqueous dispersions of PTFE resins may be compounded with polyesters without any substantial reduction in the physical properties of the polyester. Moreover, the amount of PTFE resin in the form of an aqueous colloidal dispersion needed to obtain a satisfactory performance in the Underwriters Laboratories test may be considerably reduced from the preferred range stated in British Patent 1,360,121.

Process Conditions: There is provided a method of producing a thermoplastic polyester composition comprising intimately blending a normally flammable thermoplastic polyester, an effective amount of fire-retardant additives and PTFE in the form of an aqueous colloidal dispersion, subjecting the mixture to conditions under which the polyester becomes molten and removing volatile materials from the melt.

Using the method of the process, compositions can be produced which exhibit markedly improved resistance to dripping under burning conditions. Thus, whereas commercially available glass-filled thermoplastic polyester compositions containing about 30% glass and sufficient fire-retardant additives to make them self-extinguishing, are normally classified as nondripping when tested on ⅛" samples using the Underwriters Laboratories UL94 test, the inclusion of 0.1% by weight of PTFE added in the form of an aqueous dispersion by this method improves the nondripping characteristics of the composition so that samples are classified as nondripping by this standard test when tested at thicknesses as low as 1/32".

On the more severe tests, such as the CEE 10 test of the International Commission and Rules for the Approval of Electrical Equipment, higher levels of aqueous PTFE dispersion are required to give a nondripping performance, but, nevertheless, this test also demonstrates that lower levels of aqueous PTFE improve the dripping performance by comparison with samples not containing the aqueous PTFE.

The ingredients of the composition may be mixed together, for example, by tumbling prior to feeding the mixture to the mixing apparatus or the ingredients may be added in turn to the mixing apparatus. The intimately mixed composition in the extruder is normally passed through an exit die to provide laces which may be cooled in water and chopped into granules. It is an advantage of the method that the aqueous dispersions of PTFE may be much more easily dispersed than the solid PTFE resins used in British Patent 1,360,121, so that the choice of apparatus for intimately blending the ingredients is much less critical.

The fire-retardant materials suitable for imparting nonburning characteristics to the compositions may be chosen from a wide variety of compounds which derive their effectiveness from the presence of a substantial proportion of elements such as halogen and phosphorus in the compound. Halogen-containing organic compounds are particularly effective with polyesters, especially those compounds containing substantial proportions of bromine. Preferred compounds are those of the formula:

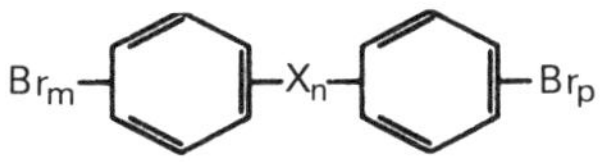

where X may be –O–, –S–, –SO–, $-SO_2-$ and $-CH_2-$; n may be 0 or 1; and each of m and p may be from 1 to 5. The fire-retardant additive will normally be effective at a concentration between 3 and 25% by weight of the polyester, although the preferred range is 5 to 15%.

Tests: The samples were tested according to both the UL94 test and to the more severe test described on page 23 of Section L of Publication 10, Part 2 of the International Commission and Rules for the Approval of Electrical Equipment (CEE 10) (October 1972). This test, which is normally used for testing electrical components, involves the application of a butane flame to the test samples for successive periods of 1 minute, 1 minute and 2 minutes. In the results below, the time periods quoted in the CEE 10 test are the cumulative times taken from the start of the first ignition.

PTFE as Polymer (% by wt of composition)	Extrusion Behavior	Melt Flow Index*	UL94	CEE10
			.... (1/16")	
8	**	Not measurable	NM	NM
4	**	4	***	NM
0.8	†	15	***	Dripped on 3rd ignition after 150 sec
0.6	†	20	***	Dripped on 3rd ignition after 165 sec
0.4	†	NM	***	Dripped on 3rd ignition after 180 sec
0.2	†	22	***	Dripped on 2nd ignition after 90 sec
0.1	†	25	***	Dripped on 2nd ignition after 60 sec
0	Smooth glossy lace, no extrusion problems	28	dripped	Dripped on 1st ignition after less than 30 sec

Note: NM is not measured.

*Measured according to ASTM method 1238-62T.
**Very high melt strength with no lumps of PTFE visible.
***VO–nondripping.
†No extrusion problems. Lace had smooth silky finish.

The compositions gave a VO performance on the UL94 test and were nondripping down to concentrations of 0.1% by weight of PTFE as polymer. The behavior on the more severe CEE 10 test varied, depending on the concentration of PTFE, although dripping occurred at all concentrations up to 0.8% PTFE.

Stable Aqueous Emulsions of Trichlorotrifluoroethane

Highlights of the Technological Achievement: A stable emulsion comprising from 70 to 98.5% by weight of trichlorotrifluoroethane, from 1 to 20% by weight of water, and from 0.5 to 10% by weight of a mixture of at least one hydrophilic amine phosphate ester selected from the neutralization product of:

(a) A phosphoric compound selected from a monophosphoric, diphosphoric, or a mixture of monophosphoric and diphosphoric compounds of (1) straight- or branched-chain aliphatic alcohols containing from 6 to 20 carbon atoms or (2) ether alcohols obtained by the ethoxylation by 1 to 5 mols of ethylene oxide of an alkylphenol or of a straight- or branched-chain aliphatic alcohol containing from 6 to 20 carbon atoms; and

(b) A hydrophilic, water-soluble amine selected from primary, secondary, or mixture of such amines; and

at least one lipophilic phosphate ester selected from the neutralization product of (a) and a lipophilic amine, the mixture of amine phosphate esters containing by weight thereof at least 5% by weight of the hydrophilic amine phosphate ester.

Reference: P. Chabert and L. Foulletier; U.S. Patent 4,191,666; March 4, 1980; assigned to Produits Chimiques Ugine Kuhlmann, France.

Background: The use of emulsions of water in trichlorotrifluoroethane for the cleaning of parts and various objects is well-known. These emulsions make it possible in effect to eliminate in a single operation greasy dirt, which is soluble in trichlorotrifluoroethane, and dirt which is soluble in water such as mineral salts and residues of the fluxes used in soldering. Such emulsions, however, are not stable and this limits their use.

Process Conditions: The trichlorotrifluoroethane (1,1,2-trichloro-1,2,2-trifluoroethane) is the major component of the emulsion comprising, as noted, from about 70 to 98.5% by weight of the emulsion.

As to the surface-active agent, it is produced from the combination in the form of salts of (a) a monophosphoric or diphosphoric ester of a mixture of monophosphoric and diphosphoric esters of straight- or branched-chain alcohols containing from 6 to 20 carbon atoms, such as octanol, 2-ethylhexanol, decanol, dodecanol stearyl alcohol, alcohols or mixtures of branched-chain alcohols obtained by the Oxo process, or the ether alcohols obtained by the addition of ethylene oxide on to the preceding alcohols or on to alkylphenols such as octylphenol, nonylphenol and containing a number of ethoxy groups of between 1 and 5; and (b) a pair of amines or mixtures of amines, each of which has been selected from one of the two following groups.

The first group comprises the secondary amines of the formula RNHR', in which R and R' are aliphatic radicals containing less than 4 carbon atoms and which may contain a primary alcohol function such as diethylamine, diethanolamine, N-methyl and N-ethyl ethanolamine, dipropylamine, diisopropylamine; and the primary amines of the general formula R_1NH_2, in which R_1 has the same meaning as the radicals R and R' above and may also contain ether functions such as 2-(2-aminoethoxy)ethanol. All these amines are very soluble in water and possess a hydrophilic character.

The second group comprises straight-chain or branched aliphatic amines containing a number of carbon atoms of between 6 and 20 such as hexylamine, octylamine, 2-ethylhexylamine, dodecylamine, stearylamine, oleylamine; commercial mixtures of fatty amines containing a number of carbon atoms of between 8 and 20; cyclic amines such as cyclohexylamine, N-methylcyclohexylamine, N-ethylcyclohexylamine, dicyclohexylamine; the diamines $R_2NHC_3H_6NH_2$ in which R_2 is an alkyl radical containing 8 to 20 carbon atoms; and commercial mixtures containing these amines, for example, oleylaminopropaneamine and stearylaminopropaneamine; and the polyamines such as diethylenetriamine, triethylenetetraamine and tetraethylenepentamine.

All these amines are sparingly soluble or insoluble in water and soluble in oils and nonpolar solvents, and possess a lipophilic character, except the polyamines which, although very soluble in water, give, in combination with the phosphoric esters, salts which are very sparingly soluble in water, but soluble in nonpolar organic solvents and, therefore, belong to this group.

Example: A commercial octylphosphoric acid (mixture of phosphates of monooctyl, 60 mol % and dioctyl, 40 mol %) is neutralized exactly (within ±1%) with diisopropylamine on the one hand and dodecylamine on the other hand, and mixtures of the resultant products used to prepare a series of 2% solutions in trichlorotrifluoroethane.

Water is gradually added to each of these solutions which are agitated mechanically so that the water is solubilized, forming a clear and colorless emulsion which is stable when at rest. The following table shows the maximum quantity of water which is solubilizable at 20°C in 100 g of 2% solution of mixture of amine octylphosphates, reckoned on the composition by weight of this mixture.

..... Mixture of Octylphosphates		Maximum Quantity of Water-Solubilizable
Diisopropylamine	Dodecylamine	
.... (wt %)		(g/100 g of solution)
100	0	<0.1
82	18	0.5
80	20	6
68	32	6
62	38	2.5
53	47	1.5
0	100	<0.1

The octylphosphoric acid used in this example and the optimum composition of the mixture of amine octylphosphates is: 68 to 80% of diisopropylamine octylphosphate and 32 to 20% of dodecylamine octylphosphate.

Hair Grooming Aid Containing Fibrillatable Polytetrafluoroethylene Resin

Highlights of the Technological Achievement: This process concerns a treatment which may be applied on practically all hair and which substantially improves the cosmetic properties thereof. More particularly, it contemplates a hair grooming aid to beautify hair, particularly living hair, which includes as a necessary ingredient a solid phase of finely divided fibrillatable polytetrafluoroethylene (PTFE) resin. The method includes a process of grooming hair comprising applying the grooming aid to hair and brushing the hair at a temperature above the transition temperature at which PTFE fibrillates to form microscopic and submicroscopic fibers in an amount sufficient to give the hair a natural appearance and to maintain its set for a long time.

Reference: G.A. Shaw; U.S. Patent 4,184,973; January 22, 1980; assigned to The Harshaw Chemical Company.

Background: Hair lacking elasticity is difficult to style and manage. Hair that has been overprocessed by being chemically penetrated for tinting or bleaching, and the like, over and over again usually is damaged and will no longer take a set. The hair just hangs. Sets are ineffective because the elasticity has been lost.

Most available so-called conditioners are ineffective to elasticize the hair, and this loss of elasticity detracts from the natural springy look of healthy hair. Particularly, with the emphasis on the dry look, it is as desirable to retain the spring seen in normally elastic hair as it is difficult to treat it to do so. Therefore, a need exists to provide a grooming aid and method to enhance the naturalness and elasticity of hair, and to bias the hair so-treated into a natural set, which has remarkable endurance. The hair grooming aid of this process is directed to fulfilling this need.

Process Conditions: The actual quantity of a colloidal aqueous dispersion of PTFE which may be in a hair grooming composition may vary widely though, in general, no more PTFE is used than is required to produce the desired cosmetic effect with a normal application. Thus, though fibrillatable PTFE is a valuable polymer, it is effective in a hair dressing in such a small proportion that it does not contribute significantly to the cost of a hair grooming aid. Particularly since hair treated with PTFE tends to retain a portion of the PTFE even after it is washed, subsequent grooming of the hair may be accomplished with a most sparing use of this grooming aid.

To this end, a typical grooming aid may include from 0.001 to about 1% PTFE based on the weight of the nonsolid phase, the specific amount used being determined largely by the specific function of the grooming aid.

An important part of this process is the discovery of means for adding the PTFE to hair to provide an even application of the PTFE particles at very low concentrations, for example, 0.01% by weight of hair, and at concentrations less than, for example, 1% by weight of hair. The higher concentrations will work very well, but will mat the hair and leave an obvious excess of white PTFE agglomerates which will brush out to a degree resembling loose dandruff. At levels somewhat below 1%, for example, 0.2% by weight of hair, this latter effect will still persist to some degree.

The optimum results of holding the hair in place and retaining or enhancing the natural appearance of the hair is typically in the range of between 0.01 and 0.05% of PTFE by weight of hair for most common types of hair.

Example: A shampoo is formulated with 5 pbw solid PTFE in 100 pbw liquid. Thus, in 100 g of shampoo, approximately 9 pbw of 60% solids T-30 colloidal aqueous dispersion of PTFE are used. A small portion of the shampoo is placed in the palm of the hand and rubbed into wetted hair until the desired sudsing or shampooing action is obtained. The hair is thereafter washed in a conventional manner, recognizing that some, and possibly a major proportion of the PTFE solids in the hair, is washed away leaving an effective concentration of PTFE.

The washed hair is then damp-dried with a towel, combed, set in the desired style and permitted to dry at room temperature in the usual manner. After the hair dries, it is warmed by a heat lamp and brushed repeatedly with a soft bristle brush until the desired body and obedience is obtained. Generally, fibrillation of the PTFE particles is evidenced by a slight, but noticeable drag on the bristles of the brush. The same effect may be obtained when the hair is still damp or nearly dry, except that damp hair is preferably brushed at an elevated temperature. It is found that the treated, brushed hair has been imbued with a desirable luster and softness, and enhanced body and elasticity such as is a characteristic of natural, healthy hair; yet, treated hair, unlike untreated hair, maintains its set, stays acceptably clean longer and can be repositioned with brushing.

Fluorocarbon Polymer-Polymeric Adjunct Coating Compositions Stabilized Against Discoloration

Highlights of the Technological Achievement: The discoloration of finishes derived from coating compositions containing (a) fluorocarbon polymer, along with (b) a polymeric adjunct, and (c) a liquid carrier, can be significantly reduced by adding to the composition, before it is applied, a compound which decomposes to an oxide or hydroxide of a particular metal.

Reference: E. Vassiliou and W. Van Hoeven, Jr.; U.S. Patent 4,147,683; April 3, 1979; assigned to E.I. DuPont de Nemours and Company.

Background: Cookware, coated with fluorocarbon polymers of various sorts, has come into widespread use. Most housewives prefer to use such cookware in their kitchen because food is less likely to stick to it than to uncoated cookware and because it is so easy to clean.

These fluorocarbon polymer coatings are ordinarily provided in various colors. The darker ones of these are generally satisfactory, but the lighter colors, especially white and unpigmented, tend to be discolored or gray because of carbonaceous residues of the various adjuncts and surfactants originally present in the compositions from which the coatings are derived.

Process Conditions: This discoloration or grayness can be significantly reduced and the purity of the color enhanced by using as the coating composition:

(a) About 25 to 96% by weight of the total of (a) and (b) of a fluorocarbon polymer of monoethylenically unsaturated hydrocarbon monomers completely substituted with fluorine atoms or a combination of fluorine atoms and chlorine atoms, the polymer having a number average molecular weight of at least 20,000;

(b) About 2 to 75% by weight of the total of (a), (b) and (c) of a polymer of ethylenically unsaturated monomers which depolymerizes and whose depolymerization products vaporize in the temperature range of about 150°C below the fusion temperature of the fluorocarbon polymer used to about the fluorocarbon polymer's decomposition temperature, the amounts of (a) and (b) totaling 100%;

(c) Enough of at least one compound of at least one of the metals: cobalt, iron, bismuth, cerium and manganese, which decomposes in the temperature range of about 100° to 500°C to give at least 0.2% by weight of the metal in the compounds of an oxide or hydroxide to provide at least about 0.005 pbw of metal per 100 pbw of (a); and

(d) a liquid carrier.

Representative of the monoethylenically unsaturated acid units are those derived from acrylic acid, methacrylic acid, fumaric acid, itaconic acid and maleic acid (or anhydride). Acrylic acid and methacrylic acid are preferred because of their availability. The acid units of these polymers can optionally be esterified with glycidyl esters of 4 to 14 carbon atoms (total).

The glycidyl ester preferred for its availability is Cardura E ester (Shell Chemical Company). This is a mixed ester of the general formula:

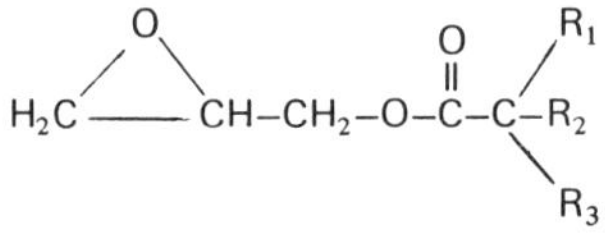

where R_1 is CH_3 and R_2 and R_3 are lower alkyl groups, R_1, R_2 and R_3 containing a total of 7 to 9 carbon atoms.

The polymeric adjuncts preferred for use for the better coalescence they confer to films during the curing stage are:

(1) Methyl methacrylate/2-ethylhexyl acrylate/methacrylic acid terpolymers, preferably the 35-50/40-60/1-15 (weight ratio) terpolymers;

(2) Butyl acrylate/methyl methacrylate/Cardura E/acrylic acid quadripolymers, preferably the 23-27/33-37/21-25/15-19 quadripolymers;

(3) Styrene/methyl methacrylate/Cardura E/acrylic acid quadripolymers, preferably the 28-32/28-32/21-25/15-19 quadripolymers; and

(4) Methyl methacrylate/ethyl acrylate/methacrylic acid terpolymers, preferably the 30-45/45-60/1-15 terpolymers.

FORMALDEHYDE PRODUCTS

PROCESSING

Preparation of Cationically Modified Formaldehyde Resins of Prolonged Stability

Highlights of the Technological Achievement: Cationic resins soluble in water and stable in solution, based upon melamine and formaldehyde, primary or secondary aliphatic amines, alkanolamines or a polyamine are prepared by reacting melamine with formaldehyde in a methylolation reaction in methanol in the presence of one or more of the aforementioned amines.

Reference: G. Tesson; U.S. Patent 4,182,839; January 8, 1980; assigned to Manufacture de Produits Chimiques Protex Sociétè Anonyme, France.

Background: The new resins prepared according to the process are principally intended to improve the humidity resistance of paper and they have also been found to be particularly desirable as adhesion-promoting agents for varnish or other coatings applied to protect solid supports.

Process Conditions: According to the process, the resin solution is prepared by the steps of:

(a) Carrying out a methylolation reaction of melamine with formaldehyde in a methanol solution and in the presence of one or more of the aforementioned amines by heating the solution to a temperature of 60° to 85°C for 15 minutes to 2 hours at an alkaline pH of 7.5 to 10;

(b) Carrying out a methylation-condensation reaction with the solution resulting from step (a) at a pH of 5 to 6, preferably 5.4 to 5.7, until a minimum viscosity of 17 seconds is obtained, as measured at 20°C with a consistometric head of the Afnor No. 6 type; and

(c) Terminating the methylation reaction by the addition of aqueous formaldehyde and continuing the condensation for a period of

30 minutes to 2 hours at 60° to 90°C at a pH of 5.5 to 7, preferably a pH of 6 to 6.5

It has been found to be convenient to effect only a partial methylation of the methylol groups of the resin in order to retain, to a significant extent, the good reactivity of the latter.

Methylation Reaction: According to another feature of the process, the methylation reaction [step (b)] is followed during the course of the reaction by infrared spectrography of samples of the reaction system. The characteristic absorption band of the methoxy group is about 1,080 cm^{-1} as the methoxy group is formed during condensation in an acid phase and in the presence of melamine and formaldehyde in a methanol solution containing a salt of an amine, alkanolamine, or a polyamine. When the desired methylation level is obtained, it is desirable to add a quantity of formaldehyde, in the form of an aqueous solution, to carry out the desired condensation.

It is indeed surprising to note that a very slight methylation is all that is required and that methylation can be carried out until a peak is discovered in the absorption curve of infrared spectrography at the frequency of 1,080 cm^{-1}, this being sufficient to yield a resin of good stability and having excellent reactivity.

Simultaneously with the methylation reaction in the acid medium there occurs a condensation of the methylol groups of the melamine among themselves or with the hydrogens of the amines of melamine and the amines (primary, secondary, alkanolamines or polyamines) introduced into the system. A condensation of the methylol groups of melamine with those of trialkanolamine also can occur. This condensation is characterized by an increase in the viscosity in the course of the reaction.

The molar ratio of formaldehyde to melamine during the methylation step should be between 3 and 9, preferably 6. After the methylation-condensation reaction it is preferable to add an excess of formaldehyde in the form of an aqueous solution. In practice it has been found that there is no theoretical limit to the molar ratio of formaldehyde to melamine which should be supplied in this step but it has been found to be useless to exceed a ratio of 25.

Amines: It has been found to be advantageous to provide 0.1 to 2.5, preferably 0.5 to 2, mols of an aliphatic primary amine or an aliphatic secondary amine, of an alkanolamine or of a polyamine, or a combination thereof, per mol of melamine.

It has been found to be totally unimportant whether one or more of these amines are used or the amine is utilized in the form of its mineral acid or organic salt. The preferred salts are the hydrogen chloride, acetic acid, formic acid and the like. The salts and amines can be used individually or in a mixture.

As the aliphatic primary or secondary amine, one should use preferably a compound of the formula R_1-NH_2 or

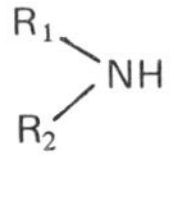

in which R_1 and R_2 are identical or different and each represents a lower alkyl group containing 1 to 4 carbon atoms.

Best results have been found with methylamine, dimethylamine, ethylamine, diethylamine, propylamine, isopropylamine, dipropylamine, butylamine.

The alkanolamines which are utilized according to the process have the formula:

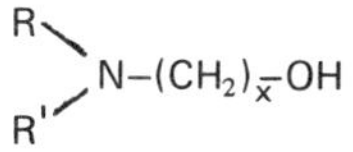

in which x = 1 to 4, and R and R' are identical or different, each representing hydrogen, an alkyl group or alkanol of 1 to 4 carbon atoms. In this case as well, the alkyl is lower alkyl and the preferred alkanolamines are monoethanolamine, diethanolamine, propanolamine, diethylethanolamine, triethanolamine.

The polyamines which can be utilized according to the invention have the formula: $H_2N-(CH_2-CH_2-NH)_xH$, with x = 1 to 5, such as ethylenediamine, diethylenetriamine, triethylenetetramine, tetramethylenepentamine.

Product: The modified melamine formaldehyde resin according to the process, which is water-soluble, is of course prepared by the reaction of melamine with formaldehyde in methanol in the presence of the aforementioned amines, alkanolamines and polyamines utilized separately or in admixture. The methylolation reaction is, as noted, carried out in an alkaline medium at a pH of 7.5 to 10 by the addition of the amine and readjustment of the pH by means of a mineral or organic acid, at a temperature between 60° and 85°C and generally with reflexing of the methanol. The methylation time should vary with temperature but in all cases lies between 15 minutes and 2 hours.

The methylation-condensation reaction is then carried out upon the reaction system previously prepared by acidifying it to a pH of 5 to 6, preferably to a pH of 5.5 to 5.8 with a mineral acid or an organic acid, preferably hydrochloric acid. In this stage the reaction stage is heated to 70° to 90°C. The reaction is terminated when the measured viscosity at 20°C using an Afnor No. 6 consistometric head reaches a minimum of 17 seconds and preferably lies in the range of 20 to 100 seconds.

The methylation-condensation reaction can also be followed spectographically as noted using an infrared spectrometer and is terminated when a peak in the absorption curve is obtained at about 1,080 cm^{-1}, which is characteristic of the methoxy group. Thereupon the methylation reaction is terminated by the addition of aqueous formaldehyde and condensation of the melamine with the formaldehyde is carried out for 30 minutes to 2 hours at 60° to 90°C and a pH of 5.5 to 7 and preferably a pH of 6 to 6.5. The solution is then cooled to 20°C. The cationic melamine-formaldehyde resin thus obtained is totally soluble in water and possesses a stability in storage such that it can be stored for periods greater than 1 year at 20°C without modification or loss of ability to treat paper.

Example: A humid pellicle of regenerated cellulose is immersed in a bath containing 6% glycerol and 0.25 and 0.5% of the cationically modified melamine-formaldehyde resin expressed as dry material to test improved adhesivity. The regenerated cellulose is obtained by acid coagulation of the cellulose xanthate.

After drying to a water level of 6 to 12%, there is applied to this support a layer of nitrocellulosic varnish to a thickness of 1 μ. After drying for 1 hour in air and another 3 minutes at 95°C, the adhesion of the varnish to the dry support is tested as is the adhesion in a humid atmosphere of 95% relative humidity. The tests are made by measuring the tear resistance expressed in g/cm of the size of the test specimen. The results are given in the table below.

Resin Concentration in Adhesion Promoting Bath (%)	Resistance to Tearing Dry	Moist
0.25	257	92
0.5	310	162

The tests demonstrate that the adhesion of a coating material to a support is increased by application of a solution of the resin according to the process. In tests applying the resin to paper, the resistance of the paper to humid environments was noted.

Amine-Modified Lignosulfonate-Extended Phenol-Formaldehyde Resins

Highlights of the Technological Achievement: An amine-modified lignosulfonate is formed from a water-soluble basic amine and a water-soluble lignosulfonate salt. The amine-modified lignosulfonate is incorporated into a phenol-formaldehyde resole resin to provide a binder particularly suitable for mineral fiber insulation mats.

Reference: P.C. Herault; U.S. Patent 4,177,313; December 4, 1979; assigned to Georgia-Pacific Corporation.

Background: Mineral fiber mats are made by passing the mineral in molten form through a fiber forming device, e.g., multiple orifice plate or spinnerette, and spraying an aqueous thermosetting binder resin onto the fibers while they are at an elevated temperature. The binder spray deposits a thin coating of resin on the fibrous surfaces. The resin-coated fibers may then be laid down in a random fashion on a traveling grate or conveyor and passed through an oven where the resin is cured and the individual fibers are bonded together to form a mat. Among the materials which are formed into mats by this basic process or variants thereof are silica, magnesia, alumina, soda, lime, slag and the like. Products of this nature are referred to in the art as glass wool, rock wool, slag wool and the like.

The aqueous binder system which has found widespread application for mineral fiber mats is based on a phenol-formaldehyde resin.

Process Conditions: The present process provides an amine-modified lignosulfonate containing from 2 to 12% by weight of a water-soluble basic amine and from 88 to 98% by weight of a water-soluble lignosulfonate salt. The amine-modified lignosulfonate is incorporated into a phenol-formaldehyde resin to provide a resin binder composition comprising an aqueous solution of a phenol-formaldehyde resole resin and from 40 to 100% by weight based on phenol-formaldehyde resin solids of a water-soluble basic amine-modified lignosulfonate.

It has been found that when an amine is blended with a water-soluble salt of lignosulfonic acid, the resulting amine-modified lignosulfonate provides desirable and advantageous properties. Liquid chromatographic analysis of unmodified lignosulfonate when compared to amine-modified lignosulfonate shows a definite change in molecular weight distribution. When an amine is blended into lignosulfonate some of the smaller molecular weight groups appear to be shifted toward higher molecular weight chains.

Particularly preferred binders for mineral fiber mats include urea in addition to the amine-modified lignosulfonate extender and phenol-formaldehyde resin. The urea functions as a formaldehyde scavenger and as an antipunking agent (it tends to suppress the smoldering of paper or the like which may be attached to or in contact with the newly formed hot mineral fiber mat). Desirably, the urea is present in amounts of from 1 to 15% by weight based on the total weight of extender and phenolic solids.

Product: The use of amine-modified lignosulfonate extenders in conjunction with phenol-formaldehyde resins according to this process results in:

(1) Binders whose cure characteristics are similar to unextended phenolic insulation resins;

(2) Binder efficiencies better than those of unextended phenolic resins and at least equal to the efficiencies of unmodified lignosulfonate extended phenolic binder systems;

(3) Tensile strengths equivalent to standard lignosulfonate extended phenolic binders;

(4) Utilization of higher levels of extender while maintaining the required tensile strengths; and

(5) Improvements in tensile strength retention when exposed to a moist atmosphere.

In addition to its utility as a binder for mineral fiber mats, the extended binder of this process is useful as a laminating resin, inter alia, for wood or paper laminates, as a binder for foundry sand, and as an adhesive.

Tensile strength test data demonstrate the following trends:

(1) Substitution of the modified lignosulfonate solutions increases the tensile strength of all binder formulations, when compared to commercially available lignosulfonate solutions;

(2) Substitution of the modified lignosulfonate solutions improves the dry strength retention in all binder formulations; whereas, the commercial unmodified lignosulfonates do not;

(3) Some salts of lignosulfonate perform somewhat better than other salts of lignosulfonate and the same is true of salts of modified lignosulfonate.

Producing Aqueous Urea-Formaldehyde

Highlights of the Technological Achievement: Aqueous urea-formaldehyde resin solutions are prepared by (a) forming a first aqueous solution containing urea and formaldehyde in a molecular ratio in the range urea:formaldehyde = 1:3 to 1:1.5, (b) reacting the first aqueous solution under neutral or alkaline conditions to form a reaction mixture containing substantial amounts of methylol urea products, (c) vacuum evaporating the reaction mixture to produce a concentrated mixture having a solids content of 55 to 80% by weight, and (d) further condensing the concentrated mixture under acid conditions to produce a resinous material. The urea may be wholly or partially replaced by melamine.

Reference: D.A. Hubbard; U.S. Patent 4,174,310; November 13, 1979; assigned to Imperial Chemical Industries Limited, England.

Background: Known methods for producing aqueous urea-formaldehyde resin solutions are as follows:

(1) The condensation of urea with standard 36% HCHO w/w formalin (e.g., 35 to 45% w/w HCHO) at about 45 to 50% w/w solids content, followed by vacuum distillation to produce a final solids content between 55 and 80% by weight.

(2) The condensation of urea with stripped formalin of about 50% HCHO w/w or more. In this method no vacuum distillation is required.

(3) The condensation of commercial Ureaform with added urea to produce a resin solution having the final solids content desired without vacuum distillation. Ureaform is essentially stripped formalin as used in method (2) above, but stabilized with urea in smaller amounts than those required for the subsequent production of urea-formaldehyde resins.

The above methods have various disadvantages depending upon the method. These are as follows.

Method (1), in use for many years, is generally rather lengthy and requires excessive plant capacity due to the unwanted water present until the final process step is carried out.

Method (2) permits the manufacture of substantially increased tonnages of final product from a resin plant of any given size but has the disadvantage that concentrated formalin is highly unstable and difficult to transport. A further problem with method (2) is that the energy cost of concentrating formalin before the reaction with urea is greater than that required to vacuum evaporate the low solids resin solution produced in method (1).

Method (3) uses Ureaform, which is a stable product and clearly avoids the stability problem of method (2). However, the higher cost of stripping weak formalin solutions still remains. Additionally the conversion of Ureaform to useful commercial urea-formaldehyde resins is more difficult than the conversions in methods (1) and (2), owing to the need for very careful monitoring of the extra urea additions required and these steps can frequently be extremely lengthy and therefore costly. This is evident from U.K. Patents 1,376,385 and 1,376,388.

Process Conditions: A process for producing aqueous urea- and/or melamine-formaldehyde resin solutions comprises the steps (a), (b), (c) and (d) given above.

The amounts of methylol urea and/or methylol melamine products in the reaction mixture comprise upwards from 80% by weight of the reactants, i.e., less than 10% by weight of the formaldehyde present is in the form of free formaldehyde.

The first aqueous solution may be formed by any conventional process using urea and/or melamine and standard commercial formalin (e.g., about 37% w/w) at a neutral or near neutral pH.

The first aqueous solution may be reacted, e.g., by leaving for 12 to 36 hours (preferably 24 hours) at ambient temperature or by heating for 10 to 60 (preferably 15 to 30) minutes at 50° to 100°C to produce the reaction mixture. The reaction mixture may then be, immediately if so desired, vacuum-evaporated from its initial solids content of approximately 45% by weight up to the solids content required within the range 55 to 80% by weight, for example 70% by weight.

The concentrated mixture thus formed is a clear nonresinous product which is further condensed, e.g., by heating to 70° to 100°C at atmospheric pressure, under mildly acid conditions which may range preferably from pH 5 to 6.5. The temperature may be slightly higher than 100°C if this step is carried out under a pressure slightly higher than atmospheric pressure.

After vacuum evaporation the temperature of the concentrated mixture produced is preferably not allowed to fall before the condensation step is commenced.

After vacuum evaporation no further evaporation is required, and at the end of the resin-forming step the product is suitably cooled and its pH increased, preferably to between 7.0 and 8.0.

Continuous Processing: The process lends itself to the production of resins on a continuous basis in that the reaction mixture can be fed through a conventional continuous evaporator and the resin-forming step can then be carried out in one or more continuously stirred tank reactors. The process may also be carried out as a batch process in which the concentrated mixture is formed immediately before the resin-forming step.

The resin-forming step may be carried out partially or completely at pressures above atmospheric pressure. This gives materials of improved water resistance. Mixed resins may be formed by resins formed in this fashion being mixed with conventionally-produced resins.

Advantages: The advantages of the process lie in the ease of evaporation of the reaction mixture (under vacuum). The energy consumption is materially lower than that required for stripping commercial formalin in a stripping column and the evaporation rates can be substantially higher than those obtained during vacuum evaporation of a viscous urea-formaldehyde resin solution because there is practically no increase in viscosity during the evaporation step. At the same time full advantage can be taken of the fast throughput in the final step equivalent to that obtained from the use of concentrated formalin solution as a starting material.

By comparison with the known processes described above it is estimated that production rates can be increased by at least 2.6 times in a single vessel used both as a reactor and an evaporator, and by at least 1.6 times in a system using a vessel for reaction only and a conventional evaporator for the concentration step.

The process is very suitable for preparing low cost resins.

Formalin Used: Formalin with a formaldehyde content of 36.5% by weight, a methanol content of 5.8% by weight, and an acidity of 0.02% by weight was used. pH adjustments were made using very small amounts of 2 N NaOH solution or 10% formic acid solution.

Example: A series of experiments were performed to measure the changes in composition occurring during the evaporation of a formalin/urea premix, and the effect of different types of thermal treatment prior to evaporation. All evaporations were performed under reduced pressure and at an evaporation rate of about 5 ml/min (except where shown). All premixes were adjusted to pH 7 (except where shown). The temperature of evaporation was as indicated; however where the premix had been heated above the evaporation temperature vacuum was applied at this temperature to assist in the cooling.

The distillate was analyzed for free formaldehyde, and a progressive but usually small amount of formaldehyde was found to be evolved. The concentration of formaldehyde in the first 10% of the distillate was found to be approaching twice that of the overall concentration of formaldehyde in the distillate. The results are presented in the following table.

Thermal Treatment Premix Temp. (°C)	Treatment Time	Evaporation Temp. (°C)	Concentration* Initial . (% by wt) . .	Final	Molar Ratio Formaldehyde: . . . Urea . . . Initial	Final	Formaldehyde in Distillate (% by wt)
–	–	47-48	52	83	2.1	1.96	5.10
23	17 hr	46-48	52	79	2.1	2.03	2.63
50	30 min	47	52	79	2.1	2.03	2.66
50	30 min	47	54	81	1.95	1.89	2.19
50	30 min**	46-47	54	80	1.95	1.89	2.40
80	30 min	66-68	53	78	2.0	1.95	1.79
80	60 min	65-68	53	78	2.0	1.96	1.57
80	120 min	65-67	53	80	2.0	1.97	1.17
90	15 min	65-67	53	78	2.0	1.95	1.97
90	60 min	64-67	53	79	2.0	1.97	1.36
Reflux (98)	15 min	66-68	53	80	2.0	1.96	1.72
Reflux	30 min	58	52	80	2.1	2.07	1.34
Reflux	30 min	47-48	54	81	1.95	1.92	0.96
Reflux	10 min	66-68	53	80	2.0	1.96	1.60
Reflux	10 min	64-68***	53	77	2.0	1.96	1.68

*Calculated concentrations of urea + formaldehyde.
**At pH 9.
***Evaporation rate = 3.3 ml/min.

APPLICATIONS

Aqueous Solutions of Etherified Melamine-Formaldehyde Resins with Long Shelf Life and Low Free Formaldehyde Content

Highlights of the Technological Achievement: An etherified melamine-formaldehyde resin is described that has prolonged shelf life, comparatively low free formaldehyde content, and is especially adapted as a binder for glass fiber mats. The melamine-formaldehyde resin is prepared in several distinct stages including condensation in a basic aqueous solution, etherification by cooling and acidification, neutralization with triethanolamine, and final stabilization. The free formaldehyde content of the resin so produced can be adjusted by reacting it with urea.

Reference: J.-P. Meunier, J. Joachim and B. Kafka; U.S. Patent 4,183,832; January 15, 1980; assigned to Saint-Gobain Industries, France.

Background: Methods for preparation of such resins are already known, notably from German Patent 2,005,166. According to this former process an alkaline condensation of melamine with formaldehyde is effected in the presence of a polyol. The formaldehyde is obtained in the form of solid paraformaldehyde. The use of paraformaldehyde presents a significant disadvantage: the price per kilogram of anhydrous formaldehyde in this form is 2 to 3 times higher than in aqueous solution form.

The purpose of the present process is a method of preparation of a solution of melamine-formaldehyde resins by condensation in alkaline solution of formaldehyde and of melamine followed by an etherification.

The present aqueous solutions yield excellent properties to the glass fiber mats such as tensile strength and suppleness.

Process Conditions: The operating conditions of the four phases of the method will be described in more detail hereafter.

First Phase – The melamine used can be of technical quality, yet standardized at 99% by weight at least in melamine, and 99% of the particles have a diameter less than 160 μ.

The formaldehyde used must be in aqueous solution at a weight concentration of 30 to 50%; and preferably one will use a 36% aqueous solution because of its wide availability. These solutions should, however, not include more than 10% maximum of methanol by weight and this content will be preferably less than 1% in order to avoid the presence of methoxyl groups in the etherified resin. In fact, to maximally increase the water dilutability, it is preferred that etherification be realized above all by the polyol or by the triethanolamine.

The reaction is conducted as follows: First mix the required quantities of the aqueous solution of formaldehyde, a polyol, and triethanolamine and this mix, while agitated, is brought to the reaction temperature, ranging between 60° and 70°C and preferably between 63° and 68°C. At temperatures lower than 60°C the dissolving of the melamine in the environment is too slow.

Then add the melamine, little by little, in 10 to 15 minutes, maintaining the temperature and the agitation for a period varying from 30 to 90 minutes.

The reaction is stopped by rapid cooling at a precise moment which is determined by the cloud point of the reacting mix that is measured and that is followed during the reaction.

This cloud point is characteristic of the degree attained by the alkaline condensation. It is measured by the successive taking of samples from the reacting environment at regular time intervals. These samples are cooled by agitating them and then the temperature is noted from which a cloud is produced. This temperature, rather low at the beginning of the reaction, rises as this is continued.

The alkaline condensation reaction is thus stopped by rapid cooling when the cloud point appears between 40° and 65°C and, preferably, between 45° and 55°C, which is produced at the end of a time period already indicated of 30 to 90 minutes.

If the reaction is stopped for a cloud point less than 40°C, the cooled reacting environment then presents a high viscosity, which makes it difficult to maintain homogeneity by agitation. Inversely, one cannot surpass a cloud point temperature higher than that of the reaction, without causing the reacting environment to opacify at the reacting temperature. The same is the case where the temperature of the reaction would be 70°C, it would be, however, very difficult to measure a cloud point temperature higher than 65°C because the separation from the reacting temperature would be too slight.

Second Phase (Etherification) – It is the etherification phase of methylolmelamines obtained in the first phase by the one or several polyols that were already introduced into the reacting environment. These polyols can be the following: ethyleneglycols, diethyleneglycol, triethyleneglycol, glycerol, saccharose, d-glucose. Yet the preferred diol, because of its price and its availability is ethyleneglycol.

One of the following acids can be used: sulfuric, hydrochloric, orthophosphoric, nitric, formic, monochloracetic.

Immediately after the alkaline condensation is cooled, as already said, and while maintaining constant the obtained temperature, the environment is acidified to a pH, the value of which should depend on the chosen temperature. It is in these conditions of temperature and pH that the etherification reaction is effected.

These conditions should range between the following limits: from 20°C for a pH of 1.5 to 40°C for a pH of 3.0, or preferably, 33°C for a pH of 1.8 to 37°C for a pH of 2.2.

Third Phase – This phase of stabilization is important because it improves two properties of glass fiber mats formed with binders prepared from aqueous solutions of the resin of the process.

In fact, it was found that an increase in the duration of this stabilization phase increases the flexibility index of the glass fiber mats as well as their stress resistance.

Fourth Phase – The percentage of free formaldehyde of the aqueous resin solutions is, after the stabilization phase, even higher than the initial formaldehyde/melamine (F/M) molecular ratio. It is on the order of 6 to 12% or even more.

Such concentrations of free formaldehyde disturb the manufacturing personnel because the vapors emitted irritate their eyes and respiratory systems.

It is thus the principal objective of this fourth phase to reduce to at least 6% the content of free formaldehyde of aqueous resin solutions. Such concentrations no longer emit to the air irritating vapors except in a quantity tolerable to the workers.

But this operation presents two further advantages: one is to increase the tensile strength of glass fiber mats; the other is to reduce the speed of increasing viscosity of stored aqueous resin solutions and thereby to contribute to prolonging this possible shelf life.

This fourth and last phase consists of adding to the aqueous solution, at the temperature of the end of the third phase, urea of a ratio of preferably 0.8 to 1.2 molecule per molecule of melamine, this quantity depending on the initial F/M ratio and consequently, on the content of free formaldehyde in the reacting environment at the end of the third phase.

The urea is used in solid form, urea in pearls, because its endothermic dissolution in the water of the environment lowers the temperature and does not diminish the dry extract of the final product as would an aqueous urea solution.

The product is thus left to cool naturally for 12 to 24 hours before storing or using it.

Production of Encapsulated Water-Dispersible Materials

Highlights of the Technological Achievement: Capsules containing water-dispersible material are described. They are formed from an admixture of low HLB emulsifier, oily water-immiscible solvent for the emulsifier, urea-formaldehyde prepolymer, water-dispersible material, and water. This admixture is agitated to form a water-in-oil emulsion to which amphiphatic acidic catalyst is added to cure the prepolymer. After curing to solid form, the capsules are separated from the remaining liquid of the admixture.

Reference: R. Golden; U.S. Patent 4,157,983; June 12, 1979; assigned to Champion International Corporation.

Background: Encapsulation, and particularly micro-encapsulation, of various materials is a common means of providing such materials with a protective environment. Encapsulation with, for example, a urea-formaldehyde resin, provides such materials with an external matrix which insulates them, partially or completely, from conditions which might lead to their loss or to loss of their desired active properties.

Typical examples of the use of encapsulation include protection of pigments or dyes to obtain improved bleed resistance, light fastness and compatibility. Water-soluble fire retardants are commonly encapsulated for use under conditions of high humidity. Pharmaceuticals, pesticides, herbicides, antifouling agents, catalysts and the like may be encapsulated in order to provide for controlled release of such materials over an extended period of time. Similarly, encapsulation of photographic emulsions provides improved characteristics; of opacifying agents such as titanium dioxide, for more efficient pigment particle spacing; and of other well-known materials for improvement in their particular uses.

In general, prior art methods suffer from one or more of a number of basic procedural drawbacks. Chief among them is the approach to curing of the encapsulating prepolymer.

Process Conditions: The term amphiphatic catalyst as used in this discussion includes acids which are hydrophilic and oleophilic in nature, in other words, these acids are soluble in both water and water-immiscible liquids. Specific examples of such acids include inorganic acids, such as hydrogen chloride, titanium tetrachloride, silicon tetrachloride, boron trihalide, and boron trihalide complexes.

By adding the catalyst directly to the water-in-oil emulsion, the onset of curing of the urea-formaldehyde prepolymer is closely controlled. Premature gellation or curing is thereby avoided. Also, polymerization cannot occur before the dispersed droplets have obtained the size (determined during emulsification) desired for the eventual product capsules.

The amount of prepolymer used is a function of the type of product capsule to be produced. Ordinarily, however, a weight ratio of between 1000:1 to 1:1 of prepolymer to water-disperisble material is utilized, with the higher amounts of prepolymer preferably being provided for the more active materials encapsulated.

The emulsifier utilized in accordance with the present process should be one having a low HLB value. This value of hydrophile-lipophile balance is ordinarily within the range of between 2 and 8, so as to insure the stability of the water-in-oil emulsion.

Suitable low HLB emulsifiers include stearic monoethanolamide, polyethylene oxide-polypropylene oxide block copolymers (such as Pluronic L122, Pluronic P123).

After formation of the emulsion, the catalyst is added. Ordinarily, the agitation is continued so as to retain control over the droplet size. This is not necessary, however, because reaction may proceed rapidly and the resultant formation of resin about the droplets is sufficient to preserve their sizes. From about 0.5 to 10% of catalyst by weight of prepolymer is ordinarily added to the emulsion.

After completion of the polymerization step, the capsules containing the water-dispersible material are separated from the remaining liquid. This may be done by simple filtration.

Catechol Modified Resorcinol-Formaldehyde Latex Cord Dip

Highlights of the Technological Achievement: An aqueous alkaline dispersion of a rubber latex and dihydroxyphenol-aldehyde resins consisting essentially of resorcinol-aldehyde resins and catechol-aldehyde resins is useful in forming an adhesive bond between reinforcing materials such as nylon and rubber compounds. It is essential that the catechol-aldehyde resin be made separately from the resorcinol-aldehyde resin with which it is later mixed in solution. The dispersions of this process result in improved aged adhesion and lower fabric stiffness than the older resorcinol-formaldehyde latex dispersions.

Reference: R.S. Bhakuni and R.F. Laske; U.S. Patent 4,132,693; January 2, 1979; assigned to The Goodyear Tire & Rubber Company.

Background: This process is concerned with catechol-aldehyde condensation products and their application in adhering polymers to reinforcing fabrics and cords. The commonly used cord-to-rubber adhesives are water dispersions of latex and resorcinol-formaldehyde resins with other ingredients such as carbon black and blocked polyisocyanates added for specific end results. The composition and process described herein have substituted catechol-formaldehyde resins for part or all of the resorcinol-formaldehyde resin.

Process Conditions: Catechol-aldehyde resins have not, in the past, been identified with any superior adhesive properties. However, it has now been discovered that they do impart certain very desirable properties to reinforced rubber acticles, namely improved aged adhesion and lower fabric stiffness.

The process of manufacturing a typical resorcinol-formaldehyde latex adhesive is generally comprised of the following steps:

(1) Dissolve resorcinol in water in the mixing tank.

(2) Add a suitable aldehyde such as formaldehyde in solution form to the solution in (1).

(3) Add sodium hydroxide (or another suitable base such as ammonium hydroxide or potassium hydroxide) slowly to the mixing tank.

(4) After the resin is formed add the latex.

(5) Age for approximately 12 to 24 hours.

Formaldehyde is the preferred aldehyde and the one most commonly employed, but any aldehyde may be used which is water-soluble to the extent of at least ½ to 5 wt % and capable of condensing with a dihydric phenol in a reasonable length of time. Examples of other suitable aldehydes are acetaldehyde, furfural, propionaldehyde, and crotonaldehyde.

When sodium hydroxide is used as the base catalyzing the resinification reaction, the mol ratio of sodium hydroxide to resorcinol is generally from 0.05:1 to 0.4:1. In the case of other bases, the gram equivalent corresponding to these ratios for sodium hydroxide would be used.

The process is usually carried out at atmospheric pressure and room temperature although a cooling means is sometimes employed to remove the heat of the resinification reaction.

This method has modified the normal process for manufacturing adhesive dip. A catechol-aldehyde resin is made by reacting catechol and an aldehyde in the presence of a base in water at a temperature of from 40° to 95°C. The mol ratio of catechol to aldehyde is in the range of 1:10 to 3:1 and a ratio between 1:3 and 1.4:1 is preferred. The mol ratio of sodium hydroxide (or gram equivalent of another base) to catechol is preferably from 0.4 to 0.8. A sufficient quantity of the resin is normally made for use in several batches of cord dip adhesive. This operation is called preforming.

This procedure (preparing the resorcinol-aldehyde and catechol-aldehyde resins separately) has three principal advantages: (1) the cord dip mixing time is independent of the relatively longer reaction time for catechol-aldehyde resin; (2) the resorcinol and catechol are not competing for the same reactive sites on the aldehyde molecules; and (3) reaction temperature can be controlled at the best setting for the catechol-aldehyde and the resorcinol-aldehyde reactions independently. If resorcinol and catechol were to be reacted in the same system with an aldehyde, the resorcinol would react much faster with the aldehyde, thus affecting the ratio of the two resins and the type of resin obtained, and the mixing time would have to be increased to accomodate the catechol. The preformed catechol-aldehyde resin is added as an ingredient to the adhesive between steps (3) and (4) of the process given previously.

Product: The cord dip adhesives can be utilized with reinforcing cords of rayon, nylon, polyester and aramid polymers. The finished reinforcing cord produced by applying these adhesives to reinforcing cord and processing wetted cord in the normal manner can be incorporated in a variety of reinforced rubber articles such as tires, hose, and conveyor belts.

Example 1: A conventional resorcinol-formaldehyde latex (RFL) cord dip adhesive was made according to Example 4 of U.S. Patent 3,268,467. Another adhesive of the same composition except that the resin was comprised of a 80:20 mixture of resorcinol-formaldehyde/catechol-formaldehyde resins was also made. The two adhesives were tested for adhesion using ASTM Test No. D2138 (U-Adhesion) on two types of nylon cord. The results appear in the following table.

Adhesion Aging—Nylon 6 (1260/3)*

	U Adhesion (Newtons)			
	Cord A.		Cord B.	
Adhesive System	**Original**	**Aged**	**Original**	**Aged**
Resorcinol dip	163	169	146	65
Catechol-resorcinol dip	127	187	158	113

*Cord aging conditions: 5 hr at 149°C and 276 kPa in airbomb.

In both cases, the aged adhesion of the cord dipped in the catechol-resorcinol adhesive showed a marked improvement in aged adhesion compared to the RFL adhesive.

Example 2: A catechol-formaldehyde (CF) resin formulation was prepared and used to replace various amounts of the resorcinol-formaldehyde (RF) portion of a production adhesive formulation similar to that of Example 1. Nylon 66 (1260/3) was processed with these dips under standard conditions and was then calendered in a natural rubber compound, vulcanized, and tested.

In addition to the U-Adhesion test several other tests were run on samples of the rubber/fabric composite, which tests included: (1) Peel Adhesion, ASTM D413; (2) Flex Strip Adhesion, ASTM D2630; and (3) stiffness as measured on the Gurley Stiffness Tester. The results have been recorded in the following table.

	Resin Ratio (RF/CF)			
	100/0*	80/20	60/40	40/60
Peel adhesion				
Adhesion force (Newtons)	191	187	209	200
Percent coverage	75	75	75	70
1" flex strip adhesion				
Unflexed (Newtons)	107	94	107	94
Percent coverage	85	80	90	90
Flexed (Newtons)	49	49	89	71
Percent coverage	80	85	85	90
U adhesion (Newtons)				
Original	165	165	165	178
Oven aged 24 hr, 121°C	67	76	94	102
Bomb aged 5 hr, 121°C 276 kPa	160	179	200	173
Stiffness, mg/end	165	152	129	123

*Control.

GELATIN AND GELATIN PRODUCTS

PROCESSING

Gelatin from Collagen

Highlights of the Technological Achievement: The process relates to the manufacture of gelatin, whereby the gelatin is obtained from fresh derms removed from scraps of hides, sinews and tendons within the three hours following the slaughter of the animal.

Reference: L. Oudem; U.S. Patent 4,176,117; November 27, 1979.

Background: The known processes are slow to carry out and require expensive installations; the raw materials are collected together at the site of treatment after having travelled in bulk under conditions which do not guarantee their preservation; the raw materials sometimes arrive in a virtually rotting state and their handling and sorting are particularly unpleasant for the personnel; in addition, the state of the raw materials consequently requires several washings.

Subsequently, the treatment includes successive cookings of the products in water, each cooking lasting 6 to 8 hours, until gelatin is obtained in the state of broth; this latter must then be concentrated, after which the broth is placed in a cold chamber to provoke gelling of the product; the gelled block is then cut into slices and possibly further reduced to granules or platelets, which are conveyed to a drier with a view of obtaining a product with low water content.

Apart from the considerable time required for carrying out the known processes, the procedures also imply a certain number of handlings of the product before gelling, and after gelling, i.e., at the stage of obtaining the final product; in fact, the gelatin block is cut into slices and possibly reduced to finer particles before passing to the drying phase, and then various operations are carried out manually by the personnel.

This process remedies these drawbacks and enables a finished product in the form of substantially dehydrated gelatin to be continuously produced in a considerably shorter period of time than that required for carrying out prior known processes.

Process Conditions: A process for manufacturing gelatin from animal waste, wherein the raw materials (constituted by hide wastes, sinews or tendons) are treated, straight after slaughter, and are subjected to the following successive operations:

(A) The raw materials (derms of hides, sinews or tendons) are washed and degreased in a solution of sodium hydroxide enabling the pores to dilate and facilitating elimination of the blood and fatty matter;

(B) The material is subjected to a chopping operation reducing it to the state of paste;

(C) The pH of the paste is adjusted in order to neutralize the sodium hydroxide resulting from initial washing phase (A) and the paste thus brought to a neutral pH is washed several times in cold water and rinsed;

(D) The paste is then subjected to filtration by pressing to express the water therefrom and conserve only the solid matter forming the cake; and

(E) The cake resulting from the preceding filtration and constituted by virtually pure collagen, is subjected to cooking, yielding a broth of gelatin of high density, which is subsequently subjected to gelling at low temperature and to drying.

The reduction of the raw materials into paste is preferably effected in a so-called cutter, of the type used for chopping meat. The high density broth is preferably poured onto a conveyor belt circulating in a low-temperature chamber, at about 5°C to obtain gelling. The belt is preferably made of stainless steel. The raw materials, before the phase of reduction into paste, are subjected to washing by immersion in a sodium hydroxide solution of normality close to 0.25 N, at a temperature of 10°C for 2 to 6 hours, the whole being subjected to stirring. The paste, before the phase of filtration by pressing, is preferably subjected to several washings in cold water, at a temperature lower than 30°C with stirring and rinsing.

The cake obtained after pressing and constituted by the dry extract of the washed paste, constituted by substantially pure collagen, is dried and stored for an indefinite period of time; this intermediate product, constituted by a collagen cake, being subsequently taken out of storage and subjected to the subsequent phase of cooking to obtain gelatin.

The collagen cake is cooked at a temperature of 50° to 60°C by immersion in a quantity of water corresponding to twice or six times the weight of collagen treated, this cooking being continued for 3 to 5 hours, the broth obtained, cooled to 40°C, then being subjected to gelling and being poured to this end on a steel conveyor belt circulating in a chamber cooled to a temperature of 5°C.

Drawing: Figure 13.1 shows a general diagram illustrating the functioning of this process.

Figure 13.1: Process for Obtaining Gelatin from Collagen

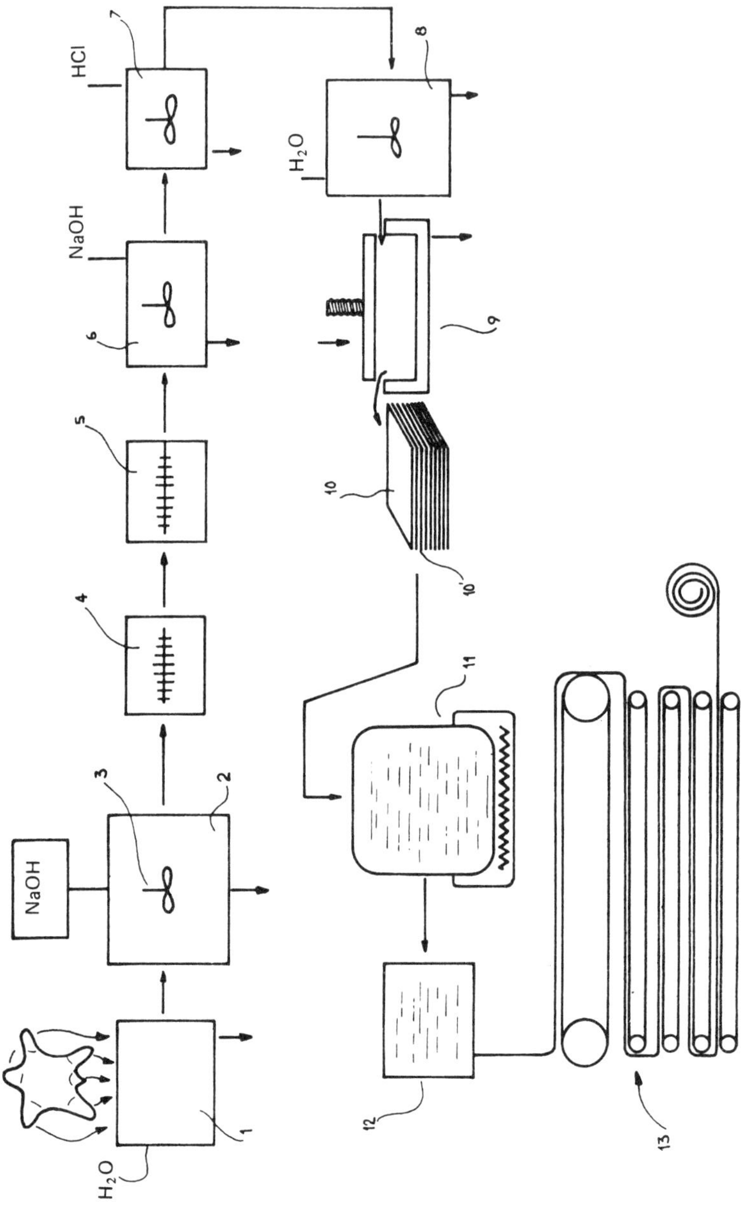

Source: U.S. Patent 4,176,117

High-Quality Gelatin Having Low Chromium Content from Chromed Waste Stock

Highlights of the Technological Achievement: A process for preparing low-chromium content gelatin from a chrome stock containing trivalent chromium which comprises treating a trivalent chromium-containing stock with a solution of a compound which forms coordination complexes with trivalent chromium present in the stock.

Reference: L.R. Smith; U.S. Patent 4,151,161; April 24, 1979; assigned to Canada Packers Limited, Canada.

Background: Although chrome waste or scrap stock has long been recognized as a potentially valuable source of gelatin, it is largely unused as a gelatin stock because it is not directly suitable for the conventional processes by which gelatin is made from traditional raw materials such as ossein, pigskin, limed hide splits and the like. If the usual techniques were applied directly to chrome stock, the gelatin produced would have a high chromium content which would render it unfit for most applications and, in particular, for edible use. It is apparent from the prior art that the problem of producing a gelatin which has a low-chromium content such as below 100 ppm, as well as a desirable bloom strength of 200 to 300 g or better, and an acceptable viscosity of from 30 to 50 mp or better from a chrome stock has not been satisfactorily achieved.

This process provides a low-chromium content edible gelatin having the desirable bloom and viscosity from a chrome stock. The process also provides for precipitation of chromium from the treatment solution to enable reuse of the treatment solution. The process is characterized by the removal of chromium prior to gelatin extraction.

Drawing: The process is schematically set forth in Figure 13.2. The figure includes not only the essential steps of the process but also illustrates certain optional steps. As is evident from the figure, the process is a multistage process which essentially includes preliminary dechroming, conditioning the resultant material to depolymerize collagen, oxidation of the conditioned stock to convert the remaining chromium to the hexavalent state, extraction of the gelatin from the oxidized material and preferably deionization to remove residual chromium.

Process Conditions: In a preferred embodiment, the preliminary dechroming is obtained by treating the splits with 3 pbw of a 10% w/v solution of sodium glycolate, having a pH of 8.0 to 9.0, for 24 hours at 40°C. This chrome-rich solution is removed and a second, identical chromium extraction is made by treating the splits with a fresh glycolate solution. In this way, most of the chromium present is displaced from the scrap and dissolved in the solution.

Example: The gelatins obtained from chromium-tanned splits had the following characteristics. Magnesium oxide (added as % by wt of stock) aided in the insolubilization of the chromium.

. . . Extraction . . .		. . . First Extract (60°C)			. . Second Extract (70°C) . . .		
MgO Used (% by wt)	pH	Chromium (ppm)	Bloom (g)	Viscosity (mpoise)	Chromium (ppm)	Bloom (g)	Viscosity (mpoise)
0.2	6.0-6.5	55	252	35.5	60	249	48.5
	7.0-7.5	66	327	42.0	50	260	55.0
	8.0-8.5	70	263	32.5	41	251	45.5
	8.0-8.5	24	284	37.0	14	235	52.5

(continued)

Conditions of ... Extraction ...		... First Extract (60°C)			.. Second Extract (70°C) ...		
MgO Used (% by wt)	pH	Chromium (ppm)	Bloom (g)	Viscosity (mpoise)	Chromium (ppm)	Bloom (g)	Viscosity (mpoise)
1.0	6.0-6.5	23	223	36.5	23	159	44.5
	7.0-7.5	15	231	40.5	25	158	55.0
	8.0-8.5	7	245	38.5	26	157	43.0

Conditions of ... Extraction....		... Third Extract (80°C)....			...Fourth Extract (95°C) ...		
MgO Used (% by wt)	pH	Chromium (ppm)	Bloom (g)	Viscosity (mpoise)	Chromium (ppm)	Bloom (g)	Viscosity (mpoise)
0.2	6.0-6.5	55	203	60.0	64	162	64.5
	7.0-7.5	64	194	65.5	73	161	81.0
	8.0-8.5	61	177	45.0	66	137	46.5
	8.0-8.5	45	180	63.5	91	127	59.5
1.0	6.0-6.5	38	131	55.5	92	78	49.5
	7.0-7.5	52	131	60.0	75	74	44.0
	8.0-8.5	63	136	63.5	99	78	37.5

Figure 13.2: High-Quality Gelatin Having Low Chromium Content from Chromed Waste Stock

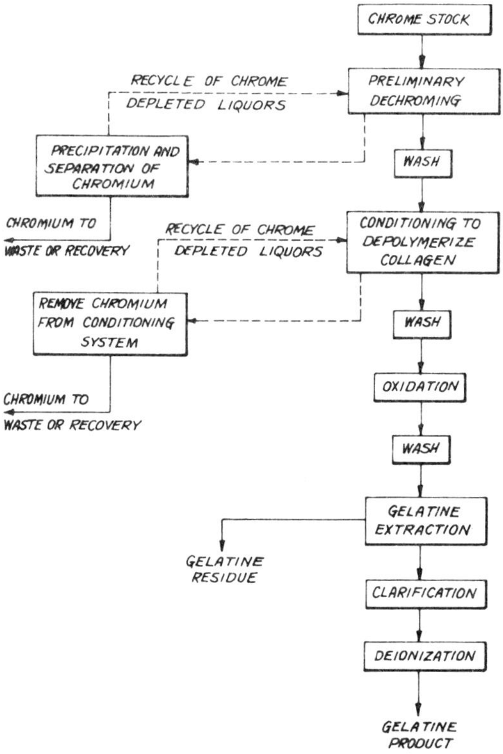

Source: U.S. Patent 4,151,161

APPLICATIONS

Gelatin Explosive Compositions

Highlights of the Technological Achievement: A gelatin explosive composition is disclosed comprising mixtures of nitric esters, o-nitrotoluene, blasting soluble nitrocotton, oxidizing salts, fuels and at least one dispersing agent. The gelatin explosive composition is markedly reduced in the quantities of liquid nitric esters and blasting soluble nitrocotton content.

Reference: B.C. Ghosh, G.M. Chopra and S.N. Sen; U.S. Patent 4,180,425; December 25, 1979; assigned to Indian Explosives Limited, India.

Background: Prior compositions were found to have the following major drawbacks:

(A) These gelatinous compositions were incapable of extrusion in small diameter down to 22 mm in the conventional cartridging machines;

(B) The sensitivity was adversely affected on storage within a period of six to nine months; and

(C) To facilitate extrusion and make the stuff more plastic, substantial quantity of wheat flour (a valuable foodstuff) had to be added.

The compositions of this process, containing a small amount of dispersing/wetting agent have not only helped to overcome all the disadvantages mentioned above but have also imparted marked improvements in a number of critical and special properties of the explosives and are superior in explosive characteristics and performance. The compositions are much safer in handling compared to the other compositions without a dispersing agent.

Tamarind seed flour (TSF) has been used as a waterproofing agent in explosive compositions. The use of waterproofing agents such as TSF is optional. Waterproofing agents such as TSF are generally used in the proportion of up to 1% w/w. However, it has additionally been found that by increasing the percentage of TSF, in addition to its waterproofing properties, its presence in explosive compositions of this process brings about still better dispersion of nitroglycerin and/or ethylene glycol dinitrate (NG/EGDN) in the explosive composition thereby enabling further reduction in the quantity of NG/EGDN and blasting soluble nitrocotton (BSNC). It has been found that increasing the percentage of TSF up to 3% achieves optimum results.

According to this process there is provided an improved gelatin explosive composition with a marked reduction in the quantities of liquid nitric esters and blasting soluble nitrocotton content, comprising essentially mixtures of nitric esters like nitroglycerin and/or nitroglycol, o-nitrotoluene, blasting soluble nitrocotton, oxidizing salts such as ammonium nitrate, fuels such as wood meal and/or waxed wood meal characterized in that the gelatin explosive composition contains at least one dispersing agent such as herein described and, if desired, a waterproofing agent such as tamarind seed flour or guar gum.

Examples 1 through 6: Explosive compositions were prepared from the components in the following table. The dispersing agent used was Cirrasol TW, which was an amide of a fatty acid.

Ingredients, %	Example No.					
	1	2	3	4	5	6
Nitroglycerin/nitroglycol	22.5	23.0	23.5	20.0	22.0	21.5
Nitrocotton (dry wt)	0.6	0.6	0.7	0.6	0.6	0.6
o-Nitrotoluene	1.0	1.0	1.5	4.0	–	–
Ammonium nitrate	55.3	65.0	68.5	65.0	47.5	35.0
Wood meal	1.0	0.5	0.8	0.5	1.0	0.3
Tamarind seed flour	1.0	1.0	1.0	1.0	0.5	–
Barytes	8.0	6.0	2.0	6.0	6.5	10.0
Waxed wood meal	1.0	1.3	1.4	1.3	1.0	0.7
China clay	1.0	1.0	–	1.0	1.0	–
Guar gum	–	–	–	–	–	0.5
Rosin	0.3	0.3	0.3	0.3	0.3	0.3
Sodium chloride	8.0	–	–	–	19.3	30.8
Cirrasol TW	0.3	0.3	0.3	0.3	0.3	0.3

Hardening Gelatin and Photographic Light-Sensitive Material

Highlights of the Technological Achievement: A photographic silver halide light-sensitive material having at least one hydrophilic colloid layer containing a hardened gelatin and/or a gelatin derivative.

Reference: H. Sera, T. Ishii, J. Yamaguchi and H. Shiraishi; U.S. Patent 4,173,481; November 6, 1979; assigned to Fuji Photo Film Co., Ltd., Japan.

Background: Gelatin is used as a binder for many photographic light-sensitive materials. For example, gelatin is used as the main component of a silver halide light-sensitive emulsion layer, an emulsion protective layer, a filter layer, an intermediate layer, an antihalation layer, a backing layer, a film support subbing layer and a baryta layer.

These photographic materials containing gelatin are treated with various aqueous solutions having different pHs and temperatures. Since a layer containing gelatin which has not been treated with a hardening agent has properties mainly dependent on the gelatin, it has poor water resistance and swells excessively in an aqueous solution, so that the mechanical strength is greatly reduced, and in an extreme case, the gelatin layer is sometimes dissolved, particularly in an aqueous solution having a high temperature of above 30°C or a highly alkaline aqueous solution. These properties are fatal defects for the physical properties of layers which make up photographic light-sensitive materials. However, most known gelatin hardeners all have one or more defects.

Process Conditions: It has been discovered that almost all of the disadvantages of known hardeners can be improved by treating a hydrophilic colloid layer containing gelatin and/or a gelatin derivative with a compound represented by the following general formula:

$$CH_2{=}CH{-}SO_2{-}(CH_2)_a{-}\left(\underset{OH}{CH}\right)_b{-}(R)_n{-}\left(\underset{OH}{CH}\right)_c{-}(CH_2)_d{-}SO_2{-}CH{=}CH_2$$

where a and d each represents 1 or 2; b and c each represents 0, 1 or 2 with the proviso that both b and c are not simultaneously 0; R represents a divalent group; and n represents 0 or 1.

In the above general formula, the divalent group represented by R includes any divalent group, but preferably a cyclic hydrocarbon group such as an arylene group having 6 to 12 carbon atoms, e.g., an m-phenylene group, etc., and an acyclic hydrocarbon group such as an alkylene group having 1 to 8 carbon atoms, e.g., a methylene group, an ethylene group, a trimethylene group, a tetramethylene group, etc. The divalent group represented by R can also be an aralkylene group having a total of 8 to 10 carbon atoms.

1 to 3 of the carbon atoms of the group defined above for R can be replaced by a hetero atom such as a nitrogen atom, a sulfur atom, an oxygen atom, etc. Suitable examples of groups containing hetero atom(s) include a group containing a $-CH_2OCH_2-$ group, a $-(CH_2CH_2O)_2CH_2CH_2-$ group, a $-CH_2CH_2OCH_2CH_2-$ group, a

group, etc. More preferably, R is a divalent branched or straight chain alkylene group having 1 to 4 carbon atoms.

Examples of compounds of the general formula which can be used in this process are given below.

$$CH_2{=}CHSO_2CH_2CH(OH)CH_2SO_2CH{=}CH_2$$

$$CH_2{=}CHSO_2CH_2CH(OH)CH(OH)CH_2SO_2CH{=}CH_2$$

$$CH_2{=}CHSO_2CH_2CH(OH)CH_2CH(OH)CH_2SO_2CH{=}CH_2$$

$$CH_2{=}CHSO_2CH_2CH(OH)CH_2CH_2SO_2CH{=}CH_2$$

$$CH_2{=}CHSO_2CH_2CH(OH)CH(OH)CH(OH)CH_2SO_2CH{=}CH_2$$

$$CH_2{=}CHSO_2{-}CH_2{-}CH(OH){-}CH_2{-}O{-}CH_2{-}CH(OH){-}CH_2{-}SO_2{-}CH{=}CH_2$$

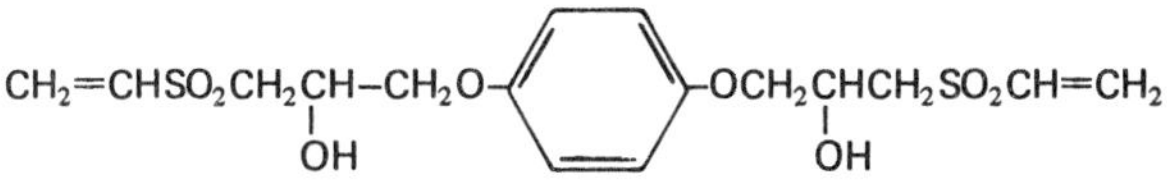

The amount of the hardener of this process used can be selected freely depending on the desired objective. The amount used generally ranges from 0.01 to 20% by weight, preferably 0.1 to 10% by weight, based on the weight of dry gelatin.

The gelatin hardener of this process may be used in the photographic light-sensitive materials in combination with a matting agent. Examples of matting agents which may be used include particles of water-insoluble organic or inorganic compounds, with an average particle size of 0.2 to 10 μ, preferably 0.3 to 5 μ.

Examples of suitable organic compounds which can be used are water-dispersible vinyl polymers such as polymethylacrylate, polyacrylonitrile, acrylonitrile-α-methylstyrene copolymers, polystyrene, styrene-divinylbenzene copolymers; cellulose derivatives such a methylcellulose, ethylcellulose, cellulose acetate, cellulose acetate propionate; starch derivatives such as carboxy starch, urea-formaldehyde-starch reaction products; and gelatin hardened with known hardeners and hollow microcapsules of hardened gelatin, which is obtained by hardening of a coacervate. Examples of inorganic compounds are silicon dioxide, titanium dioxide and calcium carbonate.

Dental Prophylactic Paste

Highlights of the Technological Achievement: A dental prophylactic paste comprising an abrasive, a humectant, gelatin alone or in combination with silica gel or a water-soluble silicate and optional flavoring and preservative ingredients having a viscosity of 1,200 to 150,000 cp at 30° to 40°C, which possesses improved flowability by which the paste can be poured through a nozzle under low pressure into individual containers as well as improved nonspatterability and cleaning efficiency in use by the dentist.

Reference: M.C. Gaffar; U.S. Patent 4,165,368; August 21, 1979; assigned to Colgate Palmolive Company.

Background: The primary object of this process is to provide a prophylactic dental paste with good flowability properties so that it can be poured into individual containers through a nozzle with a minimum of pressure, is substantially nonspattering and has improved cleaning efficiency in use by the dentist.

It has been discovered that by using gelatin instead of agar-agar in a range by weight of preferably 1 to 5%, and abrasives in a range by weight of preferably 40 to 60%, a paste is provided having a viscosity range of 1,200 to 150,000 cp at 30° to 40°C whose characteristics of flowability and nonspatterability are improved.

The gelatin which is employed in the instant prophylactic dental paste may be Type A having a gel strength of 75 to 350 Bloom g or Type B having a gel strength of 75 to 250 Bloom g as described in the publication "Gelatin" (Gelatin Manufacturers Institute of America, Inc., New York, 1973).

Examples: Table 1 contains examples of dental prophylactic pastes with and without fluoride in which medium pumice is the abrasive. Table 2 contains examples of dental prophylactic pastes containing fine pumice. Table 3 contains examples of dental prophylactic pastes with and without fluoride in which nonpumice abrasives are used.

Table 1

	Example No.					
	1	2	3	4	5	6
Paste ingredients, % w/w						
Pumice, medium	40	45	50	40	45	50
Glycerin	28	28	28	30	30	30
Gelatin (Type A, 300 Bloom g)	1	1	1	2	2	2
Sodium metasilicate	5	5	5	–	–	–
Syloid 244 (silica gel)*	–	–	–	2	2	2
Flavor	–	–	–	0.5	0.5	0.5
Sorbitol powder	–	–	–	2	2	2
Color	–	–	–	0.02	0.02	0.02
Potassium sorbate (preservative)**	–	–	–	0.05	0.05	0.05
Water, purified	26	21	16	23.43	18.43	13.43
Viscosity, cp at 35°C	1,400	1,500	6,500	1,250	5,750	24,000

*Particle size from 1 to 30 microns.
**Sodium benzoate may also be used as a preservative instead of potassium sorbate.

The viscosity was measured immediately after mixing using Brookfield Viscometer Model LVT, spindle No. 4, speed 12 rpm at 30° to 40°C, generally at 35°C and recorded in cp.

Table 2

	Example No.		
	7	8	9
Paste ingredients, % w/w			
Pumice, fine	40	45	50
Glycerin	30	30	30
Gelatin (Type A, 300 Bloom g)	2	2	2
Syloid 244 (silica gel)	2	2	2
Flavor	0.5	0.5	0.5
Sorbitol powder	2	2	2
Color	0.02	0.02	0.02
Potassium sorbate	0.05	0.05	0.05
Water, purified	23.43	18.43	13.43
Viscosity, cp at 35°C	4,400	19,000	126,000

Table 3

	Example No.		
	10	11	12
Paste ingredients, % w/w			
Silica, Ottawa #295 (abrasive)	50	60	60.00
Glycerin	30	20	8.00
Syloid 244 (silica gel)	2	2	2.00
Flavor	0.5	0.5	0.2600
Sorbitol powder	2	2	10.000
Color	0.02	0.02	0.007
Potassium sorbate	0.05	0.05	0.033
Gelatin (Type A, 300 Bloom g)	2	2	2
Water, purified	13.43	13.43	16.556
Sodium biphosphate, NF	–	–	0.418
Phosphoric acid, 85% NF	–	–	0.086
Ammonium Silicofluoride	–	–	0.6400
Viscosity, cp at 35°C	3,750	26,250	9,400

Preparation of Matted Photographic Layers Containing Gelatin

Highlights of the Technological Achievement: In a process for the preparation of matted photographic layers gelatin particles which are dispersed in a water-in-oil emulsion are hardened, separated from the emulsion, redispersed in water, added to the layer which is required to be matted, and the casting composition is applied to a substrate and dried.

Reference: B. Morcher and R. Brück; U.S. Patent 4,148,943; April 10, 1979; assigned to AGFA-Gevaert AG, Germany.

Background: It is known to produce matted photographic layers by milling gelatin powder in a volatile organic solvent such as ethyl alcohol which contains formaldehyde, displacing the solvent with another, nonpolar organic solvent such as benzene, evaporating off the nonpolar solvent and using the hardened gelatin powder for matting the photographic layers, if necessary first redispersing the powder in water as described in U.S. Patent 2,043,906. One disadvantage of this process, however, is that even the most vigorous milling can only produce relatively coarse particles which moreover vary considerably in size. The matt effect produced in photographic layers containing these substances is too coarse for most purposes.

Process Conditions: The hardeners used should be at least partly soluble in the organic solvent. Although hardeners which are insoluble in the organic solvent could conceivably also be emulsified in the water-in-oil emulsion, it is then necessary to ensure that the hardening reaction does not set in until the emulsion has been completely prepared. This can be achieved, for example, by using a slow acting hardener.

The preferred procedure comprises, as a first step, the emulsification of the gelatin solution up to the desired particle fineness, and the second step of hardening the obtained gelatin particles. This procedure can be used for emulsifying and hardening aqueous solution containing from 6 to 30% by weight of gelatin. Where it is necessary to start from more slightly concentrated gelatin solutions, the gelatin may be hydrolyzed by enzyme action.

This process makes it possible for exceptionally fine grained matted photographic gelatin layers to be obtained. The advantage of a fine grained matt effect is particularly important when the process is used for matting extremely thin layers which form constituents of multilayered photographic materials, e.g., color photographic materials. Moreover, using modified gelatin particles as matting agent has the effect that the refractive indices of the binder and matting agent are substantially identical, thus avoiding the undesirable optical effects which are particularly troublesome in photographic materials.

Another advantage of using matting agents made of modified gelatin is that their swelling properties are very similar to those of the binder used for the layer so that the matting agents in no way interfere with the penetration of the processing baths into the layers.

The roughness of a matted photographic layer can be determined by measuring the peak to valley height by means of a Hommel-Tester Type T surface measuring instrument (Hommelwerke GmbH). Such results are summarized below.

Matting Dispersion in Emulsion Binder (% by Wt.)	Peak to Valley Height in Layer Surface (μ)
Pure gelatin	0.2-0.4
5	1.0-2.0
10	1.5-2.5
20	2.0-3.0
50	2.5-3.5

Emulsion layers which contain more than 10% by weight of the matting dispersion as binder are excellent for retouching purposes.

Color Photographic Material Comprising Acid-Treated Gelatin

Highlights of the Technological Achievement: Acid-treated gelatin having a gel strength of at least 250 g (Bloom) is added to silver halide emulsion layers, ultraviolet absorber layers, etc., which are coated on a resin coated paper and which contain high boiling point solvent, whereby deterioration of physical properties of surface of the photographic paper can be prevented.

Reference: H. Arase and K. Yabuuchi; U.S. Patent 4,146,398; March 27, 1979; assigned to Mitsubishi Paper Mills, Ltd., Japan.

Background: The protected coupler type color photographic emulsion containing the high boiling point solvent is inferior to the emulsions not containing any high boiling point solvent in physical properties at the gelled state. Therefore, in the case of employing a high temperature and high speed drying which is required for promotion and simplification of coating and drying of the emulsion, the emulsion layer gelled with a cold air in a setting zone tends to result in collapse of its gel structure in the subsequent warm air drying zone (usually 30° to 60°C).

Thus, the dried emulsion layer with the gel structure collapsed tends to be overswelling at the development process to bring about the so-called reticulation phenomenon which forms creases on the surface of the layer. The emulsion layer in which the reticulation phenomenon has occurred cannot be subjected to ferrotype drying. Therefore, the creases remain as irregularities on the surface after drying, whereby uniform smoothness is lost and beautiful mirror luster cannot be obtained.

In an attempt to overcome the defects mentioned above, it has been found that when a part or all of the usual lime-treated gelatin which is a protective colloid for the high boiling point solvent-containing layer is substituted with an acid-treated gelatin having a gel strength of at least 250 g (measured by the Paggi method), the defects can be overcome effectively.

In general, the acid-treated gelatin (also called acidic gelatin) is superior in physical properties to lime-treated gelatin (also called alkali-treated gelatin or lime-dipped gelatin) and especially the gelatin having a gel strength of at least 250 g cannot be obtained by lime-treatment.

According to this process, with the use of acid-treated gelatin having a gel strength of at least 250 g, physical properties of the emulsion layer gelled with a cold air in the setting zone can be improved to prevent collapse of gel structure of the layer in the subsequent warm air zone, whereby it becomes possible to speed up the coating and drying step, to prevent reticulation of the emulsion surface during development and to obtain the smooth surface having an excellent mirror luster after drying. Gelatins having a gel strength of less than 250 g are inferior in these effects.

The effect is conspicuously high when the acid-treated gelatin is added in an amount of at least 50% of the total amount of the used gelatin in the high boiling point solvent-containing layer. When the acid-treated gelatin is used in a layer not containing any high boiling point solvent (such as an intermediate layer, protective layer, etc.), addition of preferably at least 30% per total amount of gelatin in the layer can result in sufficient effect. The amount of the acid-treated gelatin used may vary depending upon the kind and amount of the high boiling point solvent used.

The effect of this process can be attained by using the acid-treated gelatin having a gel strength of at least 250 g in at least one layer in the case of multilayer color photosensitive materials. The more the number of layers in which the acid-treated gelatin is used, the higher the effect is and the highest effect is obtained when the acid-treated gelatin is used in all layers.

When the number of layers in which the acid-treated gelatin is used is small, it is preferred that the acid-treated gelatin be used in an amount of as much as possible in the uppermost layer containing the high boiling point solvent. On the other hand, when the acid-treated gelatin is used in many layers, the amount of the acid-treated gelatin used in each layer may be relatively small.

Gelatin-Containing Photographic Layer Incorporated with Hardener

Highlights of the Technological Achievement: Compounds of the formula

$$(CH_2{=}CH{-}SO_2{-}CH_2{-}CH_2{-}CO{-}A \rightarrow_n P$$

where each A, which may be either the same or different, is selected from direct bond, –NH– and –O–, P represents an n-valent aliphatic moiety which can be interrupted by a nitrogen atom or phenylene radical, and n is 3 or 4, provided that when A is direct bond the nitrogen atom interrupting the aliphatic moiety can be connected through the direct bond, and can be effectively used as gelatin hardeners in gelatin-containing photographic layers.

Examples of the compounds include 1,1,3,3-tetrakis(vinylsulfonylpropionamido)-propane, α,α'-tetrakis(vinylsulfonylpropionamido)-p-xylene, 1,2,3-tris[β-(vinylsulfonyl)propionyloxy]propane, N,N',N"-tris[β-(vinylsulfonyl)propionyl]diethylenetriamine, and tetrakis[β-(vinylsulfonyl)propionyloxymethyl]methane.

Reference: K. Emoto, K. Yabuta, K. Yamashita and K. Futaki; U.S. Patent 4,134,770; January 16, 1979; assigned to Mitsubishi Paper Mills, Ltd., Japan.

Background: It is known that in order to be of practical use, a gelatin-containing photographic layer must be hardened. Of the many compounds proposed as gelatin hardeners, those which have been known as particularly useful are aldehydic compounds such as formaldehyde, mucochloric acid and formaldehyde addition products; nitrogen-containing six-membered heterocyclic compounds having two or more reactive chlorine atoms in the molecule; compounds having reactive vinyl groups such as divinyl sulfone, divinyl ketone, and N,N',N"-triacryloylhexahydro-1,3,5-triazine; compounds having in the molecule two or more reactive three-membered rings such as ethyleneimino and epoxy groups; and polymeric hardeners such as dialdehyde starch and a maleic half-ester of polyvinyl alcohol or starch. However, when used in gelatin-containing photographic layers, these known hardeners produce results which are not always desirable with respect to physical and photographic properties.

Process Conditions: Compounds represented by the following general formula (I) can overcome the difficulties of the conventional hardeners:

(I) $(CH_2{=}CH{-}SO_2{-}CH_2{-}CH_2{-}CO{-}A)_{\overline{n}}P$

where each A, which may be either the same or different, is selected from direct bond, –NH– and –O–, P represents an n-valent aliphatic moiety which can be interrupted by a nitrogen atom or phenylene radical, and n is 3 or 4, provided that when A is direct bond the nitrogen atom interrupting the aliphatic moiety can be connected through the direct bond.

Of the compounds having the general formula (I), those particularly useful are selected from the compounds of the following general formula (II):

(II)

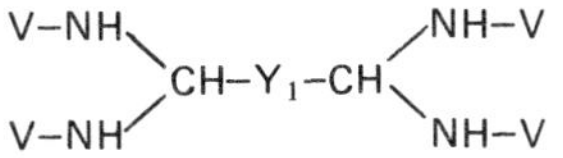

where V represents $CH_2{=}CHSO_2CH_2CH_2CO$, and Y_1 represents a phenylene radical or $-(CH_2)_{\overline{m}}$, where m is zero or an integer from 1 to 3.

Examples of compounds suitable for use are listed below. In the following examples, V represents $CH_2{=}CHSO_2CH_2CH_2CO$.

(1)

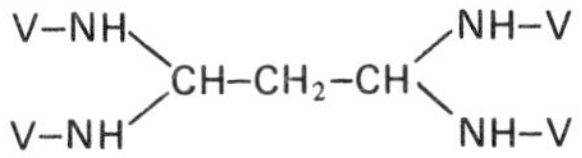

(2)

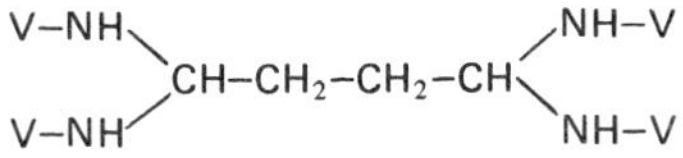

(3)

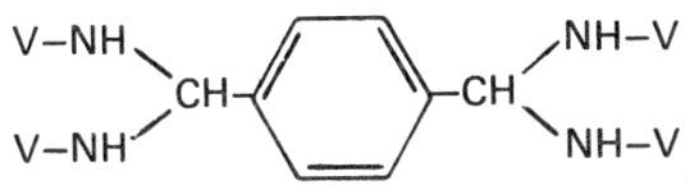

(4)

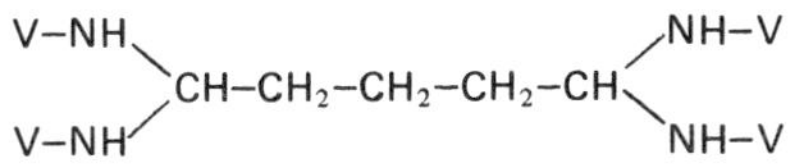

Example: Compounds (1) through (4) were each added to a gelatin composition in an amount corresponding to 2% by weight of gelatin. The resulting composition was applied onto a sheet of resin-coated paper and set by placing on ice. After having been air-dried, the gelatin layer was heated at 90°C for 1 hour and tested for hardening degree. The results obtained are shown below.

Compound No.	Hardening Degree (g)
(1)	108
(2)	95
(3)	84
(4)	65
None	<2

Hardening degree, as herein referred to, was determined as follows. Pretreatment of the test specimen was conducted by warming the test specimen and immersing it in a color developer solution (Kodak Co.) at 20°C for 3 minutes. A ballpointed needle was placed on the pretreated specimen at right angle to the surface. The ballpoint, 1 mm in diameter, was allowed to travel in contact with the surface at a rate of 1 cm/sec. The hardening degree is the minimum load, in grams, under which the ballpoint marks a scratch on the gelatin surface.

As is clear from the table, the compounds showed a remarkable hardening effect on gelatin.

Gelatin Benzimidazole Blends as Inhibitors for Carboxylic Acids

Highlights of the Technological Achievement: Compositions comprising a mixture of (a) gelatin and (b) a compound of the formula:

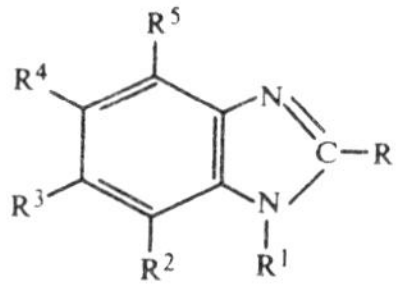

where R, R^1, R^2, R^3, R^4 and R^5 are each a group selected from the class consisting of $-H$, $-Cl$, $-CHCl_2$, $-CH_2Cl$, $-CCl_3$, $-SH$, $-SCH_3$, $-NH_2$, $-NHC{\equiv}N$, $-CH_2C{\equiv}N$, $-NHC(O)NH_2$, $-NHC(NH_2){=}NH$, $-CH_2OH$, $-NO_2$ and monovalent hydrocarbon radicals having from 1 to 18 carbon atoms; provide effective corrosion inhibition for aqueous carboxylic acid solutions in the cleaning of metal industrial equipment.

Reference: J.D. Anderson; U.S. Patent 4,209,418; June 24, 1980; assigned to Union Carbide Corporation.

Background: Acid cleaning operations are commonly employed to remove adhering substances, such as mill scale and fly ash, from the interior surface of vessels, tubes and related industrial processing equipment, particularly where such equipment is fabricated from ferrous metals.

Aqueous processing equipment, especially that operating at elevated temperatures, tends to acquire a build-up of insoluble deposits. Typically, acid cleaning is used to remove these deposits. For example, acid cleaning is employed to remove lime deposits or water scale from power plant boilers and piping systems and evaporating equipment. Likewise, acid cleaning is employed to remove scale and deposits from processing equipment in such plants as refineries, utility companies, paper mills, chemical plants, and the like.

Since cleaning acid tends to remove a portion of the basis metal of the industrial equipment with each cleaning, the use of inhibitors to minimize basis metal loss in acid cleaning can substantially extend the life of the equipment.

It has been found that gelatin together with a compound selected from a specific class of aromatic amines provides particularly effective inhibition for aqueous carboxylic acid solutions in the cleaning of metal industrial equipment.

Process Conditions: The carboxylic acid useful in the cleaning solutions and cleaning composition concentrate blends of this process can be any carboxylic acid such as, for example, citric, formic, oxalic, acetic, glycolic, pyruvic, malonic, propionic, lactic, β-hydroxypropionic, methoxyacetic, glyceric, β-iodopropionic, maleic, oxaloacetic, crotonic, vinylacetic, succinic, methylmalonic, malic, tartaric, n-butyric and isobutyric acids.

The amount of components (a) and (b) in the inhibitor blend is not critical and can vary over a wide range. Preferably, the blend will contain from 95 to 5 parts by weight of (a) and from 5 to 95 parts by weight of (b) per 100 parts by weight of (a) and (b).

This process is preferably effected at atmospheric pressure, although super- or subatmospheric pressure may be used, if desired, for some purposes. The cleaning solutions of this process are preferably employed at an elevated temperature. Although the solution temperature can vary over a wide range, suitable temperatures would include those from 150°F or lower to 220°F or higher.

To produce the inhibitor blend, crude gelatin (PCS-1 Extra Special Fine Ground Glue) (Peter Cooper Corporation) was added to the purified benzimidazole in the amounts specified in the table below. As can be seen from the table, tests were also conducted using, in addition to gelatin and benzimidazole, ethyl quinolinium iodide and 2-thio-4,6-dimethylpyrimidine hydrochloride.

The inhibiting strength of the inhibitors in the cleaning solutions was determined as follows. A strip of mild steel (1010 cold rolled steel) ½" wide by 7" long is prepared by solvent wiping with acetone on a cotton rag, then bending the strip to a suitable shape. The dried, cleaned strip was weighed, immersed in the cleaning solution at 200°F for 24½ hours, then water-rinsed, dipped in acetone, dried and reweighed.

The weight loss is converted to weight loss in $lb/ft^2/24$ hr by multiplying the weight loss in grams times 0.045. For industrial chemical cleaning, it is preferred that the weight loss not exceed 0.005 $lb/ft^2/24$ hr. The results are as follows:

Run No.	Inhibitor	Inhibitor Amount (g in 100 ml)*	Weight Loss (g)	Weight Loss ($lb/ft^2/24$ hr)
1**	Benzimidazole	5.0	0.3440	0.0156
2	Gelatin	5.0	0.2061	0.0093
3	Gelatin	5.0	0.1427	0.0065
	2-Thio-4,6-dimethyl pyrimidine hydrochloride	0.1		
4	Gelatin	2.5	0.0599	0.0027
	Benzimidazole	2.5		

(continued)

Run No.	Inhibitor	Inhibitor Amount (g in 100 ml)*	Weight Loss (g)	Weight Loss ($lb/ft^2/24$ hr)
5	Gelatin	2.5	0.0599	0.0027
	Benzimidazole	2.5		
	2-Thio-4,6-dimethyl pyrimidine hydrochloride	0.1		
6	Gelatin	1.7	0.0414	0.0019
	Benzimidazole	1.7		
	Ethyl quinolinium iodide	1.7		
7	Gelatin	1.7	0.0311	0.0014
	Benzimidazole	1.7		
	Ethyl quinolinium iodide	1.7		
	2-Thio-4,6-dimethyl pyrimidine hydrochloride	0.1		

*Of ammoniated EDTA cleaning solution.
**Run 1 was made using a cleaning solution with pH 4.5; all other runs made at pH 4.8.

As can be seen from the results given in the table, the 2-component (Runs 3 and 4), 3-component (Runs 5 and 6) and 4-component (Run 7) inhibitor blends of the process provide excellent inhibition for an ammoniated EDTA cleaning solution with a synergistic effect compared to the single component inhibitor (Runs 1 and 2).

Silver Diffusion Transfer Receiving Layer Comprising HEC and Gelatin

Highlights of the Technological Achievement: A receiving element for use in a photographic silver diffusion transfer process which comprises a transparent support carrying a layer comprising gelatin, hydroxyethylcellulose (HEC) and noble metal silver-precipitating nuclei; wherein the nuclei are present in a level of from 0.1 to 0.4 mg/ft^2, gelatin is present at a level of from 0.05 to 1.5 mg/ft^2 and HEC is present at a level of 0.1 to 7 mg/ft^2. Preferably, the noble metal is palladium. The receiving element is particularly useful in an additive color diffusion transfer process.

Reference: E.H. Land and S.M. Bloom; U.S. Patent 4,186,013; January 29, 1980; assigned to Polaroid Corporation.

Background: Procedures for preparing photographic images in silver by diffusion transfer principles are well-known in the art. For the formation of the positive silver images, a latent image contained in an exposed photosensitive silver halide emulsion is developed and almost concurrently therewith, a soluble silver complex is obtained by reaction of a silver halide solvent with the unexposed and undeveloped silver halide of the emulsion.

Preferably, the photosensitive silver halide emulsion is developed with a processing composition in a viscous condition which is spread between the photosensitive element comprising the silver halide emulsion and a print-receiving element comprising, preferably, a suitable silver-precipitating layer.

The processing composition effects development of the latent image in the emulsion and, substantially contemporaneously therewith, forms a soluble silver complex, for example, a thiosulfate or thiocyanate, with undeveloped silver halide. This soluble silver complex is, at least in part, transported in the direction of the print-receiving element and the silver thereof is largely precipitated in the silver-precipitating element to form a positive image thereon.

Process Conditions: A combination of HEC and gelatin as the binder for the nuclei has been found which provides superior photographic results. An excess of HEC is employed compared to gelatin. The HEC and gelatin comprise the sole polymeric binder matrix for the nuclei.

The binder matrix in the silver-precipitating layer of this process provides a more stable positive image. It is believed that oxidation of the positive image silver, which may result in silver being relocated behind the wrong filter screen element producing a red effect in the image, is inhibited. The silver image is also less frangible and therefore more suitable for use in a motion picture format where bending and transporting the film unit is common. Significantly greater densities are achieved compared to prior art receiving layers without deleterious effect on speed, range or D_{min}.

Preferably, a ratio of 5 to 1 of HEC to gelatin is used. As levels of HEC are increased, substantially no change in tone is observed. However, at the higher levels, in additive color film units, blue density increases at a greater rate than red and green densities and the silver image tone thus tends to become browner.

Example 1: Three solutions were prepared. Solution A contained 3,140.4 g water, 3.5 g glacial acetic acid, 3.6 g gelatin (20% solution), and 1.7 g $SnCl_2 \cdot 2H_2O$. Solution B contained 196 g water and 4.0 g HEC. Solution C contained 1,400 g water and 28.7 g $PdCl_2$ solution (80.6 g HCl and 166.6 g $PdCl_2$ per liter of solution).

Solution A was heated to between 81° and 82°C and 330 g of Solution C added with agitation. Addition was completed within 5 seconds and the solution cooled to 24°C with continuing agitation. Water lost through evaporation was replaced. 3.3 g of 10% alkyl phenoxypolyoxy ethanol (PE 120, NOPCO Chemical Division of Diamond Shamrock Company) were added. After mixing 5 minutes, 138 g of 2-propanol as a coating aid were added. To the combination of Solutions A and C (3,472 g) was added the quantity of Solution B to give the desired HEC/gel ratios.

Example 2: A film unit was prepared comprising a transparent polyester film carrying on one surface, an additive color screen of approximately 1,500 triplets per inch of red, blue and green filter screen elements in repetitive side-by-side relationship; 328 mg/ft^2 polyvinylidine chloride/polyvinyl formal protective overcoat layer; a nucleating layer comprising palladium nuclei with the coverages and polymers designated below; an interlayer formed by coating 1.9 mg/ft^2 gelatin, 2.3 mg/ft^2 acetic acid and 0.19 mg/ft^2 octylphenoxy polyethoxy ethanol surfactant; a hardened gelatin silver iodobromo emulsion (0.73 μ mean diameter grains) coated at a coverage of 124 mg/ft^2 of gelatin and 150 mg/ft^2 of silver with 7.18 mg/ft^2 propylene glycol alginate and 0.73 mg/ft^2 of nonylphenol polyglycol ether (containing 9.5 mols of ethylene oxide) panchromatically sensitized with 5,5'-dimethyl-9-ethyl-3,3'-bis(3-sulfopropyl)thiacarbocyanine triethyl ammonium salt (0.53 mg/g Ag); 5,5'-diphenyl-9-ethyl-3,3'-bis(4-sulfobutyl)oxacarbocyanine (0.75 mg/g Ag); anhydro-5,6-dichloro-1,3-diethyl-3'-(4"-sulfobutyl)benzimidazolothiacarbocyanine hydroxide (0.7 mg/g Ag); and 3-(3-sulfopropyl)-3'-ethyl-4,5-benzothia-thiacyanine betaine (1.0 mg/g Ag); red, green, green and blue sensitizers respectively; and the following antihalo top coat.

Top Coat

	Milligrams per Square Foot
Gelatin	400
Dow 620 (carboxylated styrene/butadiene copolymer latex)	204
Propylene glycol alginate	25.7
Dioctyl ester of sodium sulfosuccinate	1.2
Benzimidazole-2-thiol gold Au^{+1} complex, as gold	5
Dexad-11 (polymerized sodium salts of alkyl naphthalene sulfonic acid)	0.38
Pyridinium bis-1,5(1,3-diethyl-2-thiol-5-barbituric acid) pentamethine oxanol	5.6
4(2-Chloro-4-dimethylamino-benzaldehyde)-1-(p-phenyl carboxylic acid)-3-methyl pyrazolone-5	7

Processing Composition A

	Percent by Weight
Sodium hydroxide	9.4
Hydroxyethylcellulose (Natrosol 250HH)	0.7
Tetramethyl reductic acid	9.0
Potassium bromide	0.6
Sodium sulfite	0.8
2-Methylthiomethyl-4,6-dihydroxypyrimidine	9.0
4-Aminopyrazolo-[3,4d]-pyrimidine	0.02
N-benzyl-α-picolinium bromide (50% solution)	2.9
Water	67.6

Processing Composition B comprised Processing Composition A with 3.3% by weight of sodium tetraborate·$10H_2O$.

Film units prepared according to the above procedure were given a 16 mcs exposure with a xenon sensitometer and processed with mechanical rollers with an 0.8 mil gap disposing the processing composition between the top coat and a polyethylene terephthalate cover sheet. The film unit was held in the dark for one minute and then the cover sheet was removed, retaining the rest of the film unit together and then drying.

The following tables set forth spectral data from film units of this process employing in the receiving layer 0.2 mg/ft^2 of gelatin; 0.15 mg/ft^2 Pd and the designated coverage of HEC.

In Table 2 the silver coverage in the emulsion was 110 mg/ft^2 and the average mean diameter of the grains was 0.59 μ. Control A contained no HEC and the nuclei were formed under a blanket of nitrogen, excluding air. Control B contained no HEC.

The spectral data was obtained by reading the neutral column to red, green and blue light in an automatically recording densitometer.

Table 1

Ex. No.	Polymer Type	Polymer Coverage (mg/ft^2)	Processing Composition	D_{max} Red	D_{max} Green	D_{max} Blue
3	A	0.8	B	3.32	3.59	3.53
4	B	0.8	B	2.98	3.26	3.15
5	A	0.8	A	3.44	3.76	3.69
6	B	0.8	A	3.29	3.60	3.45
A*	–	–	A	3.03	3.35	3.24
7	A	0.8	A	3.30	3.59	3.56

*Control.

Table 2

Ex. No.	Polymer Type	Polymer Coverage (mg/ft^2)	Processing Composition	D_{max} Red	D_{max} Green	D_{max} Blue
B*	–	–	B	3.13	3.21	3.00
8	A	0.8	B	3.23	3.36	3.20
9	C	0.8	B	3.21	3.31	3.16
10	A	1.6	B	3.55	3.62	3.60
11	A	3.2	B	3.12	3.35	3.60
12	A	6.4	B	2.72	3.15	3.62
B*	–	–	B	2.97	3.12	2.98
13	A	0.8	B	3.18	3.37	3.24
14	A	1.0	B	3.29	3.39	3.30
15	A	1.2	B	3.25	3.39	3.36
16	A	1.4	B	3.35	3.39	3.42
17	A	1.6	B	3.32	3.40	3.45
B*	–	–	B	3.18	3.28	3.08
18	A	1.0	B	3.20	3.40	3.30
19	E	1.0	B	3.20	3.40	3.30
20	C	1.0	B	3.20	3.40	3.30
21	D	1.0	B	3.30	3.40	3.30
22	E	1.0	B	3.20	3.40	3.30

*Control.

The HEC polymers used were Type A, Natrosol 250HHP; Type B, Natrosol 250L; Type C, Natrosol 250H4R; Type D, Natrosol 250MH; and Type E, Natrosol 250HHR (Hercules, Inc.).

In the above tables, the increase in density obtained in film units of this process is shown. It will be seen that advantageous results, in the form of increased densities can be obtained over a relatively wide range and that the viscosity range of the particular HEC employed is not critical.

INORGANIC PRODUCTS

PROCESSING

Stable Dispersions Containing Trialkyltin Fluorides

Highlights of the Technological Achievement: The tendency of dispersions containing trialkyltin fluorides in organic liquids to agglomerate is avoided by using as the dispersion medium specified organic liquids in combination with from 0.5 to 10%, based on the weight of the dispersion, of specified inorganic compounds. The choice of both organic liquid and inorganic compounds is critical to achieving long-term stability of the resultant dispersion.

Reference: C.B. Beiter and L.A. Hafner; U.S. Patent 4,191,580; March 4, 1980; assigned to M & T Chemicals Inc.

Background: This process relates to compositions containing a dispersed form of a trialkyltin fluoride. More particularly, this process relates to stable dispersions of trialkyltin fluorides which are capable of being stored for extended periods of time without any significant increase in viscosity.

A number of trialkyltin fluorides, particularly tri-n-butyltin fluoride, effectively inhibit the attachment and growth of barnacles and other organisms responsible for fouling of submerged surfaces such as the hulls of sea-going vessels and the pilings of docks and other facilities exposed to salt water. The trialkyltin fluorides are, therefore, useful as the toxicant for antifouling coatings. A typical antifouling coating contains the toxicant, one or more pigments and a film-forming polymer. All of these components are dissolved or dispersed in an organic solvent such as xylene or toluene, optionally in combination with a ketone such as 2-butanone.

Previously, acceptable coatings containing trialkyltin fluorides, which are solid materials at ambient temperatures, have been difficult to prepare. Dispersions of trialkyltin fluorides in organic solvents exhibit a strong tendency to agglomerate in the form of large particles.

Process Conditions: This process provides a stable, thixotropic dispersion of a trialkyltin fluoride, the dispersion consisting essentially of:

(1) from 40 to 70% by weight of a trialkyltin fluoride of the formula R_3SnF, wherein R is alkyl containing from 2 to 12 carbon atoms;

(2) from 20 to 60% by weight of at least one organic liquid having a kauri butanol value of 96 or less and selected from the group consisting of alcohols containing from 4 to 12 carbon atoms, aliphatic hydrocarbons containing from 5 to 12 carbon atoms, aromatic hydrocarbons;

(3) from 0.5 to 10% by weight of a compound selected from the group consisting of:

(a) lithium and sodium salts of p-toluenesulfonic, phenylphosphonic and silicic acids;

(b) nitric acid salts of calcium, magnesium, sodium, lithium, iron and zinc;

(c) salts of p-toluenesulfonic or phenylphosphonic acid and an element selected from the group consisting of magnesium, calcium, strontium and barium;

(d) the chlorides of metallic elements selected from the group consisting of divalent copper, silver, gold and the elements in Groups 2-A, 2-B, 4-A, 4-B, 5-A, 6-B, 7-B and 8 of the periodic table, and

(e) carboxylic acid salts of lead, manganese, zirconium, barium and strontium, wherein the carboxylic acid contains from 2 to 12 carbon atoms.

Product: Typical antifouling coatings that can be prepared using the dispersed form of trialkyltin fluoride contain from 1.0 to 20.0 of the trialkyltin fluoride composition, including one or more of the aforementioned salts and organic liquids, from 10 to 50% by weight of pigments, typically titanium oxide or zinc oxide alone or in combination with colored pigments such as ferric oxide, from 10 to 50% by weight of at least one film-forming component, which typically includes vinyl chloride homopolymers and copolymers, rosin and chlorinated rubbers such as polychloroprene, and from 20 to 60% by weight of one or more organic solvents, including xylene, cyclohexanone, 2-butanone and mixtures of hydrocarbons commonly referred to as "aliphatic naphtha" and "high flash naphtha." A small amount of a viscosity modifier such as a bentonite clay is usually included to impart thixotropy to the final coating composition.

Example: A typical red formulation suitable for use with the dispersed form of tri-n-butyltin fluoride can be prepared using the following procedure. Combine the following ingredients using a high-speed stirrer to obtain a uniform dispersion: 4.89 pbw rosin (70% by weight in xylene) and a mixture containing 0.47 pbw bentonite clay and 0.14 pbw methanol. Combine this mixture with 12.00 pbw cyclohexane and add the following pigments: 13.90 pbw red iron oxide; 10.30 pbw talc; and 6.05 pbw zinc oxide.

Add 34.13 pbw of 3% by weight solution of VAGH (a film-forming vinyl chloride-partially hydrolyzed vinyl acetate copolymer) in a 2-butanone/xylene mixture. Add 17.65 pbw of a dispersed form of tri-n-butyltin fluoride. Stir the mixture at high speed until the composition exhibits a fineness of 4 to 5 on the Hegman N.S. scale.

Self-Hardening Water-Soluble Mold

Highlights of the Technological Achievement: A mold produced by adding to a molding material consisting mainly of alumina, a binder prepared by adding an inorganic water-soluble substance to an alcohol or a mixture of an alcohol and water, kneading and molding the mixture, and then drying the molded mixture has water-solubility and self-hardening properties. Further, by suitable selection and combination of the inorganic water-soluble substance and the solvent in the binder, the drying step becomes unnecessary or the molding material shows fluidity at room temperature and a self-hardening water-soluble mold requiring no ramming is obtained.

Reference: K. Koike; U.S. Patent 4,196,011; April 1, 1980; assigned to Hitachi, Ltd., Japan.

Background: Prior art water-soluble molds can be removed from the casting by dissolving or decomposing with water, but the removing speed has not necessarily been great. The mold comprising alumina as a main material and a water-soluble compound as a binder which was previously discovered to solve the abovementioned difficulty (Japanese Patents 816,642 and 816,643) may be easily disintegrated with water and it is easy to remove the mold. However, this mold has defects in that it must be dried before casting since it contains moisture and the drying time is prolonged particularly in the case of a large mold. Also, when used as an outer mold, the moisture content is high and the mold strength is not satisfactory owing to such a high moisture content. In this case, drying is still required.

Also, there is known a process wherein an inorganic water-soluble substance, the hydrate of which is stable at room temperature, and water are added to alumina, the mixture is heated to a temperature of 100°C or less to dissolve the water-soluble substance in its water of crystallization and thereby fluidize the whole mixture, the fluidized mixture poured into a pattern and cooled to solidify the mixture, and the resulting molded article dried and used as a mold for casting metals. However, this process has a defect in that narrow spaces of the pattern cannot be filled by the fluidized mixture owing to a reduction in fluidity with a decrease in temperature when the fluidized mixture is poured into the pattern.

Process Conditions: The mold can be obtained by dissolving or dispersing an inorganic water-soluble substance in an alcohol (a polyhydric alcohol and/or a monohydric alcohol), adding the resulting solution or dispersion alone or in admixture with a small amount of water to alumina, and then molding the mixture by ramming or compressing. This mold has a low moisture content and self-hardening properties. Therefore, it can be used as a mold for casting metals without forced drying or after slight surface drying for increasing the mold strength.

At temperatures from room temperature to the boiling point of the alcohol added, the strength of the mold is maintained by the binding force of the solution of the water-soluble compound in the alcohol.

As the water-soluble compound added as a binder in a molding material for this mold, any inorganic water-soluble compound may be used so long as it is alcohol-soluble or alcohol-dispersible. Examples are shown in the following table together with the main material and their uses. Examples of alcohols that may be used include monohydric alcohols such as ethyl alcohol and polyhydric alcohols such as sorbitol, glycerol, ethylene glycol, etc.

Main Material	Inorganic Water-Soluble Compound	Use
Al_2O_3	NaCl	Aluminum alloy
	KCl	
	$CaCl_2$	
	Na_2CO_3	
	K_2CO_3	
Al_2O_3	Na_3PO_4	Copper alloy, cast iron, aluminum alloy
	K_3PO_4	
Al_2O_3	$NaAlO_2$	Cast steel, copper alloy, cast iron, aluminum alloy
	$KAlO_2$	
	Li_2O	
	BaO	
	SrO	
	LiOH	
	$Ba(OH)_2$	
	$Sr(OH)_2$	
	$BaAl_2O_4$	

Note: To 100 pbw of an inorganic water-soluble compound is added 60 to 200 pbw of a mixture consisting of 2 to 50% by weight of a polyhydric alcohol (ethylene glycol, etc.), 0 to 60% by weight of a monohydric alcohol (ethyl alcohol, etc.) and 0 to 88% by weight of water, and 2 to 40 pbw of the binder thus obtained is then added to 100 pbw of alumina.

The fluid water-soluble mold of the process utilizes the hydration reaction of an inorganic water-soluble substance for fluid molding of a water-soluble mold. Thus, alumina is added to a solution of an inorganic water-soluble substance in an alcohol, and water is then added to form a slurry. Hydration reaction proceeds gradually in this state and the slurry is gradually solidified. In this case, solidification time or hardening time can be changed or controlled according to the time of pouring the slurry into a pattern and before solidification in the pattern and the time of removing the mold from the pattern by controlling the amounts of the alcohol and water added. Further, this process has advantages in that molds of a complicated and fine shape can be correctly produced, and the moisture content of the mold can be reduced, whereby the drying operation is facilitated and the drying time is shortened.

The inorganic water-soluble substance used in this mold, a hydrate of which is stable at room temperature, is exemplified by $Na_2CO_3 \cdot 10H_2O$, $Na_3PO_4 \cdot 12H_2O$, $Ba(OH)_2 \cdot 8H_2O$ and $Al_2(SO_4)_3 \cdot 18H_2O$. These compounds have a high melting point, are stable at the melting temperature of metals after dehydration, and are comparatively large in the amount of water bound as water of crystallization.

If alumina is added to a solution of these compounds in an alcohol and an amount of water to be fixed as water of crystallization or less water is then added to the mixture to form a slurry, the reactions as mentioned below occur in the mixture, water is fixed as water of crystallization of these compounds, and the slurry is solidified.

Among the inorganic water-soluble substances as mentioned in the table above, sodium carbonate (Na_2CO_3), sodium phosphate (Na_3PO_4), barium hydroxide [$Ba(OH)_2$] and barium oxide (BaO) are suitable, and hydration reaction proceeds as follows:

$$Na_3PO_4 + 12H_2O \rightarrow Na_3PO_4 \cdot 12H_2O$$

$$Ba(OH)_2 + 8H_2O \rightarrow Ba(OH)_2 \cdot 8H_2O \text{ or}$$

$$BaO + 10H_2O \rightarrow Ba(OH)_2 \cdot 8H_2O + H_2$$

This reaction proceeds rapidly in water, but gradually in a mixture of an alcohol and water. Thereby, hardening time can be controlled by varying a ratio of water to the alcohol. The alcohol which is set free has no influence on solidification if its amount is small. When its amount is large, however, an alkali metal or its chloride is previously added to fix the alcohol as alcoholate of the metal:

$$Al + 3R{-}OH \rightarrow (R{-}O)_3Al + (3/2)H_2$$

wherein R is an alkyl group. The alcohol used here is selected according to the solubility of the inorganic water-soluble compound. The abovementioned monohydric alcohols and polyhydric alcohols may be used. Also, there may be used substances which may react with water to form an alcohol such as mono- or polyethylene oxide $-\!\!\left[CH_2CH_2O\right]\!\!-_n$, etc.

Polyanion-Stabilized Aluminum Hydrogels

Highlights of the Technological Achievement: Aluminum hydrogels which are stabilized at pH below 7 with a polymer containing carboxyl groups, e.g., carboxymethylcellulose or gum arabic. The gels are useful as phosphate binding agents in the treatment of hyperphosphatemia. Such polymers are incorporated during the formation of the aluminum hydrogel, permitting a high yield to be obtained, and preventing colloidal sol formation during exchange with monovalent anions.

Reference: N.S. Mason and R.E. Sparks; U.S. Patent 4,181,718; January 1, 1980.

Background: This process relates to aluminum hydrogels formed at pH below 7 in the presence of a carboxy-containing polymer and to their use as phosphate binding agents, particularly as phosphate binding agents in the treatment of hyperphosphatemia.

Hyperphosphatemia may be treated by removal of phosphorus from a patient by binding phosphates in the intestinal tract. The prior art phosphate binding agents are antacid hydrated hydrogels, e.g., aluminum carbonate gel, which are insoluble in water or acids and remove phosphate by ion exchange. In pharmaceutically available form the hydrogels are partially dehydrated with a resultant loss in original phosphate binding capacity. Because this dehydration is irreversible, these materials cannot be rehydrated to recover this capacity. This is particularly a problem when these prior art gels are used in tablet form, wherein phosphate capacity can be irreversibly decreased by a factor of six. These prior art gels are also so astringent in taste that it is difficult for a patient to take the amount of gel required to maintain a phosphorus balance.

Process Conditions: The phosphate-binding agents of this process are aluminum hydrogels formed at pH below 7 in the presence of polymeric anions. Broadly, they may be prepared by the addition of a large anion, e.g., 1,000 to 700,000 molecular weight, particularly a high molecular weight hydrophilic anionic polymer to an aluminum salt either before neutralization of the aluminum salt or during its subsequent exchange with solutions containing monovalent ions which would otherwise solubilize most of the gel. Hence, the polyanion stabilizes the gel, permitting it to be obtained in high yield in the dilute solutions which are necessary to obtain gels having high phosphate binding capacity, by the addition of a large anion, particularly a high molecular weight anionic polymer to an aluminum salt either during its subsequent exchange with solutions containing monovalent ions which would otherwise solubilize most of the gel.

Hence, the polyanion stabilizes the gel, permitting it to be obtained in high yield in the dilute solutions which are necessary to obtain gels having high capacity.

The aluminum hydrogels of this process may be prepared by the following methods:

Methods A, B, and C can be used when the anion in the aluminum salt, e.g., SO_4^{-2} is undesirable pharmaceutically and is exchanged for a pharmaceutically acceptable anion, e.g., acetate ion.

(A) an aluminum salt, e.g., Al_2SO_4 is neutralized in an aqueous medium containing a stabilizing polyanion, e.g., gum arabic, followed by washing and separation of the resulting gel by conventional techniques, e.g., centrifugation. The anion, e.g., SO_4^{-2}, is then exchanged in an aqueous acid solution, e.g., acetic acid for the corresponding acid ion, e.g., acetate ion. After the ion exchange, conventional washing and separating techniques are used to separate the gel;

(B) The procedure of (A) is followed except that the polyanion is added before the ion exchange step of the aluminum salt and the resulting polyanion stabilized gel is then ion exchanged and recovered as described in (A);

(C) The procedure of (A) is followed except that the polyanion is added during the ion exchange step, and the resulting stabilized gel is recovered as described in (A); Methods D and E may be used when the starting material contains a pharmaceutically acceptable anion, e.g., tartrate ion, and therefore, no ion exchange step is necessary.

(D) The procedure of (A) is followed except that the ion exchange step is omitted and the gel recovered as described in (A); and

(E) The procedure of (B) is followed except that the ion exchange step is omitted and the gel recovered as described in (B).

Anionic Polymers: Among the anionic polymers that may be added during the gel formation are carboxymethylcellulose (having a molecular weight range of 5,000 to 700,000, preferably 80,000 to 700,000 and a degree of substitution of 0.4 to 1.5, preferably 0.5 to 1.2); gum arabic; gelatin (having a molecular weight range of 15,000 to 250,000, and an isoelectric point of 4.7 to 5); ethylene maleic anhydride copolymer (having a molecular weight range of 1,000 to 100,000, preferably 60,000 to 90,000); and the like. Gum arabic and carboxymethylcellulose are the preferred hydrophilic polymers.

The stabilized gel of this process may contain from about 1 to 80% of the anionic polymer based on the weight of the gel.

Example 1: The following formulation for a liquid oral suspension containing an effective amount of the active compound may be formulated using conventional methods. The formulation is useful in treating hyperphosphatemia.

Component	Grams
Gum arabic-stabilized aluminum hydrogels	100
Peppermint oil	0.01
Sodium benzoate	0.1
Water, qs	

Example 2: Tablets suitable for administration which contain the following ingredients may be prepared by conventional tabletting techniques. Such tablets are useful in treating hyperphosphatemia.

Ingredients	Weight (Milligrams)
Gum arabic-stabilized aluminum hydrogels	500
Tragacanth	10
Lactose	100
Corn starch	25
Talcum	15
Magnesium stearate	2.5

Catalyst Support Prepared by Alcohol Treatment of Hydrogels

Highlights of the Technological Achievement: A method is provided for treating a hydrogel with a saturated, aliphatic alcohol selected from the group consisting of 1-pentanol, 2-pentanol, 3-methyl-1-butanol, 4-methyl-2-pentanol, 2,2-dimethyl-1-propanol and 1-hexanol to remove water thereby producing a dried product that retains a porous structure and is particularly suitable upon impregnation with a polymerization catalyst for use in promoting the production of ethylene polymers of melt index in excess of 6.6.

Reference: J.N. Short and D.R. Witt; U.S. Patent 4,152,503; May 1, 1979; assigned to Phillips Petroleum Company.

Background: The preparation of chromium-containing polymerization catalysts supported on substrates prepared from hydrogels is well known. The preparation of hydrogels and removal of water therefrom to produce these substrates has been described in a variety of methods. The impregnation of polymerization catalysts onto supporting substrates is also well known. Previously, however, the methods for producing a chromium-containing catalyst impregnated upon a support for polymerization of ethylene polymers have not shown means by which a melt index of at least 6.6 can be obtained.

Process Conditions: The hydrogels contemplated in this process are generally the inorganic gels which can be converted into products usefully employed as catalysts or as catalyst supports in various hydrocarbon conversion processes. Examples are the hydrogels formed from the precipitation of the oxides of aluminum, germanium, iron, molybdenum, silicon, tin, titanium, tungsten, vanadium and zirconium and combinations thereof. However, the silica-containing hydrogels are of particular interest in this process. The silica content of the various cogels, tergels, etc., on a dry basis, will generally range from 80 to 99.8 wt %.

The resulting hydrogel is preferably aged for at least one hour at an elevated temperature, i.e., 60° to 95°C, after which soluble impurities such as sodium sulfate, etc. are washed out by exchange with a dilute aqueous solution of ammonium nitrate as is known in the art. Generally, the amount of interfering ion or ions is less than 0.1 wt % based on the dry weight of the gel following this treatment.

Following the alcohol treatment, the treated gel is recovered and the alcohol removed by evaporation or flashing. The addition of a chromium compound can be accomplished, if desired, by impregnation with a nonaqueous solution, for example, by a hydrocarbon solution of tert-butyl chromate.

Example: Samples were activated by calcining at 1600°F (871°C) for five hours in a fluidized bed and portions thereof were employed in a particle form polymerization process to polymerize ethylene. A stainless steel stirred reactor containing 1¼ lb (568 g) isobutane as the diluent was employed. In each instance the reaction was conducted at 230°F (110°C) and 550 psig (3,792 kPa) in the absence of hydrogen. The melt index of each recovered polymer, compared at a productivity of 5,000 g polymer per gram of catalyst, was determined in accordance with the procedure of ASTM D1238-62T, condition E. The results obtained are given in Table 1 for azeotrope distillation and Table 2 for continuous extraction.

Table 1: Melt Index Results from Azeotrope-Dried Catalyst

Run No.	Organic Compound Used	Remarks	Water-Removal* Time (min)	Polyethylene Melt Index
1	Ethyl acetate standard	control	100	6.3
2	Butyl acetate	control	31	4.3
3	Amyl acetate	control	25	3.4
4	Methyl isobutyl ketone	control	31	4.9
5	2-Pentanone	control	40	5.9
6	3-Pentanone	control	40	4.7
7	n-Butanol	control	26	4.8
8	Isobutanol	control	39	4.9
9	2-Methyl-1-butanol	control	31	6.4
10	3-Methyl-1-butanol	process	30	7.2
11	2,2-Dimethyl-1-propanol	process	41	6.7

(continued)

Table 1: (continued)

Run No.	Organic Compound Used	Remarks	Water-Removal* Time (min)	Polyethylene Melt Index
12	1-Pentanol	process	30	7.2
13	2-Pentanol	process	37	7.2
14	4-Methyl-2-pentanol	process	35-40	6.8
15	1-Hexanol	process	20	6.9
16	2-Ethyl-1-butanol	control	35	6.0
17	Cyclohexanol	control	60	5.2
18	2-Ethyl-1-hexanol	control	29	5.3

*Calculated as the time liquid started to fill the receiver until the liquid coming into the receiver was clear.

Table 2: Melt Index Results from Continuous Extracted Catalyst

Run No.	Organic Compound Used	Remarks	Water-Removal* Time (min)	Polyethylene Melt Index
19	Ethyl acetate standard	control	105	6.4
20	Isopropyl ether	control	450	1.1
21	Diethyl ketone	control	245	6.1
22	4-Methyl-2-pentanone	control	285	5.5
23	Isobutanol	control	30	5.8
24	3-Methyl-1-butanol	process	105	7.5
25	4-Methyl-2-pentanol	process	200	7.1
26	Cyclohexanol	control	60	4.6

*Calculated as the time liquid was begun to be pumped through the heated section containing the sample until the cooled effluent was clear.

APPLICATIONS

Stable Solution for Dental Remineralization

Highlights of the Technological Achievement: An aqueous solution useful for remineralizing subsurface carious lesions of dental enamel which solution contains sources of calcium ions and phosphate ions as well as fluoride ions and further includes as an agent to stabilize the solution against precipitation, an antinucleating agent such as ethylenediamine tetramethylenephosphonic acid or water-soluble salt thereof, the pH of the solution being 5 to 9, preferably close to physiological conditions, such as about 6.8 to 7.5.

Reference: M.C.S. Gaffar and A. Gaffar; U.S. Patent 4,183,915; January 15, 1980; assigned to Colgate-Palmolive Company.

Background: It is known that dental caries begins with lesions of so-called "white spots," which are demineralized areas below the surface of intact dental enamel. If unchecked, surface enamel above a subsurface lesion eventually collapses, leading to cavitation and subsequent loss of tooth structure.

In order to arrest demineralization, and, indeed, in order to remineralize "white spots" various compositions have been proposed. They are, however, not substantially permanent and precipitation can occur, particularly when fluoride ions are present.

It is an advantage of this process that a one-part stable aqueous remineralizing solution is suitably prepared as a dental mouth rinse and also can be incorporated into other dentifrice compositions such as a dental cream or gel, mouth spray, troche, chewable tablet, lozenge and the like.

Process Conditions: The antinucleating properties of the agents employed in the process appear to be effective to prevent precipitate formation from the calcium and phosphate ions of the solution particularly with the fluoride ions also present. Desirably, the antinucleating agent of the process is present in an amount of about 1 to 500 ppm of the solution, preferably about 25 to 250 ppm, such as about 225 ppm.

The antinucleating agent of the process is desirably a diamine tetramethylenephosphonic acid of the formula $(M_2O_3PH_2C)_2N(CH_2)_n(CH_2PO_3M_2)_2$ wherein n is an integer from 1 to 10 and M is hydrogen or an orally acceptable cation such as alkali metal (e.g., sodium or potassium), ammonium or C_{1-18} mono-, di- or trisubstituted ammonium (e.g., mono-, di- or triethanolammonium) salt.

Phosphonoacetic acid (PAA) and its water-soluble orally acceptable salts are also desirable antinucleating agents. They are characterized by the formula $M_2O_3PCH_2COOM$, wherein M has the meaning indicated above.

The peroxydiphosphate (PODP) and particularly the alkali metal salts thereof (e.g., potassium or sodium) are likewise desirable antinucleating agents. They are characterized by the formula $M_4P_2O_8$ wherein M has the meaning indicated above. The desirable oligomer antinucleating agents have the formula:

wherein M is hydrogen or a water-soluble orally acceptable cation (as indicated above); R_1, R_2, R_3 and R_4 are independently H, methyl or ethyl; Y is at least one hydrophilic member of the group consisting of –COOM, $-COHN_2$ and $-CH_2OH$; X is at least one hydrophobic member of the group consisting of –CN, –COOR, $-COOR_5OR$, –CONHR and $-COONHR_5COR$; R is C_{1-8} alkyl; R_5 is C_{1-4} alkylene; a is 0 to 7; and a + b is 4 to 15.

These oligomers are anionic and of relatively low and accurately regulated degree of polymerization, (in contrast to the conventional free radical redox polymerization conducted with an oxidative initiator such as hydrogen, alkyl or acyl peroxides, persulfates or hydroperoxides in relatively large amounts and a reductive activator such as $NaHSO_3$, $Na_2S_2O_4$ or sodium formaldehyde sulfoxylate in relatively low amounts generally added subsequently to the polymerization medium) are prepared by a reductive polymerization in which a much larger amount of a bisulfite salt, e.g., $NaHSO_3$ (sodium bisulfite, sodium acid sulfite), a reducing agent, is the initiator charged initially with the monomer, and an oxidizing agent is added in smaller amounts as the activator during the polymerizing or oligomerizing process.

Subscript a in the formula represents the number of mols of hydrophobic groups, and subscript b the number of mols of hydrophilic groups, in the oligomer molecule. The proportion of X (i.e., the value of a) must be small enough or even zero, to avoid the production of a too large, sticky and hydrophobic polymer molecule, and will of course be dependent for the most part in any particular instance on the identity of the X and Y groups, i.e., the hydrophobic-containing and hydrophilic-containing monomeric reactants. Mixtures of such oligomers may of course also be employed.

Examples of monomers containing hydrophilic Y group are acrylic acid, methacrylic acid, α-ethylacrylic acid, β-methylacrylic acid, α,β-dimethylacrylic acid and orally acceptable salts of these acids.

Haloplatinate Dental Composition for Preventing and Inhibiting Dental Caries

Highlights of the Technological Achievement: A dental composition comprising, as an essential ingredient, water-soluble haloplatinate and a pharmaceutical carrier or diluent, prevents and inhibits dental caries. The composition exhibits its anticarious effect by inhibiting both digestion of the soft tissue in dentin with a protease and decalcification of the hard tissue in dentin with an acid.

Reference: H. Okuda and N. Hori; U.S. Patent 4,200,626; April 29, 1980; assigned to Toyo Seiyaku Kasei Co., Ltd., Japan.

Background: To prevent dental caries, there has been used a water-soluble fluoride, such as stannous fluoride, and sodium silicofluoride. The fluoride is formulated into a composition, for example, a solution suitable for a treatment, such as painting it on the tooth surfaces or mouth washing, and such treatment is often carried out in a primary school, a kindergarten, a dental clinic or the like. Further, there is also carried out a treatment, such as internal administration of a tablet containing the above fluoride or addition of the fluoride to tap water. These treatments obtain considerable results. However, the effect of the fluoride as a medicament is not always enough to prevent or inhibit dental caries.

Process Conditions: The haloplatinate is formulated into the composition in an amount of not less than 0.01% by weight (as platinum), preferably 0.1 to 5% by weight, on the basis of the total weight of the composition. The pharmaceutical carrier or diluent to be used includes water, glycerin, propylene glycol, polyethylene glycol (macrogol), ethyl lactate or a mixture thereof.

Semisolid Water-Based Coating Compositions

Highlights of the Technological Achievement: This process relates to coating compositions which are "solid" prior to application, yet, which when applied to a substrate, flow and level in a manner similar to conventional coating compositions.

Reference: J.E. Jones, R.G. Temple, J.A. Seiner, S.L. Young and W.G. Boberski; U.S. Patent 4,196,107; April 1, 1980; assigned to PPG Industries, Inc.

Background: One major criticism of conventional coating compositions is the mess attendant in their handling and application. First, typically, the coating composition must be stirred to redisperse the solids content, and then at least a portion of the coating composition is transferred to a pan adapted to a roller or pad applicator. Splashing, dripping and spilling, prior to and during the actual application of the coating composition to a substrate are always real dangers.

Contrary to the prior art systems, in the coating compositions of the process, once the gel structure formed by the interaction of an electrolyte and a colloidal gelling agent is destroyed by the application of shearing forces, there is little or no recovery of a substantial gel structure contributed by the gelling system. Thus, the compositions of the process act as conventional paints when coated upon a substrate, providing excellent flow and leveling.

Process Conditions: The coating compositions of the process comprise a gelled coating composition, capable of application and flow on a solid substrate, comprising: (a) a water-reducible polymer, (b) water, (c) an electrolyte, and (d) a colloidal gelling agent, gelable by interaction with the electrolyte, in an amount sufficient, when interacted with the electrolyte, to gel the composition to form a stable solid or semisolid and to provide a composition reducible to a coating application viscosity upon the application of a relatively small shearing force at a relatively small shearing rate. The gelled coating composition, after being sheared to a coating application viscosity, does not recover substantial (colloidal gelling agent contributed) gel structure.

Generally, the coating compositions of the process are conventional water-based coating compositions containing, in addition to their conventional components, appropriate amounts of a gel-forming system which is a combination of an electrolyte and colloidal gelling agent, gelable by interaction with the electrolyte.

The film-forming polymer includes, but is not limited to, water-reducible inorganic and organic polymers, such as acrylic polymers, vinyl resins and chlorine substituted vinyl resins, polyphosphorylonitriles, polyethers, saturated and unsaturated polyesters, polyurethanes and polyepoxides as well as natural resins or chemically modified natural resins.

A particularly useful group of colloids are colloidal silicas, usually with high SiO_2/Na_2O ratios, where the particle surface is partially composed of silanol groups which are partially ionized and stabilized in the presence of sodium counter ions. Similar colloids are useful where the counter ion is, for example, potassium, lithium, ammonium, substituted ammonium and quaternary ammonium ions. Examples of such colloids are available as Ludox, Nalcoag and Nyacol.

In addition to silica sols, other sols which are electrolyte gelable include As_2S_3 sols, Fe_2O_3 sols and the like. An example of a useful colloidal alumina (Baymal) is a free-flowing powder consisting of clusters of minute fibrils of boehmite (AlOOH) alumina.

The gel strength and coating composition application viscosity were characterized by the use of a rheometer, i.e., a Rotovisco RV3 viscometer from Haake, Inc. using a SVIIP sensing system measured from 0 to 1,000 rpm, using a 50/500 dual measuring head with the spike being the scale reading at about 15 sec^{-1} shear rate (for about 20 rpm) and the end point at about 779 sec^{-1} (or about 1,000 rpm). (All Rotovisco viscometer values reported are measurements made using the 500 measuring head.)

Example: A solid coating composition was prepared from the following ingredients:

Ingredients	Amounts (Parts by Weight)
Carboxymethylhydroxyethylcellulose (Hercules 420H, 2% solids)	459
Polyphosphoric ester acid anhydride potassium salt (Strodex SEK-50, 50% solids)	28.0
Sodium salt of a polymeric carboxylic acid (Tamol 731, 25% solids)	48.0
Nonylphenoxypolyethoxyethanol (Triton N-57)	8.0
Sodium alkyl aryl polyether sulfonate (Triton X-200)	4.0
Silica and mineral oil defoamer (Defoamer L-475)	8.0
Phenyl mercuric acetate (30% solution)	2.0
Ethylene glycol	160.0
Phenyl ether of ethylene glycol	32.0
Diatomaceous earth (Celite 499)	42.4
Calcium-modified SiO_2 (Hi-Sil 422)	80.0
TiO_2 (Du Pont R-900)	938
TiO_2 (Du Pont R-902)	402
Barium sulfate	200
High-boiling hydrocarbon solvent (Soltrol 200)	106
Nonionic surfactant (Triton X-45)	3.0
Silica and mineral oil defoamer (Defoamer L-475)	20.0
Clay slurry (Al-Sil-Ate NC, 68% solids)	530
Water	196.7
Sodium salt of styrene-maleic anhydride copolymer (5.4% solids)	420
Latex Vehicle (Rohm and Haas AC-64)	1,123.2
Colloidal silica	312

After heat processing, the semisolid coating composition had the following viscosity properties: spike, 24.5; end, 9.5.

Several weeks later, the above solid coating composition was studied to demonstrate the lack of viscosity recovery with time. Using the Rotovisco RV3 viscometer, a first curve (Figure 14.1-initial) was obtained on the solid coating composition. The resultant reduced viscosity coating composition sample was then sealed, and after 24 hours, a second curve (Figure 14.1-sheared and 24-hour recovery) was obtained.

The low shear rate (16 sec^{-1}) spike readings show a lack of substantial viscosity recovery based on gel structure: initial spike, 30.5; end, 9.5; 24-hour recovery rerun spike, 6.0; end, 9.0. (The curves in Figure 14.1 are presented on a logarithmic scale.)

As the curves in Figure 14.1 indicated, some viscosity recovery occurred, a significant portion of which is contributed by the conventional coating formulation. As with most conventional coating compositions intended for use with brush, roller or pad applicators, a degree of thixotropy is built into the composition to aid conventional application properties.

Using the same procedure, a number of solid coating compositions were prepared. Lower levels of colloid or higher relative levels of electrolyte provided more stable compositions.

Figure 14.1: Viscosity Studies for Semisolid Water-Based Coating Compositions

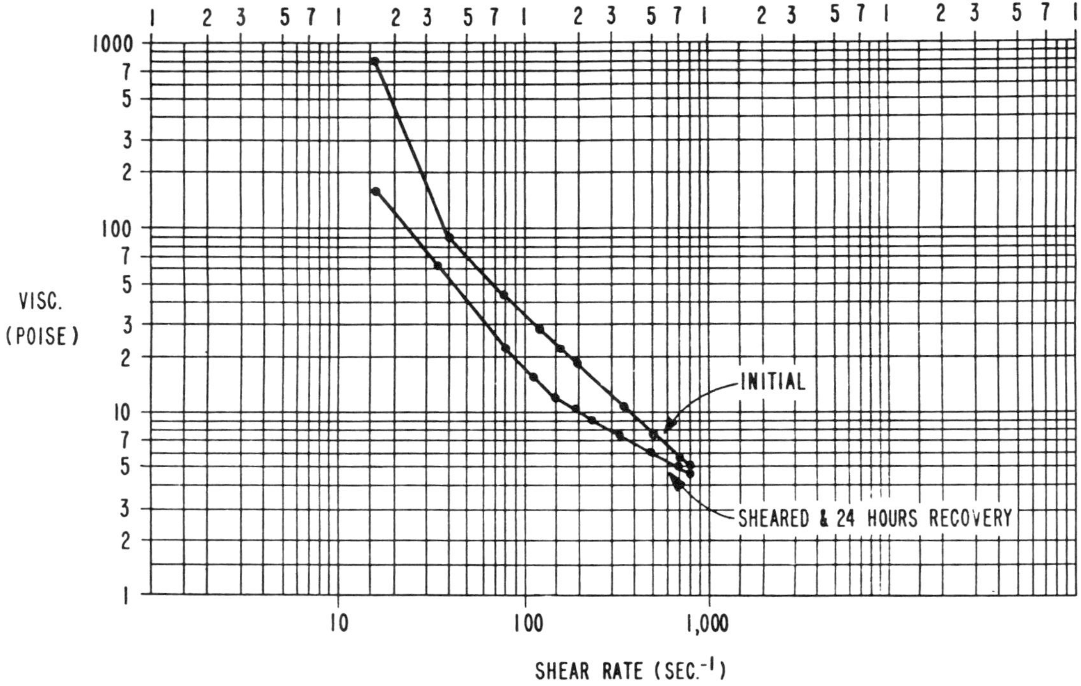

Source: U.S. Patent 4,196,107

Antifouling Paint

Highlights of the Technological Achievement: An antifouling paint for ships' hulls, comprising a film-forming copolymer and a pigment component, has the property that a rough film of the paint becomes smoother in relatively moving seawater. The film-forming copolymer contains from more than 50 up to 80% by weight of units of at least one triorganotin salt of an olefinically unsaturated carboxylic acid, the balance of the copolymer being units of at least one olefinically unsaturated comonomer. The pigment component comprises a metalliferous pigment capable of reacting with seawater to form a water-soluble metal compound and a pigment which does not react with or dissolve in seawater.

Reference: G. Hails and J.D. Symonds; U.S. Patent 4,191,579; March 4, 1980; assigned to The International Paint Company Limited, England.

Background: The growth of marine organisms on the submarine parts of a ship's hull increases the frictional resistance of the hull to passage through water, leading to increased fuel consumption and/or a reduction in the speed of the ship. Marine growths accumulate so rapidly that the remedy of cleaning and repainting as required in dry dock is too expensive to contemplate. The alternative which has been practiced with increasing efficiency over the years, is to limit the extent of fouling by applying to the hull a topcoat paint incorporating antifouling agents.

The antifouling agents are biocides which are freed from the surface of the paint over a period in concentrations lethal to marine organisms at the hull surface. The antifouling paint fails only when the concentration of biocide available at the paint surface falls below the lethal concentration, and with modern paints up to two years of useful life is expected.

Process Conditions: According to the process, an antifouling paint for ships' hulls comprises:

(a) a film-forming copolymer containing from more than 50% up to 80% by weight of units of at least one triorganotin salt of an olefinically unsaturated carboxylic acid, the balance of the copolymer being units of at least one olefinically unsaturated comonomer, and

(b) a substantially water-insoluble pigment component, part of which comprises a metalliferous pigment capable of reacting with seawater to form a water-soluble metal compound and at least 5% by weight of which comprises a pigment which does not react with or dissolve in seawater, the proportion of the metalliferous pigment capable of reacting with seawater to form a water-soluble metal compound being sufficient that a rough film of the paint becomes smoother in relatively moving seawater.

The antifouling paint according to this process exhibits the antifouling action to be expected from its ability to release triorganotin ions into seawater as well as the remarkable property of the paint described in British Patent 1,457,590 that relatively moving seawater smooths the painted surface.

Figure 14.2: Antifouling Paint

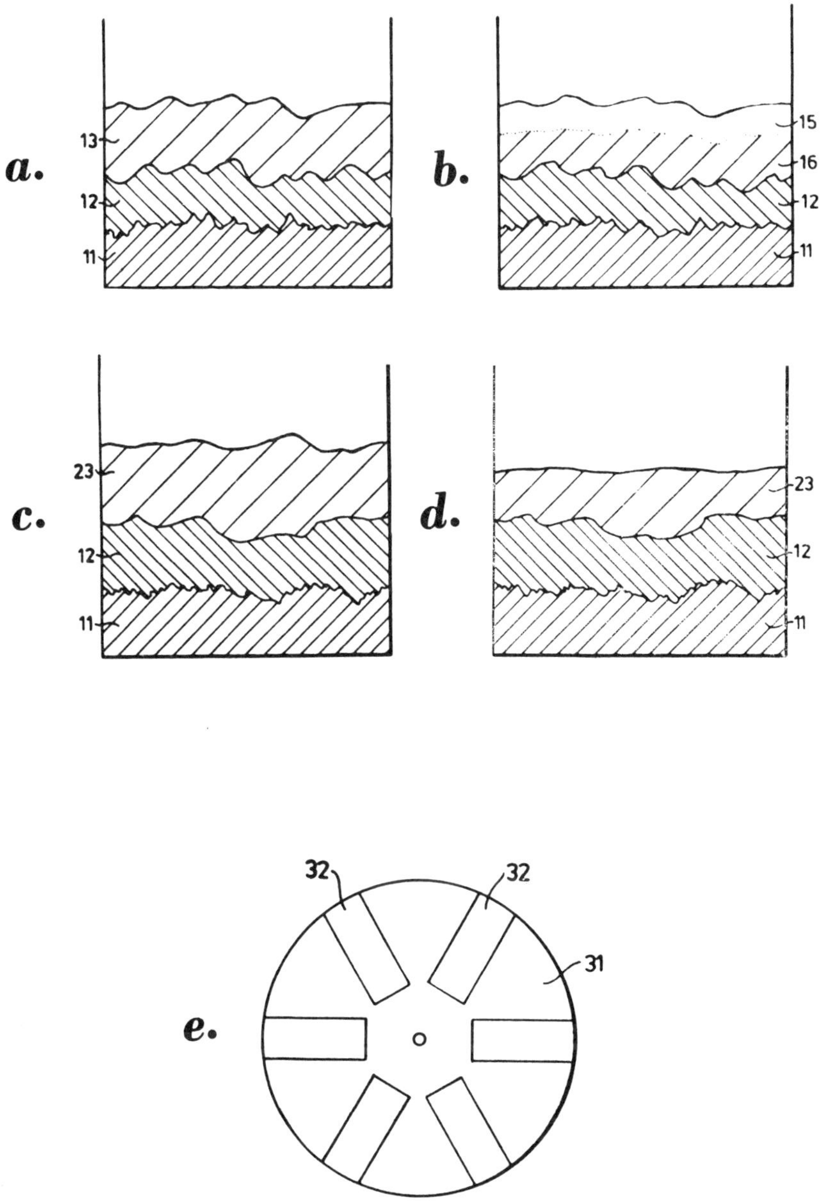

(continued)

Figure 14.2: (continued)

f.

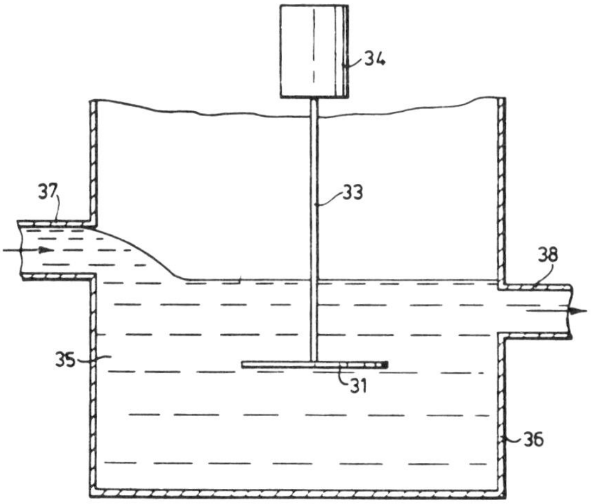

(a)-(d) Diagrammatic cross sections through films of antifouling paint on the hull of a ship
(e) Plan of a test disc used to test paints
(f) Sectional elevation of an apparatus using the test disc of Figure 4.2e

Source: U.S. Patent 4,191,579

The difference in the performance of the antifouling paint according to the process and a conventional antifouling paint is illustrated in Figure 14.2a through 14.2d of the above drawings. All of those four figures are diagrammatic cross-sectional views through part of the hull of the ship showing the steel plate of the hull **11**, the anticorrosive paint **12** and the antifouling paint **13** or **23**. Figure 14.2a represents a newly painted hull using a conventional antifouling paint **13** and Figure 14.2b shows the hull with the same paint near the end of its useful life, that is, about 18 months later.

As seen in Figure 14.2b, an antifouling agent dispersed in the film-forming matrix has been leached from the coating by seawater to a depth shown as a dotted line between the regions **16** and **15**. The outermost region **15** overlies the potentially useful antifouling region **16**, inhibiting the leaching of antifouling agent in lethal concentrations from that region to the outer surface of region **15** so that fouling ensues.

Figure 14.2c represents a hull freshly painted with a topcoat of antifouling paint **23** according to this process on an underlying anticorrosive paint. Over a period, the topcoat paint is planed by the seawater, becoming thinner and smoother as shown in Figure 14.2d, yet the antifouling capability is unchanged and no inactive surface develops to hinder the release of the antifouling agent.

The smoothing performance of paints in relatively moving seawater can be tested in the apparatus shown schematically in Figures 14.2e and 14.2f. The mild steel disc **31** shown in Figure 14.2e having a diameter of 23 cm is coated overall with a conventional anticorrosive paint and overcoated in radial stripes such as **32** with the paints under test. The stripes are applied by a standardized applicator adapted to deposit a film 20 μ thick (wet thickness).

The disc is mounted on a shaft **33** driven by an electric motor **34** and immersed in flowing seawater **35** contained in a vessel **36** having an inlet **37** and overflow **38**. The peripheral speed of the disc is 33 knots.

Organotin: Upon contact with seawater, the copolymer in paints according to the process generates a water-soluble residue which essentially is a copolymer salt formed as a result of the depletion in the copolymer organotin content caused by exchange of tin ions for metal ions present in the seawater. It has been found that the copolymer must contain more than 50% by weight of organotin salt units in order to generate the water-soluble residue at a sufficient rate. If the organotin content of the copolymer is less than about 50%, the organotin ion may still be exchanged with a metal ion from seawater, but the residual polymer either separates from the ship's hull in a way which does not lead to a smoother surface, or remains upon the hull. In the latter case the paint behaves similarly to the known antifouling paints as described with reference to Figures 14.2a and 14.2b.

The more the organotin salt content of the copolymer exceeds 50% by weight, the greater the rate of ion exchange with seawater and the quicker the copolymer becomes water-soluble enough to be swept away from the paint surface by the seawater flowing over the surface.

The triorganotin salt of the olefinically unsaturated carboxylic acid incorporated in the copolymer suitably has the general formula:

$$R_3SnOOCCR'{=}CHR''$$

where the three groups R are the same or different alkyl radicals containing up to 8 carbon atoms or aryl or aralkyl radicals, R' is H or methyl and R'' is H or $-COOSnR_3$, where R is defined as above. Thus the cation of the salt, R_3Sn^+, is exemplified by tributyltin, tripropyltin, triethyltin, tribenzyltin, diethylbutyltin, tribromophenyltin, diphenyltolyltin, tritolyltin, diethylphenyltin, ethyldiphenyltin, octyldiphenyltin and diethyloctyltin. The preferred cations are those having three identical organic groups attached to the tin atom. The anion of the salt, $-OOCCR'{=}CHR''$, may be, for example, acrylate, methacrylate, maleate or fumarate. Examples of the salt come from pairing any listed anion with any listed cation.

Mineral-Stabilized Resin Emulsion

Highlights of the Technological Achievement: A mineral-stabilized resin emulsion comprising a low to zero acid number hydrocarbon resin modified with an organic acid to the extent necessary for emulsification, a mineral stabilizer and a complexing agent such as an inorganic acid, a polyvalent metal ion or ammonium ion compound.

Reference: F.H. Haney; U.S. Patent 4,190,568; February 26, 1980.

Background: The drawbacks associated with prior asphalt emulsion and acrylic coating materials have led to the development of the mineral-stabilized resin emulsion which is similar to the prior asphalt emulsions in appearance but can be easily pigmented and used either by itself as a decorative weather coating, as an outer protective coating applied to an underlying asphalt emulsion, or applied as a coating to roofing paper and optionally surmounted with a granular material or other suitable surfacing agent. Further, the mineral-stabilized resin emulsion produces a dry emulsified film which is water resistant and more resistant to flow at elevated temperature.

Process Conditions: The resin to be emulsified may be any hydrocarbon resin, wax or the like having an acid number too low to be emulsified by conventional emulsification methods, and plasticized with a suitable process oil, if necessary, whereby the resin has a softening point of 80° to 150°F and a penetration of 40 to 190 dmm at 77°F (ASTM D5).

Referring to the mineral colloid emulsification system, the anions derived from the organic acid modified resin do not react with the anions derived from the mineral stabilizer slurry in the system unless they are promoted by a source of cations.

Generally, the resin is heated above its softening temperature but below its flash point, for example, to about 180° to 280°F and the selected organic acid for modifying the resin is stirred into the resin, by the action of a mechanical blender such as a pony mixer, ribbon mixer, Charlotte colloid mill, Lancaster colloid mill or the like.

Depending on the softening point of the resin, the clay slip may be heated to 140° to 160°F; however, if about ⅓ to ½ of the clay slip is put into the mixing container there is usually enough heat from the resin to bring about emulsification as the parts are mixed together. The balance of the ingredients are added proportionately over a short time period, e.g., 5 to 15 minutes. It is preferable to stir and cool the finished emulsion to keep the particle size from growing. Whipping the cold emulsion can reduce the particle size. The usual water content of a finished emulsion runs from about 35 to 55%.

The mineral stabilizer is sifted into a portion of water contained in a mechanical mixer and allowed to form a slurry. The complexing agent is added to the slurry in the form of a solution made with the remaining quantity of water, thereby forming the emulsifier. The melted acid-modified resin mixture is slowly poured into approximately one-half of the emulsifier during continued stirring. The balance of the emulsifier and acidified resin are then slowly added over a period of about 5 minutes until the entire amounts are added and emulsified. The emulsion is stirred and allowed to cool to room temperature, and subsequently dried.

Polyphenylene Sulfide Resin Coating Composition

Highlights of the Technological Achievement: A composition comprising a polyphenylene sulfide resin powder and an aluminum oxide powder of colloidal size. The composition is useful as a powder coating composition. The composition exhibits improved adhesion to a substrate, even in hot water.

Reference: M. Takahashi, O. Ishii, M. Naito, Y. Kusuhara and N. Imahigshi;

U.S. Patent 4,183,840; January 15, 1980; assigned to Sekisui Kagaku Kogyo Kabushiki Kaisha, Japan.

Background: Polyphenylene sulfide resin has recently been much studied for its capacity for forming coated films having good thermal stability, chemical resistance and mechanical properties. Coated films of the polyphenylene sulfide resin are generally considered to have better adhesion to substrates than those of other resins. In practical application, however, their adhesion is not sufficient, and particularly in hot water, tends to decrease markedly.

A general method known to improve the adhesion of resinous coated films involves blending or graft-copolymerizing a compound having a polar group such as an epoxy resin or acrylic acid with a resin. Since a paint film of the polyphenylene sulfide resin requires high temperatures and long periods of time for baking, when a paint of the polyphenylene sulfide modified by this method is used, the polar organic compound blended or grafted with it will decompose during baking at high temperatures. It is extremely difficult therefore to improve the adhesion of the polyphenylene sulfide by this general method.

Process Conditions: The polyphenylene sulfide is a resin whose main chain is composed mainly of the recurring unit:

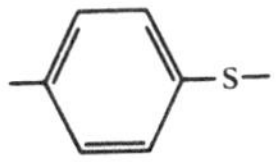

in which the phenylene group is optionally substituted by halogen or an alkyl group with 1 to 12 carbon atoms. The resin may have a partially oxidized, branched or crosslinked structure. The polyphenylene sulfide resin and a method for its production are disclosed, for example, in U.S. Patent 3,354,129. To be suitable as a paint, the resin generally has a melting point of 200° to 350°C, preferably 250° to 300°C, and a melt flow (ASTM D1238, under a load of 5 kg at 600°F) of 30 to 6,000 g/10 min, preferably 100 to 3,000 g/10 min. The particle diameter of the resin powder is not particularly limited, and may be the one convenient for spray coating, for example, 1 to 150 μ, preferably 10 to 100 μ. The polyphenylene sulfide powder is commercially available under the designation Ryton PPS. In the process, such a commercially available product can be used.

Product: The powder paint has the following advantages over dispersion paints:

(a) In the case of the powder paint, a thick coating can be formed by one cycle of coating operation. In the case of the dispersion paint, a thick coating cannot be formed by one coating operation, and the coating must be carried out a number of times until the desired thickness is obtained. Hence, the dispersion paint has poor operability.

(b) When the dispersion paint is used, pinholes tend to form in the coated film by the volatilization of the liquid medium. These pinholes cause deterioration of the properties of the coated film, especially hot water resistance. The occurrence of pinholes could be prevented by carefully controlling the speed of drying. In practice, however, it is difficult to exercise good control all the time. The powder paint, on the other hand, can give superior coated films without involving such a defect or difficulty.

(c) When the dispersion paint is used to coat the inner surface of a tubular product, especially a narrow pipe, a coating of uniform thickness is considerably difficult to obtain because the slurry flows. In contrast, when the powder paint is used, a coating of uniform thickness can be easily obtained in such a case.

Rendering Surfaces Permanently Water-Wettable

Highlights of the Technological Achievement: The surfaces of articles of manufacture fabricated from aluminum and other materials which are not permanently water-wettable are so rendered by coating the surface with a continuous film containing colloidal alumina of a water-soluble form of basic film-forming polymer which is curable to water-insolubility; curing the polymer to water-insolubility; and hydrolyzing the surface portion only of the cured polymer.

Reference: M.E. Rickert, Jr.; U.S. Patent 4,181,773; January 1, 1980; assigned to General Electric Company.

Background: This process relates to a process for rendering surfaces permanently water-wettable and to water-wettable articles of manufacture thus produced.

It is sometimes desirable that a surface of an article of manufacture exposed to water be water-wettable. The term "water-wettable" as used herein means the ability of the surface to retain a substantially unbroken film of water thereon, preferably for at least 10 seconds and more preferably for at least 30 seconds after removal from water, e.g., after being immersed in water or flushed with a stream of water. This property is desirable because a water-wettable surface exposed to water condensation has less of a tendency to drip droplets of water therefrom than a surface which is not water-wettable.

Process Conditions: A process for rendering permanently water-wettable a surface of an article of manufacture which normally is not permanently water-wettable, which comprises the steps of:

(a) coating the surface with a continuous film containing colloidal alumina of a water-soluble form of a film-forming basic polymer which is curable to water-insolubility;

(b) curing the coated polymer on the substrate surface to water-insolubility; and

(c) thereafter hydrolyzing substantially only the surface portion of the cured polymer.

Product: If the article of manufacture is formed of or coated with a polymer, necessarily the polymer must be one which can be coated with a water-soluble basic polymer which is stable in the curing step. Preferred substrates are those formed from a base metal, especially those formed from aluminum or an aluminum alloy, e.g., of high Si content, or a composite of aluminum and steel because, inter alia, the aluminum is protected from the staining which occurs on unprotected aluminum, as well as being rendered permanently water-wettable.

The article of manufacture can be in any solid shaped form, including rigid and flexible sheets, films, foils and extruded, cast, stamped and machined three-dimensional-shaped articles, including water pipes and the heat exchanger surfaces of air conditioners and evaporators and the frames of refrigerator doors, drain troughs, fresh food liner tops and other refrigerated areas which are susceptible to sweating. Preferred are those forming the heat exchanger surface of evaporators, especially aluminum refrigerator evaporators, and other fabricated metal articles of manufacture.

The basic polymers used to coat the starting substrate are those which: (a) form water-soluble salts with acids; (b) will form a continuous coherent film on the selected substrate; and (c) are curable to water-insolubility.

These properties, rather than the specific chemical structure of a basic polymer, determine its suitability for use in this process. Therefore, a wide variety of structural types of film-forming basic polymers, i.e., having a basic nitrogen atom, can be employed.

Preferred basic polymers are curable basic resins which can be deposited as a film and from an aqueous dispersion thereof, onto an article by cationic electrodeposition. Examples of such polymers are copolymers derived from: (1) 3 to 50% by weight of a 5- or 6-membered, heterocyclic N-vinyl amine, e.g., N-vinylpyrazole and N-vinylindole; (2) 3 to 30% of an amide of acrylic and/or methacrylic acid; (3) 20 to 94% by weight of one or more of an alkyl ester of acrylic or methacrylic acid, styrene and alkylvinylbenzene; and (4) 0 to 30% by weight of another ethylenically unsaturated monomer.

The basic polymer is preferably in the form of an organic acid salt thereof. Although salts of strong acids, e.g., phosphoric, are also operable, they are not preferred.

Preferably, the polymer coating applied to the substrate has a thickness of 3×10^{-3} to 40×10^{-3} mm, more preferably 9×10^{-3} mm. The thickness of the coating can be regulated by the concentration of the polymer in the solution, the viscosity of the solution, the number of immersions, the spraying time and/or the final voltage when the coating is applied electrolytically.

NATURAL GUMS

PROCESSING

Producing Heteropolysaccharides by Bacterial Fermentation of the Genus *Xanthomonas*

Highlights of the Technological Achievement: A process for producing modified heteropolysaccharides derived from a bacteria of the genus *Xanthomonas* comprises the steps:

(a) Preparing an aqueous saline heteropolysaccharide solution;

(b) Heating the saline heteropolysaccharide solution to a temperature of at least 100°C;

(c) Maintaining the saline heteropolysaccharide solution at a temperature of at least 100°C for a period of time sufficient to increase the injectivity and filterability characteristics of the heteropolysaccharide, but for an insufficient time to impair the viscosity-imparting properties of the heteropolysaccharide; and

(d) Removing or separating the proteinaceous materials and/or residual whole bacterial cells or other cell debris from the saline- and heat-treated heteropolysaccharide product.

Reference: L.A. Naslund and A.I. Laskin; U.S. Patent 4,182,860; January 8, 1980.

Background: A genuine need exists for a method for producing a *Xanthomonas* heteropolysaccharide which can be readily filtered and separated from other debris in the solution, such as bacterial cells and proteinaceous material. Moreover, a need exists for a heteropolysaccharide which can be injected into porous media, such as oil-bearing sandstone formations, and which can be added to foodstuffs or pharmaceutical preparations without discoloring or markedly changing the clarity of such preparations.

Process Conditions: A more detailed preparation of the readily filtered heteropolysaccharide solution is given below:

(a) Preparing an aqueous solution containing from 200 to 30,000 ppm by weight of a heteropolysaccharide fermentation product produced by the action of bacteria of the genus *Xanthomonas* and at least 0.5 wt % of at least one salt, preferably from 0.5 to 10 wt % inorganic salts selected from the group consisting of sodium chloride, calcium chloride and mixtures thereof, to obtain a saline heteropolysaccharide solution;

(b) Heating the saline heteropolysaccharide solution to a temperature of at least 100°C, preferably in the range from 100° to 180°C; and

(c) Maintaining the saline heteropolysaccharide solution at a temperature of at least about 100°C, preferably for a period of time sufficient to improve the filterability characteristics, but not so long that the viscosity-imparting properties of the saline- and heat-treated heteropolysaccharide are substantially impaired, e.g., from 1 to 30 minutes.

Oil Recovery: A process for recovering oil from a subterranean formation penetrated by at least one injection well and at least one production well by injecting into the injection well and into the subterranean formation an aqueous solution containing a heteropolysaccharide produced by the action of bacteria of the genus *Xanthomonas*, wherein the heteropolysaccharide has been modified such that the modified heteropolysaccharide is capable of imparting a viscosity of at least 4.0 cp to an aqueous test solution containing 2 wt % NaCl and 0.2 wt % $CaCl_2$ when the modified heteropolysaccharide is added to the test solution at a concentration of approximately 600 ppm by weight, as measured on a Brookfield viscosimeter with a UL adapter at 60 rpm at 25°C.

The modified heteropolysaccharide is further capable of imparting a filterability such that more than 1,000 ml of a different aqueous test solution containing 8.8 wt % salt comprised of NaCl and $CaCl_2$ at a 10:1 weight ratio and approximately 600 ppm concentration by weight of the modified heteropolysaccharide will pass through a 13 mm diameter Millipore filter having a pore size of 5 μ at a constant pressure drop of about 1.55 psig without plugging.

***Xanthomonas* Bacteria:** The heteropolysaccharides which are used in the practice of this process are generally fermentation products produced by the action of bacteria of the genus *Xanthomonas* upon carbohydrates. Representative species of these bacteria include *Xanthomonas campestris, Xanthomonas phaseoli, Xanthomonas malvacearum, Xanthomonas carotae, Xanthomonas translucens, Xanthomonas hederae, Xanthomonas papavericola, Xanthomonas begoniae* and *Xanthomonas incanae.*

Studies have shown that the production of the heteropolysaccharides is a characteristic trait of members of the genus *Xanthomonas*, but that certain xanthomonads synthesize such heteropolysaccharides with particular efficiency and are, therefore, more attractive for purposes of the process than are others. *Xanthomonas campestris, Xanthomonas begoniae* and *Xanthomonas incanae* are outstanding in this respect and, hence, are preferred species.

Commercially Available Heteropolysaccharides: Heat treating in a saline solution with subsequent filtration can be incorporated in the process steps for producing commercial grade heteropolysaccharide or, alternatively, commercial grade heteropolysaccharide can be produced and then subjected to the treating process of this method. For example, commercially available heteropolysaccharide can be obtained from a number of sources, one example being Kelzan XC (Kelco Company).

The commercial grade heteropolysaccharide is dissolved in water in a concentration of 200 to 30,000 ppm in distilled water with the addition of from 0.5 to 10 wt % salts. This solution is heated to a temperature ranging from 100° to 180°C for a period of time ranging from 1 to 300 minutes. The heteropolysaccharide solution can then be subjected to mechanical separation to remove all entrained residual material such as clumps of incompletely solubilized heteropolysaccharide residual proteinaceous material and residual whole bacterial cells or other cellular debris.

Example: A series of tests were carried out in order to determine the effect of the saline heat treatment on a variety of commercially available *Xanthomonas* heteropolysaccharide samples. Duplicate flasks with each set containing 3,000 ppm of one of six commercially available polymers were prepared. The samples used were as follows:

(A) Kelzan—the standard industrial grade heteropolysaccharide suitable for most industrial applications;

(B) Kelzan XC—an industrial grade heteropolysaccharide of particular value as an additive to oil well drilling mud;

(C) Kelzan MF—an industrial grade heteropolysaccharide for secondary and tertiary oil recovery;

(D) Kelzan—same as (A), but a different batch;

(E) SS-3192-laboratory sample of highly purified heteropolysaccharide (Kelco Company); and

(F) XB-23—*Xanthomonas* heteropolysaccharide (General Mills Chemical Inc.).

One set of flasks was heated at 121°C for 3 hours and then cooled to room temperature; the second set received no heat treatment. In the cases of samples (E) and (F), a third set of flasks was treated for 1 hour instead of 3 hours. The 3,000 ppm concentrates were sheared in a Waring blender at full power for 3 minutes and were used to prepare 600 ppm solutions in 8.8% brine.

The results of viscosity and filtration tests in 8.8% brine are shown in the table and the method can be applied to a wide variety of *Xanthomonas* heteropolysaccharide samples, ranging from rather crude to very pure preparations and from different manufacturers.

In the tests, saline heat treatment was done using 3,000 ppm heteropolysaccharide concentrates in 2.2% brine; the filterability and viscosity measurements were done on the 600 ppm heteropolysaccharide solutions in 8.8% brine; and the flow rates are shown for the points at which 100 or 1,000 ml of solution has passed through the filter.

Effect of Saline Heat Treatment

	Viscosity (cp)	Filterability: Flow Rate (ml/min) 100 ml	1,000 ml	Remarks
Sample A				
Untreated	5.85	–	–	Filter plugged at 35 ml
Treated 3 hr	5.35	16.0	5.6	More than 1,000 ml throughput without plugging
Sample B				
Untreated	5.35	–	–	Filter plugged at 50 ml
Treated 3 hr	4.95	15.5	2.7	More than 1,000 ml throughput without plugging
Sample C				
Untreated	5.75	–	–	Filter plugged at 44 ml
Treated 3 hr	5.60	12.5	2.5	More than 1,000 ml throughput without plugging
Sample D				
Untreated	6.05	–	–	Filter plugged at 30 ml
Treated 3 hr	5.50	14.5	2.4	More than 1,000 ml throughput without plugging
Sample E				
Untreated	4.65	–	–	Filter plugged at 54 ml
Treated 3 hr	3.40	17.0	13.0	More than 1,000 ml throughput without plugging
Treated 1 hr	4.65	13.8	12.0	More than 1,000 ml throughput without plugging
Sample F				
Untreated	5.35	–	–	Filter plugged at 12 ml
Treated 3 hr	3.40	12.8	2.7	More than 1,000 ml throughput without plugging
Treated 1 hr	4.65	7.0	0.1	More than 1,000 ml throughput without plugging

Amphoteric Graft Copolymers of *Xanthomonas* Hydrophilic Colloid and Partially N-Aminomethylated Acrylamide

Highlights of the Technological Achievement: A series of amphoteric graft copolymers of *Xanthomonas* hydrophilic colloid and partially N-aminoalkylated acrylamide has been prepared by reacting the corresponding acrylamide copolymer with an alkyl amine and an active aldehyde. The resulting copolymer is cationized via acidification or quaternization, and is reverted to its anionic form via basification.

Reference: J.S. Racciato, I.W. Cottrell and J.L. Shim; U.S. Patent 4,155,885; assigned to Merck & Co., Inc.

Background: To create the desired, easily ionizable copolymers, the amino functions of the acrylamides present in the *Xanthomonas* colloid-acrylamide copolymers are partially converted to aminoalkylamines via Mannich reaction.

Although the method has been applied to polyacrylamide, it has not been used in modifying any of the *Xanthomonas*-acrylamide copolymers which incorporate all the *Xanthomonas* heteropolysaccharide functional groups such as hydroxyl, carboxyl, ester and ketal. Surprisingly, these active functional groups have not interfered with the aminoalkylation.

Process Conditions: The amphoteric copolymers of this process are graft copolymers of *Xanthomonas* hydrophilic colloid and partially N-aminoalkylated acrylamide in which the weight ratio of *Xanthomonas* colloid to the acrylamide (ZN:AM) ranges between 1:1 to 1:10 and where in each of the polymers, the percentage of the N-aminoalkylated acrylamide (NAM%) groups ranges from 30 to 85% of the total available acrylamide groups.

The N-aminoalkylated acrylamide function is defined in formula (1):

(1)

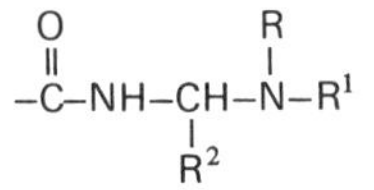

which is cationized via acidification to form an acid addition salt of formula (2):

(2)

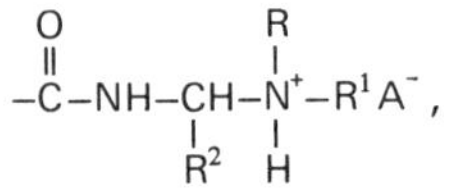

or quaternization to form the quaternary salt of formula (3):

(3)

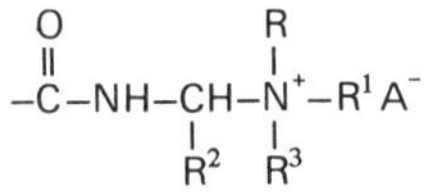

where R and R^1 are independently lower alkyl, especially C_{1-6} alkyl; R^2 is hydrogen, lower alkyl especially C_{1-6} alkyl, or benzyl; R^3 is lower alkyl especially C_{1-6} alkyl, or benzyl; and A^- is an anion of a strong acid such as hydrochloric acid, hydrobromic acid, hydroiodic acid, sulfuric acid, phosphoric acid, trichloroacetic acid, trifluoroacetic acid, p-toluenesulfonic acid, 2,4-dinitrophenol sulfonic acid, or the like.

First, with regard to the weight ratio of the *Xanthomonas* colloid to the acrylamide (ZN:AM), the preferred form of the copolymers of this process is that where the weight ratio ZN:AM is from 1:1 to 1:5, especially from 1:1 to 1:3 for higher aqueous viscosity.

Second, concerning the percentage of the N-aminoalkylated acrylamide of the total available acrylamide groups (NAM%), the preferred form of the copolymers of this process is that where the percentage (NAM%) ranges from 45 to 55%.

Third, regarding the N-aminoalkylated acrylamide function, the preferred form of the copolymers of the process is that where R and R^1 represent independently C_{1-3} alkyl such as methyl, ethyl or propyl; R^2 represents hydrogen, methyl or benzyl; R^3 represents C_{1-4} alkyl such as methyl, ethyl, propyl or butyl; and A^- is a halide anion such as chloride, bromide, or iodide.

A still more preferred version of this process is that wherein R and R^1 are both methyl; R^2 is hydrogen; R^3 is methyl; and A^- is chloride.

Aminoalkylation: The aminoalkylation is carried out in an inert solvent such as water, a lower alkanol such as methanol, ethanol, a di(lower alkyl)ketone such as acetone, methyl ethyl ketone or diethylketone, a lower alkyl lower alkanoate such as methyl or ethyl acetate or the like, or mixtures thereof. The process comprises the treatment of a suitable copolymer with a dialkylamine of the formula:

(4)

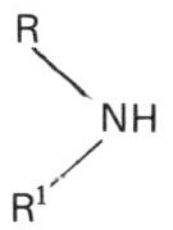

where R and R^1 are as previously defined, at preferably 65° to 75°C for a sufficient length of time, usually 10 to 60 minutes, followed by subsequent treatment with an aldehyde of the formula R^2CHO, where R^2 is as previously defined, at preferably 65° to 75°C until the reaction is complete, usually about 15 minutes. If desired, the resulting product is then cationized by:

(a) Addition of a strong acid of formula HA, where the anion thereof, A^-, is as previously defined to form the acid addition salt of formula (2); or

(b) Quaternization with an appropriate reagent such as an alkyl halide of the formula R^3A where A is selected from chloride, bromide or iodide.

Again if desired, the resulting quaternary halide of formula (2) is basified with a strong base such as sodium hydroxide, potassium hydroxide, ammonium hydroxide or the like, to form the corresponding quaternary hydroxide followed by salt formation with a strong acid HA to form the quaternary salt of formula (3).

Similarly, upon basification with an inorganic base such as sodium hydroxide, sodium bicarbonate, potassium carbonate and ammonium hydroxide or the like, the ammonium salts of the copolymers of formula (2) are reverted to their anionic forms containing the N-acrylamidoalkylamine function as shown in forula (1); while in the meantime the glucuronic acid molecules contained in the *Xanthomonas* colloid are converted to carboxylate salts.

Product: The copolymers are used as retarding or leveling agents in dyeing textiles such as acrylic fabrics. Usually a 0.5 to 1.5% solution of the copolymer, based on the weight of the fabric to be dyed, containing 0.25 to 1.0% of the dye, also based on the weight of the fabric to be dyed, in a polar solvent such as water, methanol, ethanol, acetone or aqueous solutions thereof, is adjusted to a pH value required by the dye used, with a weak acid such as acetic acid, propionic acid or the like. The fabric is then immersed in this solution and the entire mixture is heated to reflux until the dyeing process is completed, usually about 1 to 2 hours.

Other uses of the copolymers include flocculants or pigment retention aids in paper manufacturing or coagulant aids or agglomerating agents in water treatment.

Example 1: To a solution of 8.0 g of *Xanthomonas* hydrophilic colloid-acrylamide (1:3) copolymer in 300 ml of water at 70°C are added 19.0 g of 40% aqueous dimethylamine. The reaction solution is stirred for about 5 minutes before 5.8 g of paraformaldehyde are added slowly.

After the addition is completed, the reaction mixture is stirred for an additional 20 to 40 minutes and then cooled to 55°C. While stirring, 6.0 g of ammonium chloride and 8.4 g of concentrated hydrochloric acid are added and the mixture is stirred for an additional 30 minutes before it is cooled to room temperature and diluted with 600 ml of acetone to precipitate the product. After filtration, it is dried in vacuo at 50°C and then milled to 11.3 g (65%) of a 40 mesh powder which was insoluble in water, showed IR bands at 3,050 cm^{-1} (NH_4^+), 1,670 cm^{-1} and 1,550 cm^{-1} [–C(O)NH–]. Based on the percent weight add-on (41.3%), it is estimated that 35.3% of the original acrylamide function has been aminomethylated.

Example 2: In the similar manner as described in Example 1, 12.0 g of *Xanthomonas* hydrophilic colloid-acrylamide (1:1) copolymer are treated with 19 g of 40% dimethylamine and 5.8 g of paraformaldehyde to form 15 g (70.4%) of a copolymer which is insoluble in water. Based on the percent weight add-on (25%), it is estimated that 32.3% of the original acrylamide function has been aminomethylated.

Example 3: In the similar manner as described in Example 1, 7.2 g of *Xanthomonas* hydrophilic colloid-acrylamide (1:5) copolymer are treated with 40% aqueous dimethylamine (19 g) and paraformaldehyde (5.8 g) to afford 14.9 g (89.9%) of a water-soluble copolymer in which 82.8% of the existing acrylamide is dimethylaminomethylated. The viscosity of a 1% aqueous solution of the product was 25 cp.

Treatment of Xanthan Gum to Improve Clarity

Highlights of the Technological Achievement: A xanthan gum fermentation beer containing at least 1% gum is filtered at a temperature of at least 112°C. The gum subsequently recovered from the filtrate has improved clarity.

Reference: F.E. Corley and J.B. Richmon; U.S. Patent 4,135,979; January 23, 1979; assigned to Merck & Co., Inc.

Background: U.S. Patent 3,355,447 discloses a process for improving the clarity and stability of *Xanthomonas* hydrophilic colloids by heating a fermentation beer to a temperature from 66° to 77°C, holding the beer at this temperature for at least 20 minutes, cooling to a temperature in the range of 4° to 38°C, diluting to a concentration not above 1% and filtering. This process, however, is relatively time consuming and costly because of the requirement that the *Xanthomonas* hydrophilic colloid be diluted and filtered.

Raw Materials: The starting material for the conduct of the process is a *Xanthomonas* hydrophilic colloid solution and preferably a crude solution thereof known as a beer. A *Xanthomonas* hydrophilic colloid that is particularly suitable for use in accordance with the process is such a colloid prepared by the bacterium *Xanthomonas campestris.*

The *Xanthomonas campestris* colloid is a polymer containing mannose, glucose, and potassium glucuronate. In such a colloid, the potassium portion can be replaced by several other cations without substantial changes in the property of the material for the purpose. The colloid, which is a high molecular weight exocellular material, may be prepared by the bacterium *Xanthomonas campestris* by the whole culture fermentation of a medium containing 2 to 5% commercial glucose, an organic nitrogen source, dipotassium hydrogen phosphate and appropriate trace elements. The incubation time of the final medium is approximately 96 hours at 30°C under aerobic conditions.

In preparing the colloid as aforementioned, it is convenient to use corn steep liquor or distillers' dry solubles as an organic nitrogen source. It is expedient to grow the culture in two intermediate stages prior to the final inoculation in order to encourage vigorous growth of the bacteria. These stages may be carried out in media having a pH of about 7.

In the first stage, a transfer from an agar slant to a dilute glucose broth may be made and the bacteria cultured for 24 hours under vigorous agitation and aeration at a temperature of about 30°C. The culture so produced may then be used to inoculate a higher glucose (3%) content broth of larger volume in a second intermediate stage. In this stage, the reaction may be permitted to continue for 24 hours under the same conditions as the first stage. The culture so acclimated for use with glucose by the aforementioned first and second stages is then added to the final glucose medium.

In the aforementioned method of preparing *Xanthomonas campestris* hydrophilic colloid, a loopful of organism from the agar slant is adequate for the first stage comprising 290 ml of the glucose medium. In the second stage, the material resulting from the first stage may be used together with nine times its volume of 3% glucose medium.

In the final stage, the material produced in the second stage may be admixed with nineteen times its volume of the final medium. A good final medium may contain 3% glucose, 0.5% distillers' solubles, 0.5% dipotassium phosphate, 0.1% magnesium sulfate having 7 molecules of water of crystallization, and water. The reaction in the final stage may be satisfactorily carried out for 96 hours at 30°C with vigorous agitation and aeration. A *Xanthomonas* colloidal solution in the aforementioned stage of preparation is known as a beer. This is a preferred starting material for practicing the method, i.e., the beer resulting directly from the fermentation may be used as starting material.

Process Conditions: An aqueous solution containing at least about 1 wt % of *Xanthomonas* hydrophilic colloid (xanthan gum) to be clarified, for example, a fermentation beer, is filtered at a temperature of at least about 112°C up to a temperature beyond which thermal degradation of the gum occurs. Preferably, the temperature is from 127° to 149°C. It is preferred not to maintain the beer for prolonged times, e.g., beyond 30 minutes, at elevated temperature in order to avoid degradation of the gum. The cooling may take place rapidly in a few minutes or slowly over a period of hours. For economical operation, however, it is preferred that the fermentation beer contain at least 1.5 wt % xanthan gum. While the upper limit of xanthan gum is subject to variation, as a practical matter, solutions containing above 4.0% are too viscous to be filtered.

Product: The filtrate obtained according to the process is a clear solution, substantially free of fermentation insolubles. The gum separated from this filtrate by conventional means, e.g., alcohol precipitation, spray drying or drum drying, is a highly clarified product suitable for all uses of xanthan gum and particularly foodstuff and pharmaceutical use as well as other uses where a clarified xanthan gum is required, e.g., in secondary and tertiary oil well recovery operations. The clarified fermentation broth itself which results from the hot filtration process may also be used directly in such oil well recovery operations after appropriate dilution.

Production of Polygalactomannan Ethers

Highlights of the Technological Achievement: This process provides an improved method for producing alkyl ethers of polygalactomannan gum by contacting polygalactomannan gum with alkyl halide in the presence of a metal hydroxide catalyst and a quaternary ammonium phase-transfer agent.

Reference: A.A. DeGuia, R.W. Stackman and A.B. Conciatori; U.S. Patent 4,169,945; October 2, 1979; assigned to Celanese Corporation.

Background: Polygalactomannan gums swell readily in cold water and can be dissolved in hot water to yield solutions which, characteristically, have a high viscosity even at a concentration of 1 to 1.5%. Guar gum and locust bean gum, as supplied commercially, usually have a viscosity (at 1% concentration) of 1,000 to 4,000 cp at 25°C using a Brookfield Viscometer, Model LVF, spindle No. 2, at 6 rpm.

For a number of commercial applications, it is advantageous to have a gum that provides a solution viscosity lower than that imparted by the same gum in the form in which it is ordinarily sold commercially. Thus, for various food products where a polygalactomannan is incorporated as a thickener or is used in a gel- or jelly-type product, it is desirable to use a gum which has a 1% solution viscosity between 1,000 and 2,000 cp at 25°C. Solutions having lower viscosity than those produced with conventional commercial gum are also advantageous in paper making procedures, either in the beater or when used as a sizing. Low viscosity solutions of polygalactomannan gums also find special application in the sizing and printing of textiles.

Process Conditions: A process for preparing alkyl ethers of polygalactomannan gum which comprises reacting polygalactomannan gum with an alkyl halide in the presence of (a) at least a stoichiometric quantity of a strongly basic metal hydroxide, based on the weight of alkyl halide, and (b) between 0.05 and 5 wt % of a water-soluble quaternary ammonium halide phase-transfer agent, based on the weight of polygalactomannan gum in an aqueous reaction medium at a temperature between 10° and 100°C for a reaction period sufficient to achieve a degree of substitution by alkyl ether groups between about 0.01 and 3.0.

Example 1: *Purification of Guar Gum* – Guar gum is extracted with methanol to remove methanol-soluble oils. The guar gum so treated is wetted with isopropanol, then sufficient water is added slowly to form a 0.5% solution. After standing overnight, the solution is centrifuged at 8,000 rpm for 30 minutes. The clear supernatant is decanted from the insoluble residue and filtered through glass fiber filter paper. The filtrate solution is diluted with ethanol to precipitate the guar gum. The precipitate is filtered, dried and ground in a Wiley mill through a 40 mesh screen.

The purified guar gum powder has less than 0.1% nitrogen content, and about 0.48% ash content. For the purposes of the process, a polygalactomannan alkyl ether can be prepared from either the purified or unpurified guar gum.

Example 2: *Preparation of Polygalactomannan Alkyl Ethers –*

	Formulations A	Formulations B	Formulations C	Formulations D
Water, g	115	115	105	105
Gum splits, g	100	100	100	100
50% sodium hydroxide, g	20	20	40	40
$[(CH_3)_4N]^+Cl^-$, g	–	0.1	–	0.1
Methyl bromide, g	25	25	50	50
Degree of substitution	0.21	0.26	0.30	0.54
Efficiency of methyl bromide utilization, %	44	56	32	57

The guar splits are slurried in the water medium in a pressure reactor, heated to 50°C and purged with nitrogen for 1 hour. The sodium hydroxide solution and phase-transfer agent (when used) are added to the mixture and stirred for 10 to 20 minutes.

The methyl bromide is slowly metered into the reaction over a period of several hours while maintaining the reactor pressure below 20 psi. The reactor is cooled and the mixture neutralized with acetic acid. The solid product is washed with 50% methanol, and then is ground in the presence of methanol in a Waring blender. The polygalactomannan methyl ether thus prepared is recovered and dried.

The data above illustrates the increased alkyl halide utilization which is achieved by using a phase-transfer agent in accordance with the process.

Example 3: *Hydration of Guar Methyl Ether Gum in a Salt Solution* – The viscosity (cp) of a 1% solution of guar gum methyl ether (0.3 DS) was measured in comparison with guar gum:

	Guar Methyl Ether	Guar
Water	4,540	4,600
Calcium chloride (10%)	4,680	4,300

Guar methyl ether gum hydrates to a greater degree than does guar gum in a salt solution. It has also been found that guar gum does not hydrate in 50% aqueous methanol, while under the same conditions guar gum methyl ether hydrates and the solution develops higher viscosity.

Improving Water-Resistance of Pullulan

Highlights of the Technological Achievement: A method for improving the water-resistance of pullulan, which is a linear high polymer having the structural formula shown on the following page and is useful as films, sheets and fibers.

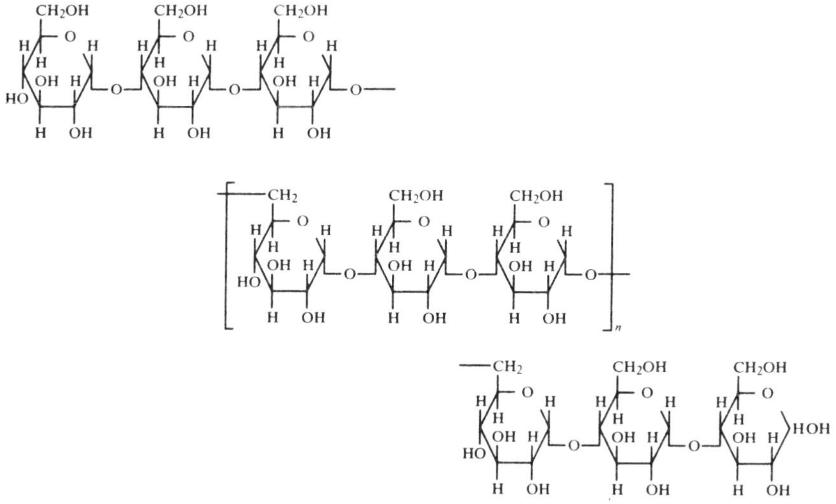

Reference: M. Fujimoto, K. Fukami, K. Tsuji and T. Nagase; U.S. Patent 4,186,024; January 29, 1980; assigned to Sumitomo Chemical Company, Limited and Hayashibara Biochemical Laboratories, Inc., both of Japan.

Background: Pullulan is easily soluble in cold or hot water and forms an aqueous solution which has a very low viscosity and is very stable without gelation or retrograde on a long-term standing.

Pullulan also has an excellent film-forming ability and the resulting film is superior in transparency, hardness and strength, and has a very low permeability to gases such as oxygen and nitrogen.

Further, pullulan is nontoxic and is decomposed spontaneously in soil or water. Thus, the pullulan is usable as films, sheets, fibers, binders, adhesives and paper-converting agents.

The application of pullulan is limited, however, to areas where a high resistance against water or humidity is not required because the solubility of pullulan in water is too high and, hence, the products made of pullulan are not resistant to water or humidity and are liable to become tacky in a humid atmosphere.

As the result of a study, it has been found that the water-resistance of pullulan can be improved without loss of the excellent properties of pullulan by the addition of dialdehyde polysaccharides to pullulan.

Process Conditions: The molecular weight of the pullulan used in the process is not particularly limited, but an average molecular weight of 1.0×10^4 to 2.0×10^6 is preferred.

The dialdehyde polysaccharide used in this process is one which is obtained by oxidation of polysaccharides with aqueous periodic acid or sodium periodate, and contains some dialdehyde parts of the following formula in its molecule.

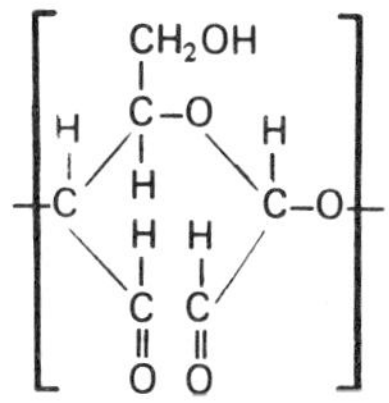

The polysaccharides may be, for example, starches (e.g., wheat starch and potato starch), celluloses, dextran, dextrin, pullulan and the like. The oxidation degree is not particularly limited if the resulting dialdhyde polysaccharides are soluble in cold or hot water. A desirable dialdehyde polysaccharide is obtained when 50 to 90% of the glucose units in the molecule are oxidized into the dialdehyde as above, particularly in the case of starches.

As to the mixing ratio of the dialdehyde polysaccharide to pullulan, 0.5-50:100 by weight is generally used, but 1-30:100 by weight is preferable from the practical point of view. The thus obtained mixture of pullulan and dialdehyde polysaccharides is then mixed with water so as to become uniform. Generally, 1 to 20 times by weight of water is conveniently added to the mixture. In this process, various compounds may be added in order to accelerate the process wherein pullulan is made waterproof.

Accelerating Agents: As the accelerating agents, there may be exemplified mineral acids such as sulfuric acid, hydrochloric acid and phosphoric acid; organic acids such as p-toluenesulfonic acid, β-naphthalenesulfonic acid and acetic acid.

Example 1: 100 pbw of each mixture comprising pullulan having an average molecular weight of 150,000 and dialdehyde starch [Karudasu No. 5 (Nihon Karitto Co., Ltd.)] in a mixing ratio as set forth in the following table were dissolved in 500 pbw of water at 90°C to obtain a uniform solution. This solution was coated on a glass plate and dried at room temperature to obtain a film of 30 μ in thickness. The table shows the tackiness of this film at a relative humidity of 90% and the rates of dissolution in water at 20° and 80°C.

	. . . . Composition			Rate of Dissolution	
Sample	Pullulan (parts)	Dialdehyde Starch (parts)	Tackiness	20°C	80°C
A	95	5	Hardly observed	~1 hr	1 min
B	90	10	Not observed	3 hr	2 min
C	70	30	Not observed	8 hr	3 min
Control	100	0	Observed	<1 min	<1 min

Example 2: 80 pbw of pullulan having an average molecular weight of 350,000 and 20 pbw of dialdehyde starch (Karudasu No. 5) were dissolved in 500 pbw of water at 90°C. One-half of the resulting solution was adjusted to pH 4.8 with acetic acid. Both solutions were each coated on a glass plate and dried at 100°C for 5 hours to obtain a film of 30 μ in thickness. The results are shown in the following table.

Sample	Composition: Pullulan (pbw)	Composition: Dialdehyde Starch (pbw)	pH of Aqueous Solution	Rate of Dissolution of Film*
A	80	20	6.7	about 20 hr
B	80	20	4.8**	insoluble
Control	100	0	4.8**	within 1 min

*Test in 50°C water.
**Adjusted with acetic acid.

Ionic Pullulan Gels

Highlights of the Technological Achievement: A process for producing ionic pullulan gels which comprises gelation of pullulan by crosslinking and introduction of ionic groups into the resulting pullulan gel, the former comprising reacting pullulan with a bifunctional compound in a solvent in the presence of an alkaline substance, and the latter comprising reacting the resulting pullulan gel with a compound having a carboxyl, sulfonic, phosphoric or amino group in a solvent in the presence of an alkaline substance.

Reference: F. Fujita, K. Fukami, M. Fujimoto and T. Nagase; U.S. Patent 4,174,440; November 13, 1979; assigned to Sumitomo Chemical Company, Limited, Japan.

Background: As the well-known, water-swellable gels having an ionic group, there may be mentioned dextran, cellulose and starch substituted with an ionic group such as a diethylaminoethyl or carboxylmethyl group. Dextran is a glucose polymer and its structure is such that glucose units are linked one after another via α-1,6 linkages and the highly reactive hydroxyl group at the C-6 position of the glucose unit takes part in the ether linkage. Consequently, dextran is not well suited for the production of gels having a high level of ion exchange capacity.

Cellulose and starch are high in crystallinity and very poor in stability in water so that the reaction does not proceed rapidly. Therefore, gels having a high level of ion exchange capacity are hardly obtainable. Further, mechanical strengths of the gels are low upon swelling in water. Accordingly, there is a strong demand for hydrophilic gels having a high ion exchange capacity and high mechanical strengths even upon swelling in water.

On the other hand, pullulan used in the method is a linear polymer of maltotriose wherein maltotriose units, a trimer of glucose, are linked one after another via α-1,6 linkage which is different from that of maltotriose. Pullulan has, unlike dextran, a large number of highly reactive hydroxyl groups of primary alcohol at the C-6 positions.

Although pullulan has a glucose unit, its property is different from that of starch or cellulose. That is, pullulan is noncrystalline and is very easily soluble in both cold water and hot water, producing aqueous solutions of extremely low viscosity. Consequently, pullulan reacts with compounds having an ionic group in aqueous solutions more rapidly and with a higher degree of substitution than dextran, cellulose and starch. As a result, there are obtained reaction products having a high level of ion exchange capacity.

Further, pullulan has many desirable properties as a hydrophilic gel, both in the form of aqueous solution and solid membrane. The aqueous solution is very stable for a long time without gelation and ageing, and the membrane is very superior in transparency and mechanical strength. In addition, pullulan has a superior membrane-forming ability, no toxicity and good suitability to the living body.

For obtaining ionic pullulan from pullulan, however, it is necessary to react pullulan with a compound having an ionic group in an aqueous medium and precipitate the resulting ionic pullulan with a water-miscible organic solvent, followed by purification by the reprecipitation method. This process is, therefore, troublesome in operation and uneconomical since organic solvents are required. Further, in the subsequent process for obtaining ionic pullulan gels by crosslinking of the ionic pullulan, the following difficulties are encountered:

(a) When the ionic pullulan is, for example, a cationic pullulan containing amino groups, there is a fear of the amino groups losing uniformity of basicity since they take part in the crosslinking reaction; and

(b) When the ionic pullulan has anionic groups such as carboxymethyl groups, the viscosity of its aqueous solution becomes extremely high and its concentration in the aqueous solution becomes difficult to increase, as a result of which gels having a low water regain upon swelling are difficult to obtain.

Process Conditions: A process for producing ionic pullulan gels which comprises reacting pullulan, which is a water-soluble high polymer, with a compound of the formula X_1-R_1-Y, where X_1 and Y are each a halogen atom or an epoxy group, and R_1 is an aliphatic group having 1 to 30 carbon atoms which may have 1 to 2 hydroxyl groups or a hetero-atom such as an oxygen atom in the carbon chain, in the presence of an alkaline substance, and then reacting the resulting crosslinked water-swellable pullulan gel with a compound of the formula X_2-R_2-Z, where X_2 is a halogen atom or an epoxy group; Z is a carboxyl, sulfonic, phosphoric group or salt thereof, or an amino group of the formula:

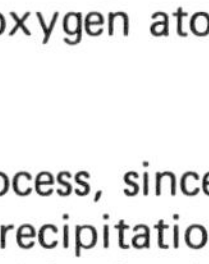

where R_3 and R_4 are each a hydrogen atom or a methyl, ethyl, hydroxyethyl, or phenyl group or salt thereof; and R_2 is a C_{1-20} hydrocarbon group containing an aliphatic group and/or an aromatic group which may contain one or more hydroxyl groups or one or more oxygen atoms in the carbon chain, in the presence of an alkaline substance.

Advantages: According to the process, since the synthesized ionic pullulan gels can be used as they are without precipitation with organic solvents and purification by reprecipitation, the method is simple and economical. In addition, the following advantages are obtained:

(a) Since there is no necessity of further chemical treatment after ionic groups have been introduced, homogeneous ionic pullulan gels are obtained; and

(b) The viscosity of aqueous pullulan solutions which are a starting material is so low that the water regain upon swelling can freely be controlled.

It was also found that the ionic pullulan gels of the process combine superior mechanical strengths upon swelling and superior performance as materials for separation and purification. As to the former, the gels of the method are superior to other water-swellable gels having ionic groups. As to the latter, they can widely be used for separation and purification of compounds ranging from low molecular weight to high molecular weight.

Culture of *Pullularia pullulans*: Pullulan can be isolated and recovered as a tacky substance secreted in a culture of a strain belonging to the genus *Pullularia.* A strain of the species *Pullularia pullulans* is subjected to the shaking culture at 24°C for 5 days in a medium containing 10% starch syrup, 0.5% K_2HPO_4, 0.1% NaCl, 0.02% $MgSO_4{\cdot}7H_2O$, 0.06% $(NH_4)_2SO_4$ and 0.04% yeast extract, or to the culture in a medium containing glucose as a carbon source, whereby pullulan is obtained as a tacky substance secreted from the cells into the culture liquor. The molecular weight of pullulan used in the process is not particularly limited, but preferably it is within the range of 1×10^4 to 100×10^4.

Pullulan Gels: As examples of the bifunctional substance used for gelation of pullulan, there may be given compounds of the formula $X_1{-}R_1{-}Y$, where X_1, R_1 and Y are as defined above, for example, epichlorohydrin, epibromohydrin, dichlorohydrin, dibromohydrin and 1,2,3,4-diepoxybutane.

As to the pullulan gels thus obtained, both their water regain and particle size are not particularly limited; but, for producing the ionic pullulan gels of the process, the water regain, which is expressed in the gram number of water absorbed per gram of dry gel, is generally in the range of 1 to 100 g/g, preferably in the range of 1 to 50 g/g and the particle size is preferably a spherical form of 10 to 500 μ in diameter.

Ionic Groups: Introduction of ionic groups into pullulan gels is carried out in the presence of an alkaline compound. This reaction may be expressed as follows:

$$\text{pullulan gel–OH} + X_2R_2Z \xrightarrow{\text{alkali}} \text{pullulan gel–OR}_5Z + X_2$$

where X_2, R_2 and Z are as defined above and R_5 is a hydrocarbon residue.

When X_2 is an epoxy group, no anionic ion is produced. From the equation, it is clear that when X_2 is a halogen atom, the alkali is consumed; while when X_2 is an epoxy group, the alkali acts only as a catalyst. Further, when X_2 is an epoxy group, the ring-form linkage of the epoxy compounds is cleft by the action of the hydrogen from the hydroxyl groups of pullulan gels to form a linkage of:

$$-CH_2-\underset{\displaystyle OH}{\underset{|}{CH}}-$$

As a result, hydroxyl-containing groups $-R_5Z$ are linked to pullulan gels.

The compounds which are reacted with pullulan gels for introduction of ionic groups are represented by the formula X_2R_2Z in which X_2, R_2 and Z are as defined above. As examples of the compounds where Z is an amino group, there may be given 2-dimethylaminoethyl chloride, 3-amino-1,2-epoxypropane, N,N-(2,3-epoxypropyl)methylaniline, and inorganic acid salts or alkyl halide thereof. As examples of the compounds where Z is a carboxyl group, there may be given chloroacetic acid, bromoacetic acid, chloropropionic acid and salts thereof. As examples of the compounds where Z is a sulfonic acid group, there may be given chloromethanesulfonic acid, bromoethanesulfonic acid, chloroethanesulfonic acid and salts thereof.

Temperature: The reaction conditions are also not particularly restricted. Although a reaction temperature of 200°C or less is generally suitable, undesirable side reactions would take place in some cases at a temperature exceeding 100°C. A reaction temperature of 5° to 100°C is, therefore, more desirable.

Example: 100 pbw of pullulan having a mean molecular weight of 50,000 were dissolved in 200 pbw of water and 100 pbw of a 5 N aqueous sodium hydroxide solution were added thereto to make a homogeneous solution.

This solution was gradually added to a dispersion medium comprising 15 pbw of polyvinyl acetate and 800 pbw of toluene to disperse the solution in a droplet form. In this operation, a stirrer equipped on the reaction vessel was kept at a rate of 800 rpm. One hour after addition of the aqueous solution, 25 pbw of epichlorohydrin were added and the reaction was carried out at 50°C for 3 hours. After the reaction was finished, the reaction mixture was cooled to room temperature and treated as follows.

The produced gel was deposited at the bottom of the reaction vessel, followed by decantation of the supernatant liquor. The gel was dispersed in 500 pbw of toluene and filtered. This operation was repeated three times to remove polyvinyl acetate. Thereafter, the gel was dispersed in methanol, washed and filtered. The gel was then dispersed in water and neutralized with dilute hydrochloric acid, followed by decantation and filtration. This process was repeated several times for purification. The gel was dispersed again in water, contacted as it was with ethanol and filtered. The gel was then dried at 70°C for 24 hours under reduced pressure to obtain 73 pbw of a dry gel. This gel had exactly a spherical particle form and its water regain was 6.0 g/g.

Pullulan Aminoalkyl Ether

Highlights of the Technological Achievement: Pullulan aminoalkyl ether is obtained by reacting pullulan with an aminoalkylating agent in a solvent in the presence of an alkali. This ether may be used as a cationic polymer of considerable utility in industries. For instance, when used in a paper-coating material, it gives desirable effects on gloss, RI dry pick, and RI wet pick; furthermore, as compared with pullulan, markedly improves physical properties of the coated paper, such as tearing strength, tensile strength and wrinkle recovery. It also improves retention of loading materials such as clay and calcium carbonate.

Reference: F. Fujita; K. Fukami and M. Fujimoto; U.S. Patent 4,167,623; September 11, 1979; assigned to Sumitomo Chemical Company, Limited and Hayashibara Biochemical Laboratories, Inc., both of Japan.

Background: Pullulan is a neutral substance, as is apparent from its chemical structure. Among water-soluble polymers there are many such as, for example, carboxymethylcellulose and polyethyleneimine, which have an electric charge in an aqueous solution and are industrially utilized to make the most of their anionic or cationic characteristics in aqueous solutions. Therefore, it is considered to be of significance to render pullulan positively charged in an aqueous solution, its inherent characteristics being retained.

Process Conditions: Pullulan aminoalkyl ether provided by the process can be represented by the general formula:

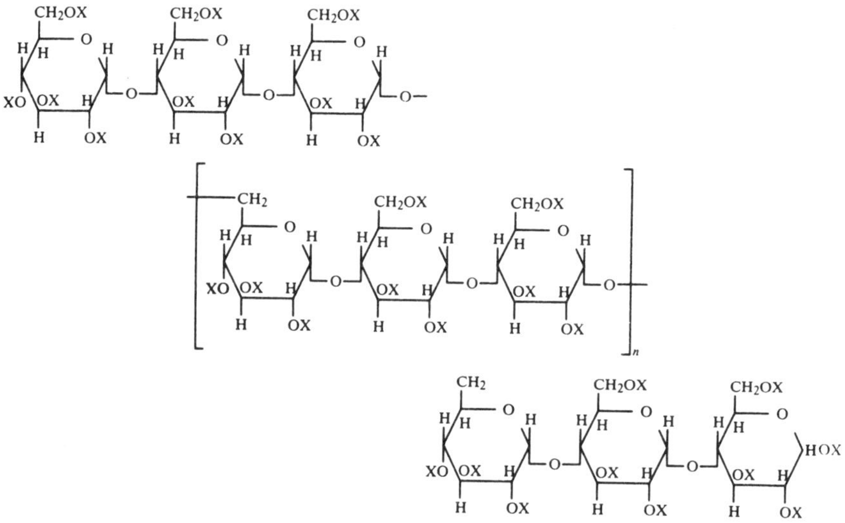

where n is an integer from 20 to 4,000 and X is H or a group represented by the general formula $-R_1-NR_2R_3$, where R_1 is an alkylene group having 1 to 5 carbon atoms or a hydroxyalkylene group having 1 to 4 carbon atoms and R_2 and R_3 are each hydrogen or an alkyl group having 1 to 4 carbon atoms, the number ratio of hydrogen atom to the group $-R_1-NR_2R_3$ in X in one molecule being 0.001 to 2.0.

The salts of pullulan aminoalkyl ether of the process include inorganic acid salts (e.g., hydrochloric acid salt, hydrobromic acid salt, sulfuric acid salt, nitric acid salt and the like), organic acid salts (e.g., acetic acid salt, oxalic acid salt and the like), and quaternary alkylamine salts thereof.

Pullulan aminoalkyl ether of the process can be prepared by reacting pullulan with an amine compound of the formula:

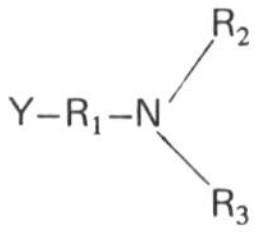

where Y is a halogen atom and R_2 and R_3 are each as defined above, in the presence of an alkali (e.g., sodium hydroxide, potassium hydroxide, etc.).

Pullulan aminoalkyl ether of the process where $-R_1-$ is propylene can also be prepared by using instead of the above formula a 3-amino-1,2-epoxypropane derivative of the formula:

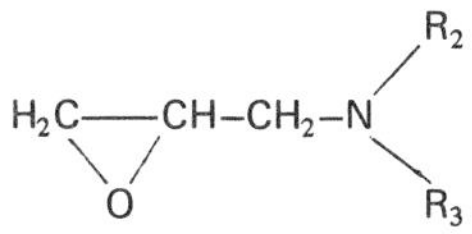

where R_2 and R_3 are each as defined above. Moreover, pullulan aminoalkyl ether of the process where R_2 and R_3 are hydrogen atom can be prepared by using alkyleneimines such as ethyleneimine and propyleneimine.

The amount of the amines to be reacted with pullulan is 0.001 to 10, preferably 0.01 to 10 mols per glucose unit of pullulan, depending on the amine content of the intended pullulan aminoalkyl ether.

The solvent used in the reaction is not limited to any type, but a solvent which dissolves one or both of the pullulan and the aminoalkylating agent is preferred. For example, water, dimethyl sulfoxide, dimethyl formaldehyde and dimethyl acetamide are used. The reaction conditions are also not particularly limited, but the reaction temperature is preferably 0° to 200°C.

Hydrophilic Crosslinked Pullulans

Highlights of the Technological Achievement: Crosslinked pullulans are obtained by the reaction of pullulan with a crosslinking agent selected from the group consisting of polycarboxylic acids, halogenated dicarboxylic acids, polycarboxylic anhydrides, aldehyde compounds, N-methylol compounds, isocyanate compounds, metaphosphoric acid salts, divinyl compounds, bisaziridine and bifunctional compounds of the formula X–R–Z, where X and Z are each a halogen atom or an epoxy group and R is an aliphatic residue having 1 to 30 carbon atoms in a solvent and are useful as gels or beds for chromatography or in various fields requiring high degrees of water-absorbency and retention.

Reference: T. Nagase, K. Tsuji, M. Fujimoto and F. Masuko; U.S. Patent 4,152,170; May 1, 1979; assigned to Sumitomo Chemical Company, Ltd. and Hayashibara Biochemical Laboratories, Inc., both of Japan.

Background: Pullulan is a linear, high polymer where units of maltotriose, which is a trimer of glucose, have been repeatedly bonded through the α-1,6-linkages which are different from the linkages of the trimer.

Although pullulan used in the process contains glucose units in its molecule, it is entirely different in its molecular structure from the well-known polysaccharides, for example, starch, oxidized starch, enzymated starch, etherified starch, cationic starch, aminated starch, cellulose, alkylcellulose, hydroxyalkylcellulose, carboxymethylcellulose, gum arabic and the like. Therefore, the properties of pullulan are also very different from that of those substances. For example, pullulan is easily soluble in cold water and forms an aqueous solution which is very stable with neither gelation nor retrogradation, even on a long-term standing.

Pullulan is different in properties from those of water-soluble synthetic high polymers such as polyvinyl alcohol, polyacrylamide, polyacrylic acid, sodium polyacrylate and polyvinyl pyrrolidone.

Synthetic resins such as polyethylene, polypropylene, polystyrene, polyvinyl chloride and polymethyl methacrylate are widely used, but many problems arise when the resins are treated as waste. That is, the resins as waste cause pollution of the rivers and seas; polyvinyl chloride resins generate a poisonous hydrogen chloride gas on burning and resins such as polyethylene, polypropylene and polystyrene generate a large quantity of heat on burning, thus shortening the life of burning furnaces.

Process Conditions: The method provides

(a) Hydrophilic gel materials, which are hardly soluble in water, but water-swellable, prepared by crosslinking pullulan with a crosslinking agent selected from the group consisting of saturated or unsaturated polycarboxylic acids, polycarboxylic anhydrides, aldehyde compounds, N-methylol compounds, isocyanate compounds, metaphosphoric acid salts, divinyl compounds and bisaziridine;

(b) Hydrophilic gel materials having a water-absorption (the number of grams of water absorbed by 1 g of a dry gel) of 1-100 g/g, which are prepared by reacting pullulan and a bifunctional compound of the formula X–R–Z, where X and Z are each a halogen atom or an epoxy group and R is an aliphatic residue having 1 to 30 carbon atoms;

(c) A method for the preparation of spherical hydrophilic gel materials by the reaction between pullulan and a bifunctional compound of the formula X–R–Z, where X, Z and R are as defined above, which comprises conducting the reaction by dispersing an aqueous pullulan solution in a liquid dispersion medium which is immiscible with the aqueous pullulan solution and contains a dispersion stabilizer; and

(d) Crosslinked pullulans, which are water-resistant, prepared by reacting pullulan with polycarboxylic acids, polycarboxylic anhydrides or halogenated dicarboxylic acids and a catalyst.

The molecular weight of the pullulan used in the process is not particularly limited, but it is preferably within a range from 1.0×10^4 to 1×10^6, and more preferably 3.0×10^4 to 1×10^6.

The crosslinked pullulans of the process are obtained by reacting pullulan with a suitable crosslinking agent in a suitable solvent and, if necessary, in the presence of a suitable catalyst.

Crosslinking Agents: The crosslinking agents include polycarboxylic acids, halogenated dicarboxylic acids, polycarboxylic anhydrides, aldehyde compounds, N-methylol compounds, isocyanate compounds, metaphosphoric acid salts, divinyl compounds, bisaziridine and bifunctional compounds of the formula X–R–Z, where X, Z and R are as defined above. The crosslinking agents are used in an amount of 0.1 to 100 pbw, based on 100 pbw of pullulan.

The solvents are used in an amount of 30 to 2,000 pbw, based on 100 pbw of pullulan. The reaction between pullulan and the crosslinking agent is generally carried out at a temperature of 15° to 150°C.

Product: The hydrophilic gel materials thus obtained have a water-absorption of more than 0.1 g/g, preferably more than 0.2 g/g, at the swelling equilibrium in water. The water-absorption is controlled by properly selecting the molecular weight of pullulan, the kind and amount of the crosslinking agent and the crosslinking conditions.

The reaction comes to an end in 1 to 24 hours in general, but it is desirable to complete the reaction in 2 to 5 hours. The crosslinked products are sometimes produced within 10 minutes, depending upon the reaction conditions. When crosslinked products are obtained by such a reaction, it is sometimes desirable to subject them to an after-curing at temperatures somewhat higher than the initial ones. In general, the after-curing temperature is higher by about 10° to 20°C than the temperature of the preceding reaction. The after-curing is sufficiently achieved within about 7 hours in general, and it is desirable to complete the curing within 2 to 3 hours. After completion of the after-curing, the resulting products are neutralized, washed and dried.

Since the hydrophilic gel materials of the process have such an excellent swellability, they exhibit a molecular sieve effect on various compounds which are dissolved not only in aqueous solvents, but also in organic polar solvents such as dimethylformamide, and thus they are expected to be useful in the fields of separation and purification such as desalting, wastewater treatment, and the like. Preferred shapes of the crosslinked products used for such purposes are spheres having a diameter of 10 to 500 μ.

Example: 10 pbw of pullulan having an average molecular weight of 100,000 were dissolved in 100 pbw of N,N'-dimethylformamide at 70°C. After adding 5 pbw of adipic acid and 1 pbw of p-toluenesulfonic acid as a catalyst to the solution, stirring of the mixture was continued at 70°C for 10 hours. The reaction mixture was coated on a glass plate and heat-treated at 130°C for 5 hours. The film-like crosslinked product thus obtained was insoluble in water and had a water-absorption (the weight in grams of water or a specified liquid which is absorbed by one gram of the crosslinked pullulan when the pullulan is dipped in water or the liquid) of 3 g/g.

APPLICATIONS

Polishing Slurry of Xanthan Gum and a Dispersing Agent

Highlights of the Technological Achievement: A solution for use in suspending 0.05 to 15 μ ceria or alumina particles to produce a slurry for polishing lenses and semiconductor wafers is disclosed.

Reference: G.B. Regan; U.S. Patent 4,161,394; July 17, 1979.

Background: Lenses, both glass and plastic, and semiconductor wafers are polished before use. Polishing is generally accomplished by bringing the workpiece into rubbing contact with a pad while circulating a slurry of finely-divided abrasive particles in an aqueous medium into the area of contact between the pad and the workpiece. The slurry drains from the workpiece surface into a sump from which it is recirculated to further contact with the workpiece by a pump.

Polishing slurries commonly encounter three problems in greater or lesser degree. One problem is the tendency of the solid particles of the slurry to settle rapidly especially when the polishing action is interrupted with the result that the settled material must be redispersed before its use in polishing can be continued. A second problem is that polishing slurries commonly have little capacity to hold the debris removed by the workpiece during polishing action in suspension with the result that the useful life of the polishing slurry is short. A third problem is the tendency of polishing slurries to foam during the course of use with the result that recirculation by a pump is inefficient and even impossible in serious foaming situations.

Process Conditions: The slurries of the process comprise four components. The first and major component is water which constitutes 80 to 90% by weight of the total composition. The second component is xanthan gum. Xanthan gums are high molecular weight (~1,000,000) polysaccharide gums produced by fermenting hexoses with the organism *Xanthomonas campestris*. The fermentation product gum may be modified by acylation if desired. A suitable xanthan gum is Keltrol (Kelco Company). The quantity of xanthan gum contained in the composition is small, in the range of 0.02 to 0.25% by weight, based on the water content of the composition. The xanthan gum functions as a suspending agent and to a small degree because of its low concentration as a viscosity-increasing agent.

The third component of the composition is a dispersing agent. While a large number of dispersing agents are known, only two have been found which, in combination with the xanthan gum component of the composition, are effective to produce a stable slurry having a large capacity to hold debris removed from the workpiece in suspension. One of the dispersing agents is the condensate of naphthalene monosulfonic acid with formaldehyde. The chemical character of the condensate is shown by the following structural formula:

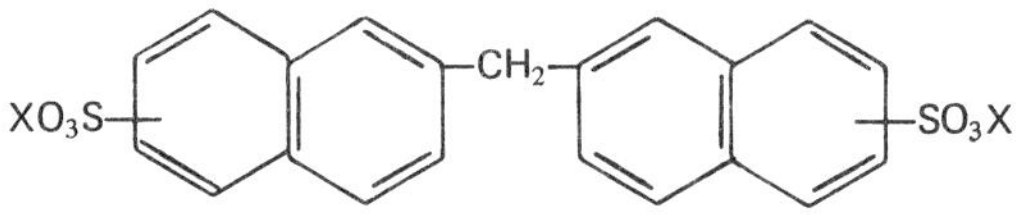

where X is hydrogen, ammonium or alkali metal.

The second dispersing agent found effective for use in the composition is polymethacrylic acid. Polymethacrylic acids having molecular weights in the range 8,000 to 18,000 and preferably in the range 10,000 to 16,000, are effective. The dispersing agent is present in the composition in amounts ranging from 0.2 to 1.2% by weight. The two dispersing agents may be used in the acid form or in the form of their ammonium and alkali metal salts, the salt forms being preferred.

The fourth component of the slurry is a finely-divided abrasive which is preferably alumina, ceria or zirconia. The abrasive particles are small, being in the range 0.05 to 15 μ in diameter. The particle size of the abrasive should be selected with a view to the character of the material to be polished. For polishing plastic lenses, it is preferred to use abrasive particles ranging in size from 0.05 to 1 μ. For polishing glass lenses, it is preferred to use abrasive particles ranging from 1 to 2 μ in size. For polishing semiconductor wafers, it is preferred to use abrasive particles about 7 μ in diameter. The abrasive particles are present in the polishing compositions in amounts ranging from 10 to 20% by weight of the total composition.

Product: Two practical considerations relating to the use of the polishing slurries make it desirable in many instances to supply the user with a concentrated solution of the xanthan gum and the dispersing agent in water and permit the user to disperse abrasive particles of the size appropriate for the intended use in the concentrate diluted with water all within several days of the time of intended use.

While the slurries are very stable, the particles do settle out after about 7 days standing and it must be redispersed before the slurry is used. Also, since the particle size of the abrasive material is varied as indicated above depending on the character of the material to be polished, supplying the user with a concentrated solution of the xanthan gum and dispersing agent in water permits the user to dilute the concentrate and disperse abrasive particles of the appropriate size in the diluted concentrate just prior to use. A suitable concentrate consists of water containing from 0.06 to 0.75% by weight of xanthan gum and from 0.5 to 5.0% by weight of the dispersing agent.

1 part of this concentrate is mixed with 2 parts by volume of water and then the abrasive particles of appropriate size are added in amounts ranging from 10 to 20% by weight of the diluted concentrate and dispersed in it.

Example: A solution for use in suspending finely-divided ceria or alumina, or zirconia to produce a slurry for polishing lenses and semiconductor wafers was prepared. The solution consisted of 10 ℓ water, 30 g xanthan gum and 300 g dispersing agent. The dispersing agent was the sodium salt of condensate of formaldehyde with naphthalene monosulfonic acid having the structure shown above, where X is hydrogen.

A functionally equivalent solution is obtained by using polymethacrylic acid having a molecular weight in the range 8,000 to 18,000 or the ammonium or alkali metal salts of the polyacid as the dispersing agent.

Crosslinkable Polymer Gums for the Recovery of Oil

Highlights of the Technological Achievement: This process relates to the recovery of oil from underground deposits. The method relates to gel compositions consisting of water, a polysaccharide polymer, an acid generating salt and a melamine resin. The method also relates to the process of utilizing the composition to recover underground oil.

Reference: G.T. Colegrove; U.S. Patent 4,157,322; June 5, 1979; assigned to Merck & Co., Inc.

Background: Methods for plugging off, diverting or reducing the rate of undesired fluid movement in porous media make up a substantial amount of technology, including the in-place formation of gels. Such gels are used to plug highly permeable zones in the formation, thus diverting the water or other fluid through the less permeable zones, thereby improving sweep efficiency and providing greater oil recovery. These prior art gels are normally prepared on the surface and are then pumped underground into the formations.

A disadvantage of these prior art methods is that the gelled compositions are subjected to shear degradation during the pumping operation as they are pumped through pipes, perforations and fractures. This results in the breakdown of crosslinks and the loss of the gel's ability to plug and maintain impermeability.

Process Conditions: Crosslinkable polymers that may be utilized in this process include synthetic polymers such as partially hydrolyzed polyacrylamides. Also included among the polymers that can be utilized are cellulose and cellulose derivatives such as carboxymethylcellulose, methylcellulose, methylhydroxypropylcellulose, hydroxyethylcellulose or polysaccharides such as xanthan gums, guar gums, locust bean gums, gum tragacanths, alginates and their derivatives.

The polymer may be utilized over a rather broad range of concentrations from 0.05% by weight based on the water to 2.0%. However, a more preferred range is from 0.5 to 1.5%.

The gelation reaction is catalyzed by a salt which is acid-generating upon the application of heat. The gelation reaction occurs under acidic conditions. Another important advantage of this process is that the solution pH is close to neutral when made up, but when the heat from underground decomposes the salt, an acid is formed in situ, lowering the pH and causing the gelling reaction to occur. The acid-generating salt can be utilized over a range from about 0.5 to 1.0% by weight of water. A more preferred range is from about 0.1 to 0.5%.

The melamine resin that is utilized in this process can be a commercial product such as the reaction product of melamine and formaldehyde, i.e., Cymel 373, Cymel 370 and Cymel 380 (American Cyanamid). The resin concentration by weight based on water is from 0.3 to 1.25%.

Gelation Rate: The gelation rate of the composition depends on the amount of each of the components and the temperature at which the reaction is conducted. Thus, the gel rate and gel strength of the compositions can be tailored by adjusting the amount of the acid-generating salt, the resin amount and the temperature. The higher the temperature at given concentrations of resin, acid-generating salt and polymer will result in a faster gelation time. If a more gelled composition is desired, the polymer and resin concentrations at a given temperature are increased.

Example: A simulated core model, packed with coarse gravel, is utilized to demonstrate the water-diverting characteristics of the process composition. When plain water is passed through the core model packed with coarse gravel, a high flow rate indicates a permeability of about 6 darcies.

An aqueous solution containing 0.5% xanthan gum (Kelzan), 0.5% partially methylated melamine-formaldehyde resin (Cymel 373), 0.25% monoammonium phosphate and 0.2% formalin by weight of composition is then injected into the core model containing coarse gravel and the solution allowed to gel for 8 hours at 130°F. An attempt to pass water through the core after the gelled composition indicates a flow rate effectively reduced to about zero which indicates complete plugging of the porous channels.

Mixed Xanthan Gum and Locust Bean Gum Therapeutic Compositions

Highlights of the Technological Achievement: Compositions comprising xanthan gum, locust bean gum and a pharmaceutical agent in a pharmaceutically acceptable liquid vehicle. More specifically, these compositions are useful in drug delivery systems, particularly for ophthalmic drugs.

Reference: M.K. Pramoda and S.-L. Lin; U.S. Patent 4,136,173; January 23, 1979; assigned to American Home Products Corp.

Background: This process relates to a pharmaceutical delivery system and a method for the controlled administration of drug to a patient. This process also relates to a pharmaceutical delivery system possessing the property of being liquid at a pH of less than 3.5 to 5 and a gel at a pH of greater than 5. In preferred embodiments, the process relates to bioerodable drug delivery systems for the controlled administration of drug to a mammalian patient over a prolonged period of time, especially to the eye of such patient.

In another aspect, the process relates to the reversibility of the system from liquid to gel and gel to liquid, thereby enabling varied methods of application suited to the different biological environments encountered in a mammalian patient.

Process Conditions: Xanthan gum means *Xanthomonas* hydrophilic colloid produced by the bacteria of the genus *Xanthomonas.* Illustrative of such a colloid is the xanthan gum Keltrol (Kelco Co.), which may be described as a high molecular weight natural carbohydrate or, more specifically, polysaccharide. The generic term "xanthan gum" defines exocellular biopolysaccharides which are produced in a pure culture fermentation process by the microorganism *Xanthomonas campestris.*

Three different monosaccharides are found in the basic structure: mannose, glucose and glucuronic acid (as a mixed potassium, sodium and calcium salt). Each repeating block within the polymer contains sixteen sugar units. The linear portion consists of five 1→4-linked D-glucose units, four 1→4-linked D-mannose units and four 1→2-linked D-glucuronic acid units. At C-3 of two glucose units there are mannose branches and at some other position, as yet undetermined, in the repeating unit is a side chain derived from pyruvic acid and glucose, i.e., 4,6-O-(carboxyethylidine)-D-glucose. Each repeating unit contains an average of 3.4 acetyl groups which are believed to be connected at C-6 of mannose.

Locust bean gum is a high molecular weight polysaccharide derived from *Ceratonia siliqua.* Chemically, locust bean gum is a galactomannan best illustrated with galactose units located on every fourth mannose unit, with smaller amounts of pentoglycan, protein, cellulose and ash. A detailed description of the composition, physical and chemical properties, preparation, etc. of locust bean gum is given in *Industrial Gums*, "Polysaccharides and Their Derivatives", Academic Press, pp. 361-376 (1959).

The process relates to a composition of matter comprising a solution of plant seed gum and microbial gum in a liquid vehicle where the pH of the solution is adjusted, depending on the application, to a pH of less than 3.5 to 5 for liquid administration, and to a pH of greater than about 5 for gel administration.

Product: The formulations of this process are administered topically to the eye in the form of ophthalmic solutions. The concentrations of ingredients in these formulations are expressed as percent weight by volume, unless noted otherwise. In preferred compositions, the plant seed gum is locust bean gum and the microbial gum is xanthan gum; generally, the total gum concentration in solution is from 0.01 to 1%. It is preferred to keep the total gum level in solution at 0.1 to 0.5%.

In these gelable compositions, furthermore, the proportions of the respective xanthan and locust bean gums are important. The percent of xanthan gum in the mixed gum system ranges from 6 to 94% of the total gum content, the remainder being locust bean gum.

Suitable drugs for use in therapy with the drug delivery systems of the process include:

Protein drugs such as insulin;

Desensitizing agents such as ragweed pollen antigens, hay fever pollens, dust antigen and milk antigen;

Vaccines such as smallpox, yellow fever, distemper, hog cholera, fowl pox, antivenom, scarlet fever, diphtheria toxoid, tetanus toxoid, pigeon pox, whooping cough, influenzae, rabies, mumps, measles, poliomyelitis, Newcastle disease, etc.;

Antiinfectives such as antibiotics, including penicillin, tetracycline, chlortetracycline, bacitracin, nystatin, streptomycin, etc.; sulfonamides including sulfacetamide, sulfamethizole, etc.; antivirals including idoxuridine; and other antiinfectives including nitrofurazone and sodium propionate;

Antiallergenics such as antazoline, methapyrilene, chlorpheniramine, etc.;

Antiinflammatories such as hydrocortisone, cortisone, prednisolone, etc.;

Decongestants such as phenylephrine, naphazoline and tetrahydrozoline;

Sympathomimetics such as epinephrine;

Sedatives and hypnotics such as phenobarbital, secobarbital sodium, codeine, etc.;

Psychic energizers such as 3-(2-aminopropyl)indole acetate and 3-(2-aminobutyl)indole acetate;

Tranquilizers such as reserpine, chlorpromazine, and thiopropazate;

Androgenic steroids such as methyltestosterone and fluoxymesterone;

Estrogens such as estrone, 17β-estradiol and diethylstilbestrol;

Progestational agents such as progesterone, megestrol, melengestrol, etc.;

Hormonal agents such as the prostaglandins, e.g., PGE_1, PGE_2;

Antipyretics such as aspirin, sodium salicylate and salicylamide;

Antispasmodics such as methantheline, papaverine and methscopolamine bromide;

Antimalarials such as the 4-amino-quinolines, chloroquine and pyrimethamine;

Antihistamines such as diphenhydramide, tripelennamine, perphenazine, etc.;

Cardioactive agents such as hydrochlorothiazide, chlorothiazide and aminotrate;

Nutritional agents such as vitamins, essential amino acids and essential fats;

Anti-Parkinsonism agents such as L-dopa (1,3,4-dihydroxyphenylalanine); and

Investigative antihypotensive agents such as dopamine [4-(2-aminoethyl)pyrocatechol].

Gelling Compositions for Food Products

Highlights of the Technological Achievement: A gelling composition used as a food gelling agent, in water or milk, and comprising a gelling mixture consisting of a first gelling agent, which is a galactomannan, and a second gelling agent, which is an agar and/or xanthan, where the galactomannan has undergone a depolymerization treatment so that its solution in water at 1% has a viscosity in the range of 10 and 1,000 cp at 25°C.

Reference: G. Brigand, H. Kragen and R. Rizzotti; U.S. Patent 4,200,661; April 29, 1980; assigned to Societe CECA SA, France.

Background: The process concerns the use, as a food gelling agent in water and milk, of a mixture of two gelling agents, the first being a galactomannan having undergone a depolymerization treatment so that its solution in water at 1% has a viscosity in the range of 10 and 1,000 cp at 25°C, and the second comprising agar and/or xanthan, eventually completed by another polysaccharide such as carrageenan, furcellaran or pectin. In the same measuring conditions, the underpolymerized galactomannans form high viscosity solutions (locust bean gum, 1,200 to 3,600 cp).

The gelling agents normally used to gel milk are sulfated polysaccharides such as carrageenans, furcellarans and agars. The products give good results with a milk whose pH is close to 6. The use of an acidified milk whose pH is often close to 4 brings about a flocculation of the casein which deteriorates the aspect and quality of the product and makes it difficult to be eaten. This phenomenon is accentuated by the presence of sulfated polysaccharides which, chemically, precipitate with the casein in these pH conditions.

The addition of a galactomannan, such as locust bean gum, to agar appears to avoid the precipitation of the casein, doubtlessly by a protecting colloid effect and to improve the texture, but the gels obtained are stiff and do not have the organoleptic qualities designated by the term "mouth feel". On using a quantity of galactomannan, such as locust bean gum, which is larger than that of agar, this disadvantage is partially remedied, but the cohesiveness is still great and the gel has a disagreeable rubbery texture.

The first gelling agent is depolymerized galactomannan, for example, locust bean gum, the latter having undergone a depolymerization treatment such that its solution in water at 1% has a viscosity in the range of 10 and 1,000 cp at 25°C and the second gelling agent is constituted by agar and/or xanthan, eventually completed by a polysaccharide such as carrageenan, furcellaran or pectin. This composition is used as a food gelling agent of water or milk either at a pH close to 7 or after acidification; this acidification may be more particularly obtained by adding a fruit juice. In the case of milk, this acidification may be obtained by microbic action leading to a food product (yogurt).

Yogurt is a product having a certain gellified texture obtained by adding to milk, which is usually pasteurized, certain ferments which acidify the milk through a coagulation of the casein. The transformation is made at a temperature close to 40°C.

If the yogurt is preserved at an ambient temperature, the ferments continue their action, the acidity continues to develop, the product loses its qualities and its preservation is thereby limited.

This heating (thermization) is nevertheless limited to approximately 70°C with common gels. Beyond this temperature, the protecting effect shades off and the casein undergoes a dehydration process leading to a heterogeneous texture which is characteristically sandy and very disagreeable, which can lead to a total separation if the temperature rises too much.

It has been found, and this constitutes an advantage of the method, that, due to the gelling agent which is the subject of this method, yogurt can be treated at very much higher temperatures, while preserving the product's complete homogeneity. This thermization at a higher temperature, i.e., higher than 70°C, allows a longer preservation period by assuring a better destruction of the ferments.

Example: Solutions of more or less depolymerized locust bean gum were coagulated with an agar solution to obtain a final mixture containing 25% of agar and 75% of locust bean gum. The depolymerization of the locust bean gum was measured by taking a sample of its solution and coagulating it; after drying, the viscosity at 1% of this locust bean extract was measured. The complexes obtained were used at 1% to make water gels by dissolution on boiling and cooling of the solution.

The characteristics of the obtained gels are measured in two ways. On the one hand, the gel stiffness is measured by the force required for sinking a 1.1 cm diameter piston to a depth of 4 mm into the gel; on the other hand, the gel cohesiveness is measured by the force required to crush the unmolded gel 6 cm in diameter and 4 cm high with a 10 cm diameter plate.

The results are shown in the table and the corresponding curves are shown in the figure.

Viscosity* (cp)	Gel Stiffness (g)	Gel Cohesiveness (g)
60	58	120
150	64	350
360	69	620
900	68	900
1,170	69	2,100
2,000	69	2,500
3,000	70	2,400

*At 1% of the locust bean extract contained in the complex.

Figure 15.1: Agar-Locust Bean Gum Gel

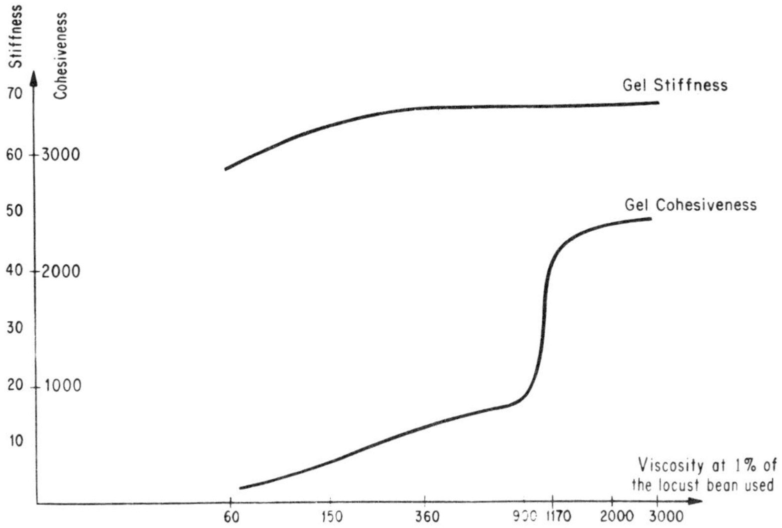

Source: U.S. Patent 4,200,661

These results show that in an agar-locust bean gum gel, the gel stiffness varies little with the depolymerization rate of the locust bean used and the gel cohesiveness varies with the depolymerization rate of the locust bean gum.

In addition, it has been established, for different proportions of agar and locust bean gum, that (a) the locust bean gums having viscosities higher than 1,000 cp give a very great cohesiveness which is expressed on tasting by the rather disagreeable sensation of an elastic block; (b) the locust bean gums having viscosities in the range of 300 to 800 cp give less cohesiveness and more melting on tasting; and (c) the locust bean gums having a viscosity lower than 300 cp give even more melting on tasting, but the gels sometimes become more difficult to unmold.

The stability of the depolymerized agar-locust bean gum complex in an acid medium makes it particularly interesting in acid milk gels. In addition, below the isoelectric point of the milk (of the casein), the utilization of the complex avoids casein precipitation.

On the one hand, the depolymerized agar and locust bean gum, being practically neutral, do not coprecipitate with the milk casein, as opposed to the carrageenans or the furcellaran which are the usual dairy gelling agents, at least in a neutral medium. On the other hand, the depolymerized agar and locust bean gum prevent casein precipitation on playing the role of the protecting colloid.

Compositions for Stabilizing Soft Serve and Hard Frozen Yogurt

Highlights of the Technological Achievement: This process relates to a stabilizer for soft-serve and hard-frozen yogurt. The stabilizer blend consists of propylene glycol alginate, sodium alginate, guar, carrageenan and an emulsifier.

Reference: R.S. Igoe; U.S. Patent 4,178,390; December 11, 1979; assigned to Merck & Co., Inc.

Background: Most prior art gums and blends have been found to react with the milk protein during the processing procedures, resulting in yogurts which are coarse-bodied, grainy and which exhibit whey-off. (The term "whey-off" refers to the separation of fluid from solid material.)

Process Conditions: The process is directed to compositions of matter which act as a stabilizer for soft-serve and hard-frozen yogurt. The stabilizer is a combination of ingredients having the following composition: 45 to 60% propylene glycol alginate, 15 to 35% sodium alginate, 10 to 20% guar, and 1 to 10% carrageenan.

The stabilizer composition can, in turn, be utilized in combination with an emulsifier and is referred to as a blend. The stabilizer blend, therefore, is a mixture of the stabilizer as described above with an emulsifier. The stabilizer may constitute a range of 55 to 75% by weight of the stabilizer blend and the emulsifier may constitute 45 to 25% by weight of the stabilizer blend. The amount of emulsifier may vary, however, depending on which particular one is chosen. The stabilizer and stabilizer blend components are mixed by conventional methods including dry-blending techniques which are known to those skilled in this art.

The stabilizer can be utilized in the soft-serve or hard-frozen yogurt at a range of 0.1 to 0.4% by weight. A preferred range for stabilizer would be 0.2% by weight. The stabilizer blend can be utilized in the soft-serve or hard-frozen yogurt at a range of 0.15 to 0.60% by weight. A preferred range for stabilizer blend would be 0.3% by weight.

Typical emulsifiers that may be utilized include monoglycerides, diglycerides, lecithin, polysorbate 80 and polysorbate 65. These emulsifiers may be used alone or in combination with each other. For example, the polysorbates may be utilized with the monoglycerides or diglycerides, and the like.

Example 1: The following soft yogurt formulation is a typical utilization of the stabilizer blend. The formulation is based on 100% total ingredients by weight.

Ingredient	Parts by Weight
Milk, 3.5% fat	54.0
Milk solids, nonfat	6.0
Sugar	11.0
Corn syrup solids, D.E. 36	8.0
Stabilizer blend*	0.27
Flavor concentrate	0.40
Color	Trace
Water	20.33

*The stabilizer blend is made up of the following:

Propylene glycol alginate	0.096
Sodium alginate	0.046
Guar	0.028
Carrageenan	0.010
Mono-/diglyceride	0.072
Polysorbate 80	0.018

Under medium to high speed agitation, the stabilizer blend and milk solids (nonfat) are added to the milk. The milk solution is pasteurized at 180° to 185°F for 30 minutes and then homogenized at 2,000 psi. After cooling the milk to 90°F, the yogurt culture is then added. The inoculated milk is incubated at 90°F for 15 hours to obtain a yogurt of pH 4.0, which is then stirred to a smooth consistency.

With mild stirring of the yogurt, the sugar, flavor concentrate, color and water are added. The yogurt is then homogenized at zero pressure to reduce the viscosity. The yogurt is then cooled over cooling coils at 40°F and then frozen on a Sweden single barrel freezer. The product is drawn at 19° to 20°F from the freezer to obtain a desired 50% overrun.

A formulation is run repeating the above ingredients and amounts, except that the emulsifier is omitted. The product obtained is of the same texture, appearance and smoothness as the product in which the emulsifier is utilized.

Examples 2 through 6: The procedure utilized in preparing Examples 2 through 6 is the same as Example 1, except that the amounts of ingredients in the formulation are changed as indicated.

	Example No. (pbw)				
Ingredients	2	3	4	5	6
Milk, 3.5% fat	68.6	57.1	54.3	71.4	57.1
Milk solids nonfat	4.1	5.54	6.0	5.1	7.04
Sugar	11.0	6.0	11.0	10.0	6.6
Corn syrup solids, D.E. 36	4.0	7.0	8.0	9.0	8.8
Stabilizer/emulsifier*	0.30/0.15	0.23/0.12	0.20/0.10	0.13/0.07	0.13/0.07
Flavor concentrate	0.9	0.5	0.8	1.0	—
Flavor puree	—	14.0	—	—	20.26

(continued)

Ingredients	Example No. (pbw) 2	3	4	5	6
Color	trace				
Water	10.96	9.51	19.6	3.3	–

*The stabilizer/emulsifier is made up of the following ingredients.

Propylene glycol alginate	0.135	0.1175	0.106	0.078	0.078
Sodium alginate	0.105	0.0575	0.05	0.0247	0.0195
Guar	0.045	0.0391	0.032	0.0143	0.0208
Carrageenan	0.015	0.0161	0.012	0.011	0.0117
Mono-/diglyceride	0.12	0.096	0.08	0.056	0.056
Polysorbate 80	0.03	0.024	0.02	0.014	0.014

The products obtained in the above examples have a smooth, grain-free characteristic when prepared and obtained in the desired 50% overrun.

Methyl Ether of Polygalactomannan Gums for Use in Hydraulic Well-Treating Fluids

Highlights of the Technological Achievement: This process provides a hydraulic fluid composition adapted for fracturing of subterranean formations, which comprises an aqueous medium, methyl ether of polygalactomannan gum as a gelling agent, and a breaker additive for subsequent reduction of fluid viscosity.

Reference: R.N. DeMartino; U.S. Patent 4,169,798; October 2, 1979; assigned to Celanese Corporation.

Background: Hydraulic fracturing fluids commonly use a viscosity-increasing agent such as a water-soluble polymer to improve the flow characteristics and the suspending ability of the fluids. Carboxymethylcellulose, polyacrylamide, polyvinylpyrrolidone, guar, tragacanth, Irish moss and modified starch have been used as water-soluble or water-dispersible additives to provide pumpable gels having solids-suspending properties.

The disadvantage of using natural gum as a gelling agent in hydraulic fracturing fluids is the lessening of fluid loss control caused by plugging of apertures, which reduces the permeability of a fractured formation. In the case where a natural resin such as guar gum is used as a gelling agent in combination with an enzyme breaker additive in a hydraulic fracturing fluid, the hydraulic system is unsatisfactory because the hydrolysis residue of the natural gum is present in sufficient quantity to create a plugging effect and reduces the permeability of a fractured formation.

Accordingly, it is a main object of the process to provide natural gum derivatives having improved viscosity and suspending properties for application as gelling agents in hydraulic well-treating fluid compositions.

Process Conditions: Polygalactomannan gums are hydrocolloid polysaccharides composed principally of galactose and mannose units and are usually found in the endosperm of leguminous seeds, such as guar, locust bean, honey locust, flame tree, and the like. Guar flour, for example, is composed mostly of a galactomannan which is essentially a straight-chain mannan with single-membered galactose branches. The mannose units are linked in a 1→4 β-glycosidic linkage and the galactose branching takes place by means of a 1→6 linkage on alternate mannose units. The ratio of galactose to mannose in the guar polymer is, therefore, 1:2. Guar gum has a molecular weight of 220,000.

Locust bean gum is also a polygalactomannan gum of similar molecular structure in which the ratio of galactose to mannose is 1:4. Guar and locust bean gum are the preferred sources of the polygalactomannans, principally because of the commercial availability thereof.

Polygalactomannan gums swell readily in cold water and can be dissolved in hot water to yield solutions which characteristically have a high viscosity even at a concentration of 1 to 1.5%. Guar gum and locust bean gum as supplied commercially, usually have a viscosity (at 1% concentration) of 1,000 to 4,000 cp at 25°C using a Brookfield Viscometer, Model LVF, spindle No. 2, at 6 rpm.

Breaker Additive: The breaker additive is preferably an enzyme which, under formation fracturing conditions, autonomously degrades the polygalactomannan methyl ether gum gelling agent so as to reduce the viscosity of hydraulic fluid which is under hydrostatic pressure.

The enzyme breaker additive can be used in a quantity between 0.01 and 5 wt %, based on the weight of polygalactomannan methyl ether gum in a hydraulic fluid composition. Hemicellulase enzyme is illustrative of a suitable breaker additive for hydraulic fluid compositions containing methylated guar gum.

The polygalactomannan methyl ether gum component of the process hydraulic fluid compositions is produced by contacting the gum with a methylated reagent in the presence of a basic compound. The reaction proceeds readily at a temperature between 0° and 100°C.

Example 1: *Preparation of Methyl Ethers of Polygalactomannan Gum –*

	A	A′	B	B′
Isopropanol (50%), ml	1,800	1,800	1,800	1,800
Guar gum, g	300	300	–	–
Locust bean gum, g	–	–	300	300
50% sodium hydroxide, g	25	50	25	50
Methyl bromide, g	25	50	25	50
Degree of substitution	0.2	0.3	–	–
Viscosity, 1% solution, cp	4,310	4,540	–	–

The polygalactomman gum is slurried in the isopropanol solution in a pressure reactor, heated to 50°C and purged for 1 hour. The sodium hydroxide solution is added to the slurry and the mixture is stirred for 10 minutes.

The methyl bromide is slowly metered into the reactor over a period of several hours, while maintaining the reactor pressure below about 20 psi. The reactor is allowed to cool and the reaction mixture is neutralized to a pH of 7 with acetic acid, filtered, washed with 50% isopropanol, and then with 100% isopropanol. The polygalactomannan methyl ether product is recovered and air-dried.

Example 2: *Hydration of Guar Methyl Ether Gum in a Salt Solution* – The viscosity (cp) of a 1% solution of guar gum methyl ether (DS of 0.3) was measured in comparison with guar gum as follows.

	Guar Methyl Ether	Guar
Water	4,540	4,600
Calcium chloride, 10%	4,680	4,300

Guar methyl ether gum hydrates to a greater degree than does guar gum in a salt solution. It has also been found that guar gum does not hydrate in 50% aqueous methanol, while under the same conditions, guar gum methyl ether hydrates and the solution develops higher viscosity.

Dialkylacrylamide Ether Adducts of Polygalactomannan Gums

Highlights of the Technological Achievement: This process provides a dialkylacrylamide ether adduct of polygalactomannan gum compositions. This process further provides a hydraulic fluid composition adapted for fracturing of subterranean formations, which composition contains dialkylacrylamide ether adduct of polygalactomannan gum as a gelling agent.

Reference: R.N. DeMartino and A.B. Conciatori; U.S. Patent 4,137,400; January 30, 1979; assigned to Celanese Corporation.

Background: The productivity of oil and gas wells can be improved by increasing the area of communication within a selected producing zone. The drainage area can be increased by hydraulic fracturing of the producing zone to provide fractures and channels emanating from the well base area into the contiguous subterranean formations.

The hydraulic fracturing process is accomplished by rapid pumping of an aqueous fluid medium down a well which penetrates the subterranean formation where fracturing is desired. The rapid pumping of the aqueous fluid creates a hydrostatic pressure which energizes splitting forces in the confined zone. Pressures as high as 10,000 psi are used to effect formation fracturing.

Process Conditions: Illustrative of a preferred class of polygalactomannan ether adduct gum is one represented by the formula:

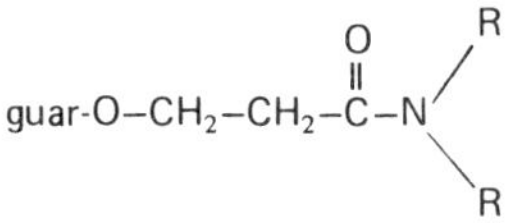

where R is an alkyl group containing between 1 and about 4 carbon atoms.

The N,N-dialkylacrylamide ether adduct of polygalactomannan gum is produced by contacting natural polygalactomannan gum with N,N-dialkylacrylamide (or N,N-dialkylmethacrylamide) in the presence of a basic compound. The etherification reaction proceeds readily at a temperature between about 0° and 100°C.

The N,N-dialkylacrylamide reactant is used in a quantity sufficient to provide the desired degree of substitution of hydroxyl groups with oxypropionamido groups, e.g., a DS between 0.01 and 3.0, and preferably between 0.1 and 1.5.

The enzyme breaker additive can be used in a quantity between 0.01 and 5 wt %, based on the weight of polygalactomannan ether adduct gum in a hydraulic fluid composition. Hemicellulase enzyme is illustrative of a suitable breaker additive for hydraulic fluid compositions containing polygalactomannan ether adduct gum.

The quantity of the N,N-dialkylacrylamide ether adduct of polygalactomannan gum incorporated in a process hydraulic composition can vary in the range between 0.05 and 5 wt %, based on the weight of the water component. A preferred range is between 0.1 and 2 wt % of gum, based on the weight of water.

Product: It was also found that the N,N-dialkylacrylamide ether adduct of polygalactomannan gum provided by the process has superior properties in comparison with the corresponding acrylamide ether adduct polygalactomannan gum. Particularly significant for the purposes of the process is the greater resistance to hydrolysis exhibited by the N,N-dialkylacrylamide moiety as compared with the corresponding acrylamide moiety of the gum derivatives. Acrylamide tends to hydrolyze to carboxylic acid under the etherification reaction conditions:

$$\text{gum-O-CH}_2\text{-CH}_2\text{-}\overset{\overset{\displaystyle O}{\|}}{C}\text{-NH}_2 \longrightarrow \text{gum-O-CH}_2\text{-CH}_2\text{-}\overset{\overset{\displaystyle O}{\|}}{C}\text{-OH}$$

The presence of carboxylic acid groups in the gum derivatives is detrimental in the applications contemplated by the process.

Example: *Preparation of Dimethylacrylamide Adducts of Polygalactomannan Gum –*

	Formulation Reference					
	A	A′	B	B′	C	C′
Acetone (50%), ml	1,800	1,800	1,800	1,800	1,800	1,800
Guar gum, g	300	–	300	–	300	–
Locust bean gum, g	–	300	–	300	–	300
Sodium hydroxide (50%), g	30	30	30	30	30	30
N,N-dimethylacrylamide, g	75	75	150	150	300	300
Approximate DS	0.11	0.11	0.25	0.25	0.5	0.5

The formulations that are listed above in the table are reacted in the following manner. The galactomannan gum is slurried in the acetone solution, then heated to 40°C and purged for 1 hour with nitrogen. The caustic solution is added to the slurry, and the mixture is stirred for 10 minutes.

The N,N-dimethylacrylamide reactant is added to the mixture, and the etherification reaction is conducted at 40°C over a period of 4 hours. The reaction mixture is neutralized with acetic acid, filtered, washed twice with 50% acetone, and once with 100% acetone. The polygalactomannan ether adduct product is recovered and air-dried.

In comparison with the corresponding unmodified natural polygalactomannan gum starting materials, the N,N-dimethylacrylamide ether adduct derivatives hydrate faster and reach peak viscosity more quickly. Also, the higher the degree of substitution of N,N-dimethylacrylamido groups in the polygalactomannan ether adduct derivatives, the greater the stability under varying conditions of pH and temperature. The polygalactomannan ether adduct derivatives of the process have the further advantages of improved solution stability and resistance to bacterial degradation.

The polygalactomannan ether adduct derivatives also exhibit greater stability in alcohol solution than the corresponding unmodified natural gum.

	..... 1% Solution, cp	
	50% Methanol	75% Methanol
N,N-dimethylacrylamide adduct of guar		
DS, 0.11	5,000	0
DS, 0.25	3,700	1,200
DS, 0.5	3,100	2,500
Guar	0	0

N,N-dimethylacrylamide ether adduct of guar gum (DS 0.26) exhibits long-term aqueous solution stability in comparison with guar gum. For example, a 5,250 cp ether adduct viscosity (1% solution) remains substantially unchanged over a period of 20 days. Unmodified guar gum (1% solution, 5,000 cp) loses measurable viscosity within 2 to 3 days.

Thickening Agent Containing a Polygalactomannan Gum and a Copolymer of an Olefinically-Unsaturated Dicarboxylic Acid Anhydride

Highlights of the Technological Achievement: This process provides a dry-blend composition adapted for application as a thickening agent in aqueous solutions. The composition consists essentially of an admixture of a polygalactomannan gum and a copolymer containing carboxylate groups. The process composition contributes exceptionally high viscosity to an aqueous solution due to a synergistic viscosity effect derived from the interaction of the polygalactomannan and copolymer components.

Reference: R.N. DeMartino: U.S. Patent 4,143,007; March 6, 1979; assigned to Celanese Corporation.

Background: Hydrophilic vegetable gums have been known and utilized for a variety of industrial applications. They are readily utilized in aqueous solutions in many manufacturing processes such as textile, cosmetic, foodstuff, and the drilling and secondary production of subterranean wells. Of particular importance is their ability to form viscous colloidal dispersions in aqueous environments.

Process Conditions: The dry-blend composition adapted for application as a thickening agent in aqueous solutions comprises (a) a polygalactomannan gum and (2) a copolymer of an α,β-olefinically-unsaturated dicarboxylic acid anhydride and a comonomer selected from α-olefinically-unsaturated hydrocarbons and alkyl vinyl ethers. The process further provides an aqueous composition having a pH in the range between 4 and 13.

Polygalactomannan Gum Component -- Preferred types of polygalactomannan gums for application as a component of the process thickener composition include guar gum and locust bean gum. Another preferred type of polygalactomannan gum is the hydroxyalkyl ether derivatives which have a degree of substitution between about 0.1 and 3, and which have an alkyl group containing between 1 and about 6 carbon atoms. Etherification of polygalactomannan gums is generally accomplished with polygalactomannan gums in the form of finely-divided powders or in the form of gum splits.

Guar gum and other polygalactomannan hydrocolloids are derived from certain seeds of the plant family leguminosae. The seeds are composed of a pair of tough, nonbrittle endosperm sections referred to as splits, between which is sandwiched a brittle embryo layer. The entire structure is enclosed in a tough seed coat. This renders them difficult to reduce into a finely-divided state. One method of separating the endosperm splits is described in U.S. Patent 3,132,681. Methods of reducing endosperm splits into finely-divided powder are described in U.S. Patents 2,891,050 and 3,455,899 and references cited therein.

Hydroxyalkyl ethers of polygalactomannan gums are readily prepared by the reaction of a polygalactomannan gum with an alkylene oxide containing between 1 and about 6 carbon atoms, e.g., ethylene oxide, propylene oxide, butylene oxide, hexylene oxide, and the like.

The reaction of a polygalactomannan gum and an alkylene oxide is conveniently conducted at a temperature between 20° and 100°C and a pressure between 15 and 150 psi, in a solvent medium in the presence of an alkaline catalyst such as sodium hydroxide or calcium hydroxide. Suitable solvent media include water, cyclic and acyclic alkyl ethers, alkanones, hydrocarbons, dialkylformamide and the like, and mixtures thereof.

Copolymer Component – The preferred types of copolymers for application as a component of the process thickener composition are those which are composed of an α,β-olefinically-unsaturated dicarboxylic acid anhydride and a comonomer selected from α-olefinically-unsaturated hydrocarbons and alkyl vinyl ethers.

Illustrative of α,β-olefinically-unsaturated dicarboxylic acid anhydride comonomers are those represented by the formula:

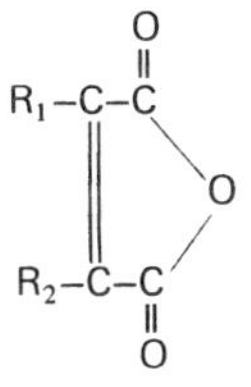

where R_1 and R_2 are independently selected from the group consisting of hydrogen, halogen, cyano, alkyl, aryl, alkaryl, aralkyl, and cycloaliphatic.

Illustrative of a suitable copolymer is Gantrez AN-179 (GAF) which corresponds to the structural formula:

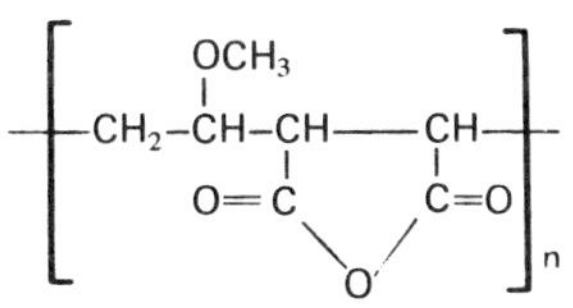

where n is an integer between 11 and 700.

The polygalactomannan gum and copolymer components are used in the thickener composition in a ratio between 0.05 and 15 pbw polygalactomannan gum per part by weight of copolymer.

The aqueous composition of the process is prepared by adding the thickener composition to an aqueous medium in a quantity between 0.2 and 5 wt %, based on the total aqueous composition weight. On the average, the preferred quantity of added thickener composition will be in the range between 0.3 and 2 wt %, based on the total aqueous composition weight.

Example: This example illustrates the preparation of dry-blend and aqueous compositions of the process. A dry-blend composition is prepared by the simple expediency of admixing the respective composition components. It is not required that the admixture be blended until homogeneous, since the dry-blend composition components dissolve individually when the composition is added to an aqueous medium as a thickening agent.

A typical dry-blend composition is prepared by admixing 10 g of purified guar gum with 10 g of a copolymer of maleic anhydride and isobutylene [Isobam HH (Kuraray Co. Ltd., Japan)] or a copolymer of maleic anhydride and methyl vinyl ether [Gantrez AN-179 (GAF)].

An aqueous composition is prepared by adding a dry blend composition as described above to an aqueous medium as a thickening agent. Dissolution of the copolymer component (e.g., Isobam HH or Gantrez AN-179) is aided by adjusting the pH of the aqueous medium into the range of 7 to 11 with sodium hydroxide.

As an alternative method of preparation of the aqueous compositions, the polygalactomannan gum and the copolymer components are first dissolved individually in water to form separate aqueous solutions. An aliquot of each solution is then combined to produce one of the desired high-viscosity aqueous compositions of the process. The following procedure and data are illustrative of this alternative preparative method.

Guar gum and hydroxypropyl ether of guar gum are individually dissolved in water to form separate 1% by weight solutions. Isobam HH is dissolved in water in the form of the sodium salts to produce a 1% by weight solution. Gantrez AN-179 is dissolved in water in the form of the sodium salt to produce a 3% by weight solution. All of the solutions are allowed to stand overnight.

Various aliquots of the polygalactomannan gum solutions and the copolymer solutions are blended, and the viscosities of the resulting solutions are measured after 1 hour, using a Brookfield RVF Viscometer, spindle No. 4, at 20 rpm. The viscosity data are summarized in the following table.

Guar gum, wt %	100	90	80	70	60	50	40	30	20	10	0
15% Isobam HH, wt %	0	10	20	30	40	50	60	70	80	90	100
Viscosity, cp	4,400	6,200	6,850	7,200	7,500	7,650	7,300	6,800	6,200	5,400	5,350
Hydroxypropyl guar gum, wt %	100	90	80	70	60	50	40	30	20	10	0
1% Isobam HH, wt %	0	10	20	30	40	50	60	70	80	90	100
Viscosity, cp	4,200	7,000	7,600	7,800	7,900	7,750	7,500	6,850	6,900	5,850	5,100
Guar gum, wt %	100	90	80	70	60	50	40	30	20	10	0
Gantrez AN-179, wt %	0	10	20	30	40	50	60	70	80	90	100
Viscosity, cp	4,700	8,650	9,300	9,000	7,900	6,850	5,800	4,500	3,600	2,500	1,750

Hair Shampoo and Conditioning Lotion

Highlights of the Technological Achievement: A process for shampooing and/or conditioning the hair by applying to the hair a lotion containing a shampoo and/or conditioning agent including an aqueous or aqueous-alcoholic solution, emulsion or gel containing a water-soluble salt of chitosan.

Reference: P. Gross, E. Konrad and H. Mager; U.S. Patent 4,202,881; May 13, 1980; assigned to Wella AG, Germany.

Background: Hair shampoos customarily contain anionic surface-active agents, such as, for instance, fatty alcohol sulfates, fatty alcohol ether sulfates, fatty acid monoglyceride sulfates, alkylaryl sulfates and sulfosuccinates. Special hair shampoos are compositions which contain nonionic, cationic, ampholytic surface-active agents or mixtures of these agents as active wash components. Among the nonionic surface-active agents there may, for instance, be mentioned nonylphenolpolyglycol ethers, sorbitan fatty acid esters, sorbitan polyoxyalkylene fatty acid esters, fatty acid ethoxylates and fatty alcohol ethoxylates.

Representative of the cationic surface-active agents are, for instance, alkyldimethylbenzylammonium chlorides, pentaoxyethylammonium chlorides, alkyltrimethylammonium chlorides, alkyldimethylammonium saccharinates and alkyldimethyldichlorobenzylammonium chlorides.

The method, therefore, has the object to provide for a process by which an improved conditioning of the hair is possible during or without shampooing the hair and, if desired, together with a dyeing of the hair.

Process Conditions: The objective of the process is accomplished by adding to the hair conditioner and/or hair shampoo a water-soluble salt of a chitosan. As a result of this addition, the conditioning effect is substantially improved and particularly the combing properties and the gloss and touch of the hair are improved, even in damaged hair.

The chitosan is a high-polymer amine and adapted to form salts with acids. It is made by alkaline deacetylation of chitin. The complete deacetylation is difficult since the alkali during the reaction penetrates the chitin particles only imperfectly. A virtually acetyl-free chitin, that is pure chitosan, can be obtained only by repeated alkali treatment or by fractionating. The chitosan commercially available, therefore, constitutes a more or less deacylated product with a chitosan content of about 70 to 90% by weight. These commercial products as well as products with a higher content of chitosan can be used for the purposes of the process equally well.

By neutralization of the free amino groups of the chitosan with acids, the corresponding salts can be obtained. According to the process, however, the salts can only be used if they are soluble in water. Suitable acids for making the salts are, for instance, hydrochloric acid, formic acid and acetic acid.

The amount of acid preferably is chosen to obtain sufficient neutralization of the free amino groups of the initial chitosan. Thus, for instance, in order to form an aqueous solution of 1 g of chitosan (having 90% of free amino groups), 3.36 g of 10% acetic acid are necessary.

The compounds of the process are aqueous solutions, emulsions or gels which are characterized by their contents of water-soluble salts of chitosan. The concentration of the salts in the composition should, preferably, be between 0.5 and 6.0% by weight. However, depending on the total composition, there may also occasionally be used higher or lower amounts of chitosan salt.

In general, the application of the compositions in the process is effected by applying the solution, etc., to the hair, permitting it to act on the hair for 3 to 10 minutes and then removing it again by rinsing.

Example 1: *Hair Conditioner* – A gel is prepared consisting of 0.6 g hydroxypropylmethylcellulose, 2.1 g of chitosan (having 90% free amino groups), 5.3 g formic acid (10% conc), 0.5 g lauryl pyridinium chloride and 91.5 g water (fully desalted).

Example 2: *Shampoo* – A solution is prepared consisting of 50.0 g dimethylcarboxymethylenepropyleneamidostearate betaine (30% conc), 2.0 g chitosan (having 90% free amino groups), 2.2 g lactic acid (90% conc) and 45.8 g water (fully desalted).

Solid Oral Pharmaceutical Product with Increased Efficacy and Predetermined Steady State of Solubility

Highlights of the Technological Achievement: The process relates to a solid oral pharmaceutical preparation with protracted release of the active ingredient consisting of discrete solid granules containing the active ingredient soluble in the stomach and auxiliary agents and of an equally solid external phase surrounding the granules.

Reference: L. Gyarmati, I. Rácz, P. Szentmiklósi and J. Plachy; U.S. Patent 4,199,560; April 22, 1980; assigned to Novex Talalmanyfejleszto Es Ertekesito Kulkereskedelmi Rt, Hungary.

Background: The method gives a solution of a very important problem in the field of antacid therapy. Though it is well-known that hyperacidity does not cause ulcers, it can be a condition of their formation; ulcers form only on surfaces exposed to the effect of hydrochloric acid. The presence of excess hydrochloric acid prevents the curing of ulcers already formed, so the therapy requires a proper diet and in a given case, besides observing other instructions, the neutralization of the excess acid to a proper degree. Thus, the acid-binding medicine for the given purpose must satisfy the following requirements:

(a) The initial neutralizing effect must be quick and must keep its efficacy during the normal digestion time of the stomach;

(b) It must bind the required amount of acid;

(c) It must raise the pH value of the gastric acid to a level at which the pepsin activity is of reduced value, but not fully inhibited;

(d) It may not cause a higher rise in the pH value of the gastric acid than desired; and

(e) It should not cause a systemic alkalosis, even when applied repeatedly and for a long time.

Process Conditions: The process relates to controlled-release two-phase solid oral pharmaceutical compositions (tablets, coated pills, capsules) with protracted action and comprises the granulation of a powder mixture which contains the active ingredient, a solid carrier, and, optionally, pharmaceutical auxiliary agents with a liquid consisting of an aqueous emulsion which contains hydrophobic and hydrophilic components.

Owing to its specific physico-chemical properties, the release time of the active ingredient is further increased to a significant extent, thereby augmenting the protracted biological utilization of the active ingredient to a level approaching the theoretical optimum and pressing the granules thus obtained into tablets or filling the same into capsules. The special granulating emulsion contains as hydrophobic component stearic or palmitic acid, and as hydrophilic components, two types of nonionic surface-active agents, namely Tween-type emulsifiers and, optionally, a mucous-forming substance, e.g., sodium carboxymethylcellulose. By appropriate selection of the components of the emulsion and changing their quantitative ratio, the rate of release of the active ingredient could be modified.

The process is based on the recognition that acid-binding metal compounds, being generally insoluble in water in neutral medium (like basic bismuth nitrate, aluminum hydroxide, magnesium hydroxide, magnesium trisilicate and other substances of amphoteric character used as active ingredients in antacidic preparations) enter into molecular interaction at an approximately neutral pH interval with gel-forming substances of vegetal or animal origin, e.g., tragacanth, albumin, casein, pectin, etc., and form a sticky homogeneous mass of gel-like structure and strongly increased viscosity from which the basic metal compound is released extremely slowly and uniformly even in acidic medium.

The following table shows the macroscopically-observable behavior of a suspension containing 10% basic bismuth nitrate in 1% tragacanth gel at different pH values of the medium.

	Time of Standing (hr)				
pH	**Directly After Preparation**	**8**	**16**	**24**	**48**
2-3	Homogenous suspension	Slightly sedimented suspension	Progressed sedimentation	*	*
4	Homogenous suspension	Thick-flowing, homogenous suspension	Gelatinous, thick, mucilaginous suspension	**	***
5-6	Homogenous suspension	Thick-flowing homogenous suspension	Homogenous, gelatinous gel	**	***
7	Homogenous suspension	Thick-flowing, homogenous suspension	Gelatinizing, thick, mucilaginous suspension	** **	*** ***
8	Homogenous suspension	Slightly sedimented suspension	Progressed sedimentation	†	†
9	Homogenous suspension	Slightly sedimented suspension	Progressed sedimentation	*	*

*Completely sedimented suspensions; the sedimented part is not gelatinous, it is powder-like; the supernatant liquid is of mucilaginous consistency.

**Homogenous gels of gelatinous consistency.

***Gelatinous gels with starting syneresis, the separated liquid is not of mucilaginous consistency.

†Completely sedimented suspensions; the sedimented part is slightly gelatinous, the supernatant liquid is of mucilaginous consistency.

Example: For the preparation of tablets containing a spasmolytic as active ingredient, the following substances are used, calculated for one tablet: 0.15 g 3,4,6,7-tetraethoxy-1-benzal-3,4-dihydroisoquinoline hydrochloride; 0.20 g tragacanth; 0.12 g sodium alginate; 0.05 g aluminum hydroxide and 0.04 g sodium carboxymethylcellulose.

Adhesive Compositions Consisting of Guar Gum, Locust Bean Gum, Pectin and Gum Karaya

Highlights of the Technological Achievement: An adhesive composition adapted for use with an ostomy appliance consisting of a mixture of a hydrocolloid gum, a pressure-sensitive adhesive and a cohesive strengthening agent. The cohesive strengthening agent is either an inert natural or synthetic fibrous material, finely-divided cellulose, a crosslinked dextran, a crosslinked carboxymethylcellulose or a starch-acrylonitrile graft copolymer.

Reference: J.L. Chen, R.D. Cilento, J.A. Hill and A.L. La Via; U.S. Patent 4,192,785; March 11, 1980; assigned to E.R. Squibb & Sons, Inc.

Background: Major abdominal surgery for a number of diseases involving different parts of the gastro-intestinal and urinary tract can result in a patient being left with an abdominal stoma. The three most common types of abdominal stoma are the colostomy, the ileostomy, and the ileal conduit. In the case of an ileostomy, ileal conduit and many colostomy operations, the patient is unable to control the passage of bodily waste material and must rely upon an appliance attached to their body to collect this material.

Several systems have been used for this purpose. Colostomates normally use a disposable appliance which is either a one-piece or two-piece unit. The term "disposable" refers to the fact that the pouch or bag portion of the unit can be disposed of after a single use. A closed end pouch sealed along all of its edges with only an opening for the stoma is suitable for this purpose. The two-piece disposable unit consists of such a bag suspended from a reusable mounting ring that is supported by a belt.

Process Conditions: The hydrocolloid gum component of the adhesive compound should have a large capacity to absorb moisture, should provide wet adhesion and should also hydrate and swell at a relatively slow rate so as not to cause disintegration of the adhesive composition. Guar gum, locust bean gum and mixtures thereof have been found to be suitable with guar being preferred and such gums can be present at from about 15 to 40% by weight of the adhesive composition.

An additional gum substance having soothing or healing properties can be included within the adhesive composition. Pectin, gum karaya and mixtures thereof have been found to be suitable with pectin being preferred and can be present at 0 to 25% by weight of the adhesive composition provided that the total amount of gums within the adhesive composition is from 25 to 55% by weight.

Product: The adhesive composition of this process can be sterilized by means of gamma radiation. Preferred adhesive compositions within the scope of this process comprise on a weight basis from 2 to 12% cotton or from 10 to 30%

finely-divided, purified wood cellulose as the cohesive strengthening agent; from 15 to 40% guar gum, from 0 to 25% pectin provided that when the cohesive strengthening agent is cotton, the guar gum and pectin combined are from 35 to 55% and when the cohesive strengthening agent is wood cellulose, the guar gum and pectin combined are from 20 to 45%; and from 45 to 55% of the mixture of low molecular weight polyisobutylene and butyl rubber where the ratio of low molecular weight polyisobutylene to butyl rubber is about 4:1 on a weight basis.

The following example is illustrative of the process. Other suitable adhesive compositions can be obtained by minor variations in the amounts of ingredients used.

Example: This example is directed to preparing an adhesive having the following composition:

Ingredient	Percent by Weight
Polyisobutylene, viscosity average molecular weight (Flory) of 36,000 to 45,000*	40
Butyl rubber of a viscosity average molecular weight (Flory) of 425,000**	10
Guar gum of high-grade, extra-fine powder***	30
Finely-divided, purified wood cellulose†	20

*Vistanex LM-MS (Exxon)
**Exxon, Grade 077
***Jaguar A-40-F (Stein Hall Co.)
†Solka Floc BW-100 (Brown Co.)

2 kg of butyl rubber are broken down by mixing in a kneader mixer for 2 to 5 minutes. 4 kg of the low molecular weight polyisobutylene are added and mixed with the butyl rubber for 2 to 5 minutes. 6 kg of guar gum and 4 kg of the finely-divided, purified wood cellulose are combined in a powder mixer and the resulting powder is added to the polyisobutylene-butyl rubber mixture. Mixing of the ingredients is continued until a homogeneous mass is formed with the polyisobutylene and butyl rubber completely interdispersed (about 10 to 20 minutes). The remaining 4 kg of the low molecular weight polyisobutylene are added and mixing is continued until a homogeneous dough is formed (about 10 to 20 minutes).

This dough mass, while hot and soft, is extruded and flattened. A sheet of polyethylene of 1.5 mils thickness is pressed over one side and silicone-coated release paper on the other. The resultant mat is cut into the desired shape.

Acidic Agar Gel Electrochromatography of Hemoglobins

Highlights of the Technological Achievement: The common hemoglobin variants A, A_2, F, S and C are discretely separated by electrochromatography under acidic conditions on an agar gel which utilizes an agar gel having a wet thickness of from 0.1 to 0.5 mm and a potential of from 60 to 110 V.

Reference: B.I. Bluestein, C.A. Lepp and R.D. Mason; U.S. Patent 4,209,373; June 24, 1980; assigned to Corning Glass Works.

Background: Interest in population screening programs for abnormal hemoglobins in the U.S. has increased primarily because of the National Sickle Cell Disease Program developed by the Department of Health, Education and Welfare. Thus, screening for hemoglobinopathies has become very important. For example, many hospitals screen all incoming surgical patients for such abnormalities. Because electrophoresis on cellulose acetate is the primary procedure in the Federal Sickle Cell Disease Program, cellulose acetate electrophoresis at pH 8.4 to 9.2 is the most common electrophoretic screening method for hemoglobinopathies. It has been found, however, that cellulose acetate cannot adequately separate certain combinations of hemoglobin variants.

The hemoglobin variants separated by the citrate agar electrochromatography procedure can be estimated visually, scanned by a densitometer to determine the relative amounts of each variant present, or stained to enhance visualization or densitometry readings of the hemoglobin variants. Typical staining procedures, however, involve hemoglobin-specific dyes such as benzidine or o-dianisidine. Benzidine is a known carcinogen and has been removed from the market by the FDA. The o-dianisidine is a benzidine derivative and is suspected to be carcinogenic as well. Thus, such compound also may be removed from the market. In addition, these stains are light-sensitive and darken upon storage. Furthermore, the prior art agar gels must be stored wet and thus can be bulky and difficult to store.

Process Conditions: Accordingly, the process provides, in a method for discretely separating the common hemoglobin variants A, A_2, F, S and C by electrochromatography on an agar gel, under acidic conditions utilizing citrate gel and well buffers, each of which has a citrate concentration of approximately 0.05 M and a pH of approximately 6.0 and a potential of 50 to 90 V, the improvement which comprises using: (a) an agar gel having a wet thickness of from about 0.1 to 0.5 mm; and (b) a potential of from 60 to 110 V.

The separated hemoglobin variants are visualized by means of a nonspecific protein stain such as amido black. Such stain is used in place of the prior art hemoglobin-specific stains, which stain is nontoxic, not sensitive to light, and highly sensitive to the protein content of the hemoglobin variants.

The method requires the use of an agar gel having a wet thickness of 0.1 to 0.5 mm. The nature of the agar is not critical. Suitable agar preparations are readily available commercially and have been found to be suitable in the method. The amount of agar typically used is about 1% by weight per volume. It will be apparent to those having ordinary skill in the art that greater or lesser amounts of agar can be used, provided that appropriate adjustments in the other parameters are made.

Optionally, the gel can contain minor amounts, e.g., less than about 10% weight per volume, of one or more humectants. Examples of suitable humectants include sucrose, hydroxyethylcellulose, glycerin and sorbitol.

Finally, it should be noted that the gel can contain up to about 1% weight per volume of agarose. Such use of agarose, however, is optional. If used, a preferred amount of agarose is 0.5% weight per volume. In addition, the gel can contain small amounts of preservatives, such as sodium azide and disodium ethylenediaminetetraacetate.

The citrate gel buffer is prepared from sodium citrate and citric acid in accordance with well-known procedures. The concentration of citrate in the buffer can range from 0.04 to 0.065 M. A particularly useful concentration is 0.055 M, in which case the buffer is readily prepared by dissolving 16.17 g of sodium citrate dihydrate in 990 ml of deionized distilled water, after which the pH is adjusted as desired with 30% weight per volume citric acid solution. The gel buffer pH typically can vary from 6.0 to 6.5, with a pH of 6 being preferred. In the example just given, the concentration of sodium citrate is 0.055 M, while the total citrate concentration, which includes both citrate and citric acid, is about 0.06 M.

Product: As already pointed out, the gel must have a thickness of 0.1 to 0.5 mm. A particularly preferred wet thickness is from 0.3 to 0.4 mm. The citrate well buffer similarly is prepared in accordance with known procedures. In general, the citrate concentration will be in the range of 0.055 to 0.065 M, with a concentration of 0.06 M being preferred. The pH of the well buffer can vary from 6.0 to 6.5, with a pH of 6.0 being preferred.

The electrophoresis is carried out in accordance with well-known procedures, using a potential of 60 to 110 V. The preferred potential is 90 V.

Air-Treating Gel Composition

Highlights of the Technological Achievement: An improved air-treating gel of significantly enhanced stability, primarily thermal stability, comprising in an aqueous system a weight ratio of 0.3:1 to 5:1 carrageenan to stearate salt, a volatile active agent, a stearate solubility enhancer and water.

Reference: A.L. Streit and E. Sansanelli; U.S. Patent 4,178,264; December 11, 1979; assigned to Airwick Industries, Inc.

Background: A number of air-treating compositions have been marketed with a reasonable degree of success. These compositions, which are primarily based on carrageenan, release perfume or odor counteractants continuously over a period of time. Although carrageenan adequately forms gels and is capable of entrapping essential oils and aromatics for slow release, the physical properties of the carrageenan gel create a number of problems which must be closely watched during production and formulation.

A major disadvantage of such carrageenan systems is their lack of stability when such gels are exposed to the extreme temperatures frequently encountered during storage and/or shipping in the summer and winter months. Such lack of stability is exhibited by one or more of the following: product weight loss, deterioration of the product appearance, loss of effectiveness and the like. Specifically, such carrageenan-based compositions show poor freeze-thaw stability in that samples which return to ambient temperatures from temperatures below -20°C exhibit a significant fluid loss and corresponding reduction in size and appearance.

Other gel compositions are prepared utilizing stearate salt gelling agents. Although stearates also adequately form gels and entrap volatiles they too exhibit poor temperature stability, in this instance, at high temperatures. Thus, samples which are returned to ambient temperatures after having been exposed to temperatures above 48°C lose their integrity completely and return to their liquid state.

It should be recognized that this difficulty is a meaningful one in that such temperatures can be reached in warehouses and the like during hot summer months or in tropical climates.

Process Conditions: The air-treating gels of the process comprise from 1.5 to 15% by weight of a 0.3:1 to 5:1 weight ratio of carrageenan and stearate salt, from 0.5 to 6.0% by weight of essential oils and aromatics, from 1.0 to 20.0% by weight of a solvent component, with the balance being water.

As noted, the carrageenan and the stearate salt are present as the gel ingredient in the weight ratio of 0.3:1 to 5:1, and preferably 1:1 to 3:1, in order to provide the substantially improved thermal stability described herein. Concentrations which reflect additional amounts of either ingredient beyond the indicated range result in gels which tend to exhibit the disadvantages noted with gels prepared from the individual gelling agents.

In preparing air-treating gels in accordance with the process, the carrageenan is initially slowly added to water maintained at a temperature of 45° to 70°C. The solution is heated to a temperature of 80° to 95°C for complete hydration of the powder and then cooled to a temperature of 65° to 90°C. Thereafter, either (a) the glycol and stearate are individually dispersed in the carrageenan solution; or (b) the solvent is heated to above 85°C in a separate vessel in which the stearate is dispersed and the hot solution is added, under agitation, to the carrageenan solution.

Finally, a clear, premixed solution of active agent and any optional emulsifiers are added to the hot gel solution whereupon the gel is cooled to about 60° to 85°C, poured into a mold and allowed to cool to produce the formed gel masses.

Example: This example illustrates the preparation of a typical gel composition of the process as well as the substantially improved freeze-thaw stability thereof. The following formulations were prepared by adding the carrageenan to water at 49°C, heating the solution to 82°C and cooling to 71°C. Thereafter, the glycol, stearate and perfume were individually dispersed in the hot carrageenan solution. The resulting gel composition was then cooled.

	Formulations (pbw)				
	1	2	3	4	5
Carrageenan	5.1	–	–	3.1	3.1
Sodium stearate	–	5.1	–	2.0	–
Potassium stearate	–	–	5.1	–	2.0
Propylene glycol	6.0	6.0	6.0	6.0	6.0
Perfume	3.0	3.0	3.0	3.0	3.0
Tap water	balance to 100				
Physical state at RT	Solid	Solid	Liquid	Solid	Solid
Physical state at 54°C	Solid	Liquid	–	Solid	Solid

Formulations 1, 2, 4 and 5 were then subjected to the following freeze-thaw test procedure. A preweighed gel unit was placed in a freezer at a temperature of -18°C. The gel was removed when it reached -18°C (18 hours). The gel unit was then thawed under ambient conditions until it stabilized at room temperature (8 hours). All excessive syneresis was removed and the unit reweighed to determine weight loss (%). The following weight loss (%) values were obtained:

formulation 1, 2.4; formulation 2, 1.8; formulation 4, 0.2; and formulation 5, 1.0.

The results thus indicate the excellent physical characteristics of the gels of this process as well as the synergistic properties derived from utilizing the carrageenan-stearate blend as the primary gelling agent.

Stabilized Spoonable Ice Cream

Highlights of the Technological Achievement: The properties of ice creams, scoopable at deep-freeze temperatures, are improved by incorporating stabilizer mixtures comprising (a) locust bean gum and/or tara gum; and (b) kappa-carrageenan and/or xanthan gum and/or agar-agar.

Reference: I.C.M. Dea and D.J. Finney; U.S. Patent 4,145,454; March 20, 1979; assigned to Thomas J. Lipton, Inc.

Background: A problem with conventional ice creams is that at deep-freeze temperatures, e.g., -20°C, they cannot be served or eaten as readily as when they are at normal eating temperatures, e.g., -10°C. The consumer cannot treat them even approximately in the normal manner immediately when taken from the deep freeze. In some cases, conventional ice creams cannot even be scooped out with a spoon at -20°C, i.e., they are not spoonable.

Reformulation to ensure that such properties, e.g., spoonability at deep-freeze temperatures, are approximately those expected at normal eating temperatures is comparatively simple. The difficulty is that such reformulation leads to products that do not have acceptable properties, in particular stability, at normal eating temperatures. It has seemed impossible to get an ice cream that has, at both deep-freeze and normal eating temperatures, even approximately the serving and eating properties conventionally expected at normal eating temperatures and that is sufficiently stable. In particular, it is difficult to get adequate stability without the ice cream having unacceptably cloying mouth-feel. The process provides stabilizer systems with which this can be achieved.

Process Conditions: For an ice cream to be spoonable at -20°C, its log C (C is its penetrometer value) should be less than 2.8, particularly preferably less than 2.5; a correlation exists between spoonability and log C.

The process, therefore, provides an ice cream stabilized with a stabilizer mixture comprising locust bean gum and/or tara gum with other stabilizers and prepared by aerating and freezing and then hardening to a temperature below -20°C an ice cream mix containing the stabilizer mixture characterized in that the ice cream is an ice cream that, after it has been hardened, has a log C at -20°C of less than 2.8 and the stabilizer mixture comprises (a) locust bean gum and/or tara gum; and (b) xanthan gum and/or kappa-carrageenan and/or agar-agar.

As a general guidance to achieve a stabilized ice cream with desired log Cs at -20°C the following limits apply; in exceptional circumstances, stabilized ice creams with the required log C at -20°C can be achieved outside these limits. The general limits are that the amount of (a) should be greater than 0.07% by weight of the ice cream mix, the amount of (b) should be greater than 0.02% by weight of the ice cream mix; and the amount of total stabilizers should be such that the viscosity of the mix prior to aerating and freezing is not greater than 4 poises at 20 reciprocal seconds at 5°C. Use of ice cream mixes with viscosities over 4 poises lends to problems in processing and in mouth-feel of the hardened ice cream.

Substances such as low dextrose equivalent (DE) maltodextrins and modified starches can be added to improve mouth-feel, although preferably not in amounts such that the viscosity of the ice cream mix is more than 4 poises at 20 reciprocal seconds at 5°C. Thus, low DE maltodextrin can advantageously be present to improve the mouth-feel. By low DE is meant a maltodextrin with a DE of less than 20. The maltodextrin should be soluble. The lower limit for DE at which maltodextrins become insoluble depends particularly on the other ingredients in an ice cream mix, but whether a given maltodextrin is soluble in any such mix can be determined readily by experiment and in particular by noting whether it leads to the desired effect.

Maltodextrins can, for example, be obtained by the mild hydrolysis of starch. Enzymatic hydrolysis, optionally under acidic conditions, of the starch can be used; the conditions are so mild that negligible repolymerization occurs. This is in contrast to dextrins which typically are made from starch by hydrolysis and repolymerization using high temperature and pressure.

Stability: Because of the particularly good properties, e.g., stability, which are obtained, (b) is preferably xanthan gum. When (b) is xanthan gum, the weight ratio of (a) to (b) is preferably 7:1 to 1:7, particularly preferably 4:1 to 1:2, when (a) is locust bean gum, and preferably 7:1 to 1:2, particularly preferably 4:1 to 1:1, when (a) is tara gum. With a low excess of (a) to (b), i.e., weight ratio 4:1 to 1:1, aeration of the ice cream and the texture of the ice cream are particularly good. Total (a) plus (b) is preferably in the range 0.15 to 1%, particularly preferably 0.25 to 0.5%. Preferably, the amount of (a) is at least 0.1% and is less than 0.4%. Preferably, the amount of xanthan gum is at least 0.05%.

Use of kappa-carrageenan as (b) also is a preferred form of the method. Improved properties, in particular stability, are obtained. When (b) is kappa-carrageenan, the weight ratio of (a) to (b) is preferably in the range 1:1 to 7:1. The total of (a) and (b) is preferably in the range 0.15 to 1%, particularly preferably 0.15 to 0.5%. The ratio of (a) to (b) is particulary preferably in the range 3:1 to 7:1. The amount of (a) is preferably greater than 0.1%.

Example: Ice cream was prepared from the following formulation.

Ingredient	Percent by Weight
Made-up skimmed milk, 32.5% solids	30
Sucrose, as liquid syrup	25*
Stabilizers	
Locust bean gum	0.2
Kappa-carrageenan	0.03
Butter oil	6**
Glycerol	2
Palm monoglycerides, emulsifier	0.5
Flavor and color	0.04
Water to 100% by weight	

*15% solids plus 10% water
**Equivalent to 7.5% butter

Pimiento Paste

Highlights of the Technological Achievement: The process provides a method for producing pimiento paste for use in stuffing olives. The method involves sorting and chopping pimiento, preserving it in brine, washing the preserved pimiento to desalt it, mixing the desalted pimiento with sodium alginate, guar gum, potassium sorbate and water, and homogenizing the resulting mass until a gelatinous mixture is obtained from which any entrapped air is thereafter removed.

Reference: J.M. Gonzalez and C.L. Cortes; U.S. Patent 4,168,325; September 18, 1979; assigned to Medina Garvey Aceitunas, SA, Spain.

Background: The paste for stuffing olives produced by this process is a substitute for the pastes being used which are made from oven-peeled pimientos. This process guarantees a healthy, high quality product for the consumer since the paste contains the same type of pimiento mentioned above and certain nontoxic agglutinates described in detail below. The quality and quantity of these substances have been approved by the FAO/WHO Joint Program for Food Standards.

Among the advantages of using this pimiento paste are the following:

(a) Uniform color and texture of product in comparison with the same qualities in the natural oven-peeled pimiento being used;

(b) More efficient use of material since the substantial reduction of by-products produced makes it possible to shape this paste into very long strips which are the correct size for stuffing—these conditions are very limited in natural pimiento; and

(c) Finally, the increase of production by the stuffing machines which can produce at a higher rate when provided with a more homogeneous product.

Process Conditions: The process for producing pimiento paste comprises the following stages: (1) preparation of the raw material; (2) homogenization of the different ingredients; and (3) shaping the paste.

In the first stage of the process, that is, in the preparation of the raw material, natural oven-peeled pimiento is preferably used as the main ingredient. In this stage, the pimiento is first sorted and chopped, a procedure which varies according to the original condition of the pimiento. This step eliminates the remaining seeds, pieces of burnt skin, insects or other impurities that might affect the quality of the end product. The quality of the pimiento, which is preserved in brine, is then improved by successive washings with water, as by maceration or by continuous agitation, to desalt the pimiento. Microbiological control during this operation insures an aseptic, nontoxic product, uncontaminated by diseases affecting the pimiento. This preparatory stage may be considerably reduced through the use of natural oven-peeled pimiento preserved by heat treatment, that is, by sterilization.

In the second stage of the process, homogenization of the different ingredients of the pimiento paste is carried out, as by a crusher-mixer, to homogenize the following ingredients in the proportions, by weight, indicated: desalted pimiento, 50.00%; sodium alginate, 7.00%; guar gum, 1.00%; potassium sorbate, 0.02%; and water, enough to make up 100%.

Once a gelatinous mixture has been produced, the air trapped in the mixture is removed before proceeding to the third stage of the process. This air removal is carried out either by producing a vacuum or through compression and filtration. In this way, a semiliquid, completely uniform paste is produced. The purpose of this air removal is to avoid subsequent oxidation-reduction reactions that would alter the protein-vitamin content of the pimiento.

Product: Because the basic ingredient of the final product is natural pimiento, and since the composition of this ingredient, like all agricultural products, is not always the same due to the effect of climate, fertilizer, soil quality, etc., the percentage of the abovementioned mixture cannot always be maintained exactly. The composition of the paste is adjusted with water in order to maintain the same characteristics which, as closely as possible, resemble a specific model in color, texture, elasticity, refractive index, etc.

POLYAMIDES AND POLYIMIDES

PROCESSING

Fibrids from Polyamideimide Resins

Highlights of the Technological Achievement: Fibrids are prepared by introducing a solution of a polymer into a liquid precipitation medium while exposing it to shearing forces. A polyamideimide resin solution is admixed with a solvent which is miscible with the resin solvent but which is a nonsolvent for the resin itself.

Reference: B. Sander, K. Scherling, R. Steinberger and E. Ropte; U.S. Patent 4,187,143; February 5, 1980; assigned to BASF AG, Germany.

Background: The manufacture of fibrids from polymers, by dissolving the latter in a suitable organic solvent and introducing the resulting solution into a liquid precipitation medium while at the same time exposing it to shearing forces, has been disclosed.

However, if the conventional process is applied to solutions of polyamideimide resins, finely divided pulverulent particles, showing no tendency to form a web, are obtained in place of discrete fibrids. The particles obtained in most cases have a high solvent content, so that their subsequent working up is difficult and expensive.

It is an object of the process to provide a method in which solutions of polyamideimide resins are passed in a liquid precipitation medium while exposed to shearing forces, whereby discrete fibrids exhibiting a high degree of fibrillation and a high tendency to form a web are obtained directly.

Process Conditions: Fibrids of polyamideimide resins are 0.1 to 5 mm long and from 5 to 200 μm thick and have a freeness of from 20° to 90° SR and a specific surface area from 1 to 80 m^2/g. For the purposes of this method, fibrids are fibrous particles of synthetic polymers which morphologically, in size and shape, and in their properties resemble cellulose fibers.

It has proved particularly appropriate to use water as the precipitation medium. In order to increase the viscosity of the precipitation medium the water may be entirely or partially replaced by, for example, ethylene glycol, diethylene glycol or glycerol. Dispersing assistants may also be added to the precipitation medium.

In this instance, i.e., when using a polyamideimide resin as the synthetic polymer, 1-methyl-2-pyrrolidone, N,N-dimethylformamide and N,N-dimethylacetamide have proved particularly suitable organic solvents for the polymers. The polyamideimide resin solutions employed in general have a solids content of preferably 15 to 25 wt %, and their viscosity is preferably 15,000 to 60,000 cp, measured by means of a falling ball viscometer.

Before being introduced into the precipitation medium, the solution of a polyamideimide resin in one of the above organic solvents is mixed with a liquid which is a nonsolvent for the polyamideimide resin but is miscible in all proportions with the organic solvent and the precipitation medium. The preferred organic liquid is an organic cyclic ether, e.g., tetrahydrofuran or 1,4-dioxane. The volume ratio of solvent to cyclic ether is from 1:1.5-1:5, preferably from 1:2.5-1:4. The cyclic ethers or other nonsolvents for the polymer may be added to the polymer solution until the polymer starts to precipitate. Within the above limits, the most advantageous volume ratio can in each case be determined by a simple laboratory experiment.

Polyamideimide resins are polymers which contain recurring units of amide groups and imide groups, for example, as shown in the general formula:

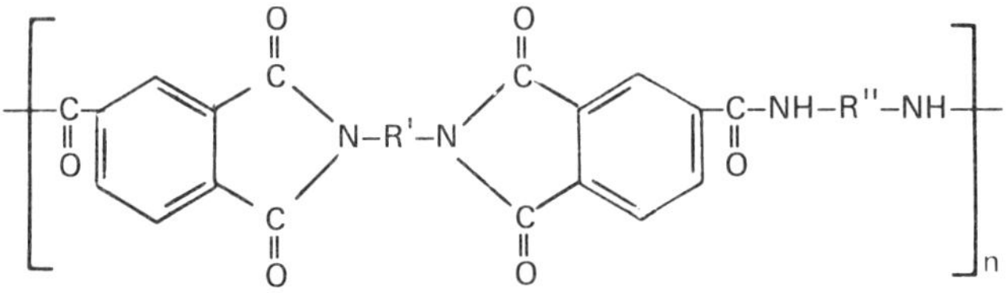

where n is an integer, R' and R'' may or may not be equal, R' and R'' are m- or p-phenylene or:

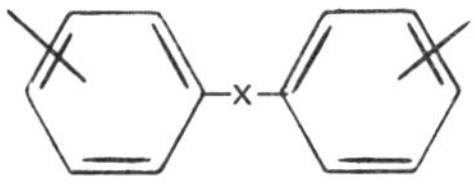

and x is CH_2, O, S, CO or SO_2, and where the polymer may or may not contain additional amide linkages of the structure: –OC–R'–CONH–R''–NH–.

Drawings: Devices useful for creating the shearing forces in the production of the subject fibrids are illustrated in Figure 16.1. Figure 16.1a is a diametric section of a rotor-stator type apparatus for creating the shearing forces mechanically; Figure 16.1b is a side elevation, partly in diametric section, of another apparatus for creating the shearing force hydraulically; and Figure 16.1c is a section view taken on section plane **A-B** of Figure 16.1a.

If the process is carried out continuously it is possible to use, for example, the following apparatus (Figure 16).

Figure 16.1: Roto-Stator Apparatus for Creating Shearing Forces Mechanically

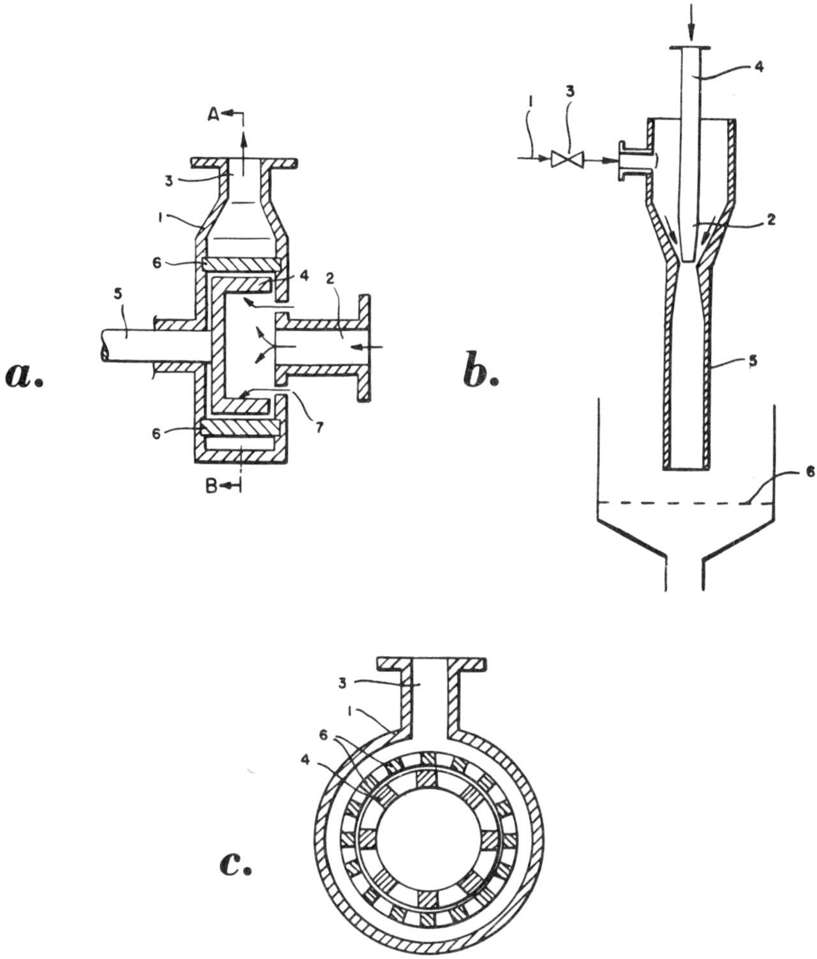

Source: U.S. Patent 4,187,143

Example 1: 49.5 parts of 4,4'-diaminodiphenylmethane were dissolved in 627 parts of 1-methyl-2-pyrrolidone and 96 parts of trimellitic anhydride were added. After adding 125 parts of toluene, the theoretical amount of water was removed at 140° to 150°C, and the toluene was recovered. On cooling the reaction mixture, the diimidedicarboxylic acid precipitated in a well-crystallized form. For the further conversion to the polyamideimide, the reaction batch was brought to 80°C and 62.5 parts of 4,4'-diphenylmethanediisocyanate were added. The temperature was then raised to 190°C at the rate of 10°C/30 min. The mixture was then cooled to 120°C and 1.25 parts of 4,4'-diphenylmethanediisocyanate were added twice more, at intervals of 1 hr. After 2 hr, a viscous polyamide-imide solution having a solids content of 22% and a solution viscosity of 34,600 cp (measured by means of a falling ball viscometer at 20°C) was produced. The intrinsic viscosity of the polymer was 0.81 dl/g (measured as an 0.5% strength solution in 1-methyl-2-pyrrolidone at 30°C).

Example 2: 49.5 parts of 4,4'-diaminodiphenylmethane and 96 parts of trimellitic anhydride were reacted in 627 parts of 1-methyl-2-pyrrolidone to give the diimidedicarboxylic acid, as described in Example 1. After the reaction solution had cooled, the diimidedicarboxylic acid which had crystallized out was filtered off and converted to its acid chloride by boiling with excess thionyl chloride or by reaction with phosgene in a chloroform suspension.

145.5 parts of this acid chloride and 49.5 parts of 4,4'-diaminodiphenylmethane were reacted in 627 parts of 1-methyl-2-pyrrolidone at 0° to 5°C, to give the polyamideimide. After stirring for 30 min, a viscous solution was obtained, which had a solids content of 22%, a solution viscosity of 55,000 cp and an intrinsic viscosity of 1.00 dl/g.

Production of Linear, Segmented Polyamideureas and Their Aqueous Dispersions

Highlights of the Technological Achievement: A process for the production of polyureas wherein solutions of organic polyisocyanates are reacted with primary and/or secondary organic polyamines in the presence of tetracarboxylic acid dianhydrides. The polyisocyanates used are preferably difunctional linear prepolymers containing terminal isocyanate groups, while the amines used are preferably diamines. The resultant product, which contains carboxyl groups, may then be neutralized and dispersed in water.

Reference: O. Lorenz; U.S. Patent 4,160,753; July 10, 1979; assigned to Bayer AG, Germany.

Background: It is known that polyurethane ionomers can be produced from prepolymers. Anionic segmented polyurethane ionomers can thus be produced by reacting NCO-terminated prepolymers with bifunctional anionic chain extenders such as diamino- or dihydroxy- carboxylic acids in accordance with the following equation.

$$\text{OCN}\sim\text{NCO} + \underset{\text{COOH}}{\text{H}_2\text{N}\text{———}\text{NH}_2} \longrightarrow \text{CO–NH}\sim\text{NH–CO–NH}\underset{\text{COOH}}{\text{———}}\text{NH–}$$

Unfortunately, this reaction is difficult to control because the free carboxyl groups are able to react with the isocyanate groups. In some cases, insoluble internal salts are even formed. Accordingly, the anionic component is preferably used in the form of the sodium or potassium salt. The chain extension of NCO-prepolymers with salts of aliphatic diaminocarboxylic acids has proved to be the most favorable, in this case, polyaddition to form the macromolecule taking place over a short period.

Accordingly, the object of the instant process is to provide a simplified process for the production of polyureas which results in polyadducts which contain free carboxyl groups and which may be converted into polyureas dispersible in water by simply forming salts from these free carboxyl groups by treatment with suitable bases.

Process Conditions: In cases where difunctional diisocyanates, especially difunctional NCO-prepolymers are used, the process of the instant method results in the formation of linear, segmented polyureas having carboxyl groups which are directly situated on the main chain and are nonstatistically arranged. The polyadducts synthesized in this way show extremely favorable elastic properties which may be commercially exploited. In particular, high-quality coatings can be readily obtained from these plastics. The number of reactive hydrogen atoms may readily be selected in such a way that buildup of the corresponding macromolecule results in a substantially linear end product, which is often ideal for optimum elastomeric behavior.

The reaction is accompanied by the formation through ring cleavage, of carboxyl groups which can be converted into the corresponding salts by the addition of bases. In order to better illustrate the instant process, the reaction of an NCO-terminated prepolymer with a diamine containing two primary amino groups in the presence of a pyromellitic acid dianhydride may be described as follows.

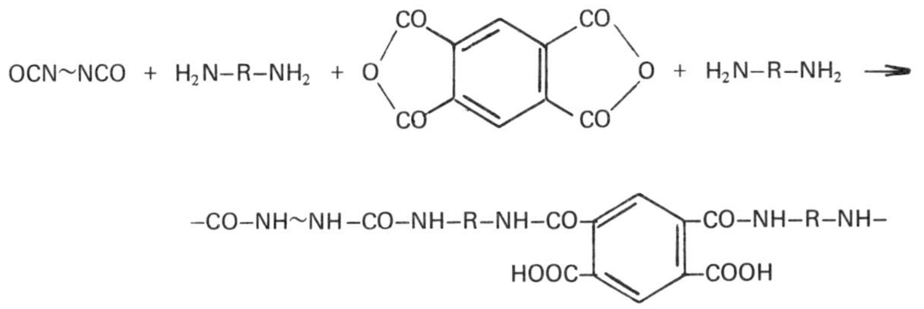

It is thus possible by virtue of the process according to this method to readily obtain linear, segmented polyamideureas having carboxyl groups which are situated directly on the main chain and which are nonstatistically arranged, and which may be converted into the corresponding carboxylate anions by a simple neutralization reaction.

The following compounds are particularly suitable starting compounds for use in this process.

(a) NCO-prepolymers having preferably two terminal isocyanate groups and molecular weights of preferably 800 to 4,000.

(b) Primary or secondary diamines of the formula $(R)_n(NHR')_2$, in which R represents a divalent aliphatic hydrocarbon radical with 2 to 18 carbon atoms, a divalent aromatic hydrocarbon radical with 6 to 15 carbon atoms, a divalent cycloaliphatic hydrocarbon radical with 5 to 15 carbon atoms or a divalent araliphatic hydrocarbon radical with 8 to 18 carbon atoms, and R' represents hydrogen or an aliphatic hydrocarbon radical with 1 to 4 carbon atoms, and n is 0 or 1.

(c) Tetracarboxylic acid dianhydrides: (1) pyromellitic acid dianhydride (1,2,4,5-benzene tetracarboxylic acid dianhydride); (2) bicyclo[2.2.2] oct-7-ene-2,3,5,6-tetracarboxylic acid-2,3:5,6-dianhydride; (3) perylene-3,4,9,10-tetracarboxylic acid dianhydride; (4) naphthalene tetracarboxylic acid dianhydride; (5) tetracarboxylic acid anhydrides of the following formula:

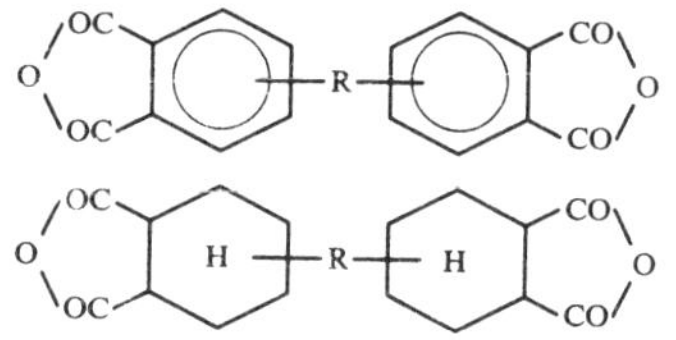

in which R represents a C_{1-12}-alkylene, –CO–, –O–, –S–, $-SO_2$-bridge, the alkylene bridge optionally also containing heteroatoms. In other words, the alkylene group may be attached to the rings through ester groups. Pyromellitic acid anhydride, benzophenone tetracarboxylic acid dianhydride and esters of trimellitic acid anhydride with simple glycols of the type mentioned in (a) are preferred.

(d) Inorganic or organic bases for optionally converting the free carboxyl groups into carboxylate anions including monofunctional primary, secondary or tertiary amines such as, methylamine and triethylamine or inorganic bases such as ammonia, or sodium hydroxide.

To carry out the method of this process, the polyisocyanate (a) is preferably dissolved together with the tetracarboxylic acid dianhydride (c) in a suitable solvent. The polyamine (b), preferably also dissolved, is added dropwise with stirring to the resulting solution at 10° to 100°C, preferably at 20° to 60°C. The quantitative ratios in which the reactants are used are selected in such a way that from 0.05 to 0.6, preferably from 0.1 to 0.5, acid anhydride groups and from 0.4 to 4 preferably from 0.8 to 2.5, NCO-groups are used per amino group during the reaction. Suitable solvents include any solvents which are inert under the reaction conditions and which boil at temperatures in the range of 30° to 200°C, preferably from 40° to 80°C.

Example 1: 47.6 g of 2,2,4-trimethyl hexamethylene diisocyanate are added with stirring at 80°C to 250 g of an adipic acid-hexane diol-neopentyl glycol polyester of molecular weight 2,000 (OH-number 56) which had previously been dehydrated in vacuo at 120°C. The melt is heated to 115° to 120°C and kept at that temperature for a period of 2 hr. The reaction mixture is stirred with 400 ml of acetone under heat, resulting in the formation of a clear solution. 7.37 g of pyromellitic acid dianhydride in 170 ml of acetone are added, and then 4.05 g of ethylene diamine dissolved in 125 ml of acetone are slowly added dropwise, the temperature in the reaction vessel being 56°C.

The reaction mixture is then stirred for approximately 20 min, after which a solution of 4.37 g of KOH in 417 g of deionized water are added with intensive stirring at 55° to 60°C. This results in the formation of a stable dispersion which is partly freed from acetone in vacuo at 45°C, so that a latex having a viscosity of about 40 cp (20°C) is ultimately obtained.

Example 2: The procedure is as in Example 1, except that 38.0 g of hexamethylene diisocyanate are used instead of 2,2,4-trimethyl-hexamethylene diisocyanate. The polyurethane latex formed has a viscosity of about 60 cp at 20°C.

Example 3: The procedure is as in Example 2, except that 16.38 g of 2,2-bis-(cyclohexylamine)-propane are used instead of 4.05 g of ethylene diamine. The dispersion obtained has a viscosity of about 150 cp at 20°C.

Mechanical properties of films prepared from the dispersions obtained in accordance with Examples 1 through 3 are set out in the following table.

Mechanical Properties of Films

Ex. No.	Strain at 300% Elongation (kp/cm²) 1st Strength	6th Strength	Permanent Elongation (%)	Breaking Strength (kp/cm²)	Breaking Elongation (%)
1	13.9	11.6	27.5	154	1,125
2	37.0	32.8	12.5	209	770
3	44.7	35.7	29.4	140	705

Water-Solubilized Resin Composition with a Volatile Nitrogenous Base Salt of a Polyimide Precursor

Highlights of the Technological Achievement: A water-solubilized resin composition for forming a thermally stable polyimide polymer prepared by mixing: (a) a nitrogenous base salt of a polyimide precursor containing an amino group at both terminals of the molecule and having a residual acid value ratio of 50 to 5% obtained by reacting 1,2,3,4-butanetetracarboxylic acid (BTC) and a molar excess of a diamine in water, or in a water-soluble solvent or in the presence of both a water-soluble solvent and water with (b) a nitrogenous base salt of a 3,3',4,4'-benzophenonetetracarboxylic acid diester (BTDE) whose ester linkage is positioned at either one of the adjoining two carboxyl groups in the same benzene nucleus thereof.

Reference: Y. Noda, M. Yoshihara and T. Takigawa; U.S. Patent 4,144,210; March 13, 1979; assigned to Nitto Electric Industrial Co., Ltd., Japan.

Background: Because of superior thermal stability, chemical resistance and electrical characteristics, imide-type polymers have heretofore found many applications, for example, as films, adhesives, paints and molded articles used at high temperatures. Thermally stable polyimides are obtained generally by polymerizing aromatic tetracarboxylic acid derivatives such as pyromellitic dianhydride, and diamines in organic polar solvents to form soluble high-molecular-weight polyamide acids, shaping them, and then chemically or thermally treating the shaped products to induce intramolecular cyclization.

Polyamide acids, which are polyimide precursors, readily cyclize during storage to form insoluble infusible polyimides. However, since an expensive basic solvent such as N-methyl-2-pyrrolidone, N,N-dimethylacetamide or N,N-dimethylformamide is used in a large quantity, the polymer solution itself is expensive, and a large quantity of solvent is evolved in the process of manufacturing the final products. The evolved solvent tends to pollute the air and water, and poses a problem of environmental pollution.

Process Conditions: The water-solubilized resin solution (a) is prepared by blending 1 mol of BTC or a mixture of BTC with a minor proportion of a derivative thereof (e.g., a monoanhydride, dianhydride, ester, amide, etc.) with 1.1 to 2.0 mols of a diamine, subjecting the mixture to amidation and imidization to form a polyimide precursor containing an amino group at both terminals of the molecule and having a residual acid value of 50 to 5% in water, or if desired, in a water-soluble solvent or in the presence of both a water-soluble solvent and water, and then reacting the polyimide precursor with a nitrogenous base to form a carboxylic acid salt.

One of the important characteristics of this process is that the polyimide precursor containing an amino group at both terminals of the molecule and having a residual acid value ratio of 50 to 5% obtained by the above operation exhibits very good solubility in water as a result of reaction with a nitrogenous base.

Raw Materials: Examples of water-soluble solvents which can be used in this process include compounds expressed by the following general formulas: monohydric alcohols of the formula (a), R_1–OH, wherein R_1 is a monovalent residue of an aliphatic or alicyclic compound having 3 to 8 carbon atoms; glycols of the following formula (b), HO–R_2–OH, wherein R_2 is a divalent residue of an aliphatic or alicyclic compound having 3 to 8 carbon atoms, or a divalent residue of the formula, $-(CHR_3CH_2O)_{\overline{n}}CHR_3CH_2-$, in which R_3 is a hydrogen atom or a methyl group, and n is an integer of 0 to 5; glycol ethers of the formula (c), $R_4O-(CH_2CH_2O)_{\overline{m}}H$, wherein m is an integer of 1 to 3 and R_4 is a lower alkyl group having 1 to 4 carbon atoms; and trihydric alcohols expressed by the following formula (d), $R_5(OH)_3$, wherein R_5 is a residue of glycerin or trimethylolpropane.

Examples of suitable compounds of the formula (a) are isopropyl alcohol, n-butyl alcohol and t-butyl alcohol. Examples of suitable compounds of the formula (b) are ethylene glycol, diethylene glycol and triethylene glycol. Examples of suitable compounds of the formula (c) include ethylene glycol monoalkyl (monomethyl, monoethyl, monoisopropyl or monobutyl) ethers. Examples of compounds of the formula (d) are glycerin and trimethylolpropane.

3,3′,4,4′-Benzophenonetetracarboxylic Acid Diester: A suitable temperature at which the polyimide precursor can be converted to the salt is 0° to 200°C, and temperatures of 40° to 120°C are preferred. Thus, the water-solubilized resin solution [1] can be obtained easily.

The water-solubilized compound [2] which results in the production of a high-molecular-weight thermally stable polymer or reaction with the terminal amino groups of the water-solubilized resin solution [1] is a derivative of a 3,3',4,4'-benzophenonetetracarboxylic acid diester (BTDE).

BTDE is a diester in which each ester linkage is positioned at either one of the two carboxyl groups present in the same benzene nucleus, that is, one of the two ester linkages is positioned at the 3-position or the 4-position of the benzophenone nucleus and the other ester linkage is positioned at the 3'-position or the 4'-position of the benzophenone nucleus.

The nitrogenous base salt of BTDE used in this process is characterized by the position of the ester linkage. Moreover, it can be more easily water-solubilized than similar salts derived from other aromatic tetracarboxylic acids such as pyromellitic acid.

In addition, organic metal salts such as tin, zinc, manganese, iron, cobalt, lead or a like salt of octenoic acid or naphthenic acid which are not water-soluble in nature but are useful as a reaction accelerator for the reaction between the water-solubilized resin [1] and the water-solubilized compound [2] can be also mixed with the composition prior to processing as long as the amount used is in the range of from 0.01 to 0.1 wt %.

Products: *Laminate* – The characteristics of a laminate of the process and the results are shown below.

Tensile strength, 20°C	30 kg/mm^2
Flexural strength, 20°C	60-65 kg/mm^2
Flexural strength, 200°C	48-50 kg/mm^2
Insulation resistance, initial stage	2 x 10^{14} ohms
Insulation resistance, D-4/100*	3 x 10^{13} ohms
Volume resistivity	2 x 10^{15} ohms-cm
Dielectric constant, 1 kHz	4.0
Dielectric loss tangent, 1 kHz	0.3
Water absorption, D-24/23*	0.12 %
Thermal stability test, E-24/180*	Acceptable
Heat loss, 250°C x 10 days	1.0 %

*Represents numbers of ASTM

Enamelled Wire – The characteristics of an enamelled copper wire using a coating of the process are shown below.

Thickness of coated film	0.040 mm
Flexibility, self-diameter winding	
after 10% stretch	Good
after 15% stretch	Good
after 20% stretch	Good
Heat softening temperature, 2 kg load	490°C
Reciprocating abrasion resistance,	
600 g load	130 times
Slit twist, 20 cm span	93 counts
Dielectric breakdown voltage	
normal conditions	12.1 kV
after heating at 260°C for 168 hr	11.1 kV
Burn-out property 42A	
20 sec on, 10 sec off (120 V)	136 cycles

APPLICATIONS

Water-Soluble Polyesterimide Resin Wire Coating Process

Highlights of the Technological Achievement: Water soluble polyesterimide prepolymers are prepared: (a) having 5 to 60%, preferably at least 35% of imide moieties; (b) an excess hydroxyl content in excess of 60% and preferably more than 100%; and (c) a number average molecular weight of 600 to 1,300. To water solubilize the prepolymers formed, there are added amines, preferably tertiary amines. The products are useful as wire enamels. To enhance the thermoplastic flow of the enamel, there are added organic titanates.

Reference: D. Laganis and P.M. Begley; U.S. Patent 4,206,261; June 3, 1980; assigned to Schenectady Chemicals, Inc.

Background: Polyesterimide wire coatings thinned in conventional cresylic acid-aromatic hydrocarbon solvent blends have been used commercially for a number of years. Taking the base polymer one can solubilize it in water along with an amine. In attempting to use the same curatives as are employed in the non-aqueous or organic solvent system previously mentioned it was soon discovered that they were insoluble in water. This necessitated a search for crosslinkers that were soluble in water, and those were mainly organic titanates, such as Tyzor TE and LA which are titanium chelates of triethanolamine and the ammonium lactate salt, respectively.

Process Conditions: As the imide forming components, there can be used (a) anhydrides, such as trimellitic anhydride, etc.; and (b) polyamines, preferably aromatic amines, such as methylene dianiline, oxydianiline, and phenylene diamines.

Reactants (a) and (b) are usually employed in an amount of approximately 2 mols of (a) per mol of (b) to form the diimide-diacid. This reaction product may be represented as follows:

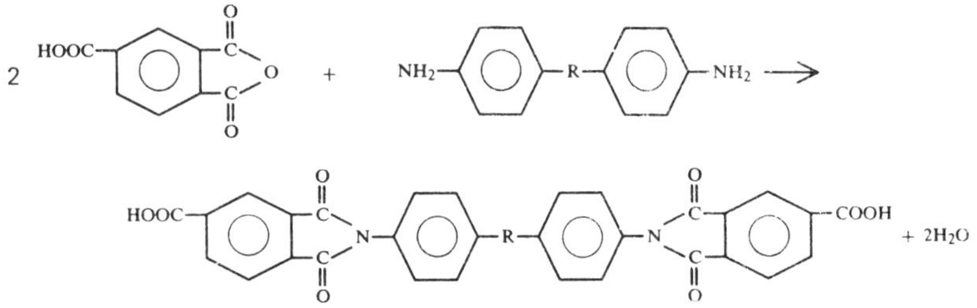

where R is CH_2 in the case of methylene dianiline or the oxygen atom (O) in the case of oxydianiline.

Then in the formulation of the polymer, this diimide-diacid is considered as a portion of the total dicarboxylic acid content which contains predominately aromatic diacids with or without a minor quantity of aliphatic diacids. The preferred major diacid is terephthalic acid. When aliphatic carboxylic acids are included, e.g., in an amount up to 50 eq % of the total acid, there can be used, for example, adipic acid, sebacic acid, maleic acid (or its anhydride), azelaic acid, and glutaric acid.

These diacids may then be subsequently esterified by polyols, such as tris(2-hydroxyethyl) isocyanurate (THEIC) alone, or in admixture with dihydric alcohols, such as ethylene glycol, neopentyl glycol, propylene glycol, diethylene glycol, 1,3-bis(hydroxyethyl)-5,5'-dimethylhydantoin and others in various ratios.

The preparation of the polyesterimide is not restricted to any specific sequence of reactions. All the reactants may be charged to the reactor initially, heated to about 400° to 460°F and reacted in situ.

With respect to the determination of diimide-diacid content, earlier mention was made of its formation by the reaction of approximately 2 mols of an aromatic anhydride with 1 mol of an aromatic diamine. The combination of the molar content of this product and an aromatic diacid or aromatic-aliphatic diacid mixture is used to calculate the percentage of each. An example of this would be as follows.

	Mols		Mols
Trimellitic anhydride	1.00	Diimide diacid	0.50
Methylenedianiline	0.50		
Terephthalic acid	0.50	Aromatic diacid	0.50

Thus, on a mol percent basis the diimide diacid content would be 50%. In the preparation of the polyester an excess of alcoholic groups (OH) over acid groups (COOH) is preferred in order to get better flow on wire due to the formation of a lower molecular weight polymer with a lower softening point range. This may be expressed as an OH/COOH ratio range of preferably 2.20:1-2.50:1. In other words, an excess of OH groups over COOH groups of 80 to 150% is required to get the flow and other properties needed.

There is no need to employ solvents during the preparation of the polyesterimide prepolymer. Solvents of the polar type may be subsequently employed as cosolvents with distilled water in the preparation of aqueous solutions. Conventional cresylic acid-aromatic hydrocarbon (e.g., xylene) blends may also be used to make solvent-type wire enamels that perform equally as well as the aqueous types. A comparison of wire properties of solvent versus water-based enamels using the same polymer shows only a slight superiority of the former.

To solubilize these inherently water-insoluble resinous prepolymers in water, various amines may be employed that react with the free carboxyl groups or amic acid groups available to form the salts that are soluble in water. These amines may be of the alkyl, alkanolamine, or morpholine types. In general, the tertiary amines work best from the standpoint of fast cure, and least moisture sensitivity in the resultant baked film.

Example: *(A) Preparation of Polymer –*

Reactants	(g)	Mols	Mol % of Total Diacids	Mol % of Total Reactants
(A) Ethylene glycol	273	4.40	–	36.55
(B) Tris(hydroxyethyl)isocyanurate (THEIC)	935	3.58	–	29.73
(C) Trimellitic anhydride (TMA)*	603	3.14	–	–
(D) Methylene dianiline (MDA)*	310	1.56	–	–
(E) Terephthalic acid	415	2.50	61.58	20.76

*Or as diimide diacid OH/COOH = 2.40/1, 1.56 mols; mol % of total diacid 38.42 and mol % of total reactants 12.96.

Materials A, B, C, D and E were charged into a 5 ℓ, 3-necked flask equipped with an agitator driven by an electronically-controlled motor, gas inlet tube, thermometers for flask and distilling head, 3-bubblecap Snyder fractionating column and water-cooled condenser. A blanket of nitrogen was applied to the charge, and the temperature increased gradually to about 450° to 460°F, and held there until the desired physicals were attained. At a viscosity of U ¼ as measured at 30% solids in cresylic acid the reaction was terminated, and the melt discharged into a pan to solidify.

(B) Preparation of Aqueous Wire Enamel – The hard resin was then fractured into small chunks, and 400 g of the polymer along with 50 g of N-methyl pyrrolidone (NMP) was charged to a 3 ℓ, 3-neck roundbottom flask, and heated to about 250° to 270°F until the polymer was fluid and dissolved. At a temperature of 240° to 250°F a mixture of 50 g each of distilled water and dimethylethanolamine (DMEA) were added to the flask carefully. Additional water, amine, and solvent were added to reduce the viscosity to Z (on the Gardner-Holdt scale), obtain a pH of 7.5 to 8.5, and solids of 53.3%. Then on a solids-to-solids basis 4.5% Tyzor TE was added to the solution to make the finished enamel having a viscosity of X ¼, a pH of 8, and a solids content of 54.1%.

The effect of titanate content was studied with wire enamels prepared with the polymer. The principal effect of increased titanate content was the large rise in the dissipation factor. The other wire properties were all quite comparable with those of solvent based enamels.

Salts of a Basic Nitrogenous Compound as Stabilizers for Neutralized Polyamide-Acid Resin

Highlights of the Technological Achievement: A method for stabilizing a water-soluble resin composition which comprises adding, as a stabilizing agent, a carbonate and/or an organic carboxylate of a basic nitrogenous compound to the water-soluble composition, the water-soluble composition, being prepared by condensation-reacting an acid component consisting of 1,2,3,4-butane tetracarboxylic acid alone or a polycarboxylic acid component containing at least 50 mol % of 1,2,3,4-butane tetracarboxylic acid and a diamine in the presence of a water-soluble solvent.

Reference: K. Yamaguchi, Y. Suzuki, M. Komima and H. Shibata; U.S. Patent 4,183,835; assigned to Nitto Electric Industrial Co., Ltd., Japan.

Background: Heretofore, various organic solvent-type resin solutions have been developed and are used in many fields. However, organic solvent-type resin solutions involve the release of a large amount of solvent in the production steps, e.g., heating and/or drying steps, and recently have become a problem from the standpoints of environmental pollution and environmental safety. Specifically, solvents used in the fields of electrical insulating varnishes, etc., are mainly acidic or basic solvents such as cresol, xylenol, phenols, N-methyl-pyrrolidone or the like, and there are further great problems such that the deposition of the resin solution becomes inflamed.

In order to solve the above defects, attempts have been made to make the solvent for the resin solution safer, i.e., to use water as the main component of the solvent, by employing a water solubilization method, an emulsion method, etc.

Process Conditions: A method for stabilizing a water-soluble resin composition which comprises adding, as a stabilizing agent, a carbonate and/or an organic carboxylate of a basic nitrogenous compound to the water-soluble composition in an amount of 0.01 to 10 pbw based on 100 pbw of the resin content in the water-soluble composition, the water-soluble composition being prepared by condensation-reacting an acid component consisting of 1,2,3,4-butane tetracarboxylic acid alone or a polybasic acid component containing at least 50 mol % of 1,2,3,4-butane tetracarboxylic acid and a diamine in the presence of a water-soluble solvent or, if desired, in the presence of the water-soluble solvent and water to obtain a condensation product having a residual acid value ratio of 3 to 50% of the acid value before condensation followed by formation of a salt of the condensation product with a basic nitrogenous compound.

It is preferred for the reaction temperature to range from 60°C to the boiling point of the reaction solution under normal pressure.

The water-soluble solvents which can be used in this process can be represented by the following formulas: (a) R–OH wherein R represents a monovalent organic residue of an aliphatic or alicyclic compound having 3 to 8 carbon atoms; (b) $HO{-}R_1{-}OH$ wherein R_1 represents (1) a divalent organic residue of an aliphatic or alicyclic compound having 3 to 8 carbon atoms, or (2) a divalent moiety having the formula $-(CHR_2CH_2O)_{\overline{n}}-CHR_2CH_2-$ wherein n represents 0 to 5 and R_2 represents a hydrogen atom or a methyl group; (c) $R_3O-(CH_2CH_2O)_{\overline{m}}R_4$ wherein m represents 1 to 3, R_3 represents a lower alkyl group having 1 to 4 carbon atoms, and R_4 represents a hydrogen atom or a $-OOCCH_3$ group, or R_3 and R_4 are both methyl groups; or (d) $R_5(OH)_3$ wherein R_5 represents a glycerin or trimethylolpropane residue.

The temperature at the formation of a salt of the condensation product ranges from 0° to 200°C, preferably room temperature (e.g., 20° to 30°C) to 120°C.

Product: A characteristic feature of this process is to prevent an increase in viscosity of the water-soluble composition during storage thereby to stabilize the water-soluble composition by adding as a stabilizing agent a carbonate and/or an organic carboxylate of the basic nitrogenous compound to the water-soluble composition in an amount of 0.01 to 10 pbw based on 100 pbw of the resin content in the water-soluble composition.

Example: A water-soluble composition was prepared and to the resulting compositon various kinds of salts of basic nitrogenous compounds as shown in the following table were added to observe the effect of preventing increases in viscosity.

Sample No.	Salt Added*	Initial Viscosity (poises at 30°C)	Viscosity During Storage at 50°C Under Sealed Conditions (poises at 30°C) After 3 Days	After 5 Days	After 10 Days	After 20 Days
1	None	20	60	150	500	gel state
3	Ammonium carbonate	20	20	22	24.3	28
6	Ammonium formate	20	21.5	25.3	29.8	39.3
7	Ammonium acetate	20	24.5	32.3	45.8	67.3
8	Ammonium propionate	20	30	41.5	68.5	110
9	Triethanolamine carbonate	20	21.3	22.5	25.8	30.3

*Amount added: 1 part per 100 parts of resin content.

Cationic Aqueous Emulsion Systems

Highlights of the Technological Achievement: Water-based coatings have been developed comprising water-containing dispersed liquid droplets of water-immiscible, hydroxy functional organic material, which are emulsified in the water, the water having dispersed therein an amine salt constituted by a salt of a solubilizing acid with a polyamide having a plurality of terminal primary amine groups.

Reference: A.T. Spencer; U.S. Patent 4,169,080; September 25, 1979; assigned to Mobil Oil Corporation.

Background: Most aqueous cationic systems are of the solution or colloidal dispersion type. In contrast, this process relates to cationic aqueous emulsion systems which cure by the unblocking of isocyanate groups blocked for stability in the emulsion by a volatile blocking agent.

Emulsion coating systems have various advantages. In conventional coating application, the aqueous systems in which the resins are dissolved or colloidally dispersed are very viscous whenever the resin solids content approaches desirable levels. Emulsion systems possess lower viscosity at higher solids content. Emulsion systems in electrocoat application tend to possess better throwing power and superior flow in the deposited film. This superior flow is the result of the water insoluble organic solvents which are present in the discontinuous phase of the emulsion and which are, hence, electrodeposited with the emulsion droplets. As a result, there are advantages in electrodepositing an emulsion even though the factor of lower viscosity at high solids content is not a factor in an electrocoat system which is normally operated at low solids content.

Process Conditions: An aqueous emulsion is provided in which water has emulsified therein liquid droplets of water-immiscible, hydroxy-functional organic material, and the water has dispersed therein an amine salt constituted by a salt of a solubilizing acid with a polyamide having a plurality of terminal primary amine groups, about one of which is unsubstituted, and at least 70% of the isocyanate-reactive amino hydrogen atoms in the polyamide other than in the unsubstituted primary amine group being reacted with one isocyanate group of an organic polyisocyanate.

The remaining isocyanate groups of the polyisocyanate are blocked by a volatile blocking agent. This provides blocked isocyanate groups which are nonreactive at room temperature, but when the isocyanate groups unblock at elevated temperature, these unblocked isocyanate groups are reactive with the hydroxy-functionality of the organic hydroxy-functional material to cure the system.

The preferred polyhydric polyethers are diglycidyl ethers of a bisphenol, such as bisphenol A, having an average molecular weight in the range of 350 to 3,800, which have been reacted to consume all of the epoxy functionality. In particularly preferred practice, the epoxy resin is reacted with at least 1 mol of a bisphenol per epoxide equivalent to provide a bisphenol-terminated bisphenol-epichlorohydrin polymer having a molecular weight in the range of 600 up to 4,000.

The polyamides which are used are selected to have a plurality of terminal primary amine groups. These are conveniently provided by esterifying a polycarboxylic acid, which is preferably dicarboxylic, with 1 mol per equivalent of carboxyl of a polyamine containing a plurality of terminal primary amine groups. The polycarboxylic acid may vary considerably to include such diverse acids as succinic acid, adipic acid, phthalic acid or mellitic acid, but dimer fatty acids are preferred.

The polyamines which are used are preferably diprimary amines illustrated by alkylene diamine, like ethylene diamine. Most preferably, all of the isocyanate-reactive amino hydrogen atoms except the one primary amine group are reacted, and the reaction is a simple addition reaction which proceeds easily at moderate temperature (50° to 100°C). The reaction of amino hydrogen with isocyanate is itself well known.

Example: *Preparation of Half-Blocked Diisocyanate* – A reaction flask is charged with 2,592 g of tolylene diisocyanate (commercial mixture containing mostly the 2,4-isomer) and heated to 45°C under a nitrogen blanket. After removal of the heat source, 1,408 g of 2-ethoxyethanol are added over 2 hr. Cooling is supplied as needed to maintain 59°C±5°C. The temperature is then allowed to rise to 70°C and this temperature ±2°C is maintained for 2 hr.

Hydrophilic Styrene-Maleic Anhydride Copolymer Fibers for Use in Papermaking

Highlights of the Technological Achievement: Hydrophilic fibers may be prepared by discharging a heated and pressurized dispersion of a styrene-maleic anhydride copolymer into a zone of reduced temperature and pressure, and then modifying the fibers so produced by treatment with an aqueous admixture of selected cationic and anionic water-soluble, nitrogen-containing polymers. Blends of the hydrophilic fibers with wood pulp provide paper products having improved physical properties.

Reference: T.W. Rave; U.S. Patent 4,158,595; June 19, 1979; assigned to Hercules Incorporated.

Background: A considerable amount of effort has been expended in the development of fibrous synthetic pulps having hydrophilic properties. Much of the effort has been with polyolefins. For example, according to U.S. Patent 3,743,570 polyolefin fibers having a high surface area are treated with a hydrophilic colloidal polymeric additive composed of a cationic polymer such as melamine formaldehyde and an anionic polymer such as carboxymethylcellulose. Hydrophilic polyolefin pulps also have been prepared by dispersing a polyolefin and a hydrophilic additive, such as polyvinyl alcohol or a hydrophilic clay, in a liquid which is not a solvent for either component at its normal boiling point, heating the resulting dispersion at superatmospheric pressure to dissolve the polymer and any solvent-soluble additive, and spurting the resulting composition into a zone of reduced temperature and pressure to form a fibrous product.

Using the spurting procedure of German Patent 2,413,922, anionic pulps from mixtures of polyolefins and copolymers of olefinic compounds with maleic anhydride or acrylic or methacrylic acids have been prepared. Blends of these pulps with wood pulp have provided paper with better tensile strength than paper made with the copolymer component.

Process Conditions: In accordance with this process, it has been found that paper having further improved dry-strength properties can be prepared by forming a spurted fibrous anionic pulp from a styrene-maleic anhydride copolymer and then modifying this fibrous product by stirring a suspension of the fibers in a dilute aqueous solution or dispersion of a blend of a certain type of cationic, water-soluble, nitrogen-containing polymer and a certain type of anionic, water-soluble, nitrogen-containing polymer. The fiber modifying step of the process of this method results in deposition of the blend of cationic and anionic nitrogen-containing polymers on the spurted fibers, and the originally anionic fibers are converted into modified fibers which are capable of bonding to the cellulosic fibers of wood pulp. It is a significant feature of the process here described that the cationic nitrogen-containing polymer used in the fiber modifying step of the process is one which imparts no substantial amount of wet strength to the paper product and thus permits the reworking of broke readily to take place.

As an example of the process, a styrene-maleic anhydride copolymer is dispersed in a mixture of hexane and water, and the dispersion is heated in a closed system to a temperature of 220°C. Under these conditions, the pressure generated by the hexane and water vapors is of the order of 500 psi. After introducing nitrogen to increase the vapor pressure of the system to a pressure of about 1,000 psi, the contents of the system are vented to the atmosphere through an orifice, resulting in evaporation of the water and the hexane solvent and formation of the fiber product.

The fiber product then is suspended in an aqueous medium formed by blending a dilute aqueous solution of, for example, the reaction product of ammonia with epichlorohydrin-modified poly(diethylenetriamineadipic acid) with a dilute aqueous solution of, for example, glyoxal-modified poly(acrylamide-coacrylic acid), and the components of the resulting suspension are brought into intimate contact with each other by stirring. The treated fibers may then be isolated and stored in wet cake form, or the suspension containing the fibers may be used directly in a papermaking process.

Other Cationic Polymers: Another group of preferred cationic polymers useful in accordance with this method is that in which the polymers are poly(diallyldialkylammonium chloride)s. These are linear polymers having units of the formula:

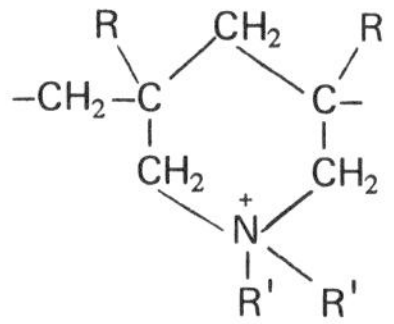

where R is hydrogen or lower alkyl and R' is alkyl or a substituted alkyl group. Polymers having units of the above formula are obtained by polymerizing quaternary ammonium chloride salt monomers in which the quaternary ammonium cation is represented by the formula:

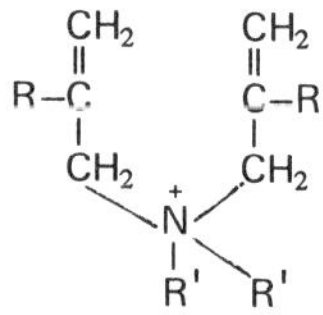

in which R and R' are as indicated above, in the presence of a free radical catalyst. In both of the above formulas, each R can be the same or different, and as stated, can be hydrogen or lower alkyl.

A final group of effective cationic polymers useful in accordance with this process is that in which the polymers are homopolymers, or certain copolymers, of acrylate and methacrylate alkyl esters containing quaternary ammonium groups. These monomers may be copolymerized with an acrylamide having the formula:

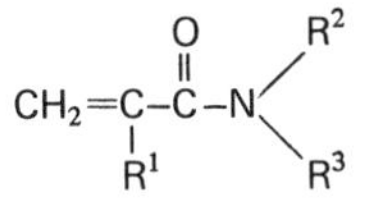

wherein R^1, R^2 and R^3 may each be hydrogen or a lower alkyl group having one to four carbon atoms. Representative compounds of the above formula are acrylamide, methacrylamide and N-isopropylacrylamide, with acrylamide being preferred. The acrylamide monomers may be used in amounts up to 75 mol % when copolymerized with the acrylate and methacrylate esters containing quaternary ammonium groups.

Hydrophilic Polyolefin Fibers for Use in Papermaking

Highlights of the Technological Achievement: Hydrophilic polyolefin fibers may be prepared, for example, by discharging a heated and pressurized solution of an anionic polyolefin composition containing carboxylic functionality into a zone of reduced temperature and pressure, and then modifying the fibers so produced by treatment with an aqueous admixture of selected cationic and anionic water-soluble, nitrogen-containing polymers.

Reference: T.W. Rave; U.S. Patent 4,154,647; May 15, 1979; assigned to Hercules Incorporated.

Background: A significant deficiency of hydrophilic polyolefin pulps has been that, when they have been blended with wood pulp, the resulting paper products have exhibited considerably less strength than that of a paper prepared from wood pulp alone. However, some improvement in the strength of paper made from blends of polyolefin pulps and wood pulp has been realized by imparting an anionic character to the polyolefin pulp.

Process Conditions: As an example of the process of this method, polypropylene and an ethylene-acrylic acid copolymer are dispersed in a solvent such as methylene chloride, and the dispersion is heated in a closed system to a temperature of 190°C to dissolve the polymer components in the solvent. Under these conditions, the pressure generated by the methylene chloride vapors is of the order of 600 psi. After introducing nitrogen to increase the vapor pressure of the system to a pressure of 1,000 psi, the resulting solution is vented to the atmosphere through an orifice, resulting in evaporation of the methylene chloride solvent and formation of the fiber product.

The fiber product then is suspended in an aqueous medium formed by blending a dilute aqueous solution of, for example, the reaction product of ammonia with epichlorohydrin-modified poly(diethylenetriamineadipic acid) with a dilute aqueous solution of, for example, glyoxal-modified poly(acrylamide-coacrylic acid), and the components of the resulting suspension are brought into intimate contact with each other by stirring. The treated fibers may then be isolated and stored in wet cake form, or the suspension containing the fibers may be used directly in a papermaking process.

The temperatures employed for carrying out the reaction between the dicarboxylic acid and the polyalkylenepolyamine may vary from 110° to 250°C or higher at atmospheric pressure. For most purposes, however, temperatures between 160° and 210°C have been found satisfactory and are preferred. When reduced pressures are employed, somewhat lower temperatures may be utilized.

In carrying out the reaction, it is preferred to use an amount of dicarboxylic acid sufficient to react substantially completely with the primary amine groups of the polyalkylene polyamine but insufficient to react with the secondary and/or tertiary amine groups to any substantial extent. This will usually require a mol ratio of polyalkylenepolyamine to dicarboxylic acid of from 0.9:1-1.2:1. However, mol ratios of from 0.8:1-1.4:1 may be used with quite satisfactory results. Mol ratios outside of these ranges are generally unsatisfactory. Thus, mol ratios below about 0.8:1 result in a gelled product or one having a pronounced tendency to gel while mol ratios above 1.4:1 result in low molecular weight polyamides.

In converting the aminopolyamide, formed as above described, to a cationic resin, it is reacted with epichlorohydrin at a temperature from about 45° to 100°C and preferably between 45° and 70°C until the viscosity of a 20% solids solution at 25°C has reached about C or higher on the Gardner scale. This reaction is preferably carried out in aqueous solution to moderate the reaction. A pH adjustment is usually not necessary. However, since the pH decreases during the polymerization phase of the reaction, it may be desirable, in some cases, to add alkali to combine with at least some of the acid formed.

Crosslinked Starch-Polyamide-Polyamine Interpolymers as Paper Strength Additives

Highlights of the Technological Achievement: Crosslinked starch-polyamide-polyamine interpolymers are prepared and described. Wet- and dry-tensile, dry-burst, and concora crush strengths of paper products, prepared from both acid and alkaline pulp furnishes, are significantly increased by the wet-end addition of the interpolymers.

Reference: G.E. Hamerstrand and M.E. Carr; U.S. Patent 4,152,199; May 1, 1979; assigned to U.S. Secretary of Agriculture.

Background: Among the more important of the strength improving chemical additives are the synthetic or starch-derived cationic polymers, oxidatively crosslinked starch xanthanes (U.S. Patent 3,160,552) and starch polyethyleniminothiourethane (U.S. Patent 3,436,305) which is made by reacting starch xanthate with polyethylenimine. There are several strength factors, which must be considered when producing paper products, that include wet- and dry-tensile and dry-burst strengths, crush resistance, tear factor, fold endurance, and pick resistance. Most of the prior art additives will improve either the wet strength or certain of these dry strength properties, sometimes at the expense of other properties. One problem that often occurs is low retention of the additives, which then end up in the white water creating a removal problem.

Process Conditions: Sodium starch xanthate is a derivative of starch having the following formula:

$$\text{starch}\left[\text{O}\overset{\overset{\displaystyle S}{\|}}{\text{C}}\text{S}^{-}\ \text{Na}^{+}\right]_{DS}$$

where degree of substitution (DS) is the number of xanthate groups per anhydroglucose units (AGU) in starch and where starch is unmodified starch. Sodium starch xanthates, having DS values of 0.01 to 3, have been prepared and are useful as starting materials for preparation of the compounds of the instant process. However, xanthate with DS values of from 0.05 to 0.5 are most readily prepared and are, therefore, preferred.

The polyamide-polyamine-epichlorohydrin resin (PAE) starting material can be any of the products described in U.S. Patent 2,926,154. The PAE used in the examples was shown by nuclear magnetic resonance (NMR) and elemental analyses to be a polymer having the following repeating unit:

$$\left[-(CH_2)_2NH\overset{\overset{\displaystyle O}{\|}}{C}(CH_2)_4\overset{\overset{\displaystyle O}{\|}}{C}NH(CH_2)_2R- \right]_n$$

where R equals the following substituents in the approximate ratios of 3:1:1:

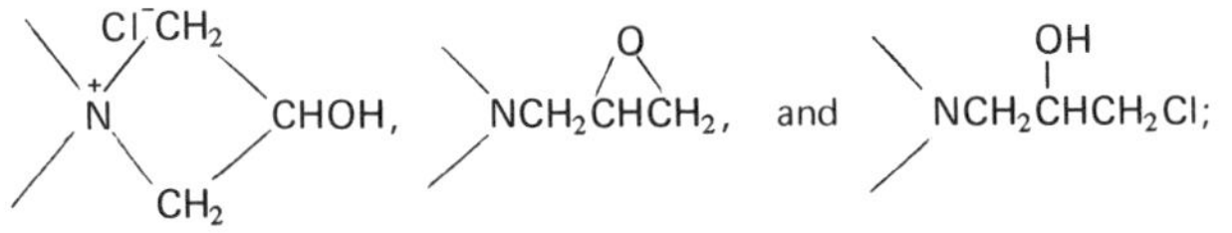

and n equals 65 to 130 based on a unit weight of 300 and a molecular weight range of 20,000 to 40,000. The chain length and the number of –R– groups in the PAE repeating unit above will vary according to the dicarboxylic acids and the polyalkylamines used in the preparation of the PAE.

Reactions between sodium starch xanthate and PAE result in two different products depending on reaction conditions. The most important of these conditions are reaction time and the ratio of PAE repeating units to starch xanthate groups which are present in the reaction mixture.

When aqueous solutions of PAE are added dropwise to aqueous solutions of sodium starch xanthate, precipitation occurs at PAE to sodium starch xanthate weight ratios of from 0.24:1-0.67:1, depending primarily on pH of the reaction mixture, concentrations of the aqueous solutions of reactants, and DS of sodium starch xanthate. Products from these reactions have a ratio of PAE repeating units to xanthate groups of 1:1-3:1. The above ratio will be known herein as PAE:Stx and is defined as the ratio of the number of PAE repeating units to the number of starch xanthate groups.

Sodium starch xanthate (0.13 DS, 0.50 wt % solution) reacted with PAE (0.01 wt % solution) at pH 5 to form a precipitate having a PAE:Stx of about 1:1. Reactions conducted in the same manner at pH 9 resulted in products having a PAE:Stx of about 1.5:1. In other words, increasing pH increases the amount of PAE per sodium starch xanthate required to form the precipitate. Decreasing sodium starch xanthate concentration has the same effect. A 0.13 DS sodium starch xanthate at a concentration of 0.01% reacted with PAE (0.01 wt % solution) at pH 5 to give a precipitate having a PAE:Stx of about 1.9:1. Precipitates were produced in good yield at pH values from 5 to 10. Reaction temperatures of 5° to 45°C had little effect on reaction efficiency.

In the above reaction the polysalts precipitate in good yields as long as the proper amounts of reactants are completely combined within less than 10 min.

Product: In actual paper machine runs, additions can be made at any wet-end position including the headbox. Papers produced by these procedures from pulp furnishes in which contact time for PAE and sodium starch xanthate were from 2 to 30 min, exhibited essentially no differences in strength properties.

When furnishes were treated with only sodium starch xanthate, almost no starch xanthate was retained in the paper. Retention in linerboard as high as 95% occurred when sodium starch xanthate was added at the 1% O.D. level to a pH 5 furnish containing 0.5% O.D. PAE. Newsprint, prepared from pH 7 pulp furnishes containing 0.5% PAE and 0.5% sodium starch xanthate, retained 98% of the latter component. Using the preferred method of addition and levels of addition of 0.125 to 0.5% O.D. PAE and 0.25 to 0.5% O.D. sodium starch xanthate, newsprint was prepared which had strength properties that were as good or better than high-test grade, commercially prepared newsprint.

POLYESTERS

PROCESSING

Phosphate Ester Coating on Inorganic Fillers for Polyester Resins

Highlights of the Technological Achievement: An improved process for dispersing inorganic fillers, such as alumina trihydrate, in a polyester resin is described wherein the improvement comprises significantly lowering the viscosity of the polyester-filler dispersion by coating the filler with 0.05 to 1.0%, based on the weight of the filler, of a polar phosphate ester surfactant containing acid groups and polar ether groups, prior to adding the filler to the polyester resin.

Reference: J.J. Koenig and H.P. Hsieh; U.S. Patent 4,183,843; January 15, 1980; assigned to Aluminum Company of America.

Background: It is well known that the presence of alumina trihydrate, as is the case with other inorganic fillers or pigments, has the adverse effect of increasing the dispersion viscosity of the fillers in plastics, particularly polyester resins. Such increased viscosity is proportional to the amount of inorganic filler in the dispersion. Although loading the mixture with more than 50% filler or pigment may be desired, such a mixture exhibits an unworkable viscosity. Therefore, prior to this process, the loading had been limited by viscosity.

Process Conditions: Inorganic filler polyester resin dispersions are used primarily in two areas, in polyester spraying applications and in the area of molding compounds, including sheet molding compounds and bulk molding compounds. Spraying, or spray-up as it is commonly called, is a simple and economic molding process for fiberglass reinforced polyester products, such as bathtubs and shower stalls. In such applications, the alumina trihydrate is used in part for flame retardancy and smoke suppression that is typically required to meet the standards of various building codes.

For polyester spraying applications, resin, filler and fiberglass are typically used in the following concentrations.

Material	Concentration
Polyester resin	50% by wt of resin and filler
Inorganic filler	50% by wt of resin and filler
Fiberglass	10-20% by wt of resin, filler and fiberglass

The spray-up polyester resin, such as Koppers B304-60, a typical resin, when used alone, i.e., without the inorganic filler, has a viscosity of about 300 cp at 20 rpm and 25°C as measured on the Brookfield RVT Model Viscometer. As the filler loading is increased, using alumina trihydrate as an example, the alumina trihydrate-polyester dispersion exhibits a significant increase in viscosity as shown in the following representative table:

Alumina Trihydrate	Hydrate (%)	Brookfield Viscosity (cp)*
None	0	300
A (8.6 μm)**	45	1,300
	50	1,845
	60	5,120
	65	15,100
B (5.1 μm)**	45	1,530
	50	2,500
	60	8,080
	65	24,750
C (3.5 μm)**	45	1,695
	50	2,505
	60	12,000
	65	64,600

*At 20 rpm, spindle No. 3 at 25°C.
**Particle size median by Sedigraph.

Note also the significant increase in dispersion viscosity as the hydrate loading increases above 50%. As dispersion viscosities approach 2,000 cp at 20 rpm, the dispersions become difficult to mix and handle. It will also be appreciated from the following table that small increases in hydrate loading, in the 50 to 65% range result in significant increases in flame retardancy.

Polyester Resins

Alumina Trihydrate (%)	Oxygen Index (% O_2)
0	18
30	22
40	25
50	28
60	38
65	53

The amount of coating material that is to be applied to the inorganic filler is from about 0.05 to 1.0%, and most preferably about 0.25%, based on the weight of the filler. The coating is applied by dry coating or wet coating. In dry coating, the filler in the form of a dry powder with a particle size median of less than 15 μ by Sedigraph, is coated with the surfactant before being mixed with the polyester resin. It has been found that the wet-coated trihydrates gave even lower dispersion viscosity than the dry-coated trihydrates. See the following representative table.

Method of Coating*	Triton QS-30 Coating Level** (%)	Brookfield Viscosity*** (cp)
–	–	1,750
Dry	0.05	1,500
Wet	0.05	650

*50% hydrate loading in Koppers B304-60 polyester resin.
**Coating level based on weight of trihydrate.
***At 20 rpm, spindle No. 3 at 25°C.

Product: The polyester compound of the process comprises a product of a dispersion of −15 μ in inorganic filler, such as alumina trihydrate, coated with 0.05 to 1.0%, based on the weight of the filler, of a polar phosphate ester surfactant containing acid groups and polar ether groups, that is dispersed in a polyester resin. Although the process comprehends the use of any polyester resin, exemplary resins include Koppers polyester resin B304-60 and Ashland polyester resin Aropol 8349 T-09. B304-60 polyester resin contains 45 to 50% monomer (styrene) and a small amount of cobalt naphthenate promoter to effect a rapid cure at room temperature. Aropol 8349 T-09 polyester resin contains about 45% monomer and some promoter.

To determine that the effective polar phosphate ester coating agents containing acid groups and polar ether groups would reduce the viscosity of dispersions of other inorganic pigments in polyester resins, other pigments were tested. Coating of calcium carbonate, titanium dioxide, clay and silica with such agents prior to adding such pigments to a polyester resin exhibited significant viscosity reductions although not as drastic as in the case of the alumina trihydrate.

Water-Soluble Polyesterimide Prepolymers

Highlights of the Technological Achievement: This patent provides essentially the same information as U.S. Patent 4,206,261 on p 347 by the same authors.

Reference: D. Laganis and P.M. Begley; U.S. Patent 4,179,423; December 18, 1979; assigned to Schenectady Chemicals, Inc.

Aqueous Coating Compositions Containing Polyester with Carboxyl Groups

Highlights of the Technological Achievement: A carboxyl-group-containing polyester having an acid number of 30 to 150, a hydroxyl number of 20 to 150 and a Patton alkyd constant of 0.9 to 1.2 which contains in cocondensed form

(a) Saturated alcohols selected from the group consisting of dihydric aliphatic, polyhydric aliphatic, dihydric cycloaliphatic, and polyhydric cycloaliphatic alcohols,

(b) Dicarboxylic acids selected from the group consisting of aliphatic, aromatic and cycloaliphatic dicarboxylic acids,

(c) Cyclic carboxylic acids selected from the group consisting of tribasic and polybasic cyclic carboxylic, characterized in that of the cocondensed compounds (c), which are only attached to the polyester through a carboxyl group, at least one other carboxyl group is condensed with a compound selected from the group consisting of aliphatic monoalcohols having from 3 to 20 carbon atoms, cycloaliphatic monoalcohols having from 3 to 20 carbon atoms and secondary amines containing from 6 to 36 carbon atoms.

Reference: H.-P. Patzschke; U.S. Patent 4,172,822; October 30, 1979; assigned to Herberts GmbH.

Background: The use of synthetic resins which can be diluted with water is of considerable importance in the lacquer industry because the resulting reduction in the organic solvent content not only complies with pollution control requirements but also lessens the danger of fire and explosions.

It has been found that resistance to aging is seriously affected by hydrolyzed fatty acids or polycarboxylic acids, such as phthalic acid, tetrahydrophthalic acid and trimellitic acid, which are attached to the chain structure in an extremely labile manner by their addition as semiesters, and which are soluble in water subsequent to hydrolysis. It has now surprisingly been found that these disadvantages can largely be obviated by using the polyesters of the present process.

Process Conditions: Since all alkyd resins contain ester groups which are sensitive to hydrolysis, it was surprising to find that a stable, high-grade stoving lacquer can be obtained from a combination of polyester modified with monoalcohols, masked polyisocynanates and/or formaldehyde condensation resins as crosslinking agents and, optionally, modifying agents containing OH-groups, providing the individual components and the quantitative ratios are carefully selected and also providing a special production process is applied. By incorporating special dicarboxylic acids, it is possible to build up a chain structure which is more stable to hydrolysis, and to coordinate the suitable acid numbers and hydroxyl numbers with the required average molecular weights, molecular weights of from 1,000 to 8,000, particularly from 3,000 to 6,000 being preferred for the polyester.

The polyesters themselves preferably contain from 0.3 to 3.0 and, with particular preference, from 0.5 to 2.5 meq of aliphatic, aromatic and/or cycloaliphatic dicarboxylic acids condensed into them per gram of resin. Of the tribasic or polybasic cyclic carboxylic acids, from 0.8 to 2.0 and, most preferably from 1.1 to 1.5 mmol/g of resin is attached to the polyester through only one carboxyl group.

Stages: The following procedure carried out in stages may be used, and it is preferably adapted to avoid any clouding and gel formation:

(1) The esterification of dicarboxylic acids, preferably aromatic and cycloaliphatic dicarboxylic acids, which are unable to form an intramolecular anhydride, with dialcohols, preferably dialcohols which contain either secondary OH-groups or primary OH-groups which are sterically hindered by β-substitution, a polyester containing OH-groups being formed through an excess of alcohol.

(2) In order to ensure solubility in the aqueous alkali medium, carboxyl groups have additionally to be introduced into the OH-

containing polyesters. This may be done by reaction at temperatures below 190°C with an aromatic or cycloaliphatic dicarboxylic acid formed from a polycarboxylic acid containing more than 2 carboxyl groups, such as trimesic acid.

The molar ratios used in the overall process are selected in such a way that a viscosity suitable for the particular application envisaged is obtained for the required acid number of from 30 to 150.

In the context of the process, etherified formaldehyde crosslinking resins are aminoplasts, phenolplasts or methacrylamide copolymers of which the methylol groups may be etherified with lower alcohols.

The solution polymerization is carried out at a solids content of 70 to 90% by weight in the presence of radical initiators. It is preferred to use lower monoalkyl glycol ethers, such as ethyl glycol or butyl glycol as solvent.

Phenolic Resins: Suitable thermally reactive phenolic resins are standard commercial grade aryl, alkyl or aryl-alkyl phenolic resins of the resol type which contain a considerable number of methylol groups of methylol ether groups.

The phenolic resins may have very low molecular weights, such as methylol phenol ether corresponding to the formula:

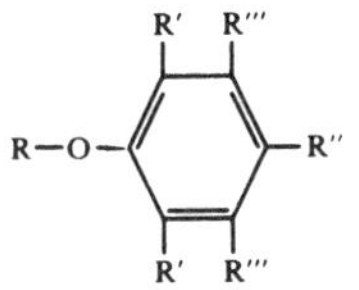

wherein R is optionally halogenated alkyl or alkenyl (preferably chlorinated); R' is H or $-CH_2OH$; R''' is H or alkyl; and R'' is $-CH_2OH$,

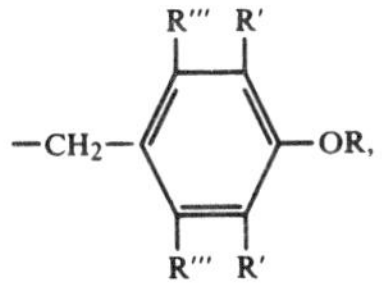

or

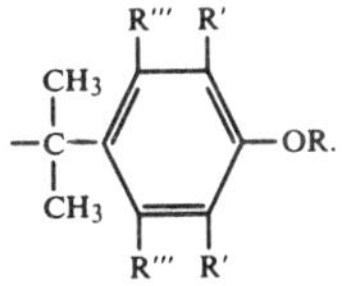

The alkyl or alkenyl groups in the above formula have from 1 to 6 and preferably from 1 to 4 carbon atoms.

High Solids Coating Compositions

Highlights of the Technological Achievement: This process is directed to a high solids coating composition and more specifically to a coating comprising low molecular weight hydroxy-containing polyesters, an epoxy resin and at least one crosslinking agent consisting of an isocyanate, an alkoxy alkyl melamine, or an amino condensation product.

Reference: W.J. Yapp and R.E. McDonald; U.S. Patent 4,169,825; October 2, 1979; assigned to The Sherwin-Williams Company.

Background: This process relates to high solids polyester-containing coating compositions and more specifically to nonvolatile coating compositions particularly useful in the architectural and appliance industry. More specifically, this process relates to high solids coating compositions and to a process for applying a polyester-containing coating on various substrates and particularly on metal substrates.

In preparing coatings, e.g., for the appliance industry, etc., it is important that the compositions remain stable at temperatures ranging up to 250°F or higher and are capable of being applied at these temperatures by conventional methods, on a variety of surfaces. It is important also that these coating compositions have a maximum cure temperature of about 400°F, are flexible, have good adhesion characteristics, have a gloss ranging from 85 to 90 and are substantially stain resistant.

Recently, the trend has been toward strict controls and legislation which prohibits the emission of solvents, e.g., hydrocarbons into the environment particularly since it is known that some hydrocarbons are photochemically reactive and, therefore, enhance the problems related to emissions. Thus to minimize the pollution problems, the coating industry has made a considerable effort toward eliminating or at least reducing the amount of emissions by resorting to water-based coatings or to the use of the coatings which are substantially free of organic solvents.

Process Conditions: It has been found in accordance with this process that a nonvolatile, high solids coating particularly useful in the appliance industry can be obtained by using a comparatively low molecular weight polyester having a hydroxyl content ranging up to 12% by wt in combination with an epoxy resin. The polyester and epoxy resin mixture is crosslinked at increased temperatures, e.g., with an amino condensation product and, if needed, blended further with a lesser amount of styrene copolymer.

More specifically, this process relates to a high solids coating, e.g., containing less than 20% by wt of a volatile solvent which comprises from 50 to 80 pbw of at least one polyester having an average molecular weight ranging up to 1,200, e.g., 200 to 1,000 and a hydroxyl content ranging preferably from 3.0 to 6.0 pbw of an epoxy resin or polyepoxide, and preferably from 20 to 30 pbw of at least one crosslinking agent selected from the class consisting of capped isocyanates, low molecular weight alkoxy methyl melamines and condensation products of an aldehyde with an aminotriazine, an aldehyde with urea, an aldehyde with guanamine and the alkylolates of these condensation products.

The high solids compositions of this process may be coated onto various substrates and particularly on metal substrates at temperatures ranging from room or ambient temperatures up to 250°F, depending on the viscosity of the composition and the amount of solvent, if any is necessary, and then subsequently cured at temperatures ranging up to 400°F. The exact temperature, however, and amount of time necessary for the coating to cure will depend not only on the particular composition, i.e., the amount and type of polyester, epoxy resin and crosslinking agent, but also on the presence of a curing catalyst.

High Solids Coating: A particularly high solids coating of this process (e.g., approximately 91% of nonvolatile material or 86% nonvolatile volume) can be reduced to application viscosity by heating the composition, e.g., up to 250°F and with the addition of a comparatively small amount of solvent or by both heating and the addition of solvents. The high temperature application alone permits the composition to be applied at solid levels above 80% nonvolatile volume. The high temperature ICI cone and plate viscometer readings indicate that the 86% NVV compositions reach application viscosity (less than one poise) at about 180°F. Successful applications of the coating at this temperature have been completed and typical ICI viscosity measurements for some compositions are indicated below.

Temperature (°F)	Viscosity (poise)
122	4.0
167	1.2
212	0.6

In preparing the alcohol blocked isocyanates, various aliphatic, cycloaliphatic or aromatic alcohols may be used as blocking agents. The blocked isocyanates are formed by reacting sufficient quantities of the blocking agent with the isocyanate under conditions to insure that there are substantially no free or unreacted isocyanate groups. This reaction is exothermic and, therefore, the polyisocyanate and the blocking agent are preferably mixed at temperatures below about 75°C and more preferably below about 50°C.

The polyesters crosslinked with the abovementioned crosslinking agents are characterized as esters derived from carboxylic acids and polyols and as having a hydroxyl content ranging from 2.0 to 12% by wt and average molecular weights not in excess of 1,200 and preferably ranging from 200 to 1,000. More specifically, the polyesters, for purposes of this process, are derived from aromatic and/or aliphatic carboxylic acids, the anhydrides thereof and various aliphatic and/or aromatic polyols.

The hydroxyalkyl acrylates, for purposes of this process, may be characterized by the formula:

$$CH_2{=}\overset{\displaystyle R_2\atop|}{C}{-}COOR_3$$

wherein R_2 is either hydrogen, a halogen, e.g., chlorine or bromine, an alkyl or substituted alkyl group, e.g., halogen-substituted alkyl group wherein the alkyl group has 1 to 2 carbon atoms and R_3 is a mono- or polyhydroxyalkyl or substituted alkyl, e.g., halogen-substituted hydroxyalkyl group having up to 12 and preferably 2 to 8 aliphatic carbon atoms.

The styrene copolymerized with the hydroxyalkyl acrylate may be characterized by the formula:

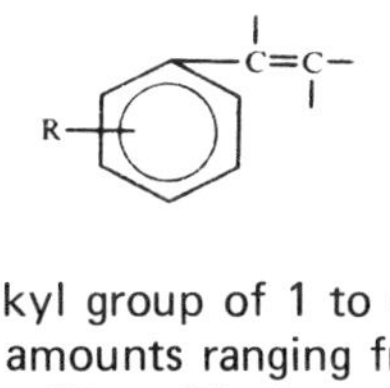

wherein R is hydrogen or an alkyl group of 1 to 4 carbon atoms, e.g., tertiary-butyl styrene and is present in amounts ranging from about 10 to 60% and preferably in amounts ranging from 25 to 30 pbw of the total reactants. The alkyl-substituted styrene may be either an ortho-, meta- or para-alkyl-substituted styrene such as para-substituted tertiary-butyl styrene.

The following example illustrates the preparation of the hydroxy-containing polyesters.

Example:

Reactants	Parts by Weight
2,2,4-Trimethyl-1,3-pentane diol	561.17
Adipic acid	132.05
Isophthalic acid	210.43
Dibutyl tin oxide	4.54
Cellosolve acetate	91.81

The above reactants were added to the reactor equipped with a fractionating column and a thermometer. The diol, adipic acid, part of the isophthalic acid and the dibutyl tin oxide were added to the reactor and the mixture was heated to about 250°F. After melting, agitation was started and the temperature was increased to 360°F. At this temperature, the remaining portion of the isophthalic acid was added to the reactor which was then held at 360°F for a period of about 2½ hours. The reaction mixture was then heated to 385°F at which point the reaction product appeared clear and acid values were taken. When the acid value fell below 10, the temperature of the reaction mixture was lowered to 250°F and the Cellosolve acetate was added to the mixture. The reaction mixture was then cooled and a polyester having an average molecular weight of 600 was obtained.

The hydroxy-containing polyester of the example, having as an average molecular weight 600, was used to prepare a coating composition having a low viscosity.

Water-Soluble Branched Polyesters

Highlights of the Technological Achievement: Water-soluble polyester having an apparent molecular weight between 600 and 5,000 and a SO_3M group content between 5 and 40 mol % based on the number of mols of dicarboxylic and/or polycarboxylic acids esterified in the polyester, and an at least trifunctional alcohol and/or an at least trifunctional carboxylic acid content between 1 and 40 mol % based on the number of mols of dicarboxylic and/or polycarboxylic acids esterified in the polyester so that the polyester is correspondingly branched and M denotes a cation of an alkali metal, the ammonium ion or the cationic radical of an organic amine. The improved polyester is particularly suited as a leveling agent in the disperse dyeing of polyester fibers or the like.

Reference: F. Engelhardt, K. Hintermeier, J. Ribka and H. Beutler; U.S. Patent 4,167,395; September 11, 1979; assigned to Cassella AG, Germany.

Background: Preferred water-soluble polyesters of the present process have an apparent molecular weight ranging from 700 to 4,000, or have from 5 to 40 mol % of the acid and alcohol that is at least trifunctional, or have a SO_3M content from about 10 to 30 mol %, or have two or three features together.

Raw Materials: Dicarboxylic acids suitable for the manufacture of the polyesters of the present process are, in the first place, all dicarboxylic acids

$$\mathrm{HO{-}\overset{\overset{\displaystyle O}{\|}}{C}{-}R^1{-}\overset{\overset{\displaystyle O}{\|}}{C}{-}OH}$$

or corresponding acid halides, anhydrides or esters thereof which supply to the final polyester the radical

$$-\overset{\overset{\displaystyle O}{\|}}{C}-R^1-\overset{\overset{\displaystyle O}{\|}}{C}-$$

wherein R^1 represents a direct bond or a divalent, aliphatic, cycloaliphatic or aromatic radical and this radical can also carry a SO_3M group. R^1 preferably denotes a divalent, aliphatic radical having 2 to 10 carbon atoms, a divalent, cycloaliphatic radical having 6 to 8 carbon atoms, or a divalent, aromatic radical having 6 to 12 carbon atoms.

The polyesters of the present process preferably contain radicals of the formula

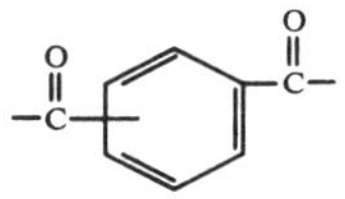

in an amount of at least 40 mol % of all the dicarboxylic acid groups in the polyester.

These acids, as indicated above can be introduced into the reaction mixture as their esters, anhydrides or acid chlorides. Particularly good results are obtained if at least 40 mol % of all the dicarboxylic acid groups have the formula

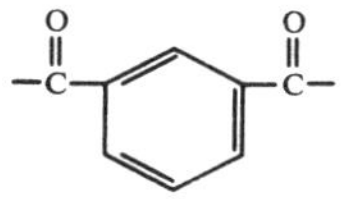

Instead of the dicarboxylic acids and diols it is also possible to use hydroxycarboxylic acids or derivatives, especially methyl or ethyl esters thereof, which can for example introduce into the final polyester the structure

$$-O-R^3-\overset{\overset{\displaystyle O}{\|}}{C}-$$

wherein R^3 represents a divalent, aliphatic, cycloaliphatic or araliphatic radical which can, in addition, also contain a SO_3M group. Particularly desired R^3 groups are divalent aliphatic radicals having 1 to 10 carbon atoms, divalent cycloaliphatic radicals having 6 to 11 carbon atoms or divalent araliphatic radicals having 8 to 12 carbon atoms. In the araliphatic radical the aliphatic part can be interrupted or separated from the aromatic part by a hetero atom such as oxygen. The following are examples of suitable hydroxycarboxylic acids: glycollic acid, lactic acid, 3-hydroxypropionic acid, 4-hydroxybutyric acid, 5-hydroxy-3-pentenoic acid, mandelic acid, 3-hydroxymethylcyclohexanecarboxylic acid, 4-hydroxymethylcyclohexanecarboxylic acid and 6-hydroxymethyldecalin-2-carboxylic acid, citric acid, malic acid, tartaric acid,

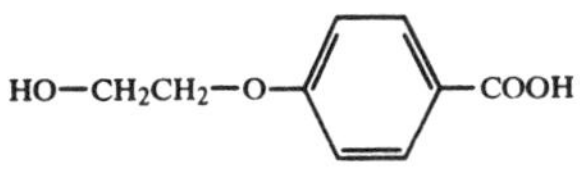

and

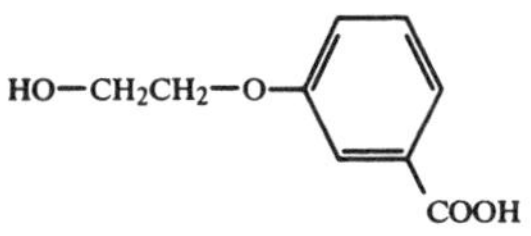

The following should be mentioned as examples of hydroxycarboxylic acid esters which carry SO_3M groups:

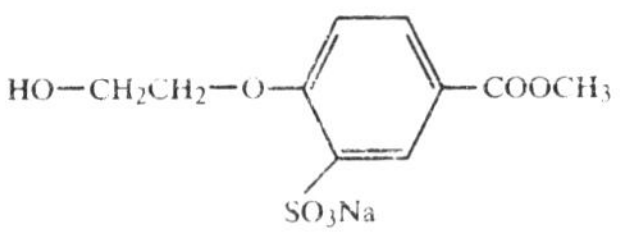

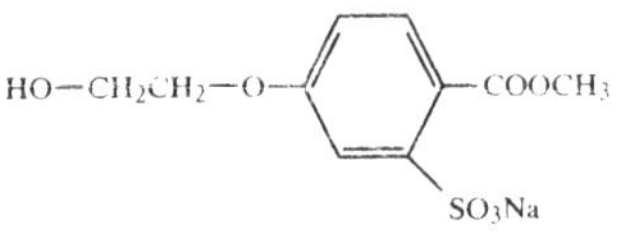

and

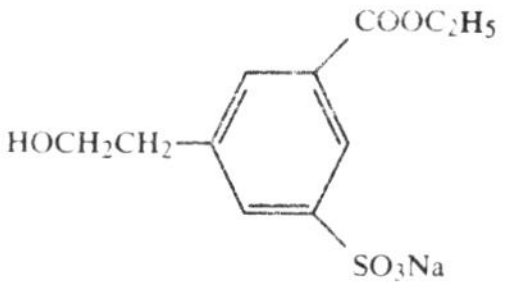

Process Conditions: The polyester of the present process is readily made by interesterifying a mixture of the acids and the alcohols generally by heating to temperatures of preferably 150° to 230°C, and best in an inert atmosphere such as under nitrogen or carbon dioxide.

In the condensation reaction, all the reactants can be subjected to the polycondensation reaction from the start, that is to say they can all be initially mixed together.

The condensation reaction can be carried out as an esterification, transesterification or acylation, or as any combination of these techniques.

If in carrying out the polycondensation reaction, at least one reactant possesses a SO_3M group, the polyesters according to the present process can be obtained directly.

In order to manufacture the polyesters of the process, 80 to 140 mol % of the alcoholic component consisting of diols and branching polyhydroxy compounds are normally employed relative to 100 mol % of the carboxylic acid component consisting of the dicarboxylic acids and branching polycarboxylic acids.

Product: The polyesters of the process are mostly produced in the manufacturing process at a pH value of 2 to 5, so that it is frequently suitable to increase the pH value to 6 to 7 by stirring in an aqueous solution of a base. For this purpose normally aqueous solutions of alkali hydroxides, especially sodium and potassium hydroxide or of aqueous ammonia are used. Normally, such base is employed as provides the cation already contained as M in the SO_3M groups.

When used as leveling auxiliaries in polyester dyeing, the water-soluble polyesters of the process exhibit a certain partial carrier action. A reversible addition reaction of the disperse dyestuff with the molecules of the auxiliary probably takes place, and this reaction is discontinued in favor of the final adsorption of the dye onto the fiber.

APPLICATIONS

Water-Soluble Insulating Varnish

Highlights of the Technological Achievement: Water-dispersible oil-modified polyesters and oil-free polyesters are formed by the reaction of aromatic dicarboxylic acids, aliphatic dicarboxylic acids and polyols to form the polyester. In oil-free polyesters, oils of the respective fatty acids are excluded but the other reactants are similar to those employed in the preparation of oil-modified polyesters.

Reference: D. Laganis; U.S. Patent 4,179,420; December 18, 1979; assigned to Schenectady Chemicals, Inc.

Background: It is generally known that maleic anhydride or trimellitic anhydride can be used to form polyesters which can become water-soluble. The maleic anhydride or trimellitic anhydride acts to provide carboxylic sites for aqueous solubilization. The polyesters produced from these reactants are rendered water-soluble by the aid of an amine, a metal oxide, hydroxide or carbonate.

The process provides new polyesters which can be formed from compositions containing considerable amounts of monocarboxylic acids; alternatively, the polyesters of the process can be formed from components with the exclusion of monocarboxylic acids altogether. Some of these polyesters are commonly known as oil-free polyesters.

The polyesters of the process may be used in improved coating systems for a variety of metal and nonmetal substrates and are characterized by superior heat and chemical resistance properties.

Process Conditions: Vegetable oils of the drying and nondrying types, or their respective fatty acids, are reacted with a combination of (1) specific aromatic dicarboxylic acids set forth hereinafter; (2) trimellitic anhydride, aliphatic dicarboxylic acids and their anhydrides and cycloaliphatic dicarboxylic acids and their anhydrides; and (3) polyols to form the alkyd or the oil-modified polyester. In the oil-free polyesters, oils or their respective fatty acids are excluded but the reactants used to form the oil-free polyesters are similar to those employed in the preparation of the alkyds.

Another aspect of the method is directed to the process for producing the oil-free and oil-modified polyesters. It has been discovered that if the aromatic dicarboxylic acid is reacted initially with the polyols (and/or diols), to form an intermediate, and subsequently the intermediate is reacted with the aliphatic (or cycloaliphatic) dicarboxylic acids or anhydrides, or trimellitic anhydride, the process to form the polyesters occurs more readily and at a lower temperature to allow better control of polymer formation.

The aromatic dibasic acids which are used include isophthalic acid, terephthalic acid, 4,4'-benzophenonedicarboxylic acid and mixtures thereof.

Oil-Free Polyesters – The oil-free polyesters of the process contain preferably 60 to 70%, based on the total mols of the dibasic acid (or anhydride thereof) used to form the polyester, of an aromatic dibasic acid of the class defined above; and preferably 20 to 40 mol % of an aliphatic dibasic acid (or anhydride thereof) or trimellitic anhydride based on the total mols of dibasic acid used to form the polyester; 20 to 55 mol % of a diol based on all reactants; and 0 to 35 mol % of a higher polyol based on total mols of reactants. From 0 to 50 mol % or more of the aliphatic dibasic acid can be replaced by trimellitic anhydride. The trimellitic anhydride, for example can be up to 20 mol % based on all reactants.

The OH:COOH ratio of the oil-free polyesters may range from 1:1 to 1.3:1.

Oil- or Fatty Acid-Modified Polyester – The oil- or fatty acid-modified polyesters are formed from

(1) An unsaturated or saturated fatty acid of considerable chain length, (e.g., having 9 to 18 carbon atoms) wherein the percent of fatty acid in the composition may range from 5 to 60 wt %, based on all reactant compositions; or 5 to 65% by wt of vegetable oil;

(2) 30 to 50 % by wt of the aromatic dicarboxylic acid;

(3) 3 to 30 % by wt of the aliphatic (or cycloaliphatic) dicarboxylic acid;

(4) 10 to 27% by wt of a polyol having at least three hydroxyl groups and optionally, 0 to 41% by wt of diol, usually 0 to 27% of diol.

The OH:COOH ratio of oil-modified polyesters in accordance with the process ranges from 1:1 to 1.5:1.

Method – In accordance with the process, the aromatic dicarboxylic acid is reacted initially with the polyols (and fatty acids if employed). This reaction step obviates the problem of the lesser solubility of the aromatic dicarboxylic acids in the reaction medium, particularly isophthalic or terephthalic acid, and their slower reactivity which results in the necessity for higher processing temperatures. In the subsequent step, the aliphatic (or cycloaliphatic) dicarboxylic acid or trimellitic anhydride is reacted with the intermediate formed in the first step, at lower temperatures than temperatures of the first step. The lower temperatures of the second step, preferably at 320° to 375°F, allow for better control of polymer formation.

As set forth above, the aliphatic or cycloaliphatic dicarboxylic acids are advantageously employed to provide easily solubilizible sites in the form of free carboxyl groups that are subsequently reacted with amines or metal hydroxides to form salts that are soluble in water.

In one embodiment of the process, the polyol of at least three hydroxyl groups and/or the diol are heated to elevated temperatures of at least 300°F and then the aromatic dicarboxylic acid is added thereto. Heating of the diol, polyol with at least three hydroxyl groups and aromatic dicarboxylic acid is continued to the point at which distillation commences. The esterification reaction is continued until an acid number of 10 to 20 is attained at 100% solids.

Thereafter, the contents are cooled to a temperature of 300°F. The aliphatic dicarboxylic acid, cycloaliphatic dicarboxylic acid or trimellitic anhydride is then added to the contents of the reaction mixture which are at a temperature of about 300°F. Heating is resumed until an acid value range of from 25 to 90 is attained on a sample diluted to 80% solids. This procedure is critical and it is the addition of the trimellitic anhydride, aliphatic (or cycloaliphatic) dibasic acid or anhydride in the second stage of the process which appears to be the crux of the method.

The advantage of the process is that it results in the possibility of producing polyesters from aromatic dicarboxylic acids which polyesters are reacted at an acid value range of 25 to 90 and which polyesters are characterized by being capable of forming an aqueous solution that is clear. Moreover, the polyesters produced in accordance with the process are characterized by package and storage stability.

The Varnish: The varnishes of the process are formed from compositions which contain 25 to 50% by wt of a polyester concentrate (of at least 30% solids content, usually about 80% solids content). The solubilizing agent, including amines, metal oxides, hydroxides or carbonates is present from 1.5 to 4% by wt. The triazine component is added in amounts of 4 to 25% by wt of the varnish composition. Water generally forms from 30 to 55% by wt of the total varnish composition.

Fibrous Article Sized with a Branched Water-Dispersible Polyester

Highlights of the Technological Achievement: A polyester size composition for synthetic fibers useful in protecting the fibers from abrasion during the weaving process or to improve the adhesion to the fibers of conventional size compositions and especially useful with polyester fibers such as those based upon polyethylene terephthalate.

Reference: R.B. Login; U.S. Patent 4,158,083; June 12, 1979; assigned to BASF Wyandotte Corporation.

Background: It is common in the textile industry to use as sizing materials easily removed high molecular weight polymers such as polyvinyl alcohol and starch but with the advent of synthetic fibers, especially polyester fibers based upon polyethylene terephthalate, yarns made of these materials are not suitably sized with these conventional sizing materials but require a synthetic sizing material having suitable adhesion to the synthetic fiber.

The polymers of the prior art can be distinguished structurally from those of this process and also on the basis that certain polymers of the prior art are hydrophobic and, therefore, would lack the water solubility required for use as a sizing material. The water-disperisble polyester sizes of the prior art contain sulfonate groups attached to the backbone of the polyester by direct substitution on an arylene or an alkylene group of the polymer backbone and, therefore, would differ structurally from those polyesters described herein in which the sulfonate group is attached to an aromatic nucleus through an aliphatic or cycloaliphatic hydrocarbon chain and ester linkage.

Process Conditions: In preparing the polyester size compositions of this process, the necessary reactants can be added to a single reaction vessel equipped with condensing equipment to remove the water evolved during the formation of the polyester. The conditions of heating are generally at a temperature of 150° to 200°C over a time of 1 to 8 hours.

The polyester size compositions of the process generally contain, based upon the sum of the theoretical mol percentages of the other acid components of the polyester, 5 to 50 mol % of at least one carboxylic acid monomer containing the $-SO_3M$ group or mixtures thereof; 95 to 50 mol % of at least one aromatic or aliphatic dicarboxylic acid or mixtures thereof and 100 mol % of at least one diol or polyglycol or mixtures thereof. The ease of dispersibility of the polyester size of the process is directly related to the amount of the carboxylic acid or ester derivative of a monomer or monomers containing the $-SO_3M$ group.

Suitable hydroxysulfonic metal salts can be prepared from the appropriate acids as described hereinafter and correspond to the general formula $HO{-}Q{-}SO_3M$, wherein Q is a bivalent aliphatic or cycloaliphatic saturated or unsaturated hydrocarbon radical having 2 to 8 carbon atoms and its valence bonds on different carbon atoms and which can contain halogen substitutents and in which M is an alkali metal. Representative examples of hydroxy aliphatic and cycloaliphatic hydrocarbon sulfonic acids are 2-hydroxyethanesulfonic acid (isethionic acid) and 2-hydroxy-1-propanesulfonic acid.

Useful metals are sodium, lithium and potassium.

The hydroxysulfonic acids of the process are preferably reacted with a polyfunctional aromatic carboxylic acid anhydride or mixtures thereof. A proportion of the hydroxysulfonic acid is preferably utilized in the reaction so as to leave the sulfonate-group-containing polyfunctional compound at least difunctional in order that the reaction product can be utilized as a monomer ester reactant in the preparation of a polyester which will function as a chain extender rather than a chain terminator as would be the case when the reaction product is monofunctional.

The diol or polyglycol used in preparing the polyester size compositions of the process can be a polyethylene glycol having the generalized formula:

$$HO(OCH_2CH_2)_nOH$$

or a polymethylene glycol having the generalized formula $HO(CH_2)_nOH$, wherein in each formula n is an integer of from 2 to 10.

Aqueous Binder Composition Comprising a Polyvinyl Acetate and a Polyester Useful for Glass Fiber Mats

Highlights of the Technological Achievement: A glass mat, the fibers of which are sized with a size comprising a polyester-compatible material, the fibers of the mat being bonded with an emulsion bonding agent comprising a polyvinyl acetate and a polyester.

Reference: S.G. Dunbar and J.L. Antle; U.S. Patent 4,156,668; May 29, 1979; assigned to Owens-Corning Fiberglas Corporation.

Background: The use of mats comprised of mineral fibers is well known. Such mats, generally known as chopped strand mats, are typically used in contact molding processes as reinforcements in the manufacture of boats, automotive parts and the like. In these uses, mats comprising mineral strands, most frequently glass strands coated with a size, are formed by bonding the strands together with a powdered polyester binder, the binder being soluble in the solvent contained in room-temperature-curable resin into which the mats are incorporated for strength purposes.

This process provides a chopped strand mat binder comprising a polyvinyl acetate-polyester emulsion, the emulsion being employed in conjunction with the glass strands sized with a polyester-compatible size. This combination of size and binder provides a glass mat having low stiffness and excellent dry comformability which characteristics provide good drape properties; that is, the mat can be easily pressed into a mold and made to conform to the mold shape.

Process Conditions: The size composition of this process will be comprised of a water-soluble epoxide in an amount within the range of 1 to 6 wt %. In the preferred embodiment, the water-soluble epoxide will be present in an amount of about 3.50 wt %.

Any suitable water-soluble epoxide can be employed. A particularly suitable water-soluble epoxide is represented by the formula:

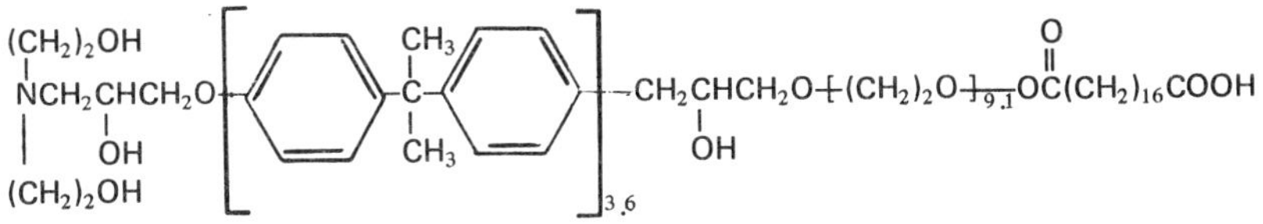

The size composition also comprises an organosilane coupling agent in an amount within the range of 0.01 to 1.00 wt %. In the preferred embodiment, the organosilane coupling agent will be present in an amount of 0.1 wt %.

The aqueous emulsion mat binder of this process will contain a polyvinyl acetate emulsion in an amount within the range of 4 to 6 wt % and preferably in an amount of about 5.2 wt %.

The emulsion mat binder of this process will contain a polyester which is the reaction product of tetraethylene glycol and adipic acid employed in equimolar quantities and which has an acid value within the range of 10 to 20. The polyester will be used in an amount within the range of 1 to 3 wt % and preferably in an amount of 2 wt %. A suitable polyester can be obtained by usual reaction techniques which involve heating the tetraethylene glycol to 200°F, adding the adipic acid and heating the mixture up to 340°F while removing distillate and, finally, heating the reaction mixture up to 440°F and maintaining the reaction mixture at a temperature for a period sufficient to produce a polyester having an acid value within the above range, a 140 viscosity within the range of 115 to 170 cp at 350°F and a refractive index within the range of 1.4735 to 1.4745.

The mat binder will also preferably contain a water-dispersible antifoaming agent. One suitable material is commercially available as Antifoam B. The antifoaming agent will be present in an amount within the range of 0.010 to 0.015 wt %. Preferably, an amount of 0.012 wt % will be employed.

Water will comprise the balance of the mat binder, being present in an amount of about 93 wt % such that the binder contains an amount of solids within the range of 3.0 to 9.0 and preferably 7.2 wt %.

Aqueous Polyester Coatings

Highlights of the Technological Achievement: Aqueous heat-curable coating compositions comprise polyester resin, aminoplast resin, and an allyl ether of a diol or triol wherein the allyl ether alcohol functions as a conventional cosolvent for aiding in dispersing the polyester resin in water and under curing conditions becomes an integral part of the cured coating.

Reference: E.T. Turpin; U.S. Patent 4,195,102; March 25, 1980; assigned to SCM Corporation. (Other Reference: U.S. Patent 4,156,667)

Background: This process substantially suppresses solvent popping in the cured film and can obviate the necessity of using volatile organic cosolvents by providing a unique blend of a polyester resin with disclosed allyl ether alcohols which not only function as conventional cosolvents but also are crosslinkable into the coating upon subseuqent curing operations. Higher nonvolatile solids content, reduced pollution, and superior performance coatings are thereby obtained.

Process Conditions: The allyl ether alcohols of this process broadly are monoallyl ethers of diols or triols and diallyl ethers of triols represented by the following general structure:

$$[CH_2{=}CH{-}CH_2O\text{-}]_m R_2 \text{-}[(O{-}CH_2{-}\underset{\displaystyle R_1}{\underset{|}{C}H}\text{-})_x{-}OH]_n \qquad (1)$$

R_1 is a hydrogen or a methyl group; and

R_2 is a divalent or trivalent hydrocarbon group and m is 1 or 2, n is 1 or 2, (n)(x) is 1 to 18, inclusive, or;

R_2 is a divalent polyoxyalkylene radical represented by the general structure

$$-CH_2{-}\underset{\displaystyle R_4}{\underset{|}{C}H}\text{-}(O{-}CH_2{-}\underset{\displaystyle R_3}{\underset{|}{C}H}\text{-})_y{-}$$

R_3 and R_4 independently are hydrogen or a methyl group, m is 1, n is 1, and x + y is 2 to 18, inclusive,

and where the molecular weight of the allyl ether alcohol ranges up to 1,200, preferably between 300 and 700.

The allyl ether alcohol represented in formula (1) can be prepared by the reaction of an alkylene oxide with allyl alcohol at temperatures of about 20° to 100°C in the presence of a Lewis acid catalyst.

Referring now to the polyester resin, such resin is the product obtained by condensing (A) polyol or mixture of polyols including hydroxyl precursors such as oxiranes or the like, and (B) polybasic acid or corresponding anhydride having at least two carboxyl groups for providing good water dispersibility of the polyester resin. The basic constraints placed on the polyester resins are that they are water-dispersible by forming water-soluble salts from their carboxyl groups and they contain sufficient hydroxyl and/or carboxyl groups for crosslinking with the aminoplast crosslinking resins for cure of the coating.

The polyester resin can be made by conventional polyesterification condensation techniques which are well known in the art. While the ratios of the starting materials can vary greatly, generally the molar ratio of (A) to (B) is preferably between 1 and 1.5. The ratio of hydroxyl to carboxyl groups in the starting materials most often is between about 1 and 1.5, but these figures can vary substantially. The acid number of the polyester resin should be preferably between 30 and 70 for providing good water-dispersibility of the resin and assisting in catalyzing the aminoplast cure of the coating.

Reduced Solvent Emission: In practicing this process, the allyl ether alcohol is blended with the polyester resin for providing the requisite viscosity of polyester resin for incorporation in water. The weight ratio of polyester resin to allyl ether alcohol broadly ranges from about 0.5 to 15. Within this range, there is substantially no volatile organic cosolvent required for reducing the polyester in water nor for providing storage or can stability of the coating.

As used herein, substantially no volatile organic cosolvent (or substantially devoid of volatile organic cosolvent) means that the coating composition contains not in excess of about 5% by wt of the cosolvent. The weight ratio of polyester resin to allyl ether alcohol in this embodiment advantageously is about 1:1 to 10:1 and preferably about 2:1 to 7:1. Reduced solvent emission during curing operations is a decided benefit of the process when the coating composition contains substantially no volatile organic cosolvent.

A measure of such solvent volatilization is known as the "relative solvent emission" (RSE), and is defined as the percentage of organic volatiles emitted from the coating during curing based on the weight of the coating resin solids which form the cured coating. The following formula shows the calculation of the RSE as will be used in this case:

$$\text{RSE} = 100\% \times \frac{\text{Aminoplast volatiles} + \text{amine agent volatiles} + \text{organic cosolvent (by wt)}}{\text{Film-forming solids of the coating (by wt)}}$$

As displayed in the foregoing formula, the organic volatiles normally found in the coating are the amine neutralizing agent used to neutralize polyester resin, organic cosolvents used in the composition, and a portion of the aminoplast resin which is displaced from the aminoplast during the curing and is volatilized thereafter. For example, when the aminoplast resin is hexamethoxymethylmelamine, it is assumed that approximately 20% by wt of the aminoplast in the form of chemically displaced methanol occurs and such loss forms part of the solvent emission for calculating the RSE of a coating. The coating composition of this process can provide RSE values as low as 7 to 8% while maintaining superior performance properties of the cured coating.

Solvent Popping: Though reduced organic cosolvent and low RSE values are decided features of the process, the specified allyl ether alcohols perform another valuable function in the coating even when conventional amounts of volatile organic cosolvent are used in the coating. Thus, when organic cosolvent is the principal ingredient employed for reducing the polyester in water, such cosolvent can cause solvent popping in the cured coating. Solvent popping is the result of heat curing the coating whereby the cosolvent is volatilized from the film to leave pops or blisters which are imperfections in the cured film detracting from appearance and effective protection for the coated substrate. Use of the allyl ether alcohols in such instances, wherein the weight ratio of polyester resin to allyl ether alcohol is less than about 9:1, substantially precludes solvent popping in the cured film even though substantial amounts of volatile organic cosolvent are contained in the coating. Desirably, then, such weight ratio should be preferably between 5:1 and 9:1 when electrostatic spraying techniques are used to apply the coating on a substrate.

POLYETHYLENE GLYCOL (PEG)

PROCESSING

Polybasic Acids Formed from Polyethylene Glycol

Highlights of the Technological Achievement: Polyamides and polyesters having greatly improved resistance to static electricity build-up are obtained by the incorporation of 0.1 to 20% by weight poly(alkylene ether) esters of polybasic acid. The antistatic properties of these polymers are significantly improved when poly(ethylene oxide) esters of dimer and trimer acids are employed.

Reference: I.S. Schlossman and R.E. Wiman; U.S. Patent 4,165,303; Aug. 21, 1979; assigned to Emery Industries, Inc.

Background: In an effort to provide permanent antistatic protection for fibers and filaments, poly(alkylene ethers) have been blended with the polymers prior to melt-spinning. Ester-terminated poly(alkylene ethers) have been incorporated in polyolefins and polyesters, also. Ethoxylated triglycerides of hydroxystearic acid have also been suggested for use in polyamides to give rapid dissipation of electrostatic charges. None of these antistatic agents have been completely acceptable, however. Either the antistats are not resistant to repeated washing and dry-cleaning, they significantly decrease the antisoiling properties and/or dyeability of the polyester or they significantly reduce other important chemical or physical properties of the polymer.

Process Conditions: The antistatic agents of this process are mixed esters of polybasic acids and poly(oxyalkylene) glycols and contain one or more radicals of the formula

$$-\overset{\displaystyle O}{\overset{\|}{C}}-\!\left(OCHR'CH_2\right)_x\!-O-$$

where R' is hydrogen, a methyl or ethyl group and x indicates the number of repeating oxyalkylene units and ranges from 4 to 450. These radicals may be located internally or in a terminal position of the ester molecule, or both. The mixed ester compounds correspond to the general formula

(1) $$A-(\overset{O}{\overset{\|}{C}}-B)_m$$

where A represents a hydrocarbon radical containing from 24 to 64 carbon atoms, m is 2, 3 or 4, and B is hydroxyl, $-\!\!(OCHR'CH_2)_xOH$ or

$$-\!\!(OCHR'CH_2)_x\!\!-O\overset{O}{\overset{\|}{C}}-A\!\!\left(\overset{O}{\overset{\|}{C}}-B_1\right)_n$$

where n is 1, 2 or 3, B_1 is a radical as defined for B and A, R' and x are the same as previously defined, with the proviso that at least one of the B groups is $-\!\!(OCHR'CH_2)_xOH$ or

$$-\!\!(OCHR'CH_2)_x\!\!-O\overset{O}{\overset{\|}{C}}-A\!\!\left(\overset{O}{\overset{\|}{C}}-B_1\right)_n$$

The hydrocarbon radical A may be branched, straight-chain or cyclic and, while essentially saturated, small amounts of ethylenic unsaturation may be present. Also, the number of valences of the hydrocarbon radical A will vary with the value of m, i.e., when m = 2, A will be divalent and when m = 3, A is trivalent. The molecular weights of the esters may vary widely depending primarily on the molecular weight of the poly(oxyalkylene) glycol employed and the extent to which the esterification reaction is driven.

In a preferred embodiment of this process the polybasic acid is a dimer or trimer acid, or mixture thereof, and the poly(alkylene glycol) is a poly(ethylene glycol) of molecular weight 200 to 10,000.

In terms of the designations already employed for formula (1), these preferred compositions would have m = 2 or 3, R' = hydrogen, x = 4 to 225, n = 1 or 2 and A would be a bivalent or trivalent radical containing from 34 to 51 carbon atoms. In the case where dimer (dibasic) acid is employed the mixed ester products will have the general formula

$$R_1O-\overset{O}{\overset{\|}{C}}\!\!\left[A_1-\overset{O}{\overset{\|}{C}}\!\!-\!\!(OCHR'CH_2)_y\!\!-O\overset{O}{\overset{\|}{C}}\right]_z\!\!A_1-\overset{O}{\overset{\|}{C}}\!\!-\!\!(OCHR'CH_2)_y\!\!-OH$$

where A_1 represents a bivalent hydrocarbon radical, preferably containing 34 carbon atoms, z is an integer from 0 to 3, R_1 is hydrogen or $-\!\!(CHR'CH_2O)_y\!\!-H$, R' is hydrogen and y is a number from 4 to 225.

Especially useful ester compositions of this process are derived from poly(ethylene glycols) of molecular weight 400 to 6,000 and dimer acids containing 36 carbon atoms, the poly(ethylene glycol) and dimer acid being reacted at a molar ratio of 1.4 to 2.0:1 employing conventional esterification procedures.

The poly(alkylene ether) esters of polybasic acids are obtained employing standard esterification procedures. The molar ratio of the ether glycol to dibasic acid may range from 2.5:1 to 1.4:1, and more preferably from 1.4 to 2.0:1, but if trimer or tetramer acids are employed the ratio may be 5:1 or higher.

Example 1: A poly(ethylene glycol) ester of dimer acid was prepared by the reaction of 960 grams commercially available poly(ethylene glycol) having a molecular weight of 1,000 with 348 grams C_{36} dibasic acid (Empol 1010). The molar ratio of these reactants was 1.6:1. The reaction was carried out at temperatures up to 200°C employing 0.2% acid catalyst until the theoretical amount of water (21.8 ml) was recovered. At this point the reaction product had an acid value of 8.8 (mg KOH/g sample).

A copolyester of dimethyl terephthalate, ethylene glycol and the poly(ethylene glycol) dimerate was then prepared employing a modification of No. 68, *Preparative Methods of Polymer Chemistry*, Sorenson and Campbell, Second Edition (1968).

The copolyester was prepared by reacting 40 grams bis(2-hydroxyethyl) terephthalate obtained by the transesterification of ethylene glycol and dimethyl terephthalate and 0.8 gram of the poly(ethylene glycol) dimerate in the presence of 0.01 gram antimony trioxide and 0.1 gram calcium acetate. The copolyester contained 2.8% bound poly(ethylene glycol) dimerate. A similar preparation was conducted to obtain 5% bound poly(ethylene glycol) dimerate.

When the poly(ethylene glycol) ester of dimer acid was melt-blended into poly(ethylene terephthalate) at a 2% level, similar improvement in the antistatic properties was observed over unmodified polyester.

Example 2: A series of poly(ethylene glycol) esters of various acids were prepared in accordance with the procedure described in Example 1. The reactants, mol ratio of reactants and acid value for the resulting product are set forth in the table which follows.

Sample No.	Poly(ethylene Glycol) (MW)	Polybasic Acid	Molar Ratio Poly(ethylene Glycol; Polybasic Acid	Acid Value
A	400	C_{36} dibasic	2:1	3.40
B	600	C_{36} dibasic	2:1	5.90
C	1,000	C_{36} dibasic	2:1	6.40
D	1,000	C_{36} dibasic	1.6:1	8.80
E	1,000	95% C_{36} dibasic-5% C_{54} tribasic*	2:1	4.0
F	1,000	87% C_{36} dibasic-13% C_{54} tribasic**	2:1	3.60
G	1,000	75% C_{36} dibasic-25% C_{54} tribasic***	1.6:1	17.4

*Empol 1014 dimer acid.
**Empol 1016 dimer acid.
***Empol 1024 dimer acid.

Hydrophobic Enhancing and Hydrophilic Enhancing Monomers

Highlights of the Technological Achievement: A water-reducible resin composition, a water-reducible admixture of such resin with solvents and plasticizers, a water reduced form of such admixture suitable for a coating, and a coating prepared from such water reduced mixture. The resin is required to be prepared from a balance of hydrophobic enhancing and hydrophilic enhancing monomers.

Generally, such hydrophobic enhancing monomers are required to be a balance of monomers promoting hard and soft segments of the resin.

Reference: G.L. Burroway and M.J. Maximovich; U.S. Patent 4,153,592; May 8, 1979; assigned to The Goodyear Tire & Rubber Company.

Background: A serious disadvantage of organic solvent solutions is their inherent potential toxicity, flammability and environmental pollution. As a result, a commercially useful water dispersion or solution is highly desirable.

Process Conditions: In accordance with this process, a resin, suitable for a water-reducible composition, having a ring and ball softening point of 100° to 300°C is prepared by the method which comprises free radical aqueous emulsion polymerizing, in an aqueous medium having a pH of 2 to 7, free radical organic solution polymerizing a monomer mixture which comprises, based on 100% by weight of monomers (A) preferably 70 to 85% by weight of at least one hard segment hydrophobic enhancing monomer selected from styrene, α-methyl styrene, acrylonitrile, vinyltoluene, methyl methacrylate, vinyl chloride and vinylidene chloride; (B) preferably 15 to 25% by weight of at least one soft segment hydrophobic enhancing monomer selected from at least one acrylate, one vinyl ether and/or one diene provided that the dienes are not copolymerized with vinyl or vinylidene chloride; and (C) preferably 3 to 10% by weight of at least one hydrophilic enhancing organic acid selected from acrylic, methacrylic, fumaric, itaconic and maleic acid.

The water-reducible composition comprises an admixture or solution of 100 pbw of resin with 50 to 100 pbw of coalescing, preferably water-soluble, solvent therefor, selected from ethylene glycol monomethyl ether, ethylene glycol monoethyl ether, ethylene glycol monobutyl ether, diethylene glycol monobutyl ether, diethylene glycol monoethyl ether acetate, diethylene glycol diethyl ether, ethylene glycol monomethyl ether acetate, methyl ethyl ketone, acetone, methyl propyl ketone and diacetone alcohol, and 5 to 70 pbw of at least one resin-compatible plasticizer having a MP of -40°-25°C, a BP of at least 95°C, and solubility parameter of 8-16.

Coating: In further accordance with this process, a coating composition is provided which is prepared by mixing and reacting sufficient amount of a volatile amine with the water-reducible mixture and mixing with water to form a stable dispersion or solution of the composition having a pH of 8 to 14, where the amine is selected from primary, secondary and tertiary amines having a melting point of -40° to 25°C and a boiling point of 50° to 150°C.

Example: Various resins suitable for preparing water-reducible compositions were prepared by the following method and labeled herein as Experiments A through G.

To a reactor was charged water, suitable emulsifiers, free radical-generating catalyst and sufficient amount of ammonium hydroxide with potassium pyrophosphate to achieve a pH of 3.1 to 6. The mixture was adjusted to a temperature of 30°C. With stirring, monomers were added to the mixture and allowed to polymerize at a temperature of 45°C for 8 hours.

The resulting latex was then coagulated and dried to form a particulate resin having a ring and ball softening point within a range of 100° to 300°C. The ring and ball softening point is of the type determinable by ASTM Method E28-58T.

Various monomer mixtures were used which are summarized in the following table. The columns A through G represent Experiments A through G with their corresponding amounts of monomers in parts by weight.

Monomers	A	B	C	D	E	F	G
(A) Hard segment							
Styrene	55	56	45	45	65	65	70
α-Methyl styrene	14	–	–	–	–	–	–
Acrylonitrile	–	–	25	–	–	–	–
Vinyl toluene	–	–	–	25	–	–	–
Methyl methacrylate	–	15	–	–	–	–	–
(B) Soft segment							
2-Ethylhexyl acrylate	25	–	–	–	–	20	–
Butyl acrylate	–	25	–	25	25	–	25
Ethyl acrylate	–	–	25	–	–	–	–
(C) Hydrophilic enhancing monomer							
Acrylic acid	–	–	–	–	10	15	–
Methacrylate acid	6	4	5	5	–	–	5

Crosslinked Hydrophilic Foams

Highlights of the Technological Achievement: The process disclosed is for improved hydrophilic crosslinked polyurethane foams prepared by reacting a blend of a monomeric polyol and polyoxyalkylene glycol (the –OH groups of the blend being capped with a polyisocyanate) with large amounts of an aqueous reactant. The resultant foams may be characterized with a broad spectrum of improved properties including hydrolytic stability and tensile strength.

Reference: L.L. Wood and K.C. Frisch; U.S. Patent 4,137,200; January 30, 1979; assigned to W.R. Grace & Co.

Background: Numerous attempts have been made in the prior art to produce hydrophilic polyurethane foams. Typically, these attempts have been based on (1) inclusion of a separate hydrophilic additive into a hydrophobic polyisocyanate either during or after foaming; or (2) use of reactants such as polyoxyethylene polyol and a polyisocyanate with low, i.e., near stoichiometric amounts of water, and a catalyst; and (3) foams based on noncatalytic reactions using linear polyoxyethylene diols, diisocyanate and varying amounts of water.

Process Conditions: It has been found, however, that improved hydrophilic crosslinked polyurethane foams may be prepared simply by reacting a particular isocyanate-capped polyoxyethylene polyol with large amounts of an aqueous reactant. The thus generated foams are typically characterized by having a crosslinked, i.e., nonlinear, molecular network which provides the product foams with a broad spectrum of improved properties.

Polyoxyethylene polyol used as a reactant in preparing the capped product to be foamed may have a weight average molecular weight of preferably between 600 and 6,000, with a hydroxyl functionality of preferably 2 to 8.

One method for preparing the prepolymer is by reacting polyoxyethylene glycol having a reactive functionality equal to 2 with a molar excess of a diisocyanate which leads to an isocyanate-capped polyurethane product (A) having an isocyanate functionality of two. A polyol such as pentaerythritol having a reactive functionality equal to 4 is reacted with a large molar excess of a diisocyanate to form an isocyanate-capped polyurethane intermediate product (B) having an isocyanate functionality of four.

By blending the two isocyanate-capped products thus prepared, i.e., products (A) and (B), in various molar proportions, the resulting product mixture has an average isocyanate functionality greater than two and on treatment with aqueous reactants will lead to improved hydrophilic crosslinked polyurethane foams of this process.

In addition, other monomeric or polymeric polyisocyanate crosslinking agents may be substituted for the tetraisocyanate product (B). Tolylene-2,4,6-triisocyanate having a reactive functionality of 3 is an example of a simple monomeric triisocyanate which may be usefully employed to achieve the same objective of imparting to the system an average isocyanate functionality greater than two.

The dramatic way in which the addition of water influences practice of this process is seen by consideration of the Water Index Value which is defined as equivalents of H_2O x 100/equivalents of NCO.

Here one must keep in mind that in polyurethane foaming reactions 1 mol of water ultimately consumes two NCO groups, i.e., 1.0 mol H_2O = 2 equivalents –OH which react with 2 equivalents of NCO. A Water Index Value of 100 indicates the equivalents of water and equivalents of isocyanate are balanced. An Index of 95 indicates that there is a 5% shortage of water equivalents while an Index of 105 indicates a 5% surplus of water equivalents. A slight shortage of water equivalents (i.e., a slight excess of isocyanate), usually 3 to 5%, is common practice in the prior art, particularly with flexible foams.

In this process the amount of water employed should exceed 6.5 mols H_2O/mol of NCO groups (H_2O Index Value of 1,300). The water employed can range up to 390 mols H_2O/mol NCO groups (H_2O Index Value 78,000). Thus, the available water content in the aqueous reactant is at least 6.5 and can fall within a range from 6.5 to 390 mols H_2O/mol of NCO groups, i.e., an H_2O Index Value of 1,300 to 78,000 and desirably from 4,000 to 40,000, i.e., from 20 to 200 mols H_2O/mol of NCO groups.

Example: A prepolymer was prepared by admixing 2 molar equivalents of polyethylene glycol having an average molecular weight of 1,000 and 1 molar equivalent of trimethylolpropane (TMOP). The admixture was dried at 100° to 110°C under a pressure of 5 to 15 torrs to remove water. The resulting dried mixture was slowly added over a period of about 1 hour to a vessel containing 6.65 molar equivalents of toluene diisocyanate (TDI) while stirring the TDI and polyol mixture. The temperature was maintained at 60°C. The mixture was maintained at 60°C with stirring for 3 additional hours. Then an additional 1.05 molar equivalents of TDI were added with stirring over a period of about 1 hour while maintaining the temperature at 60°C.

The final reaction mixture contained a 10% molar excess of TDI. All hydroxyl groups were capped with isocyanate and some chain extension occurred because of crosslinking of the polyols with TDI.

Hydrophilic foams have been prepared from the above prepolymer using large amounts of water as described previously. These foams exhibit good physical properties, and various materials as described above can be incorporated into the aqueous phase when preparing the foams.

APPLICATIONS

Melt Extrudable Cold Water-Soluble Films

Highlights of the Technological Achievement: Films, 0.5 to 10 mils thick, which are rapidly and completely soluble in cold water, and which are suitable for use as packaging film in automatic packaging equipment, can be prepared by conventional melt extrusion processes when prepared from a composition consisting essentially of 5 to 20 parts by weight of a polyethylene glycol (having an average molecular weight between 325 and 550) in 100 parts by weight of a partially hydrolyzed low molecular weight polyvinyl alcohol.

Reference: R.D. Wysong; U.S. Patent 4,206,101; June 3, 1980; assigned to E.I. DuPont de Nemours and Company.

Background: Many commonly used chemicals are produced and sold in pulverulent form, but are then slurried, dispersed, or dissolved in water when used by the consumer. Examples of these materials are pesticides (specifically insecticides, herbicides, nematicides, and fungicides, and the like) which are applied as a water spray, cleaning products (such as laundry detergents, bleaches, and caustic cleansers) which are dissolved in wash water, process chemicals (such as carbon black and activated charcoal) which can be slurried in water, and pigments and dyes which are dissolved or dispersed.

There are several problems associated with the use of these types of products. The first of these problems is exposure of the user and his immediate surroundings to the chemical. The second problem in using common pulverulent chemicals is accuracy of measurement. Use of too strong a solution of the same material may damage some desired plants as well as the undesired weeds. It is frequently extremely difficult to accurately measure materials which have become compacted and/or lumpy and to transfer powdery materials in areas exposed to wind.

Finally, after the chemical has been used, the user is faced with the problem of disposing of the package in which the chemical was delivered. It may contain residual amounts of a material which is a pollution problem, potentially hazardous to humans, harmful to plants and animals, or merely unpleasant and unsightly.

Water-soluble films and packages possess one or more of a variety of deficiencies such as the following:

(A) Many films which are characterized as water-soluble dissolve slowly or dissolve incompletely in cold water, resulting in the formation of gel-like particles.

(B) Many of the compositions utilized in the preparation of such films possess physical properties which are such that large, expensive, highly energy consuming and technologically complex process equipment for film manufacture, e.g., solvent-casting (or doctor knife or band-casting) is required.

(C) Many water-soluble films of the prior art possess physical properties which make the film unsuitable for use in packaging quantities of pulverulent materials in the range of one-half to ten pounds (e.g., yield strength, tear resistance, toughness, flexibility). Such films are also unsuitable for use in automatic packaging machinery.

(D) The constituents of many of the water-soluble packaging films are numerous, difficult to obtain, and/or expensive.

Process Conditions: In characterizing the polyethylene glycol to be used in the compositions and films of this process, an alternative means of identification is by the average degree of polymerization. That is, in characterizing a polyethylene glycol of the formula $HO(CH_2CH_2O)_nH$, n is the degree of polymerization.

The most preferred plasticizers because of their commercial availability are polyethylene glycols having an average molecular weight of 380 to 420, for example, Carbowax 400. Other polyethylene glycols believed to be equivalent to Carbowax 400 are Poly G400 and Gafanol E400.

Commercially available polyvinyl alcohols, having a viscosity of between 3 and 10 cp and a degree of hydrolysis of 85 to 90 mol percent, are suitable for this process. These include Elvanol 51-05 (viscosity 4 to 6 cp, 87.7 to 89.7 mol percent hydrolyzed), Gohsenol GL-05 (viscosity 4.8 to 5.8 cp, 87 to 89 mol percent hydrolyzed) and Gelvatol 20-30 (viscosity 4 to 6 cp, 85.5 to 88.7 mol percent hydrolyzed).

The preferred relative amounts of the two primary constituents are from 12 to 17 parts by weight of the polyethylene glycol in 100 parts by weight of the polyvinyl alcohol, with the most preferred embodiment being 15 parts by weight of the polyethylene glycol to 100 parts by weight of the polyvinyl alcohol.

In addition, small quantities of water-soluble polymers which are partially compatible with the polyvinyl alcohols of this process can be advantageously included as coplasticizers, provided the partially compatible polymer constitutes no more than 2 parts by weight based on 100 parts by weight polyvinyl alcohol. Examples of such partially compatible polymers include polyethers of the formula $-[CH_2CH_2O]_n-$ with a molecular weight of at least 600 which are sometimes referred to as polyethylene glycols and sometimes as polyethylene oxides, polypropylene glycols or polypropylene oxide/ethylene oxide copolymers of at least 400 molecular weight. Polyethylene oxides of molecular weight 500,000 to 1,000,000 are most preferred.

Clear-Rinse Agent for Mechanical Dishwashers

Highlights of the Technological Achievement: A clear-rinse agent concentrate for mechanical dishwashers containing a sheeting component, from 10 to 80% by weight of a reaction product obtained by reacting under acid catalysis conditions (1) an aldose selected from the group consisting of a reducing monosaccharide having 5 to 6 carbon atoms and an oligosaccharide thereof having from 2 to 4 monosaccharide units, with (2) a polyglycol ether having an average molecular weight of from 190 to 450, in a molar ratio of from 1:0.4 to 1:1.5, as well as the method of clear-rinsing employing the agent.

Reference: C.-P. Herold, D. Grodau and T. Altenschopfer; U.S. Patent 4,187,121; February 5, 1980; assigned to Henkel Kommanditgesellschaft auf Atkien, Germany.

Background: In mechanical dishwashing generally two cleaning cycles, usually separated by intermediate rinsing cycles with pure water are used. In the two cleaning cycles, different products are utilized. In the first or true cleaning cycle, alkaline-reacting agents are employed for the loosening and emulsifying of the food residues. In the after-rinsing or clear-rinsing bath, on the other hand, special clear-rinsing agents are employed. The latter should possess a good wetting power and be able to reduce the surface tension of the after-rinsing water to such a degree that it drains in a filmlike manner from the dishes and leaves no visible deposits, such as lime spots or other impurities, and completely clear dry dishes are obtained. This is called sheeting and the clear-rinsing agents concentrates are often stated to have a sheeting component.

Because of the violent agitation of the liquor in the dishwasher, these clear-rinsing agents have to be as low-foaming as possible. The customary anionic wetting agents, however, such as higher-molecular weight alkyl sulfates or alkyl sulfonates or aralkyl sulfonates are not generally usable because they foam too much.

In practice, therefore, mostly nonionic tensides based on ethylene-oxide adducts to fatty alcohols, alkylphenols, or polypropylene glycols of higher molecular weights are employed. These products, however, were also found in actual practice to be not sufficiently low-foaming in the concentration range required for a sufficient wetting effect.

These adducts have been found to cause disturbances due to excessive foam formation particularly in commercial dishwashing machines which have a very high rate of water circulation and a very high return rate of the clear-rinsing liquor into the main rinsing cycle. The same difficulties may also arise in home dishwashing machines.

Process Conditions: The polyglycol ethers used may be polyoxyethylene glycols with molecular weights of from 190 to 450, monopropoxylated derivatives of polyoxyethylene glycols with molecular weights up to 300 and ethoxylated and partially still monopropoxylated alkane diols having from 2 to 6 carbon atoms, alkoxyalkane diols having from 4 to 6 carbon atoms and alkane triols having 3 to 6 carbon atoms such as butane-1,3-diol, butane-1,4-diol, dipropylene glycol or glycerol, all having ethoxylation units and perhaps one propoxylation unit, up to a total molecular weight of 450.

Example: *Reaction of Monosaccharides* – 0.75 mol (based on the average molecular weight) of a polyglycol and concentrated sulfuric acid (1% by weight, based on the quantity of monosaccharide used) was introduced into a flask and heated to 100°C in an oil bath with stirring. 1 mol of a monosaccharide was then added. Any water of crystallization in the monosaccharide and water produced in the reaction was distilled off within 4 hours by application of a vacuum. The vacuum was adjusted to prevent foaming over.

The end of the reaction was found by determining the residual content of reducing sugar. The syrup reaction product was either dissolved in water, neutralized with a basic ion exchange resin, clarified with active charcoal and concentrated by evaporation under vacuum or it was neutralized while still hot by the addition of concentrated sodium carbonate or sodium hydroxide solution and bleached with 35% hydrogen peroxide solution at 80°C.

A product having an exceptionally low hydrogen peroxide content can be obtained by adding a suitable quantity of acetic acid anhydride after bleaching. The remaining H_2O_2 is converted into peracetic acid which rapidly decomposes under these conditions.

Reaction of Oligosaccharides – 1.4 mols of a polyglycol and concentrated sulfuric acid (1% by weight based on the quantity of oligosaccharide put into the process) were introduced into a flask and heated to 100°C in an oil bath with stirring. A concentrated aqueous solution of an oligosaccharide was then added.

The quantity of oligosaccharide was calculated to correspond to 1 mol of free reducing sugar. Water was then distilled off under vacuum. If there is any difficulty in dissolving the oligosaccharide in the polyglycol, the following procedure may be adopted. 1 mol of an aqueous oligosaccharide solution (calculated as reducing sugar) is introduced into a flask and heated to 110°C in an oil bath. A mixture of polyglycol (1.4 mols) and concentrated sulfuric acid (1% by weight, based on the oligosaccharide) is then added with stirring and at the same time water is distilled off. The polyglycol is added at such a rate that only a slight turbidity occurs in the flask. The reaction, which takes 7 to 8 hours, is assisted by the addition of a small quantity of the end product.

The reaction is completed when the reducing sugar is found to have virtually disappeared. The reaction product is worked up as indicated above.

The sugar derivatives prepared by this process, which were used as sheeting components according to this process, are summarized in the table below. They are identified by their hydroxyl numbers and the results of the investigations into their biological degradability, determined by the GF-Test according to W.K. Fischer [see *Fette-Seifen-Anstrichmittel* 65, p 37 (1963)]. The turbidity points of all of the sugar derivatives determined according to DIN 53 917 were above 80°C.

.... Starting Compound				Residual Content of Reducing Sugar	OH	Biological Degradation GF Test
... Sugar ...		.... Polyglycol				
Type	Mols	Type	Mols	(%)	No.	GF (%)
Glucose	1	Tetraethylene glycol	0.75	0.85	735	ND
Glucose	1	Tetraethylene glycol	0.6	0.52	759	ND
Glucose	1	Polyethylene glycol 200	0.75	0.94	772	ND
Glucose	1	Polyethylene glycol 300	0.75	0.65	615	81
Glucose	1	Polyethylene glycol 300	1.0	0.41	588	ND

(continued)

Starting Compound: Sugar Type	Sugar Mols	Polyglycol Type	Polyglycol Mols	Residual Content of Reducing Sugar (%)	OH No.	Biological Degradation GF Test GF (%)
Glucose	1	Polyethylene glycol 400	0.75	1.5	513	69-72
Glucose	1	Polyethylene glycol 200 + 1PO	0.75	3.4	651	69
Glucose	1	Polyethylene glycol	0.75	1.8	555	67
Glucose	1	Glycerol + 6EO	0.75	1.03	625	51
Glucose	1	Glycerol + 8EO	0.75	0.26	541	ND
Glucose	1	Glycerol + 6EO + 1PO	0.75	0.45	582	ND
Glucose	1	1,4-Butylene glycol + 3EO	0.750	0.75	721	68-72
Glucose	1	1,4-Butylene glycol + 4EO	0.75	0.66	663	68-69
Glucose	1	1,4-Butylene glycol + 6EO	0.75	0.98	553	73
Glucose	1	1,3-Butylene glycol + 4EO	0.75	0.62	643	74-75
Xylose	1	Polyethylene glycol 200	0.75	0.74	631	ND
Glucose-syrup DE 38	1	Polyethylene glycol 200	1.4	3.2	814	ND
Glucose-syrup DE 38	1	Polyethylene glycol 400	1.4	1.7	724	ND
Maltose	1	Polyethylene glycol 200	1.4	1.25	736	ND

Note: EO is ethylene oxide; PO is propylene oxide and ND is not determined.

Talc Compositions

Highlights of the Technological Achievement: Talc compositions with improved fragrance stability and other desirable characteristics comprising talc, fragrance and polyethylene glycol having a molecular weight of from 100 to 1,500.

Reference: V. Zeitz; U.S. Patent 4,185,086; January 22, 1980; assigned to Johnson & Johnson.

Background: High grade talcs have for many years been used in the form of powder for application to the skin. The talc in its finely divided powdered form is well suited for this purpose. Finely divided talc has thus found wide application in the cosmetic industry and as a powder for treating tender skins, such as those of infants and children, to prevent chafing or other irritation as would occur from diapers or wet clothing.

Numerous attempts have been made to improve fragrance retention in talc compositions but few of these attempts have solved the problems in a practical manner. Many of the attempts have resulted in improved fragrance stability but have led to other undesirable characteristics in the compositions.

Process Conditions: A preferred embodiment of this process comprises a talc composition consisting of talc, from 0.1 to 1.0% of polyethylene glycol and from 0.01 to 1.0% of perfume. The polyethylene glycol which is useful in this process should have a molecular weight of preferably 200 to 400, and should be in a fluid form for ease of handling. If a polyethylene glycol of molecular weight greater than 1,500 is utilized, it would be in solid form and could lead to processing problems.

The polyethylene glycol is utilized in 0.1 to 1.0% by weight of the total composition, preferably from 0.2 to 0.3%. If less than 0.1% by weight of the total composition is utilized, stability problems occur; and if greater than 1% is utilized, stickiness, caking and dispersion problems may occur.

Example: A talc composition is prepared as follows. 99 g of fragrance are premixed with 112.5 g of polyethylene glycol having a molecular weight of 400 in a suitable beaker and stirred until the solution turns clear. To a twin shell blender fitted with an intensifier bar to ease processing is charged 44.7885 kg of talc and the blender motor is turned on for 2 minutes at which point the intensifier drive motor is turned on for 2 minutes.

The polyethylene glycol/fragrance solution is then slowly added over a period of 4 to 5 minutes and mixing is continued for an additional 8 to 10 minutes. The resulting perfumed talc composition has the following composition in percent by weight: 99.53% talc; 0.22% perfume; and 0.25% polyethylene glycol (molecular weight 400).

Drilling Fluids Containing an Admixture of Polyethoxylated, Sulfurized Fatty Acids and Polyalkylene Glycols

Highlights of the Technological Achievement: Aqueous drilling fluids containing an admixture of a polyethoxylated sulfurized fatty acid and polyalkylene glycol. Such fluids are especially useful where reduced torque drilling fluids are needed. Another embodiment of this process relates to a method of drilling utilizing the abovedescribed fluids.

Reference: T.O. Walker; U.S. Patent 4,172,800; October 30, 1979; assigned to Texaco Inc.

Background: In a drilling operation, such as in a rotary drilling operation, a drilling fluid is forced down the drill string, about the drill bit at the bottom of the borehole and then back up to the surface. The drilling fluid employed in such a drilling operation usually is an aqueous drilling fluid and is compounded of various materials in order to impart certain desirable physical and chemical properties to the drilling fluid. For example, there is usually incorporated in an aqueous drilling fluid a hydratable clay material, such as a bentonite clay, to impart desirable viscosity and gel strength properties to the drilling fluid so as to better enable the drilling fluid to carry away the drilling cuttings from the bottom of the borehole.

Other materials such as weighting agents, e.g., barium sulfate, are employed to increase the density of the drilling fluid so as to make the drilling operation more effective and safer by overcoming the fluid pressure within the formation being drilled. Other materials such as water loss improving agents, e.g., carboxymethylcellulose, hydrolyzed starch, etc., are added to reduce the loss of fluid from the drilling fluid into the formation during the drilling operation. Still other materials such as corrosion inhibitors, bactericides and drill bit lubricants are incorporated in the drilling fluid in order to improve the drilling operation and the drilling fluid.

In accordance with this process, an improved aqueous drilling fluid is provided by incorporating therein a minor amount of a water-soluble or water-dispersible admixture of a polyethoxylated sulfurized fatty acid and a polyalkylene glycol as hereinafter defined.

Polyethylene Glycol Component: A water-soluble polyalkylene glycol having an average molecular weight of preferably between 400 and 1,000, preferably a polyethylene glycol.

A drilling fluid of this process may be prepared by the addition of a water-soluble polyethoxylated sulfurized fatty acid and a polyalkylene glycol described above to an aqueous drilling fluid or to an aqueous drilling fluid containing a clay dispersed therein.

The polyalkylene glycol component of the admixture is used in a weight ratio of 2 up to 4 parts per part of the sulfurized polyethoxylated fatty acid component. In general it has been found necessary to maintain the glycol-fatty acid ratio at 2 or more parts by weight of the glycol per part by weight of the fatty acid component so as to maintain the rheology of the drilling fluid within acceptable limits.

Drilling Fluids: An aqueous drilling fluid, the Base Mud in the following tables, was prepared using a top hole mud from the West Cote Blanche Bay Field, St. Mary Parish, La., which was treated with 6 pounds per barrel of the ferrochrome lignosulfonate dispersant Q-Broxin, its pH raised to 9.5 with caustic, and aged overnight at 150°F before use.

There was added thereto varying amounts of the admixture of the water-soluble polyethoxylated sulfurized fatty acid and the polyalkylene glycol. The physical properties of the resultant drilling fluids in terms of lubricity and drilling fluid rheology were measured using a lubricity tester.

In the following tables P.V. is plastic viscosity and Y.P. is yield point.

Table 1

Ex. No.	Conc. (lb/bbl)	Additives*	S (%)	Reduction in Torque	P.V.	Y.P.	pH
**	–	–	–	–	19.5	0	9.6
1	0.5	A	–	0	–	–	–
2	1.0	A	–	9.9	–	–	–
3	0.5	A	2	17.3	–	–	–
4	1.0	A	2	25.0	22.5	26	9.6
5	0.5	A	4	26.2	–	–	–
6	1.0	A	4	30.0	23.5	27	9.7
7	0.5	A	6	13.5	–	–	–
8	1.0	B	6	17.9	–	–	–
9	0.5	B	–	0	–	–	–
10	1.0	B	–	0	–	–	–
11	0.5	B	2	0	–	–	–
12	1.0	B	2	0	–	–	–
13	0.5	B	4	8.0	–	–	–
14	1.0	B	4	19.0	19.5	32	9.6
15	0.5	B	6	3.0	–	–	–
16	1.0	B	6	9.1	–	–	–

*A is Emerest 2646 polyethoxylated oleic acid with 10 ethylene oxide groups; and B is Emerest 2660 polyethoxylated oleic acid with 14 ethylene oxide groups.
**Base Mud

The data presented in Table 1 show that ethoxylated fatty acids alone (Examples 1, 2, 9 and 10) are ineffective as lubricity additives or (Example 2) less than 10% effective, as measured by the percent reduction in torque.

The data obtained with sulfurized ethoxylated fatty acids (Examples 4 through 8) indicate some improvement in lubricity but concomitantly the rheology of the drilling fluids have been substantially impaired. These data indicate the sulfurized ethoxylated fatty acid materials are likewise unsatisfactory as lubricity additives.

The data in Table 1 also show the criticality of the ethylene oxide content of the fatty acid, both unsulfurized and sulfurized (Examples 9 through 16) using the 14 mol ethylene oxide oleic acid material Emerest 2660, resulted in little or no beneficial effects in terms of lubricity or satisfactory rheology properties, except for Example 14.

Table 2

Ex. No.	Conc. (lb/bbl)	Additive*	S (%)	. . . PEG. . . (lb/bbl)	MW	Reduction in Torque	P.V.	Y.P.	pH
**	–	A	–	–	–	–	10.5	1	9.5
17	0.5	A	2	–	–	–	12.5	20	9.6
18	0.5	A	2	1	400	19.5	13.5	13	9.5
19	0.5	A	2	1	600	–	4.5	11	9.5
20	0.5	A	2	1	1,000	–	13	10	9.6
21	1.0	A	2	–	–	–	16	23	9.5
22	1.0	A	2	2	400	22.5	18.5	15.5	9.5
23	1.0	A	2	2	600	–	15.5	11.5	9.6
24	1.0	A	2	2	1,000	–	17	11.5	9.6
25	0.5	A	4	–	–	–	14	20	9.6
26	0.5	A	4	1	400	23.9	15	14	9.8
27	0.5	A	4	1	600	–	15.5	14.5	9.5
28	0.5	A	4	1	1,000	–	15	13.5	9.6
29	1.0	A	4	–	–	–	16	23	9.6
30	1.0	A	4	2	400	31.5	16	22	9.6
31	1.0	A	4	2	600	–	14	9.6	9.6

*Emerest 2646 polyethoxylated oleic acid with 10 ethylene oxide groups.
**Base Mud.

The data in Table 2 show that satisfactory lubricity values and drilling fluid rheology are obtained by the additive admixtures of this process. In all instances, except for Examples 17, 21, 25 and 29, the drilling fluid rheology was within acceptable limits in terms of Plastic Viscosity and Yield Point values and the lubricity properties obtained were good, ranging from 19 to 31.5% reduction in torque (Examples 18, 22, 26 and 30).

Polyglycolamine Corrosion Inhibitors

Highlights of the Technological Achievement: Polyglycolamines such as aminated polypropylene or polybutylene glycols and amine-initiated polypropylene and polybutylene glycols such as oxypropylated or oxybutylated lower alkanolamines inhibit the corrosion of ferrous metals by a nonoxidizing acid such as hydrochloric acid. Polyglycolamines of this class having an average molecular weight in the appropriate range of 300 to 5,000 are effective inhibitors in 5 to 15% hydrochloric acid at moderately elevated temperatures.

Reference: C.W. Martin; U.S. Patent 4,171,279; October 16, 1979; assigned to The Dow Chemical Company.

Background: It is known that higher alkylamines are acid corrosion inhibitors. It is also known that triethanolamine is a moderately good inhibitor in some acid systems. These compounds, however, are fully effective only under somewhat limited conditions and are particularly prone to lose their corrosion inhibiting properties as the temperature or the acid concentration increases.

Process Conditions: It has been found that certain poly(lower)alkyleneglycolamines are highly effective to inhibit the corrosion of ferrous metals by hydrochloric acid and other nonoxidizing acids such as dilute sulfuric acid even when the acid is in moderately high concentration and the temperature is moderately elevated.

The class of polyglycolamines is defined as those having an average molecular weight of 300 to 5,000 wherein the alkylene moieties are at least half, and preferably all or substantially all, of 3 to 4 carbon atoms. This class includes polypropylene glycolamines, polybutylene glycolamines, and oxyalkylated amines of several kinds, particularly the oxyalkylated lower alkanolamines which are the condensation products of an alkanolamine with propylene oxide or butylene oxide or mixture thereof, and the physical and chemical mixtures possible under the above definition.

The polyglycolamines of this process fall into two subclasses; oxyalkylated amines, particularly oxyalkylated alkanolamines, and the aminated polyglycols. In the first group which comprises oxypropylated or oxybutylated lower alkanolamines, particularly preferred compounds are those obtained by condensing butylene oxide with a trialkanolamine such as triethanolamine. These products are triols having a tertiary amine nitrogen atom in their structure.

The second group, the aminated polypropylene and polybutylene glycols, are primarily polyoxyalkylene diamines mixed with a smaller amount of the corresponding monoamine wherein only one hydroxyl group of the polyglycol molecule has been replaced by an amino group. These mixtures can be prepared by an amination method, for example, by reacting a polyglycol disulfonate or a corresponding monosulfonated polyglycol chloride with ammonia as shown in U.S. Patent 3,236,895. This group of polyglycolamines overlaps the first-described group in that the monoamines can be prepared by either method, i.e., by oxyalkylating a monoalkanolamine or by the amination reaction such as that of the cited patent.

The polyglycolamines of this process thus have certain unique advantages as corrosion inhibitors in addition to the low concentrations required for effective inhibition. Because of their molecular structure and the way in which they are made, these polyglycolamines are readily tailored to fit a particular use by adjusting the polyglycol chain length or the identity or proportion of the oxyalkylene units which form the chain. In this way, corrosion inhibition efficiency and solubility characteristics can be optimized for the conditions of acid concentration, temperature and type or types of ferrous metal encountered in a particular situation. Solubility characteristics of these compounds are particularly advantageous in that solubility generally appears to decrease with increasing temperature, causing the amine to cling to a metal surface more tightly rather than being more rapidly leached off.

Corrosion Test Procedure: SA 1010 carbon steel coupons with rounded edges and a surface area of 6.44 in^2 were used to determine inhibition of corrosion. The test coupons were cleaned by rinsing twice in 1,1,1-trichloroethane and once in acetone, then immersing in 18 to 20% HCl at room temperature for 15 minutes, rinsing with water, scrubbing with pumice soap and a brush, and finally rinsing with hot water followed by an acetone rinse. The cleaned and dried coupons were stored in a desiccator for use.

In the corrosion tests, weighted coupons were put in 150 g of aqueous HCl containing the desired amount of amine and control coupons were put in 150 g of uninhibited HCl of the same concentration, then both containers were allowed to stand at the test temperature for the desired length of time. Both sets of coupons were then removed from the acid solutions and cleaned by rinsing with water, scrubbing with pumice soap, and rinsing once with hot water and twice with acetone.

The cleaned and dried coupons were weighed to determine weight losses. The average weight losses for the two sets of coupons were used to compute the percent inhibition for the test polyglycolamine by the following formula:

$$\%\ \text{Inhibition} = \frac{\text{wt loss uninhibited} - \text{wt loss inhibited}}{\text{wt loss uninhibited}} \times 100$$

Identification of Test Amines: The following abbreviations were used to identify the amines used in the corrosion inhibition tests:

Abbreviation	Identification
MEA	Monoethanolamine
DEA	Diethanolamine
TEA	Triethanolamine
TE-3	Triethanolamine + 3 mols ethylene oxide
TE-6	Triethanolamine + 6 mols ethylene oxide
TE-10	Triethanolamine + 10 mols ethylene oxide
MSBA	Monosec-butanolamine
DSBA	Disec-butanolamine
TSBA	Trisec-butanolamine
TB-3	Triethanolamine + 3 mols butylene oxide
TB-6	Triethanolamine + 6 mols butylene oxide
TB-10	Triethanolamine + 10 mols butylene oxide
TB-16	Triethanolamine + 16 mols butylene oxide
TBM	Equal mixture of TB-3, 6 and 10
TP-3	Triethanolamine + 3 mols propylene oxide
TP-6	Triethanolamine + 6 mols propylene oxide
TP-10	Triethanolamine + 10 mols propylene oxide
AP-400	Aminated polypropylene glycol of 400 mol weight
AP-2000	Aminated polypropylene glycol of 2000 mol weight
APG-3000	Aminated polypropylene glycol of 3000 mol weight initiated with glycerol

Example 1: This corrosion test was run in 5% HCl at room temperature for 4 days according to the procedure described above. A number of alkanolamines and oxyethylated triethanolamines were also tested in order to show the superior inhibiting properties of oxypropylated and oxybutylated triethanolamines of the process. In each case, the concentration of amine was 0.2% by weight.

Amine	Inhibition (%)
TP-3	94.85
TP-6	94.91
TP-10	92.43
TB-3	94.55
TB-6	95.32
TB-10	96.79
MEA	11.21
DEA	21.77
TEA	93.13
MSBA	51.33
DSBA	35.59
TSBA	58.71
TE-3	94.74
TE-6	91.12
TE-10	92.74

Example 2: Using the same general procedure, corrosion tests were run at 175°F for 5 hours in 5% HCl. A range of amine concentrations was covered and amines TE-3 and TE-10 were run for purpose of comparison.

Amine	Concentration (%)	Inhibition (%)
TE-3	0.1	85.91
TE-10	0.1	85.68
TP-3	0.1	93.52
	0.0677	80.88
	0.0333	76.91
TP-10	0.1	90.48
	0.0677	86.48
TB-3	0.1	94.07
	0.0677	92.50
	0.0333	81.45
TB-6	0.1	95.64
	0.0677	93.53
	0.0333	91.72
TB-10	0.1	97.37
	0.0677	96.90
	0.0333	96.63
	0.0200	95.54
TB-16	0.0333	96.97
	0.0200	95.71
TBM	0.0677	94.27
	0.0333	95.86

Example 3: Several polypropylene glycolamines of this process were tested by the procedure described in Example 2 as follows.

Amine	Concentration (%)	Inhibition (%)
AP-400	0.100	95.53
	0.0333	93.66
AP-2000	0.100	96.39
	0.0333	95.06
APG-3000	0.100	96.48
	0.0333	95.63

Example 4: The oxybutylated triethanolamines and polypropylene glycolamines of Examples 3 and 4 were tested by the general corrosion inhibition test procedure outlined above but using stronger acid, 5 hours in 15% HCl at 150°F.

Amine	Concentration (%)	Inhibition (%)
TB-3	0.100	55.19
TB-6	0.100	71.68
TB-10	0.100	84.83
	0.050	79.63
	0.020	71.52
TB-16	0.100	96.22
	0.050	96.49
	0.020	96.46
	0.010	93.94
	0.005	92.55
AP-400	0.100	77.81
AP-2000	0.100	96.38
APG-3000	0.100	97.28
Tallowamine	0.100	29.20

Detergent Compositions Containing Starch

Highlights of the Technological Achievement: Laundry detergent compositions comprising an organic surface-active agent and low concentrations of substantially water-insoluble starch, the starch being added to the compositions as an intimate comixture with polyethylene glycol having a molecular weight of from 400 to 50,000, are disclosed. These compositions impart antiwrinkling and ease-of-ironing benefits to fabrics laundered with them.

Reference: J.A. Sagel and E.J. Wandstrat; U.S. Patent 4,169,064; September 25, 1979; assigned to The Procter & Gamble Company.

Background: This process relates to laundry detergent compositions of the type disclosed in U.S. Patent 3,892,681, Edwards and Diehl, issued July 1, 1975. The dry starches disclosed in that patent are, in general, quite dusty and difficult to handle, posing a potential dust explosion hazard, in the processing of the detergent compositions.

The detergent compositions described in U.S. Patent 4,116,854, Temple et al, issued Sept. 26, 1978, utilizes water-soluble starch which contains from 0.01 to 5% by weight of a substantially water-insoluble organic liquid as a means of minimizing these dust/handling problems, while simultaneously providing the fabric care benefits described in U.S. Patent 3,892,681. These compositions, however, may leave laundered fabrics with difficult to remove creases, if those fabrics are not removed from the laundry dryer promptly at the conclusion of the heated drying cycle. By utilizing the compositions described herein, it is possible to minimize both the dustiness of the compositions described in U.S. Patent 3,892,681 and the creasing properties of the compositions described in U.S. Patent 4,116,854, while also providing laundered fabrics with antiwrinkling and ease-of-ironing benefits.

Process Conditions: A particularly preferred polyethylene glycol component is PEG 6000 (Union Carbide Corporation). The starch and polyethylene glycol components are included in the comixture in a ratio of starch to polyethylene glycol of preferably from 3:1-1:5, especially 2:3. The preferred starch for use in the compositions of this process is cornstarch.

The starch/polyethylene glycol comixture may be incorporated into the detergent compositions in ways which are conventionally known and used in the detergency arts. For example, the starch and polyethylene glycol components may be mixed together and either sprayed onto or agglomerated with the remainder of the detergent composition, in granular or particulate form.

The polyethylene glycol component may also be sprayed onto the starch, with the resulting product being admixed with the remainder of the detergent composition. In these processes it is preferred that the polyethylene glycol used have a relatively low molecular weight (i.e., from 400 to 3,000) in order to give the mixture sufficient fluidity to facilitate the spraying process.

The starch/polyethylene glycol mixture may also be incorporated into the detergent compositions by premixing molten polyethylene glycol with the starch and forming either a prill or a coflake from the mixture, using conventionally known techniques, and admixing these prills or coflakes with the remainder of the detergent composition. If the prills or coflakes are admixed with the granules immediately following the spray-drying operation, the granules are relatively hot (170° to 200°F), and the prills or coflakes fuse to the granule surfaces.

Example: Three spray-dried detergent compositions were prepared and their antiwrinkling performance tested in the following manner. Composition A was prepared by combining and spray-drying the ingredients given below. Composition B, having basically the same formula as that given below except that 0.6% cornstarch replaced an equal amount of sodium sulfate, was prepared by combining the ingredients, except for the cornstarch, and spray-drying them to form a granule which was then admixed with the dry cornstarch. The handling, preparation and admixing of this composition resulted in the presence of cornstarch dust in the air.

Composition C, having the same formulation as Composition B, was prepared by forming a prill of the cornstarch and polyethylene glycol components, and admixing this prill with a spray-dried detergent granule formed with the remaining components.

Component	Percent by Weight
Sodium C_{13} linear alkylbenzene sulfonate	16.8
Sodium tripolyphosphate	32.8
Sodium sulfate	33.5
Sodium silicate solids (2.0 ratio)	6.0

(continued)

Component	Percent by Weight
Polyethylene glycol 6000	0.9
Condensation product of coconut alkyl alcohol with 6 mols of ethylene oxide	0.5
Moisture and minors (including perfume and brightener)	Balance to 100

Substantially equivalent results are also obtained where the polyethylene glycol used in the above compositions is replaced by an equivalent amount of polyethylene glycol having a molecular weight of 1,000, polyethylene glycol having an average molecular weight of 3,500, polyethylene glycol having a molecular weight of 7,500, or polyethylene glycol having a molecular weight of 9,000.

Excellent performance is also obtained when the sodium tripolyphosphate builder used in the compositions above, is replaced with an equivalent amount of sodium nitrilotriacetate, sodium melitate, sodium citrate, sodium carbonate, or mixtures thereof.

Stable Polymer Gels

Highlights of the Technological Achievement: This process is directed to gels which are stable in storage, comprising a polymer matrix, a liquid dispersing agent embedded therein and optionally fillers or additives, characterized in that the polymer matrix is a polycondensation product of aldehydes and aminoplast-formers, at least 30% by weight of the aminoplast-formers having a molecular weight of between 200 and 10,000 and containing at least one ether, thioether, acetal, ester, urethane and/or carbonate group.

Reference: D. Schäpel, K. Wagner, H.P. Müller and M. Dahm; U.S. Patent 4,160,754; July 10, 1979; assigned to Bayer AG, Germany.

Background: In substances which are termed gels within the meanings of this process, there are three main components: the high molecular weight matrix which forms the skeleton for the gel in the form of a polymer network; the liquid (hereinafter referred to as dispersing agent) embedded in the matrix; and fillers or additives dissolved and/or dispersed and/or suspended in the dispersing agent.

Gels of the type described above can be prepared from various starting materials. They may be divided approximately into natural gels and synthetic gels. The natural gels are generally based on proteins or carbohydrates. Typical examples include the gels based on alginates, gelatin, pectin, agar-agar, karaya gums and tragacanth gums, starch and carob bean flour. Carboxymethyl cellulose and the polyurethane polyureas described in German Offenlegungsschrift 2,347,299, which can be converted to foam gels as has been described in German Offenlegungsschrift 2,521,277, are examples of synthetic gel-forming substances.

For natural gels, the best dispersing agent, i.e., the liquid immobilized in the gel, is invariably water. Some of the synthetic gel-formers, particularly the polyurethane polyureas, can be so widely varied in their structure that both polar liquids such as water, alcohols or glycols and nonpolar liquids such as petroleum hydrocarbons or benzene can be embedded in them.

It is very difficult to embed in such a gel any substances which react with isocyanate groups. This applies particularly to the hydrogels which are commercially very interesting.

For many potential applications of the gels, it is advantageous if the solutions or dispersions of the starting materials from which the gels are prepared have a low viscosity. This is particularly advantageous if the gels are mixed with solid substances such as sand, animal waste or vegetable waste. In that case, loose piles of such solid material can be solidified as gel simply by pouring the gel-forming substance over them. This is of particular interest for solidifying sand or other loose terrain.

For preparing gels, it is therefore desirable to use starting materials which, when dissolved or dispersed in the dispersing agents, give rise to low viscosity solutions or dispersions which are highly stable in storage and from which temperature-stable gels can be prepared but which are still highly adaptable in their high molecular weight matrix to the selected dispersing agent, as in the case with polyurethane polyureas.

Process Conditions: Preferred modified ureas correspond to the following formula

$$(1) \qquad R(O{-}CO{-}NH{-}R'{-}NH{-}CO{-}NH_2)_n$$

in which R represents a residue of valency n which contains at least one ether, thioether, acetal, ester, ester amide and/or carbonate group and has a molecular weight of between 44 and 10,000, R' represents the residue of an organic diisocyanate, and n represents an integer of from 2 to 8, preferably 2 or 3, and most preferably 2.

In spite of their small number of functional groups, these modified α,ω-bisureas are comparable in their condensation velocities in an aqueous medium to known aminoplasts. They can be converted into the N-methylol compounds in the same way as urea.

Suitable low molecular weight diols include, for example, diethylene glycol, triethylene glycol, tetraethylene glycol, dipropylene glycol, tripropylene glycol, dibutylene glycol, 1,4-phenylene-bis(β-hydroxy ethyl ether), ester diols represented by the following general formulas:

$$HO{-}(CH_2)_x{-}CO{-}O{-}(CH_2)_y{-}OH \text{ and}$$

$$HO{-}(CH_2)_x{-}O{-}CO{-}R{-}CO{-}O{-}(CH_2)_x{-}OH$$

where R represents an alkylene or arylene group having 1 to 10, preferably 2 to 6, carbon atoms, x = 2 to 6, and y = 3 to 5. Examples include δ-hydroxybutyl-ϵ-hydroxy caproic acid ester; ω-hydroxyhexyl-γ-hydroxy butyric acid ester; adipic acid-bis(β-hydroxyethyl)-ester and terephthalic acid-bis(β-hydroxyethyl)-ester. Also suitable are diol urethanes represented by the following general formula:

$$HO{-}(CH_2)_x{-}O{-}CO{-}NH{-}R'{-}NH{-}CO{-}O{-}(CH_2)_x{-}OH$$

where R' represents an alkylene, cycloalkylene or arylene group having from 2 to 15, and preferably 2 to 6, carbon atoms, and x represents a number between 2 and 6. Examples include, e.g., 1,6-hexamethylene-bis(β-hydroxyethyl urethane) or 4,4'-diphenylmethane-bis(δ-hydroxybutyl urethane).

Another preferred class of aminoplast-formers are compounds with terminal carbamate groups which have urethane, ether, thioether, acetal and/or ester groups in addition to the urethane end groups. Particularly preferred carbamates correspond to the following formula:

$$R-(O-\overset{\overset{\displaystyle O}{\|}}{C}-NH_2)_n \qquad (2)$$

in which R and n have the same meaning as in formula (1) above.

Another special variation of the process for preparing the gels of this process comprises reacting the aminoplast-formers described above, or conventional aminoplast-formers such as urea or melamine, with formaldehyde and one of the polyols described above, preferably a polyether polyol. This reaction gives rise to polymers which have recurrent O-N-acetal structural units, preferably those represented by the following formula:

$$-\underset{\underset{\displaystyle O}{\|}}{C}-NH-CH_2-O-\underbrace{\sim\sim\sim\sim\sim\sim}_{\text{Polyether residue}}$$

Product: The gels obtained can be used for various purposes. Owing to their high elasticity, they can be used as cushioning or padding or shock-absorbing parts, while water gels can be used as substrates for hydrocultures. They are also excellent binders for solid and fibrous fillers. Hydrogels, for example, are excellent flame-resistant insulating materials and also provide a simple means for solidifying sandy and unstable terrain.

Nonionic Shampoo

Highlights of the Technological Achievement: An all-nonionic liquid shampoo which includes an amine oxide, a polyoxyethylene hexitan monohigher fatty acid ester, and at least one of a higher alkoxy polyoxyethylene ethanol, an alkyl glycoside and a mixture of a higher fatty acid lower alkanolamide and polyacrylamide.

Reference: D.B. Kenkare and C.R. Robbins; U.S. Patent 4,154,706; May 15, 1979; assigned to Colgate-Palmolive Company.

Background: Prior art shampoos have usually been based on anionic detergents but have sometimes included nonionic and amphoteric components, too, and sometimes have even included cationics. Also, shampoos based largely or primarily on nonionic detergents have been suggested but these have usually included ionic materials, such as salts, acidic or basic materials or materials which may be considered as being anionic or cationic.

It is known that anionic materials are capable of breaking salt linkages of keratins and various ionic surface active agents have been shown to be capable of rupturing the keratinic disulfide bond and thereby solubilizing proteinaceous material from the keratin. The isoelectric point of hair varies among individuals but in the absence of added electrolyte the keratin does not significantly react with acids or alkalies near the neutral point, e.g., 6.5 to 7.5, and thus remains in desired isoelectric condition.

Process Conditions: In accordance with this process, a nonionic shampoo comprises:

(A) 3 to 25 or 30% of a dilower alkyl or hydroxy lower alkyl monohigher alkyl or higher acylamido lower alkyl amine oxide or a mixture thereof wherein the lower alkyls are of 1 to 4 carbon atoms and the higher alkyl is of 10 to 20 carbon atoms;

(B) 2 to 20 or 30% of a polyoxyethylene hexitan monohigher fatty acid ester having 10 to 20 carbon atoms in the higher fatty acyl thereof and 5 or 10 to 100 mols of ethylene oxide per mol;

(C) 6 to 30% of a nonionic detergent selected from the group consisting of higher alkoxy polyoxyethylene ethanols wherein the higher alkoxy is of 10 to 20 carbon atoms and the ethylene oxide content, including the CH_2CH_2O- of the ethanol, is 6 to 20 mols per mol, and higher alkyl glycosides, the alkyl of which is of 10 to 20 carbon atoms, and mixtures thereof; and

(D) 25 to 89% of water.

Usually such percentages are 4 to 25%, 4 to 15%, 10 to 25%, and 35 to 82%, respectively.

In another aspect of this process, it has been found that component (C) may be replaced by a combination of 1 to 6% of a higher fatty acid lower alkanolamide wherein the higher fatty acid is of 10 to 20 carbon atoms and the alkanolamide is a monoalkanolamide or dialkanolamide, the alkanol of which is of 2 to 3 carbon atoms, and 0.05 to 1% of polyacrylamide. Usually the percentages of such components will be within the ranges of 1.5 to 4 and 0.1 to 0.5, respectively.

Despite the presence of the alkanolamide and the polyacrylamide in the shampoos, it is very often desirable to further improve the viscosity and pour properties of the liquid shampoos and to accomplish this it has been found to be useful to include in those compositions a polyethylene glycol higher fatty acid ester of an average molecular weight of most preferably 6,000, wherein the fatty acid moieties thereof are most preferably of 18 carbon atoms. It is preferred that the polyethylene glycol ester be a diester and most preferred of these compounds is the polyethylene glycol distearate of an average molecular weight of 6,000.

In addition to the previously mentioned constituents of the liquid shampoo, one may also employ normal and conventional adjuvants, providing that they are nonionic. Thus, there may be used various coloring agents and perfumes, thickeners, such as hydroxypropylmethyl cellulose, methyl cellulose, polyvinyl alcohol and polyvinyl pyrrolidone.

Example 1:

	Parts by Weight
Polyoxyethylene (80) sorbitan monolaurate	19.0
Dimethyl myristyl amine oxide (30% active ingredient)	25.0
Cocomonoethanolamide (85% active ingredient)	2.5
Polyacrylamide	0.2
Polyethylene glycol 6,000 distearate	2.0
Formalin	0.1
Perfume	1.0
Deionized water	50.2

When the above components are combined in the manner previously described, a very thick liquid shampoo is obtained which is of good foaming characteristics and is mild to the conjunctiva and the hair, while leaving the hair in excellent condition after shampooing. As measured by a Raymond flowmeter, the rate of flow of the shampoo is among the lowest of this process products (in the higher portion of the given viscosity range). The shampoo, and others of this process, is smarting to the eyes but is nonirritating to the conjunctiva.

When in place of the polyoxyethylene (80) sorbitan monolaurate there is employed a corresponding polyoxyethylene (80) sorbitan monopalmitate, having 80 mols of ethylene oxide per mol or polyoxyethylene (44) sorbitan monolaurate, having 44 mols of ethylene oxide per mol, the same general results are obtained.

Example 2:

	Parts by Weight
Polyoxyethylene (20) sorbitan monolaurate	19.0
Dihydroxyethyl coconut derived alkyl amine oxide (40% active ingredient)	18.8
Cocomonoethanolamide	3.5
Polyethylene glycol 6,000 distearate	2.0
Polyacrylamide	0.2
Formalin	0.1
Perfume	1.0
Deionized water	55.4

The above product, made by the manufacturing methods previously described, is a shampoo of good flow properties and thickness (comparatively high) and is also of good foaming, cleaning and hair treating characteristics.

Example 3:

	Parts by Weight
Polyoxyethylene (20) sorbitan monolaurate	19.0

(continued)

	Parts by Weight
Cocoamidopropyl dimethyl amine oxide (30% active ingredient)	10.0
Dimethyl myristyl amine oxide oxide (30% active ingredient)	15.0
Polyacrylamide	0.2
Cocomonoethanolamide	2.5
Polyethylene glycol 6,000 distearate	2.0
Formalin	0.1
Perfume	1.0
Deionized water	50.2

The above product is an excellent shampoo of good foaming and flow characteristics, which is mild to the hair and nonirritating to the eyes.

POLYETHYLENEIMINE (PEI) AND RELATED PRODUCTS

PROCESSING

Oleophilic Amidopolyethylenepolyamines

Highlights of the Technological Achievement: Polyamines are disclosed which consist essentially of linear chains of the formula:

$$-\!\left(CH_2CH_2NR \right)_n\!-$$

wherein R is H or the acyl group of a fatty acid or an alkylsulfonic acid and n is a number 6 to 22. Of the n units in the chain, in 0 to 50% R is an acyl group of 20 to 150 carbon atoms, in 5 to 75% R is an acyl group of 1 to 18 carbon atoms and in 10 to 90% R is H. They are useful as surfactants, especially as detergents in lubricating oils.

Reference: J.S. Kelyman and G.A. Paul; U.S. Patent 4,152,342; May 1, 1979; assigned to The Dow Chemical Company.

Background: A great variety of acylated polyalkylenepolyamines is known. Strang, U.S. Patent 3,298,955 discloses acylated polyamines of the formula:

$$\begin{matrix} (R'CO)_x \\ R(CO)_y \end{matrix}\left[\begin{matrix} | \\ -N-(CH_2)_n- \end{matrix} \right]_z$$

wherein R is an olefinic polymer chain of 40 to 700 carbon atoms, R' is an alkyl radical of 1 to 8 carbon atoms, n is 1 to 8, x is at least 2, y is 0 or 1 and z is at least 3 and preferably is 3 to 8. The amines used to make these products were the usual polyalkylenepolyamines such as triethylenetetramine, tetraethylenepentamine, pentaethylenehexamine and the corresponding propylene and butyleneamines. These are known to be highly branched in structure and of very limited range in chain length (z is 3 to 8). The compounds are said to be useful as lubricating oil detergents.

Norman and LeSeur, U.S. Patent 3,444,170 disclose acylated amines similar to

Strang's except that the acyl groups are all of the same size. Preferred polyalkyleneamine for use in their process is tetraethylenepentamine.

Dickson and Jenkins, U.S. Patent 3,347,789, disclose amides made by acylating polyalkyleneimines of molecular weight at least 800 with acyl groups of 6 to 40, and preferably 8 to 30, carbon atoms. Among other uses they are said to be useful as lubricating oil additives.

Process Conditions: The process relates to linear polymers consisting essentially of chains of the formula $-(CH_2CH_2NR)_{\overline{n}}-$ wherein n is a number about 8 to 22 and each R independently is H or the acyl group of a fatty acid or an alkylsulfonic acid, 10 to 95% of the R groups being H, 0 to 50% being acyl groups of 20 to 150 carbon atoms and 5 to 90% being acyl groups of 1 to 18 carbon atoms.

The polymers of the process are conveniently made from linear poly-2-oxazolines wherein the 2-position bears an alkyl substituent of 1 to 17 carbon atoms and the degree of polymerization is about 8 to 22. Such polymers are known and are readily made by the cationically catalyzed ring-opening polymerization of the oxazoline. Such polymers are acylated polyethyleneimines or, as they are sometimes called, polyethyleneamines.

The acylated polyethyleneimines are partially hydrolyzed to remove the desired proportion of acyl groups and replace them with hydrogen. It was previously known to completely hydrolyze such a polymer but apparently was not known or appreciated that the reaction could be monitored and stopped at a desired point to produce a partially hydrolyzed product having unique properties and uses. For the purpose of the process, preferably 15 to 60% of the acyl groups should be removed. This is conveniently done by heating the acylated polyethyleneimine with aqueous acid or alkali. The alkali is suitably NaOH, KOH or Na_2CO_3, and its concentration is suitably 5 to 25% by weight, based on the aqueous alkali solution.

The proportion of alkali to polyethyleneimine is not critical so long as at least one equivalent of alkali is provided for each acyl group to be removed from the polymer. Acid catalysts can be used similarly, HCl and H_2SO_4 being suitable and convenient in concentrations of about 5 to 20%, based on the water used. The reaction is suitably conducted at an elevated temperature, such as 50° to 100°C, and for a time sufficient to effect the desired degree of reaction. The reaction can be monitored by consumption of alkali or other convenient means and the reaction is stopped at the desired point by neutralization of the alkali, cooling of the reaction mixture, or other suitable means.

In the final step of the process, the partially hydrolyzed polyethyleneimine is acylated with a long-chain acylating agent to an extent such that about 5 to 50% of ethyleneimine groups in the polymer bear long-chain acyl groups.

The acyl group of the acylating agent has the formula RCOX or RSO_2X wherein R is an acyclic hydrocarbon radical of about 19 to 150 carbon atoms, preferably alkyl, and X is OH, Cl, OCH_3 or other readily replaceable group. The preferred acylating agents are those of the formula RCOX in which R is an alkyl group of 50 to 125 carbon atoms and X is OH.

The acylation reaction is suitably conducted by heating the partially hydrolyzed polyethyleneimine with the acylating agent.

(A) Preparation of N-Propionyl Polyethyleneimines: 2-ethyl-2-oxazoline was polymerized by heating a mixture of 23 g of dried strong acid ion exchange resin (Dowex 50-X20) and 200 g of 2-ethyl-2-oxazoline for 4 hours at 126°C. It was then cooled and diluted with 600 ml of water, after which the ion exchange resin was removed by filtration. The water was removed from the filtrate in a rotary vacuum evaporator and the product was further dried in a vacuum oven at 115°C. The molecular weight of the product was 750 as determined by Microgel permeation chromatography using a ⅛-inch by 15-foot column packed with Waters Micro-Styrogel resin of mixed pore size and a Waters differential refractometer, the standard being polystyrene of known molecular weight.

(B) Partial Hydrolysis of the Polyethyleneimines: The polymers prepared in (A) hydrolyzed by refluxing with aqueous NaOH or HCl as follows:

Base Hydrolysis – A solution containing 25 g of poly(N-propionyl)ethyleneimine in 300 ml of water and a solution of 40 g of sodium hydroxide in 100 ml of water were combined and heated under reflux for 3 hours. The water was removed by evaporation and the residual material was taken up in 200 ml of methylene chloride. The insoluble salts were removed by filtration. Removal of the methylene chloride left a residual solid. By NMR analysis, 10% of the propionyl groups had been removed by hydrolysis. The percent hydrolysis can be controlled by the length of heating period.

Hydrolysis by Acid – To a solution of 180 g of poly(N-propionyl)ethyleneimine (having a molecular weight of 750) in 445 g of water was added 95 g of 38% aqueous HCl. This mixture was allowed to reflux at 100°C for about 15 minutes. The solution was cooled in an ice bath, then neutralized with 1.5 kg of Dowex-Strong Base Resin-OH form to a pH of about 11. The resin was filtered and washed with water. The water was distilled off the filtrate to obtain 149 g of product. Nuclear magnetic resonance showed 30% hydrolysis of the polymer.

(C) Partial Acylation of Hydrolyzed Polyethyleneimines: In a typical experiment, 7 g of partially hydrolyzed polyethyleneimine prepared as described above and containing 18.69 meq of –NH– groups, corresponding to 23% hydrolysis, and 17.29 g (9.48 meq) of a hydrogenated carboxy-terminated polybutene having an average of about 125 carbon atoms per carboxyl group were heated at 200°C and stirred under a nitrogen atmosphere for 22 hours. After being cooled to 50°C the mixture was diluted with 100 ml of hexane, filtered, and the hexane was distilled to leave the product as a waxy solid. Infrared analysis showed only residual traces of carboxyl.

(D) Evaluation as Sludge Dispersant in a Petroleum-Based Motor Oil: The following laboratory test was used to screen the compounds for dispersant activity:

Spot Dispersancy Test – To a 5-dram bottle, a quantity of test additive is added to 5 grams of heavily sludged automobile crankcase petroleum-based motor oil. The bottle was then capped, heated to 300°F and shaken for one hour to insure proper dissolution and mixing of the components. The mixture was then heated overnight in the closed container at 320°F. After the heat treatment, the bottle was removed from the oven and allowed to cool to room temperature. Six drops of oil were dropped onto the center of a 4" x 5" piece of standard white blotter paper. After 24 hours, the diameter of the sludge spot and oil spot are measured. Dispersancy is reflected by the ability of the oil to keep the sludge in suspension. In this test, dispersancy is reflected by the difference in diameters of the sludge and oil spots.

$$SDT = (D\ \text{sludge spot}/D\ \text{oil spot}) \times 100$$

A heavily sludged oil (no test additive) will yield an SDT rating of 40 or less. An excellent dispersant will yield SDT ratings of 70 to 80 when added at 1% to 3% by weight of the active compound.

Examples 1 through 23: The table below summarizes the results of some typical experiments conducted essentially as described above. In the table, the column headed "Mers" shows the number of monomer units in the acylated polyethyleneimine used to make the compound of that example as described in (A) above, "% Hydrolysis" refers to the hydrolysis described in (B), "% Free NH Sites Reacted" refers to the partial acylation as described in (C), and "Acid" describes the acid used in the partial esterification. Examples 8 to 10 were run identically except that in the partial esterification step Example 8 was heated 30 minutes at 100°C, Example 9 was heated an additional 20 hours at 140°C, and Example 10 was heated 30 minutes at 100°C, then an additional 20 hours at 140°C and finally, another 7 hours at 190°C.

Ex. No.	Mers	Hydrolysis (%)	Free NH Sites Reacted (%)	Acid*	SDT Active Additive (%) 1	2	3
1	13	12	80	C_{100}TPB	41	89	91
2	10	23	50	C_{100}TPB	46	87	90
3	10	23	75	C_{100}TPB	43	90	90
4	10	23	100	C_{100}TPB	45	89	84
5	10	25	70	C_{50}TPB	42	44	54
6	9	27	75	C_{110}TPB	43	61	79
7	12	50	25	C_{110}TPB	69	82	78
8	11	48	25	C_{110}TPB	40	43	61
9	11	48	25	C_{110}TPB	41	68	83
10	11	48	25	C_{110}TPB	45	67	73
11	13	48	15	C_{100}TPB	41	46	89
12	13	48	25	C_{100}TPB	40	64	89
13	13	48	50	C_{100}TPB	39	64	90
14	13	48	75	C_{100}TPB	41	48	56
15	13	60	15	C_{100}TPB	41	46	59
16	13	60	25	C_{100}TPB	38	44	56
17	13	60	50	C_{100}TPB	39	41	52
18	13	60	50	C_{100}TPB	39	44	51
19	13	60	75	C_{100}TPB	39	41	46
20	10	12	80	HCTPB	44	84	83
21	10	23	75	HCTPB	47	91	66
22	10	23	50	HCTPB	43	77	88
23	11	48	25	HCTPB	44	52	72

*CTPB is carboxy-terminated polybutene. The subscript number indicates the number of carbon atoms in the acid. HCTPB is hydrogenated carboxy-terminated polybutene having an average of about 125 carbon atoms per molecule.

APPLICATIONS

Removal of Color from Paper Mill Wastewaters

Highlights of the Technological Achievement: Color is removed from paper mill wastewater by a precipitation method using relatively small amounts of polyethyleneimine as a complexing agent and precipitant for color bodies in such wastewaters.

Reference: J.J. Svarz; U.S. Patent 4,179,329; December 18, 1979; assigned to Nalco Chemical Company.

Background: It is generally recognized that the amount of organic material in the form of BOD (biochemical oxygen demand), COD (chemical oxygen demand), and color bodies currently being discharged into public water supplies from pulp and paper mills is a problem of increasing importance in water pollution.

In the manufacture of paper, cellulose from both hard woods and soft woods is converted to pulp by various processes such as the sulfite process and the sulfate process and the pulp is subjected to various treatments particularly bleaching which involve the use of chemical reagents such as chlorine, chlorine dioxide, sodium hypochlorite and caustic alkalis. As a result of these treatments, the papermaking process produces highly colored effluents. The bleaching of the pulp, in particular the caustic extract effluent, is a major contributor to pollution problems.

Lignin is a primary source of color in the pulp. Chlorination of the pulp during the bleaching operation results in the formation of color bodies which are leached from the pulp by caustic alkali solutions. Thus, the caustic extract effluent contains a major proportion of the color bodies and other organic materials which have to be disposed of in wastewater treatment.

Several processes are used or advocated for use in removing color from paper mill waste. One of these is the so-called "massive lime process." This process utilizes lime in quantities ranging from 5,000 to 20,000 ppm to remove by coagulation and sedimentation techniques the organic colored bodies from paper mill wastes.

Another process which is disclosed in U.S. Patent 3,412,018 involves the use of certain types of water-insoluble amines in combination with a diluting agent which is nonmiscible in water and which has a dielectric constant, measured at 20°C, of less than 4.8 in order to form substantially water-insoluble organophilic amine complexes which are extracted by the organic phase.

Process Conditions: In accordance with the process it has been found that color can be removed from paper mill wastewaters by adding to such wastewaters, preferably at a pH within the range of 2 to 7, and in most instances within the range of 2 to 5, a relatively small amount of a polyethyleneimine having a molecular weight of at least 300, the amount used being sufficient to form a complex solid of the imine with the color bodies in the wastewaters which can be separated by settling or filtration. The amount of the imine can be as low as 10 ppm with respect to the wastewater or bleachery effluent and, in most cases, will not exceed 2,000 ppm, depending upon the particular imine employed.

It is essential for the purpose of the process that the pH of this effluent be reduced to a pH within the range of 2 to 7 and preferably 2.5 to 5.0. In a typical paper mill plant the wastewaters from the whole mill might total 25,000,000 to 30,000,000 gallons per day and the bleachery wastewaters would constitute a smaller proportion of the total, 1,000,000 to 6,000,000 gallons per day. The process can be used to treat all of the wastewaters but it is usually more practical and more essential to treat the bleachery wastewaters in order to remove or to reduce to a minimum the color.

The process is based on the discovery that relatively small amounts of imines of the type previously mentioned will combine with the color bodies in the aforementioned wastewaters under certain pH conditions to form a water-insoluble solid and that the resultant imine-color body solids can be removed by settling or filtration.

PEI: It appears to be essential for the pupose of the process that the imines employed have an average molecular weight of at least 300. The polyethyleneimines are available commercially as PEI 3, PEI 6, and PEI 18 with average molecular weights of 300, 600 and 1,800, respectively. They are normally water-soluble and can be used as aqueous solutions. Higher molecular weight imines can be used. Inasmuch as the reaction with the color body apparently occurs at the imino

$$\left(-\underset{|}{\mathrm{N}}\mathrm{H}-\right)$$

group, it is deemed to be immaterial whether the imine itself is water-soluble or water-insoluble.

In carrying out the process the pH can be achieved by blending acidic chlorination effluent from the bleachery process with caustic extraction effluent or the caustic extraction effluent can be acidified with sulfuric acid or waste liquor containing sulfuric acid, or hydrochloric acid.

The process has been evaluated by treating the wastewaters with the imine, removing the resultant solids and comparing the clarified effluent or filtrate with the wastewater prior to treatment. This comparison was carried out in two ways: (1) using a Gardner-Holdt colorimeter and observing the readings, and (2) obtaining color readings in terms of APHA units. The treated samples were centrifuged or filtered through No. 40 filter paper to remove floc and the filtrates or centrifugates were adjusted to pH 7 before color readings were taken. The raw effluents used in the test were adjusted to pH 7 and then filtered through a millipore filter before color readings were made to get a baseline reading on original color.

The treatment of the wastewater with the imine can be supplemented by a subsequent addition of a water-insoluble amine containing at least 8 carbon atoms, preferably 8 to 24 carbon atoms, in the hydrocarbon group linked to the amino group, e.g., oleylamine.

The process will be further illustrated but is not limited by the following examples in which the quantities are given by weight unless otherwise indicated.

Example 1: A soft wood bleachery caustic alkali extract was acidified with hydrochloric acid to give a pH of approximately 3 and then treated with 400 ppm of polyethyleneimine having an average molecular weight of 300 (PEI 3). The Gardner color was reduced from 18 to 12. The temperature used in this test was 40°C. Other tests were carried out at ambient temperatures of 20° to 25°C with generally similar results in color reduction. Subsequent additions of 300 ppm of oleylamine reduced the Gardner color to 5.

Example 2: Samples of various types of paper mill wastewaters were tested using 500-ml samples to which a polyethyleneimine was added at various pH's and in various proportions. The samples were stirred at ambient temperatures at 100 to 120 rpm for 2 minutes and at 40 rpm for 2 minutes. Color and clarity on the APHA color scale on a Hellige instrument were determined. Readings were made after cup centrifuging at the treated pH and then again at pH 7.0 after adjustment. Usually color values go up 20 APHA units at pH 7.0. The additives were all added as solutions in water diluted to a 1% concentration of the polyethyleneimine.

In one series of tests the bleachery caustic extract had an APHA value of 500 and a pH of 9.1. The addition of polyethyleneimine having an average molecular weight of 1,800 (PEI-18 Dow) at a dosage of 50 ppm reduced the APHA color to 20. At dosages of 30 ppm and 10 ppm there was no reduction in the APHA color.

The same polyethyleneimine at dosages of 50 ppm, 30 ppm and 10 ppm and at a pH of 5 in each case reduced the APHA color of the sample from 500 to 20, 30 and 120, respectively.

The same polyethyleneimine at dosages of 50 ppm, 30 ppm and 10 ppm and at a pH of 7 reduced the APHA color from 500 to 60, and 200 APHA in the case of the 50 ppm dosage and the 30 ppm dosage, respectively. A dosage of 10 ppm at pH 7 produced no reduction in the APHA color and no floc formation.

The same polyethyleneimine at dosages of 50 ppm, 30 ppm and 10 ppm and at a pH of 3 reduced the APHA color from 500 to 40, 50 and 60, respectively.

Fabricating Wood Composite Panels Having Enhanced Fire Retardancy

Highlights of the Technological Achievement: What is disclosed is a method of fabricating wood composite panels having enhanced fire retardancy. The panels so fabricated have hard, durable finished surfaces which allow machine finishing without the concurrent loss in fire retardancy due to the finishing operation.

Reference: J. Maine; U.S. Patent 4,165,409; August 21, 1979.

Background: The major objective of this process is to produce wood composite structures that have a combination of aesthetically appearing durable finished surfaces and fire retardancy. This particular combination has long been sought and many methods have been described to achieve such an effect.

U.S. Patent 3,817,822 describes the preimpregnation of two panels and their subsequent gluing together via the use of a polyethylene based glue to form a composite which is fire retardant. The patent describes the added advantage of having a glue which is not affected by the use of fire retardants in the wood panels. The patent does not describe the use of fire retardants with wood that requires fine finishing nor does it describe what one needs to do to acquire woods having both fire retardancy and the ability to be finished. Thus, the problem of having fire retardancy in a wood product which can be fine finished and still retain the fire retardant property, has not been resolved.

Process Conditions: This process consists of a method of enhancing the fire retardancy of fabricated wood panel systems which method consists of:

(A) impregnating a first or outer wood member with an aqueous based fire retardant;

(B) impregnating the product of (A) with a material selected from the group consisting of:

  (1) curable tertiary butylstyrene (TBS) and

  (2) mixtures of curable tertiary butylstyrene and a solvent-based flame retardant different than the flame retardant in (A);

(C) impregnating a second or core wood member with a fire retardant as in (A) and drying;

(D) forming a composite structure from the product of (A) and (B) and the product of (C) by gluing the wood members together; and

(E) attaching a third or balance wood member to the remaining core face opposite the first wood member, whereby a fire retardant composite fabricated wood panel is obtained.

Example: Several panels measuring about 23⅞ inches in length, 3⅞ inches wide and thicknesses as shown below were prepared in the following manner:

The outer or first wood members were fashioned from $\frac{1}{40}$-inch-thick walnut wood and some were impregnated with a composition which was an aqueous solution of 12 pbw of polyethyleneimine having an average molecular weight of from 40,000 to 60,000 and a Brookfield viscosity (at 5% solids) of 28 cp at 25°C and 88 pbw of a 40/60 weight percent mixture of mono/diammonium phosphate. The aqueous solution was 20% solids. The treatment was a vacuum-pressure treatment. After the walnut members were dried, some were then subjected to an impregnation using a mixture of tertiary-butylstyrene monomer, a carrier for the monomer, a glycol crosslinker, a catalyst and in some cases Fyrol CEF.

The fire-retardant containing TBS contained 100 pounds TBS, 7 pounds of glycol and 0.75 pound of catalyst and 20 pounds of Fyrol CEF which was 50% in methylene chloride solvent. The crosslinkers useful in this process are acrylate functional glycols or polyglycols such as a dimethacrylate or diacrylate of ethylene glycol. In the above example, the crosslinker was ethylene glycol dimethacrylate.

The walnut members were then cured at 160°F and then dried and allowed to cool. Core members, comprising ⅛-inch-thick luan or basswood were then glued to the walnut members. Some of the core members were first impregnated as were the walnut members in the first step of the process, using the polyalkyleneimine phosphorus aqueous solution. A final panel, which was a balance member of aspen of 1/32-inch thickness was then glued to the backside of the core member. These balance members were not impregnated. The entire sandwich was then compressed and allowed to dry.

The tests were carried out on a two-foot Monsanto tunnel test to determine flame retardancy. Details of this test can be found in *"Journal of Paint Technology,"* Volume 39, No. 511, August 1967. The values reported herein were calculated using the formulas found in that publication and correlated with ASTM E84-61 (the 25-foot tunnel test). Results from this test can be roughly extrapolated to indicate the performance of the panels in the 25-foot tunnel test ASTM No. E84-61. The results are reported as the average of 3 runs.

	. . . . Face Panel	. . . .		Core Panel	Composite
Run No.	Impregnated with Polyalkyleneimine*	TBS** Only	Fyrol plus TBS	Treated with Polyalkyleneimine	Flame Spread Rating Result
1	yes	–	yes	yes	39.2
2	yes	yes	no	yes	43.0
3	no	yes	no	no	T***
4	yes	no	yes	no	99.4
5	no	no	yes	yes	T
6	no	yes	no	yes	122.5
7	no	no	no	no	T
8	†	no	no	no	T
9	†	no	no	yes	117.1
10	no	yes	no	yes	122.0

*Fire retardant.
**TBS is tertiary-butylstyrene.
***T is flame out of the tunnel.
†Lacquer coat.

Run No. 1 is one aspect of this process and has the best value of 39.2. The Run No. 2 is the other aspect of this process and has a good value of 43. Comparisons can be made between these two values and the other treatments listed thereunder. Thus, this data shows that the TBS treatment outlined in U.S. Patents 3,790,401 and 3,560,225 is not a fire-retardant product. Run No. 4 shows that treatment of the face panel with fire retardant and the subsequent treatment with fire retardant and TBS, i.e., steps (A) and (B) of this process, do not suffice to give an enhanced flame retardancy; the core must also be fire retardant. Run No. 5 shows that core impregnation and fire retardant plus TBS on the face panel does not give enhanced fire retardancy. Run No. 6 shows that a fire retardant core with TBS-treated face is not effective either.

A duplicate of this run can be found in Run No. 10. A plain composite was run in Run No. 7. The same type of plain composite with a lacquer coat was run in Run No. 8 and the same thing was run in Run No. 9 except that the core was fire retardant.

Polysaccharide-Containing Adsorbent

Highlights of the Technological Achievement: A polysaccharide-containing adsorbent for the removal of heavy metal ions from an effluent which adsorbent has been manufactured from:

(1) a polysaccharide-containing material,

(2) a basic, nitrogen-containing polymer which can be acylated,

(3) a crosslinking compound, which is different from component (4), containing at least two reactive substituents and, optionally,

(4) an aliphatic or araliphatic carboxylic acid containing a mobile substituent or a multiple bond which is capable of undergoing addition.

Reference: Q. Bowes, J. Haase and R.F. Wurster; U.S. Patent 4,144,190; Mar. 13, 1979; assigned to Ciba-Geigy AG, Switzerland.

Background: The new adsorbents can be manufactured by reacting or mixing and reacting the three or four different components in any desired sequence. However, these products are preferably manufactured by mixing: (1) the polysaccharide-containing material with a previously obtained reaction product of (2) basic, nitrogen-containing polymers which can be acylated and (4) an aliphatic or araliphatic carboxylic acid containing mobile substituents or a multiple bond which is capable of undergoing addition, and then crosslinking the mixture with component (3).

Process Conditions: In principle, polymers which contain amino groups which can be acylated, that is to say, which contain primary or secondary amino groups, are suitable as basic, nitrogen-containing polymeric compounds (2) for the manufacture of the adsorbents according to the process.

Suitable basic polymers are, above all, the polyalkyleneimines which appropriately have an average molecular weight (MW) of 5,000 to 200,000, preferably 10,000 to 100,000. In general, these polymers possess a Brookfield viscosity at 20°C of 500 to 20,000 cp. The polyalkyleneimines are preferably derived from alkyleneimines with 2 to 4 carbon atoms. Suitable alkyleneimines are, in particular, ethyleneimine, propyleneimine, 1,2-butyleneimine and 2,3-butyleneimine. Of all the polyalkyleneimines, polyethyleneimine is preferably used. Polyethyleneimines which have an average molecular weight of 10,000 to 100,000, in particular 30,000 to 40,000, are of particular practical interest.

A typical representative of the polyamidepolyamines is obtained from polymerized linoleic acid/linolenic acid and triethylenetetraamine and has an amine value of 350 to 400 mg of KOH/g.

Monomeric or polymeric compounds which possess crosslinking properties and which can form a chemical bond either with the polysaccharide-containing material (1) or with the basic, polymeric compound (2), or with both, can be used as component (3). These compounds, containing at least two functional groups, owe their reactivity to a multiple bond which is capable of addition, an epoxy group or aziridine group, a substituent which can be split off as a tertiary amine or, preferably, a substituent which can be split off as an anion; a combination of these factors can also be present. They contain, for example, the grouping:

$$-CO-C=C,\ -CO-C\equiv C-\ \text{or}\ -SO_2-C=C,$$

or also the isocyanate or isothiocyanate group as a multiple bond which can undergo addition. Component (3) contains substituents with a radical which can be split off as an anion, preferably with mobile halogen atoms, as a preferred, reactive group. Radicals of this type which can be split off as an anion owe their mobility, for example, to the influence of electrophilic groups, such as the –CO– or $-SO_2-$ group in saturated, aliphatic radicals. They can also owe their mobility to the influence of a quaternary nitrogen atom, such as in the group:

$$-\overset{|}{\underset{|}{N^+}}-CH_2CH_2Cl,$$

or, in aromatic radicals, to the influence of electrophilic groups in the o-position and p-position, for example, nitro groups, hydrocarbon-sulfonyl groups or hydrocarbon-carbonyl groups, or to the bond to a ring carbon atom which is adjacent to a tertiary ring nitrogen atom, such as in halogenotriazine radicals or halogenopyrimidine radicals.

Product: Effluents to be treated with the adsorbents according to the process include various industrial effluents which contain ions of metals with a standard potential of, as a rule, more than -1.5 V, preferably between -1.0 to +0.85 V, at 25°C. Such metal ions are, for example, Cd^{++}, and Cu^{++}, Ni^{++}, Sn^{++}, Pb^{++}, Hg^{++}, Ag^{+}, Fe^{+++}, Cr^{+++} and Co^{+++}.

The removal of the metal ions is appropriately carried out at 10° to 100°C. However, it is preferably carried out between 20° and 70°C. If desired, the purification of the effluents can also be carried out under pressure or under a vacuum. The pH value of the effluents can vary within wide limits, for example, between 2 and 12. However, depending on the nature of the modified polysaccharide material used as the adsorbent, pH correction, for example, to a value of 2 to 9, in particular 3 to 6, can facilitate or accelerate the purification process.

POLYSILOXANES AND RELATED PRODUCTS

PROCESSING

Stable Alkylhydridopolysiloxane Emulsions

Highlights of the Technological Achievement: An aqueous alkylhydridopolysiloxane emulsion is stabilized against loss of hydrogen by addition thereto of 0.1 to 5% by weight of an emulsifier containing perfluoroalkyl groups.

Reference: H.-H. Steinbach, K. Schnurrbusch, M. Rieder and O. Weiden; U.S. Patent 4,179,426; December 18, 1979; assigned to Bayer AG, Germany.

Background: Aqueous emulsions of organopolysiloxanes, which also contain alkylhydridopolysiloxanes, have found varied use for a relatively long time, in particular for impregnating fiber materials of all kinds, in which they are also often combined with crease-proofing agents. It is usual to impregnate textiles with them and then to heat the textiles in order to give them a water-repellent surface, the reactivity of the HSi group promoting the durability of the impregnation.

Because of the evolution of hydrogen and the demixing associated with this, this reactivity makes it difficult to store such emulsions and to transport them over relatively long routes. It is already known to reduce these disadvantages by adding a protective colloid, such as polyvinyl alcohol, and increasing the viscosity of the emulsions. Self-emulsifying solutions of the polysiloxanes, with which attempts are made to deal with the problem, are also known. However, such measures are frequently not satisfactory since they restrict the range of use, for example, they impair the ease of combination with crease-proofing agents, and since the "liquors," i.e., the emulsions prepared at the place of application by high dilution with water, become unstable.

Process Conditions: It has been shown, surprisingly, that by adding emulsifiers containing perfluoroalkyl groups to polysiloxane emulsions of this type, a very good stabilization of the Si–H bond is achieved, coupled with simultaneous stability of the emulsion.

The siloxane oils to be emulsified, which in most cases are mixtures of customary diorganosiloxane polymers and alkylhydridopolysiloxanes, wherein the alkyl radical may have up to 6 carbon atoms, such as, for example, methylhydridopolysiloxanes, to which organic solvents, such as, for example, toluene or bromobenzene, can also additionally be added, have viscosities between about 5 cp and 500 cp (20°C). The proportions should be chosen so that the sum of the organopolysiloxanes is, in total, up to about 40% by weight of the storable emulsion.

Raw Materials: Suitable emulsifiers containing perfluoroalkyl groups are, for example, perfluoroalkyl polyethers of the general formula:

$$C_xF_{2x+1}SO_2N(CH_3)CO-(OCH_2-CH_2)_n-(OCHCH_3CH_2)_m-OC_4H_9$$

wherein x is about 6 to 12, n is 2 to about 40 and m is 2 to about 40. Compounds wherein x is 8, n is 20 to 30 and m is 20 to 30 are preferred. Longer-chain perfluorinated alkanesulfonic acid salts of the composition:

$$C_xF_{2x+1}SO_3N(C_nH_{2n+1})_4$$

wherein x is 6 to 12, n is 1 to 3, such as, for example, $C_8F_{17}SO_3N(C_2H_5)_4$, are also suitable.

In certain circumstances it can be appropriate also to additionally add fluoroalkyl-substituted siloxanes to the emulsions, e.g., about 0.1 to 5% by weight. These can favor the stability of the emulsion, in particular when emulsifiers having a slightly alkaline action, such as, for example, $C_8F_{17}SO_3K$ or $C_8F_{17}SO_3N(CH_3)_4$, are used. For this purpose, organopolysiloxanes, containing perfluoroalkyl groups, of the general formula:

$$R_3SiO\left[\begin{array}{c}R\\|\\-Si-O-\\|\\R\end{array}\right]_a\left[\begin{array}{c}(CH_2)_2R_F\\|\\-Si-O-\\|\\R\end{array}\right]_b-SiR_3$$

wherein the R's independently of one another represent an alkyl radical with 1 to 6 carbon atoms or a phenyl radical, R_F represents a perfluoroalkyl radical with 4 to 12 carbon atoms, a represents a number from about 5 to 70 and b represents a number from about 5 to 70, are suitable.

An example which should be mentioned of this type of substance is:

$$(CH_3)_3Si[OSi(C_8F_{17}CH_2CH_2)(CH_3)]_{36}[OSi(CH_3)_2]_{64}OSi(CH_3)_3$$

Product: The concentration, in the emulsion, of the emulsifier containing perfluoroalkyl groups should be about 0.1 to 5% by weight (relative to the total amount of emulsion), preferably 0.2 to 1.5% by weight.

It is, of course, also possible to stabilize the emulsions mentioned with emulsifier mixtures or mixed emulsifiers. In this procedure, both fluorine-containing and fluorine-free mixtures of emulsifiers can be used, provided that the conditions according to the process are observed.

Example: 35 parts by weight of methylhydridopolysiloxane (viscosity 17 cs at 20°C) are added to a solution of 0.5 pbw of:

$$C_8F_{17}SO_2\underset{\displaystyle CH_3}{\underset{|}{N}}-\overset{\displaystyle O}{\overset{\|}{C}}-(OC_2H_4)_{30}-(OC_3H_6)_{30}-OC_4H_9$$

in 64.5 pbw of water and the entire mixture is emulsified in a homogenizing device (Gann machine). An extremely stable emulsion is obtained, which can be combined with zirconium soaps or used as a crosslinking agent, mixed with other organopolysiloxane emulsions.

When an emulsion (200 ml) prepared in the manner described was continuously stirred intensively for 24 hours at 40°C, only 2 cm^3 of hydrogen were evolved. On standing undisturbed at room temperature still no gas could be observed after 4 days.

Contacting an Organosilicon Compound with an Inorganic Hydrogel to Form an Inorganic Xerogel

Highlights of the Technological Achievement: An inorganic xerogel is produced by contacting an inorganic hydrogel with a contacting system which comprises at least one organosilicon compound and thereafter heating the resulting hydrogel to remove essentially all the liquid components therein.

Reference: M.P. McDaniel; U.S. Patent 4,190,457; February 26, 1980; assigned to Phillips Petroleum Co.

Background: Inorganic gels are well known and have long been used for various purposes, for example, for the adsorption of condensible vapors from gaseous carriers and for catalytic purposes either as a catalyst itself or as a component thereof or as a carrier. The most widely used of these inorganic gels in the dried condition is that of silica with or without the addition of other gelatinous materials such as alumina.

The manner in which a catalyst or support to be utilized in a catalytic composition is prepared can influence the average pore diameter of the support, and hence, the melt index of the polymer prepared in the presence of the catalyst composition.

In the preparation of an inorganic catalyst support (xerogel) from a hydrogel, e.g., a silica-containing hydrogel, an important step is the removal of water from the hydrogel or the drying of the hydrogel to produce the xerogel. The step is important in that the manner in which the water is removed has a great bearing upon the size of the pores of the catalyst support (xerogel). One method of drying hydrogel is to heat the hydrogel at a high temperature in order to evaporate the water. A problem, however, is that the dried hydrogel (xerogel) has relatively small pores.

The prior art has somewhat overcome this problem by removing the water from the hydrogel through repeated washings with an organic compound or by azeotropic distillation. Use of these organic compounds requires many time-consuming steps to recover a dried gel. Azeotropic distillation involves not only additional processing but also additional expense.

An object of the process, therefore, is to provide a simple process to produce an improved inorganic xerogel.

Process Conditions: The contacting system of the process comprises at least one organosilicon compound having the structure: R_nSiA_{4-n}, wherein n is an integer of 2 or 3, and wherein each R is a saturated or unsaturated hydrocarbon group wherein each R can be the same or different and wherein A is selected from the group consisting of hydroxy radicals, halides and alkoxy radicals in which the alkyl group therein contains from 1 to about 10 carbon atoms.

Preferably, R is selected from the group consisting of alkyl radicals of from 4 to 12 carbon atoms, alicyclic radicals of from 4 to 12 carbon atoms, aryl radicals of from 6 to 24 carbon atoms, and hydrocarbyl-substituted aryl radicals such as alkylaryl and cycloalkylaryl of from 6 to about 24 carbon atoms.

Some examples of suitable organosilicon compounds which can be employed in the instant process include tri-t-butylsilanol, tri-t-octylsilanol, tricyclohexylsilanol.

The preferred organosilicon compounds are triphenylsilanol, diphenylsilanediol and triphenylchlorosilane because of ready availability and relatively low cost and because they have been used with good results.

In functional terms the amount of the at least one organosilicon compound employed in the contacting step is an amount sufficient to provide an increase in pore volume of the contacted and heated inorganic hydrogel as compared to the pore volume of the heated inorganic hydrogel not contacted with the organosilicon compound. Generally, the amount of the organosilicon compound is from 0.1 to 20 weight percent of the hydrogel. The contacting system may additionally contain at least one water-soluble oxygen-containing organic liquid.

Product: The hydrogels contemplated in this process are generally the inorganic gels formed from the precipitation of the oxides of aluminum, germanium, iron, molybdenum, silicon, tin, titanium, tungsten, vanadium, zirconium and combinations of any two or more thereof. Preferred is the hydrogel comprising silica because of its wide use in industry and good performance in accordance with the process.

The time necessary for contacting the hydrogel is not believed to be critical. Normally the time employed is the amount of time sufficient to obtain penetration of the organosilicon compound into the pores of the gel which is generally less than one minute, but most any length of time can be used. Long contacting times also can be used without impaired results. Thus a contacting time within a range of about 30 seconds to about 10 hours or more can be used.

The hydrogel composition is treated at an elevated temperature by any suitable means such that essentially all liquid components contained therein are removed. Generally, this can be achieved by the application of sufficient heat. Frequently the hydrogel composition which has been contacted with an organosilicon compound and a water-soluble oxygen-containing organic liquid is heated in vacuo or in air, e.g., in an oven or with a heat lamp. Preferably, the temperature used in the treating step is within the range of 20° to 300°C. A temperature of 150°C has been used with good results.

The inorganic xerogels obtained in accordance with this process have a pore volume of at least 1.75 cc. Inorganic xerogels have been produced that have pore volumes within the range of 1.75 to 2.20 cc.

Xerogels were prepared from a silica hydrogel utilizing the organosilicon compounds listed in the table. 60 ml of isopropanol was added to each compound with the exception of Run 1. The pore volumes (determined by Innes method) of the resultant calcined xerogels are listed in the table.

Organosilicon Compound

Run No.	. . . . . . Description (wt %)		Pore Volume (cc)
1	Control, no isopropanol	–	1.46
2	Control, 60 ml isopropanol	–	1.74
3	Triphenylsilanol	1	1.92
4	Triphenylsilanol	8	1.83
5	Diphenylsilanediol	4	1.78
6	Triphenylchlorosilane	4	1.90
7	Triethylsilanol (control)	1.2	1.48
8	Triphenylsilanol	4	2.19
9	Triphenylsilanol	4	1.92
10	Triphenylsilanol	4	1.86

The results presented in the table demonstrate the effect of the various organosilicon compounds on the pore volume of each calcined xerogel. In control run, 1, where no isopropanol or organosilicon compound is employed, the silica zerogel has a pore volume of 1.46 cc. In control run 2, when isopropanol is employed as the sole additive to the hydrogel during drying, a pore volume of 1.74 cc results. However, process runs 3 to 6 and 8 to 10 clearly demonstrate a pore volume noticeably greater than that of control run 1 as well as control run 2 which used isopropanol alone.

Use of Water-Soluble Monomers for the Production of Polysiloxane Hydrogels

Highlights of the Technological Achievement: A water-insoluble hydrophilic gel comprising: (A) 20 to 90% by weight of a hydrophilic (a) polymer of identical or different water-soluble monoolefinic monomers, or (b) copolymer of the water-soluble monomers with 1 to 80% (of total monomers) of water-insoluble, identical or different monoolefinic monomers; which polymer or copolymer is crosslinked with (B) 10 to 80% by weight of a terminal polyolefinic siloxane macromer having a molecular weight from 400 to 8,500.

Reference: K.F. Mueller and E.K. Kleiner; U.S. Patent 4,136,250; January 23, 1979; assigned to Ciba-Geigy Corporation.

Background: In the pharmaceutical field the main interest lies in the slow and controllable release of drugs from such hydrogels. Drug-containing hydrogel preparations have been described as being in the form of bandages, subcutaneous implants, buccal devices, intrauterine devices, and eye inserts. They are made by complicated fabrication procedures which usually involve casting the monomer solution into a suitable mold and polymerizing in the presence of a free radical generating initiator.

It has now been discovered, that an unexpectedly ideal combination of useful properties can be obtained by synthesizing a hydrogel containing a major proportion of a polymeric polysiloxane based crosslinking agent. This combination of properties includes: hydrophilicity, oxygen permeability, flexibility and toughness in the dry and swollen state, biocompatibility, polishability and machineability in general, clarity, ease and economy of synthesis.

Process Conditions: This process pertains to a water-insoluble hydrophilic gel comprising the crosslinked copolymerization product of: (A) about 20 to about 90% by weight of the gel of (a) a water-soluble monoolefinic monomer, or mixture of the monomers, or (b) a water-soluble monomer, or mixture of the monomers, with 1 to 80% by weight of total monomers of a water-insoluble monoolefinic monomer, or mixture of the water-insoluble monomers, with (B) about 10 to about 80% by weight of the gel of a terminal polyolefinic hydrophobic siloxane macromer having a molecular weight from about 400 to about 8,500, the macromer having the formula:

$$\text{(B)} \qquad R_3CH{=}\overset{\displaystyle R_2}{\overset{|}{C}}{-}X{-}Y{-}R_1{-}\left[Y{-}X{-}\overset{\displaystyle R_2}{\overset{|}{C}}{=}CHR_3\right]_z$$

wherein R_1 is a polysiloxane chain having a molecular weight from about 200 to about 8,000, and is of the formula:

$$(CH_3)_3SiO{-}[SiO(CH_3)_2]_x{-}\left[\begin{matrix}SiO(CH_3)\\ |\\ R_6-\end{matrix}\right]_y{-}Si(CH_3)_3 \quad \text{or} \quad -R_6{-}[SiO(CH_3)_2]_x{-}SiO(CH_3)_2{-}R_6{-},$$

wherein R_2 is hydrogen or methyl, R_3 is hydrogen or $-COOR_4$, R_4 is hydrogen or alkyl of 1 to 10 carbon atoms, and with the proviso that at least one of R_2 and R_3 is hydrogen, X is oxa, $-COO-$ or $-COR_5-$, R_5 is hydrogen or alkyl of 1 to 5 carbon atoms, R_6 is a branched or linear alkylene of 1 to 7 carbon atoms or $(CH_2CH_2O)_n$ where n is 1 to 20, x is an integer of 3 to 120, y is an integer of 2 to 3, z is an integer of 1 to 2, Y is the direct bond, the diradical:

$$-R_7{-}Z_1{-}CONH{-}R_8{-}NHCO{-}Z_2{-}, \quad \text{or} \quad -R_7OCOCH_2CH_2S-,$$

R_7 is a linear or branched alkylene of 2 to 5 carbon atoms connected to X, R_8 is a diradical obtained by removing the NCO groups from an aliphatic, cycloaliphatic or aromatic diisocyanate, Z_1 is a oxa or $-NR_5-$, Z_2 is Z_1 or thia and is connected to R_6, with the proviso that if X is oxa, Y is different from a direct bond and R_2 and R_3 are hydrogen.

The hydrogels of this process are synthesized by free radical copolymerization, either in solution or in bulk, of a water-soluble monomer, (As) capable of forming a water-soluble or water-swellable polymer, i.e., hydrophilic polymer (Ap), with a di- or tri-olefinic siloxane derived macromer (B), for example, a compound consisting of a long, linear polysiloxane chain, terminated on both ends with a polymerizable vinyl group. In this way a three-dimensional macromolecular network is formed composed of two types of segments, each segment contributing its distinct physical properties to the system. The A-segment contributes water-solubility; the siloxane B-segment forms flexible crosslinks and improves oxygen permeability and tissue compatibility.

The water-soluble monomers, As, are preferably acrylic and/or methacrylic acid or the water-soluble derivatives thereof, such as hydroxyalkyl esters where alkyl is 2 to 4 carbon atoms, e.g., 2-hydroxyethyl, 3-hydroxypropyl, 2-hydroxypropyl or 2,3-dihydroxypropyl esters; also ethoxylated and polyethoxylated hydroxyalkyl esters such as esters of alcohols of the formula:

$$HO{-}C_mH_{2m}{-}O{-}(CH_2CH_2{-}O)_n{-}H$$

where m represents 2 to 5 and n represents 1 to 20; or esters of analogous alcohols, wherein a part of the ethylene oxide units is replaced by propylene oxide units.

Water-soluble comonomers, which do not contain hydroxy groups are: acrylic and methacrylic acid and alkyl ethers of polyethoxylated hydroxyalkyl esters thereof, such as esters of alcohols of the formula:

$$HO{-}C_mH_{2m}O{-}(CH_2CH_2{-}O)_n{-}CH_3$$

where m is 2 to 5 and n is 4 to 20; dialkyl aminoalkyl esters and amides, such as 2-(dimethylamino)ethyl; or 2-(diethylamino)ethyl acrylate and methacrylate.

Hydrogel Compositions: The hydrogel compositions may vary from 20 to 90% polymer (A_p) from monomer (A) and 10 to 80% siloxane macromer (B), corresponding to a typical degree of swelling of about 10 to 250%. The preferred compositions contain from about 15 to 60% macromer (B) and about 40 to 85% polymer (A_p), and range in their degree of swelling from 15 to 120%.

The following examples illustrate the process. In the example, the Degree of Swelling, DS, is defined as:

$$DS = \frac{\text{weight of swollen sample} - \text{weight of dry sample}}{\text{weight of dry sample}} \times 100$$

The monomers and the diisocyanates useful in this process are widely available as items of commerce. The polydimethyl siloxane intermediates used to prepare the siloxane macromers are commercially available from the Dow Corning Chemical Company. Suitable siloxane intermediates are: Q4-3667, molecular weight about 2,400; X2-8024, molecular weight about 3,006; Q4-3557, molecular weight about 800; and 1248, molecular weight about 6,000.

Example 1: *Preparation of Siloxane Macromer* – To a 1,000 ml, 3-necked flask fitted with a mechanical stirrer, thermometer, condenser and nitrogen inlet were charged 480 g (0.2 mol) of the polydimethyl siloxane diol, Q4-3667, molecular weight 2,400, and 88.8 g (0.4 mol) of isophorone diisocyanate. After adding 0.2 g of triethylamine catalyst, the mixture was stirred under nitrogen at 60°C for five hours till the isocyanate (NCO) content reached 2.78% (by titration, 2.95% theoretical).

500 g of this polydimethyl siloxane diisocyanate prepolymer was diluted with 125 g of 2-hydroxyethyl methacrylate to give an 87.3% solution of the siloxane prepolymer endcapped with methacrylate moieties forming the macromer in 2-hydroxyethyl methacrylate. This solution was stirred at room temperature till the isocyanate (NCO) band disappeared from the infrared spectra taken of the solution.

Example 2: To 100 g of the siloxane macromer solution prepared in Example 1 was added 0.1 g of tert-butyl peroctoate. The macromer solution in 2-hydroxyethyl methacrylate was degassed for 30 minutes and then cast between Teflon coated aluminum sheets fitted with a 1/16 inch (1.588 mm) diameter silicon cord spacer. The sheet mold was cured at 80°C for three hours with a one hour postcure period at 100°C. After cooling, a translucent, very flexible polymer sheet of 1.5 mm thickness was obtained. The DS of this polymer was 31.5% in water and 106% in ethanol.

APPLICATIONS

Aqueous Polysiloxane Emulsions with Mica Treated with Organosilicon Compound

Highlights of the Technological Achievement: Aqueous diorganopolysiloxane emulsions containing dispersed inorganic silicates which have been treated with an organosilicon compound to impart hydrophobic properties thereto. The resultant dispersions may be employed as release agents especially in molding tires.

Reference: P. Huber and J. Meusel; U.S. Patent 4,184,880; January 22, 1980; assigned to Wacker-Chemie GmbH, Germany.

Background: Aqueous diorganopolysiloxane emulsions having inorganic silicates dispersed therein and their use as release agents for molding tires have been described in U.S. Patent 3,872,038. In contrast to the aqueous diorganopolysiloxane emulsions having inorganic silicates dispersed therein, the dispersions of this process have several advantages. For example, the dispersions of this process have a lower viscosity at the same solids concentration. Moreover, the dispersions of this process filter much more rapidly than prior dispersions. In addition, the dispersions of this process can be easily sprayed with either the so-called "airless" equipment or with compressed air equipment. Furthermore, these sprayed dispersions provide a more uniform and finely dispersed coating than the dispersions used heretofore.

Process Conditions: The diorganopolysiloxane emulsions employed in this process are the same diorganopolysiloxanes which have been used or could have been used in the aqueous diorganopolysiloxane emulsions employed heretofore. In addition to the diorganosiloxane units, the diorganopolysiloxanes may also contain other siloxane units, such as triorganosiloxane units which can be located either at the end and/or along a chain consisting of diorganosiloxane units.

Other units which may be present are monoorganosiloxane units and/or $SiO_{4/2}$ units. Monoorganosiloxane units and/or $SiO_{4/2}$ units may, however, be present in addition to the diorganosiloxane units, regardless of whether or not the triorganosiloxane units are present. It is preferred that the emulsified diorganopolysiloxanes contain 1.9 to 2.1 SiC-bonded organic radicals per silicon atom. It is, however, preferred that at least 90 mol percent of the siloxane units in the emulsified diorganopolysiloxanes be diorganosiloxane units.

It is preferred that the SiC-bonded organic radicals in the emulsified diorganopolysiloxanes be hydrocarbon radicals having from 1 to 18 carbon atoms. Also, these hydrocarbon radicals may be halogenated hydrocarbon radicals. Examples of suitable hydrocarbon radicals having from 1 to 18 carbon atoms are alkyl radicals; cycloalkyl radicals such as the cyclohexyl radical; alkenyl radicals such as vinyl and allyl radicals; aryl radicals such as the phenyl radical; alkaryl radicals such as the tolyl radical; and aralkyl radicals such as the benzyl radical. Examples of halogenated hydrocarbon radicals having from 1 to 18 carbon atoms are the 3,3,3-trifluoropropyl radical and the o-, p- and m-chlorophenyl radicals. Because they are readily available, it is preferred that at least 50% of the number of SiC-bonded organic radicals in the emulsified diorganopolysiloxanes be methyl radicals. Also, it is preferred that the emulsified diorganopolysiloxanes have a viscosity of at least 1,000 cp at 25°C. Their plasticity, as measured with the parallel-plate-plastometer test, in accordance with ASTM-D 926-67, should not exceed 120 mm.

It is preferred that the inorganic silicate employed in the dispersions of this process have a particle size of 0.02 to 0.25 mm. Mica is the preferred silicate.

Any organosilicon compound capable of imparting hydrophobic properties to inorganic substances may be used to treat the inorganic silicates of this process. Organosilicon compounds which impart hydrophobic properties to inorganic substances are those having the general formula: $(R_3Si)_aZ$ or $R_{a'}SiZ'_{4-a'}$ wherein R which may be the same or different represents monovalent hydrocarbon radicals and halogenated hydrocarbon radicals, Z and Z' represent halogen, hydrogen or a radical of the formula: –OH, –OR', –NR'X, –ONR', –SR' or –OOCR' and when a is 2, then Z may also represent –O–, –NX– or –S–; R' is a hydrocarbon radical or a substituted hydrocarbon radical, preferably an alkyl radical having from 1 to 4 carbon atoms; X is hydrogen or has the same meaning as R'; a is 1 or 2 and a' is 1, 2, or 3. Other organosilicon compounds which may be employed are octamethylcyclotetrasiloxane and/or dimethylpolysiloxanes having from 2 to 12 siloxane units per molecule in which the dimethylpolysiloxanes have an Si-bonded hydroxyl group in each of their terminal units.

It is preferred that organosilanes having the general formula: $R^1_{a'}SiOR^2_{4-a'}$ wherein R^1 which may be the same or different represents hydrocarbon radicals having from 1 to 18 carbon atoms, R^2 which may be the same or different represents alkyl radicals having from 1 to 4 carbon atoms or methoxyethylene radicals and a' is 1, 2 or 3, be reacted with the inorganic silicates because they do not release any by-products which may have a detrimental effect on the dispersions of this process.

Examples of thickening agents which may be employed are magnesium aluminum silicate, carboxyethylcellulose, carboxymethylcellulose, sodium carboxymethylcellulose, sodium carboxymethylhydroxyethylcellulose, hydroxyethylcellulose, methylcellulose, methylhydroxypropylcellulose, ethylhydroxyethylcellulose, methylethylcellulose, methylhydroxyethylcellulose, polyvinyl alcohol and bentonite. It is preferred that the thickening agents be employed in an amount of from 0.01 to 1.5% by weight based on the total weight of the dispersion.

In the following, all parts are by weight unless otherwise specified.

Example: The following ingredients are placed in a mixing device which operates at 1,000 rpm in the sequence indicated: 300.0 parts of water; 0.1 part sodium carboxymethylcellulose; 8.0 parts lecithin; 50.0 parts of a copolymer consisting of ethylene oxide and propylene oxide containing 70 mol percent of ethylene oxide units and 30 mol percent of propylene oxide units and having an average molecular weight of 3,000; 100.0 parts of an aqueous diorganopolysiloxane emulsion; 4.0 parts of a diorganopolysiloxane-based commercially available antifoaming agent (SE47, (Wacker-Chemie GmbH); and 350.0 parts of hydrophobic mica (wet-ground muscovite which has been treated with 5% by weight of dimethyldiethoxysilane based on the weight of the mica).

The mica is added to the mixture comprising the other components over a period of approximately 5 minutes. Mixing at 1,000 rpm is then continued for an additional 5 minutes.

Comparison Example: The procedure described in the above example is repeated, except that the wet-ground muscovite has not been treated with an organosilane.

The following table illustrates the properties of the resultant dispersions.

Dispersion Prepared as per	Dispersion Viscosity (cp)*	Sifting Duration (hr)**	Settled . After . . 3 Days . (wt %) .	12 Days	Appearance	Sprayed Coating Characteristics***
Example	5,000	2	4.3	5.6	Fine dispersion	1.7 g weight of coating; uniformly smooth, fine grained
Comparison Ex.	7,000	5	5.5	8.1	Lumpy dispersion	1.2 g weight of coating; rough, coarse grained.

*Brookfield viscosimeter, RVF, spindle No. 2, 2rpm.
**Through 250 micron sieve, equipped with rotating stripper.
***10 seconds with DeVilbiss paint spray gun, 1.0 mm nozzle; compressed air 6 bars, at a distance of 35 cm onto a rotating disk with a diameter of about 22 cm.

Craze-Resistant Polysiloxane Resin Coatings

Highlights of the Technological Achievement: Polysiloxane resin coatings containing a dispersed phase of particulate organic modifier which is tougher than the polysiloxane and which has an adequately high coefficient of friction have improved craze resistance permitting the use of thicker coatings with greater durability.

Reference: E. Vassiliou; U.S. Patent 4,183,998; January 15, 1980; assigned to E.I. Du Pont de Nemours and Company.

Background: Polysiloxane homopolymer resins and more costly copolymer resins such as those reacted with epoxies are sometimes used as nondurable food release coatings in combination with substantial proportions of silicone oils. The oils are fluid which facilitate application of the coatings and improve release of burned-on food residues. Such coatings are often formulated for relatively low temperature cure.

It would be desirable to have a more durable polysiloxane-resin based coating which could be applied in thick layers without being subject to craze cracking and without the necessity of using multiple thin layers.

Process Conditions: The process provides a coating composition consisting essentially of a crosslinkable polysiloxane resin-forming precondensate and 5 to 30%, preferably 10 to 20%, by weight of a particulate organic material, referred to herein as a modifier since it modifies the characteristics of the coating to give improved craze resistance, wherein the modifier has an average particle size in the range of 0.1 to 15 μm, preferably 0.5 to 5 μm, and wherein the modifier is tougher than the crosslinked coating made from the coating composition, and the modifier has a higher static coefficient of friction against the polysiloxane resin than does polytetrafluoroethylene. Percentages herein are based on the combined weight of the polysiloxane resin or resin-forming precondensate and the modifier, except where stated otherwise.

Preferably, the polysiloxane resin-forming precondensate is a polysiloxane resin-forming precondensate having one or more of the units:

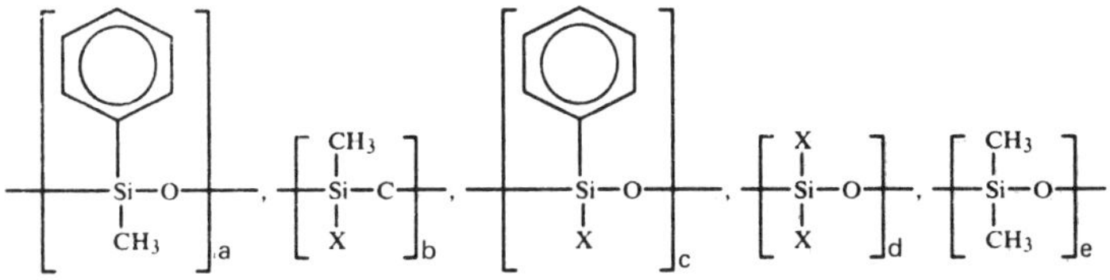

wherein X is a functional group which allows crosslinking at its site, and a, b, c, d and e are of a magnitude and in proportions which provide the resin with a degree of substitution, measured as the ratio of phenyl plus methyl groups to silicon atoms, of at least about 1.0, and a ratio of phenyl groups to silicon atoms of at least about 0.3.

Preferably, X is the hydroxyl group and the silicon-bonded hydroxyl content is at least about 3%, more preferably at least about 4%, based on the weight of the precondensate. Precondensates having a random arrangement of the repeating units in any order are, of course, included.

The modifier is preferably a polyarylene sulfide (more preferably polyphenylene sulfide), a polyarylene sulfone, a polyamide-imide resin, or a poly-para-hydroxybenzoic acid.

It is thought that the process is broadly applicable to the use of polysiloxane resins. Such resins are provided in coating compositions in the form of resin-forming precondensates of varying molecular weight containing functionality, generally silanol functionality, which causes crosslinking upon curing to form a rigid three-dimensional macropolymer matrix. It is within this matrix that the relatively tough and adherent particulate material used in the process is dispersed. The precondensates will generally be of relatively low molecular weight, such as 1,000 or 1,100, and enough of the precondensate molecules will have three or more functional groups to cause the formation of a resin with a rigid, crosslinked, three-dimensional matrix upon condensation or curing. Preferably, the matrix will be hard and have a low degree of elongation, such as 2 to 4%, in contrast with elastomers which are soft and have a high degree of elongation, such as 30% or more.

Crosslinking of the polysiloxanes can be accomplished by catalysis and heat. Crosslinking can take place at hydroxyl groups, halogens, hydrocarbyloxy groups, and acyloxy groups. If a, b, c and e are each zero and the structure is made entirely of:

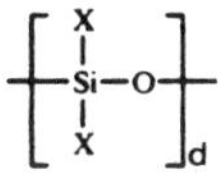

the resulting composition would be hydrated silicon dioxide which is not a resin-forming precondensate. Also, if b, c and d were all zero and the resin were made entirely of:

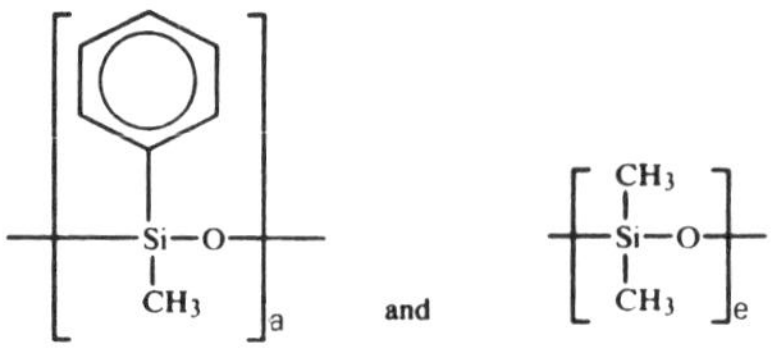

the result would be an elastomer rather than a resin.

In the following compositions, shown in the table below, the indicated materials were mixed and pebble-milled for 24 hours, producing an average particle size of the modifier generally on the order of 2 to 3 μm. The particles after milling tend to be rather rounded with fairly uniform particle sizes. The resulting formulations were filtered through a 100-mesh screen (U.S. Standard Sieve Size) and applied over grit-blasted aluminum substrates to a dry film thickness of 25 to 50 μm. The coated substrates were air-dried and then placed in a 425°C (800°F) oven until the temperature of the metal substrate reached 425°C (800°F).

In all cases, the coatings had a satisfactory appearance. They did not stress craze or form cracks and could be used satisfactorily in cookware, exhibiting substantial durability. In the compositions, all parts, percentages and proportions are given by weight.

Compositions

	Example No.							
	1	2	3	4	5	6	7	8
Polymethylphenylsiloxane								
DC 801	48.75	–	–	–	–	–	–	–
DC 803	–	58.50	–	–	–	–	–	–
DC 805	–	–	58.50	–	–	–	–	–
DC 806A	–	–	–	58.50	–	–	–	–
DC 840	–	–	–	–	48.75	–	–	–
SR 112	–	–	–	–	–	58.50	–	–
SR 82	–	–	–	–	–	–	48.75	–
SR 240	–	–	–	–	–	–	–	58.50
Xylene	26.00	25.64	25.64	25.64	26.00	25.64	26.00	25.64
VMP naphtha	3.61	3.61	3.61	3.61	3.61	3.61	3.61	3.61
8% zinc octoate in VMP naphtha	1.83	1.83	1.83	1.83	1.83	1.83	1.83	1.83
Titanium dioxide pigment	7.32	7.32	7.32	7.32	7.32	7.32	7.32	7.32
Alumina pigment	7.33	7.33	7.33	7.33	7.33	7.33	7.33	7.33
Polyphenylene sulfide Ryton V-1	5.16	5.16	5.16	5.16	5.16	5.16	5.16	5.16

Coated Polymeric Substrates

Highlights of the Technological Achievement: Abrasion resistant coatings of silica and a partially polymerized organic silanol of the formula $RSi(OH)_3$ having rapid cure times are obtained when the pH of the composition is maintained at greater than 6.0, and preferably 6.2 or greater.

Reference: R.W. Ubersax; U.S. Patent 4,177,315; December 4, 1979; assigned to E.I. Du Pont de Nemours and Company.

Background: A variety of polymeric materials are used in structural and laminating applications which require optical clarity. For example, methyl methacrylate polymers and polycarbonates are used increasingly in glazing applications and polyethylene terephthalate films can be incorporated into automotive windshields. While these clear plastic substrates perform their structural function for extended periods of time, they are subject to surface scratching which mars their appearance and optical clarity. Accordingly, coatings have been applied in the past to increase the abrasion resistance of these substrates. A need exists for coating compositions which provide better long-term abrasion resistance and permit rapid application and curing consistent with high-speed commercial applications.

Process Conditions: The process provides improved coating compositions for clear polymeric substrates which exhibit excellent abrasion resistance and curing times markedly lower than heretofore realized with earlier compositions.

Specifically, there is provided a rapidly-curing coating composition comprising:

(a) 5 to 50 weight percent solids, the solids comprising 10 to 70 weight percent of a partially polymerized organic silanol of the general formula $RSi(OH)_3$, wherein R is selected from methyl and up to 40% of a radical selected from the group consisting of vinyl, phenyl, γ-glycidoxypropyl, and γ-methacryloxypropyl, and

(b) 95 to 50 weight percent solvent, the solvent comprising 10 to 90 weight percent water and 90 to 10 weight percent lower aliphatic alcohol,

the coating composition having a pH of greater than 6.0 and less than 8.0.

The coating compositions, after application to the substrate, can be cured at a temperature of 20° to 150°C. The cure times used for these coatings are significantly lower than those previously used in the art at comparable temperatures. While ambient temperatures of 20°C require cure times of several days, elevated temperatures of 150°C will cure the coatings in several seconds. For convenience in commercial operations, a curing temperature of 100°C is preferred, at which cure times of about one minute are realized.

The process is further illustrated by the following specific examples, in which parts and percentages are by weight unless otherwise indicated.

Examples 1 through 4 and Comparative Examples A through D: Coating solutions were prepared by the cohydrolysis of tetraethylorthosilicate and methyltrialkoxysilanes. In Examples A, B, 1 and 2, a mixture of 104 g of tetraethylorthosilicate and 136 g of methyltrimethoxysilane was hydrolyzed by adding a solution of 110 g of water and 5.0 ml of 0.1 N hydrochloric acid solution. An exothermic reaction immediately ensued after which the mixture became homogeneous. The resulting solution was stirred at room temperature for an additional 4 hours.

In Examples C, D, 3, and 4, a mixture of tetraethylorthosilicate (104 g) and methyltriethoxysilane (178 g) was cohydrolyzed by reaction with 110 g of water and 5.0 ml of 0.1 N hydrochloric acid solution, and stirred at room temperature for 5 hours.

To 60 g portions of each of the above two base solutions was added 40 g of an isopropyl alcohol/water mixture (25 g isopropyl alcohol/15 g water), 0.5 ml of 10% acetic acid solution and 0.4 N sodium hydroxide to adjust the pH to various levels. The resulting solutions were 16.4% solids, of which 31% was SiO_2 and 69% $CH_3SiO_{1.5}$. Samples of primed 7-mil polyethylene terephthalate film were dip coated and immediately cured for one minute at 100°C. Abrasion resistance of the samples was tested by rubbing with steel wool and measuring the increase in haze and the results are summarized in the table below.

Example	pH	Abrasion Resistance (% ΔH)
A	5.5	1.6
B	6.0	0.4
1	6.4	0.3
2	6.8	0
C	5.5	4.3
D	6.0	1.2
3	6.4	0.1
4	6.8	0.1

Organothiol-Containing Siloxane Resins as Adhesion Promoters for Siloxane Resins

Highlights of the Technological Achievement: A composition of matter which is an aqueous dispersion of an organothiolsilsesquioxane and colloidal silica is used as an adhesion additive for siloxane resins which have a low degree of substitution.

Reference: R.H. Baney and L.A. Harris; U.S. Patent 4,177,175; December 4, 1979; assigned to Dow Corning Corporation.

Background: There are silicone resins which have been utilized to coat various substrates for a variety of purposes and whenever there were desirable properties of these resins that those skilled in the art wished to impart to the substrate, there were always ways to gain the necessary adhesion. For example, silicone resins are used on metal substrates for encapsulation purposes but some metals utilized in electrical systems simply would not adhere to the silicone resins. Some of the common approaches to adhesion in these cases was to prime the metal surface with functional silanes such as:

$$(CH_3O)_3SiCH_2CH_2CH_2OCH_2CH\overset{O}{\text{—}}CH_2$$

or

$$(CH_3O)_3SiCH_2CH_2CH_2NHCH_2CH_2NH_2,$$

allowing an air-dry and then applying the resin or, these silanes were added directly to the resin and then the resin was applied to a cleaned metal substrate. Either way, the abovementioned silanes seemed to provide good adhesion, presumably due to their compatibility in (or with) the silicone resins having a high degree of organic substitution.

When one scans the prior art surrounding the adhesion of silicone resins to various substrates, one continuous thread winds its way through the publications; there is no universal primer or adhesion additive and therefore, cautionary statements are always made in those publications that the substrates and the adhesion promoters have to be matched for the best results. Thus, it is not totally unexpected that the ordinary adhesion promoters discussed above would not function well with the siloxane resins having a low degree of organic substitution.

Process Conditions: The process provides a pigment-free aqueous coating composition comprising a mixture of (A) and (B), in which component (A) is a dispersion of colloidal silica in a lower aliphatic alcohol-water solution of the partial condensate of a silanol of the formula $RSi(OH)_3$, in which R is a C_{1-3} alkyl radical, the vinyl radical, the 3,3,3-trifluoropropyl radical, the γ-glycidoxypropyl radical, the γ-methacryloxypropyl radical and the phenyl radical, at least 70 wt % of the silanol being $CH_3Si(OH)_3$.

Component (B) is an aqueous composition comprising a dispersion of colloidal silica in lower aliphatic alcohol-water solution of a partial condensate of a silanol of the formula $R'Si(OH)_3$, in which R' is HSR"– or $(HS)_2R$"– and R" is selected from a group consisting of a divalent aliphatic hydrocarbon radical having 1 to 6 carbon atoms and phenylene.

Component (B) is prepared essentially in the same manner as is component (A). The difference between the two materials is that the partial condensate of (A) is primarily $CH_3Si(OH)_3$ while the partial condensate of (B) is derived from a mercaptofunctional substituted silanol.

Aqueous colloidal silica dispersions having a particle size in the range of 5 to 150 mμ in diameter are required.

Example 1: *Preparation of Component (A)* – A mixture of 10.5 g of glacial acetic acid and 304.5 g of methyltrimethoxysilane was slowly added to 444.1 g of a commercially available aqueous dispersion of colloidal silica having an initial pH of 3.1 containing approximately 34% SiO_2 of approximately 22 mμ particle size and having an Na_2O content of less than 0.01 wt %. The acidified dispersion was stirred while it was slightly cooled by an external ice bath to keep the temperature below about 10°C. The dispersion was stirred for about 1 hour beyond the end of the full addition of the silane while generating methanol and methyltrisilanol. Upon standing for a while, the pH stabilized at 4.5 and the material was stripped to approximately 60% solids and diluted to 35% solids using alcohol.

Example 2: *Preparation of Component (B)* – Component (B) is 50 weight percent $HSCH_2CH_2CH_2Si(OH)_3$ and 50 wt % SiO_2. A solution (142.9 g) of 48.6 g of the same colloidal silica as was used in Example 1, 94.3 g of water and 10 g of glacial acetic acid was placed in a glass reaction flask and there was added thereto 77.2 g of mercaptopropyltrimethoxysilane with moderate stirring to produce a hydrolyzate containing $HSCH_2CH_2CH_2Si(OH)_3$. 134.6 g of isopropanol was added, the mixture stirred and then stripped to 48.2% solids and then rediluted with alcohol to 35% solids.

Example 3: Component (A) from Example 1 was further diluted to 22.5 wt % solids using alcohol. Various quantities of component (B) were then added to component (A) so as to give 5 wt % of the solids of component (B) based on the total solids content of (A) and (B), in the coating composition.

Prior to coating, the test panels were cleaned in the following manner. The panels were washed with isopropanol and then hexane and then rubbed gently with a dust-free cloth containing heptane. The panels were then rinsed with isopropanol, air-dried and heated at 125°C for 2 hours, allowed to cool to room temperature and then coated immediately.

The abrasion resistance was measured according to ASTM Method D1044-56 using a 500 g test load and CS-10F abrasive wheel for 500 revolutions. The percent change in haze from an unabraded surface to an abraded surface is reported. Haze is defined as that percentage of transmitted light which in passing through the specimen deviates from the incident beam by forward scattering according to ASTM Method D1003-61.

To determine haze, the amount of diffused light is measured, divided by the amount of transmitted light and multiplied by 100. NA indicates no adhesion. The results of % change in haze are shown in the tables below:

Cure Time (hr)	.......Air Dry Time After Coating....... ...5% Additive Level; RH 65%/Cure 125°C... ½ Hour	1½ Hours	3 Hours
½	NA	NA	NA
1½	0.7	1.0	3.0
3	1.5	2.5	2.8
	... 0% Additive Level*; RH 65%/Cure 125°C....		
½	NA	NA	NA
1½	NA	NA	NA
3	NA	NA	NA
	....5% Additive Level; RH 35%/Cure 125°C		
½	NA	NA	0.8
1½	0.4	1.0	1.6
3	1.5	1.2	–
	....0% Additive Level*; RH 35%/Cure 125°C....		
½	NA	NA	NA
1½	NA	NA	NA
3	NA	NA	NA

*0% additive level is component (A) without component (B).

The data show that enhanced adhesiveness is obtained by the use of this process and that the incorporation of the component (B) in component (A) shortens the cure time of component (A).

Emulsion-Polymerized Polydimethylsiloxanes in Durable Press Resins

Highlights of the Technological Achievement: Silicone-containing durable press resin compositions are prepared by polymerizing a polydimethylsiloxane which has been emulsified in an aqueous solution of dimethylolethylene urea or dimethyloldihydroxyethylene urea. The resulting homogeneous aqueous compositions have improved resistance to separation after freeze-thaw cycling. Advantageously, more highly concentrated aqueous compositions comprising a silicone-containing durable press resin are available by the method of this process. The silicone-containing durable press resin compositions are useful for providing wrinkle-recovery and permanent press to cellulose-containing fabrics.

Reference: H.E. Griffin; U.S. Patent 4,170,581; October 9, 1979; assigned to Dow Corning Corporation.

Background: Durable press resins and silicone compositions have been individually used for a long time to provide improved properties to textiles. In addition, to further improve the physical properties of textiles treated with durable press resins, silicone materials have been combined with durable press resins to provide improved hand, tear strength, and abrasion resistance to the treated textile.

One disadvantage of prior methods is the tendency of the silicone emulsion and the blend of silicone emulsion with durable press resin solution to separate, such as by creaming or occasionally by breaking, after being subjected to one or more freeze-thaw cycles. Although the blends are easily reemulsified, problems with storage of these materials are introduced by this disadvantage.

Another disadvantage of the above methods is the necessity of processing, shipping and storing relatively large quantities of water as a component of the silicone emulsion, the durable press resin solution and their blends that are used to prepare the textile-treating bath. Additional economies could be realized if the quantity of water in these compositions that are subsequently used to prepare the textile-treating bath could be reduced.

This method comprising emulsion polymerization of lower molecular weight polydimethylsiloxanes in a solution of durable press resin provides a means for overcoming the abovementioned disadvantages. The resulting homogeneous aqueous composition obtained therefrom has improved stability to separation after freeze-thaw cycling. In addition, since the water of the homogeneous aqueous composition of this process serves a dual role in dissolving the durable press resin and functioning as the continuous phase during the emulsion polymerization of the polydimethylsiloxane, more highly concentrated compositions are made available, greater economies are realized and compositions of greater utility are obtained than those of the prior art.

Process Conditions: This process relates to a method for preparing a homogeneous aqueous composition comprising a polydimethylsiloxane and a durable press resin, and to the compositions obtained therefrom, the method comprising:

(A) preparing an emulsion consisting essentially of:

  (1) a polydimethylsiloxane having a viscosity of less than 100 mPa-sec at 25°C and being selected from the group consisting of cyclopolydimethylsiloxanes and hydroxy-endblocked linear polydimethylsiloxanes,

  (2) a surface-active siloxane-polymerization catalyst selected from aliphatic hydrogen sulfates and hydrocarbonsulfonic acids, and

  (3) an aqueous solution of a durable press resin selected from dimethylolethylene urea and dimethyloldihydroxyethylene urea,

(B) allowing the emulsified polydimethylsiloxane to polymerize until a desired increase in molecular aggregation has been obtained, and thereafter

(C) neutralizing the surface-active catalyst.

Component (3) of the emulsions that are prepared in the method of this process is an aqueous solution of a durable press resin selected from dimethylolethylene urea and dimethyloldihydroxyethylene urea. These durable press resin solutions, conveniently designated in the art as DMEU resin and DMDHEU resin, respectively, consist essentially of monomeric molecules of 1,3-bis(hydroxymethyl)-2-imidazolidinone and 1,3-bis(hydroxymethyl)-4,5-dihydroxy-2-imidazolidinone, respectively, and may further contain trace to small amounts of precursor materials such as formaldehyde, urea, ammonia and glyoxal and higher molecular weight, condensed ethylene urea.

Tests: Each fabric sample was washed five times (AATCC Test Method 124-1975) and tested for flat appearance (AATCC Test Method 124-1975) and shrinkage (AATCC Test Method 135-1973). The samples treated with textile-treating baths prepared by the method of this process had essentially equivalent shrinkage and slightly inferior flat appearance compared to the sample treated per the prior art as noted in the table below.

Composition Number	Flat Appearance	. . .Shrinkage (%) . . .	
		Warp	Fill
1	3.5	0.5	0.8
2	3.4	0.6	0.8
3	3.3	0.7	0.5
4	3.4	0.8	0.6
5*	3.8	0.5	0.6

*Prior art composition for comparison purposes only.

Aqueous Emulsion of a Hydroxyl-Endblocked Polydimethylsiloxane for Textile Treatment

Highlights of the Technological Achievement: A homogeneous blend of a concentrated aqueous solution of a durable press resin and a concentrated aqueous emulsion of a hydroxyl-endblocked polydimethylsiloxane was found to be storage-stable and to possess improved resistance to skin-over during atmospheric exposure. The preparation of the homogeneous blend and its subsequent dilution with water comprises an improved method for preparing a textile-treating composition.

Reference: R.J. Rooks; U.S. Patent 4,167,501; September 11, 1979; assigned to Dow Corning Corporation.

Background: Herein, the concentrated aqueous solution of durable press resin may also be denoted as the concentrated resin solution, and the concentrated aqueous emulsion of hydroxyl-endblocked polydimethylsiloxane may also be denoted as the concentrated silicone emulsion. A mixture of the concentrated resin solution and the concentrated silicone emulsion may also be denoted herein as the concentrated homogeneous resin-silicone blend.

Preparation of a textile-treating composition comprising a silicone component and a durable press resin is usually done immediately prior to its intended use using concentrated forms of the durable press resin and silicone component. Thus, a common method of preparing a textile-treating composition comprises adding a concentrated aqueous solution of a water-soluble durable press resin and a concentrated aqueous emulsion of a hydroxyl-endblocked polydimethylsiloxane, either sequentially or simultaneously, to a portion of water, preferably containing a surfactant, to provide a composition having suitable stability and proper dilution for applying to a textile. Preparing a textile-treating composition immediately prior to use, for example, within 8 to 10 hours of its intended use, is necessitated by the fact that the completed bath also usually comprises catalysts and curing agents for the durable press resin and the silicone component, thereby rendering the textile-treating composition unstable to long term storage. However, there are some disadvantages with this method.

A principal disadvantage with the above method for preparing a textile-treating composition is the tendency of the concentrated silicone emulsion to skin-over when it is exposed to the evaporating action of the open atmosphere.

A principal disadvantage of this skin-over is the tendency of this layer of semisolid or solid polymer to be incorporated into the subsequently prepared textile-treating compositions as noncolloidal particles which interfere with the proper use of the textile-treating composition.

Another disadvantage with the above method for preparing a textile-treating composition is the need to prepare, ship, store, inventory or otherwise handle the several individual components that are used in preparing the textile-treating compositions. It would be desirable to combine two or more of these individual components to form stable concentrated blends to facilitate and simplify the handling.

Yet another disadvantage with the abovementioned method for preparing a textile-treating composition is the possibility for error and batch-to-batch variation when admixing several components "in the field;" that is, in the relatively less sophisticated environment of the end-user. This possibility for error and batch-to-batch variation can be reduced by doing some of the mixing of components at the component manufacturer's and/or supplier's facility.

Process Conditions: This process provides an improved method for preparing a textile-treating composition comprising a water-soluble durable press resin and a hydroxyl-endblocked polydimethylsiloxane by premixing the appropriate amounts of a concentrated aqueous solution of a durable press resin and a concentrated aqueous emulsion of a hydroxyl-endblocked polydimethylsiloxane prior to preparing the textile-treating composition. The composition obtained by the premixing was found to have unexpected resistance to skin-over.

Because of the well-known tendency of concentrated hydroxyl-endblocked polydimethylsiloxane emulsions to break when mixed with polar materials such as alcohols and aqueous solutions of salts, and because of the long-standing practice in the textile-treating industry to mix the concentrated resin solutions and concentrated silicone emulsion in additional water, preferably containing a surfactant, to avoid breaking the silicone emulsion, it was surprising that a concentrated resin solution and a concentrated silicone emulsion could be mixed to provide a storage-stable homogeneous resin-silicone blend.

It was also surprising to find that a concentrated silicone emulsion could be stabilized against skin-over during atmospheric exposure by admixing therewith a concentrated resin solution to provide a concentrated homogeneous resin-silicone blend having storage stability.

Of particular interest for this process are the durable press resins based on 1,3-bis(hydroxymethyl)-2-imidazolidinone, sometimes designated as DMEU resins or dimethylolethylene urea resins; 1,3-bis(hydroxymethyl)-4,5-dihydroxy-2-imidazolidinone, sometimes designated as DMDHEU resins or dimethyloldihydroxyethylene urea resins; and N,N-bis(hydroxymethyl)methoxyethyl carbamate resins. These durable press resins are of particular interest because of their ability to form concentrated homogeneous resin-silicone blends with any of the concentrated silicone emulsions hereinafter delineated.

The durable press resins which are suitable for use in the process are sufficiently water-soluble to permit the preparation of concentrated aqueous solutions thereof. By concentrated aqueous solutions it is meant, solutions which contain at least 20% by weight, preferably from 35 to 50% by weight, of the durable press resin, based on the total weight of the concentrated resin solution. Durable press resins are commercially available as concentrated resin solutions or as solids, pastes and creams from which the concentrated resin solution may be prepared.

By hydroxyl-endblocked polydimethylsiloxanes it is meant, polymers having the general formula: $HO[(CH_3)_2SiO]_xH$ wherein x has an average value of 10 to 1,500 and corresponding to a polymer viscosity at 25°C of approximately 0.03 to approximately 250 pascal-seconds (Pa-sec). Hydroxyl-endblocked polydimethylsiloxanes having a viscosity of at least 10 preferably at least 50 Pa-sec are of particular interest in the method of this process because they provide a pleasing "hand" to fabrics treated therewith and because their concentrated aqueous emulsions are more likely to need stabilization against skin-over.

The concentrated homogeneous resin-silicone blend of this process may be used immediately to prepare a textile-treating composition or it may be transported and/or stored for up to 12 months or more as desired and subsequently used to prepare a textile-treating composition. In any event, it is no longer necessary to store concentrated silicone emulsion for preparing these textile-treating compositions. In the following examples, all parts, percentages and ratios are by weight unless otherwise specified. Viscosity was measured in centipoises and was converted to pascal-seconds by dividing by 1,000.

Example 1: A 40% emulsion of hydroxyl-endblocked polydimethylsiloxane was prepared by heating a homogenized mixture of 1,070 parts of water, 71.2 parts of dodecylbenzenesulfonic acid and 950 parts of cyclopolydimethylsiloxanes at 80°C for 4 hours. After being cooled, the emulsion was neutralized with 43.3 parts of triethanolamine in 83.6 parts of water and further mixed with 116 parts of octylphenoxypolyethoxy (40) ethanol in 30 parts of water. The emulsion had a pH of 7.7 and the hydroxyl-endblocked polydimethylsiloxane had a viscosity of 96.6 Pa-sec at 25°C.

This polydimethylsiloxane emulsion was thoroughly mixed, in various weight ratios, with a 40% solution of 1,3-bis(hydroxymethyl)-4,5-dihydroxy-2-imidazolidinone (DMDHEU) in water. Certain of the resulting mixtures also contained additional surfactant, as indicated in Table 1 below.

The resulting concentrated homogeneous resin-silicone blends, as well as the concentrated silicone emulsion only and the concentrated resin solution, were placed in separate open containers and exposed to the atmosphere for 5 days. At the end of 5 days the presence or absence of a skin on the surface of the mixture and buildup of solid deposit on the walls of the container were noted. The results, summarized in Table 1, show the advantageous resistance of the compositions of this process against skin-over. In addition, none of the tabulated mixtures of silicone emulsion and resin solution showed any separation at room temperature in a closed container.

Table 1

Blend Number	Ratio of Resin Solution to Silicone Emulsion	. . Atmospheric Exposure Results . . Skin-Over	Sidewall Deposit
1	5	None	Very little
2	2	None	Light to moderate
3	1	None	Heavy
4*	0.5	Thick	Heavy
5**	5	None	Trace
6**	2	None	Moderate
7**	1	None	Moderate to heavy
8*,**	0.5	Thick	Heavy
9*	***	Thick	Heavy
10*	†	None	Trace

*For comparison purposes only.
**Contains an additional 0.1 part octylphenoxypolyethoxy(40)ethanol/1 part silicone emulsion.
***Silicone emulsion only.
†Resin solution only.

Example 2: Six concentrated aqueous solutions of various durable press resins were each mixed with one of three concentrated aqueous emulsions of various hydroxyl-endblocked polydimethylsiloxanes in a ratio of 5 parts of durable press resin solution to 1 part of polydimethylsiloxane emulsion. 50 ml of each mixture were placed in 2 ounce containers and the closed containers were allowed to stand at room temperature for one week. The results of a visual clarity and compatibility test on these week-old samples are summarized in Table 2. Note that not all combinations provide homogeneous resin-silicone blends. This example illustrates a simple test that can be used to determine whether concentrated resin-silicone blends are homogeneous and, therefore, are within the scope of this process.

In Table 2 the resin solutions and silicone emulsions are as follows:

Resin WNM–40% solution of DMDHEU resin solution (Burlington Industries)

Protorez EU-50–DMEU resin solution (Procter Chemical)

Stanset C-90–dimethylolmethoxyethyl carbamate resin solution (Standard Chemical)

Protorez CHD–dimethylolethyl carbamate resin solution (Procter Chemical)

Aerotex 23 Special–triazine-formaldehyde resin solution (American Cyanamid)

Resin MW–melamine-formaldehyde resin solution (American Cyanamid)

DC 1111–40% emulsion of a hydroxyl-endblocked polydimethylsiloxane emulsion polymer having a viscosity of approximately 100 Pa-sec (Dow Corning)

DC 347–60% emulsion of emulsified hydroxy-endblocked polydimethylsiloxane having a viscosity of 3 to 5 Pa-sec (Dow Corning)

DC 27116–20% emulsion of emulsified hydroxyl-endblocked polydimethylsiloxane having a viscosity of 60 to 70 mPa-sec (Dow Corning)

UCC TE-24–20% emulsion of hydroxy-endblocked polydimethylsiloxane (Union Carbide)

Table 2

Resin Solution	Silicone Emulsion* DC 1111	DC 347	DC 27116	UCC TE-24
Resin WNM	H, hazy	H, cloudy	H, hazy	H, cloudy
Protorez EU-50	H, hazy	H, hazy	H, hazy	H, hazy
Stanset C-90	H, cloudy	H, cloudy	H, cloudy	H, cloudy
Protorez CHD	H, cloudy	Incompatible	Incompatible	Incompatible
Aerotex 23 Special	H, hazy	Incompatible	Incompatible	Incompatible
Resin MW	H, cloudy	Incompatible	Incompatible	Incompatible

*H denotes homogeneous and indicates a concentrated homogeneous resin-silicone blend of this process.

Aqueous Polysiloxane Preparations for Shrink-Proofing of Wool

Highlights of the Technological Achievement: A method for producing a stable, aqueous preparation of compounds which are organosiloxanes having amino groups attached to the silicon atoms which are used for the shrink-proofing of wool which preparations contain silanes which may be aminodi- or trialkoxysilanes is disclosed. Wool treated with the preparation exhibits excellent decreased shrinkage as well as an absence of felting.

Reference: G. Koerner and H.-J. Patzke; U.S. Patent 4,137,179; January 30, 1979; assigned to Th. Goldschmidt AG, Germany.

Background: It is well known that wool, in the untreated state, shrinks and felts on being washed in aqueous liquor. In order to counteract this shrinkage and felting, chemical treatments have already been recommended in which the structure of the wool is changed, or finishes have been used which contain resin that deposits on the surface of the wool fibers and envelopes the fibers. With both processes, however, products are obtained whose so-called "hand" or aesthetic feel is unpleasant to the consumer.

One of the major disadvantages of these known processes is that the organosilicon compounds must be applied as a dispersion or as a solution in a volatile organic carrier. Organic solvents, such as benzene, hexane and perchloroethylene are mentioned as suitable carriers. In many mills, however, as well as in the case of a later finishing treatment in the home, the use of organic solvents is not possible. In addition, for reasons of environmental protection, it is desirable to avoid the use of organic solvents wherever possible.

Process Conditions: Adding aminodi- and/or trialkoxysilanes to an emulsion of a dimethylpolysiloxanediol of a viscosity from 1,000 to 100,000 cp at 20°C in such amounts that in every case, there are 25 to 1,000 dimethylsiloxy units per mol of aminosilane.

Especially preferred is a preparation in which 1 mol of aminosilane corresponds to every 40 to 160 dimethylsiloxy units. A preparation in which 1 mol of aminosilane corresponds to every 60 to 100 dimethylsiloxy units is particularly preferred.

The production of the emulsion of dimethylpolysiloxanediol, which preferably contains 5 to 30 wt % of dimethylpolysiloxanediol, can take place according to various processes of the state of the art. One can either start with a dimethylpolysiloxanediol of the desired viscosity, mix this with emulsifiers and add it to water with vigorous stirring or work water into the siloxane-emulsifier mixture. It is recommended that the emulsion so obtained, be homogenized mechanically.

As emulsifiers, the well-known nonionic emulsifiers based on ethylene oxide derivatives and anionic or cationic emulsifiers can be used.

Instead of the cyclic dimethylsiloxanes, dialkoxysilanes whose alkoxy residues are linear or branched and have 1 to 4 carbon atoms may also be used. The emulsion polymerization is preferably carried out at elevated temperatures up to 95°C.

Before use, the preparations may be diluted and adjusted to the pH range desired. The concentration of the liquor and the dwell time of the textile in the liquor are selected such that after squeezing off the excess treatment fluid on the fibers to be treated, 0.1 to 10 and especially, 0.1 to 5% of the organosiloxane compounds, based on the weight of the fiber, remain.

Example of the Use of the Emulsion of the Process: A knitted material of fine wool was treated with an emulsion prepared according to the process in such a way that, after simply drying the impregnated knitted material, a solids coating of 1% resulted.

The finished material, as well as the untreated material, was laundered ten times in a domestic washing machine and dried in a tumble dryer between launderings. After 10 launderings and dryings, the control material showed a felting area shrinkage of 44% and a strongly felted surface. On the other hand, the treated sample had an average felting area shrinkage of 5% and a completely unchanged appearance. In addition, the hand of the treated material after laundering was significantly softer than the hand of the untreated material after laundering and corresponded to the hand before laundering.

The calculation of felting area shrinkage was carried out according to the following formula:

$$\text{Felting area shrinkage} = L + \%W - \%L \times \%W/100$$

wherein the % L = percent longitudinal shrinkage and the % W = percent width shrinkage.

POLYURETHANE PRODUCTS

PROCESSING

Water Swellable, Lightly Crosslinked Polyurethane Hydrogels

Highlights of the Technological Achievement: Continuous process for producing hydrogels by charging preselected feeds containing poly(oxyalkylene) polyols, organic diisocyanate and catalyst to a reaction zone followed by crosslinking by exposure to atmospheric humidity. Solvents are not used and the hydrogels are shaped in desired configuration as they exit the extruder.

Reference: K.L. Hoy, A.P. Jones and R.J. Knopf; U.S. Patent 4,209,605; June 24, 1980; assigned to Union Carbide Corporation.

Background: The use of fibrous materials as absorbents for moisture is a well known, widespread practice. In this application many of the natural and synthetic fibers have been used and extensive efforts have been made to improve their absorption properties. One of the major deficiencies of the natural and synthetic materials previously used has been their tendency to release the absorbed moisture when pressure has been applied to the moisture containing material. The fact that pressure causes the absorbed fluid to be expelled from the fibers is known as reversible absorption. For many applications, however, irreversible absorption is desired, for example, in surgical dressings, diapers, bed pads, catamenials, and the like, whereby the absorbed moisture is retained in the absorbent material under an applied pressure.

Within the past decade innovations have resulted in the production of hydrogels, which have such irreversible absorption properties, in powder or particulate form and even, in some instances, in film form using processes which require the use of solvents. An especially interesting characteristic of the hydrogel polymers is that when in contact with water they absorb it and swell to a certain point and then stop further absorption; however, the final swollen polymer is still similar in shape to its initial unswollen shape. Thus, many of the hydrogels have the ability to absorb many times their original weight in water without becoming soggy or deformed.

Process Conditions: The water swellable, lightly crosslinked, polyurethane hydrogel polymers that are produced are the reaction products of an isocyanato terminated prepolymer comprising the reaction product of (1) poly(alkyleneoxy) polyols having an average molecular weight up to about 25,000 with (2) organic diisocyanates, wherein the prepolymer is subsequently lightly crosslinked with (3) a crosslinking agent that is water or an organic polyamine, the amount of crosslinking agent used being an equivalent amount based on the number of equivalents of isocyanato groups present in the prepolymer. In the processes of this method the ambient atmospheric moisture has been found eminently suitable; as is known, this can also be controlled to establish any relative humidity concentration desired.

The initial reactants have low Brookfield viscosities of 500 to 2,500 cp, generally from about 1,000 to 1,500 cp, and upon completion of the reaction the products have a high Brookfield viscosity of 1,000,000 to 10,000,000, or greater.

The reactants are introduced in the fluid form at temperatures of preferably 60° to 90°C, and are then reacted at reaction temperatures sufficient to cause rapid reaction, preferably 120° to 150°C.

The residence time in the reaction zone of the reaction means is an important factor in this process. This residence time varies from preferably 45 seconds to 2 minutes at the reaction temperature.

Example: *Feed 1* – A mixture of 3,000 g of a poly(oxyethylene) diol having an average hydroxyl number of 13.42 and 18.56 g of a poly(oxyethylene) triol having an average hydroxyl number of 141.5 was azeotropically distilled for 2 hours using trichloroethylene, to remove 6.2 g of water. The trichloroethylene was stripped at reduced pressure of 18 mm Hg to a kettle temperature of 130°C and the contents were then cooled to 95°C and 1.5093 g of triethylene diamine were added. After thorough mixing, the contents were transferred to a feed tank under dry nitrogen atmosphere and maintained at 80°C.

Feed 2 – A 1,000 g portion of the same diol used in Feed 1 was azeotropically distilled using 200 g of trichloroethylene to remove 4.2 g of water over a period of 1.5 hours; then the trichloroethylene was stripped. The contents of the reaction was cooled to 75°C and 1,000 g of methylenebis(4-phenylisocyanate) were added and thoroughly mixed. The contents were transferred to a second feed tank under dry nitrogen atmosphere and maintained at 80°C.

Hydrogel Polymer A – In this example 39.2323 g of Feed 1 and 3.2447 g of Feed 2 were pumped using gear type metering pumps and joined before entering an in-line turbine blade mixer in which the reactants are thoroughly mixed and which are attached to and discharged into the extruder reactor. The turbine blades of the mixer were rotating at about 625 rpm and the temperature in the mixer was maintained at 90°C. A total of 42.48 g of the reaction mixture was being fed to the extruder reactor every minute having an NCO:OH equivalent ratio of 1.3:1 and a Brookfield viscosity of about 1,500 cp at 70°C.

During the reaction the first zone of the extruder was maintained at 90°C and the second reaction zone at 120°C. The residence time in the second reaction zone was 46.6 seconds. The isocyanato terminated prepolymer produced was highly viscous (Brookfield viscosity >6,000,000 cp at 120°C) and was extruded onto glass plates and rolled with a forming blade to film form. As the film extruded and was rolled it was crosslinked to the hydrogel polymer.

Hydrogel Polymer B – Following the procedure described for Hydrogel Polymer A, 38.2381 g of Feed 1 and 3.4155 g of Feed 2 were reacted to produce a polyurethane hydrogel polymer. In this instance 41.65 g/min were fed to the extruder reactor and the NCO:OH equivalent ratio was 1.4:1. In both instances the films were tough, strong and translucent. The water uptake and solubility are tabulated below.

	......Hydrogel Polymer......	
	A	B
Water absorption, times	32.0; 29.5	13.5; 10.2
Water solubility, (%)	32.5; 33.0	22.9; 15.0

Nonionic, Water-Dispersible Polyurethanes

Highlights of the Technological Achievement: The process relates to nonionic, water-dispersible polyurethanes having a substantially linear molecular structure and lateral polyalkylene oxide polyether chains containing ethylene oxide units.

Reference: K. Noll and J. Pedain; U.S. Patent 4,190,566; February 26, 1980; assigned to Bayer AG, Germany.

Background: The nonionic polyurethane dispersions described in these prior publications are eminently suitable for the production of mechanically strong, highly elastic coatings. Both nonflexible substrates, such as wood and metals, and also flexible substrates, such as textiles or leather may be coated with the dispersions.

One preferred field of application is the coating of textiles. However, it has been found that in certain cases, especially in cases where high-gloss transfer papers are used, coated textiles produced using these polyurethanes show a phenomenon which may seriously restrict their commercial value. This phenomenon is the appearance of a coating which becomes increasingly dull in the event of prolonged storage. High-gloss, dark-pigmented or colored coatings in particular show a speckled, matt-grey appearance after from 3 to 6 weeks which seriously affects the optical appearance of the sheet-form material. Although this undesirable phenomenon may be eliminated by wiping with a damp cloth, it reappears after a certain time.

Process Conditions: The process relates to nonionic, water-dispersible polyurethanes having a substantially linear molecular structure and lateral polyalkylene oxide polyether chains containing ethylene oxide units which are responsible for dispersibility, distinguished by the fact that, based on the number of alkylene oxide units, from 40 to 95% of the alkylene oxide units of the lateral polyalkylene oxide polyether chains consist of ethylene oxide units and 5 to 60% of propylene oxide, butylene oxide or styrene oxide units.

The process also relates to the preferred process for producing these water-dispersible polyurethanes by reacting organic diisocyanates with difunctional organic compounds ("difunctional" in the context of the isocyanate polyaddition reaction) containing terminal isocyanate-reactive hydrogen atoms and having a molecular weight of 300 to 6,000 in the presence of synthesis components containing hydrophilic groups which guarantee the dispersibility of the polyurethanes, and optionally in the presence of the chain extenders known in polyurethane chemistry having a molecular weight below 300.

Hydrophilic synthesis components having lateral hydrophilic chains containing ethylene oxide units corresponding to the following general formula:

(1)

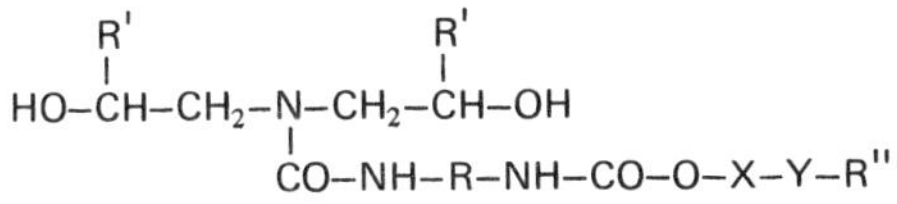

and/or to the general formula:

(2)

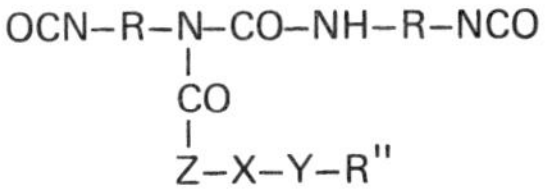

are used in the process.

Particularly preferred synthesis components are those corresponding to general formula (1) above. In general formula (1) and (2) above,

> R represents a difunctional radical of the type obtained by removing the isocyanate groups from a diisocyanate corresponding to the general formula $R(NCO)_2$ of the type mentioned above;
>
> R' represents a hydrogen atom or a monofunctional hydrocarbon radical having from 1 to 8 carbon atoms, preferably a hydrogen atom or a methyl group;
>
> R'' represents a monofunctional hydrocarbon radical having from 1 to 12 carbon atoms, preferably an unsubstituted alkyl radical having from 1 to 4 carbon atoms;
>
> X represents a difunctional radical obtained by removing the terminal oxygen atom from a polyalkylene oxide radical having from about 5 to 90 and preferably from about 20 to 70 chain members, of which from about 40 to 95% and preferably from about 60 to 90% consists of ethylene oxide units and from about 5 to 60%, preferably from about 10 to 40%, of propylene oxide, butylene oxide or styrene oxide units, of which propylene oxide units are preferred;
>
> Y represents oxygen or –NR'''– wherein R''' has the same definition as R''; and
>
> Z represents a radical which has the same definition as Y.

The process for producing the self-dispersible polyurethanes may be carried out in accordance with the methods known in polyurethane chemistry both by the one-stage process and also by the two-stage process (prepolymer process).

The lateral polyalkylene oxide chain which contains the ethylene oxide units essential to the process is preferably attached through groups: (a) corresponding to the following general formula:

$$\begin{array}{l} -N- \\ \;| \\ CO-NH-R-NH-CO-O-X-Y-R'' \end{array}$$

or (b) corresponding to the following general formula:

$$\begin{array}{l} -N- \\ \;| \\ CO-Z-X-Y-R'' \end{array}$$

wherein R, R'', X, Y and Z are as defined above.

The process as described above merely represents the preferred method, but by no means the only method of obtaining the polyurethanes according to the process. Another method of obtaining the polyurethanes according to the process is, for example, to introduce the nonionic lateral hydrophilic groups into a preferably linear polyurethane elastomer by reacting this elastomer with hydrophilic monoisocyanates corresponding to the following general formula:

$$OCN-R-NH-CO-O-X-Y-R''$$

wherein R, X, Y and R'' are as defined above.

Thermally Linkable Aqueous Polyurethane Dispersions

Highlights of the Technological Achievement: An improvement in a process for the preparation of a solvent-free exclusively aqueous dispersion of a thermally linkable, storage stable, light stable polyurethane obtained by the reaction of a hydroxy group containing polyester composition and a diisocyanate in a ketonic solvent at 10° to 50°C employing the hydroxyl-containing polyester and diisocyanate in amounts such that there are 1.5 to 2.0 NCO equivalents per OH group.

The reaction product is thereafter reacted with a primary diamine, secondary diamine or diol in such an amount that 30 to 70% of the existing free NCO groups on the reaction product are reacted and the resultant reaction product is dispersed in water and the ketonic solvent is removed. The diisocyanate used in this process is the uretdione of the 3-isocyanatomethyl-3,5,5-trimethylcyclohexylisocyanate.

Reference: W. Flakus and J. Disteldorf; U.S. Patent 4,184,989; January 22, 1980; assigned to Veba-Chemie AG, Germany.

Background: The chemical resistance and weather stability of dispersion films drying exclusively physically at room temperature, as well as also at higher temperature, can still be improved in comparison with polyurethane reaction lacquers containing solvents.

In the case of cationic polyurethane dispersions whose technical preparation requires a number of process steps, acid catalyzed after-linkages can be conducted with hydroxyl-containing hardeners. However, the precursors of the resulting films cannot be considered totally satisfactory.

Anionic polyurethane dispersions, which are more easily accessible commercially, cause some difficulties with regard to a later after-linkage since in the alkaline field smoothly proceeding reactions without discolorations in tolerable temperature ranges are not yet known.

Process Conditions: It has been discovered that one can produce on the basis of hydroxy-containing polyethers or oxy esters of linear as well as also branched structure with derivatives of the isophoronediisocyanate (3-isocyanatomethyl-3,5,5-trimethylcyclohexyl-isocyanate/IPDI) partially capped by NCO in first of all organic uretdione-isocyanate adducts, which can be converted in the simplest manner into aqueous dispersions and can be thermally linked.

The isophoronediisocyanate derivative advantageously used for the reaction is an addition product of isophoronediisocyanate with itself, the corresponding uretdione, which is prepared according to German Patent 1,934,763. It has the following formula

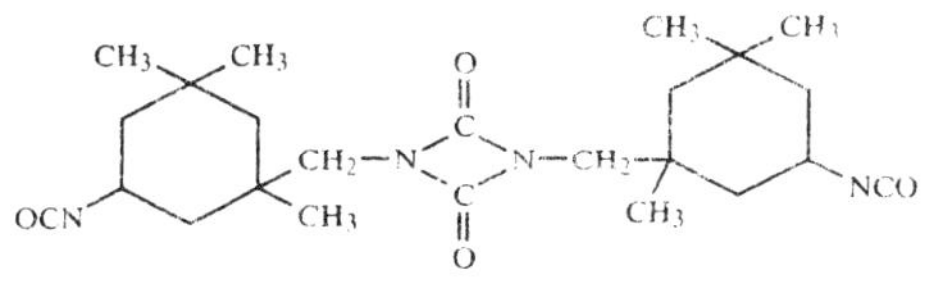

N,N'-bis(1,3,3-trimethyl-5-isocyanatomethylcyclohexyl)-1,3-diaza-2,4-dioxo-cyclobutane.

The average chemical composition of commercially produced uretdione is as follows: <1 part IPDI, 85 parts dimeric IPDI (uretdione), 15 parts trimeric IPDI (triisocyanurate).

By reaction of hydroxyl-containing polyethers or polyesters (especially those containing tin) of average molecular weights of preferably those within the range of 1,000 to 2,000, with IPDI-uretdione of commercial purity or mixtures of IPDI with IPDI-uretdione in ketonic solvents such as acetone, methyl ethyl ketone and others at room temperature, one obtains storageable adduct solutions which can also be added to a later subsequent treatment.

For the production of the abovementioned adducts, the reaction partners are reacted in the equivalence ratio of OH/NCO = 1/1.5 to 1/2, preferably of 1/2. The adducts now prepared in this manner still contain free as well as capped or terminated NCO-substituents.

Finally, the ketonic adduct solutions are submitted to a chain-extension with diols or diamines, especially 3-aminomethyl-3,5,5-trimethylcyclohexylamine and dispersed in a manner known per se with any anion-acting tenside surface acting substance, i.e., for instance, all salts of monoaminoacids known for this purpose in the simplest case with sodium aminocaproate as well as water. In this manner, all the free NCO substituents are converted into corresponding urethane or urea compounds.

Product: As properties of the polyurethane dispersions prepared according to the process and especially of the resulting coatings the following should be mentioned:

(1) Extensive stability against especially polar solvents, acids and alkali solutions.

(2) Surface hardnesses which should be classified within the area

of the solvent-containing PUR-reaction lacquers, (i.e., according to DIN 53 157).

(3) High blocking temperatures, as a rule 150° to 200°C.

(4) High stability against hot water as well as good sterilibility (sterilizing ability) (2 hours, 121°C).

To a special degree these dispersions are suitable for being used as solvent-free systems for appropriate metal annealing lacquerings. However, these dispersions can also be used for the coating of other substrates which permit a sufficient temperature requirement, i.e., of 10 minutes 150° to 180°C. As such there should still be mentioned pressboard, PVC, polyurethane and similar materials.

Example: *Adduct Formation* – 1,000 g oxy ester on the basis of 50 pbw adipic acid, 21 pbw hexanediol-1,6 and 30 pbw neopentyl glycol (OH number: 144 mg KOH/g) were reacted with 1.220 g uretdione of the IPDI (commercial quality) and 952 g acetone while stirring at room temperature within 4 to 5 hours into a 70% adduct solution. NCO number of the solution: 3.4%.

Dispersion – 400 g of the above-prepared adduct were reacted after dilution with 300 g acetone with 14 g 3-aminomethyl-3,5,5-trimethylcyclohexyl amine (isophoronediamine) and then dispersed while stirring with 41 g of a 30% aqueous solution of the sodium salt of the aminocaproic acid, as well as 27 g trimethylolpropane and 380 g of water, then stripped free of acetone. The resulting dispersion has a solid content of about 45%. The properties after hardening within 15 to 10 minutes at 150° to 180°C are listed below:

Impact stress according to Konig	141 sec.
Hardness according to Bucholz	91
Erichsen cupping	11.2 mm
Grid section	0
Blocking temperature	165°C
Tensile strength	409 kp/cm^2
Tension	8 %
Foil thickness	49 μ

Solutions or Dispersions of Polyisocyanate Polyaddition Products Containing Urethane and/or Urea Groups

Highlights of the Technological Achievement: This method relates to a process for the production of nonfoamed urethane and/or urea group-containing polyisocyanate polyaddition products which contain both cationic and also anionic groups chemically bound into the same molecule of the polyisocyanate polyaddition products, to the polyaddition products obtainable by this process and to the use thereof for coating flexible and nonflexible sheet-form materials.

Reference: A. Reischl and W. Wenzel; U.S. Patent 4,182,828; January 8, 1980; assigned to Bayer AG, Germany.

Background: It is not possible to obtain the required combination of properties using the ionic polyurethanes containing differently charged ion centers, preferably dispersed in water, which are described in the literature. The reason for this would appear to be the intensive interaction of the differently charged ion groups within the betaine structure which partly mask the properties associated with the type of ion. In general, a solution to this problem is regarded as basically impossible because, for example, the stable formation combination of cationic and anionic aqueous dispersions itself encounters insurmountable difficulties and may only be achieved in some cases with the assistance of water-compatible organic solvents.

Process Conditions: A process for the production of nonfoamed urethane and/or urea group-containing polyisocyanate polyaddition products, in which chemically fixed anionic and chemically fixed cationic groups are simultaneously present, by reacting:

(a) Organic polyisocyanates, optionally together with organic monoisocyanates; with

(b) Organic compounds containing at least two isocyanate-reactive groups in the context of the isocyanate addition reaction, optionally in the presence of organic compounds containing one isocyanate-reactive group in the context of the isocyanate addition reaction; compounds containing ionic groups and/or groups which may be converted into ionic groups being used as component (a) and/or (b), characterized in that the compounds containing ionic groups or groups convertible into ionic groups which are used in accordance with the process are both:

  (1) Compounds containing anionic groups and/or groups convertible into anionic groups; and

  (2) Compounds containing cationic groups and/or groups convertible into cationic groups.

Any groups convertible into ionic groups which are present in the polyaddition product on completion of its synthesis are at least partly converted into ionic groups in known manner by neutralization or quaternization. Finally the type and quantities of synthesis components used and, optionally, the degree of neutralization and quaternization are selected in such a way that an equivalent ratio between cationic and anionic groups of 5:1 to 1:10 prevails in the end product ultimately obtained for a total content of ionic groups of 2 to 300 meq/100 g of solids. At least one urethane or urea group is present between every cationic and anionic group which is chemically fixed in the end product.

The process also relates to nonfoamed polyisocyanate polyaddition products containing urethane and/or urea groups, characterized by a content of from 2 to 300 meq/100 g of ionic groups chemically bound into the polyisocyanate polyaddition adducts which are made up of cationic and anionic groups in an equivalent ratio of 5:1 to 1:10, the cationic groups being separated from the anionic groups by at least one urethane and/or urea group.

Particularly preferred (potential) ionic synthesis components include N-alkyl dialkanolamines, such as N-methyl diethanolamine, N-ethyl diethanolamine or N-propyl dipropanolamine, diaminosulfonates of the type described in Canadian Patent 928,323, such as the sodium salt of N-(2-aminoethyl)-2-aminoethane sulfonic acid, dimethylol propionic acid and sulfonate diols corresponding to the following general formula:

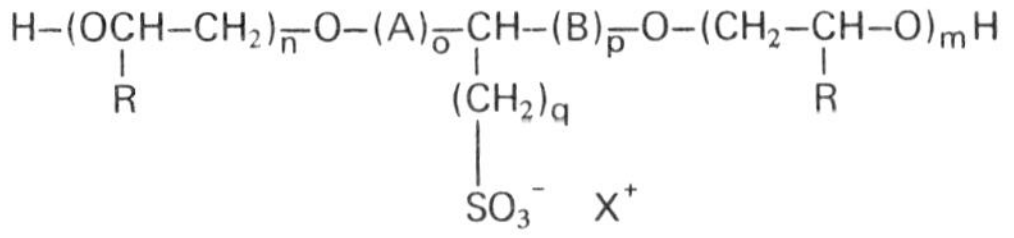

wherein A and B, which may be the same or different, each represents a difunctional aliphatic hydrocarbon radical containing from 1 to 6 carbon atoms;

R represents hydrogen, an aliphatic hydrocarbon radical containing from 1 to 4 carbon atoms or a phenyl radical;

X^+ represents an alkali metal cation or an optionally substituted ammonium group;

n and m each represents the same or different number of from 0 to 30;

o and p each represents 0 or 1; and

q represents an integer of from 0 to 2.

The potential ionic groups, if any, initially incorporated in the polyaddition product may be converted into ionic groups in a known manner by neutralizing the potential anionic and cationic groups or by quaternizing tertiary amine nitrogen atoms or tertiary phosphine phosphorus atoms which may be present in the polyaddition products.

The preferred ionic centers that may be present in the polyaddition products according to the process are the quaternary ammonium group, $-CO_2^-$ and $-SO_3^-$. The quaternary ammonium group preferably is incorporated in the main chain of the polyaddition product although lateral ammonium groups chemically bound into the polyaddition products are also possible, of the type which may be formed for example, by reacting labile halogen atoms incorporated in the chain, which in turn have been incorporated by using polyisocyanates or polyhydroxyl compounds containing labile halogen atoms, with tertiary amines, such as triethylamine, or by incorporating, for example, glycols containing lateral tertiary amine nitrogen atoms, such as 2-ethyl-2-dimethylaminomethyl-1,3-propane diol, and subsequently quaternizing the tertiary nitrogen incorporated.

Example: *Dispersion in Water/Dimethyl Formamide* – Preliminary stage: 32.00 kg of phthalic acid-adipic acid-ethylene glycol polyester (OH-number 64) (PAA); 10.58 kg of 1,6-hexamethylene diisocyanate (H); 0.945 kg of N-methyl diethanolamine (N-MDA); 44.48 kg of dimethyl formamide (DMF); and 0.995 kg of dimethyl sulfate (DMS).

Method: The polyester (PAA) was dehydrated in vacuo for 30 minutes at 130°C. 7.6 kg of diisocyanate (H) were added at 100°C. After a reaction time of 1 hr at from 100° to 130°C, the mixture was cooled to 63°C and an NCO-value of 5.69% was observed.

N-MDA was then added, followed by stirring for 1 hour at about 65°C. The mixture was then diluted with 23 kg of DMF. After 10 minutes of stirring, DMS in 1.5 kg of DMF was added and stirring continued for 30 minutes at 65°C. Another 17 kg of DMF were then stirred in. An NCO-value of 1.97% is observed. 2.98 kg of diisocyanate (H) in 2.98 kg of DMF are stirred into this prepolymer. An NCO-value of 4.0% is observed.

Extension of the preliminary stage with amines in a double flow mixer: Mixture – 500 g/min of the preliminary stage in DMF; X g/min of diethylene triamine in water (9.2%) (DAT); Y g/min of sodium ethylene diamine-2-ethane sulfonate (46.5% in water) (AAS); and 220 g/min of a DMF/water mixture (93:7).

Method – The ampholyte dispersion is prepared as follows in a double flow mixer (mixer 1, 5,000 ml volume; mixer 2, 250 ml volume, each equipped with a toothed stirrer). In mixer 1 the prepolymer of the preliminary stage is mixed with the DMF/water mixture and the DAT in water at about 3,000 rpm. The dispersion formed is extended using AAS in mixer 2. Both mixers are jacket-cooled with brine (approximately -10°C) to dissipate the heat of reaction. The final dispersion has a pH value of 7. The Ford cup viscosity amounts to 10 seconds for a solids content of 29.8% (orifice number 6). The following cation:anion ratios were adjusted in different settings:

Dispersion	DAT Cation (g)	AAS Anion (g)	Cation:Anion Ratio	Viscosity (sec)
a*	139	4.7	~4:1	10
b**	131	9.4	~2:1	11
c***	114	18.8	~1:1	15

*Contains 16.7 meq of cation and 4.3 meq of anion.
**Contains 16.7 meq of cation and 8.6 meq of anion.
***Contains 16.5 meq of cation and 17 meq of anion.

Polyurethane Hydrogels Having Enhanced Wetting Rates

Highlights of the Technological Achievement: The wetting rates of polyurethane hydrogels comprising lightly crosslinked polymers of isocyanate terminated prepolymers which are the reaction products of polyalkyleneoxypolyols with organic diisocyanates that have been lightly crosslinked with water or an organic polyamine are enhanced by treating the surface of the solid hydrogel with certain ketones, alcohols, organic amines, aromatic hydrocarbons or aqueous alkali metal hydroxide solutions.

Reference: A.P. Jones and R.J. Knopf; U.S. Patent 4,182,827; January 8, 1980; assigned to Union Carbide Corporation.

Background: The polyalkyleneoxypolyols that are used in producing the polyurethane hydrogel polymers that are treated in accord with the process are those having a molecular weight up to 25,000. These polyols can be diols, triols or tetrols, with the molecular weight of the polyol varying depending upon which is used.

Process Conditions: The suitable diols are the polyethyleneoxy glycols which have a molecular weight of preferably 6,000 to 20,000. The triols and tetrols that can be used are those having a molecular weight of from about 92 to 5,000, preferably from about 500 to 1,500. These can be the polyalkyleneoxypolyols wherein the alkyleneoxy group contains 2 to 4 carbon atoms and they can be homopolymers or block or random copolymers having 3 or 4 reactive hydroxyl groups. One can also use the aliphatic polyhydroxyl compounds of the formula $C_nH_{2n+2-m}(OH)_m$ wherein n is an integer having a value of from 3 to 6 and m has a value of 3 or 4.

The hydrogels can be produced by reacting the polyalkyleneoxydiol with an organic diisocyanate to form an isocyanato terminated prepolymer which is then lightly crosslinked with a crosslinking agent that is a mixture of an organic diamine and an organic triamine.

In another embodiment, the hydrogels can be produced by reacting a mixture of polyalkyleneoxydiols and polyalkyleneoxytriols and/or tetrols with an organic diisocyanate to form the prepolymer which is then lightly crosslinked with a crosslinking agent that is water, an organic diamine, or a mixture thereof. When a mixture of polyols is used in producing the hydrogels, the mol ratio of the diol to the higher polyols is at least about 6:1 and can be as high as about 40:1. Preferably this mol ratio is from 20:1 to 25:1. It has been observed that the mol ratio of diol to higher polyol has an effect on water uptake; the higher the mol ratio, the higher the water uptake.

The surface treating compounds are (a) liquid ketones of the general formula: RCOR' wherein R and R' are alkyl groups having from 1 to 6 carbon atoms, and wherein R' can also be a phenyl group; (b) a liquid alkanol having from 1 to 6 carbon atoms, preferably 1 to 3 carbon atoms; or (c) an organic amine having up to 12 carbon atoms, preferably up to 3 carbon atoms, which can be aliphatic, cycloaliphatic or aromatic; or (d) an aromatic hydrocarbon having from 6 to 10 carbon atoms, preferably 6 to 8 carbon atoms; or (e) a cyclic oxide having from 2 to 4 ring carbon atoms; or (f) an aqueous solution of an alkali metal hydroxide.

Product: A polyurethane hydrogel was produced for treatment in accordance with the process. A mixture was prepared containing 1,350 g of polyoxyethylenediol having an average molecular weight range of from 6,000 to 7,500, 9.08 grams of polyoxyethylenetriol having an average molecular weight range of from 950 to 1,050 and 2,632 g of trichloroethylene and heated to 76° to 91°C while refluxing to remove 2.4 ml of water. The solution was cooled to 86°C and 58.35 grams of methylenebis(4-phenylisocyanate) were added to the reactor contents.

The mixture was stirred at 86° to 90°C for 2 hours after adding about 10 drops of dibutyltin dilaurate catalyst. The isocyanato/hydroxyl ratio was 1.42/1. The prepolymer solution had a total solids content of 35%, a Brookfield viscosity of 101,520 cp at room temperature and an equivalent weight of 45,470. It was diluted with trichloroethylene to a 13.5% total solids content and then impregnated with 2,270 g of naphtha to give a prepolymer solution of 11.1% total solids having a Brookfield viscosity of 460 cp at 22°C.

A coagulating bath was prepared containing naphtha that was approximately 0.0001 N with ethylene diamine as the crosslinking agent. The prepolymer solution was extruded through a manifold head having 20 holes, each 0.43 mm in diameter, in the form of droplets which were allowed to fall through an ambient air space of about 30 cm and then into the coagulating bath and onto a conveyer belt. The pellets that form are carried through the bath a distance of about 90 cm during which time the trichloroethylene solvent is extracted by the naphtha and crosslinking takes place.

After pellets exit the bath they are dried at 40°C; they had an average size of about 4.75 mm (Pellets A). A portion of the pellets was ground and the resulting granules had an average size of about 0.76 mm (Granules A).

The Wetting Rates of the pellets and granules were determined by averaging the results of four runs of each; they were: Pellets A, 455.5 seconds or 7.6 minutes; Granules A, 203.8 seconds or 3.4 minutes. It was noted that the variation in time of each run was greater for the granules. This was attributed to their irregular particle size and their fractured skin surfaces.

Example 1: In this example Pellets A and Granules A were dipped into the treating composition rather than sprayed with it and the treated hydrogel articles were then dried at 40°C under a 50.8 cm vacuum. The treating compositions employed and the Wetting Rates of the treated hydrogels are set forth in the following table; the Controls are untreated Pellets A and Granules A.

The rate of water absorption was determined by using a random selection of the shaped hydrogel articles, e.g., the pellets, spaced on a glass microscopic slide. A drop of distilled water was placed onto the individual article or pellet while observing under a microscope. The Wetting Rate was determined visually using a stopwatch to ascertain the time required for the entire article or pellet to become transparent, an indication that water was absorbed completely throughout its cross section. The procedure is repeated from 3 to 6 times for each sample and the average of the results is reported as the Wetting Rate. For comparative purposes the Wetting Rate of the untreated hydrogel polymer from the same batch is also determined.

Treating Compositions	.... Wetting Rate (sec)	
	Pellets A	Granules A
Controls	455	204
Acetone	376	121
Acetone + 2% Surfactant 1*	216	123
Acetone + 5% Surfactant 1	216	121
Acetone + 2% Surfactant 2**	215	121
Acetone + 5% Surfactant 2	216	114
Acetone + 2% Surfactant 3***	208	121
Acetone + 5% Surfactant 3	192	117
Acetone + 2% KOH solution†	204	112
Acetone + 5% KOH solution	199	110

*Trimethylnonyl polyethylene glycol ether with average of 10 oxyethylene groups (nonionic).

**Nonylphenyl polyethylene glycol ether with average of 10.5 oxyethylene groups (nonionic).

***Dioctyl ester of sodium sulfosuccinic acid (anionic).

†10% aqueous solution.

The results clearly show the improvement in Wetting Rates achieved when the hydrogel articles are treated in accordance with the process.

Example 2: A series of experiments were carried out whereby Pellets A and Granules A were treated with aromatic hydrocarbons, ketones, alkanols, cyclic oxides and amines. The hydrogel particles were dipped into the treating composition and then dried at 40°C under a 50.8 cm vacuum. Surfactant 2 of Example 1 was added in some instances. The controls are untreated Pellets A and Granules A.

Treating Compositions	.... Wetting Rate (sec)	
	Pellets A	Granules A
Controls	455	204
Benzene	350	120

(continued)

Treating Compositions	.... Wetting Rate (sec) .... Pellets A	Granules A
Benzene + 0.5% Surfactant 2	353	117
Toluene	355	119
Toluene + 0.5% Surfactant 2	333	114
Acetophenone	348	116
Acetophenone + 0.5% Surfactant	350	116
Methyl isobutyl ketone	345	112
Methyl isobutyl ketone + 0.5% Surfactant 2	350	114
Methanol	347	115
Methanol + 0.5% Surfactant 2	345	114
Ethanol	325	117
Ethanol + 0.5% Surfactant 2	208	–
Ethanol + 1% Surfactant 2	206	105
Ethanol + 2% Surfactant 2	212	104
Propylene oxide	355	120
Propylene oxide + 0.5% Surfactant 2	352	116
Tetrahydrofuran	351	122
Tetrahydrofuran + 0.5% Surfactant 2	344	119
Pyridine	354	118
Pyridine + 0.5% Surfactant 2	360	119

APPLICATIONS

Aqueous Dispersion of Urethane Diamine Having Storage Stability

Highlights of the Technological Achievement: An aqueous coating composition which will give coatings of high adhesion strength and good weatherability can be prepared from the aqueous urethane diamine dispersion and an aqueous dispersion of a diepoxy compound.

Reference: T. Kinoshita and T. Izumi; U.S. Patent 4,208,311; June 17, 1980; assigned to Mitsui Petrochemical Industries Ltd., Japan.

Background: The process also relates to a room temperature-curable coating composition consisting essentially of the aqueous dispersion of a urethane diamine and a minor amount of an aqueous dispersion of a diepoxy compound, which exhibits superior film characteristics typified by high adhesion strength and good weatherability as well as superior coatability in such applications as paints, waterproof paints, and adhesives.

Process Conditions: The process specifically pertains to an aqueous dispersion having a pH value of 7 to 10.5 and comprising

(A) A urethane diamine having a molecular weight of about 5,500 to about 25,000 and expressed by the formula

$$H_2N(R'NH\underset{\substack{\| \\ O}}{C}ORO\underset{\substack{\| \\ O}}{C}HN)_nR'NH_2$$

wherein R is a residual moiety of a long-chain diol having a molecular weight of 500 to 15,000, R' is a divalent hydrocarbon radical, and n is an integer of 1 to 25;

(B) 0.5 to 20 pbw/100 pbw of the urethane diamine (A) of a polyoxyethylene ether as an emulsifier having a hydrophilic-lipophilic balance (HLB) of 11 to 16 and selected from the group consisting of polyoxyethylene aryl ethers and polyoxyethylene lauryl ether;

(C) 50 to 300 pbw/100 pbw of the urethane diamine (A), of water;

(D) An acid as a pH adjusting agent in an amount sufficient to adjust the pH of the aqueous dispersion to about 7 to about 10.5; and

(E) A filler in an amount which is within the range of 0 to 300 pbw/100 pbw of the urethane diamine (A) and which does not cause phase separation of the aqueous dispersion. It also pertains to an aqueous coating composition consisting essentially of (1) the aqueous dispersion of a urethane diamine and (2) an aqueous dispersion of a diepoxy compound.

The urethane diamine (A) which constitutes the aqueous dispersion of the process is known, and a method for its preparation is also known. Any urethane diamines corresponding to the abovementioned formula and having a molecular weight of preferably 7,000 to 20,000, can be used.

The aqueous dispersion having a pH value of about 7 to about 10.5 composed of the ingredients (A) to (E) may further contain (F) a curing accelerator, (G) a viscosity adjusting agent, etc. The amount of such an additive can be varied properly. For example, the amount of the curing accelerator (F) is 0 to 10 pbw/100 pbw of the urethane diamine (A), and the amount of the viscosity adjusting agent (G) is 0 to 100 pbw/100 pbw of the urethane diamine (A).

Example 1 and Comparative Example 1: A flask equipped with 3 propeller blades was charged with 100 pbw of water and 0.2 N aqueous solution of HCl as a pH adjuster. A mixture of 100 pbw of a urethane diamine having a number average molecular weight of 8,500 (corresponding to the formula given hereinabove in which R is a residue of polyoxypropylene glycol with a molecular weight of 3,000 and R' is a hexamethylene group) and 2 pbw of polyoxyethylene nonylphenyl ether of varying HLB values was added.

The contents were stirred at a speed of 500 rpm to form an aqueous dispersion. The stability of the aqueous dispersion was examined two days later by observing the state of phase separation. The results are shown in the table.

HLB of Emulsifier	pH of Aqueous Dispersion	HCl Salt of Urethanediamine (%)	Stability of the Aqueous Dispersion
13.0	10.5	0	No phase separation
13.0	10.0	5	No phase separation
13.0	9.5	10	No phase separation
13.0	8.9	13	No phase separation
13.0	8.5	18	No phase separation
13.0	8.0	45	No phase separation

(continued)

HLB of Emulsifier	pH of Aqueous Dispersion	HCl Salt of Urethanediamine (%)	Stability of the Aqueous Dispersion
13.0	7.0	95	No phase separation
12.4	8.9	13	No phase separation
14.5	8.9	13	No phase separation
10.0*	8.9	13	Phase separation occurred
18.0*	8.9	13	Phase separation occurred
13.0*	6.5	100	Phase separation occurred

*Comparative.

Similar results were obtained when the above procedure was repeated except that a 0.2 N aqueous solution of sulfuric acid was used instead of the 0.2 N HCl aqueous solution.

Example 2: An aqueous dispersion (A) was prepared from 100 pbw of a urethane diamine having a molecular weight of 9,500, 80 pbw of calcium carbonate, 78 pbw of water, 22 pbw of a 0.2 N aqueous solution of HCl and 2 pbw of polyoxyethylene nonylphenyl ether having an HLB of 13.

Separately, a flask fitted with a stirrer having 3 propeller blades was charged with 100 pbw of an epoxy resin (Epomix R-140, Mitsui Petrochemical Industries, Ltd.) and 3 parts of an emulsifier (Nissan Pronon 208, Nihon Yushi KK), and the mixture was stirred for 5 minutes at 60°C at a speed of 500 rpm. With continued stirring at 500 rpm, 10 pbw of a 5% aqueous solution of polyvinyl alcohol (KH-17, Nihon Gosei Kagaku Kogyo KK) was added dropwise. Subsequently, 87 pbw of water was added in small portions over the course of about 10 min. The mixture was stirred further for 10 minutes, and cooled to form an aqueous dispersion of epoxy resin (B). The resulting aqueous dispersion (B) remained stable for more than 6 months.

100 pbw of (A) and 5.7 pbw of (B) were mixed, coated on a glass sheet, and cured at room temperature for 2 weeks to form a sheet. The resulting cured sheet had the following properties: tensile strength (JIS A6021); 23 kg/cm^2; elongation (JIS A6021), 400%; shore A hardness, 58; and tensile shear adhesion strength (JIS K6850, adherend, an iron plate; no primer applied), 9 kg/cm^2.

Example 3: A mixture of the aqueous dispersions (A) and (B) prepared in Example 2 was applied to each of the various adherends shown in the table without applying a primer. The adhesion strength of each coated film was measured by a 180° peeling test at a pulling speed of 50 mm/min. The sample had a thickness of 1.5 mm, and the area of adhesion was 25 mm x 150 mm. The results are shown in the table below.

Adherend	Adhesion Strength (kg/cm)	Remarks
Mortar	2.9	Cohesive destruction
Slate	2.6	Cohesive destruction
Untreated aluminum	2.5	Cohesive destruction
Polyurethane-coated aluminum	2.6	Cohesive destruction
Iron	2.6	Cohesive destruction
Glass	2.4	Cohesive destruction
Plywood	*	—

*The adherend was destroyed.

Aqueous Polyurethane Dispersions Resistant to Sedimentation

Highlights of the Technological Achievement: This process relates to coating compositions based on aqueous polyurethane dispersions and to the use of these coating compositions for the production of topcoats or finishes on PVC. The coating compositions according to the process contain a mixture of two aqueous polyurethane dispersions which have particular chemico-physical properties.

Reference: W. Wenzel, Schröer, M. Preuss and H.-J. Koch; U.S. Patent 4,206,255; June 3, 1980; assigned to Bayer AG, Germany.

Background: The production of polyurethane coatings on PVC is already known in principle. Methods of coating PVC using one-component or two-component polyurethane systems dissolved in organic solvents have been described, for example, in U.S. Patents 3,262,805; 4,017,493 and 4,045,600. The use of coatings containing solvents, however, has certain practical disadvantages, particularly the fact that for economical and ecological reasons it is necessary to recover the organic solvent after the coating process. There is, therefore, a general demand for solvent-free coating compositions.

It has surprisingly been found that coating compositions which fulfill the abovementioned requirement can be obtained from aqueous polyurethane dispersions by mixing two different polyurethane dispersions each of which must fulfill different chemico-physical criteria.

Process Conditions: This process relates to coating compositions based on aqueous, sedimentation-resistant polyurethane dispersions having a solid content below 60% by weight, which are characterized in that they contain, based on the polyurethane solid content:

(A) Preferably from 5 to 50% by weight of a polyurethane which contains:

   (1) Preferably from 8 to 40 meq or pseudo meq in 100 g of polyurethane solids of anionic and/or cationic groups and/or ethylene oxide units and

   (2) At least 4% by weight, most preferably from 20 to 50% by weight, based on the polyurethane solid content, of linear alkylene groups having from 4 to 6 carbon atoms, which polyurethane results in films which

   (3) Have a softening point most preferably from 140° to 180°C, and

   (4) A Shore hardness A preferably from 30 to 95; and

(B) Preferably from 50 to 95% by weight of a polyurethane which contains:

   (1) Preferably from 8 to 20 meq or pseudo meq of anionic and/or cationic groups and/or ethylene oxide sequences in 100 g of polyurethane solids and results in films which

   (2) Melt most preferably above 220°C and

   (3) Have a Shore hardness D of most preferably 60 to 80.

The procedure generally consists of first preparing a prepolymer having isocyanate end groups from a higher molecular weight polyhydroxyl compound, optionally a low molecular weight mono, di-, or triol and an excess of polyisocyanate, and then converting this prepolymer into a polyurethane dispersion by a process accompanied by chain-lengthening, for example, by a process analogous to those described in U.S. Patents 3,479,310 and 3,756,992.

Any higher molecular weight polyhydroxyl compounds may, in principle, be used for the preparation of component (A) of the coating compounds according to the process.

Whereas the type (A) dispersions should contain from about 5 to 100, preferably from about 8 to 40 pseudo meq of hydrophilic groups per about 100 g of solid content, as already mentioned, the type (B) dispersions should contain from about 5 to 30, preferably from about 8 to 20 pseudo meq of hydrophilic groups per about 100 g of solid content.

By "pseudo meq" is meant, in the context of this process, that quantity of ethylene oxide units in the polyurethane, preferably in a polyether side chain to the main chain of the polyurethane molecule, which corresponds in its hydrophilic character to 1 meq of ionic groups. According to this definition, 0.5% by weight of $-CH_2CH_2O-$ units corresponds to 1 pseudo meq in 100 g of polyurethane solids.

Coating Compositions: The coating compositions are particularly suitable for use as topcoats or finishes for coating textiles with PVC. Three methods of applying the compositions according to the process are described below:

(1) The dispersion composition is applied (e.g., by knife-coating, spraying or printing) to a compact PVC layer which has been applied to a substrate such as a textile. The coating is generally dried at from about 70° to 190°C in a heating channel. The artificial leather obtained in this way may then be machine embossed and is then rolled up.

(2) The polyurethane dispersion composition is applied by knife-coating to a separating paper or some other substrate, e.g., a strip of steel sheet, and then dried in a heating channel. A degasified or foamed PVC paste is then applied to the dried PUR layer by knife-coating. The textile which is to be coated is then applied to the PVC paste. The PVC is gelled using a known method in a heating channel. After cooling, the separation paper or other substrate may be stripped off and the artificial leather rolled up.

(3) The composition is applied by knife-coating to a separating paper or some other substrate, e.g., a strip of steel sheet, and dried in a heating channel.

Temperatures of 170° to 200°C are generally used for gelling PVC pastes. The melting point of a topcoat on PVC should therefore be above 200°C. PVC is a synthetic material which is extremely resistant to hydrolysis.

The following table demonstrates the properties of the top film and its adherence to the PVC foam.

Finish from Dispersion	SP (°C)	. . . . Swelling . . . . . Dodecyl . Phthalate	Dioctyl (vol %) . .	Tear* Resistance (%)	(wk)	Adherence to PVC	. . Surface Feel . . . Direct Coating	Reversal Process
A	160	19	19	18	4	Good	Blocking	Sticky
B_1	230	6	6	100	6	US	D,H	D,H
B_2	230	7	7	81	6	US	D,H	D,H
$A:B_1 = 2:8$	230	19	19	83	6	Good	D,F	D,F
$A:B_2 = 2:8$	230	19	19	73	6	Good	D,F	D,F

Note: US is unsatisfactory, D is dry, H is hard and F is flexible.

*After x week's subjection to the Tropical test (70°C, 95% relative humidity).

Delivery Means for Biologically Active Agents Comprising Hydrophilic Polyurethane

Highlights of the Technological Achievement: Sustained release delivery means comprising (1) a biologically active agent, (2) a linear hydrophilic block polyoxyalkylene-polyurethane copolymer, and (3) optionally containing a buffer. Included among the biologically active agents are abortifacient compounds.

Reference: F.J.T. Fildes and F.G. Hutchinson; U.S. Patent 4,202,880; May 13, 1980; assigned to Imperial Chemical Industries, Ltd., England.

Background: The process involves a hydrophilic block copolymer of a particular type which previously has not been used in connection with any biologically active agent, and whose use in such connection is not obvious. This type of copolymer has the following advantageous properties:

(1) Because it is linear in structure, the copolymer is soluble in common organic solvents; this is important for biological applications because the copolymer can be purified, and solvent casting techniques can be used for fabrication purposes. Also, the copolymer is thermoplastic and easily molded and extruded.

(2) It is substantially insoluble and nonerodable in body fluids.

(3) It is compatible with delicate body tissues and causes no toxicity or related problems.

(4) The degree of hydrophilicity, and therefore of water-swellability, of the copolymer can be predetermined by its composition. In use, the delivery means of the process is permeated by the body fluids, which results in a diffusion path being established from the delivery means to the adjacent body tissue. As a result, there is a sustained release of the biologically active agent.

(5) It is a soft elastomer when swollen in body fluids, and therefore it can conform to body cavities and is nonirritant.

(6) A wide range of biologically active agents, as far as water-solubility is concerned, can be incorporated, in solution or in suspension, in the copolymer.

Process Conditions: The hydrophilic regions are composed of one or more polyoxyalkylenes containing the repeating unit $-(CH_2)_2O-$, $-CH_2CH(CH_3)O-$, $-(CH_2)_3O-$ and/or $-(CH_2)_4O-$, and the hydrophobic regions are composed of a polyurethane which is obtainable in known general manner from a diisocyanate and one or more dihydroxy compounds.

The copolymers which characterize this process fall within the class known as amphipathic materials. The copolymers are linear block copolymers of the type ABABAB. . ., wherein A represents the hydrophilic regions and B the hydrophobic regions. A suitable polyoxyalkylene is, for example, polyoxyethylene of molecular weight 400 to 20,000. The hydrophobic regions B are composed of a polyurethane which is obtainable in known general manner from a diisocyanate, for example, an aromatic or aliphatic diisocyanate, for example, 4,4'-diphenylmethane diisocyanate.

Preferred copolymers contain 30 to 70% by weight of hydrophilic regions A and 70 to 30% by weight of hydrophobic regions B. A preferred polyoxyalkylene is polyoxyethylene of molecular weight 600 to 6,000. A preferred diisocyanate is 4,4'-diphenylmethane diisocyanate, because it reacts at a convenient rate without the need for polymerization catalysts which may be leachable and toxic. A preferred dihydroxy compound for use in block B is 1,1'-isopropylidenebis-p-phenyleneoxydipropanol-2.

One particularly useful embodiment of this process consists of an intravaginal device comprising an abortifacient compound. A suitable abortifacient compound is, for example, an abortifacient prostaglandin derivative, for example, fluprostenol sodium [racemic (9S,11R,15R)-9,11,15-trihydroxy-16-(3-trifluoromethylphenoxy)-17,18,19,20-tetranor-5-cis,13-trans-prostadienoic acid sodium salt].

A particularly preferred embodiment consists of an intravaginal device comprising fluprostenol sodium, a biologically acceptable buffer, for example, a phosphate buffer, and a hydrophilic copolymer obtained as described above and containing 60% by weight of polyoxyethylene of molecular weight 4,000, and 40% by weight of hydrophobic polyurethane blocks obtained from approximately equal parts by weight of 4,4'-diphenylmethane diisocyanate and 1,1'-isopropylidenebis-p-phenyleneoxydipropanol-2.

The rate of release in vivo of the abortifacient drug from the intravaginal device of this method is determined by the composition of the copolymer, the size, shape and thickness of the device, the nature and amount of the abortifacient compound, and the nature and amount of any buffer, in the device. The intravaginal device can be in the form of a disc, ring, annulus or slab, or any other shape which can be retained in the vagina.

Aqueous Polyurethane-Phenolic-Formaldehyde Resin Emulsions for Use as Adhesives, Primers and Surface Coatings

Highlights of the Technological Achievement: Stable aqueous emulsions useful as coatings, primers for promoting vinyl plastisol adhesion to metal, and adhesives for bonding vinyl to metal with high environmental resistance are described.

Reference: D.J. Damico; U.S. Patent 4,197,219; April 8, 1980; assigned to Lord Corporation.

Background: An enormous demand has developed in the art for means for successfully adhesively bonding resin films, and particularly vinyl chloride-containing polymers, to metal substrates in order to obtain a combination sheet or article having the appearance, feel and corrosion resistance of the resin film and the structural strength, fire-resistance and other desirable properties of the underlying metal. While a wide variety of adhesives, adhesive systems and processes have been developed for this purpose, a continuing need exists for improved means for bonding vinyl to metal and particularly for an adhesive means for achieving a bond having improved resistance to environmental conditions, including mechanical distortion of the laminate on exposure to moisture.

In view of the increasing awareness of the hazards of organic solvents and consequent legislative pressures to limit their use, it would be especially desirable to provide a water-based adhesive for these purposes, particularly one providing improved environmental resistance. There is also a need in the art for similar coatings and, especially for primers for promoting vinyl plastisol adhesion to metals.

Process Conditions: New aqueous compositions in the form of stable emulsions or dispersions which contain:

(a) An aqueous polyurethane latex, containing from 0.02 to 1% of the dispersed phase of a salt type group selected from the class consisting of –COO–, $–SO_2O–$, >POO–, and –O–, and having a solids content of 30 to 70% by weight.

(b) An emulsion selected from the group consisting of aqueous casein cocondensed phenolic-formaldehyde resin emulsions, such emulsions having a solids content of 40 to 60% by weight; and

(c) A nonionic fluorocarbon surfactant;

the total composition containing 25 to 99% (a), 1 to 50% (b), 1 to 10,000 ppm (c), and having a total solids content of 15 to 30%, all by weight of the total composition. The stable aqueous dispersions are curable by crosslinking with glyoxal or any material acting as a formaldehyde donor.

The polyurethanes are preferably predominantly linear and are produced from polyhydroxyl compounds which have a molecular weight preferably of 300 to 10,000, organic polyisocyanates and, if desired, chain lengthening agents. In the production of the polyurethane dispersions, a component having at least one salt type group is included. This may be a compound with active hydrogen atoms as determined by the Zerewitinoff method and thus capable of reacting with an isocyanato group or it can be an isocyanate.

Any suitable polyacetal may be used, but it is preferred to use the water-soluble types, for example, those from 1,6-hexanediol and formaldehyde, from 4,4'-dihydroxyethoxydiphenyldimethylmethane and formaldehyde.

Any suitable polyester may be used such as, for example, those obtained from polyhydric alcohols and polycarboxylic acids. Any suitable polyhydric alcohol, but preferably a dihydroxy alcohol such as, for example, diethylene glycol.

After the formation of the preadduct, an organic solvent is preferably added and the reaction is preferably continued at a relatively low temperature of 20° to 100°C especially where very reactive chain lengthening agents are used. Thus, the polyurethane preadduct remains of a low viscosity so that it can be easily stirred into the aqueous phase.

Example: 250 g of an adipic acid-hexanediol-neopentyl glycol-polyester are dried in vacuo at about 120°C and reacted for 2 hours with 38.6 g 1,6-hexane diisocyanate. 150 ml acetone (water content about 0.17%) are added to the viscous melt at 80°C and when the solution has been effected, 3.75 g tartaric acid in 110 ml acetone are added and the mixture is diluted with 600 ml acetone after 3 hours. After 4 hours stirring at 55°C the mixture is diluted with 300 ml acetone and reacted with 2.82 g triisopropylamine in 30 ml acetone. 250 cc water are then stirred in and approximately half the acetone contained therein is distilled off. 2 g casein lactate in 10% aqueous solution are added as protective colloid to the thin liquid, milky white latex, and the remaining acetone is removed by further distillation.

400 g of the 27% solution obtained are adjusted to pH 7 with dilute ammonia solution, 22 cc water are added dropwise and the acetone is distilled off. A 42% latex is obtained which after the addition of some casein that has been opened up by ammonia may be further concentrated. On drying, a sticky, rubbery mass remains behind which can be cured with polyisocyanates or with substances giving off formaldehyde.

Microencapsulation with Modified Aliphatic Polyisocyanates

Highlights of the Technological Achievement: Microcapsules and a process for the production of these microcapsules, the walls of which consist of polycondensates of a film-forming aliphatic polyisocyanate containing at least one biuret group or polyaddition products thereof with a chain-extending agent.

Reference: G. Baatz, W. Schäfer and M. Dahm; U.S. Patent 4,193,889; March 18, 1980; assigned to Bayer AG, Germany.

Background: In a practical embodiment initially a stable emulsion is prepared of which the disperse phase consists of the material to be encapsulated and the first component, optionally dissolved or emulsified in an additional solvent, and of which the continuous phase is formed by a liquid immiscible with the disperse phase.

Encapsulation processes of this type have some decisive disadvantages. Thus, it is necessary to prepare a stable emulsion which does not disintegrate, even when the second component is added. The process cannot be varied to any significant extent because only few suitable reaction components are available. For practical application, only few and very similar polymers can be produced in the manner described. Furthermore, the low molecular weight reaction components are substantially nonfilm-forming. Accordingly, only little core material can be encapsulated in a large quantity of shell-forming material.

Process Conditions: In the polyisocyanates used in accordance with the process, some of the isocyanate groups can be modified by dimerization or trimerization or by conversion into carbodiimide groups, and a part is converted into biuret groups, for example, by reaction with water, tertiary butanol, formic acid or amines. Aliphatic polyisocyanates modified in this way usually have viscosities of from 8,000 to 40,000 cp at 25°C.

Reactants for these modified polyisocyanates during the microencapsulation process are low molecular weight chain-extending agents, such as water, diols or polyols, but especially amines. The microcapsules according to the process show adequate crosslinking density and, therefore, are only slightly permeable to easily volatile encapsulated substances.

The process has decisive advantages. The reaction of the modified isocyanates with the reactants is quick enough for continuous operation. High volume/time yields are obtained.

For carrying out the process, which is an encapsulation process by polyreaction at the organic phase interface in a dispersion, the wall-forming polyisocyanate can be dissolved or emulsified in the core material either as such or in the form of a solution in an inert solvent or solvent mixture.

In order to obtain better emulsification and stabilization of the dispersion, emulsification aids are preferably added to the aqueous phase. Examples are protective colloids such as polyvinyl alcohol, gelatin, carboxymethylcellulose and polyacrylates.

Encapsulation can be carried out continuously or in batches. The degree of turbulence generated during mixture determines the diameters of the microcapsules obtained. The diameter of the microcapsules can be from about 5 to about 2,000 μm according to the mixing conditions. The ratio by weight of core material to shell material in the finished microcapsules normally is from 60-90:40-10.

Example: *Production of the Isocyanate Component (Biuret Polyisocyanate)* – 1,000 pbw of hexamethylene-1,6-diisocyanate (5.952 mols) and 50 pbw of tert-butyl alcohol (0.676 mol) are mixed at room temperature. The molar ratio corresponds to 8.8 mols of diisocyanate to 1 mol of tert-butanol. The temperature of the reaction mixture is increased to 160°C over a period of about 30 minutes. A vigorous evolution of carbon dioxide and isobutylene begins at that temperature.

The temperature is slowly increased to 185°C over a period of another 30 min. After about 3 hr at 185°C, gas is no longer evolved hence the reaction is over. The reaction product is then freed from monomeric hexamethylene diisocyanate in a thin layer evaporator under a pressure of 0.2 torr and at a temperature of 160°C. Approximately 284 pbw of a viscous biuret polyisocyanate are obtained. NCO-content: 21.3% by weight. The product has a viscosity of 10,500 cp at 20°C.

Encapsulation – 25 ml of trichloroethyl phosphate (TCAP) are mixed with 5 g of the biuret polyisocyanate described above by means of a laboratory disperser of the Ultra-Turrax type (Jahnke and Kunkel AG).

300 ml of water containing 1 g of polyvinyl alcohol in solution (Moviol 50/98, Hoechst AG) are initially introduced and the mixture of TCAP and biuret polyisocyanate is emulsified therein immediately after its preparation by means of a laboratory stirrer of the Lenart-Rapid type (500 rpm). After about 1 minute, a solution of 14 g of ethylene diamine in 56 g of water is added to the emulsion. The mixture is then heated to 60°C and stirred for approximately 1 hour under the same conditions to harden the capsules and is then neutralized. The diameter of the capsules produced amounts to between 300 and 1,200 μm.

Water-Soluble Cationic Oligourethane Resins for the Treatment of Pelts or Leather

Highlights of the Technological Achievement: This process relates to water-soluble, cationic hydrophilic oligourethanes free from isocyanate groups and containing methylol groups and to the use thereof for the tanning, pretanning or retanning of pelts or pretanned leather and as dyeing auxiliaries.

Reference: H. Träubel, D. Dieterich and H. Heinze; U.S. Patent 4,187,074; February 5, 1980; assigned to Bayer AG, Germany.

Background: The aqueous oligourethane solutions according to the process have the following advantages over solutions known in the art: (1) They have excellent resistance to electrolytes in spite of the cationic character thereof; (2) the anionic character of retanned chrome leather is overcompensated; (3) the products have a soft tanning effect; (4) the products may be used for pretanning without deleteriously affecting the subsequent chrome tanning; and (5) the cationic oligourethane solutions improve the consumption of the anionic dyes and produce a more brilliant and more intense color without deleteriously affecting the leveling of the dyeing.

Process Conditions: The process thus relates to water-soluble oligourethanes which contain methylol groups and are free from isocyanate groups and have a molecular weight ($\overline{M}n$) of between about 600 and 5,000 based on:

(a) Polyethylene glycol ethers having a molecular weight ($\overline{M}n$) between about 400 and 1,500;

(b) Polyisocyanates; and, optionally,

(c) Compounds which are mono- or poly-functional towards isocyanates, preferably alcohols, having a molecular weight below about 300; and

(d) Formaldehyde,

characterized in that the proportion of component (a) in the oligourethane is from 40 to 90% by weight, preferably from 60 to 80% by weight, and in that component (a) and/or (c) together contain a total of preferably 30 to 100 meq of tertiary nitrogen atoms per 100 g of oligourethane solids content, preferably 60 to 90% thereof, which have been converted into quaternary ammonium nitrogen or into the ammonium salt by treatment with alkylating agents and/or acids.

The process also relates to a process for tanning or retanning pelts or leather, characterized in that the pelts or leather are treated with an aqueous solution of the oligourethanes which solution is at a pH most preferably between 7 and 8.

The polyethylene glycol ethers used in the preparation of the oligourethane resins may be polyethers having an average functionality of 2 to 3, and a molecular weight ($\overline{M}n$) of 400 to 1,500. These polyethers may be obtained in known manner by the addition of ethylene oxide to starting components which contain reactive hydrogen atoms, such as alcohols, amines or water.

The total quantity of tertiary and quaternary nitrogen and nitrogen in the ammonium salt form present in the oligourethane resin should be preferably from 0.35 to 1.8% by weight.

C_1 to C_{20} monohydric alcohols are advantageously used as chain-breaking agents in the preparation of the oligourethanes according to the process. The following examples serve to illustrate the process. (Figures given for the parts denote parts by weight or percentages by weight, unless otherwise indicated.)

Examples 1 through 7: The following mixture of hydroxy-functional compounds is introduced into a reaction vessel at room temperature: 492 g (0.8 mol) of a polyoxyethylene glycol having a molecular weight ($\overline{M}n$) of 615, 23.8 g (0.2 mol) of N-methyldiethanolamine and 19.2 g (0.6 mol) of methanol.

The quantity of 1,6-diisocyanatohexane indicated in the table is added and the mixture is heated, if necessary, to 80°C until no more free isocyanate groups may be detected. A solution of 3.3 g of crude commercial glutaric acid in 200 g of water and 65 g of a 37% aqueous formaldehyde solution (0.8 mol) are then added and the mixture is finally diluted with water to a solids content (including formaldehyde) of 55%.

	Example No.						
	1	2	3	4	5	6	7
1,6-Diisocyanatohexane							
Grams	117.6	134.4	151.2	168.0	184.8	201.6	215.0
Mols	0.7	0.8	0.9	1.0	1.1	1.2	1.28
pH	7.6	7.8	7.3	7.3	7.3	7.5	7.6

Comparison Experiment: The procedure is the same as in Example 2, but with the addition of 36 g (0.6 mol) of urea instead of methanol (pH = 7.4). The products from the following examples were tested analogously and the following shrinkage temperatures were found:

Ex. No.	Dry Substance* (%)	Shrinkage Temperature (°C)
1	10	86
2	10	87
3	10	86
4	10	85
5	10	84
6	10	85
7	10	85
**	10	75

*Based on the weight of pelt.
**Comparison experiment.

Polyurethanes Prepared from Alcohols and Hydrocarbon Polyisocyanates Used in Textile Wet Treatment Processes

Highlights of the Technological Achievement: Disclosed is a textile wet treatment process wherein, to increase the lubricity of the substrate, certain specified polyurethanes are employed in the treatment liquor.

Reference: J. Canela and H. Gerber; U.S. Patent 4,186,119; January 29, 1980; assigned to Sandoz Ltd., Switzerland.

Background: In the wet processing of synthetic or semisynthetic textile materials by modern processing methods, especially in winch-beck and jet dyeing machines, the undesirable formation of creases in the material represents a problem. Such crease formation, for example, can lead to unlevel dyeing.

It has been discovered that the tendency of the material to creasing can be considerably reduced or even eliminated if the wet processing is effected in the presence of certain water-soluble or -dispersible polyurethanes having the property of increasing the fiber/fiber lubricity and which can be named lubricating agents.

Process Conditions: The process relates to such lubricating agents and their use in the wet processing of textile materials, and, more particularly, provides a wet treatment process for synthetic or semisynthetic textile material comprising employing a liquor containing a polyurethane. The polyurethane is a reaction product of a compound or mixture of compounds 1, the, or each, compound 1 being a derivative of ammonia, a polyamine; urea; a polyamide; and aminoamide; an aminoalcohol; or a polyol containing up to 10 carbon atoms and up to 6 hydroxyl groups per molecule; in which at least three of the active hydrogen atoms of the amino, hydroxyl and/or amido moieties present are replaced by the same or different groups of formula 1a,

$$-\!\!\left(\text{alkylene-O}\right)_{n}\!\!-\text{H}$$

in which each alkylene independently, is straight or branched chain (C_{2-4})-alkylene, unsubstituted or substituted by a phenyl group, and n is an integer 1 to 100, and a compound or mixture of compounds 2, each compound 2 being a hydrocarbon polyisocyanate, in a molar ratio of compound(s) 1:compound(s) 2 of $\Sigma p:\Sigma r$, p being the mols of the, or each, compound 1 and r being the mols of the, or each, compound 2 reacted. Σp and Σr are so chosen that $\Sigma(p \cdot q)/\Sigma(r \cdot m) = 1.5$ to 21 and $\Sigma p \leqslant \Sigma(r \cdot m)$, q being the number of groups of formula 1a per molecule of the, or each, compound 1 and m being the number of isocyanate groups per molecule of the, or each, compound 2.

The polyurethane is water-soluble or -dispersible to a degree of at least 1 g/ℓ at 20°C, and this 1 g/ℓ aqueous solution or dispersion at 20°C should exhibit a viscosity of no more than 100 cp.

Thus the relative amounts of the reacting species used to produce the polyurethane employed in the process are such that the total number of molecules of the compound(s) 1 does not exceed the total number of NCO groups in the compound(s) 2, and the polyurethane reaction product has a ratio of free hydroxyl groups present in groups of formula 1a to urethane groups of from 0.5:1 to 20:1, it preferably being from 0.5:1 to 10:1, preferably in the upper region of the latter range when the product is of relatively high molecular weight.

By the expression, "water-dispersibility of at least 1 g/ℓ at 20°C," used above is meant that displayed when an aqueous dispersion of the polyurethane product containing at least 1 g/ℓ of the product remains stable, i.e., does not show any visible signs of deposition of the dispersed product, after standing at 20°C for at least 12 hours.

Preferably the polyurethane product is water-soluble rather than merely water-dispersible, and has a water-solubility of at least 10 g/ℓ, more preferably at least

25 g/ℓ, at 20°C. The solidification temperature of the polyurethanes used in the process preferably does not exceed 150°C.

The polyurethane products may still contain a certain amount of untreated free isocyanate groups, but for the purposes of the process, up to 1% of the unreacted isocyanate groups originally present in the starting compound(s) 2 in the product can be tolerated.

In the compounds 1, there are preferably present at least as many alkyleneoxy units as the number of active amino, hydroxy and/or amino hydrogen atoms in the original ammonia, polyamine, urea, polyamide, aminoamide, amino-alcohol or polyol.

The polyurethane lubricating agents are preferably present in the wet treatment liquor employed at temperatures between 50° and 150°C, and, independently, at a concentration between preferably 0.3 to 1 g of polyurethane reaction product (active substance) per liter of treatment liquor.

The polyurethane lubricating agents are preferably added to the treatment liquor before the wet treatment process is begun in the form of an aqueous concentrate. Such concentrates preferably contain at least 25% by weight of polyurethane reaction product (active substance), or more preferably at least 50% by weight. The viscosity of such aqueous concentrates preferably does not exceed 100,000 cp at 20°C.

Modified Aqueous Synthetic Resin Dispersions Containing Organic Diisocyanates

Highlights of the Technological Achievement: A process for preparing modified aqueous synthetic resin dispersions, comprising introducing organic diisocyanates which are liquid at room temperature into polyurethane-containing, nonsedimenting, aqueous synthetic resin dispersions.

These modified dispersions are useful as water-resistant surface coatings and leather dressings which are abrasion resistant when wet.

Reference: A. Reischl, W. Wenzel and D. Dieterich; U.S. Patent 4,186,118; January 29, 1980; assigned to Bayer AG, Germany.

Background: It has been found that the properties useful for the practical application of stable, aqueous polyurethane dispersions which are preferably free from emulsifiers and may contain water-dispersible polymers may be considerably improved by incorporating specially selected diisocyanates in the dispersions by a particular method to be described in detail below. This addition of diisocyanates causes the dispersed latex particles to be enveloped with polyurea formed from the diisocyanates.

Process Conditions: The aqueous synthetic resin dispersions used in the process are preferably emulsifier-free, nonsedimenting dispersions having a solids content of from 10 to 60% by weight, preferably from 20 to 50% by weight, although mixtures of such polyurethane dispersions with dispersions of polymerization products which optionally contain emulsifiers may also be used in the process. If such mixtures are used, the total solids content thereof may be from 10 to 60% by weight, preferably from 20 to 50% by weight, and the proportion of other polymerization products, based on the total solids content, may be up to 70% by weight, preferably up to 50% by weight. By "nonsedimenting" dispersions are meant, in this context, synthetic resin dispersions which have formed no visible precipitate after 4 weeks' storage at 20°C.

Particularly preferred polymer latices are those based on polyacrylonitrile, on butadiene-acrylonitrile copolymers or on grafted copolymers based on acrylonitrile, butadiene and styrene, as well as those based on polymethacrylates.

Diisocyanates suitable for the process are organic diisocyanates which are liquid at room temperature and have a maximum average molecular weight of 400, preferably 300.

To carry out the process, the synthetic resin dispersion which is to be modified is introduced into a reaction vessel which may be heat controlled and which is equipped with a stirrer or other mixing device and the diisocyanate is introduced at such a temperature preferably in the range of 10° to 30°C, at which foam formation is not yet observed. This temperature at which foam formation is not yet observed depends on the quantity of diisocyanate as well as on the reactivity of the diisocyanate which is used. The temperature is generally within the ranges indicated and can easily be determined by a preliminary test.

The introduction of diisocyanate is accompanied by mixing of the components. These conditions are maintained after complete addition of the diisocyanate until at least 50% of the isocyanate groups in the diisocyanate introduced have been used up. If desired, the reaction may then be completed by heating to temperatures of up to 100°C.

The quantity of diisocyanate to be used in the process is preferably from 3 to 50% by weight, based on the total solids content in the modified synthetic resin dispersion. In the case of hexamethylene diisocyanate, the preferred quantity is from 1.5 to 5% by weight.

Water-Based Compositions from Isocyanate-Terminated, Carboxyl Group-Containing Prepolymer

Highlights of the Technological Achievement: A water-based polyurethane dispersion is prepared by mixing an aliphatic diisocyanate with three critical active hydrogen compounds, to form a prepolymer containing carboxyl groups and free isocyanate groups, then dispersing the prepolymer in an aqueous medium containing a tertiary amine, which converts the carboxyl groups to carboxylate ions, and a diamine, which extends the prepolymer through its –NCO groups. These aqueous dispersions are storage-stable for at least several months and are useful in many coating applications, among others, for coating textile materials.

Reference: H.W. Wolfe, Jr.; U.S. Patent 4,183,836; January 15, 1980; assigned to E.I. DuPont de Nemours and Co.

Background: Polyurethane coating compositions are important commercial products. They provide a tough, flexible, abrasion-resistant finish for a variety of industrial articles made of wood, metal, glass, ceramic materials, and fabrics. Because of the difficulty of dispersing polyurethanes in water, the usual commercial polyurethane compositions are solutions or dispersions in organic solvents.

However, these solvent-based products have several shortcomings, especially high cost, high flammability, and toxicity of solvent vapors. There is, therefore, a great need for polyurethane coating compositions based on water dispersions.

Process Conditions: A composition obtained by mixing an isocyanate-terminated, carboxyl group-containing prepolymer made by reaction of (1) 1 equivalent of an aliphatic diisocyanate with (2) total 0.6 to 0.8 equivalent of the following active hydrogen compounds:

(a) α,α-dimethylolacetic acid or a homologue thereof,

(b) A polymeric glycol selected from polyalkyleneoxy glycols and polyester glycols having a weight average molecular weight of about 400 to 3,000; and

(c) N,N'-dicyanoethylpoly(propyleneoxy)diamine having a number average molecular weight of about 200 to 2,000;

the amount of component (a) being 0.08 to 0.25 hydroxyl equivalent per 100 g of total weight of all the reactants (1) and (2), and the relative mol proportions of compounds (b) and (c) being 30:70 to 70:30, with an aqueous solution of a sufficient amount of a tertiary amine to convert substantially all the carboxyl groups of the prepolymer to carboxylate ions, then chain-extending the prepolymer with about stoichiometric amount, based on free isocyanate groups, of a diamine having at least one hydrogen attached to each nitrogen atom.

Aliphatic diisocyanates suitable in the process include, among others, hexamethylene diisocyanate, 1-isocyanato-2-isocyanatomethyl-3,5,5-trimethylcyclohexane and methylenebis(cyclohexyl isocyanate), which is preferred.

It is recommended to add the acid (a) after the reaction of the diamine (c) with the diisocyanate (1) is complete, to avoid premature salt formation. With this reservation, the reactants can be added in any order. Yet, it is preferred to run this reaction in steps, allowing the diisocyanate to react first with N,N'-dicyanoethylpoly(propyleneoxy)diamine. The resulting polyurea catalyzes the subsequent reactions of the less reactive hydroxyl groups with additional isocyanate groups. Other catalysts, e.g., organotin compounds or tertiary amines can also be present in the reaction medium.

Molecular Weight: The prepolymer is dispersed in an aqueous solution of tertiary amine. The tertiary amine must not carry any substituents reactive with isocyanate groups. Otherwise, there are no restrictions on its molecular weight or substituent groups. The lower molecular weight aliphatic tertiary amines, such as triethylamine, tripropylamine, or tributylamine, are preferred because they present less potential health hazard than aromatic amines, are stronger bases capable of readily forming salts with carboxyl groups, and have sufficient volatility for the practical operation of the process. This step of dispersing the prepolymer in the presence of tertiary amine is carried out at ambient temperature to avoid hydrolyzing the isocyanate groups of the prepolymer. The dispersion is stable at room temperature for several hours.

The chain-extending diamine is then added at ambient temperature, and the resulting mixture is well agitated. The diamines can be any diprimary, disecondary, or primary-secondary diamine.

Test Methods: The following test methods are employed in the following examples: Sward Hardness, described in *Physical and Chemical Examination of Paints, Varnishes, Lacquers, Colors,* Gardner and Sward, 11th ed., 1950, pp 164-166; Modulus at 100% Elongation, (M_{100}), ASTM D412; Tensile Strength at Break, (T_B), ASTM D412; and Elongation at Break, (E_B), ASTM D412.

Example 1: To a mixture of 132 g (1.0 equivalent) of 4,4'-methylenebis(cyclohexyl isocyanate) containing 65% cis,trans isomer, 20% trans,trans isomer and 15% cis,cis isomer, 0.05 g of dibutyltin dilaurate, and 74 g of methyl ethyl ketone is added with mixing 42 g (0.25 equivalent) of N,N'-di-β-cyanoethylpoly(propyleneoxy)diamine having a molecular weight of about 236. After 15 min, 100 g (0.2 equivalent) of polypropyleneoxy glycol having a molecular weight of about 1,000 is added with mixing. The mixture is heated to 80°C and maintained at that temperature for 2 hours, after which 20.1 g (0.3 equivalent) of α,α-dimethylolpropionic acid is added, and heating at 80°C is continued for another hour. The resulting prepolymer is cooled to room temperature and held for at least 24 hours.

To an agitated mixture of 1.9 g of triethylamine and 70 g of water at 25°C is added 50 g of the above prepolymer. Agitation is continued for 15 minutes to insure complete dispersion. To this dispersion is added 1.2 g of ethylenediamine, and agitation is continued for 30 minutes. The resulting dispersion is quite stable and can be stored for an extended period. The dispersion is an opalescent liquid having the following properties: Viscosity at 22°C, 0.40 Pa-sec; Solids Content, 32%; pH, 7.6; and Density, 1.01 g/ml.

Draw films of 0.254 mm wet thickness are formed on plate glass. The film dries to a thickness of about 0.0762 mm in 35 to 45 minutes at 25°C. The clear dry film is removed from the glass plate for physcial testing. The following properties are obtained: Sward Hardness, 30; T_B, 30 MPa; E_B, 230%; and M_{100}, 16.1 MPa.

Example 2: Substantially identical results are obtained when Example 1 is repeated using 100 g of polyethylene adipate glycol having a molecular weight of about 1,000 in place of 100 g of polypropyleneoxy glycol of 1,000 molecular weight.

Examples 3 through 7: A series of prepolymers are prepared by the procedure of Example 1 employing the following ingredients in the amounts tabulated below.

	. . . . Example Number				
	3	4	5	6	7
4,4'-Methylenebis(cyclohexylisocyanate), equivalents	1.0	1.0	1.0	1.0	1.0
N,N'-di-β-cyanoethylpoly(propyleneoxy)diamine, MW 340, equivalents	0.25	0.15	0.1	0.1	–
N,N'-di-β-cyanoethylpoly(propyleneoxy)diamine, MW 2,000, equivalents	–	–	–	0.075	0.1
Poly(propyleneoxy) glycol, MW 1,000, equivalents	0.2	0.2	0.2	–	–
Poly(propyleneoxy) glycol, MW 2,000, equivalents	–	–	–	0.075	0.1
α,α-Dimethylolpropionic acid, equivalents	0.3	0.4	0.4	0.45	0.5

Portions of each of the resulting prepolymers are converted to dispersions employing the procedure and amounts used in Example 1. Clear films are prepared from each formulation and have the following properties:

	Example Number				
	3	4	5	6	7
Sward hardness	30	38	22	18	4
M_{100}, MPa	16.9	14.7	13.5	15.6	9.7
T_B, MPa	26.3	21.9	16.3	22.5	19.4
E_B, %	250	240	180	316	430

POLYVINYL ALCOHOL (PVA) AND RELATED PRODUCTS

PROCESSING

Recovery of Modified Polyvinyl Alcohols from Aqueous Solutions Containing Them.

Highlights of the Technological Achievement: Recovery of partially hydrolyzed copolymers of vinyl acetate and unsaturated carboxylic or sulfonic acid from the aqueous solution dissolving them by adding an acid to the solution at a temperature of at least 40°C and then separating the deposited copolymers. This process can be advantageously applied to wastewater dissolving the above hydrolyzed copolymers from desizing and washing of textile. 70 to 90% of the hydrolyzed copolymers dissolved in the solution can be recovered, and the recovered copolymers can be reused as a sizing agent for fibers.

Reference: H. Kato; U.S. Patent 4,205,147; May 27, 1980; assigned to Nippon Gohsei Kagaku Kogyo KK, Japan.

Background: Various polyvinyl alcohols are widely employed as sizing agents for fibers, and the wastewater from scouring inevitably dissolves these polyvinyl alcohol sizing agents. The recovery of the polyvinyl alcohol sizing agents from such a wastewater is of advantage to the reduction of cost for the sizing agent and the cleaning of wastewater.

As a conventional process for removing polyvinyl alcohols from the wastewater of desizing, there are known a salting out method, a flocculent sedimentation method using polyvalent metal salts, a bubble separation method, an oxidation method, an absorption method, an insolubilization method by formalization and a biochemical treatment method. However, these known methods have the disadvantages that the degree of removal is generally low and moreover polyvinyl alcohols are decomposed or, even if recovered, the recovered polyvinyl alcohols contain other additives and cannot be reused as sizing agents.

Process Conditions: This process can effectively be applied to a wastewater from a scouring process, dissolving the modified polyvinyl alcohols. The modified polyvinyl alcohols dissolved in the wastewater can be readily deposited by adding an acid and separated in high yields. The separated deposit can be redissolved by converting the recovered modified polyvinyl alcohols into the salt type with an alkali, and can be used repeatedly as a sizing agent.

The modified polyvinyl alcohols in this process are alkali-hydrolyzed copolymers of vinyl acetate and an unsaturated carboxylic or sulfonic acid, particularly those having a degree of hydrolysis in the vinyl acetate units of 50 to 80 mol %. Examples of the unsaturated carboxylic acid are acrylic acid and methacrylic acid.

Although the content of the unsaturated carboxylic or sulfonic acid in the modified polyvinyl alcohols is not particularly limited, the modified polyvinyl alcohols containing 0.1 to 2 mol %, of the unsaturated acid units are suitably employed as a sizing agent for fibers. When the content is less than 0.05 mol %, the polymers are deposited under the temperature condition for sizing, and when the content is more than 10 mol %, the polymers have excess hygroscopicity and tackiness and are inferior in sizing property.

The modified polyvinyl alcohols are prepared by hydrolyzing the copolymers of vinyl acetate and an unsaturated carboxylic or sulfonic acid in the presence of an alkali catalyst such as sodium hydroxide, potassium hydroxide or alkali metal alkoxides, and preferably in an alcoholic medium such as methanol or ethanol.

When the degree of hydrolysis is less than 50 mol %, the desizing property is not good, and when the degree of hydrolysis is more than 80 mol %, the adhesiveness to a hydrophobic fiber is poor. Also, the modified polyvinyl alcohols having the degree of hydrolysis of more than 80 mol % is almost impossible to recover from wastewater of desizing and washing.

In order to separate the abovementioned particular modified polyvinyl alcohols from wastewater dissolving them, it is necessary to conduct both procedures of (a) adding an acid to the wastewater and (b) maintaining the wastewater at a temperature of not less than 40°C, preferably not less than 50°C. These procedures can be conducted in arbitrary order. Usually after heating the wastewater to over 40°C, an acid is added to the wastewater with agitation. The modified polyvinyl alcohols cannot be deposited using only one procedure, either (a) or (b), and even if deposited, the deposition is extremely insufficient.

Typical examples of the acid employed in this process are hydrochloric acid, sulfuric acid and acetic acid. Others may be used, but these three are preferred. The acid is usually employed in excess in an amount of 5 to 20 times the equivalent or more than 20 times the equivalent. In many cases, the wastewater is below pH 7, usually pH 6 to 2, when the acid is added.

This process is more specifically described and explained by means of the following example, in which all percents are percents by weight unless otherwise stated.

Example: By employing, as a modified polyvinyl alcohol, hydrolyzed vinyl acetate-sodium monomethyl maleate copolymer having a degree of hydrolysis of 70.0 mol % which was obtained by hydrolyzing a copolymer of vinyl acetate and 0.5 mol % of monomethyl maleate in methanol in the presence of sodium hydroxide, procedures of deposition-recovery-redissolution were repeated on a laboratory scale.

The deposition was carried out by adding five equivalents of hydrochloric acid per equivalent of monomethyl maleate to an aqueous solution of the modified polyvinyl alcohol at a temperature of 60°C. After heating to 60°C, the acid was added. The solution was agitated at 600 rpm by a propeller type agitator, by which the deposited polymer rose to the surface in the form of cream. The deposited polymer could be readily recovered by merely skimming. When either the acid addition or the heating to more than 40°C was omitted, no polymer was deposited.

The polymer content of the recovered creamy matter fell within the range of 12 to 16%. The creamy matter was neutralized with sodium hydroxide by employing BTB indicator to redissolve the polymer. About 1.7 to 2.1 equivalents of sodium hydroxide to monomethyl maleate was required in the neutralization, depending on the amount of hydrochloric acid present in the creamy matter.

Polymer concentration of the aqueous solution to be treated, recovery, and viscosity and degree of hydrolysis of the recovered polymer are shown in the following table. The viscosity of the recovered polymer was measured at a temperature of 20°C with respect to the aqueous solution thereof having a concentration of 4%.

	Number of Repetitions						
	0	1	2	3	4	5	6
Polymer concentration, %	–	0.87	0.62	0.54	0.48	0.43	0.40
Recovery, %	–	72	84	87	89	87	88
Viscosity, cp	9.5	11.0	12.0	12.2	12.6	12.9	13.0
Degree of hydrolysis, mol %	70.0	70.0	69.6	69.7	69.6	69.4	69.3

As seen from the above table, the recovery and the viscosity of the recovered modified polyvinyl alcohol were stable after the third repetition, though they varied to some extent before the second repetition.

Aqueous Solution of a Mixed Salt of an Interpolymer of Styrene/Maleic Anhydride/Vinyl Acetate

Highlights of the Technological Achievement: A composition for sizing textiles which employs as the sizing agent a mixed salt comprising the reaction product of an interpolymer with a fixed alkali base and a fugitive alkali base; the latter subsequently being volatilized away during processing of the sized textile.

Reference: C.R. Williams; U.S. Patent 4,189,416; February 19, 1980; assigned to Monsanto Company.

Background: In commercial textile weaving operations, the breakage of individual warp fibers has long been a costly problem. When breakage occurs, the loom automatically stops and the fractured strand must be tied or spliced by hand before the process can continue. These warp stops can be decreased by coating the warp threads with sizes which improve abrasion resistance and add tensile strength to the fibers. It has been found, however, that yarns treated with many of the sizes of the prior art lose a substantial part of the size film when subjected to the friction and abrasion concomitant with the weaving operation. When this dusting is excessive, as often happens, the warps lose their stiffness character and weaving efficiency is adversely affected.

Process Conditions: This process relates to a textile material sizing and finishing composition which comprises an aqueous solution of a mixed salt of an interpolymer, wherein the interpolymer comprises:

(a) 1.0 molar proportion of at least one carboxylic monomer selected from the group consisting of maleic anhydride, maleic acid, fumaric acid, and partial esters of up to the 100% half-ester of maleic or fumaric acid with monohydric alcohols of from 1 to 8 carbon atoms;

(b) from 0.1 to 3.0 molar proportion of at least one monovinylidene aromatic monomer containing from 8 to 12 carbon atoms; and

(c) from zero to X molar proportion of at least one monomer selected from the group consisting of alpha-olefins of up to 8 carbon atoms wherein X is 2, vinyl esters of from 1 to 4 carbon atoms in the acyl group wherein X is 8, alpha-beta-ethylenically unsaturated nitriles wherein X is 2, acrylic and methacrylic esters of from 1 to 8 carbon atoms in the alkyl group wherein X is 0 to 5 and vinyl alkyl esters of from 1 to 4 carbon atoms in the alkyl group wherein X is 6.

The total of monomers (b) and (c) is at least 1.0 molar proportion and, in addition:

(a') from 10 to 70% of the free carboxyl groups of the interpolymer are combined with at least one fixed base selected from the group consisting of nonvolatile monobasic amines and alkali metal bases; and

(b') from 5 to 90% of the free carboxyl groups of the interpolymer are combined with at least one fugitive base selected from the group consisting of ammonia, ammonium hydroxide and volatile monobasic amines.

This is such that a total of from 20 to 100% of the free carboxyl groups are so combined. The solution is applied in an amount sufficient to provide 0.5 to 20 wt % of the salt based on the weight of the yarn. The treated textile materials are heated at a temperature high enough to volatilize the fugitive base. Advantageously this heating may be accomplished during the drying operation.

Example: 1 mol of a 1:1 copolymer of styrene and maleic anhydride is slurried in water. To the slurry is added 1 mol (i.e., 50% neutralization) of sodium hydroxide whereby the copolymer is dissolved yielding a solution containing 4.7% solids and having a pH of 6.7. The solution (Solution A) has a viscosity of 14,000 cp at 89°C, a viscosity much too high for use as a textile size. Sufficient concentrated ammonium hydroxide (29% NH_3) is added to a portion of this solution to increase the pH to 7.5 (i.e., 7.5% neutralization to a total of 57.5% neutralization).

Since a very small amount of ammonia is used, the solids content of the solution (Solution B) is still essentially 4.7%. The viscosity decreases markedly and is found to be 3,450 cp at 65°C. Films of Solution A and B are cast and permitted to dry at room temperature. During the drying, the odor of ammonia being emitted from Solution B is apparent. When dry the tensile strength of a film prepared from the sodium salt of this example is 3,050 psi at 72% RH and 20°C. The tensile strength of the mixed salt prepared in this example is 3,160 psi under the same conditions.

Textile yarn (warp 5,900 ends of 65/35 polyester/cotton yarn 30/1) is sized with the mixed salt sizing solution prepared above and is run on a commercial 9 can slasher having an internal steam pressure of 35 psi. The size box temperature is maintained at 180°F and the squeeze roll is held at a total pressure of 1,200 lb. The sized warp is observed for slasher performance and tested for weaving performance.

The sized yarn having an applied total solids of about 10 wt % based on the weight of the yarn, is cleanly and easily separated at the split rods with little or no shedding. The sizing solution in the size box remains clear and homogeneous throughout the operation and exhibits no tendency to form a surface skin. The sized yarn is subsequently woven on commercial looms. The looms operate at a high weaving efficiency with very few warp stops.

Aqueous Vinyl Chloride/Vinyl Acetate/Ethylene Copolymers

Highlights of the Technological Achievement: The process relates to aqueous copolymer dispersions containing high proportions of vinyl chloride units in the copolymer, together with vinyl acetate units and ethylene units, which dispersions contain only polyvinyl alcohol as protective colloid.

Reference: H. Eck, C. Kemenater, W. Kaiser and M. Hannebaum; U.S. Patent 4,189,415; February 19, 1980; assigned to Wacker-Chemie GmbH, Germany.

Background: Aqueous copolymer dispersions that contain vinyl chloride units, vinyl acetate units and ethylene units are known and described in German Patents 21 19 549 and 22 52 285, corresponding to U.S. Patents 3,639,326 and 3,830,761 respectively.

To obtain stable dispersions from these monomers, it was hitherto always necessary to take special steps regarding the protective colloid/emulsifier system. Further, it was often absolutely necessary to adhere to critical amounts of ionic or, alternatively, nonionic emulsifiers. Not only did these special colloid stabilization systems comprising emulsifiers, protective colloids and auxiliary monomers have to be reproducibly adhered to under relatively complicated industrial conditions, but sometimes they also had disadvantageous effects in various uses of the dispersions, such as, for example, in adhesives or concrete additives. Adhesives containing emulsifiers or auxiliary stabilizing monomers generally have a relatively low water-resistance.

Process Conditions: A stable aqueous vinyl chloride copolymer dispersion has a solids content of from 30 to 70 wt %, with the dispersion containing a copolymer consisting of: 50 to 85 wt % vinyl chloride units; 5 to 35 wt % of vinyl acetate units; 5 to 30 wt % of ethylene units; and 0 to 5 wt % of monoolefinically unsaturated N-methylolamide units, and as protective colloid, from 2 to 15 wt % based on the dispersion weight of at least one polyvinyl alcohol having a saponification number of between 20 and 240 and a polymerization degree of between 300 and 2,000.

A process is developed for the production of the above stable, aqueous vinyl chloride copolymer dispersion comprising the steps of polymerizing monomers in such amounts to give the polymer, in a sufficient amount of aqueous medium, a final solids content of 30 to 70 wt % at a temperature of +10° to + 85°C in the presence of 0.01 to 3 wt % of the monomer charge of a redox catalyst system while stirring, at a constant ethylene pressure of between 20 and 150 bars and in the presence of 2 to 15 wt % of the dispersion of at least one polyvinyl alcohol having a saponification number of between 20 and 240 and a polymerization degree of between 300 and 2,000, while maintaining the monomer concentration of the liquid monomers between 5 and 20 wt % by metering in the monomers, and postpolymerizing by the addition of a further 0.005 to 0.1 wt %, based on the monomer charge, of the respective components of the redox catalyst system.

Uses: (a) Binders in hydraulically-setting building materials, (b) adhesives in combination with acidic hardeners, and (c) as coating agents in combination with aluminum nitrate, chromium nitrate, acidic phosphates and phosphoric acid as hardeners.

Example: 400 g of polyvinyl alcohol having a saponification number of 60 and a Höppler viscosity of 13 mPa-sec (of a 4% aqueous solution at 20°C) were dissolved, together with 40 g of potassium persulfate, in 5.4 ℓ of water. A mixture of 4.2 kg of vinyl chloride and 0.8 kg of vinyl acetate were continuously metered in, while stirring at a polymerization temperature of 30°C. During the whole of the polymerization period, the ethylene pressure is held constant at 50 bars. The polymerization was started by the addition of a 1% strength sodium formaldehydesulfoxylate solution and controlled by the further addition of reducing agent. Altogether the consumption of reducing agent of the concentration was 770 g (7.7 g of reducing agent, 100% strength). After 15 hr, polymerization was complete.

In the postpolymerization phase, the following were added: 2.67 g of Rongalite C, dissolved in 15 ml of water (0.02% based on the dispersion); and 2.67 g of t-butyl hydroperoxide (0.02% based on the dispersion).

The dispersion obtained had a solids content of 51.8 wt %, a viscosity, measured in the Epprecht Rheometer stage C III, of 5,900 mPa-sec and a minimum film-forming temperature of 20°C. The dispersion had a polymeric composition of 16.7 ethylene, 70% vinyl chloride, and 13.3% vinyl acetate. The dispersion is freeze-resistant, pigment-stable and shear-stable and has good strength values. After the addition of 2 wt % of butoxyethoxyethyl acetate to the dispersion, the tear strength of the film was approximately 12.5 N/mm^2 with an elongation-at-break of approximately 350%.

The examination of wood pasted together with a mixture of the dispersion with 5 wt % of a 40% strength aqueous solution of chromium(III) nitrate, produced from 70 pbw of the hydrated ($9H_2O$) salt in 30 pbw of water, yielded according to DIN 68603, the following adhesive strengths: storage sequence No. 5, 2.7 N/mm^2; and storage sequence No. 9, 4.0 N/mm^2.

When using the dispersion as concrete additive with the co-use of a customary antifoaming agent in a quantity of 0.1 wt %, calculated on the dispersion, the values shown below are obtained. The antifoaming agent consists of 94 wt % of dimethylpolysiloxane having a viscosity of more than 20 cs at 25°C, and

6 wt % of finely divided silica. With a plastics/cement factor of 0.05 and a water/cement factor of 0.48, the following values were obtained.

	Bending Tensile Strength	Pressure Resistance
	..............N/mm^2	....
28 days normal climate	9.7	46.5
28 days under water	9.2	48.0
7 days under water and 21 days normal climate	9.0	57.8
7 days normal climate and 21 days under water	6.5	36.0

Aqueous Microemulsions of Vinyl Resins Having Carboxylic or Sulfonic Groups

Highlights of the Technological Achievement: Vinyl resins containing carboxyl or sulfonic acid functionalities can be formulated as stable aqueous microemulsions suitable for preparing coatings by converting them to ionomers in a mixture of solvents and water.

Reference: Y.-L. Fan; U.S. Patent 4,182,699; January 8, 1980; assigned to Union Carbide Corporation.

Background: Vinyl resin lacquers are known in the art and have enjoyed commercial success for coating various substrates with clear vinyl coatings. The coating operation, however, involves the removal of large quantities of organic solvents in which the vinyl resins are dissolved. For ecological and antipollution considerations solvent based coatings are undesirable. This has led to consideration of a water-borne resin system for coating applications. It has not been found feasible to simply substitute a water-borne system for the vinyl resin lacquers for several reasons.

While vinyl resin latexes may be made by emulsion polymerization, these polymerization techniques require the presence of components in the polymerization recipe which have a deleterious effect on the final coating. This is particularly true in can coatings where such criteria as blush resistance and water resistance, particularly pasteurization temperatures are required. The presence of surfactants required to maintain stable emulsions increases the susceptibility of resin coatings on substrates to break down in contact with water.

Process Conditions: Stable aqueous microemulsions of vinyl resins having an average particle size of 0.02 to 0.14 μ are prepared by vigorously blending vinyl resins having carboxylic or sulfonic acid groups therein with a base, water, at least one organic, water-miscible solvent which has an affinity for the resins and a boiling point below 160°C, and an organic, water-miscible cosolvent which is a poor solvent for the resins but is miscible with the organic solvents and has a boiling point above that of the organic solvents. It is preferred to prepare these microemulsions by blending:

(A) A normally solid vinyl resin containing carboxylic or sulfonic acid groups and having the following moieties copolymerized herein:

(1)

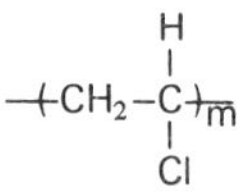

(2) 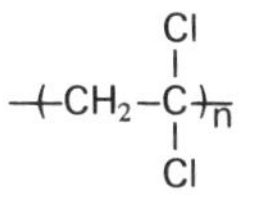

(3) 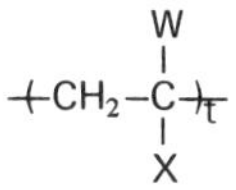

(4) 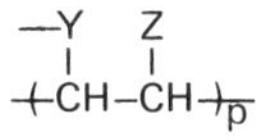

(5) 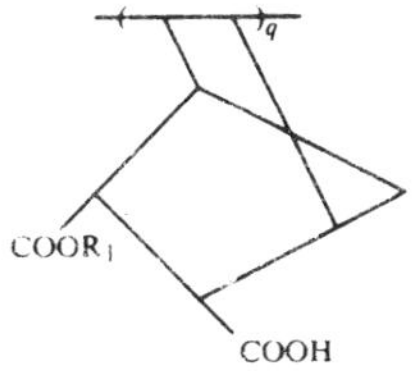

(B) Sufficient water-miscible base to neutralize about 10 to 100% of the carboxylic or sulfonic acid groups.

(C) 1 to 200 parts per 100 pbw of vinyl resin, of a water-miscible normally liquid solvent for the vinyl resin having a boiling point of up to 160°C, selected from the group consisting of lower aliphatic ketones, esters or ethers having 3 to 6 carbon atoms and cycloaliphatic ketones or ethers having 4 to 6 carbon atoms.

(D) 1 to 200 parts per 100 pbw of vinyl resin, of a normally liquid cosolvent which is miscible with water and solvent (C), having a boiling point greater than that of solvent (C).

(E) Sufficient water to provide an aqueous microemulsion having a total solids content of up to 60 wt %.

In the formulas under (A), m and n are percentages each having a value of 0 to 99%, t is a percentage having a value of 0 to 59%; p is a percentage having a value of 1 to 30% when q is 0; and q is a percentage having a value of 1 to 30% when p is 0, with the proviso that $m + n + t$ is $\geqslant 70$ and $\leqslant 99\%$, that $m + n + t + p + q = 100\%$, that when $m = 0$, n is at least 1 and that when $n = 0$, m is at least 1.

X is a monovalent radical selected from the groups consisting of –H and lower alkyls having 1 to 4 carbons. W is a monovalent radical selected from the group consisting of –H, lower alkyls, aryl having 6 to 9 carbons,

$$-\overset{\overset{\displaystyle O}{\|}}{C}-OR, \qquad -O-\overset{\overset{\displaystyle O}{\|}}{C}-R, \qquad -\overset{\overset{\displaystyle O}{\|}}{C}-O\!\left(CH_2\right)_a OH$$

wherein a is an integer having values of 1 to 3,

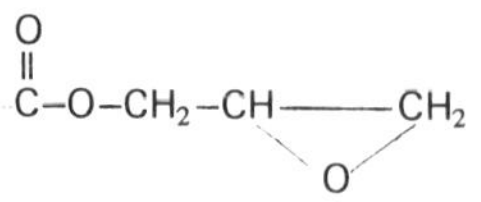

$-OR$ wherein R is a C_{1-18} alkyl, $-OH$, or

$$-\overset{O}{\overset{\|}{C}}-N(R_1)_2$$

wherein R_1 is a monovalent radical selected from the group consisting of $-H$, methyl or ethyl, and $-CN$.

Y is a monovalent radical selected from the group consisting of $-H$, methyl, $+CH_2+_bCOOR$, or $+CH_2+_bCOOH$, wherein b is an integer having values of 0 to 4. Z is a monovalent radical selected from the group consisting of $+CH_2+_bCOOH$, $-SO_3H$ and $-C_6H_4-SO_3H$, with the proviso that Y and Z are never $-COOH$ and $-SO_3H$ at the same time.

The vinyl resin of (D) is selected from:

(1) glycol monoalkyl ethers having the formula:

$$HO+CHR''CHR'''O+_rR''''$$

where each of R'', R''' is H or CH_3, r is an integer having values of 1 to 3 and R'''' is an alkyl group having 1 to 6 carbon atoms;

(2) aliphatic acidic ethers having the formula:

$$R'''''O+CH_2+_rCOOH$$

wherein R''''' is an alkyl group having 1 to 4 carbon atoms;

(3) amino esters having the formula:

$$R'''''-\overset{O}{\overset{\|}{C}}-O+CH_2+_rNH_2;$$

(4) amino ketones having the formula:

$$R'''''-\overset{O}{\overset{\|}{C}}+CH_2+_rNH_2;$$

(5) aliphatic alcohols having the formula: R_2OH, where R_2 is alkyl having 1 to 5 carbons;

(6) aliphatic carboxylic acids having the formula: R_3COOH where R_3 is H or alkyl having 1 to 4 carbons;

(7) aliphatic amines having the formula: $R_4NR_5R_6$ where R_4 and R_5 are H or alkyl having 1 to 6 carbons and R_6 is alkyl having 1 to 6 carbons with the proviso that the total number of carbons in the sum of $R_4 + R_5 + R_6 \leqslant 6$;

(8) aliphatic amino ethers having the formula: $[R_7O(CH_2)_g]_rNH_{3-r}$ where R_7 is methyl or ethyl, and g has values of 1 to 4;

(9) aliphatic dialkyl amides having the formula: $R_8CON(R_7)_2$ where R_8 is H or alkyl having 1 to 5 carbons;

(10) N-(hydroxymethyl)acrylamide and N-(hydroxymethyl)methacrylamide having the formula: $CH_2{=}CR_9CONHCH_2OH$ where R_9 is H or $-CH_3$;

(11) cycloaliphatic ether alcohols having the formula:

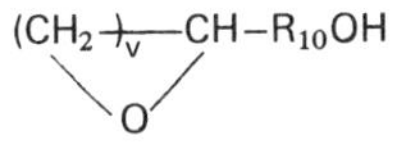

where R_{10} is alkylene having 1 to 3 carbons and v is an integer having values of 1 to 5;

(12) hydroxy esters having the formula: $R_8COO(CH_2)_rOH$; and

(13) hydroxy ketones having the formula: $R_8CO(CH_2)_rOH$.

Example 1: *Preparation of Ammonium Ionomer Microemulsion Mixed with Hydrolyzed Polyvinyl Acetate and Hexamethoxymethylmelamine* – An aqueous solution containing 6 pbw of Gelvatol 20/30 (85% hydrolyzed polyvinyl acetate), 3 pbw of Cymel-301 (hexamethoxymethylmelamine), 3 pbw of ammonium hydroxide (58%) and 244 parts of water was mixed thoroughly with a varnish consisting of 60 pbw of Bakelite VMCC vinyl resin, 30 pbw of Bakelite VERR vinyl resin, 118 pbw of MEK and 20 pbw of butyl Cellosolve.

A high-speed, high-shear blender was used. The resultant microemulsion having a solids content of 21% was concentrated under vacuum affording a microemulsion containing solids of 40%, 55% water, 4% butyl Cellosolve and about 1% ammonium hydroxide. This microemulsion was very uniform and milky, showed Tyndall scattering, had a Brookfield viscosity at 26.5°C of 75 cp, had a flow time in a No. 4 Ford cup of 15 sec at 26.5°C, and had a pH of 7.5. The average particle size, determined by the electromicroscopy, was below 0.15 μ.

These microemulsions were stable on storage at room temperature for more than 6 months. They provide protective coatings when deposited on suitable substrates and baked.

Example 2: *Preparation of Ammonium Ionomer Microemulsions Mixed with Hydrolyzed Polyvinyl Acetate and Urea-Formaldehyde Resin* – Following the procedure described in Example 1 with the exception that an equivalent weight of Beetle 60 (normally liquid butylated urea-formaldehyde resin), was substituted for Cymel-301, a microemulsion was obtained containing total solids of 40%, 55% water, 4% butyl Cellosolve and about 1% ammonium hydroxide. This microemulsion was milky and uniform, had a pH of 7.2 and a Brookfield viscosity at 27°C of 196 cp. This microemulsion provides a protective coating on substrates when deposited thereon and baked.

Preparation of a Sulfate Ester of a Polyhydroxy Polymer

Highlights of the Technological Achievement: A process for preparing a sulfate ester of polyhydroxy polymer which is a partially substituted polysaccharide or a polyvinyl alcohol which contains ether groups, ester groups other than sulfate, or a mixture of ether groups and ester groups other than sulfate.

Reference: R.G. Schweiger; U.S. Patent 4,177,345; December 4, 1979.

Background: Ester derivatives of polyhydroxy polymers are known and have been described extensively in the prior art literature. The chemical and physical properties of such ester derivatives depend to a large extent on the particular nature of the polymer, its molecular weight, the type of ester substituent group, and the degree of substitution of the polymer (hereinafter referred to as DS). Due to the manner in which ester derivatives of polyhydroxy polymers have previously been prepared, the DS of the resulting ester derivatives has not been relatively uniform. This has produced ester derivatives whose properties, e.g., water solubility and compatibility with various metallic ions, have not been generally satisfactory and has restricted the use areas for the ester derivatives.

Process Conditions: Due to substantially uniform distribution of ester groups among the polymer units in the products of the process, the products have, in general, unusual solubility characteristics. Also, solutions of the products have unusual compatibilities with metallic ions. For example, cellulose sulfate products provided according to this process which have a DS of about 0.3 to 1 are soluble in water. This is quite an unexpected result, since the prior literature reports that colloidal cellulose sulfate must have a DS in excess of 1 in order to be water soluble.

Typical of known polyhydroxy polymers which may be utilized as starting materials are the polysaccharides such as cellulose, starch, hemicellulose, guar gum, locust bean gum, gum arabic and the mannans; the polyuronic acids typified by alginic and pectic acids, and the synthetic polyhydroxy polymers such as polyvinyl alcohol. In accord with one aspect of this process, the partially etherified or esterfied polyhydroxy polymers, as described previously, are used as starting materials in the preparation of nitrate esters of sulfate esters of the partially substituted polyhydroxy polymers.

The starting polyhydroxy polymer is suspended in a suitable solvent which includes a swelling or solubilizing agent for the polymeric reaction product and a proton acceptor. Solvents which have been found suitable in serving both as a proton acceptor and a swelling or solubilizing agent include weak tertiary amine bases as typified by pyridine, quinoline and isoquinoline and also N,N-dialkyl acylamides such as N,N-dimethylformamide and N,N-dimethylacetamide (hereinafter referred to as DMF and DMAC respectively), and mixtures thereof.

Sufficient solvent should be used to avoid difficulty in handling the resulting viscous mixture. In general, it has been found that a minimum solvent to polymer weight ratio is 3:1, i.e., 3 pbw of solvent for each part of polymer.

By controlling the degree of substitution of the nitrile ester of cellulose employed as the starting material, the degree of substitution of the cellulose sulfate product may be likewise controlled. Thus, when the nitrite ester of cellulose has a degree of substitution of 2 to 3, the cellulose sulfate ester which is produced in accordance with the process has a degree of substitution up to 1.1. However, when the nitrite ester of cellulose has a degree of substitution which is less than 2, the cellulose sulfate ester has a degree of substitution greater than 1.1. In each case, the sum of the degree of substitution of the cellulose sulfate ester and the degree of substitution of the nitrite ester is equal to about 3.0.

First Step: The first step of a process in accordance with one embodiment of this method comprises nitrosating a polyhydroxy polymer suspended in a suitable solvent with dinitrogentetroxide, nitrosyl chloride or mixtures thereof to obtain the corresponding polyhydroxy polymer nitrite ester.

Second Step: In forming a nitrite ester product, the second step of the process comprises heating the polymer nitrite ester solution in the presence of nitric acid with agitation at a temperature of 60° to 110°C for 15 min to about 2 hr to obtain the corresponding polymer nitrate ester.

Although the polymeric nitrite ester solution used in Step 2 is generally the reaction mixture formed in Step 1, in which DMF or DMAC has been used as the proton acceptor and dinitrogentetroxide as the reagent in an amount sufficient for nitrosation to about the maximum DS, the polymeric nitrite ester may be isolated after Step 1 and then redissolved in one of the solvents, as stated above, and the corresponding amount of anhydrous nitric acid added. The use of the reaction mixture of Step 1 for Step 2 obviates the need for isolation of the polymeric nitrite ester. Also, the addition of nitric acid during Step 2 is not necessary in this case since, during nitrosation with dinitrogentetroxide in DMF or DMAC, enough nitric acid is formed for the subsequent nitration.

Example: *Preparation of Polyvinyl Nitrite Ester from Polyvinyl Alcohol* – To a dry 500 ml three-neck round bottom flask, 10 g of finely ground polyvinyl alcohol having a degree of saponification greater than 90% was suspended in 100 ml of N,N-dimethylformamide. The mixture was mechanically stirred and under exclusion of moisture, dinitrogentetroxide gas was introduced. The vessel was cooled with cold water to keep the temperature at approximately 25°C. A clear viscous solution was obtained upon the addition of 20 g of dinitrogentetroxide gas.

Upon analysis, the resulting product was identified as polyvinyl nitrite ester having a degree of substitution of 0.8.

It was found that when the polyvinyl alcohol starting material had a degree of saponification less than 90%, for example, a product containing about 50% acetyl groups and 50% free hydroxyl groups, substantially the same results could be produced as obtained above, however, the amount of dinitrogentetroxide gas required to solubilize the starting material was less.

Separation of Polyvinyl Alcohol from Aqueous Solutions

Highlights of the Technological Achievement: In order to protect the environment from pollution, it is desirable to keep the polyvinyl alcohol content in wastewaters as low as possible. Polyvinyl alcohol may be removed from wastewaters by means of microorganisms on the one hand, and on the other according to chemical/mechanical methods by salting-out and precipitation and subsequent separation of the solids.

Advantageous is the precipitation in the form of a polyvinyl alcohol/boric acid/didiol complex, especially by means of alkaline earth metal hydroxides in an alkaline medium. Calcium hydroxide is a particularly suitable alkaline earth metal hydroxide.

Reference: W. Zimmermann and H. Schindler; U.S. Patent 4,166,033; August 28, 1979; assigned to Hoechst AG, Germany.

Background: Although polyvinyl alcohol (PVA) has no noxious effect on either plants or animals, it increases the chemical oxygen demand (COD) and the biological oxygen demand (BOD) of the wastewater, and thus causes an undesirable environmental strain. Therefore, the PVA content of the wastewater should be kept as low as possible.

Although it is known that certain adapted microorganisms, for example, pseudomonas, are capable of degrading PVA in the biological phase of a sewage purification plant, they refuse to do so when their concentration is relatively low, as is the case in normal water courses. Since, furthermore, factories using PVA are seldom equipped with sewage purification plants having a biological degradation phase, it is necessary to develop processes for separating PVA from aqueous solutions, in order to keep the sewage system as free as possible from pollution.

It has been proposed to separate polyvinyl alcohol from aqueous solutions by adding to the solutions a coagulation agent which consists of a salt of boric acid as gelation agent and an alkali metal salt or alkaline earth metal salt of an organic acid as salting-out agent (see German Patent 2,557,443). These processes, however, have the disadvantage of requiring generally a high concentration of salt, which makes them very uneconomic; moreover, a quantitative removal of the PVA cannot be obtained by this method. There are, furthermore, technological difficulties, since the precipitate is in the form of a gel and thus is difficult to separate. Finally, the processes described allow flocculation only in the case of high PVA concentration.

It has now been found that the cited disadvantages can be overcome when the precipitation of PVA is carried out in the form of a PVA boric acid/didiol complex.

Process Conditions: This process relates to a method for the separation of polyvinyl alcohol from an aqueous solution by precipitation of the polyvinyl alcohol in the presence of orthoborate ions and alkaline earth metal ions, in an alkaline medium, which comprises either adding first a saturated boric acid solution to the polyvinyl alcohol solution and then adjusting the latter one to a pH above 8 by means of an alkaline earth metal hydroxide solution; or adjusting first the polyvinyl alcohol solution to a pH of 11 by adding an alkaline earth metal hydroxide solution and then adding a boric acid solution neutralized by an alkaline earth metal hydroxide; and subsequently separating the precipitated polyvinyl alcohol.

It has been described that boric acid forms complexes in an alkaline medium. In such a PVA/boric acid/didiol complex, at least 2 PVA chains are linked with each other via a boron atom to form a very stable six-membered ring configuration.

Simultaneously, with the formation of such a complex, an increase of the viscosity of the PVA solution is observed which depends on the concentration: highly concentrated PVA solutions gelatinize, while the consistency of dilute solutions, for example, wastewater containing less than 1% of PVA, is practically not altered. However, separation of PVA from aqueous solutions by formation of these didiol complexes could not be obtained hitherto.

For the process according to the method for separating PVA in the form of alkaline earth metal salt of the PVA/boric acid/didiol complex, there are suitable the following three technologically simple operation modes.

(1) The PVA solution is adjusted to pH 11 by means of an alkaline earth metal hydroxide, and an aqueous boric acid solution neutralized with alkaline earth metal hydroxide is then added to this solution.

(2) A saturated aqueous boric acid solution is added to the PVA solution, which is then adjusted to a pH above 8 by means of alkaline earth metal hydroxide.

(3) A saturated boric acid solution is added to the PVA solution which is then rendered alkaline by means of alkali metal hydroxide, and subsequently, the didiol complex is precipitated by adding water-soluble alkaline earth metal salts, for example, calcium chloride.

Degree of Purification: The efficiency or the degree of purification of the process of this method is advantageously determined by calculating the chemical oxygen demand (COD) according to the method of ASTM D 1252-58T on the basis of the following equation.

$$100 - \frac{COD_{\text{purified solution}}}{COD_{\text{original solution}}} \times 100 = \text{\% of efficiency or degree of purification}$$

By PVA in accordance with this method, there is to be understood a completely or partially saponified PVA which has a viscosity of 3 to 200 mPa-sec in a 4% aqueous solution, and an ester number of 0 to 400 mg KOH/g.

The concentration of the PVA solutions to be treated in accordance with this process may be in principle in all ranges which occur in the industrial processing of PVA. However, the PVA concentration is preferably in a range which is normal for wastewaters, that is, from 0.001 to 2 wt % of PVA.

Example 1: 100 parts of a 1.0% solution of a PVA, the 4% aqueous solution of which has a viscosity of 4 mPa-sec and which has an ester number of 140 mg KOH/g, are adjusted to pH 12 by addition of 0.5 part of calcium hydroxide. Subsequently, 250 parts of a precipitating solution prepared by neutralization of 45 parts of boric acid and 5 parts of calcium hydroxide in 100 parts of water are added slowly with agitation. On this dropwise addition of the precipitation agent, the mixture becomes turbid at first, and after about 5 min after addition of the total precipitating solution, a precipitate forms which settles well. The COD value of the 1% starting solution is 19,500 mg O_2/ℓ. The filtrate of the solution so treated has a COD value of 650 mg O_2/ℓ, the dilution of the PVA solution by the inorganic precipitating agent being already considered in the calculation. The efficiency or degree of purification of this reaction is represented as: 100 - 650/19,500 x 100 = 97%.

Examples 2 through 8: *Purification of PVA Solutions in Highly Dilute State* – According to the method as indicated in Example 1, 0.5 part each of calcium hydroxide is added to 250 parts each of 0.01% aqueous solutions of different types of PVA, the viscosity of a 4% aqueous solution of which is determined (column A) as well as their ester number (column B), so that a pH of 12 is adjusted, and subsequently the solutions are treated with 75 parts each of the precipitating solution consisting of saturated boric acid neutralized with calcium hydroxide, as indicated in Example 1.

The filtrate of the precipitate of the PVA complex formed which can be easily filtered has a corrected COD value (column D) as compared to the COD of the starting solution (column C). The purification degree of the solution so treated is listed in column E of the following table.

Ex. No.	A (mPa-sec)	B (mg KOH/g)	C (mg O_2/ℓ)	D (corrected) (mg O_2/ℓ)	E (%)
2	3	190	190	4	~98
3	4	140	195	2	~99
4	18	140	190	~1	~99
5	26	140	200	~1	~99
6	4	20	200	1	100
7	20	20	205	1	100
8	28	5	185	1	100

Dewatering Polyvinyl Alcohol Stabilized Polymer Slurries

Highlights of the Technological Achievement: Dewatering aqueous polymer slurries containing polyvinyl alcohol stabilizer by forming an insoluble organic destabilizer in situ from a soluble parent compound comprising ionizable groups. The slurry may then be filtered at 50°C or higher to yield a soft cake of beads essentially free of stabilizer.

Reference: W. Braun; U.S. Patent 4,154,923; May 15, 1979; assigned to Dulux Australia Limited, Australia.

Background: A number of proposals have been made to use polymeric beads, for example, spheroidal granules of synthetic polymer with a diameter of from 1 to 50 μm, as matting and opacifying agents in products such as paint, plastics and paper.

One particularly useful class of bead is that made from unsaturated polyester resin, which is cured by crosslinking it with styrene or a similar unsaturated monomer.

Beads of this type are commonly made by a process which yields a slurry of polymer beads in water containing a polyvinyl alcohol dispersion stabilizer and are described in, for example, Australian Patent 445,277. The slurries typically have a bead content of the order of 30 wt %.

When such polymer beads are to be used in an aqueous product, for example, in a latex paint, they can often be blended directly with the other components as an aqueous slurry, without any intermediate dewatering process. However, if the end product is to be essentially anhydrous, the requirement to first dewater beads which have been made as an aqueous slurry can raise serious economic and technical problems.

For example, the energy required to dry a 30 wt % slurry of beads in water by direct evaporation in an acceptable time, can add an intolerable cost penalty to the beads. Furthermore, it was found that heating a bead slurry to evaporate the water can cause hard aggregates of beads to form. These aggregates may be difficult, if not impossible, to break up satisfactorily in subsequent dispersion processes. This problem appears to result from the presence in the slurry of the polyvinyl alcohol dispersion stabilizer.

If a slurry of polymer beads in water containing a polyvinyl alcohol dispersion stabilizer is first destabilized in the manner described below, the slurry can then be dewatered by conventional means to a soft, stabilizer-free cake.

Process Conditions: The destabilizing process involves precipitating in the slurry a selected destabilizer, which causes the polymer beads to flocculate and also enables the polyvinyl alcohol stabilizer to be flushed from them with the effluent liquid, when dewatering is carried out. The slurry can then be dewatered by conventional physical means, for example, by filtration or by centrifuging, to yield a soft, essentially stabilizer-free cake of polymer beads together with some residual water. If required, this cake can then be dried, for example, by circulating heated air at a temperature lower than the softening point of the polymer beads, over and through the cake.

This dual action of the selected destabilizer is somewhat unexpected, in view of the fact that some other well-known materials, notably certain polymeric flocculants used commercially in water treatment processes, will flocculate such polymer bead slurries but will not destabilize them to the point where the slurries can readily be dewatered. It appears that flocculation of the beads alone is not sufficient to overcome the dewatering problem.

Effective destabilizers are organic materials which may, but need not be, polymeric. They must, however, have a molecular weight of at least 200. In their soluble form, the destabilizers comprise solubilizing ionizable groups, the elimination of which renders the molecule as a whole insoluble in the aqueous phase of the slurry.

The destabilizer may be, for example, a carboxylic acid. Suitable carboxylic acids are, for example, stearic, 12-hydroxy stearic, abietic, oleic and dimerized fatty acids. For the best results, the carboxylic acid should preferably have an acid value of at least 100 mg KOH/g.

Thus, the amine may, for example, be a primary amine of the structure $R-NH_2$ in which R is an alkyl group with a carbon chain length of 12 to 18 atoms. The carbon chain may be a fatty acid moiety derived from, for example, lauric, myristic, stearic and oleic acids and mixtures of such acids as they are derived from natural products. Suitable destabilizers of this type are tallow, coconut and soya amines.

The amine destabilizer may be a secondary amine of the structure R–NH–R where R is an alkyl group of the type discussed above. Alternatively, the molecule may comprise both primary and secondary amine groups.

The amine may also be a diamine with the general structure $R{-}NH{-}R_1{-}NH_2$ where R and R_1 are alkyl groups. Suitable compounds of this type are, for example, oleyl propylene and lauryl propylene diamine. The amine may be a zwitterion, for example, it can be a compound of the structure:

$$R{-}NH{-}\underset{\displaystyle CH_3}{\underset{|}{CH}}{-}CH_2{-}\underset{\displaystyle O}{\underset{\|}{C}}{-}OH$$

in which R is an alkyl group, e.g., a lauryl chain. The final cake can be readily washed with water to remove remaining traces of polyvinyl alcohol stabilizer. The solids content of the final cake is typically 60 to 70 wt %, not allowing for any water which may be present as an internal occlusion in the dispersed particles.

Example: *Effect of Temperature on the Filtration Rate of a Destabilized Bead Slurry Containing Polyvinyl Alcohol* – The beads content was 37 wt % of the slurry. To 1,350 parts of bead slurry (500 parts of beads) was added 350 parts of water and the pH adjusted with aqueous ammonia solution to 8 to 9.

300 parts of a 5 wt % solution of sodium stearate in water was then added with stirring to the batch, followed by an adjustment to a pH of 4 to 5 with 10 wt % acetic acid solution in water. The beads flocculated. The batch was then subjected to filtering trials, but using a series of processing temperatures which did not exceed that at which the polyvinyl alcohol precipitates from aqueous solution. The time taken to filter equal parts of slurry at the different temperatures and the appearance of the filtrate is shown in the following table.

Filtration Temp. (°C)	Time Taken	Condition of Filtrate
30	7 hr	Milky
50	1 hr 20 min	Slightly cloudy
70	11 min	Turbid
80	8 min	Clear

It appears from the results that complete destabilization is not being achieved at temperatures below 50°C, at which temperature and above there is a substantial increase in filter throughput rate.

Although a slight carry-over of very fine beads was observed at 50°C, becoming less noticeable as the temperature rose further, the efficiency of destabilization and rate of filtration was assessed to be at an acceptable practical level provided that minimum temperature was maintained.

Modified Partially Acetalized Polyvinyl Alcohol Films

Highlights of the Technological Achievement: A process for adjusting the adhesive force or strength of a plasticizer-containing, partially acetalized polyvinyl alcohol film to inorganic glasses to the pummel adhesion values required for a given application involves the addition of at least one silane while the resin is processed into a film by conventional methods in predetermined amounts.

Reference: R. Beckmann and W. Knackstedt; U.S. Patent 4,144,376; March 13, 1979; assigned to Dynamit Nobel AG, Germany.

Background: Composite window panes are known which consist of two or more glass panes and an adhesive film, for example, of a partially acetalized polyvinyl alcohol, which bonds the glass panes together.

For the production of composite safety panes wherein for safety reasons a particularly shatterproof bond is required, the films which have proved to be especially suitable are plasticized, partially butyralized polyvinyl alcohol films having a moisture content of 0.2 to 0.9 wt %, free OH-groups (calculated as vinyl alcohol) of preferably 16 to 23 wt %, and a plasticizer content of preferably 33 to 44 pbw per 100 pbw of partially butyralized PVA.

Composite safety glass panes, consisting of two glass plates and an adhesive film bonding the glass plates together, the film being made preferably of partially butyralized PVA, are utilized especially as windshields in automotive vehicles wherein optionally one of the glass plates can be replaced by a synthetic resin plate, such as a plate of an amorphous polyamide. Such silicate glass/silicate glass or silicate glass/synthetic resin composites are also employed in the construction field, for example, as window panes or as partitions; depending on the particular use, for example, as compound safety glasses. Multiple composites are also utilized, i.e., composite elements consisting of more than two load-bearing layers.

The partially acetalized polyvinyl alcohol resins, called PVA resins, are processed into films together with plasticizers, optionally stabilizers, as well as, if desired, further additives in devices known for film production, for example, in extruders. To adjust the films to the moisture content required for composite safety glasses, they are after-treated, if necessary, in a climatic chamber.

The problem of gaining control over the adhesion of a glass plate to a partially acetalized PVA film plays an important part in the manufacture of a windshield sector of an automobile. It is known that plasticized, partially butyralized PVA films generally exhibit very high adhesion to glass. This property is, however, undesirable for the windshield sector. Pummel adhesion values of 2 to 4 are desirable for windshields in the automobile construction.

Process Conditions: This process relates to a method for adjusting or setting the adhesive force between plasticizer-containing, partially acetalized polyvinyl alcohol films and inorganic glasses to the pummel adhesion values required for the respective purpose of the intended application, which process is characterized in that there are added to the partially acetalized polyvinyl alcohol resins, before or while it is processed into a film in accordance with conventional procedures, predetermined amounts of any one of the following materials: (a) a silicon-functional silane or a mixture of various silicon-functional silanes to reduce the adhesive force; or (b) a silicon and organofunctional silane or a mixture of silicon and organofunctional silanes to increase the adhesive force; or (c) a mixture of at least one silicon-functional silane with at least one silicon and organofunctional silane to regulate or adjust the adhesive force.

The process furthermore relates to composite safety glass panes produced with the use of such modified, partially acetalized polyvinyl alcohol films.

According to the process, preferably used as the silicon-functional silanes are those of the general formula: $R_n{-}Si{-}R'_{4-n}$ wherein R represents alike or different, saturated, straight-chain or branched alkyl groups of 1 to 18 carbon atoms, R' represents halogen, preferably Cl, or identical or different saturated alkoxy groups of 1 to 8 carbon atoms, or oxy-alkoxy groups, as, e.g.:

$$-OCH_2-CH_2-O-CH_3 \quad \text{or} \quad -OCH_2-CH_2-O-CH_2-CH_3$$

and n is 0 to 3. An example in this connection is propyltriethoxysilane. According to this process, preferred silicon-organofunctional silanes which increase the adhesive force are those of the general formula:

$$\begin{array}{c} R^1_{n'}Si{-}A_p{-}Z \\ | \\ R'_{3-n'} \end{array}$$

wherein R^1 represents a hydrolyzable group, e.g., as Cl, and $-OR''$ ($R'' = C_{1-18}$ alkyl or oxy-alkyl groups, as above). R' represents an alkyl group of 1 to 18 carbon atoms, A represents a bivalent alkylene group of 1 to 10 carbon atoms, which can optionally be branched. Z represents a functional group, e.g.:

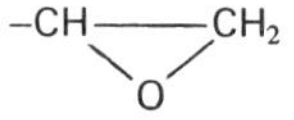

or $-CH{=}CH_2$ or $NR'''R''''$, wherein R''' represents H or aminoalkyl of 2 to 8 carbon atoms in the alkyl residue and R'''' is H or R''', and n' is an integer from 1 to 3 and, p is 1 or, in the case Z is $-CH{=}CH_2$, is 0.

Such silicon-organofunctional silanes are accordingly bifunctional. In addition to at least one hydrolyzable group reacting in the composite with the glass surface, at least one reactive group must furthermore be present which can react with the partially acetalized PVA film. Examples of such silanes include vinyltrichlorosilane, methacryloxypropyltrimethoxysilane, γ-glycidyloxypropyltrimethoxysilane, γ-aminopropyltriethoxysilane, and the like.

According to this process, isobutyltrimethoxysilane is preferably employed as the silicon-functional silane, and γ-glycidyloxypropyltrimethoxysilane is utilized preferably as the silicon-organofunctional silane.

In the process of this method, the starting compounds are partially acetalized polyvinyl alcohol resins which still contain OH-groups. Preferably, those substances are utilized containing 10 to 25 wt %, preferably 17 to 23 wt % of the vinyl alcohol residues. Partially butyralized polyvinyl alcohol is preferably employed.

The adhesive strength of a glass pane when bonded to a plasticizer-containing, partially acetalized polyvinyl alcohol film is determined in accordance with the so-called pummel adhesion test. A composite glass pane (glass/partially acetalized PVA film/glass) having a size of about 150 x 300 mm is cooled for about 2 to 8 hr at -18°C±0.5°C. The pane is then placed on a metal block inclined by 45° and pounded or pummeled with a flat-head hammer until the glass is pulverized. The testing area is of a size of about 100 x 150 mm. The adhesion is evaluated according to a scale from 0 to 10. These values are as follows.

% Free Film Surface	Pummel Adhesion Value
100	0
95	1
90	2
85	3
60	4
40	5
20	6
10	7
5	8
2	9
0	10

Examples 1 through 10: The penetration data mentioned in the following examples are obtained by the falling ball test according to German Industrial Standard 52 306 (February 1973).

The change in the pummel values due to the silane addition as demonstrated in the following tables was obtained with composite safety glass panes produced in the following manner. The silane was added to the partially butyralized polyvinyl alcohol PVB resin (content of OH-groups, calculated as the vinyl alcohol, 18.5 wt %) during the PVB film production in the extruder. The PVB resin providing the data for Table 1 originated from another batch than that resin providing the data for Table 2.

In the examples, composite panes were tested containing two glass plates of a thickness of 3 mm, as well as an interposed PVB film having a thickness of 0.76 mm. The composite panes were produced at a temperature of 130° to 135°C, a pressing force of 10 to 15 kp/cm^2, and a pressing time of 10 to 30 min.

As the plasticizer, triethylene glycol di(2-ethylbutyrate) was used in the amount of 35 pbw per 100 pbw of the partially butyralized PVA resin.

Table 1

Ex. No.	Silane* (% by wt)	Pummel Adhesion Values	Falling Ball Height** (m)	Water in Film (% by wt)	Film Thickness (mm)
1	0	10	1.50	0.5	0.76
2	0.3	7	2.50	0.5	0.76
3	0.4	5	4.00	0.5	0.76
4	0.5	3	5.50	0.5	0.76

*Isobutyltrimethoxysilane.
**Height required to penetrate composite.

The data in Table 1 illustrate that the addition of isobutyltrimethoxysilane, a silicon functional-silane, reduces the pummel values and increases the penetration resistance of the composite.

As demonstrated by the following examples, the adhesive strength is considerably increased with the addition of a silicon-organofunctional silane.

Table 2

Ex. No.	Silane* (% by wt)	Pummel Adhesion Values	Water in Film (% by wt)
5	0	1	0.5
6	0.005	3	0.5
7	0.01	5	0.5
8	0.1	10	0.5

*γ-Glycidyloxypropyltrimethoxysilane.

Examples 9 and 10 relate to the use of waste film together with fresh, partially butyralized polyvinyl alcohol resin. The pummel value is undefined due to the varying composition of the film material. The quantitative data in the Tables 1 to 3 relate to 100 pbw of film composition. The inserted residues as well as the fresh materials consist of partially butyralized PVA.

Table 3

Ex. No.	Silanes (% by wt)*	Pummel Adhesion Values
9	0.1 Silane 1** + 0.45 Silane 2***	5
10	0.2 Silane 1** + 0.6 Silane 2***	2

*Based on the total quantity.
**Silane 1 is γ-glycidyloxypropyltrimethoxysilane.
***Silane 2 is isobutyltrimethoxysilane.

Polychloroprene-Polyvinyl Alcohol Latex

Highlights of the Technological Achievement: An aqueous latex, having high colloidal stability against coagulation upon addition of ionic materials and having excellent adhesive characteristics for bonding metal foil to paper to form laminates.

Reference: D.F. Brizzolara and W. Honsberg; U.S. Patent 4,141,875; February 27, 1979; assigned to E.I. Du Pont de Nemours and Company.

Background: In the prior art, difficulty has been encountered in making, by aqueous emulsion polymerization, aqueous latexes of polychloroprene which are stable against coagulation upon the addition of electrolytes and other additives commonly included in latex formulations. Zinc oxide, for instance, is a generally used constituent of compounded latexes. The addition of dry zinc oxide to chloroprene polymer latexes prepared by aqueous alkaline polymerization of the corresponding monomer results in coagulation of the latex.

Latex adhesives for making paper-metal film laminates, in order to have optimum utility, have heretofore required the incorporation of a protein such as casein into the latex. This entails extra expense for the casein and adds another component to an already complicated composition. There is a need in the art for a paper-metal foil adhesive not requiring casein or other protein as a component.

Process Conditions: It has been found that the foregoing disadvantages of the prior art can be overcome and that compositions and processes are provided by processes comprising polymerizing in an emulsion substantially free of carboxylate soaps (1) an aqueous suspension of chloroprene monomer, (2) in the presence of (a) 3 to 10 parts of PVA and (b) 0.3 to 2.0 parts of an organic, sulfur-containing chain-transfer agent of the group consisting of dialkyl xanthogen disulfides and alkyl mercaptans, both (a) and (b) being expressed in parts by weight based on 100 pbw of monomer (1), whereby a latex is obtained of a polymer which is 10 to 95% gel polymer and in which PVA is present in a form such that it is not completely recoverable by extraction with water.

Polyvinyl Alcohol – The polyvinyl alcohols are commercially available materials. Any polyvinyl alcohol can be used which will dissolve in the aqueous polymerization system at the temperature of the polymerization. These are, in general, those polyvinyl alcohols having a mol percent hydrolysis of 80 to 100% and a degree of polymerization of 350 to 2,500. The preferred PVA, because of its solubility characteristics, has a degree of polymerization of 500 to 600 and a mol percent hydrolysis of 87 to 89%.

The amount of polyvinyl alcohol to be used in the polymerization system will, in general, range from 3 to 10 parts per 100 parts of monomer, with 4 to 6 parts being the preferred range. No more than 10 parts is required, and larger amounts result in excessive emulsion viscosities. At least 3 parts are required for adequate emulsion stability during the polymerization.

Colloidal Stabilizers – The polymerizations of this process are carried out in an aqueous suspension wherein the polyvinyl alcohol is used as the colloidal stabilizer. It is an essential feature that the polymerization system be substantially free of emulsifying agents containing carboxylate groups, such as the soaps of long-chain acids and rosin acids. The system may, however, contain from 0.01 to 1 wt %, based on monomers, of an acid-stable anionic surface active agent.

Copolymerizable Monomers – In a preferred embodiment of this process the aqueous suspension of chloroprene to be polymerized contains 0.4 to 10 mol % of an alpha,beta-unsaturated carboxylic acid, based on total monomer content.

The alpha,beta-unsaturated short-chain carboxylic acids which are used as comonomers in preparing the copolymers of this process may be represented by the following structure:

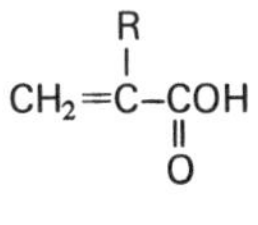

wherein R is hydrogen, an alkyl radical containing 1 to 4 carbon atoms, or a carboxymethyl radical. The preferred acid is methacrylic acid because of its reactivity and ready availability.

Inclusion of the carboxyl-containing comonomer is advantageous because it contributes to the colloidal stability of the latex. The preferred proportion is 1 to 5 mol %, 2 to 4% being particularly preferred.

Chain-Transfer Agent – The presence of a chain-transfer agent is essential in the polymerization system to control the molecular weight and to produce latexes having utility as adhesives. The operable range is 0.3 to 2 pbw per 100 parts of monomers. The preferred range is 0.5 to 1.5 parts.

The dialkyl xanthogen disulfides can be represented by the structure:

$$RO-\underset{\underset{S}{\|}}{C}-S-S-\underset{\underset{S}{\|}}{C}-OR'$$

wherein R and R' are alkyl radicals having 1 to 8 carbon atoms. The preferred dialkyl xanthogen disulfides are those in which each alkyl radical has 1 to 4 carbon atoms, especially diisopropyl xanthogen disulfide.

Polymerization – The polymerization is carried out by forming an aqueous emulsion of the monomers, the chain-transfer agent, the polyvinyl alcohol, and if used, the acid-stable surface active agent. The concentration of monomer in the emulsion is not critical but will usually range from 40 to 55%. The pH of the polymerization emulsion is on the acid side. The preferred pH is 2 to 4.

Latex Produced – The latexes prepared as described above differ from conventional neoprene latexes in several important respects. The polyvinyl alcohol is not present merely as a conventional emulsifying agent but is at least partially incorporated into the polymer. This is evidenced by the fact that the PVA is not completely recoverable from the polymer by extraction with water. The exact nature of the process has not been established, but the possible reactions involve grafting of chloroprene polymer chains at activated sites on the PVA or esterification or salt formation between the hydroxyl groups of the PVA and, when present, carboxyl groups in the chloroprene copolymers. Probably all of these reactions occur to some extent.

The polymer is not completely soluble in solvents such as benzene, toluene, or tetrahydrofuran. The amount of the solvent-insoluble (gel) component varies depending on a number of factors. The amount of gel will increase as the temperature of polymerization increases and tends to decrease with an increasing amount of chain-transfer agent. The gel content will increase as the acid latex ages. The formation of gel after polymerization can be inhibited by neutralizing the latex and providing a buffering system to minimize changes in pH of the latex during storage. In general, the amount of gel will range from 10 to 95 wt % of the polymer composition.

Procedure for Determining Gel in the Polymer: A film of latex is cast on a surface coated with polytetrafluoroethylene and allowed to dry for at least 24 hr. Two discs 1" in diameter are died out of each film and weighed.

The thickness of the film is such that the discs weigh between 0.1 and 0.3 g. This weight is the original weight [O]. The discs are swollen in tetrahydrofuran in a closed jar for at least 24 hr. The swollen polymer is removed from the solvent, and excess solvent is removed by blotting. The swollen film is then weighed in a closed weighing disk. This gives the swollen weight [S]. The swollen polymer is redried and weighed to give the swollen weight [G]. The percent gel is calculated as follows.

$$\%\ \text{gel} = \frac{[G]}{[O]} \times 100$$

Example: In this example, all parts and percentages are by weight unless otherwise indicated. Two samples are prepared using the following recipes.

	Parts A	Parts B
Chloroprene	98	95
Methacrylic acid	2	5
Diisopropyl xanthogen disulfide	1.2	1.2
Water	100	100
PVA	5	5

The redox catalyst system is cumene hydroperoxide and sodium hydrosulfite, which are added as required to initiate and maintain polymerization. The polymerization is carried out to full monomer conversion (about 98%). Temperature of polymerization is 40° to 45°C in the case of A and 40°C in the case of B.

At the end of the polymerization, an emulsion containing about 0.01 part each phenothiazine and 4-tert-butylpyrocatechol is added as a stabilizer against any further polymerization.

The latexes are compounded with 5 parts of zinc oxide and 2 parts of an antioxidant [4,4'-thiobis(6-tert-butyl-m-cresol)] per 100 parts of latex solids and are tested as foil-laminating adhesives. They show excellent adhesion with paper failure. Portland cement can be added to the latex or the latex can be added to Portland cement in any proportion without coagulating the latex. Under the same conditions, conventional chloroprene polymer latexes coagulate almost immediately.

In further experiments, Portland cement and sand mixtures are prepared and enough water is added to give a workable mass. Latex is added to this mixture without coagulating the latex. Typical amounts involved are 25 parts of Portland cement, 75 parts of sand, and the following amounts of other materials.

Latex Solids Parts	Total Water Parts	Percent Latex Solids Based on Cement
3.7	11-15	15
6.3	15-16	25

The samples of concrete are immersed in 10% aqueous HCl for 24 hr. Weight loss of the samples ranges from 8 to 16% as compared with 42% in the case of the cement not containing the latex. This shows that the latex can be used advantageously in concrete to improve its stability against attack by acid. While this application is not a new one for chloroprene polymer latexes, the use of these latexes has heretofore been made difficult by the problems of incorporating them into the cement.

Aqueous Ethylene Copolymer Dispersions With High Viscosities

Highlights of the Technological Achievement: A process for the preparation of an aqueous ethylene-containing copolymer dispersion with viscosities above 20 Pa-sec (measured with a Brookfield viscosimeter at 10 rpm).

Reference: C. Kemenater and G. Bade; U.S. Patent 4,133,791; January 9, 1979; assigned to Wacker-Chemie GmbH, Germany.

Background: The preparation of high viscosity polymer dispersions in autoclaves equipped with stirrers frequently represents a major problem, since it is more and more difficult with increasing viscosity to remove the reaction energy liberated in the polymerization process. In the case of vinyl acetate homopolymerization, the problem of the heat transport of the heat of reaction can be solved by returning the volatilizing monomer into the polymerization process after cooling. This possibility, however, is not present in the case of ethylene copolymerization in pressure autoclaves.

Process Conditions: A process for the preparation of an aqueous ethylene-containing copolymer dispersion with viscosities above 20 Pa-sec (measured with a Brookfield viscosimeter at 10 rpm) consists essentially of:

(1) polymerizing an aqueous monomer mixture of ethylene and at least one monomer copolymerizable with ethylene under conventional ethylene copolymerization conditions in the presence of from 1 to 15 wt % of polyvinyl alcohol with a degree of hydrolyzation of 70 to 100%, to a solids content of from 30 to 60 wt %, to give a copolymer dispersion,

(2) evenly distributing from 1 to 8 wt % of water-soluble polyalkylene glycols having molecular weights of between 6,000 and 100,000 selected from the group consisting of polyethylene glycol and mixtures of polyethylene glycol/polypropylene glycol, in the copolymer dispersion, and

(3) recovering the aqueous ethylene-containing copolymer dispersion with viscosities above 20 Pa-sec.

The process is preferably directed to the preparation of high viscosity ethylene/vinyl acetate copolymer dispersions having an ethylene content of preferably 10 to 25% of the copolymer.

The viscosity measurements were made on a Brookfield viscosimeter at 10 rpm and reported in pascal-seconds, one pascal-second being equivalent to 1,000 cp.

The polymerization of the ethylene-containing monomer dispersions is executed, for example, in such a way that protective colloid solutions and/or aqueous emulsifying agents, reducing agents, in some cases also heavy metal salts and vinyl esters (or vinyl chloride) are introduced and heated to the polymerization temperature with stirring. After the addition of ethylene, the addition of measured amounts of peroxide is begun. The course of the polymerization is regulated by the dosage of the peroxide catalyst. In some cases, a short term postpolymerization follows the end of the catalyst dosage. The dosage of the peroxide catalyst is most frequently done in the form of dilute solutions. The polymerization temperature is preferably 10° to 60°C. The pressure of the ethylene is preferably 5 to 100 atm.

Addition of polyethylene glycol (PEG) to an ethylene/vinyl acetate copolymer dispersion with 48 wt % solids content, 5 wt % of polyvinyl alcohol with a degree of hydrolysis of 90, and 18 wt % of ethylene in the copolymerizate.

PEG (% by wt)**	Viscosity* PEG MW 20,000 Flakes	Melt	Aqueous Solution 50%	20%	PEG MW 6,000 Flakes	50% Aqueous Solution
0	7.0	7.0	7.0	7.0	7.0	7.0
1	21.2	20.8	20.4	11.7	–	–
1.5	35.0	–	30.5	20.0	–	–
2	45.6	50.5	38.5	20.8	20.4	19.8
3	70.8	69.5	54.0	24.8	42.5	30.0
4	–	–	67.2	26.7	57.2	40.5
5	–	–	–	–	68.0	47.5
6	–	–	–	–	–	53.5

*Pa-sec = 1,000 cp, measured according to Brookfield 10 rpm.
**In 100 parts.

Figure 22.1a is a graph of the above values where the PEG has a molecular weight of 20,000 and Figure 22.1b, that where the PEG has a molecular weight of 6,000.

Figure 22.1: Increase in Viscosity of Ethylene/Vinyl Acetate Copolymer Dispersion on Addition of PEG

a.

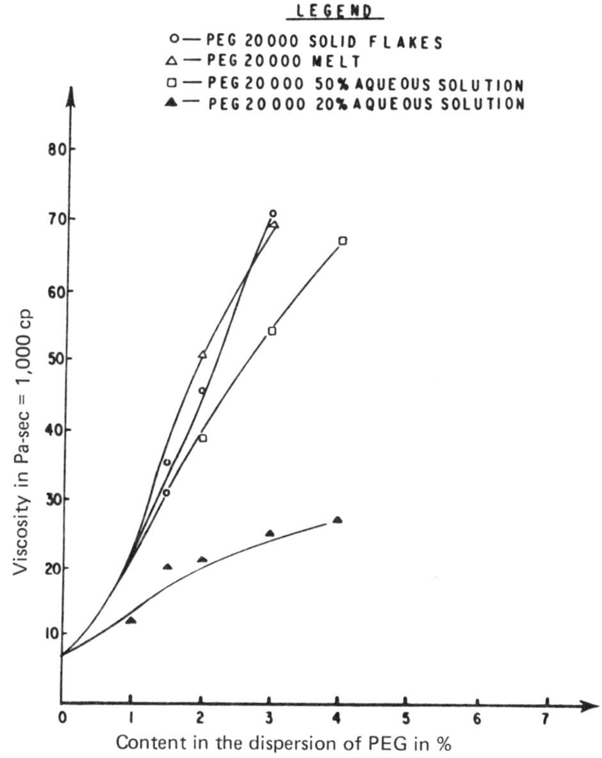

(continued)

Figure 22.1: (continued)

b.

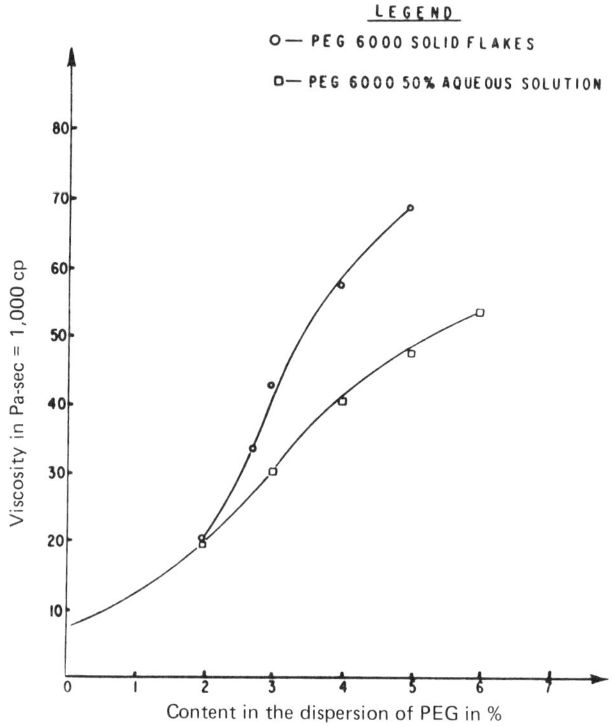

(a) Addition of PEG with MW 20,000
(b) Addition of PEG with MW 6,000

Source: U.S. Patent 4,133,791

APPLICATIONS

Silver Diffusion Transfer Receiving Layer Comprising Gelatin and Polyvinyl Alcohol

Highlights of the Technological Achievement: A receiving element for use in a photographic silver diffusion transfer process which comprises a transparent support carrying a layer comprising gelatin, polyvinyl alcohol and noble metal silver-precipitating nuclei; wherein the nuclei are present in a level of 0.1 to 0.4 mg/ft^2, gelatin is present at a level of 0.05 to 1.5 mg/ft^2 and polyvinyl alcohol is present at a level of 0.1 to 5 mg/ft^2. Preferably, the noble metal is palladium. The receiving element is particularly useful in an additive color diffusion transfer process.

Reference: E.H. Land and S.M. Bloom; U.S. Patent 4,186,015; January 29, 1980; assigned to Polaroid Corporation.

Background: Additive color reproduction may be produced by exposing a photosensitive silver halide emulsion through an additive color screen having filter media or screen elements each of an individual additive color, such as red or green or blue, and by viewing the reversed or positive silver image formed by transfer to a transparent print-receiving element through the same or a similar screen which is suitably registered with the reversed positive image carried by the print-receiving layer.

Process Conditions: A combination of polyvinyl alcohol and gelatin as the binder for the nuclei has now been found which provides superior photographic results. An excess of polyvinyl alcohol is employed compared to gelatin.

The silver-precipitating layer of this process provides a more stable positive image, particularly with respect to high humidity conditions, as well as an increase in film speed.

In a preferred embodiment a ratio of 4 to 1 polyvinyl alcohol to gelatin is employed. In a particularly preferred embodiment, the silver-precipitating layer comprises 0.15 mg/ft^2 palladium, 0.2 mg/ft^2 gelatin and 0.8 mg/ft^2 of polyvinyl alcohol.

Sizes for Textile Fibers

Highlights of the Technological Achievement: A size for textile fiber which comprises a modified polyvinyl alcohol type resin obtainable by the alkali saponification of a copolymer comprising a vinyl ester and an ethylenically unsaturated dicarboxylic acid substantially free from the monoester, diester and anhydride as essential and predominant copolymeric units. The size has excellent adhesive and cohesive strengths with respect to the textile fibers as well as superior desizing properties and workability.

Reference: K. Kajitani, T. Moritani, K. Moritani and M. Shiraishi; U.S. Patent 4,172,930; October 30, 1979; assigned to Kuraray Co., Ltd., Japan.

Background: Whereas starch, polyvinyl alcohol, acrylic copolymers, etc. are conventional textile sizes, those sizes have their own disadvantages and cannot be used alone successfully with different kinds of textile fibers. This has compelled the use of two or more of such sizes in various formulations. The problems encountered when two or more different sizes are used in combination include the instability of the resultant composition size due to the poor compatibility of component sizes, the poor desizing efficiency arising from the dissimilar optimum desizing conditions for component sizes, the complexity of treatment of desizing effluents, etc. Recently, in particular, the disposal of wastewater has become an important consideration from the standpoint of environmental protection.

The continued use of mixed sizes notwithstanding the above problems is because no single known size is able to discharge the multiple functions required of sizes for textile fibers. The characteristics of the representative commercial textile sizes heretofore in use are set forth in the following table.

It will be apparent that while starch is the oldest, commonest and least expensive size of all, it ranks very low in functional terms, for it gives only a harsh and brittle film. Often called synthetic starch, polyvinyl alcohol is a water-soluble synthetic high molecular weight substance containing a large number of hydroxyl radicals just as does starch. Because it yields a considerably tougher film than starch, with excellent cohesive and protective properties with respect to fiber, polyvinyl alcohol has rapidly come into use, replacing starch, as a textile size over the past decade or so.

Characteristics	Starch	Polyvinyl Alcohol	Acrylate Size
Cost	y	O	x
Size film properties	x	y	O
Adhesion to synthetic fiber	x	x	O
Stripping of size on weaving	x	O	O
Gumming up tendency	O	O	x
Odor of size solution	O	O	x
Stability of size solution	x	O	O
Treatment of desizing effluent	O	O	x

Note: y is very desirable, O is desirable, and x is undesirable.

Nonetheless, polyvinyl alcohol sizes have one disadvantage, i.e., an inadequate bonding or adhesive affinity for synthetic fiber. The acrylate size, on the other hand, has a property compensating for the above disadvantage of polyvinyl alcohol, that is an excellent adhesive affinity for synthetic fiber. However, this size is not only expensive but, because of the several disadvantages mentioned above, is rarely employed alone. Thus, acrylate size has been added as a modifier to starch or polyvinyl alcohol sizes for the purpose of increasing their adhesive affinity for synthetic fiber. Furthermore, both starch and polyvinyl alcohol are able to be microbiologically degraded and, as such, lend themselves well to the treatment of desizing effluents. In the case of acrylate sizes, no suitable strain of microorganism has been discovered that would degrade them and, therefore, in consideration of effluent disposal, it is preferable not to add an acrylate size.

In view of the current status of textile sizes, it was thought that if it were possible to overcome the disadvantage of polyvinyl alcohol size, i.e., its low adhesive affinity for synthetic fiber, so that the size could be utilized without the addition of an acrylate size, a considerable commercial advance could be materialized. From the above point of view, several proposals have been made as to the improvement of sizing characteristics by the modification of polyvinyl alcohol by way of copolymerization.

Process Conditions: A modified polyvinyl alcohol resin obtainable by an alkali saponification of a copolymer containing a vinyl ester and an ethylenically unsaturated dicarboxylic acid as essential and predominant copolymeric units has an adhesive affinity as high as 10 to 100 times that of ordinary polyvinyl alcohol resin and even surpassing the adhesive power of acrylate sizes, as well as unusually desirable desizing properties.

The ethylenically unsaturated dicarboxylic acid which is employed as a copolymeric component in the production of the modified polyvinyl alcohol resin for the textile size according to this process must be substantially free from the corresponding monoester, diester and anhydride. It is also desirable to preclude, as far as possible, the presence of unsaturated carboxylic acid esters as an impurity or third component. This is because the ester structure in the product resin would undergo a transesterification reaction in the course of drying or handling to produce intermolecular crosslinking, thus adversely affecting the water solubility of the resin and considerably impairing the sizing properties of the product.

The proportion of the ethylenically unsaturated dicarboxylic acid, especially itaconic acid, in the copolymer should be selected with reference to the substrate fiber to be sized and the specific object of sizing, and it is normally desirable to employ this component in the range of 0.1 to 10 mol %, and for still better results, in the range of 0.5 to 5 mol %.

When the fiber to be sized is a staple fiber, the cohesion characteristics of the size such as a fluff-binding property and the desizing, gumup and other performances of the size at the weaving step must be important considerations. In view of these considerations and the compatibility of the size with starch which is commonly used in combination with polyvinyl alcohol sizes, it is appropriate to select a degree of saponification within the range of 70 to 100 mol %. When the selected degree of saponification is below 70 mol %, the size film will be too soft, thus causing the gumup problem. Moreover, the compatibility with starch is rather adversely affected when the degree of saponification is too low.

It will be seen that the properties of yarn sized by the size of this process are at least comparable to the properties of similar yarn sized with the commercial acrylate size.

For the process size the following were used: principal size 100 parts itaconic acid modified polyvinyl alcohol (degree of modification = 3); lubricant 5 parts (PX-7, higher aliphatic sulfate); and antifoam 0.3 part (Pionin K-16). For the acrylate size the following were used: principal size 100 parts polyvinyl alcohol (degree of polymerization 550, degree of saponification 88.0 mol %); 40 parts Plus-size T-823 acrylic copolymer; and lubricant 5 parts (Sitex T-190).

Size	Trial No.	No. of Yarn Breaks	Gum-Up at Dropper	Removal of Size at Dropper	Pickup of Size by Velvet Cloth
Process	1	0	None	Little	Little
	2	0	None	Moderate	Moderate
	3	1	None	Little	Little
Acrylate	1	1	None	Little	Moderate
	2	2	Little	Moderate	Moderate
	3	0	None	Little	Moderate

Then, using this sized yarn, a plain-weave fabric was constructed on a test weaving machine and a desizing test was performed.

Thus, the woven fabric (12 cm x 20 cm) was dried at 130°C for 30 min and then allowed to stand in a desiccator containing silica gel for 15 min. The dry weight (W_2) of the yarn was determined. The fabric was immersed in distilled water (80°C), the amount of which was 500 times the weight of the fabric, for a period of 10 min. It was then rinsed with about 2,000 times of its weight of water for 1 min, dried at 130°C for 30 min and allowed to stand in a desiccator for 15 min. Then, the dry weight (W_3) was determined. The percent size pickup was calculated by means of the following equation.

$$\text{Size pickup (\%)} = \frac{W_2 - W_3}{W_3} \times 100$$

Then, a $1/300$ N boric acid saturated iodine solution was dripped onto the desized fabric and, after drying the change in hue was assessed by comparison with the standard hue to obtain the percent size residue. In all cases, the results were in the range of 0.001 to 0.003%, the desizing efficiencies being not less than 99.5%.

Reactive Dyeing Systems Using Dyes with Carboxylic Acid Groups on Polyvinyl Alcohol

Highlights of the Technological Achievement: Carboxy substituted aromatic chromophores are reactively linked to fibers containing alcoholic hydroxyl groups, amino groups or thiol groups, such as cellulosic fibers, with the linkage produced in the presence of a cyanamide compound, such as cyanamide or dicyandiamide.

Reference: B.L. McConnell, R. Thornton and L.A. Graham; U.S. Patent 4,171,954; October 23, 1979; assigned to Burlington Industries, Inc.

Background: Fiber-reactive dyes are employed quite widely in coloring cellulosics and proteinaceous fibers. They, of course, exhibit excellent washfastness, resistance to rubbing, tinctorial powers, ease of application and leveling. The latter quality is a measure of uniformity and most important for long dye runs and color matching. The reaction of the dye with cellulosic fibers is basically an etherification or esterification reaction and is broadly represented as:

(1)

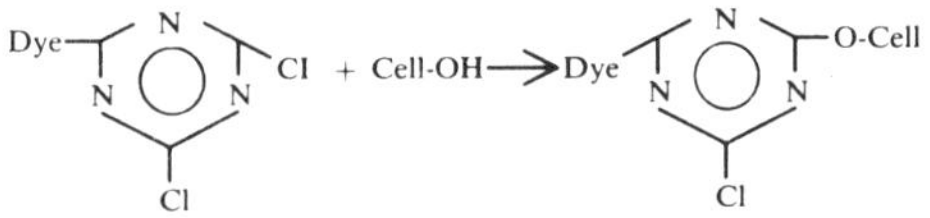

(2) $\text{Dye–}SO_2CH_2CH_2OH + \text{Cell–OH} \longrightarrow \text{Dye–}SO_2CH_2CH_2O\text{–Cell}$

Reactive dyes of Formula (1) are labile halides, and as such they are by nature of limited stability in storage and difficult to work with. Their reactions are generally effected under alkaline conditions. There are few presently available reactive dye systems which operate well under acidic conditions. An acidic system is often desirable in the dyeing of mixed fabrics such as cotton blends with polyester, wool or nylon, the latter two fibers being dyeable by acid dyes. However, acid dyes generally have no affinity for cellulose and usually only find use in dyeing of proteinaceous fibers.

Process Conditions: This process provides a method of dyeing a polymeric substrate containing certain Zerewitinoff-active hydrogen atoms, such as nonphenolic alcoholic hydroxyl groups, thiol groups, or amino groups. The substrate is contacted with a cyanamide compound such as cyanamide or dicyandiamide, and with a carboxy-substituted aromatic dye. The cyanamide and the dye carboxylic groups are present in a ratio of equivalents of at least 2:1 respectively, and in a system having a pH of 1.5 to 11. The substrate contacted with the cyanamide compound and the dyes is heated to a temperature of at least 200°F for a time sufficient to fix the dye on the substrate.

The resulting dyed textile substrates exhibit reasonably good fixation, especially when the carboxyl group is attached to the dye through the carbon atom of a phenyl ring, with the phenyl ring carrying an amino group on a carbon atom adjacent to that attached to the carboxyl group, such as in derivatives of anthranilic acid.

It is most preferred that the carboxyl group be directly attached to a ring carbon atom of an aromatic ring of the dye. In a particularly preferred embodiment, wherein especially good levels of fixation have been obtained, the carboxyl group is directly attached to such a ring carbon atom, and the next adjoining ring carbon atom carries an amino ($-NH_2$) group. Thus, the dye will be of the general formula:

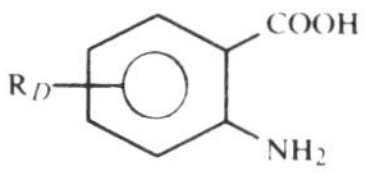

where R_D is the remaining part of a dye and is attached to one position of the aminocarboxyphenyl ring. Clearly the dye may contain one or more, generally one or two, aminocarboxyphenyl rings.

Another aspect of this process resides in the provision of a dyebath containing a tinctorial amount of, as one of the essential components therein, a dye of the general formula:

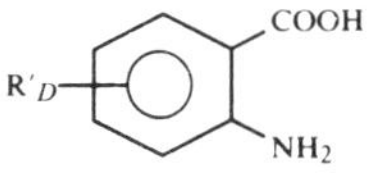

where R'_D is the remaining part of the dye and is attached through a single or, as in naphthalene or anthraquinone derivatives, two adjacent ring positions of the aminocarboxyphenyl ring.

This group of the modified dyes of this process have the general formula:

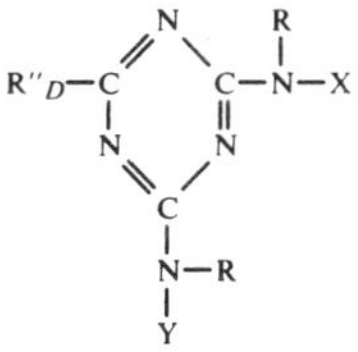

wherein R''_D is the remaining part of a dye, the linkage of which to the triazine ring is preferably through an R''–N< group, wherein R'' is H or CH_3, R is H or CH_3, X is ArCOOH wherein Ar is phenyl or naphthyl, which may be further ring-substituted by lower alkyl of 1 to 4 carbon atoms, Cl, Br, NO_2, SO_3H and COOH, Y is X, H, CH_3, or phenyl, which may be substituted by lower alkyl of 1 to 4 carbon atoms, Cl, Br, NO_2 and SO_3H.

Example: In this example fiberglass cloth sized with polyvinyl alcohol is used as the substrate. A dye of the following formula was used:

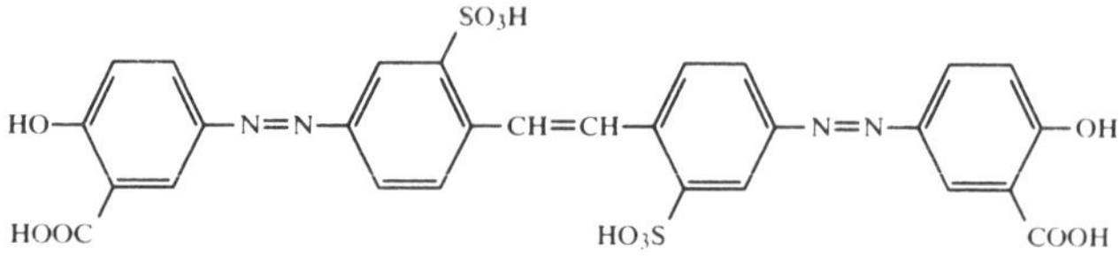

(A) As a control, an aqueous solution containing 0.5% of the above dye was prepared and padded on a swatch of fiberglass cloth sized with polyvinyl alcohol then dried and cured for 90 sec at 215°C (420°F). No cyanamide compound was used.

(B) Next an aqueous solution was prepared containing the following ingredients expressed in weight percent: 0.5% of the above dye, 4.0% dicyandiamide, 0.1% Igepal CO-710 (surfactant), 0.4% H_3PO_4, and 0.2% UCAR Z-212.

The above dyebath was adjusted to a pH of 5.5 and padded onto a swatch of PVA-sized fiberglass, dried and cured for 90 sec at 215°C (420°F).

Both swatches were scoured with a detergent solution of 1.5 g/ℓ Synthrapol SP and 1.5 g/ℓ soda ash at the boil for 2 min. Color retention was then determined by measurement with a spectrophotometer. The results were as follows. Swatch A (control) retained 7.8% and swatch B retained 55.0%.

These studies indicate that in the presence of the cyanamide compound the carboxylic dye reacts with the PVA on the fiberglass; there was no affinity for the PVA in the absence of the cyanamide-containing padding solution.

Ucar Z-212 (Union Carbide Corp.), a nonionic thickener, is reported to be a mixture of 70% hydroxyethyl cellulose, molecular weight up to 10,000 and 30% polyethylene glycol, molecular weight 4,000. Igepal CO-710 is nonylphenoxypoly(ethyleneoxy)ethanol surfactant. Synthrapol SP (ICI America, Inc.), used as a detergent and wetting agent, is a blend of nonionic and ionic surfactants.

The term retention as used herein means that percentage of the color remaining after one process rinse, the color of the fabric coming from the curing oven being taken as the base, i.e., 100%. The term endurance means the percentage of the base color remaining after the process rinse and five launderings, except where the number of launderings is noted to be otherwise. The color measurements are made on a Beckman DBG spectrophotometer.

Sealing Solution Formed from PVA and Acrylic Resin

Highlights of the Technological Achievement: The process is directed towards an improved sealing solution and in particular antigraffiti solutions. The solution is formed by combining a suitable polyvinyl alcohol with a solution or emulsion of acrylic resin to form a partially crosslinked shared polyvinyl alcohol/acrylic resin, catalyzed by exposure to heat and/or light. The properties of the solution are enhanced by the addition of a zinc complex solution containing zinc tetramine and/or an aldehyde donor such as formaldehyde or glyoxal.

Reference: R.P. Hansen; U.S. Patent 4,169,088; September 25, 1979; assigned to Ecologel Pty. Limited, Australia.

Background: The application of graffiti and other undesirable words, pictures and slogans to public buildings, memorials, sculptures and the like has become a growing problem and in the U.S. measures taken to control or rectify the graffiti cost in excess of six hundred million dollars annually. The most common method of applying graffiti is from an aerosol spray pack can, however, it has been found that markings caused by felt tip pens are the most difficult to remove.

Conventional methods of removal of graffiti include sand blasting, which removes a significant portion of the substrate and usually causes it to become more porous. This porosity then in turn accelerates damage caused by weathering. Alternatively, commercial solvents have proved unsatisfactory particularly if the graffiti has had a chance to harden. The severity of the damage is related to the depth of penetration of the marker substance especially into unpainted porous substrates such as brick, stone, concrete, etc. Removal of stains from such surfaces is extremely difficult and costly and frequently the act of cleaning off an offensive slogan may itself cause considerable physical damage. Use of solvents may worsen the problem by carrying the stain into the adjacent unmarked areas of the surface.

Process Conditions: The process requires the following reactants. The acrylic resin has a pure methylmethacrylate base, and a degree of carboxylation between 1.5 and 10%. Preferably the resin also has a glass transition temperature of at least 20°C. For low temperature ambient conditions and/or high humidity, the lower proportion of carboxyl radical in the acrylic is favored, as this gives better early weather resistance. The higher degree of carboxylation allows more complete crosslinking of the PVA/acrylic complex thus improving the degree of solvent resistance, however, this is possible only in warmer and drier ambient conditions.

The acrylic resin should comprise more than 55% of the total solids of the formulation and be prepared in the form of a water solution or as an emulsion in water. The total nonvolatiles should not be more than 35% of the complete formulation, and be prepared as a water solution or emulsion for effective penetration and sealing of the substrate. A formulation suitable for spraying onto walls with common spraying equipment requires to be of a viscosity less than 500 cp; this is the practical limiting factor on total nonvolatile content of the formulation. The concentration of the solution may be reduced if the viscosity is higher than this figure.

The quantity of PVA should be such that the number of hydroxyl groups present therein should be between 5% and more than 50% of the number of carboxyl groups present in the acrylic.

To the solution/emulsion of acrylic resin and polyvinyl alcohol is added a zinc complex solution containing zinc tetramine. This is added as a 1 to 10% solution in ammoniacal water that has been made up separately. Also an aldehyde donor, such as formaldehyde or glyoxal may be added to the solution system in one variation of the process to reduce the solubility of the PVA during initial drying and curing by covering a proportion of the hydroxyl groups to form formal derivatives.

The above solution provides a coating composition which avoids the disadvantages of previously known protective coatings at a much reduced cost. It is a paint-repellant synthetic terpolymer resin complex suitable for application to most surfaces.

It forms an inert colorless film which is substantially invisible, resistant to sunlight and prevents permanent staining by almost all substances commonly used in graffiti defacement.

Water-Soluble Films from Polyvinyl Alcohol Compositions

Highlights of the Technological Achievement: Films prepared by melt extrusion or aqueous casting from compositions consisting essentially of: (a) a resin mixture containing a low molecular weight polyvinyl alcohol and a medium molecular weight polyvinyl alcohol, and optionally, a copolymer of vinyl alcohol and an ethylenically unsaturated ester, and (b) as a plasticizer a polyethylene glycol. The films are suitable for use as packaging film in automatic packaging equipment.

Reference: R.D. Wysong; U.S. Patent 4,156,047; May 22, 1979; assigned to E.I. Du Pont de Nemours and Company.

Background: The films of this process eliminate problems associated with the use of pulverulent types of products. One such type of problem involves exposure of the user to the chemical, or chemicals, of which such products are composed. Opening a package of finely ground material, measuring an amount of the material, and transferring the measured amount from the package to equipment in which the material is contacted with water can generate air-borne dust with which the user may undesirably come in contact. Thus, a pesticide dust may be irritating to the eyes or mucous membranes of the nose or throat of the user.

Such air-borne dust can also contaminate the area in which it is generated. For example, a herbicide dust could damage plants in the area in which the packages are opened, and pigment or activated charcoal dusts present extremely difficult clean-up problems.

Process Conditions: The compositions of this process consist essentially of (a) resin mixtures of 40 to 55 pbw of a low molecular weight polyvinyl alcohol, 25 to 60 pbw of a medium molecular weight polyvinyl alcohol, and 0 to 20 pbw of a copolymer of vinyl alcohol and a comonomer containing a group having the formula:

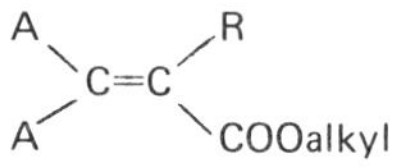

and (b) 10 to 30 parts of a polyethylene glycol per 100 parts of the resin mixture as a plasticizer therefor.

The low molecular weight polyvinyl alcohol has a viscosity in the range between 3 and 10 cp (measured on a 4% aqueous solution at 20°C by the Hoeppler falling ball method, ASTM-D 1343-56, part 8, 1958, page 486) and a degree of hydrolysis in the range between 85 and 90 mol %, dry basis, determined by saponification, e.g., KOH. The medium molecular weight polyvinyl alcohol has a viscosity in the range of 10 to 35 cp, and a degree of hydrolysis in the range of 85 to 90 mol % (both measured as aforesaid). Both the low molecular weight partially hydrolyzed polyvinyl alcohol and the medium molecular weight partially hydrolyzed polyvinyl alcohol used according to this process may be prepared by the partial acylation of a fully hydrolyzed polyvinyl alcohol, or preferably by the partial hydrolysis of a polyvinyl ester.

Whether prepared by acylation or hydrolysis, the acyl group in either case may contain up to 4 carbon atoms. The hydrolysis reaction may be acidic or basic and be performed in water or alcohol. Preferably, the polyvinyl alcohol (whether low or medium weight) is prepared by base catalyzed partial methanolysis of polyvinyl acetate.

The copolymer is made up of 90 to 98 wt % polymerized vinyl alcohol units and 10 to 2 wt % of polymerized ester units which in monomeric form have the formula:

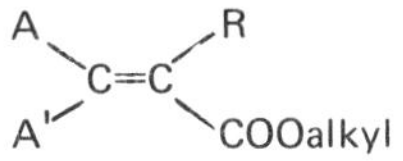

wherein A is H or $-CH_3$; A' is H or $-$COOalkyl; R is H, $-CH_3$ or $-$CHR'$-$COOalkyl; provided that when R is $-$CHR'$-$COOalkyl, both A and A' are H; R' is H or $-CH_3$; and alkyl contains 1 to 4 carbons.

Preferred monomers, because of commercial availability, are methyl methacrylate, methyl acrylate and methyl fumarate. This vinyl alcohol/unsaturated ester copolymer has a degree of hydrolysis in the range of 95 to 100% (i.e., it contains between 0 and 5 mol % of units of the vinyl ester from which it is derived, e.g., vinyl acetate, the degree of hydrolysis being measured as aforesaid), and has a viscosity between 10 and 60 cp (measured as above). These copolymers can be prepared by the process disclosed in U.S. Patent 3,689,469.

The plasticizers of this process are polyethylene glycols having an average molecular weight in the range of 285 to 420. This average molecular weight can be determined by measuring the hydroxyl value of a polyethylene glycol sample, e.g., the sample is acylated with an organic acid anhydride, such as acetic anhydride or phthalic anhydride, in pyridine solution followed by titration with a base such as alcoholic KOH or aqueous NaOH with phenolphthalein as an indicator.

As is common with commercial polymers of this nature, a polyethylene glycol with an average molecular weight within the 285 to 420 range would be expected to contain some fractions of lower and higher molecular weight (i.e., lower and higher degrees of polymerization), as typified by distribution curves.

Example: *Powder Feed Preparation for Melt Extrusion* – The following powders were fed to a ribbon blender having a capacity of 3 ft^3: 25 lb of Elvanol 51-05 G low molecular weight polyvinyl alcohol (viscosity between 4 and 6 cp, and a degree of hydrolysis between 87.7 and 89.7 mol %); 17.5 lb of Elvanol 52-22 G medium molecular weight polyvinyl alcohol (viscosity between 21 and

25 cp, and a degree of hydrolysis between 87.2 and 89.2 mol %); 7.5 lb of a vinyl alcohol/methyl methacrylate copolymer (viscosity between 20 and 35 cp and a degree of hydrolysis between 99.5 and 100 mol %).

A solution of 10 lb of Union Carbide Carbowax 400 polyethylene glycol (average molecular weight in the range of 380 to 420) and 2.5 lb of water was sprayed onto the above mixture of powders while tumbling. The water promotes ready absorption of the PEG into the polymers and increases flowability of the resulting powder blend.

Auxiliary Packaging Tests – A film prepared according to the above example gave the following results.

Elongation*	143/173%
Tensile modulus*	131,600/61,500 psi
Yield point*	3,100/3,000 psi
Tensile strength*	9,300/4,300 psi
No. of stress flex cycles*	730/860
Slip-coefficient	0.23

*Machine direction/transverse direction

Suspension Polymerization of Vinyl Chloride with Polyvinyl Alcohol (Saponification No. 300-500)

Highlights of the Technological Achievement: A process is disclosed for the suspension polymerization of vinyl chloride in which during the polymerization an agent is added preventing the forming of incrustations on the walls of the reactor.

Reference: C. Foschi, F. Fronzoni and C. Mora; U.S. Patent 4,152,500; May 1, 1979; assigned to Anic SpA, Italy.

Background: It is known that, in the suspension polymerization of vinyl chloride, either alone or in admixture with other monomers, remarkable amounts of polymer accumulate onto the walls of the autoclave. The importance of such a deposit is normally dependent on the type of stirring, on the formulations which are used (type and amount of suspending agents and of catalysts), on the impurities present in the monomers and on the number of polymerization runs carried out.

Such a deposit reduces the heat transmission through the walls of the autoclave and gives place to difficulties in controlling the reaction; moreover the product is highly contaminated by hard and vitreous crystals. As a rule the scales are removed, after each polymerization run, either manually or with high pressure water, which in any case requires 1 hr or longer times; moreover, in the case of manual operation, problems of environmental hygiene arise for the personnel attending to the descaling operation.

Process Conditions: It has been found that the suspension polymerization of vinyl chloride, either alone or in admixture with other monomers, is possible without undergoing any deposit of incrustations onto the walls of the autoclave, and without worsening of the polymer properties, by adding, during the polymerization, a mixture of polyvinyl alcohols having an average saponification number of between 300 and 500 together with a natural substance like clay or bentonite or a hydroxyl compound with colloidal behavior of an element of the Groups IIA, IIB, IIIA, IVA, VIII of the Periodic Table and, only in the latter case and at the end of the polymerization, a strong inorganic acid or a hydrolyzable salt thereof.

The polymers of vinyl chloride, produced according to this process, are granular homopolymers of vinyl chloride or copolymers of vinyl chloride containing 80% or more of vinyl chloride or 20% or less of vinylidene compounds including a terminal group $CH_2{=}C<$. These vinylidene compounds comprise, for example, vinyl halides and vinylidene chloride.

The polymerization reaction is normally carried out under pressure in the presence of a catalyst soluble in the monomer (oil) at temperatures lower than about 100°C.

Suitable oil soluble catalysts comprise, for example, lauryl peroxide, peroxydicarbonates, inclusive of diisopropylperoxydicarbonate, caprylyl peroxide, azodiisobutyronitrile or the like, normally at temperatures in the range of 40° to 60°C. The amount used is normally maintained as little as possible consistently with a good polymerization velocity; less than 1 part, such as 0.2 part of lauryl peroxide or 0.02 part of diisopropylperoxydicarbonate. The catalyst can be used as it is or directly prepared in the polymerization autoclave. The suspending agents used comprise any one amongst the well known materials useful for such a purpose, including gelatin, polyvinyl alcohol with a low saponification number, water soluble cellulose ethers, such as methylcellulose and the like.

The amount of polyvinyl alcohol having saponification numbers of between 300 and 500 varies from 0.01 to 0.5 wt % with respect to the weight of the monomer, whereas the amount of the natural substance or hydroxylated compound varies between 0.1 and 0.5 wt % with respect to the monomer or monomers fed.

Water-Soluble or Water-Dispersible Hot-Melt Adhesive Compositions Based on PVA with Residual Acetate Groups

Highlights of the Technological Achievement: Adhesive compositions which comprise polyvinyl alcohol (PVA) having 30 to 60 mol % residual acetate group content are produced by dissolving polyvinyl acetate (PVAc) having a degree of polymerization ($\bar{p}$) of 60 to 200 in methanol, the amount of the methanol ranging from about the stoichiometric amount necessary for obtaining polyvinyl alcohol having the desired degree of hydrolysis to about two times as much as the above-defined stoichiometric amount and carrying out alkaline hydrolysis. If desired, a suitable amount of a plasticizer and/or an ethylene-vinyl acetate copolymer can be added thereto.

Water soluble or water dispersible hot-melt adhesive compositions containing such polyvinyl alcohol as a main ingredient are suitable for paper making such as bookbinding or paper bag making, etc., in which products are obtained that can be reclaimed after use.

Reference: M. Sumi and J.-i. Suenaga; U.S. Patent 4,140,668; February 20, 1979; assigned to Unitika Ltd., Japan.

Background: Hot-melt adhesives have been developed as nonsolvent type adhesives which permit high speed processing and which have excellent adhesive properties suitable for mass production. Hot-melt adhesives are mainly used for processing paper goods which can be collected after use and reclaimed. However, prior hot-melt adhesives as are commonly used have serious defects from the viewpoint of the economics of collecting and reclaiming the used paper because they do not dissolve or disperse in water and, therefore, are difficult to reclaim or they deteriorate to lower the commercial value of the reclaimed paper.

As water-soluble polymers which melt by heating, vinyl acetate-vinyl pyrrolidone copolymers and water-soluble nylons are known. However, they cannot be used for bookbinding or sealing corrugated paper and cartons because they have so high a viscosity that commonly used applicators cannot be used to apply the same. Further, they have insufficient water dispersibility or low adhesive properties. Further, remoistenable hot-melt adhesive compositions containing partially hydrolyzed polyvinyl acetates having a degree of hydrolysis of 50 to 85 mol % as a main ingredient, which can be used for envelopes, postage stamps, gum tapes or wall paper have been described.

PVA obtained by acid hydrolysis has extremely low adhesive strength at a high temperature and high humidities. On the other hand, PVA obtained by known alkaline hydrolysis processes is difficult to melt, has a high melting point, a high melt viscosity and poor heat stability and undergoes foaming upon melting. In order to lower the melting point, it is necessary to add a large amount of a plasticizer or plasticizers. However, adhesive strength is deteriorated by the addition of a plasticizer(s), and thus the resultant PVA has comparatively low adhesive strength and cannot even be used as a remoistenable adhesive which does not require high adhesive strength.

Process Conditions: It is an object of this process to provide adhesive compositions having a melt viscosity of about 1,000 to 8,000 cp at about 180°C and which are not evaporated when heated.

The water-soluble or water-dispersible hot-melt adhesive compositions of this process are obtained by adding one or more plasticizers in an amount of above $(0.8\bar{p} - 10)$ wt % and below $(0.08\bar{p} + 25)$ wt % based on PVA and/or an ethylene-vinyl acetate copolymer in an amount of 0 to less than 50 wt % based on PVA.

In this process, it is important to produce PVAc having good heat stability and a low degree of polymerization. The average degree of polymerization of the PVAc required in this process is in the range of 60 to 200. The degree of polymerization is obtained by determining the viscosity of a solution of PVAc in acetone in the amount of about 5 g/l at 30°C and solving the following equation:

$$\log \bar{p} = 1.613 \log \frac{[\eta] \times 10^4}{7.94}$$

where $[\eta]$ is intrinsic viscosity.

In order to produce PVAc having such a degree of polymerization, it is preferred to polymerize vinyl acetate (hereinafter referred to as VAc) in alcohol having a chain transfer constant above 20×10^{-4} at the polymerization temperature. Examples of such alcohols include ethyl alcohol, isopropyl alcohol and isobutyl alcohol.

When VAc is polymerized in one of these alcohols or in a mixture of these alcohols, PVAc having good heat stability is obtained because lower amounts of acetaldehyde are formed by an ester interchange reaction represented by the following scheme and the amount of aldehyde groups introduced into the terminal groups of the polymer decreases.

$$\underset{\displaystyle OCOCH_3}{CH_2{=}CH} + ROH \longrightarrow CH_3COOR + CH_3CHO$$

After polymerization, the solution of PVAc is heated under reduced pressure to remove the alcohol(s) used as the polymerization solvent, whereby PVAc is obtained. PVAc available on the market is produced by the solution polymerization of VAc in methanol (the chain transfer constant thereof at 60°C is in the range of 3×10^{-4} to 5×10^{-4}). In this case, in order to decrease the degree of polymerization to below 200, the VAc concentration should be below 25 wt %. Accordingly, this process has low productivity and is not economical. Further, the resultant PVAc is colored and has low heat stability.

It is well known to use a so-called telomerization reaction which comprises polymerizing by adding a strong transfer agent to produce PVAc having a low degree of polymerization. In this case, however, the resultant PVAc easily colors and has very low heat stability and is easily gelled upon melting with heat, and the PVA resulting therefrom by hydrolysis is further colored.

Hydrolysis Conditions: Preferred hydrolysis conditions in this process are as follows: absolute methanol having a water content below 0.05 wt % is used; the concentration of PVAc is 65 to 87 wt %; sodium hydroxide, calcium hydroxide or sodium methylate is used as the alkali catalyst in an amount of about 1/1,000 to 7/1,000 molar equivalents based on PVAc; and hydrolysis is carried out at about 30° to 40°C under atmospheric pressure.

Characteristics of this hydrolysis are that the period of time from the addition of the alkali catalyst to gelation is as short as below 30 minutes, the degree of hydrolysis is easily controlled because an equilibrium value of the degree of hydrolysis can be obtained within an hour after gelation, the resultant hydrolyzed product (PVA) can easily be handled and dried because it is in a caky state, and the amount of organic solvent recovered after hydrolysis is small.

Product: To summarize, water-soluble or water-dispersible hot-melt adhesive compositions of this process which are most suitable for bookbinding and sealing of corrugated papers and cartons are preferably obtained as follows (with earlier given conditions being used unless otherwise indicated).

(a) The VAc is polymerized in ethanol or isopropyl alcohol (weight ratio of VAc:alcohol is 40:60–70:30) using azobisisobutyronitrile as a catalyst at 60° to 70°C for 3 to 6 hr under atmospheric pressure.

(b) The resultant PVAc having a degree of polymerization of 120 to 160 is dissolved in absolute methanol (water content of the methanol is below 0.1 wt %). In this case, PVAc must be dissolved in the absolute methanol which is in an amount of from the stoichiometric amount necessary to obtain PVA having a most preferred degree of hydrolysis (about 50 to 60 mol %) to 2 times as much as the stoichiometric amount.

(c) Alkaline hydrolysis is carried out by adding an alkali catalyst, e.g., NaOH, in an amount of about 1/1,000 to 5/1,000 molar equivalents based on PVAc at 30° to 40°C under atmospheric pressure. Thus, it is possible to obtain the PVA having 40 to 50 mol % residual acetate group content.

(d) The most preferred adhesive compositions of this process are obtained by melt-mixing the resultant PVA with a polyethylene glycol having a weight average molecular weight of 400 to 600 as a plasticizer, an ethylene-vinyl acetate copolymer having a melt index of 400 and a phosphoric acid compound, e.g., phosphorous acid, in an amount of 5 to 35 wt %, 15 to 35 wt % and 1 wt % based on PVA, respectively. The thus obtained adhesive compositions easily dissolve and disperse in water at 20° to 30°C. They have a softening point of from 120° to 140°C and are in the melt state at about 160°C and have a melt viscosity of from 1,000 to 6,000 cp at 180°C.

Tire Protection Composition

Highlights of the Technological Achievement: A liquid tire sealing and/or tire balancing composition containing an organic volatile amine or amine salt soluble therein, which is effective to impart corrosion resistance to the metallic belts of a belted tire is disclosed. The organic volatile amine or amine salt component is selected from the group consisting of lower alkylamines, lower cycloalkylamines, naphthylamines, morpholines, benzothiazoles, and benzoate, carbonate and nitrite salts thereof. The composition is introduced into the metallic-belted tire cavity, and the volatile amine penetrates through the rubber and lays down a protective coating on the metallic belt without harming the metal to rubber bonding.

Reference: J.B. Kent; U.S. Patent 4,137,206; January 30, 1979; assigned to New World Products, Inc. and Tri-Texas, Inc.

Background: The tire manufacture art has experienced difficulty in dealing effectively with the problem of corrosion in metallic wires of belted tires. In an attempt to avoid corrosion, substitution of noncorrosive wire such as aluminum in place of steel wire has been suggested. However, chemical pretreatment of corrosion susceptible wire belting has been excluded as a solution for the corrosion problem, because the metallic surface modification which is beneficial in inhibiting corrosion is detrimental to bonding of elastomer to the metal wiring.

Volatile amines and certain salts thereof have previously been employed for corrosion inhibition of metallic equipment and other articles of manufacture. However, the ability of certain volatile organic amines, or the nitrite, benzoate or carbonate salts thereof, upon introduction into a tire cavity, to inhibit corrosion of metallic belts enclosed in the elastomeric tire carcass, has not been previously appreciated.

Process Conditions: This process provides a liquid tire sealing and/or balancing composition which contains a volatile organic amine component. Upon introduction into the cavity of a metallic belted tire, the volatile amine migrates to the inner surface of the tire, penetrates into and through the elastomer to the metallic wire surfaces of the reinforcing belt, and forms a protective coating upon such metal surfaces. The contact of the amine with the metal is effected in situ, and within the tire carcass, subsequent to tire fabrication and hence after formation of the metal to elastomer bonding in the tire. The amine does not interfere with either polymerization, vulcanization or metal-elastomer bonding.

These compositions comprise aqueous solutions of polyvinyl alcohol and have an asbestos fibrous component. The viscosity of the compositions measured at 100°F is within the range of 1,000 to 2,200 cp as measured with a Brookfield viscometer, model LVF, No. 1 spindle, at 60 rpm. Such compositions may contain fiberglass fibers as an additional fibrous component and may also contain antifreeze components, such as ethylene or diethylene glycols as well as bactericides, coloring agents and other corrosion inhibiting compounds for the wheels such as potassium dichromate, sodium tetraborate, sodium metasilicate and potassium tripolyphosphate.

Other suitable tire sealing and balancing compositions are those comprising aqueous solutions of methylcellulose and having an asbestos fibrous component. The viscosity of these compositions measured with a Haake viscometer at 84°F generally is within the range of 1,000 to 3,000 cp. Additionally, methylcellulose formulations containing natural gums and other corrosion inhibitors such as disclosed in U.S. Patent 3,352,696 are also suitable.

Tire balancing and sealing compositions comprising polyvinyl alcohol solutions containing fibrous components are suitable delivery vehicles. In such tire balancing and sealing composition the amount of polyvinyl alcohol present is expressed, relative to the water content, by ratio of polyvinyl alcohol to water, this being from 0.5:100–5.0:100. The preferred ratio of polyvinyl alcohol to water is from 0.5:100–1.0:100.

The polyvinyl alcohol employed in the tire sealing and balancing composition may be of low, medium, or high viscosity and may be partially hydrolyzed, fully hydrolyzed or super hydrolyzed.

Polyvinyl alcohols are manufactured by polymerizing vinyl esters or vinyl acetates and hydrolyzing the polymers to the alcohol. The degree of polymerization is reflected by solution viscosity and is specified by the viscosity of a 4% aqueous solution. One preferred polyvinyl alcohol is fully hydrolyzed and of medium viscosity. An example of such a polyvinyl alcohol is Vinol 325 (Air Products). Vinol 325 is a polyvinyl alcohol from 98.0 to 98.8% hydrolyzed polyvinyl acetate, and a 4 wt % aqueous solution has a viscosity of 28 to 32 cp as measured with a Brookfield viscometer, model LVF, No. 1 spindle, 60 rpm at 20°C.

Another preferred polyvinyl alcohol is partially hydrolyzed and of medium viscosity. An example of such a polyvinyl alcohol is Vinol 523 (Air Products). Vinol 523 is a polyvinyl alcohol of 87.0 to 89.0% hydrolyzed polyvinyl acetate, and a 4 wt % aqueous solution of same has a viscosity of 28 to 32 cp as measured with a Brookfield viscometer, model LVF, No. 1 spindle, 60 rpm at 20°C.

The viscosity of the polyvinyl alcohol containing formulations is preferably in the range of 1,000 to 2,200 cp at 100°F. The viscosity of the methylcellulose containing formulations is preferably in the range of 1,000 to 3,000 cp at 84°F.

POLYVINYLPYRROLIDONE (PVP) AND RELATED PRODUCTS

PROCESSING

Increasing Molecular Weight of Vinylpyrrolidone Polymer

Highlights of the Technological Achievement: Process for increasing the molecular weight of poly-N-vinylpyrrolidone over that normally obtained by suspension polymerization of N-vinylpyrrolidone in a saturated hydrocarbon suspension medium, by adding to the suspension medium a controlled amount of water, specifically an amount between 1 and 35% by weight based on the total weight of N-vinylpyrrolidone monomer.

The polymeric products of this process are preferably water-soluble granules having a porous structure; although the process also concerns the poly-N-vinylpyrrolidone having a number average molecular weight of between 400,000 and 2,000,000, preferably between 450,000 and 1,500,000 in any physical form.

Reference: D.H. Lorenz, E.P. Williams and H.S. Schultz; U.S. Patent 4,190,718; February 26, 1980; assigned to GAF Corporation.

Background: In general, solid poly-N-vinylpyrrolidone polymers have many desirable properties including stability toward heat and light, an absence of toxicity or irritability to skin, solubility in common solvents such as alcohols, ethers, ketones, as well as water and aqueous systems. Such valuable properties have adapted moderate and higher molecular weight polymers of this type to various uses such as dispersing agents, protective coatings, laminates and films, viscosity modifiers for aqueous systems and film forming agents.

Because of these basic advantages, it is desirable to further extend the molecular weight range of poly-N-vinylpyrrolidone so that the polymer can have even wider application as, e.g., chemically inert filters, molecular sieves for organic compounds, catalyst carriers and the like and can be used to even better advantage in certain applications known such as cosmetic thickeners, pigment and dyestuff dispersants, and as oil recovery agents where such high molecular weight polymers can be employed to act as displacement agents in the recovery of oils from wells by use of water thickened with such improved polymers as hereinafter described. In addition to increasing the molecular weight of the polymer, it is also highly advantageous to produce poly-N-vinylpyrrolidone in porous granular form.

Process Conditions: By this process it is found that the molecular weight of poly-N-vinylpyrrolidone, prepared via a process of suspension polymerization, can be increased by means of a relatively simple and commercially feasible expedient.

According to this process, there is provided a suspension polymerization for N-vinylpyrrolidone which comprises polymerizing at elevated temperature a liquid suspension medium containing the N-vinylpyrrolidone monomer, a saturated aliphatic hydrocarbon solvent as the suspension medium, an alkylated N-vinylpyrrolidone or N-vinylpyrrolidone-olefin copolymer (number average molecular weight at least 8,000) as a suspending or granulating agent to assist in maintaining the suspension, a free radical initiator and a controlled amount of water, between 1 and 35% by weight based on weight of N-vinylpyrrolidone monomer.

The suspension medium undergoing reaction may additionally contain between 0.001 and 1 mol percent of a difunctional vinyl crosslinking agent for N-vinylpyrrolidone when a polymer K value above 100 is desired.

Solvents: Illustrative of those solvents suitable for this process are the saturated aliphatic hydrocarbons containing from 5 to 20 carbon atoms which include mixtures of such compounds as may be prepared for such purpose or as may be found in commercially available mixtures such as, for example, a petroleum cut of C_6 to C_{16} alkanes or other sources of hydrocarbon mixtures such as a kerosene or petroleum fraction boiling at about 175°C, e.g., Varsol, but above the temperature at which the polymerization is effected.

The preferred solvents for this process are those alkanes having boiling points between 50° and 100°C which are more easily separated from the polymeric product by boiling or evaporation. Most preferred of this group are normal and branched chain hexane.

Stabilizers: Such stabilizers or suspension aids which are found to be useful in this process are the poly-N-vinylpyrrolidone alkyl-modified polymers (e.g., N-vinylmethylpyrrolidone polymers and copolymers with methyl substituted monomer) and the N-vinylpyrrolidone C_{10} to C_{20} olefin copolymers commercially available as Antaron P-904, Antaron P-804, Antaron V-816, Antaron V-516, Ganex P-904, Ganex V-220 and Ganex V-516, which are substantially insoluble in the solvents of this process.

Generally, for this purpose, the stabilizer is employed in an amount preferably between 1 and 3% by weight based on the weight of monomer. Conveniently, the above stabilizers are polymers having a number average molecular weight between 8,000 and 20,500.

Catalyst: The catalyst or polymerization initiators found to be most effective in this suspension polymerization are those of the free radical type which are soluble in the monomer. Accordingly, suitable initiators for this process include azodiisobutyronitrile (Vazo-64).

Example 1: Into a steel autoclave, containing a double 6-bladed turbine stirrer and a thermowell, which has been rinsed with distilled water and thoroughly purged with nitrogen, is added the following mixture: 2,125 g N-vinylpyrrolidone monomer; 425 g distilled water (17% by weight based on total weight vinylpyrrolidone); 50 g Ganex V-516 (a suspending agent comprising a 50/50 graft copolymer of vinylpyrrolidone and C_{16} olefin); and 5,000 g n-heptane.

In a separate cylinder, 375 g of N-vinylpyrrolidone monomer are mixed with 1 g Vazo-64. Both the reactor and the cylinder are evacuated to 10" Hg vacuum gage and the vacuum released with dry-grade nitrogen. This procedure is repeated 12 times.

After the contents of the stirred reactor are brought to 75°C, the contents from the cylinder are charged to the reactor with continuous stirring. The reactor is then sealed and held at a temperature of between 74° and 76°C for a period of 8 hours after which it is cooled to room temperature and the contents discharged to a filter. The resulting filter cake is then tray dried at 50°C under mild vacuum to provide a dry weight of 2,375 g poly-N-vinylpyrrolidone.

The relative viscosity of the poly-N-vinylpyrrolidone is determined by preparing a 1% (weight per volume) solution of the dried polymer in chemically pure ethanol. Viscosity of the above polymer is found to be 4.43 which corresponds to K value of 91.8 and a number average molecular weight of about 370,000.

The substantially high molecular weight poly-N-vinylpyrrolidone obtained in this example can be added to a cosmetic face cream formulation in an amount of between 0.1 and 10% by weight of the formulation to produce a thickening effect therein, and consequently to promote moisturizing properties in the cream.

As a specific example, to 50 g of Albolene Cream (a cleansing cream) is added 1% of the above polyvinylpyrrolidone of 92 K value. The resulting cleansing cream is considerably thickened to the consistency of a heavier night cream and moisturizing properties of the cream are apparent.

The above suspension polymerization is repeated except that the addition of 425 g of distilled water is replaced with an equivalent amount of n-heptane so that 5,500 g of n-heptane are employed. The resulting polymer has a K value less than 83.

Examples 2 through 5: The procedure and equipment outlined in Example 1, together with the formulation designated, is repeated for Examples 2 through 5, except that varying amounts of the same solvent or different hydrocarbon solvents are employed.

The specific solvents used as the suspension medium and the concentration ratios of solvent to N-vinylpyrrolidone (VP) on a weight to weight basis are set forth in the following table.

		Suspension Characteristics		
Ex. No.	Solvent	Solvent:VP (wt:wt) 2:1	Solvent:VP (wt:wt) 1.8:1	Solvent:VP (wt:wt) 1.6:1
2	Benzene	Poor dispersion; not maintainable	Poor dispersion; not maintainable	Poor dispersion; not maintainable

(continued)

Ex. No.	Solvent	Suspension Characteristics Solvent:VP (wt:wt) 2:1	Solvent:VP (wt:wt) 1.8:1	Solvent:VP (wt:wt) 1.6:1
3	n-Hexane*	Excellent	Very good	Very good
4	n-Heptane**	Excellent	Excellent	Excellent
5	1,1-Dimethyloctane	Very good	Very good	Very good

*n-Hexane controls exotherm because of reflux temperature (68°C), is equally good with other suspension aids, e.g., Ganex V-220 and Ganex P-904 vinyl acetate graft on Ganex V-220.

**n-Heptane is equally good with the suspension aids abovementioned for n-hexane.

Preparation of Insoluble Polyvinylpyrrolidone

Highlights of the Technological Achievement: Insoluble crosslinked homopolymers of vinylpyrrolidone or copolymers of vinylpyrrolidone are prepared by reaction of the soluble homopolymer or copolymer with certain acid chlorides at a temperature between 75° and 140°C, to obtain a precipitate. Alternately, the insoluble product may be prepared by heating a partial amide chloride of a homopolymer or copolymer of vinylpyrrolidone.

Reference: K.W. Dixon; U.S. Patent 4,180,633; December 25, 1979; assigned to GAF Corporation.

Background: This process provides a method for preparing insoluble polyvinylpyrrolidones from the soluble polymers, which method is considerably easier in operation than previous methods and which is capable of producing polymers of varied and controlled crosslinked density.

Process Conditions: The preferred starting material for this process is typically a soluble homopolymer or copolymer of vinylpyrrolidone. When the substance is a copolymer, the comonomers may be, for example, vinyl acetate, methyl methacrylate, ethyl acrylate, vinyl chloride, styrene, ethylhexyl maleate, isobutylvinyl ether, or acrylonitrile. The term copolymer as used herein includes polymers of vinylpyrrolidone with one or more of the monomers specified and/or other monomers copolymerizable therewith.

The acid chloride reactant is, for example, thionyl chloride, phosgene, phosphorus trichloride, phosphorus pentachloride, oxalyl chloride, and the like. Preferred are thionyl chloride and phosgene.

Solvent: The reaction medium is preferably a solvent in which the acid chloride reactant is soluble and in which the polymer starting material may be soluble. However, it is not required that either the acid chloride or the polymer be soluble in the medium. A wide variety of solvents may be employed. The solvent chosen must be one which does not react with the vinylpyrrolidone polymer or with the acid chloride.

Suitable solvents include hydrocarbons (aliphatic and aromatic) chlorinated hydrocarbons, ketone and ethers. Examples of suitable solvents are benzene, toluene, xylene and n-heptane.

The soluble vinylpyrrolidone polymer is dissolved or suspended in the solvent, heated to reflux temperature and the acid chloride is added. The reaction mixture is kept at a temperature between 75° and 140°C, such as, for example, the reflux temperature of the solvent, until crosslinking of the vinylpyrrolidone polymer is completed. The precipitate is then separated in the usual manner.

Reaction temperature is not critical, but it has been found that a range of preferably 90° to 120°C is appropriate, owing to the speed of reaction within this range. Suitable solvents boiling in this range include the so-called high boiling alkanes and aromatics, particularly those mentioned specifically above.

Crosslinking: The ratio of polyvinylpyrrolidone polymer and acid chloride reactant varies according to the amount of crosslinking desired in the final insoluble product. In theory, the amount of acid chloride to be used will have a stoichiometric relationship to the percentage of the amide unit in the polyvinylpyrrolidone desired to be modified to amide chloride units of the intermediate.

Crosslinking has been found to occur at molar ratios of acid chloride to polymer amide units as low as 1:10. For most purposes, the preferred molar ratio of acid chloride to amide units is from 0.5:1 to 1:1. An excess of acid chloride may be used (it having been found that satisfactory crosslinking occurs at a molar ratio of acid chloride to amide units of 3:1), but it is not economically attractive.

Product: Insoluble polyvinylpyrrolidone products are useful as components of pharmaceutical compositions and in the beverage industry.

Example: To a 100 ml flask equipped with magnetic stirrer, thermometer and condenser, were added 5.0 g of the partial amide chloride of polyvinylpyrrolidone (K-90, GAF Corporation, average molecular weight 300,000) and 50 ml of dry xylene.

The slurry was heated to reflux, where evolution of hydrochloric acid began. After 2 hours, no more hydrochloric acid was evolved. The mixture was cooled to 80°C and 2 ml of water were added. After refluxing an additional one-half hour, the slurry was cooled and filtered. After drying, 3.6 g of an off-white solid were recovered. The solid was insoluble in water and other organic solvents.

Carboxylated Vinylpyrrolidone Copolymers

Highlights of the Technological Achievement: Copolymers of at least two acrylate monomers such as isobutyl methacrylate and 2-ethylhexyl acrylate, an organic acid or acid anhydride, and N-vinyl-2-pyrrolidone are prepared by emulsion polymerization. The polymer is then coagulated and dried, and dry resin is mixed with a water miscible coalescing solvent and a volatile amine. The amine-treated resin is water reducible and when used as a coating resin produces coatings of unusual water resistance and adhesion to metal surfaces.

Reference: B.B. Gross; U.S. Patent 4,137,392; January 30, 1979; assigned to The Goodyear Tire & Rubber Company.

Background: A majority of the coating resins commercially used are insoluble in water. Therefore, in general practice they have been dissolved in a suitable organic solvent or dispersed in water with the aid of emulsifying or surfactant agents in order to provide a coating composition suitable for application. A serious disadvantage of organic solvent solutions is their inherent potential toxicity, flammability, and environmental pollution.

The Clean Air Act of 1970 constitutes the basis of air pollution control regulations. It provided that the Environmental Protection Agency should issue National Ambient Air Standards, which has been done. These standards provide for limitations on particulate matter, sulfur dioxide, carbon monoxide, oxides of nitrogen, hydrocarbons and photochemical oxidants. The last two of these, arising from coating solvents, are the major concerns to the coating industry. Many state implementation plans under the Clean Air Act have followed the general framework of Rule 66, which was adopted by Los Angeles County, California in 1966 and which greatly limits these latter two contaminants.

These regulations are generally addressed to the users of coatings rather than the manufacturer. Coatings manufacture can, in general, be conducted with little or no emission of solvents into the atmosphere, but the application and curing of these coatings on finished products such as automobiles and refrigerators require the evaporation of large amounts of solvent and thus fall under these regulations.

The most common systems being developed by coatings companies to meet these regulations are waterborne, powder, electrodeposition, high solids, conforming solvents, and polymer-monomer blends for ultraviolet light cure type of coatings. It is expected that water reducible or waterborne coatings will have a major share of the coatings market in the future. Although very few waterborne coatings are devoid of organic solvents, several of the state regulations have exempted water-thinned coatings which contain 20% or less volatile nonexempt material.

Process Conditions: A resin is provided which is suitable for use in a water reducible composition, which resin comprises, based on 100% by weight of monomeric units within the polymer:

(A) From 0 to 97% of at least two acrylate monomers selected from the group consisting of methyl acrylate, ethyl acrylate, 2-ethylhexyl acrylate, lauryl acrylate, isodecylmethacrylate, 2-ethylhexyl methacrylate, butyl acrylate, butyl methacrylate, and isobutyl methacrylate;

(B) From 0 to 95% of a monomer selected from the group consisting of styrene, α-methylstyrene, acrylonitrile, vinyl toluene, methyl methacrylate, vinyl chloride, and vinylidene chloride;

(C) From 0.5 to 5% N-vinyl-2-pyrrolidone; and

(D) From 2 to 15% of an unsaturated carbonyl compound selected from the group consisting of acrylic acid, methacrylic acid, fumaric acid, itaconic acid, maleic acid and maleic anhydride, with the proviso that the total percent of monomeric units must be 100%.

Generally, harder coatings are produced from resins containing monomers from Group (B), such as styrene 20 to 30% with 60 to 75% of two monomers from Group (A). The resins of this process can be produced by reaction of the monomers in an aqueous emulsion medium.

In the synthesis of the resins of this process the monomers are mixed with water, surfactants, chelating agents, modifier, and a free radical catalyst. These ingredients are mixed and reacted at approximately 52°C, pH of from 2 to 7 (preferably 2.5 to 6), and for a time of between 8 and 16 hours.

Example 1: Polymerizations were carried out in one quart (0.95 liter) glass bottles which were clamped in a rotating wheel-polymerizer. The wheel of this polymerizer was located in a water bath held at 52°C. Each bottle was charged with the ingredients, purged with nitrogen, and capped with a perforated cap having a rubber disc seal through which a hypodermic needle could be inserted for the taking of samples. Two bottles were run for each recipe.

The percent conversion for a given polymerization time was determined by measuring total percent solids of the sample at that time and estimating conversion from a total solids-conversion straight line relationship. The total solids was determined by drying a weighed sample of the latex in a weighed aluminum dish by means of an infrared lamp until the sample appeared to be dry. The bottle polymerizations ran approximately to completion. Reaction ingredients, conditions, and final product characteristics are given in the following table.

Emulsion Polymerization of Acrylic Resins

	Sample No.		
	1	2	3
Ingredients, pbw			
Water	151.0	151.0	151.0
Emulsifier	3.0	3.0	3.0
Metal complexing agent	0.1	0.1	0.1
Sodium acid pyrophosphate	0.2	0.2	0.2
$K_2S_2O_8$ (free radical catalyst)	0.1	0.1	0.1
tert-Mercaptan	0.8	0.8	0.8
Isobutyl methacrylate	71.0	42.0	40.0
2-Ethylhexyl methacrylate	24.0	28.0	27.0
Styrene	–	25.0	25.0
N-vinyl-2-pyrrolidone	2.0	2.0	5.0
Methacrylic acid	3.0	3.0	3.0
Properties			
Final % solids	40.3	39.8	39.8
Final pH	2.7	6.1	6.0
Final viscosity, cp	22	30	25
Reaction time, hr	12	12	12

*Complex surfactant phosphate ester acid neutralized with NaOH to pH = 6.5. Surfactant used herein was Gafac RE-410.

Example 2: The resins produced in Example 1 above were separated from the reaction product mixtures by coagulation and drying. The duplicate bottles were blended together before coagulation. The coagulant in each case was a mixture of 6 g alum (aluminum sulfate), 3 g sulfuric acid (95 to 98%), and 4,000 ml of water. The temperature of the coagulation for Sample 1 is 49° to 60°C, and for Sample 2 and Sample 3, 60°C. The slurry resulting from each coagulation was filtered using a cloth-lined Buchner funnel. After filtration, the resin cake was washed twice with cold soft water. The resin was then dried at 60°C in an oven.

Film forming water reducible coating compositions can be prepared from resins such as those of Example 2 by mixing such a resin with a coalescing solvent generally at a temperature of 25° to 80°C, and generally for a period of 20 to 60 minutes. Films formed by applying the coating compositions to substrates are generally dried for a period of 8 to 24 hours.

Use of a Vinyl Monomer to Obtain an Aqueous Polymer Emulsion with Improved Stability

Highlights of the Technological Achievement: In producing an aqueous emulsion containing substantially no emulsifier, an aqueous polymer emulsion remarkably elevated in chemical stability is obtained by copolymerizing a specific amount of a particular vinyl monomer containing polyoxyethylene units as an indispensable copolymerization component.

Reference: S. Komiya; U.S. Patent 4,200,563; April 29, 1980; assigned to Japan Exlan Company Limited, Japan.

Background: In order to maintain the stability of aqueous polymer emulsions, it has been generally the practice in producing such emulsions to add a low molecular weight emulsifier upon polymerization or to compound a low molecular weight emulsifier or dispersant after polymerization. In recent years, there is an increasing demand in using aqueous polymer emulsions for paints, cement or mortar mixing, adhesives, textile treatment and paper processing. For such applications, the emulsion in intact form is usually used and not the polymer itself removed from the emulsion.

Accordingly, when the emulsion contains an emulsifier or dispersant, foaming of the emulsion will occur on account of the emulsifier contained, and this makes its handling upon processing extremely difficult. Of course, the addition of a defoaming agent is attempted to prevent foaming, but when such an emulsion containing a defoaming agent is used for the abovementioned purposes such as paint or cement, the mechanical strength or moisture-proofing of the film or shaped product will be seriously lowered by the secondary action of the emulsifier or defoaming agent, and this will constitute a great hindrance in practical use.

In order to obviate the defects attendant on such emulsifier-containing emulsions, aqueous emulsions free from low molecular weight emulsifiers have been noted in recent years, and many studies are being made of such emulsifier-free aqueous polymer emulsions or methods for producing the same.

Process Conditions: Producing an emulsifier-free aqueous polymer emulsion by copolymerizing in an aqueous medium a monomer mixture consisting of a polyoxyethylene units-containing vinyl monomer (A) represented by the following Formula (1) which forms 0.01 to 30% by weight of the total monomer mixture:

$$CH_2{=}C(R_1){-}C({=}O){-}(O{-}CH_2{-}CH_2)_n{-}OR_2 \quad (1)$$

where R_1 is H or CH_3; R_2 is H, CH_3 or $C_6H_4{-}(CH_2)_m{-}H$; n is an integer of from 4 to 30; and m is an integer of from 0 to 20, and one or more different radical-polymerizable vinyl monomers (B) which form the remainder.

The aqueous polymer emulsion obtained by the method of this process has very good chemical stability although it contains no emulsifier or other additives. Accordingly, in the case where the aqueous polymer emulsion is applied to a production process in which it is mixed with a large amount of an electrolyte, as in the case of paint production or cement or mortar kneading, the use of the aqueous polymer emulsion will produce a good mixing effect without causing any precipitation or coagulation.

Emulsion: The aqueous polymer emulsion obtained in accordance with this process is remarkably improved in its chemical stability without particularly adding any additive, for example, polyvinyl alcohol, low molecular weight anionic surface-active agent or the like. Furthermore, the aqueous polymer emulsion according to this process makes the use of a defoaming agent unnecessary because it represents substantially no foaming. Therefore, when the emulsion is used for film formation, for example, it does not undergo any adverse effect due to subsidiary materials such as additives, defoaming agents or the like, and the resistance to exfoliation, the resistance to water and the physical properties such as strength of the resulting film are not impaired at all.

In addition, the aqueous polymer emulsion according to this process is superior to the conventional emulsifier-free aqueous polymer emulsions in mechanical stability and freeze stability.

Product: Thus, since the aqueous polymer emulsion according to this process has various merits, it can be used not only for film formation, adhesives, textile treatment and paper processing, but also especially for paints and cement and mortar mixing, and, therefore, its industrial value is very important.

As preferred methods of producing the aqueous polymer emulsion of this process, there may be mentioned methods in which a monomer mixture consisting of a vinyl monomer containing polyoxyethylene units, a vinyl monomer containing a sulfonic acid group and other vinyl monomers are copolymerized in the presence or absence of a specific water-soluble polymer; or methods in which a vinyl monomer containing polyoxyethylene units and other vinyl monomers are copolymerized in the presence of a specific water-soluble polymer.

Representative of such methods is a method in which a monomer mixture composed of a vinyl monomer (A) represented by the abovementioned general formula (1) and a monomer group [vinyl monomers (B)] consisting of a vinyl monomer or vinyl monomers containing sulfonic acid or a salt thereof as represented by the following general formula (2) and/or the following general formula (3):

(2)

$$CH_2{=}\overset{\displaystyle R_1}{\overset{|}{C}}{-}\underset{\underset{\displaystyle X}{|}}{A}_x{-}(CH_2)_m{-}SO_3M$$

where R_1 is H or CH_3; A is a benzene ring; X is H or an alkyl group of C_1 to C_4; M is H, NH_4 or an alkali metal atom; x is 0 or 1 and m is an integer of from 0 to 4,

(3)

$$CH_2{=}\overset{\displaystyle R_1}{\overset{|}{C}}{-}\overset{\displaystyle O}{\overset{\|}{C}}{-}O{-}(CH_2)_n{-}SO_3M$$

where R_1 is H or CH_3; M is H, NH_4 or an alkali metal atom; and n is an integer of from 1 to 4, and at least one different copolymerizable vinyl monomer, is copolymerized without using any low molecular weight emulsifier in the presence or absence of a specific water-soluble polymer by the usual radical polymerization technique, or a method wherein a vinyl monomer represented by the abovementioned general formula (1) and at least one different copolymerizable monomer are copolymerized in the presence of a specific water-soluble polymer by the usual radical polymerization technique.

Example: 250 parts of demineralized water and 0.0063 part of ferrous chloride were supplied to a polymerization tank. With the temperature in the tank maintained at 60°C and under stirring, an aqueous solution composed of 4 parts of sulfopropyl methacrylate and 4 parts of polyethylene glycol (23 mols) monoacrylate (23 M) dissolved in 50 parts of demineralized water, a monomer mixture liquid composed of 52 parts of styrene and 40 parts of butyl acrylate, 50 parts of a 4% aqueous solution of ammonium persulfate, and 50 parts of a 6.8% aqueous solution of acid sodium sulfite were continuously added dropwise into the polymerization tank from separate supply ports, respectively, to start polymerization.

The dropping speed of these monomer and catalyst liquids was adjusted so that the dropping would be completed in 30 minutes. After the termination of the dropping, the polymerization was continued for 1½ hours under the same condition. The aqueous polymer emulsion thus obtained had a solid matter concentration of 20.8%, a pH of 2.5, a viscosity of 7 cp and an average particle diameter of 100 mμ. The emulsion was examined for its chemical stability, and was found to have excellent chemical stability which represented a value above 50%. The mechanical stability was also excellent, and no substantial foaming was observed.

On the other hand, the emulsion produced in the same way as above except that 23 M was not used was good in solid matter concentration, pH, viscosity, particle diameter and mechanical stability, but the chemical stability was very poor, being only 3%.

Producing Finely Divided Aqueous Emulsions from Materials of Limited Water Solubility

Highlights of the Technological Achievement: Process for producing finely divided aqueous emulsions from materials of limited water-solubility. These materials are diffused into preformed particles dispersed in water which are formed from vinyl compounds by emulsion polymerization under conditions which cause the preformed seed latex to include in the particles 10 to 100% oligomers having a molecular weight of 200 to 10,000. These seed particles take up 20 to 500 times their own volume of the material of limited water-solubility.

Reference: J. Ugelstad; U.S. Patent 4,186,120; January 29, 1980; assigned to SINTEF (Selskapet for industriell og teknisk forskning ved NTH), Norway.

Background: Conventional methods for making fine dispersions in water are contingent upon the use of relatively large quantities of emulsifier and/or very effective homogenization equipment. The instability of an aqueous emulsion is due to two mechanisms. The drops which come in the neighborhood of one another will be able to flocculate, i.e., pass over into greater accumulation, something which, if the viscosity is low enough, will be followed by a coalescence, i.e., an amalgamation. This type of instability will be of particular significance for relatively large drops.

With smaller drops another type of instability will be dominant, namely instability made possible by virtue of diffusion. This is due to the solubility in water of a material present in a small drop being greater than for the same material in a larger drop. This effect will exhibit itself the more the smaller the drops are. The material which is dispersed will diffuse from small drops over into the large until the small drops have disappeared. This assumes that the dispersed material has a certain solubility so that it will diffuse through the water phase from small drops to larger drops. However, this will be effected rapidly even with materials which have relatively low solubility.

Material with a solubility of about 10^{-2} g/dm^3 H_2O, drops of a size of less than 100 nm will certainly not be able to be produced by homogenization because they will disappear very quickly by diffusion. It is known that finely divided emulsions can be produced without homogenization, so-called microemulsions, which are stable, but this demands very large emulsifier concentrations, 20 to 30% based on the dispersed material against 0.1 to 2% which is the case in this process.

Process Conditions: Seed particles can be formed which have a far higher ability to swell with material of limited water-solubility. Seed particles are formed by conventional emulsion polymerization but provision is made for a drastic reduction in molecular weight of at least a portion of the material in the seed particle so that the particles include a substantial proportion of oligomers. This reduction of the molecular weight can be done, for example, by adding a chain transferer or chain stopper during the formation of the seed latex. This involves a drastic increase in the ability of the seed particles to take up the material of limited water-solubility. This concerns all materials which can diffuse through the water phase in a reasonable time which requires a solubility of greater than about 10^{-5} g/dm^3 H_2O.

In order to form a seed for use in this process any vinyl compound can be used which can be utilized in making a latex. Examples are styrene, vinyl acetate, acrylates, vinyl chloride and ethylene. The process consists of making a seed latex in the usual way by emulsion polymerization, the difference being that provision is made, for example, by chain transferers, that the latex particles have a molecular weight of less than 10,000, preferably less than 5,000.

After the formation of this seed latex, additional water and all necessary emulsifiers are added in order to give sufficient stability to the finished emulsion. After this, the material of limited water-solubility is added in an amount relative to the seed particles of 20 to 500 times by volume, depending upon how large one wishes the particles to be in the final emulsion. There can also be employed a mixture of seed latices having seed particles of different size and, in this way, there is obtained the size distribution of the swollen particles which one requires.

Product: This process is not distinguished from that which is described in Norwegian Patent 760554 as regards the polymerization. While there is formed in the Norwegian patent a preemulsion of a water-insoluble material by homogenization with the formation of drops in which the monomer diffuses in, so are these drops in this process, however, replaced by very small particles formed by conventional emulsion polymerization with the difference from normal emulsion polymerization being that one gets oligomers having a molecular weight of 200 to 10,000 instead of polymer molecules, on forming the seed latex by the addition of chain transferers or chain stoppers.

Liquid Coating Composition Containing Vinyl Fluoride-Hexafluoropropylene Resin

Highlights of the Technological Achievement: A liquid coating composition containing vinyl fluoride-hexafluoropropylene resin which has superior storage stability even in high concentrations and can form a coating layer having outstandingly superior mechanical strength properties such as tensile strength, initial Young's modulus or hardness and superior transparency and gloss.

Reference: Y. Tamura and Y. Shishido; U.S. Patent 4,183,837; Jan. 15, 1980; assigned to Teijin Limited, Japan.

Background: Generally, fluorine resins have excellent resistance to chemicals and weather, possess superior electrical properties, and have many special utilities. Because of superior weatherability in comparison to other resins, the fluorine resins are expected to be useful in the field of coating, and the development of this utility has recently attracted attention. On the other hand, the fluorine resins have the disadvantage of having poor processability and solubility.

Generally, polyvinyl fluoride and polyvinylidene fluoride have similar mechanical properties, but have extremely different solubilities. At room temperature, polyvinyl fluoride is insoluble in almost all solvents, but polyvinylidene fluoride has relatively superior solubility. It is soluble at room temperature in a solvent such as N,N-dimethylformamide or N,N-dimethyl acetamide in a concentration of up to 15% by weight. It is well known that at elevated temperatures, polyvinylidene fluoride becomes soluble in a wider variety of solvents than does polyvinyl fluoride.

Process Conditions: Extensive work on the vinyl fluoride-hexafluoropropylene copolymer had led to the discovery that a vinyl fluoride-hexafluoropropylene copolymer having an inherent viscosity of 0.2 to 2 and composed of 5 to 50 mol percent of units derived from hexafluoropropylene, 0 to 10 mol percent of units derived from a monomer selected from the group consisting of vinyl esters of aliphatic carboxylic acids having 1 to 6 carbon atoms, or 0 to 1 mol percent of units derived from a monomer selected from the group consisting of unsaturated aliphatic carboxylic acids having 3 to 6 carbon atoms, lower alkyl esters of acrylic acid having an epoxy group and lower alkyl esters of methacrylic acid having an epoxy group, and the remainder being units derived from vinyl fluoride exhibits good solubility at room temperature in inexpensive organic solvents of more kinds than in the case of the vinylidene fluoride-hexafluoropropylene copolymer.

It is also known that a solution of vinyl fluoride-hexafluoropropylene in such a solvent shows very superior storage stability, and a coated layer prepared from such a solution has improved mechanical strength such as tensile strength, initial Young's modulus and hardness and also superior transparency and gloss over those of the vinylidene fluoride-hexafluoropropylene resin liquid coating composition.

APPLICATIONS

Shaping and Stretching Aqueous Formic Acid Solutions of Polypyrrolidone

Highlights of the Technological Achievement: A method of preparing shaped polypyrrolidone articles having a high degree of tensile strength comprising forming the articles from solutions of polypyrrolidone in hydrous formic acid and drawing the articles while impregnated with the hydrous formic acid. The process also includes the step of enhancing the tenacity of shaped polypyrrolidone articles by impregnation thereof with hydrous formic acid and drawing the impregnated articles.

Reference: C.R. Chute; U.S. Patent 4,185,063; January 22, 1980.

Background: Polypyrrolidone, formed by the polymerization of 2-pyrrolidone, commonly termed nylon-4 is used for the formation of films and fibers of high strength. Methods for the preparation of polypyrrolidone are described in U.S. Patent 2,638,463.

It is an object of this process to provide a composition and method for preparing shaped polypyrrolidone articles having higher tensile strength and other more desirable properties than those formed according to prior processes.

Process Conditions: It is essential to the practice of this process that a substantially hydrous formic acid solution be employed for forming the polypyrrolidone articles. More specifically, it is critical to the practice of this process that formic acid solutions containing from 3 to 20% by weight of water be utilized to form the polypyrrolidone solutions. Preferably, formic acid solutions containing from 3 to 9% by weight of water have been found to be particularly valuable in forming articles of the highest tensile strength from polypyrrolidone.

When employing formic acid solutions containing more than 20% by weight of water, there is a tendency for the solutions to gel. When employing formic acid solutions containing less than 3% by weight of water, weaker fibers and more opaque films are produced.

In order to achieve satisfactory production of strong fibers, filaments, and films, the ratio of polypyrrolidone to hydrous formic acid solutions must be at least 20% by weight. The higher the polymer concentration in the solution, the more optimum the conditions for the formation of filaments and fibers. As the molecular weight of the polypyrrolidone employed increases, however, the less polymer is usually required for the formation of suitable fibers and filaments.

It is preferred to employ solutions comprising from 1.5 to 4 parts by weight of hydrous formic acid per part of polypyrrolidone. This range has been found suitable for the formation of strong shaped articles of polypyrrolidone from all molecular weight ranges of polypyrrolidone employing 80 to 97% hydrous formic acid.

It is generally preferred to employ polypyrrolidone having a molecular weight corresponding to a relative viscosity within the range of 2 to 10, as determined by the relative viscosity of a 1% solution thereof in metacresol. It is more specifically preferred to employ polypyrrolidone having a relative viscosity of 3 to

6. The most preferred viscosity range is from 2.5 to 4.2.

Fairly rapid dissolution of the polypyrrolidone in the hydrous formic acid may be achieved by merely mixing the two materials. However, in order to speed up solution in the case of the high molecular weight polypyrrolidone, gentle heating up to 60°C may be employed. At higher temperatures decomposition or depolymerization of the polypyrrolidone may occur.

Fibers, filaments, and films of suitable denier may be formed by conventional wet- or solution-spinning methods. It is a critical element of this process that the fibers, filaments or films formed be immediately drawn or stretched in order to achieve the highest degree of tensile strength. It has been found that the hydrous formic acid residues in the shaped articles assist in the drawing process and act to enhance the tensile strength thereof.

The precise mechanism behind the enhancement of tensile strength and other properties of the shaped polypyrrolidone articles is not known. However, the absence of hydrous formic acid residues in the shaped articles during drawing or stretching results in articles having markedly lower tensile strengths than articles cold-drawn in the presence of the hydrous formic acid.

The shaped articles may be prepared without the utilization of heat; however, warm or hot air for purposes of solvent recovery, etc., may be utilized if undertaken at less than about 92°C for short periods of time to prevent degradation. Suitable additives for modifying the properties of the polypyrrolidone may be included in the polypyrrolidone-hydrous formic acid solutions.

The articles formed from the hydrous formic acid solutions or treated therewith following formation are drawn from 100 to 800% elongation, preferably from 400 to 600% elongation.

It is to be understood that when impregnating the already formed polypyrrolidone articles, the impregnation may be accomplished utilizing either solutions of formic acid or vapors thereof. Obviously, the concentration of the formic acid and the time of immersion are mutually dependent.

Example: 1 part by weight of polypyrrolidone having a relative viscosity of 4 as determined by the relative viscosity of a 1% solution thereof in metacresol is gradually and slowly dissolved in 4 parts of hydrous formic acid (90% by weight). The solution was agitated until no polymer was visible and until all air pockets had been removed from the resulting viscous solution. Fibers of 4 denier were wet-spun from this solution and immediately drawn about 300% elongation while still impregnated with formic acid to yield fibers having 0.93 denier and 33 grams per denier tensile strength and 22% elongation at break. The procedure was carried out in air without the use of a coagulating bath.

Substantially identical fibers from an identical solution which were only partially drawn were dried and cold-drawn to the same extent yielding fibers having a tensile strength of only 5.1 grams per denier.

Fibers from identical solutions were prepared, only partially drawn, dried and then cold-drawn according to the procedure described above to yield fibers having the following characteristics before and after solution-drawing and cold-drawing.

	Tensile Strength		
	Drawing Impregnated Fibers		Cold-Drawing
Denier	(g/den)	Denier	(g/den)
1.73	16.7	1.90	3.4
2.23	19.4	1.55	4.3
1.68	24.1	0.26	4.5

Hydrophilic Polymers and Contact Lenses of High Water Content

Highlights of the Technological Achievement: Crosslinked hydrophilic interpolymers and contact lenses made therefrom comprising hydrophilic units of which the only or major proportion are derived from an N-vinyl heterocyclic monomer, hydrophobic units derived from the group consisting of styrene and ring-substituted styrenes, and crosslinking units derived from monomers containing two or more polymerizable double bonds per monomer unit.

Reference: S. Loshaek and C.M. Shen; U.S. Patent 4,182,802; January 8, 1980.

Background: In order to optimize the advantages of these hydrophilic lenses, materials with very high water content are desired. However, previously the strength of lenses with very high water contents, e.g., 60 to 95%, have been found to be low, the strength (resistance to tearing, puncturing, etc.) decreasing progressively with increase in water content. Such lenses are readily damaged during handling, in general are not very durable, and have poorer visual acuity. In some instances such contact lenses can only be inserted and removed by a professional practitioner because of their fragility. Further, the high water content lenses of the prior art cannot withstand repeated heat disinfection or sterilization without deterioration or destruction of the lens.

Polymers of the prior art which are of sufficiently high water content,to give high gas permeabilities, e.g., 60 to 95% by weight of the combined weight of polymer plus water, and especially above 70% water content, are very weak and are readily torn or otherwise physically damaged during handling. Such polymers are exemplified in British Patent 1,391,438 and U.S. Patents 3,639,524 and 3,943,045. The polymers are prepared from monomer compositions which contain a relatively high amount of crosslinking monomer, the latter being required to prevent the hydrophilic polymer from substantially dissolving in aqueous media. This excessive crosslinking frequently results in a weak polymer.

Process Conditions: The lenses can have up to 95% water content with excellent strength and optical properties and high gas permeability.

In addition, the hydrophilic polymers and contact lenses made therefrom, although sufficiently flexible to provide patient comfort are stiffer than those of the prior art for the same water content, thereby providing improved visual acuity, strength, and durability while retaining the same high level of oxygen permeability.

Briefly stated, this process comprises crosslinked interpolymers, and contact lenses made therefrom, comprising hydrophilic units derived from N-vinyl heterocyclic monomers, units derived from hydrophobic monomers taken from the group consisting of styrene and ring-substituted styrenes, and crosslinking units derived from monomers containing two or more polymerizable double bonds per monomer unit.

The preferred N-vinyl heterocyclic monomer is N-vinyl pyrrolidone and, optionally, other hydrophilic monomers can be used with the N-vinyl heterocyclic monomer as long as the N-vinyl pyrrolidone is present in major amount. Styrene is the preferred hydrophobic monomer and diallyl itaconate the preferred crosslinking monomer.

Use: Contact lenses, such as corneal contact lenses, are made by shaping the hydrophilic polymer in the dry state using conventional and well-known procedures and apparatus and thereafter hydrating the dry lens to the final configuration.

Figure 23.1: Comparison of the Hardness of Polymers of This Process (Curve A) vs Prior Art Polymers (Curve B)

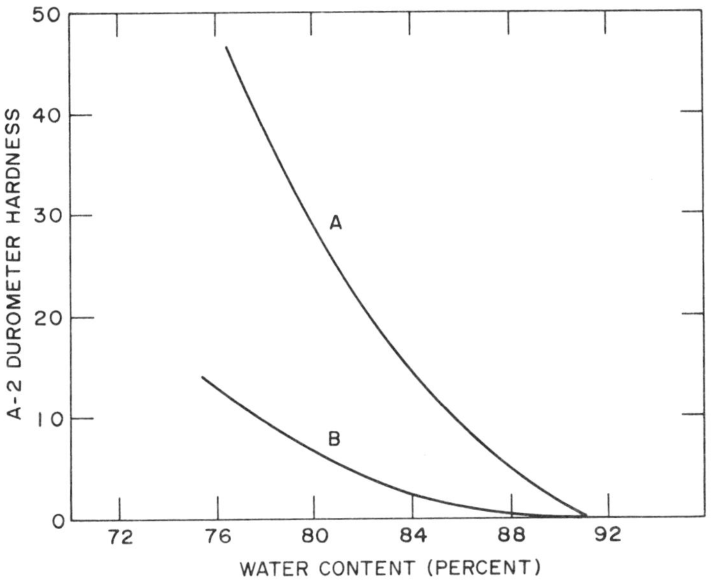

Source: U.S. Patent 4,182,802

Contact Lenses of High Water Content

Highlights of the Technological Achievement: Contact lenses comprising hydrophilic units of which the only or major proportion are derived from an N-vinyl heterocyclic monomer hydrophobic units derived from a hydrophobic monomer, and allylic crosslinking units containing more than one polymerizable double bond.

Reference: S. Loshaek and C.M. Shen; U.S. Patent 4,158,089; June 12, 1979; assigned to Wesley-Jessen Inc.

Background: It has not been possible to have hydrophilic polymers from which strong hydrophilic gel contact lenses of high water content and gas permeability can be made, which lenses can be repeatedly cleaned, disinfected, or sterilized by thermal means or by chemical means without damage to their optical or physical properties.

Process Conditions: Briefly stated, this process comprises a crosslinked hydrophilic interpolymer comprising hydrophilic units of which the only or major proportion are derived from an N-vinyl heterocyclic monomer; derived from allylic monomer crosslinking units containing more than one polymerizable double bond selected from monoallyl itaconate, diallyl itaconate, triallyl cyanurate, triallyl isocyanurate, triallyl trimellitate, N,N-diallylmelamine, multifunctional allyl monomers formed by the esterification of polybasic acids with allyl alcohol, or mixtures thereof; and optionally, hydrophobic units derived from a hydrophobic monomer; the proportions of the monomers being adjusted to give the desired water content and resistance to dissolution when the polymer is swollen to equilibrium in an aqueous environment.

This process also relates to corneal contact lenses made from such polymers. Preferably, the interpolymers, and lenses made therefrom, contain a hydrophobic monomer and for maximum strength there is used the maximum amount of hydrophobic monomer and minimum amount of crosslinking monomer consistent with the desired water content and resistance to dissolution of the polymer.

The hydrophilic polymers of this process consist of a major proportion of hydrophilic units derived from an N-vinyl heterocyclic monomer, the preferred monomer being N-vinyl pyrrolidone, optionally a minor proportion of hydrophilic units derived from other hydrophilic monomers such as hydroxy alkyl acrylates and methacrylates, acrylic and methacrylic acids, itaconic acid, maleic acid and fumaric acid and allylic crosslinking units selected from the group consisting of allyl esters derived from the esterification of polybasic acids with allyl alcohol, monoallyl itaconate, triallyl cyanurate and N,N-diallylmelamine, the preferred crosslinking monomer being diallyl itaconate which contains a polymerizable vinyl double bond in addition to two allyl bonds.

Examples 1 through 16: A series of interpolymers and lenses were made and evaluated by physical tests described in the patent. Results are given in the following table. In the table, vinyl pyrrolidone is listed as VP, methyl methacrylate as MMA, and diallyl itaconate as DAI.

Ex. No.	Water Content (%)	Composition . . (% by wt) . . . VP	MMA	DAI	Oxygen Permeability	Extractibles* (%)	Strength Rating	Extensibility (%)
1	98	100	0	0.05	66	20	1	40
2	83	100	0	1.00	33	14	2	40
3	57	100	0	5.00	12	6	5	40
4	94	95	5	0.05	54	13	1	40
5	90	90	10	0.05	48	14	2-3	160
6	90	95	5	0.20	44	10	1-2	40
7	83	80	20	0.10	33	10	4	240
8	83	85	15	0.15	34	12	2-3	80
9	80	85	15	0.50	31	12	3-4	100
10	74	70	30	0.05	23	10	8	160

(continued)

Ex. No.	Water Content (%)	Composition . . (% by wt) . . . VP	MMA	DAI	Oxygen Permeability	Extractibles* (%)	Strength Rating	Extensibility (%)
11	73	95	5	2.00	23	9	2	100
12	71	80	20	1.00	20	8	3	40
13	60	60	40	0.10	13	9	–	rigid**
14	33	40	60	0.05	4	6	–	very rigid**
15	20	25	75	0.05	1	2	–	very rigid**
16	30	Dura Soft Lens***			4	3	10	400

*Percent by weight extracted from the polymer usually after three extractions, no substantial increase in extractibles was noted between the last and next-to-last extraction.
**Lens did not break during hand pull.
***Control.

Nonionic, Water-Soluble Polymers for Improving Dry Strength of Paper

Highlights of the Technological Achievement: A nonionic copolymer composed of (in weight percent) 5 to 30% N-vinyl pyrrolidone, 15 to 60% acrylamide, and 30 to 70% methyl methacrylate is disclosed. The copolymer is added in minor amount to paper pulp, especially pulp containing black liquor contamination, to improve the paper-making operation and the dry strength of the paper.

Reference: H.S. Killam; U.S. Patent 4,167,439; September 11, 1979; assigned to Rohm and Haas Company.

Background: In recent years manufacturers of paper from unbleached kraft and other unbleached pulps have tended to utilize dirtier pulps, i.e., those containing black liquor and attendant solids, dissolved sulfates, hemicelluloses, and other contaminants. Polymeric additives for improving the dry strength of paper are normally employed in processing pulp. Such materials include natural gums and starches and synthetic organic polymers, usually composed of cationically or anionically charged particles which require a fiber deposition aid such as alum for effectiveness.

With the so-called dirtier pulps becoming more prevalent, however, the efficacy of the conventional dry strength resins is much reduced. One solution to the problem is disclosed in U.S. Patent 3,840,489 which teaches the use of an aqueous dispersion of a copolymer of at least 60% by weight of acrylamide and at least 5% of a hydrophobic monomer such as styrene, methyl methacrylate or acrylonitrile.

Process Conditions: The preferred polymers comprise linkages of 10 to 25% N-vinyl pyrrolidone, 15 to 35% acrylamide, and 45 to 70% methyl methacrylate. Specific copolymers disclosed include those having 20, 10 and 5% N-vinyl pyrrolidone, 30, 20 and 25% acrylamide and 50, 70 and 70% methyl methacrylate.

The copolymers optionally may contain minor amounts of other polymerized hydrophobic monomers, such as butyl acrylate, ethyl methacrylate, styrene, and the like; however, the amount of these linkages should be less than 5% because their presence will decrease water-solubility and effectiveness. In addition, the copolymers optionally may contain minor amounts of other polymerized hydrophilic monomers, such as N-methylol acrylamide, acrylic acid, vinyl acetate, methacrylamide, methacrylic acid, methyl acrylate and the like, the amount of the linkages not to exceed 5% in order to avoid interference with the proper deposition of the polymer on the pulp fibers and with the development of dry strength.

Emulsion Polymerization: Although the copolymers embodied herein are de scribed as water-soluble, they are more explicitly provided as dispersions or colloidal dispersions in aqueous medium of finely-divided solids of such small size, i.e., ranging from 0.05 to 0.15 μ, preferably 0.05 to 0.08 μ, which dispersions essentially perform as true solutions.

Accordingly, the copolymers are prepared by conventional emulsion polymerization techniques whereby the monomers in appropriate proportions are polymerized in an agitated aqueous medium containing a small amount of dispersing agent, such as sodium alkyl aryl sulfonate, the reaction being catalyzed by a free-radical initiator, for example, sodium persulfate. Normally, 0.1 to 1% of initiator, based on monomer weight, is used and the temperature of polymerization will range from 20° to 90°C. The reaction will generally be substantially complete after 1 to 4 hours. The amount of water in the reaction medium is usually adjusted so that the copolymer product is recovered in aqueous dispersion containing from 10 to 20% polymer solids by weight, although other solids contents can be used.

As an indication of the molecular weights of the copolymers embodied herein, the viscosities of aqueous mixtures thereof containing 10% by weight of polymer solids are in the range of 10 to 500 cp at 25°C at a pH of 4 to 8. Viscosity does not vary appreciably with changes in pH. Although the copolymer may be recovered as dried solids by separation from the aqueous carrier by conventional means, it is more convenient to retain the product in admixture with water for shipment and ultimate utilization after dilution, in paper pulp processing.

Product: It has been discovered that the hereindescribed copolymer provides marked improvements in the dry strength properties of paper when added to the pulp from which the paper is made, especially when the pulp contains significant amounts of black liquor and related contaminants. Good results are obtained using 0.15 to 0.5% of the copolymer based on the weight of dry paper fiber in the pulp. The copolymer is usually added to the pulp as a dilute dispersion (1 to 5%) in water at any convenient point such as between the fan pump and the head box.

Example: To a mixture of 700 ml water, 8 g of a 23% solution of sodium dodecylbenzene sulfonate, 30 g acrylamide, 50 g methyl methacrylate and 20 g N-vinylpyrrolidone, stirred at 20° to 22°C and blanketed by an inert stream of nitrogen, is added 0.1 ml of tert-butylhydroperoxide, then 0.1 g sodium formaldehyde sulfoxylate in 10 ml water. 1 ml of a 0.15% solution of ferrous sulfate is added to prevent coagulation of the polymer which begins to form as the temperature of reaction rises to about 33°C in 10 minutes. A temperature of 25° to 35°C is maintained for 3 hours. The product has a viscosity of 70 cp and dispersed solids content of 12.4% consisting of the copolymer: 50% methyl methacrylate, 30% acrylamide and 20% vinylpyrrolidone.

This resin dispersion is evaluated as an additive for improving the dry strength of kraft paper prepared from pulp slurry in which sulfate black liquor content, alum content, resin addition, pH and water temperature are simulated under plant conditions. The effectiveness of the resin is gauged by measuring the

Mullen burst strength (in lb/in^2) of the paper, an indication of its dry strength.

The pulp is beaten to 650 Canadian Standard Freeness (CSF) and pH is controlled at 6. The black liquor is added as a 50% solids solution and is in addition to that already present in the pulp. Water used is at room temperature. As is conventional the amounts of additives are based on the oven dry weight of paper fibers in the pulp (i.e., 1 g polymer and 100 g pulp gives 1% polymer and 2% of the 50% black liquor solution, and 100 g pulp gives 1% black liquor solids on dry fibers). Thus, in this test series, % black liquor means % black liquor solids by weight on the fibers. The data are summarized as follows:

Sheet No.	Liquor	Alum (%)	Black Liquor Solids (%)	Polymer (%)	Mullen Burst Strength (psi)
1	Basis weight = 62 g/m^2 (32 lb/3,000 ft^2)	0	1	0	31.9
2		0	1	0.25	34.3
3		2	1	0	32.6
4		2	1	0.25	36.3
5	Basis weight = 98 g/m^2 (60 lb/3,000 ft^2)	0	1	0	52.4
6		0	1	0.25	62.0
7		2	1	0	53.5
8		2	1	0.25	62.4
9	Basis weight = 98 g/m^2	0	2	0	42.5
10		0	2	0.25	47.9
11		2	2	0	43.4
12		2	2	0.25	48.6
13		4	2	0	43.6
14		4	2	0.25	51.3
15		6	2	0	42.9
16		6	2	0.25	51.7

Aqueous Size Composition for Glass Fibers Containing Epoxy Resin, Polyethylene Glycol Monostearate, PVP, a Silane and a Glucan-CMC Mixture

Highlights of the Technological Achievement: An aqueous size composition is disclosed. The composition is particularly suited for application to glass fibers, and consists essentially of an emulsified epoxy resin, a γ-methacryloxypropyltrialkoxysilane wherein the alkoxy group has from 1 to 3 carbons, a polyethylene glycol monostearate, a polyvinylpyrrolidone, a water dispersible powder of disintegrated β-1,4-glucan material and sodium carboxymethylcellulose dispersed therein and an organic acid in an amount sufficient to provide a pH from 3 to 6.

Reference: R.A. McCoy; U.S. Patent 4,159,362; June 26, 1979; assigned to Owens-Corning Fiberglas Corporation.

Background: Glass fibers for use as a reinforcing material are drawn through small openings in a melting tank from a body of molten glass. Large numbers of individual filaments are gathered together to form a roving, strand or the like. Glass is an extremely abrasive material. As a consequence, absent some sort of protective coating, the individual filaments of a roving or the like abrade one another during ordinary handling. A protective coating, usually called a size, is applied to glass fiber filaments before they are gathered into a roving or the like in order to prevent such abrasion.

Various sizing materials have been suggested for application to glass fibers to prevent or minimize abrasion, and for the further purpose of increasing adhesion between the filaments and a material, e.g., a synthetic resin or synthetic resinous composition, to be reinforced by the fibers. Various sizes for glass fibers have been suggested by the prior art.

Process Conditions: It is usually preferred that the total solids content of the size composition of this process be from 3 to 7%, that the epoxy resin solids be from 2 to 4%, the glycol monostearate solids from ½ to 1¾%, and that the dispersible glucan sodium carboxymethylcellulose solids be from 0.1 to 0.5%. The preferred silane in the composition is γ-methacryloxypropyltrimethoxysilane, and is preferably present in an amount ranging from 10 to 25%, based upon the epoxy resin solids. The terms percent and parts are used herein to refer to percent and parts by weight, unless otherwise indicated.

Use: Sized, glass fiber roving or the like is also provided according to this process. Such roving or the like is produced by applying a size, as defined above, to glass fiber filaments as they are being gathered together in the production of roving, sliver, strand or the like, and curing the size, e.g., by heating for 18 hours at a temperature of 265°F. Preferably, the cured size constitutes from ¼ to ¾%, based upon the weight of fiber and size.

STARCH, STARCH PRODUCTS AND RELATED PRODUCTS

PROCESSING

Producing Glucan Using *Elsinoe*

Highlights of the Technological Achievement: A glucan characterized by repeating units of [3)–Glc–(1 → 4)–Glc–(1 → 4)–Glc–(1 →] wherein Glc represents α-D-glucopyranose residue is produced by cultivating a microorganism of the genus *Elsinoe* capable of producing the glucan on a nutrient medium containing one or more members selected from the group consisting of various starch hydrolyzates, mannose, fructose, mannitol and xylose to produce the glucan in the medium, separating and recovering the glucan. It is designated elsinan.

Reference: K. Yokobayashi and T. Sugimoto; U.S. Patent 4,200,693; April 29, 1980; assigned to Kabushiki Kaisha Hayashibara Seibutsu Kagaku Kenkyujo, Japan. (Other References: U.S. Patents 4,202,940 and 4,202,966.)

Background: The known polysaccharides comprising α-linked D-glucose, namely, α-glucan, include starch derived from plants, glycogen derived from animals, microbial dextran and microbial pullulan. Although these α-glucans have been consumed in large amounts, their uses have been mainly directed to the food and pharmaceutical industries.

The glucose residues present in elsinan are essentially linear molecules comprising mainly α-1,4 and α-1,3 linkages in the molar ratio of 2.0 to 2.3:1.0. A very few of the glucose residues linked at the C-1 and C-3 positions with the adjacent glucose residues are branched at the C-6 position by α-1,6 linkage. Such glucose residues are, at most, one out of every 70 glucose residues.

The mean molecular weight of elsinan is freely adjustable in the range of 5,000 to 10,000,000, because the glucan is producible by either chemical or biochemical procedure and is easily hydrolyzable with hydrochloric acid, sulfuric acid, etc.

The production of the elsinan disclosed in this process is attainable by utilizing microorganisms of genus *Elsinoe.* For example, *Elsinoe leucospila* is employable for the effective production of the elsinan. The following microorganisms of genus *Elsinoe* are also employable for the production of elsinan:

Elsinoe ampelina	IFO 5263, IFO 6359
Elsinoe araliae	IFO 6166, IFO 7162
Elsinoe fawcetti	IFO 6442, IFO 8417, ATCC 13200
Elsinoe annonae	ATCC 15027
Elsinoe corni	ATCC 11189
Elsinoe haveae	ATCC 12570
Elsinoe lepagei	ATCC 13008
Elsinoe tiliae	ATCC 24510

The process for producing elsinan will be described in further detail. According to this process, elsinan is produced by cultivating abovementioned microorganisms of genus *Elsinoe* on a nutrient medium containing at least one or more members of a group consisting of starch hydrolyzates, mannose, fructose, mannitol and xylose as sugar sources and separating and recovering the formed elsinan therefrom.

Starch Hydrolyzates: Any starch hydrolyzates are employable as sugar sources for the nutrient medium in this process. For example, single compounds such as glucose, maltose, maltotriose or maltotetraose, or combinations thereof, or partial starch hydrolyzates containing at least one of these compounds are employable.

Partial starch hydrolyzates include acid and/or amylase-conversion starch syrup or starch syrup solid with a desirable degree of hydrolysis. Partial starch hydrolyzates with a degree of hydrolysis preferably 10 or more, are suitable for producing the elsinan disclosed in this process.

Besides the single compound, fructose, a sugar mixture of glucose and fructose obtained by isomerization of glucose, a sugar mixture obtained by partial or complete inversion of sucrose, and date extract containing a mixture of glucose and fructose may be effectively used as the sugar source for the nutrient medium in this process. Not only mannitol, a single compound, but also a sugar mixture of mannitol and sorbitol obtained by hydrogenation of fructose may be effectively used as sugar source.

pH and Temperature: The initial pH of the culture medium should be in the range that favors microbial growth and elsinan production. Generally, 5 to 8 is preferable. Similarly, the cultivation temperature should be in the range that favors the microbial growth and elsinan production. Generally, 20° to 30°C is preferable. Cultivation is carried out until a maximum yield of elsinan is obtained, generally 3 to 7 days.

Example: A liquid medium, consisting of 5 w/v % glucose, 0.5 w/v % defatted soybean, 0.042 w/v % Na_2HPO_4, 0.018 w/v % KH_2PO_4, and water, was sterilized at 120°C for 20 minutes and then cooled. Thereafter, the medium was inoculated with *Elsinoe leucospila* FERM-P No. 3874 at an initial pH of 6.8 and subjected to submerged culture at 24°C for 5 days.

After pasteurizing the resultant cultural broth at 85°C for 15 minutes, the broth was subjected to centrifugation (5,000 *g* for 20 minutes) to remove the cells and mycelia therefrom. Crude elsinan was obtained as a precipitate in a white plumage or gum form with the addition of 1.5 volumes of ethanol to the thus-obtained clear supernatant.

The crude elsinan was dissolved in water and subjected to centrifugation to remove insoluble substances, as described above, and then precipitation was effected by adding ethanol again to the supernatant. After the procedure was repeated 3 times, the precipitate was lyophilized. White powder of purified elsinan was obtained at an approximately 60% dry solids basis yield against the sucrose used in the medium.

The viscosity of a 3 w/w % aqueous solution of the purified elsinan determined at 30°C, using a Brookfield rotational viscometer, was 380 cp. The estimation of molecular weight distribution of the purified elsinan by the gel filtration method gave a distribution range from 10,000 to 10,000,000 or more, as illustrated in Figure 24.1.

A 5 w/w % aqueous solution of the purified elsinan was cast uniformly on a clear glass plate and air-dried. A colorless, clear, intensive, flexible and self-supporting film was formed. The excellent film formability of elsinan leads to the applications as packaging film material and coating agent.

Figure 24.1: Molecular Weight Distribution of Purified Elsinan by the Gel Filtration Method

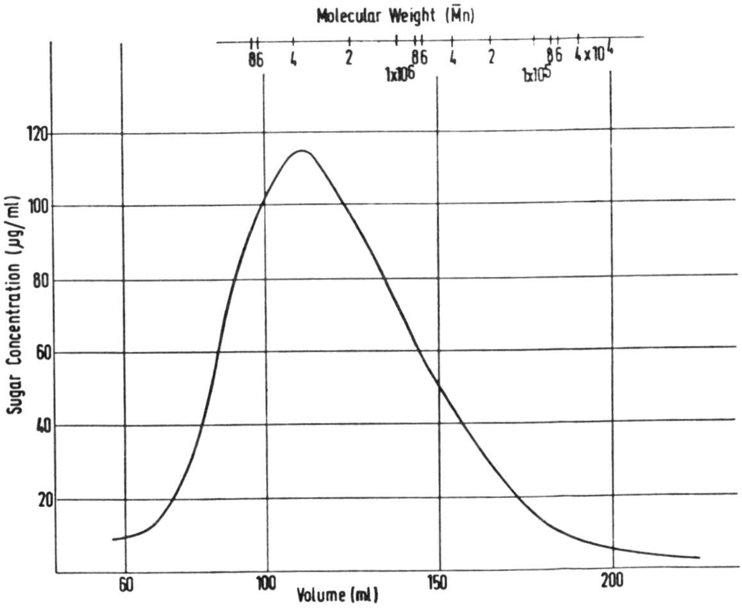

Source: U.S. Patent 4,200,693

Aqueous Heteropolysaccharide Composition Using Hydrolyzed Starch

Highlights of the Technological Achievement: A process for producing a heteropolysaccharide by a bacterial fermentation procedure in which a species of bacteria or a mutant thereof is incubated in a fermentation medium which contains a carbon source, preferably a hydrolyzed starch, a source of magnesium ions, a source of phosphorus, a source of nitrogen and water with the incubation taking place at a temperature of 28° to 35°C.

Reference: K.S. Kang, G.T. Veeder III and D.D. Richey; U.S. Patent 4,186,025; January 29, 1980; assigned to Merck & Co., Inc.

Background: It is known that heteropolysaccharides can be produced by certain microorganisms. Some of such heteropolysaccharides function as hydrophilic colloids and because of their viscosity properties and rheology have been used as thickening agents for aqueous systems. One known method of obtaining such heteropolysaccharides is by fermentation of an appropriate nutrient medium with certain *Xanthomonas* organisms (compare U.S. Patents 3,020,207 and 3,256,271).

This process pertains to a heteropolysaccharide which is produced by the action of a bacteria on a selected carbon source. In the following description, it will sometimes be referred to as Heteropolysaccharide 10 or as Heteropolysaccharide S-10 under which designation it is described and claimed in U.S. Patent 3,933,788.

Process Conditions: This compound may be prepared by fermentation of a suitable nutrient medium with a strain of *Klebsiella pneumoniae* which does not grow at 37°C in the absence of iron. A deposit of this organism was made with the American Type Culture Collection on August 11, 1971 under Accession No. ATCC 21711. The organism was identified as *Erwinia tahitica*. When the 8th Edition of this work appeared in 1974, however, a change in classification tests resulted in a change of identification of the organism from *Erwinia tahitica* to *Klebsiella pneumoniae.*

A restricted deposit of a double-blocked mutant of this organism which requires iron, insufficient of which is available in the human body, for growth at 37°C has been deposited with the American Type Culture Collection on July 25, 1977, under Accession No. ATCC 31311. Under conditions of the deposit, the mutant culture will be freely available to the public upon issuance of a United States patent on this process, and is to be released to any person authorized by the Commissioner of Patents and Trademarks to receive it under Rule 114 of the Rules of Practice in patent cases.

The bacterium identified as ATCC 21711 is a nonvirulent strain of *Klebsiella pneumoniae* isolated from a Tahitian soil sample. *Klebsiella pneumoniae* is a ubiquitous organism found in water, soil, and all manner of vegetable matter. In addition, it is found in large numbers in industrial settings such as cooling tower water and paper mill effluents. In none of these situations is there evidence for *Klebsiella pneumoniae* causing any health problem. Soil isolates such as this are generally considered to be safe.

The hydrolyzed starch was obtained by heating pearl starch (Staley Manufacturing Company) at a concentration of 30% by weight in water with an α-amylase enzyme (Tenase, supplied by Miles Laboratory) at a weight ratio of 45 to 700 parts of starch for each part of the α-amylase enzyme. After addition of the α-amylase enzyme, the solution was incubated at 60°C for 2 hours, then placed in a boiling-water bath for 30 minutes or autoclaved for 15 minutes at 15 psi at a temperature of 120°C. The pH of the E-1 medium was 7.6 to 7.8.

Sterilized carbohydrate solutions were added to a heat-sterilized basal medium in a sufficient amount to give a total carbohydrate concentration of 0.5% by weight. The pH of the basal medium was about 7 and contained a few drops of bromcresol purple, which was used as the pH indicator. Following this, the liquid growth medium was inoculated with bacteria, the test tube was closed, and the organism was incubated at a temperature of 30°C.

The following carbohydrates were fermented with production of acid and gas: D-arabinose, D-cellobiose, D-fructose, D-galactose, D-glucose, inositol, D-lactose, D-maltose, mannitol, D-mannose, D-raffinose, L-rhamnose, D-ribose, salicin, D-sucrose, D-trehalose and D-xylose. Dextrin was fermented with production of acid only. Adonitol, dulcitol, inulin, sodium alginate and starch were not fermented.

The bacteria were able to utilize a wide variety of carbohydrates for growth. In the above tests, control tests were carried out under the same conditions with the same media without the presence of the bacteria. The control test showed no change in the pH during incubation, thus indicating that the resultant pH change was caused by the bacteria.

Example: A series of experiments was conducted to determine the effect of using various carbon sources in the bacterial fermentation on the yield of Heteropolysaccharide 10. The seed for the experiments was obtained by growing *Klebsiella pneumoniae*, ATCC 21711 for 24 hours in E-1 medium containing 3% by weight of hydrolyzed starch as the carbon source.

The seed was then transferred aseptically to sterilized shake flasks containing 90 ml of basal E-1 medium and 10 ml of a 30% by weight aqueous solution of the carbon source under test. The quantity of the seed inoculum was 1% by weight of the total weight of the medium. After inoculation of the flasks, the mixture was incubated under aerobic conditions at 30°C on a rotary shaker having a controllable speed range of 160 to 400 rpm. The results of these tests are set forth in the following table.

Carbon Source (3%)	Fermentation Time (hr)	Final (pH)	Final Beer Viscosity (cp)
Glucose	43	3.45	5
Hydrolyzed starch	72	7.00	4,600
Sucrose	72	4.40	50
Dextrin-1	72	7.50	0
Maltose	72	6.70	50

As shown in the above table, the results of the fermentation vary greatly depending upon the carbon source which was employed. The use of hydrolyzed starch as the carbon source produced a satisfactory yield (1.61%) of the heteropolysaccharide as evidenced by the high viscosity of the fermentation beer. However, the use of glucose or dextrin-1 as the carbon source produced essentially no heteropolysaccharide. Similar results are obtained with *Klebsiella pneumoniae* Fer-1, ATCC 31311.

The rheological properties of a typical heteropolysaccharide are summarized in the following table.

1% by weight viscosity, cp	1,400
2% by weight viscosity, cp	7,800
Yield point, dynes/cm^2	6.6
Shear stress, dynes/cm^2	132
2% by weight Haake viscosity, cp	156,000
Thixotropy	1.08
Pseudoplasticity	
Low degree (LDP)	80
High degree (HDP)	172

In the above table, the 1% and 2% viscosities were measured on a Brookfield Synchro-Lectric Viscometer Model LVF at 60 rpm with the No. 3 and No. 4 spindles, respectively. The remaining rheological data was obtained using a Haake Rotovisco-Dual 50/500 employing a model MV 1 cup and bob.

Pseudoplasticity is the log slope of shear rate versus shear stress. The LDP is calculated using the formula

$$\left(\frac{2.000}{\log B/A} - 1\right) \times 100$$

where A is the shear stress at 0.0845 sec^{-1} and B is the shear stress at 8.45 sec^{-1}. The 2.000 is the log of the ratio of shear rates. The HDP is calculated by the formula

$$\left(\frac{2.210}{\log C/B} - 1\right) \times 100$$

where B is the shear stress at 8.45 sec^{-1}, C is the shear stress at 1,370 sec^{-1} and 2.210 is the log of the ratio of the two shear rates. The 2% Haake viscosity is taken using the MVI cup and bob and recording the reading at 0.0326 sec^{-1}. This viscosity (cp) is calculated by multiplying the reading times 11,632.

Production of Xylose by Enzymatic Hydrolysis of Xylan

Highlights of the Technological Achievement: A process for the production of xylose by enzymatic hydrolysis of xylan wherein an aqueous solution containing xylan is treated with a carrier having bonded thereto xylanase enzyme and a carrier having bonded thereto β-xylosidase and, optionally, uronic acid-splitting enzyme.

Reference: J. Puls, M. Sinner and H.-H. Dietrichs; U.S. Patent 4,200,692; April 29, 1980; assigned to Projektierung Chemische Verfahrenstechnik GmbH, Germany.

Background: It is known that the breakdown of vegetable, water-soluble, cell-wall polysaccharides to monomeric sugars involves at least two groups of enzymes, i.e., glycanases, which split the bonds within a polysaccharride at random (with the exception of the bonds at the end of a chain) and glycosidases, which break down the oligosaccharides released by the glycanases into monomeric sugars. Thus, for the breakdown of xylans β-1,4-xylanases and β-xylosidases are necessary. If xylans are present which contain as side groups 4-O-methylglucuronic acid, it is also necessary to use a previously unknown enzyme which splits uronic acid. The two groups of enzymes differ as regards their molecular weight and the conditions in which they develop their optimal activity.

Process Conditions: According to this process there is provided a method for the preparation of xylose by enzymatic hydrolysis of xylan wherein an aqueous xylan-containing solution is treated with:

(A) A carrier having bonded thereto enzymes of the xylanolytic type wherein substantially all of the enzymes are xylanase enzymes; and

(B) A carrier having bonded thereto enzymes of the xylanolytic type wherein substantially all of the enzymes are β-xylosidase and, optionally, uronic acid-splitting enzymes.

As stated above, there are uronic acid-containing xylans and xylans which contain no uronic acid. If xylans containing uronic acid are to be enzymatically split according to this process, the carrier referred to above under (B) must also contain bonded uronic acid-splitting enzyme. If the xylans contain no uronic acid, the uronic acid-splitting enzyme constituent is not required.

In a further aspect of this process there is provided a method for the production of purified enzymes bonded onto carriers, wherein a raw enzyme preparation containing xylanase, β-xylosidase and, optionally, uronic acid-splitting enzymes is separated by ultrafiltration into one fraction which contains substantially only xylanase enzymes, and a second fraction which contains substantially only β-xylosidase and, optionally, uronic acid-splitting enzymes, and wherein each of the separated fractions is bonded separately to the appropriate carrier.

The process provides a highly simple and effective way of producing the monosaccharide xylose in high yield from xylans which are available in large quantities from plant, i.e., vegetable, raw materials. Xylose is a valuable sugar which can be used per se or reduced to xylitol, which latter material is also a valuable substance previously relatively difficult to obtain in large quantities.

The conditions of xylan hydrolysis by means of carrier-bonded enzymes differ from xylan hydrolysis with free enzymes in that higher temperatures can be selected because of the greater stability of the bonded enzymes. This allows the hydrolysis to be effected more rapidly. Temperatures preferably in the range of 40° to 45°C generally yield optimal results.

Carrier-Bonded Purified Enzymes: The carrier-bonded purified enzymes used according to this process are preferably produced by removing the insoluble particles of a raw enzyme solution, conveniently by normal filtration, filtering the solution through an ultrafilter having a cut-off of preferably about 100,000 molecular weight, filtering the supernatant through an ultrafilter with a cut-off of preferably 300,000 molecular weight. The filtrate thus obtained, which principally contains β-xylosidase and possibly uronic acid-splitting enzymes is bonded onto a carrier.

The filtrate from the ultrafiltration with the separating range first referred to above is filtered through an ultrafilter with cut-off of from 10,000 to 50,000 molecular weight, preferably 30,000 molecular weight, and the filtrate thus obtained, which principally contains xylanase, is bonded onto a carrier. In order to carry out this process it is advisable to dissolve the raw enzyme in preferably 20 times the amount of water.

Example: *Carbohydrate Composition of the Aqueous and Alkaline Extracts* – Aliquot proportions of the aqueous and alkaline extracts were subjected to total hydrolysis. The quantitative determination of the individual and total sugars was carried out with the aid of a Biotronic Autoanalyzer [M. Sinner, M.H. Simatupang and H.H. Dietrichs, *Wood Science and Technology* 9, (1975) pp 307-322]. In the autoanalyzer the wood subjected to total hydrolysis was examined. Figure 24.2 shows the diagram obtained for red beech.

Figure 24.2: Analysis of Red Beech

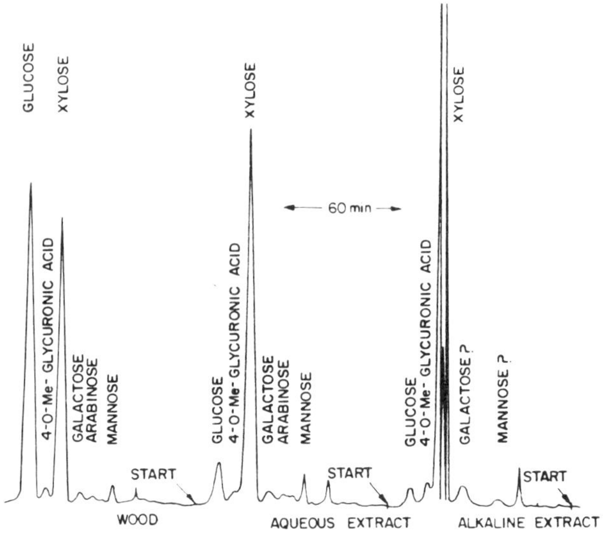

Source: U.S. Patent 4,200,692

Extract	Dissolved Carbohydrate Total (%)**	Fractions (%)* Xylose	Glucose
Red beech H_2O	13.5	69	13
NaOH	7.0	83	3
Oak H_2O	13.2	65	11
NaOH	6.8	81	5
Birch H_2O	11.2	77	8
NaOH	7.3	84	3
Poplar H_2O	8.3	76	6
NaOH	6.5	83	3

(continued)

Extract	Total (%)**	Dissolved Carbohydrate Fractions (%)* Xylose	Glucose
Eucalyptus H_2O	9.5	71	8
NaOH	5.0	80	3
Wheat H_2O	7.0	53	21
NaOH	8.3	88	3
Barley H_2O	6.1	41	25
NaOH	9.5	88	3
Oats H_2O	5.1	44	20
NaOH	4.4	88	3

*In relation to extract.
**In relation to starting material absolutely dry.

Preparing Crystalline Fructose from High-Fructose Corn Syrup

Highlights of the Technological Achievement: A process for preparing free-flowing particles of fructose with or without dextrose, which process comprises: admixing a high-fructose corn syrup with ethanol, the corn syrup and ethanol of defined moisture content, to form a clear homogeneous solution; seeding the solution with crystalline fructose and recovering free-flowing particles of crystalline fructose from the seeded solution.

Reference: B.K. Dwivedi and S.K. Raniwala; U.S. Patent 4,199,374; April 22, 1980; assigned to Chimicasa GmbH, Switzerland.

Background: Corn syrup is typically prepared by the incomplete hydrolysis of starch and generally comprises: a mixture of glucose, maltose and maltodextrans; the hydrolysis of the cornstarch to produce the corn syrup is typically effected by the action of acids or enzymes on the cornstarch.

The primary uses of corn syrup are typically in the food industry where it is employed as a sweetener having a high dextrose equivalent or as a thickener or bodying agent in soft drinks where it is employed with a low dextrose equivalent. Fructose, a naturally occurring sugar, is prepared from cornstarch by enzymatic or microbial action. Syrups containing a high concentration of fructose are commercially available. Fructose being a natural sugar, it is desirous to employ the fructose in crystalline form. Therefore, it is desirable to obtain crystalline free-flowing fructose from commercially available high-fructose corn syrup compositions.

Process Conditions: A process is provided for preparing free-flowing fructose particles from high-fructose syrup compositions, such as corn syrup and starch compositions, and which process comprises: providing the syrup composition, typically having a fructose concentration over 90% by weight of fructose; forming a homogeneous, clear, alcohol solution of the syrup composition with a food-acceptable, volatile alcohol, such as ethanol; seeding the homogeneous, clear, alcohol syrup solution with fine crystals of a polysaccharide, such as, for example, fructose or glucose; and recovering the crystalline fructose from the clear seeded solution.

In the process the ratio of syrup composition to the alcohol ranges from 1:2 to 2:1, but preferably from 1:1 on an equal volume to equal weight basis. It is important in the process that the moisture content of the syrup and the alcohol be controlled with the moisture of the syrup and alcohol, such that they satisfy the following equation:

$$Q = \frac{W_s X_{M1} + W_a X_{M2}}{W_s + W_a X_{M2}}$$

In accordance with the above equation the Q must be less than 0.25 and in the equation Ws is the weight of syrup, Wa is the weight of alcohol, and X_{M1} and X_{M2} are the weight percent of moisture in syrup and alcohol.

The mixture of the high-fructose syrup composition with the alcohol should provide for a clear, homogeneous solution. Thereafter, the solution is seeded with a fine-particle-size, food-acceptable material, typically another sweetener, such as a polysaccharide and more particularly fine crystalline fructose of glucose and the like.

After seeding, the solution may be stirred and then is allowed to stand wherein the fructose crystals are then formed and the crystals are later recovered, harvested by filtration, centrifugation, or other crystal-recovering means. The crystals are then washed with an alcohol, such as ethanol, and dried under vacuum. Such crystals as recovered are typically free-flowing and have a particle size of less than 400 mesh and typically less than 200 to 300 mesh. In one embodiment the seed crystals employed are pure crystalline fructose, such as the crystalline product obtained by the process.

The corn syrup composition employed in the process is typically a commercial syrup composition which is high in fructose, such as a high-fructose corn syrup or starch syrup composition, or if desired, may be a high-fructose composition which is prepared directly employing high fructose.

In the process where the moisture content of the high-fructose syrup is too high in that the moisture content of the syrup and the alcohol does not satisfy the equation, then the syrup composition should have the moisture removed simply by concentrating the high fructose syrup composition in a vacuum, for example, at temperatures of 30° to 90°C, e.g., 35° to 55°C, but typically should be concentrated under vacuum at a temperature at which no color is formed in the concentration. It has been found that in corn syrup compositions which are concentrated under vacuum, a formation of a yellow color typically inhibits and retards the crystallization and recovery of the free-flowing fructose particles.

In one process, high-fructose corn syrup is concentrated in a vacuum at a temperature between 35° and 55°C to a moisture content of X_{M1}. This concentrated syrup is mixed with ethyl alcohol of X_{M2} moisture content. The mixture is stirred until a homogeneous, clear solution is obtained, preferably at 30° to 55°C. The above mass is seeded with 0.5 to 5.0%, for example, 2 to 3%, of crystalline fructose having a particle size of about 0.05 to 0.50 mm and then stirred for 5 to 10 hours. Fine crystals are obtained which grow further when allowed to stand for a longer period of time, for example, 3 to 12 hours. The crystals are harvested by filtration, centrifugation or other suitable means. These crystals are then washed with alcohol and dried under vacuum.

Example: High-fructose corn syrup with the following composition on a dry matter base was used: 90% fructose; 7% dextrose; and 3% other saccharides.

1,000 g of the above syrup at room temperature were mixed with 710 ml, 200-proof alcohol. Contents were well mixed and seeded with 16 g of crystalline fructose. The resulting mass was allowed to stand for 12 hours. The crystalline mass so obtained was washed with alcohol and dried under vacuum. The yield was 36%. The particle size was smaller than 300 mesh.

Manufacture of Crystalline Fructose

Highlights of the Technological Achievement: A process for the manufacture of free-flowing crystalline fructose, which process comprises seeding an 88 to 96% by weight fructose syrup with 2 to 15% by weight of fructose seed crystals at 120° to 160°F, the seed crystals having a size not greater than 250 μ, and permitting crystallization to occur at 50° to 90°F and at a relative humidity of less than 70%.

Reference: B.K. Dwivedi and S.K. Raniwala; U.S. Patent 4,199,373; April 22, 1980; assigned to Chimicasa GmbH, Switzerland.

Background: Crystalline fructose is commercially manufactured by a very involved and time-consuming crystallization process which requires relatively sophisticated processing machinery and apparatus and careful control of process conditions, to ensure the production of a satisfactory crystalline-fructose product. Anhydrous fructose crystals may be obtained by crystallization of the fructose from both aqueous solutions and aqueous alcohol solutions, such as aqueous methanol and aqueous ethanol solutions of fructose.

The employment of solvents in fructose solutions is undesirable, both from the economic point of view and since the resulting crystals precipitated from the solution and the liquid containing the solvent must have the solvent completely removed prior to consumption or use of the crystalline fructose, while economic operations also dictate the removal of the solvent from the mother liquid for use in the process.

Process Conditions: It is essential, in the practice of this process that during crystallization, the relative humidity should be maintained below 70%, most preferably between 35 and 45%, and the surrounding temperature should be maintained either between 50° and 90°F or sequentially decreased to this range or lower. The seed size of the crystals should not be greater than 250 μ, and typically less than 200 μ.

It has been found that the heat-sensitive, water-soluble crystallization pellets of the process may be dried to a desired moisture content by cooling the pellets to a temperature below 34°F and then granulating the pellets, such as by grinding, to less than 18 mesh. Thereafter, the ground pellets may be dried in a low relative humidity, for example, less than 50%, such as 35 to 45%, at oven temperatures, for example, 110° to 160°F, to provide free powdered material with a low moisture content, for example, 0.5 to 2%.

Thus the high heat sensitivity of the crystallization pellets makes the product difficult to grind, since the product tends to melt due to heat build-up in the grinding mill. However, the pellets, when cooled to 34°F or below, may be ground, and, thereafter, may be dried in conventional ovens to a low moisture content.

Example: The term high-fructose corn syrup used in this example is used for the following commercially available product:

Dry Basis Composition	Percent
Ash (sulfated)	0.03
Carbohydrate components	
Fructose	90
Dextrose	7
Other saccharides	3
Typical analysis	
Solids	80
Moisture	20

The syrup was cooked to a moisture content of 92±0.5% and seeded with 5% seed of a size of 177 to 250 μ. The product was placed in molds and allowed to sit at 78°F and a relative humidity of 30%. The crystallization commenced in 24 hours. The resultant product was dried in a vacuum oven maintained at 29" of Hg. The temperature was gradually increased from 70° to 200°F in 16 hours. The product was worked up to the required particle size. Free-flowing, nontacky, crystalline material was obtained.

Some cooked syrup was seeded with the product obtained above and the crystallization time reduced to 8 hours.

Purifying Maltose Solution

Highlights of the Technological Achievement: From a maltose solution of a purity of 75 to 90%, a high-purity maltose solution having a maltose purity of more than 98.5% is obtained in a high yield by treating activated carbon with the aqueous solution of an organic solvent.

Reference: M. Yamada, A. Mori, Y. Sato, K. Terayama and S. Miyatani; U.S. Patent 4,182,634; January 8, 1980; assigned to Hokkaido Sugar Co., Ltd., Japan.

Background: In recent years, more people have come to favor foods that are only mildly sweet. Use of maltose as a way of decreasing sweetness and, at the same time, retaining other properties intact or having them enhanced, is gaining popularity and use of maltose is increasing. Examples of maltose-containing products available in the market include malt syrup, high-maltose syrup and plain maltose. The maltose purities in these commercial products are respectively on the order of 20 to 30%, 40 to 50%, and 75 to 90%. These products are used in confectioneries and other foodstuffs.

In addition, maltose finds utility as a medium for microorganic cultivation and as a raw material for industries in general. Further, maltose purified to an exceptionally high purity of more than 98.5% is used in medicines, especially in solutions (for injection) intended for supplementing the sugar supply of patients of diabetes mellitus.

Process Conditions: An object of this process is to provide a method for the purification of a maltose solution which is free from the various disadvantages suffered by the conventional methods for the purification of maltose, specifically a process for recovering a maltose solution of a high purity of more than 98.5%

in a high yield from a maltose solution of a purity of 75 to 90% by removing from the original maltose solution those highly obstinate oligosaccharides, such as maltotriose and maltotetraose, composed of three or more glucose units.

A process which comprises bringing activated carbon into contact with the aqueous solution of an organic solvent, adding the same organic solvent to the maltose solution under treatment until the concentration thereof equals that in the aforementioned aqueous solution of organic solvent, feeding the maltose solution onto a bed of the activated carbon already treated with the aforementioned aqueous solution of organic solvent, supplying to the bed the same aqueous solution of organic solvent as that used for the contact with the activated carbon, recovering a maltose-rich fractional eluate from the resultant eluate issuing from the bed, and separating the organic solvent from the maltose-rich fractional eluate.

In the method of purification according to this process, the activated carbon to be used is treated in advance with the aqueous solution of an organic solvent and, to the maltose solution subjected to the treatment, the same organic solvent as mentioned above is added until the concentration thereof equals that in the aforementioned aqueous solution of organic solvent, so that when the maltose solution comes into contact with the activated carbon, the affinity of the activated carbon for maltose is weaker than for those oligosaccharides, such as maltotriose and maltotetraose, composed of three or more glucose units and, as the result, these oligosaccharides are preferentially adsorbed by the activated carbon and the maltose is eluted alone. Thus, there is obtained a maltose solution of high purity.

According to the method of purification of this process, since the activated carbon adsorbs substantially no maltose, the recovery ratio of maltose is notably high.

In the graph of Figure 24.3a, the fraction numbers are indicated along the horizontal axis and Fractions 1 through 15 represent the fractions of the eluate resulting from the ion-exchanged water and Fractions 16 through 25 the fractions of the eluate resulting from the aqueous 10% ethanol solution. The vertical axis of the graph is graduated for the sugar content. It is seen from the graph that the first elution of sugar in the eluate began in Fraction 4, that the total of sugar contained in the eluate of Fractions 4 through 8 was 50.9% based on the amount supplied and that in the case of the eluate resulting from the aqueous 10% ethanol solution, the total of sugar contained in the five fractions of the highest sugar contents (Fractions 21 through 25) was 10.4%.

The combination of Fractions 4 and 5 and that of Fractions 21 and 22 were respectively assayed. The results were as shown in the following table.

	Glucose (%)	Maltose (%)	Maltotriose (%)	Others (%)
Fractions 4 and 5 combined	0.4	94.0	2.1	3.5
Fractions 21 and 22 combined	0.0	98.8	0.7	0.5

It is evident from the table that the solids present in the combination of Fractions 4 and 5 still contained 5.6% of the maltotriose, etc., as the impurities, although there was increased maltose purity. This large content of maltotriose, etc., in the eluate may be due to the fact that supplying of 70 mg of the maltose solution as a reducing sugar per ml of activated carbon was an overload.

Figure 24.3: Purifying Maltose Solution

a.

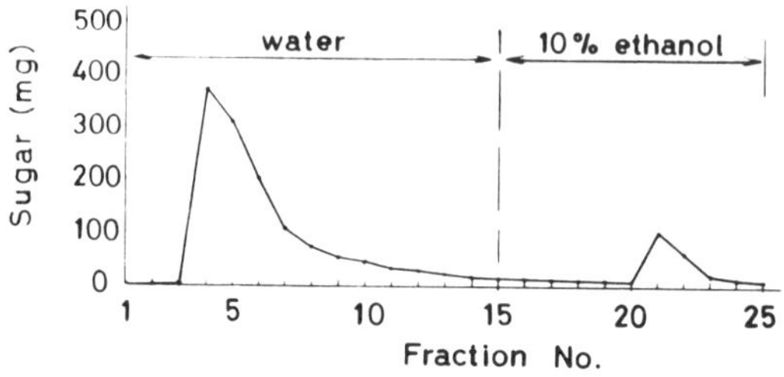

b.

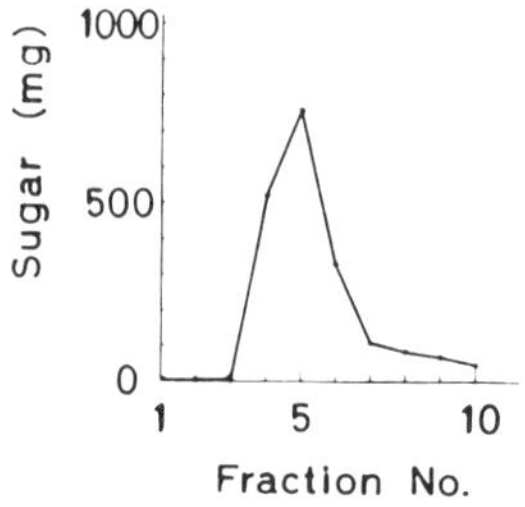

(a) Sugar contents of each of the fractional eluates with treatment according to prior art

(b) Sugar contents of each of the fractional eluates with treatment according to process

Source: U.S. Patent 4,182,634

In the graph of Figure 24.3b, fraction numbers are indicated along the horizontal axis and the sugar contents of the fractions are shown against the vertical axis. It is evident from this graph that substantially no sugar was contained in the first 30 ml of the eluate and the elution of sugar began to appear in Fraction 4.

The total amount of sugar contained in the eluate of Fractions 4 through 8 was 81.5% of the total amount of sugar added to the column. The sugar was composed of 0.3% of glucose, 99.3% of maltose, 0% of maltotriose and 0.4% of others based on the solids content.

Glucoamylase Immobilized on Cationic Colloidal Silica

Highlights of the Technological Achievement: A process for production of dextrose from starch wherein a starch hydrolyzate produced using soluble α-amylase or a combination of soluble glucoamylase and α-amylase is treated with glucoamylase immobilized on a cationic colloidal silica to produce a dextrose-containing syrup.

Reference: N.J. Mueller and D.J. Holik; U.S. Patent 4,202,939; May 13, 1980; assigned to CPC International Inc.

Background: Hydrolysis of starch to dextrose has been accomplished by means of enzymes. The principal enzyme used for such purposes was, and continues to be, glucoamylase. This enzyme effectively hydrolyzes the starch by cleaving one molecule of dextrose at a time from the starch molecule. As a practical matter, however, it is necessary first to thin the starch before subjecting it to the action of glucoamylase. This thinning step may be accomplished either by means of acid or enzyme. The starch is thinned to a dextrose equivalent of 10 to 20, then treated with glucoamylase. This two-stage process is referred to as an acid-enzyme process or an enzyme-enzyme process, depending upon the nature of the thinning step employed.

In the acid-enzyme process, starch is liquefied and hydrolyzed in an aqueous suspension containing 20 to 40% starch and an acid, such as hydrochloric acid. The suspension is then heated to a high temperature, i.e., between 70° and 160°C and at a pH between 1 and 4.5 to liquefy and partially hydrolyze the starch.

In the enzyme-enzyme process, starch is liquefied and partially hydrolyzed in an aqueous suspension containing 20 to 40% starch and a liquefying enzyme, such as bacterial α-amylase enzyme at a temperature of from 85° to 105°C. The dextrose equivalent of the liquefied and partially hydrolyzed starch is generally less than 20 and preferably less than 10. The mixture is then subjected to a temperature preferably between 110° and 150°C to insure complete starch solution. The starch hydrolyzate is then cooled to a temperature of less than 95°C and subjected to further treatment with bacterial α-amylase to hydrolyze the starch to a dextrose equivalent of up to 20.

By either process the digested starch may thereafter be converted to dextrose or dextrose-containing syrups by other enzymes such as glucoamylase.

It has been discovered that the small particles of cationic colloidal silica can be agglomerated to larger particles if they are gelled by raising the pH to 6.5, then frozen and thawed. It has been found, furthermore, that when the colloidal silica is agglomerated by this method in the presence of glucoamylase, the enzyme is immobilized in an active form. No crosslinking agent is necessary to bind the enzyme to the silica, and the enzyme does not leach out when the particles are suspended in or washed with aqueous solutions.

In contrast, if the colloidal silica is first agglomerated and the resulting particles are then contacted with the enzyme, very little of the enzyme is bound to the silica.

An important advantage of this enzyme composite is that it is relatively noncompressible. For example, it undergoes no volume change when spun in a centrifuge at 2,000 rpm for 15 minutes. This noncompressibility, coupled with good particle size makes the enzyme composite very suitable for carrying out enzymatic conversions in fixed-bed reactors with a minimum of pressure drop across the reaction zone. These properties of the composite also make it useful for enzymatic conversions in expanded bed (upflow) reactors, continuously stirred tank reactors and batch reactors.

Process Conditions: A process for the production of an insoluble, active enzyme comprises (a) contacting an enzyme solution with cationic colloidal silica, (b) converting the mixture to a gel by raising the pH to 6.5, (c) freezing the gel by holding it below -15°C for 24 hours, (d) thawing the frozen solid by allowing it to stand at 20°C, and (e) separating the resulting solid enzyme-containing particles.

In another embodiment, this process is directed to a method for producing dextrose or dextrose-containing syrups from a starch hydrolyzate which comprises the steps of (a) treating a starch hydrolyzate at a pH between 4.0 and 4.5 with an immobilized enzyme composite consisting of a starch-hydrolyzing enzyme, and (b) recovering the dextrose product.

The colloidal silica used in this process is an aqueous colloidal dispersion of silica particles, e.g., Ludox.

The glucoamylase used in this process can be any of the well-known fungal amylase preparations, particularly those derived from members of the *Aspergillus* genus, the *Endomyces* genus or the *Rhizopus* genus. A particularly preferred glucoamylase is that available from the process described in U.S. Patent 3,042,584 whereby a fungal amylase preparation is freed of undesired transglucosidase activity by treatment in an aqueous medium with a clay material.

The gelled mixture may be stabilized by the addition of crosslinking agents such as glutaraldehyde. In the case of glucoamylase and cationic colloidal silica, better results were obtained without the use of such a crosslinking agent.

The gel is frozen at any convenient temperature below the freezing point of the mixture, preferably -15° to -20°C. Time of cooling will vary with the temperature, volume of material to be frozen and other cooling conditions. The mixture should be held below the freezing point long enough so that the mixture is completely frozen.

The frozen mixture is allowed to thaw slowly, usually in a room at 15° to 25°C. However, any temperature may be employed which does not cause inactivation of the enzyme.

The pH of the substrate may vary from 3.5 to 7.0, with the optimum pH varying with the enzyme used. With glucoamylase, the preferred pH range is 4.2 to 4.5.

This process using an immobilized glucoamylase composite may be used to produce syrups in which a high percentage of the carbohydrate is dextrose. The percentage of dextrose in the product may be adjusted by varying the contact time between the hydrolyzate and the enzyme, the temperature, the concentration of the hydrolyzate used, and the composition of the hydrolyzate used. It is thus possible to produce syrups with controlled amounts of dextrose as well as those with high amounts of dextrose.

APPLICATIONS

Cold-Water-Dispersible, Gelling Starches

Highlights of the Technological Achievement: A cold-water-dispersible, modified starch characterized by forming a gel having a Bloom strength of at least 45 g is prepared by converting a tapioca starch to a water fluidity of 10 to 63, reacting the starch with a crosslinking agent to within defined Brabender viscosity limits sufficient to provide the modified starch with gelling properties, and drum-drying the converted, crosslinked starch to obtain the modified starch product. The gelling instant starch thus prepared is particularly useful in food systems of the type which gel upon standing, such as pie fillings, jellies, and puddings.

Reference: C.-W. Chiu and M.W. Rutenberg; U.S. Patent 4,207,355; June 10, 1980; assigned to National Starch and Chemical Corporation.

Background: The major drawback of these uncooked, so-called instant, food formulations is that they do not have the firm gel structure of the cooked formulations, i.e., they do not cut as cleanly with a spoon, and their texture, rather than being smooth, is generally described as grainy in both appearance and in mouth-feel characteristics. Moreover, the setting salts are not operative at low pH or in non-milk-based food systems.

Process Conditions: The above and related objects are achieved by a cold-water-dispersible, modified tapioca starch with gelling properties, prepared by drum-drying a tapioca starch which has been converted to a water fluidity of 10 to 63 and thereafter reacted with at least 0.005% by weight of a crosslinking agent, the converted, crosslinked starch being defined as having a Brabender viscosity differential, measured between 80° and 95°C, of from -20 to +30%, measured at 7% solids using a 700 cm/g cartridge, or of from -35 to -4%, measured at 10% solids using a 350 cm/g cartridge, and as having a Brabender viscosity at 80°C of at least 100 B.U., measured at 7% solids using a 700 cm/g cartridge, or of at least 400 B.U., measured at 10% solids using a 350 cm/g cartridge, the converted, crosslinked starch after drum-drying being capable of forming a gel having a Bloom strength of at least 45 g.

Example 1: This example illustrates the procedure for conversion of starch to a required water fluidity level.

A total of 2,000 g of tapioca starch was stirred into 2,500 ml of distilled water and heated in a water bath to 45°C. Then, 6.0 g of concentrated hydrochloric acid (36.5 to 38%) were added and the mixture stirred at 45°C for 16 hours. The hydrolysis was stopped by neutralizing the mixture with 3% sodium hydroxide solution to a pH of 5.5 to 6.5. The slurry was then filtered, washed and dried. The water fluidity of the thus-recovered starch was found to be 10.0, as measured by the test procedure described above.

Example 2: This example illustrates the preparation of several converted starches having a range of water fluidities of from 15.2 to 68.0.

Starch Samples 2 through 13 designated in the following table were prepared using the procedure of Example 1, varying the amounts of starch, water and acid, and the type of acid employed. Starch Samples 2, 3 and 5 were reacted at 45°C for 16, 18 and 17.5 hours, respectively, while Starch Samples 4 and 6 through 13 were reacted at 52°C for 16 hours. The water fluidity (W.F.) of each starch obtained was determined as in Example 1, and the results given in the table below. It is noted that a given degree of conversion can be obtained in shorter reaction times than those used herein when higher concentrations of acid and/or higher temperatures below the gelatinization temperature are employed.

Starch Sample No.	W.F.	Amount of Starch (grams)	Amount of Water (grams)	Type of Acid	Amount of Acid (grams)
2	15.2	2,000	2,500	H_2SO_4	4.0
3	22.0	3,000	3,750	HCl	12.0
4	25.0	2,500	3,750	H_2SO_4	6.25
5	26.7	3,000	3,750	HCl	15.0
6	37.0	2,000	3,000	H_2SO_4	8.0
7	47.8	2,000	3,000	H_2SO_4	14.0
8	51.2	5,000	3,750	H_2SO_4	35.0
9	56.2	2,000	3,000	H_2SO_4	20.0
10	60.5	2,000	3,000	H_2SO_4	25.0
11	63.0	2,000	3,000	H_2SO_4	30.0
12	65.8	2,000	3,000	H_2SO_4	36.0
13	68.0	2,000	3,000	H_2SO_4	42.0

Texturized Starch Products

Highlights of the Technological Achievement: Uniform starch particles prepared from one or more starches selected from tapioca, corn, waxy maize, potato, sago, arrowroot and cereal; and one or more gelling hydrocolloids selected from sodium alginate, sodium pectate, hydroxypropylcellulose, methylcellulose, methylhydroxypropylcellulose, methylethylcellulose, carrageenan, furcellaran, agar, gelatin, a mixture of xanthan gum and locust bean gum, and curdlan; by adding water, extruding, cutting and drying; useful in preparing retorted or aseptically packaged, tapioca-style pudding and other improved, edible texturized starch products.

Reference: H. Cheng; U.S. Patent 4,192,900; March 11, 1980; assigned to Merck & Co., Inc.

Background: An extruded starch product for use in mass-produced tapioca-style pudding processed in an automated high-temperature canning system, in which the starch particles are resistant to disintegration on cooking, is described in U.S. Patent 4,025,657. However, the process of that patent uses heat extrusion to gelatinize a portion of the starch in order to achieve a binding force, whereas this process uses a gelling hydrocolloid as binder.

Process Conditions: Prepackaged foods in single-serving sizes have experienced an increase in popularity. In order to participate in this increasing market, prepackaged puddings have had to satisfy a number of requirements regarding taste, texture and appearance. This has been particularly true for tapioca-style puddings, which are uniquely characterized by the presence of translucent pearls of gelatinized tapioca starch granules, and by a distinctive texture and taste.

In a typical aseptic packaging process for ready-to-eat puddings, condensed or dehydrated milk is employed, or else a falling film evaporator is used to concentrate milk, which is then flash-heated, followed by cooling in a tubular-type heat-exchange unit. A further development of this process is the ultra-high-temperature, short-time (UHTST) method in which the pudding ingredients (milk, flavoring, starch, etc.) are combined into a slurry that is pumped through a series of preheater units. In these preheater units the slurry temperature is raised to from 140° to 150°F, after which the slurry is removed to a cooker unit where its temperature is raised rapidly to from 285° to 305°F for a time of only 8 to 12 seconds.

From the cooker unit the slurry goes to a series of coolers where its temperature is lowered. Following this cooking and cooling, the pudding slurry is aseptically packaged, in cans or other containers, which are then sealed, ready for distribution or consumption.

Among the advantages which characterize the UHTST method are better preservation of product texture, better flavor, and better overall appearance.

In accordance with this process there is prepared an improved retorted or aseptically packaged, tapioca-style pudding wherein the improvement comprises the use therein of uniform starch particles obtained by

(a) preparing a dry homogeneous mixture of one or more starches selected from tapioca, corn, waxy maize, potato, sago, arrowroot and cereal; and one or more gelling hydrocolloids selected from sodium alginate, sodium pectate, hydroxypropylcellulose, methylcellulose, methylhydroxypropylcellulose, methylethylcellulose, carrageenan, furcellaran, agar, gelatin, a mixture of xanthan gum and locust bean gum, and curdlan;

(b) adding water and mixing until smooth;

(c) cool extruding the mixture through a 1⁄32" to ¼" hole die and cutting to produce starch particles;

(d) where the gelling hydrocolloid selected is sodium alginate or sodium pectate, soaking the starch particles in a solution of calcium, barium, or magnesium ions;

(e) drying the starch parties to a moisture content of from 1 to 12%; and

(f) screening the starch particles to pass No. 10 and be retained by No. 14, U.S. Standard Sieve Series;

the starch particles having heat and shear stability during retorting or processing for aseptic packaging, yet swelling without dissolving during the processing to produce a tapioca-style pudding of improved tapiocalike appearance and texture.

Starches: The starches employed are selected from tapioca, corn, waxy maize, potato, sago, arrowroot and cereal starches such as rice and wheat. Derivatives of these starches, such as crosslinked, etherified and esterified starches, may also be employed. Blends of two or more of these starches together may be employed. However, it is preferred, overall, to use tapioca starch in preparing the improved tapioca-style pudding of this process.

Gelling Hydrocolloids: The gelling hydrocolloids employed in this process are selected from sodium alginate, sodium pectate, hydroxypropylcellulose, methylcellulose, methylhydroxypropylcellulose, methylethylcellulose, carrageenan, furcellaran, agar, gelatin, a mixture of xanthan gum and locust bean gum, and curdlan. Two or more of these hydrocolloids may be employed together. Overall, however, it is preferred to employ sodium alginate.

Sodium alginate, the preferred hydrocolloid for use in this process, along with sodium pectate, is gelled or crosslinked by contacting it with calcium, barium, or magnesium ions. It is preferred to employ sodium alginate with a molecular weight range of from 75,000 to 100,000.

The following hydrocolloids employed in this process are cold-water-soluble but thermal gelling with hot water: hydroxypropylcellulose, methylcellulose, methylhydroxypropylcellulose, and methylethylcellulose. The following hydrocolloids employed in this process are hot-water-soluble but are gelled on cooling: carrageenan, furcellaran, agar, gelatin, a mixture of xanthan gum and locust bean gum, and curdlan. Thus, all of these hydrocolloids are gelled during the manufacturing process for the tapioca-style pudding of this process.

Examples 1 through 6: Two test methods were utilized as follows.

Test Method A (in Distilled Water) – Measure 200 ml of boiling distilled water into a beaker containing a 1" magnetic stirrer. Heat the contents of the beaker on a hot plate at 300° to 335°F and stir at 150 rpm. Add 10 g of a sample to be tested and 10 g of sugar, and heat for 25 minutes. Pour the cooked product into 2,000 ml of cold distilled water, screen through a No. 30 mesh screen, drain for 1 minute, and record the net weight of cooked product as Fish-Eye Retention Weight.

Test Method B (in Milk) – The same procedures as that for Test Method A are followed, except that 200 ml of boiling milk are substituted for the 200 ml of boiling water and the 10 g of sugar are omitted.

The samples (but not the controls) were prepared in accordance with the following procedure. Blend all dry ingredients well in a Hobart mixing bowl. Add water and mix until smooth. Cool-extrude through a 1/16" hole die on a Hobart meat grinder. Cut the extruded strands into 1/16" lengths. Soak in 3% $CaCl_2$ solution for 30 minutes and rinse with water. Dry in an air-forced oven at 110° to 130°F to a moisture content of less than 12%. Screen through No. 10 and No. 14 mesh screens. Collect products on No. 14 mesh screen as samples.

The ingredients for the various samples and the results of Test Methods A and B for these samples are set out in the table of values below.

No.	Ingredients	(%)	Fish-Eye Retention Weight (g) A	B	Comments
1*	Minute tapioca starch (through No. 10 mesh, on No. 14 mesh)		58	60	Uncooked product dissolves very easily in cold water on stirring. Cooked product is too soft and stringy, does not have good heat and shear stability.
2	Tapioca starch	57.40	24	22	Uncooked product dissolves very easily in cold water on stirring. Cooked product is extremely soft and stringy.
	Pregelatinized, modified starch**	16.30			
	Water	26.30			

(continued)

No.	Ingredients	(%)	Fish-Eye Retention Weight (g) A	B	Comments
3	Tapioca starch	58.67	66	–	Cooked product is firmer in texture and body than control; better in heat and shear stability.
	Pectin	1.33			
	Water	40.00			
4	Tapioca starch	61.82	79	–	Uncooked product does not dissolve in cold water on stirring.
	Sodium alginate	0.70			
	Water	37.48			Cooked product is much firmer in texture and body than control; has excellent heat shear stability.
5	Tapioca starch	59.9	75	–	Uncooked product does not dissolve in cold water on stirring.
	Sodium alginate	0.1			
	Pregelatinized, modified starch**	2.0			Cooked product is firmer in texture and body than control; has excellent heat and shear stability.
	Water	38.0			
6***	Same as Ex. 5		–	60	Similar to Example 5.

*Control not subject to the preparation method employed with the other Examples.
**Redisol 88 starch.
***The sample did not receive a calcium chloride treatment, calcium ions present in milk.

Foodstuffs Containing Crosslinked Starches Using STMP

Highlights of the Technological Achievement: A dual derivatizing process is disclosed for the preparation of cold-water-swelling starches and foodstuffs thickened therewith. The starch base is first subjected to a primary crosslinking reaction using sodium trimetaphosphate (STMP) according to conventional techniques, then reslurried, treated with additional STMP and drum-dried to effect a secondary crosslinking reaction in situ.

Reference: R.W. Rubens; U.S. Patent 4,183,969; January 15, 1980; assigned to National Starch and Chemical Corporation.

Background: Recent attempts to replace certain starch derivatives used in foods, e.g., epichlorohydrin-inhibited starches, with other FDA-acceptable starch derivatives, have placed new emphasis on starch derivatives formed with phosphate derivatizing reagents, especially sodium trimetaphosphate.

According to this process the granular starch base is derivatized in a primary crosslinking step with STMP at a pH of 10 to 12 and at a temperature sufficiently low to maintain the starch in an unswollen granular form. The slurry is then neutralized to a pH of 5 to 6.5, recovered and preferably washed in accordance with conventional known techniques. At this stage, however, rather than reslurrying and drum-drying as is carried out in typical STMP reaction procedures of the prior art, the starch is reslurried, adjusted to a pH of at least 7.5 and additional STMP is added to the slurry. The resulting slurry is then fed onto a drum dryer and drum-dried in accordance with conventional procedures to effect a secondary crosslinking reaction in situ.

The foodstuffs thickened with the starches resulting from this unique dual derivatizing reaction are characterized by superior viscosity-related properties as evidenced by their viscosity curves which, after peaking, maintain a high viscosity with little or no degradation or breakdown even after exposure to elevated temperatures for extended periods of time.

Figure 24.4 shows Brabender viscosity curves of the dually crosslinked starch according to this process (Curve B) compared to the control distarch phosphate ester prepared in accordance with prior art techniques (Curve A).

Figure 24.4: Brabender Viscosity Curves

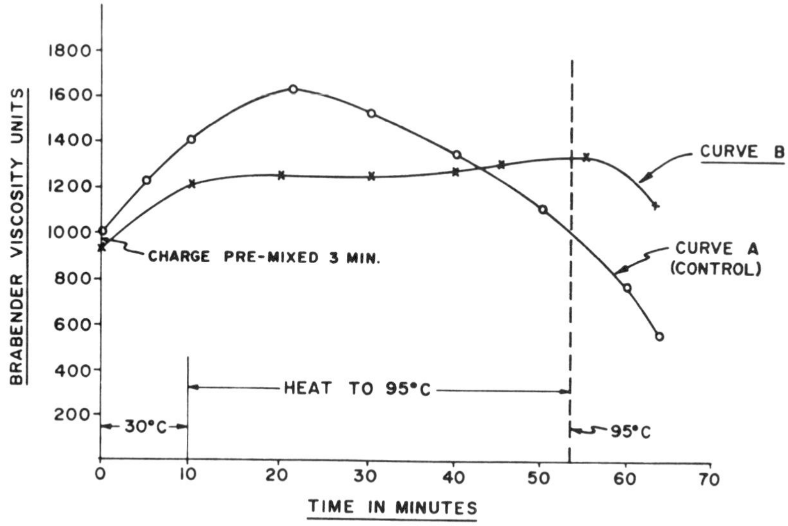

Source: U.S. Patent 4,183,969

Sterile Aqueous Solution of Isomaltooligosaccharides for Parenteral Administration to Mammals

Highlights of the Technological Achievement: A sterile aqueous solution containing 0.02 to 30 g per 100 ml solution of a mixture of isomaltooligosaccharides.

Reference: B.G. Ingelman, A.W. Richter and K.A. Granath; U.S. Patent 4,201,772; May 6, 1980; assigned to Pharmacia Aktiebolag, Sweden.

Background: Carbohydrate-containing substances consisting of individual isomaltooligosaccharides, e.g., isomaltose, isomaltotriose, isomaltotetrose, isomaltopentose, isomaltohexose, isomaltoheptose, isomaltooctose, isomaltononose or isomaltodecose, are known. They have preferably been isolated from partial hydrolysates of dextran. Mixtures of isomaltooligosaccharides are also known, e.g., in the form of partial hydrolysates of dextran. It is known that such substances may react as haptens with antibodies against dextran but no product which could be used in practice has been available.

According to this process, however, a very particularly selected mixture composed of isomaltooligosaccharides is used. This mixture has favorable properties for use in sterile aqueous solutions for parenteral administration, preferably for injection or infusion in blood vessels, and can be manufactured industrially on a large scale and can be used in practice.

The sterile aqueous solutions for parenteral administration (preferably solutions for injection or infusion) according to this process are characterized in that they contain preferably 0.03 to 30 g (especially 0.1 to 30 g, for example, 0.3 to 20 g) per 100 ml solution of a mixture of isomaltooligosaccharides, the mixture of oligosaccharides comprising:

(A) 0 to 15% by weight of isomaltose;

(B) 20 to 65% by weight, preferably 20 to 60%, of isomaltotriose, isomaltotetrose and isomaltopentose, each of these oligosaccharides being present in an amount of at least 5% by weight and at most 25% by weight;

(C) 20 to 65% by weight, preferably 20 to 60%, of isomaltohexose, isomaltoheptose, isomaltooctose and isomaltononose, each of isomaltohexose and isomaltoheptose being present in an amount of at least 5% by weight and at most 25% by weight and the total amount of isomaltooctose and isomaltononose being at least 5% by weight and at most 25% by weight; and

(D) 0 to 30% by weight, preferably 0 to 25%, of isomaltooligosaccharides of 10 to 20 glucose units, 0 to 10% by weight being isomaltooligosaccharides of 15 to 20 glucose units, the percentages being calculated on the total weight of the mixture of isomaltooligosaccharides.

The isomaltooligosaccharide mixture in the solution preferably comprises 0 to 10% by weight, for instance 2 to 10% by weight, of isomaltose.

The process also comprises a method of improving clinical dextran having a weight average molecular weight within the range of 30,000 to 80,000 with respect to side effects on parenteral administration, which method is characterized in that the clinical dextran is admixed with a mixture of isomaltooligosaccharides, the proportions between the mixture and the clinical dextran being 1 part by weight of the mixture to 0.5 to 500, preferably 1 to 300, parts by weight of the clinical dextran, e.g., 1 to 100 parts.

Producing Nitrogen-Containing Polysaccharides

Highlights of the Technological Achievement: The fungi belonging to the *Coriolus* genus are extracted by using a dilute alkaline solution having a concentration within a certain specified range whereby the low molecular weight substances with molecular weight of less than 5,000 are obtained in the extract solution. These low molecular weight substances are separated from the extract solution to obtain a nitrogen-containing polysaccharide having excellent antitumor activity.

Reference: S. Ueno, C. Yoshikumi, F. Hirose, Y. Omura, T. Wada, T. Fujii and E. Takahashi; U.S. Patent 4,202,969; May 13, 1980; assigned to Kureha Kagaku Kogyo KK, Japan.

Background: It is known that a polysaccharide having an antitumor effect can be produced by refining the extract of the Basidiomycetes with an aqueous solvent. This known method, however, has a serious effect that extraction efficiency of the active components is low and hence practical adaptability of this method to industrial production of the antitumor substances is poor. This method employs salting-out by use of ammonium sulfate, dialysis, precipitation by use of an organic solvent, or gel filtration as the extract refining means, but such refining means are extremely poor in workability, and hence such method is not of a type advantageous for removing the low molecular weight substances (with molecular weight of less than 5,000) contained in the extract.

The low molecular weight substances have also nil inhibitory activity against Sarcoma-180 solid tumors in mice in intraperitoneal administration, and further, they have bitterness and a disagreeable odor, so that presence of such substances is undesirable in utilization of the polysaccharides as medicaments.

Process Conditions: A nitrogen-containing polysaccharide having an antitumor effect and other various pharmacodynamic effects can be obtained in a high yield when a fungus belonging to genus *Coriolus* of family Polyporaceae of the class Basidiomycetes is extracted with an aqueous solution having an alkali concentration within a specified range and the obtained extract is refined by means of ultrafiltration and/or reverse osmosis.

The primary object of this process, therefore, is to provide a method of advantageously producing a nitrogen-containing polysaccharide having an excellent antitumor activity and other various pharmacodynamic effects from a fungus belonging to the *Coriolus* of the Basidiomycetes.

The concentration of the alkaline solution used for extraction of the Basidiomycetes in this process must be within the range of 0.01 to 2 N because if the concentration is less than 0.01 N, the result is not much different from that obtained from extraction with water, while if the concentration exceeds 2 N, the result may be decomposition.

Preferred extraction of the Basidiomycetes can be satisfactorily accomplished by using an alkaline solution of the abovementioned concentration range at a temperature of preferably 80° to 98°C, for a period of 20 to 600 minutes. It should be noted that an extraction temperature of lower than 50°C results in insufficient extraction of the active component, while an extraction temperature of over 100°C may invite reduction of activity of the obtained active component.

The preferred range for extraction time varies depending on the concentration and temperature of the alkaline solution used, but usually it is preferable to use an extraction time within the above-defined range, i.e., 20 to 600 minutes. It is possible to obtain a satisfactory result by one extraction, but if desired, extraction may be repeated several times.

It is considered that ultrafiltration is suited for fractionating of material with molecular weights of over 10,000 while reverse osmosis is suited for fractionating of material with molecular weight of less than 1,000.

The substance obtained in the abovedescribed manner according to this process is liver brown in color and has a nitrogen content of from 2 to 8%, in many cases 3 to 6%. It exhibits no distinct melting point and is gradually blackened and decomposed at a temperature of higher than 120°C. As for solubility of the substance of this process, it is soluble in water but almost insoluble in alcohol, pyridine, chloroform, benzene and hexane. It is also tasteless and odorless.

Various color reaction tests of the substance obtained according to the method of this process gave the results as shown in the table below.

Color Reaction Test	Color	Results
α-naphthol sulfuric acid (Molisch's reaction)	Purple	Saccharides
Indole sulfuric acid (Disch's reaction)	Brown	Saccharides
Anthrone sulfuric acid	Greenish blue	Saccharides
Phenol sulfuric acid	Brown	Saccharides
Tryptophane sulfuric acid	Purplish brown	Saccharides
Lowry-Folin process	Blue	Peptide bonds
Ninhydrin reaction (after hydrochloric acid hydrolysis)	Greenish blue	α-amino acids

The results shown in the above table imply that the substance of this process (hereinafter referred to as the process substance) is a nitrogen-containing polysaccharide. The molecular weight of the process substance, as measured according to an ultracentrifugal method, ranged from 5,000 to 300,000 and the weight average molecular weight ranged from 10,000 to 100,000. Other measuring methods, such as fractionating by use of an ultrafiltration membrane, also gave the values of 10,000 to 100,000. Therefore, it may be estimated, with high reliability, that the average molecular weight of the process substance is within the range of 10,000 to 100,000.

The nitrogen-containing polysaccharide obtained according to this process not only demonstrated a high antitumor activity with high inhibition ratio against Sarcoma-180 solid cancer in mice in intraperitoneal administration, but also proved effective in oral administration. This is indicative of very high availability of the nitrogen-containing polysaccharide of this process as an oral antitumor agent, and in fact, such effect has been confirmed in various experiments.

Oral administration of the process substance also produced an excellent effect in improvement of liver function, increase of appetite, adjustment of intestinal disorders and promotion of urination. It is also effective for treatment of leprosy.

Nitrogen-Containing Polysaccharide for Promoting Drug Sensitivity of Bacteria Resistant to Antibiotics

Highlights of the Technological Achievement: A nitrogen-containing polysaccharide, which is obtainable by extracting mycelia of Basidiomycetes belonging to the genus *Coriolus* of Polyporaceae with an aqueous solvent, is effectively used in promoting drug sensitivity of bacteria resistant to antibiotics.

Reference: C. Yoshikumi, Y. Omura and T. Hotta; U.S. Patent 4,188,384; February 12, 1980; assigned to Burroughs Wellcome Co.

Background: Advancement of the chemotherapeutical techniques and, particularly, development of a variety of antibiotics, have realized a drastic reduction of bacterial infectious diseases such as tuberculosis, dysentery, etc. However, these modern achievements have invited, on the other hand, a rapid increase of the so-called drug-resistant bacteria which defy the existing chemotherapeutants. To make the matter worse, many multiple-drug-resistant bacteria, i.e., the bacteria resistant to two or more kinds of drugs, have developed from the serious germs such as *Mycobacteria, Shigella, Staphylococci,* etc., and these pose a serious problem in the field of chemotherapy against such infectious diseases.

For such reasons, antibiotic preparations usually have a short life cycle and hence the drug manufacturers are compelled to market new kinds of antibiotics one after another in rapid succession at an enormous cost. This can result in a huge loss which cannot be ignored from the viewpoint of social economy.

Process Conditions: The nitrogen-containing polysaccharide which is used as an active component of the agent according to this process is obtainable by extracting mycelia of Basidiomycetes belonging to the genus *Coriolus* of Polyporaceae with an aqueous solvent. The term mycelia of Basidiomycetes belonging to the genus *Coriolus* of Polyporaceae as used herein is based on the classification in *Coloured Illustration of Fungi of Japan* by Rokuya Imazeki and Tsugio Hongo (Hoikusha Publishing Company).

The method of preparing the nitrogen-containing polysaccharide used as the active component in this process is briefly described as follows. The mycelia of Basidiomycetes belonging to the genus *Coriolus*, which are used as starting material for the extraction process can be produced either naturally or by artificial culture. In case of artificially culturing the fungi by using a liquid medium, it is possible to use as starting material all the culture product including not only the produced mycelia but also the eluate in the liquid medium after culture.

The thus-obtained starting material is extracted with an aqueous solvent and the extract is subjected to suitable treatments such as filtration, neutralization, concentration of the filtrate, dialysis, salting-out, etc., to remove the low molecular material (with molecular weight of lower than 5,000) from the extract, and is dried to obtain a powdery substance. The thus-obtained powdery substance is composed of a nitrogen-containing polysaccharide and can be used as the active component in this process.

The antibiotics which exhibit their effect against the drug-resistant bacteria when combined with the process substance are those derived from mold, bacteria, *Actinomyces* and the like, and typical examples of such antibiotics are penicillin G, streptomycin, kanamycin, chloramphenicol, tetracycline, erythromycin, aminobenzylpenicillin, cephaloridine and colistin.

The process substance produces the effect in promoting the drug sensitivity of the following drug-resistant bacteria to the antibiotics: *Escherichia coli, Streptococcus faecalis; Staphylococcus aureus; Enterobacter aerogenes; Salmonella enteritidis; Shigella sonnei; Klebsiella pneumoniae; Proteus mirabilis;* and *Pseudomonas aeruginosa*. Its physicochemical properties are shown in Figure 24.5.

Figure 24.5: Nitrogen-Containing Polysaccharide

a.

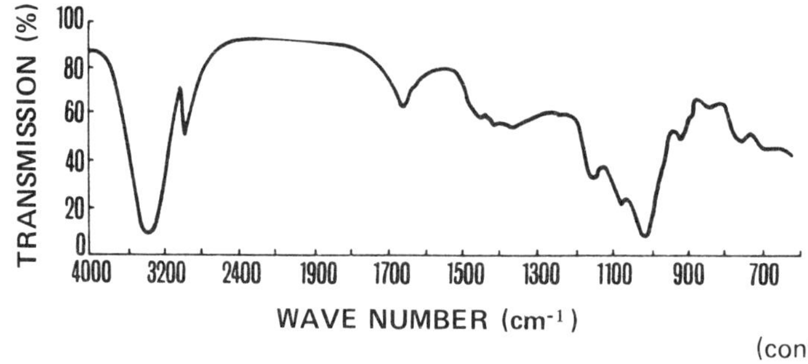

(continued)

Figure 24.5: (continued)

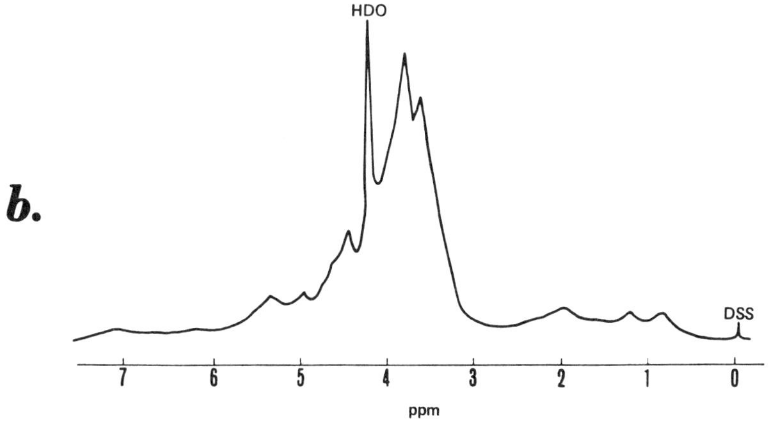

(a) Infrared absorption spectrum of the nitrogen-containing polysaccharide of the process
(b) Proton nuclear magnetic resonance absorption spectrum of the nitrogen-containing polysaccharide of the process

Source: U.S. Patent 4,188,384

Separating Hydrocarbon Pollution from a Hydroorganic Medium

Highlights of the Technological Achievement: Water, polluted with hydrocarbons, and similar pollutants in which separation occurs into an upper layer, rich in hydrocarbons and residues, and a lower layer consisting of a hydrocarbon suspension or emulsion in water, is treated to remove the pollutants from the water.

Reference: G. Sader; U.S. Patent 4,184,949; January 22, 1980.

Background: Separation of various hydrocarbons, polluted with residues to a greater or lesser degree, from an aqueous medium raises technical problems which are often difficult to resolve. In the attempt to remove crude oils spilled at sea, use has been made of either physical or mechanical processes for removing the black tide layer, or of chemical processes intended to cause coagulation and suspension of hydrocarbon particles which are then separated to a greater or lesser degree from the water by various means.

Another source of such hydroorganic media which must be separated, very frequent in practice, relates to the degreasing or dewaxing of various materials, such as, in particular, automobiles (buses, trucks and passenger cars), and machine tools. Such degreasing or dewaxing processes take place in industrial plants, such as refineries and textile dyeing mills, and in the food industry, etc.

In general, machines capable of spraying jets of water or steam at a pressure of several bars are used to clean off the film of monomer or polymer hydrocarbon compounds (included under the general term wax) which temporarily protects these materials during storage and/or transport. A small quantity of solvent is introduced into the water spray (kerosene or the like) to facilitate elimination of all traces of oil or grease.

The aforementioned processes, particularly water spraying, are indeed very effective, but their implementation gives rise to enormous quantities of aqueous liquids polluted by large fractions of organic compounds and various wastes and residues. Very often this wastewater is discharged untreated into the sewer which causes dangerous contamination and pollution of water courses, rivers, and underground water. Moreover, the treatment required to separate the hydrocarbon fraction calls upon techniques which are generally too complex and cumbersome for many industrialists.

Process Conditions: As a general rule, the hydroorganic media with which the process deals, and which result from a cleaning operation of the aforementioned type, separate into two relatively distinct layers because of their difference in density: an upper layer rich in hydrocarbons and including numerous residues, this layer generally being a dark brown color; and a lower layer with a turbid appearance and a milky white color comprised of a suspension or emulsion of a small quantity of hydrocarbons or other organic substances in water. These layers may be separated by primary decantation.

It is known that the polyoses are one of the groups in the glucide family, the other group being the ones (or monoses), i.e., nonhydrolyzable reducing sugars. Any type of polyose (or oside) may be suitable according to this process, i.e., the holosides or polyholosides such as maltose, lactose, saccharose, or the higher polyholosides such as starch, or heterosides, i.e., compounds giving oses plus other organic compounds by hydrolysis. Of these compounds, starch is particularly suitable and has the advantage of easy accessibility at a modest cost.

Any of the products used in wastewater treatment can be used as the coagulation agent, with or without the addition of polyelectrolyte. However, it is particularly advantageous to use alums, i.e., double sulfates of various metals and, of these, common alum or double aluminum-potassium sulfate, which is commercially very accessible.

The quaternary ammonium salts used to treat the supernatant layer may be composed of any one of the products already used to treat water because of their algicidal, bactericidal, germicidal or other effects. In general, these are halogenated salts of amines or quaternary ammonium hydroxides, such as alkyldimethylbenzyl ammonium chloride in which the alkyl group is C_8 to C_{16} or mixtures thereof, or corresponding chlorobenzyl derivatives.

The additive ranges will be preferably 0.5 to 5 parts of quaternary ammonium salts for 100 parts of supernatant layer (or first aforementioned layer), and 1 to 20 parts (by weight) of the polyose-coagulant mixture, the proportions of the latter to one another varying between 10 and 90 parts of the one to 90 to 10 parts of the other, for 100 parts of the turbid or milky layer (the second aforementioned layer).

Treatment of Wastewater Rich in Nutrients

Highlights of the Technological Achievement: Isolated polysaccharide assimilating yeasts are added to a wastewater rich in nutrients thereby making such yeasts assimilate nutrients, so that biochemical oxygen demand (BOD) of the wastewater is efficiently decreased. Among the polysaccharide assimilating yeasts, *ellipsoideus* YS-1, FERM-P No. 4498, in particular, can decompose and assimilate sulfurous acid, starch and protein at once, and, therefore, the BOD of the wastewater from starch production plants can be considerably decreased.

Reference: K. Yoshizawa, K. Nojiro, K. Mitsutomi, H. Hashimoto, A. Noguchi and K. Tanno; U.S. Patent 4,183,807; January 15, 1980; assigned to National Tax Administration Agency, Toho Zinc Co. Ltd., and The Hokuren Federation of Agricultural Cooperative Associations, all of Japan.

Background: The treating method of wastewater according to this process is applicable in a very wide range. For example, it is applicable very effectively to the wastewater of the production related with food such as wastewater of production of alcoholic liquors such as Japanese sake, shochu, wine and beer, wastewater of production of drink such as fruit juice and refreshing drink, etc., wastewater of production of yeasts, wastewater of bean paste production, wastewater of noodle production, wastewater of potato starch production, wastewater of treatment of fishes and meats, wastewater of soybean paste or curd, wastewater of paste-removing treatment and wastewater of dyeing treatment, and so forth.

The polysaccharide assimilating yeasts used in this process have been found in a very wide range, and these yeasts are known strains taxonomically. However, it has not previously been recognized that these yeasts have a polysaccharide assimilating property to decompose and assimilate polysaccharide in wastewater, thereby purifying the wastewater.

Process Conditions: The polysaccharide assimilating yeasts used in this process can be easily isolated and collected. The effective strains of this process and their deposition numbers are as follows:

Debaryomyces hansenii D-N, FERM-P No. 3592;
Pichia acaciae AM-37W, FERM-P No. 3593;
Hansenula anomala Y-1, FERM-P No. 3594;
Pichia nakazawae LKB-335, FERM-P No. 3595;
Pichia dispora Pd-671, FERM-P No. 3596;
Candida sp. AS-50, FERM-P No. 3597;
Endomycopsis platypodis AM-46, FERM-P No. 3598;
Saccharomyces cerevisiae Br-1, FERM-P No. 3972; and
Saccharomyces cerevisiae var. *ellipsoideus* YS-1, FERM-P No. 4498.

Example: Treatment tank containing 1.5 liters of the wastewater of potato starch production [clarifier wastewater: chemical oxygen demand (COD) 20,800 ppm] was inoculated with 10^7 cells of *Saccharomyces cerevisiae* var. *ellipsoideus* YS-1, FERM-P No. 4498 per 1 ml of wastewater, and after the cultivation the wastewater was added into the tank at a rate of 1 ml/min, and the content of the tank was stirred at a temperature of from 24° to 25°C while the overflowed wastewater treated in the tank for the dwelling time of 24 hours was continuously added into the subsequent activated sludge treating tank.

In the tank the treated water was stirred with aeration for a dwelling time of 72 hours, and then the treated wastewater was subjected to the natural sedimentation thereby separating the yeasts. The supernatant liquid was discharged, but due to the continuous treatment for 12 days COD of the treated water was 17 to 19 ppm.

High Calorie Solutions of Low Molecular Weight Glucose Polymer Mixtures for Intravenous Administration

Highlights of the Technological Achievement: High calorie solutions of low molecular weight glucose polymer mixtures useful for intravenous administration, preferably via a peripheral vein, are disclosed. Methods of preparing the solutions and their preferred rate of infusion at less than 5 mg/min/kg of patient body weight are also disclosed. The solutions can be administered concomitantly with amino acids, lipid emulsions, vitamins and electrolytes.

Reference: A.B. Ramsay, D.R. Luebke, D.T. Guzek, C.-L. Hsieh, R. Roteman and W.D. Leathem; U.S. Patent 4,182,756; January 8, 1980; assigned to Abbott Laboratories.

Background: Conventionally, aqueous solutions of glucose or fructose have been infused into patients intravenously to provide exogenous calories. Because of the osmotic relationship between human plasma and blood corpuscles, an important consideration of any solution parenterally administered to a patient is that it can cause damage to the patient's corpuscles if it is not substantially isotonic to human blood. If the patient's plasma becomes strongly hypertonic, the corpuscles shrivel. If it becomes strongly hypotonic, the corpuscles swell and may burst. Accordingly, the aqueous solutions of glucose or fructose most commonly intravenously infused into a patient are such isotonic 5% w/v solutions.

1 liter of 5% w/v glucose solution provides 50 g of glucose per each 1,000 ml of solution. Each gram of glucose provides 3.4 calories and, therefore, 1 liter of a 5% w/v glucose solution provides 170 calories per each 1,000 ml of solution. Generally, adult patients require a total of 2,000 to 5,000 calories per day. To satisfy that need, a patient would have to receive from 12 to 35 liters of 5% w/v glucose solution per day. Clearly, such an infusion of 12,000 to 35,000 ml of water would grossly overhydrate the patient.

Alternatively, hypertonic glucose or fructose 10 to 20% w/v solutions can and have been intravenously injected in order to provide the patient more calories. However, their infusion for extended periods of time is limited by the inability of the patient's peripheral veins to handle such hypertonic solutions without serious venous thrombosis or thrombophlebitis resulting.

It is well known that 25 to 50% w/v solutions of glucose or invert sugar have been intravenously administered to patients via a cannula inserted into the vena cava. While this technique provides a means by which the patient may intravenously receive high calorie solutions, it is not totally satisfactory in that insertion of the cannula is a surgical procedure not generally performed by a nurse, or many surgeons, and the dangerous possibility of septicemia or serious thrombosis in the vena cava, as well as potential damage of the heart and large vessels by the catheter tip exists.

A nontoxic metabolizable high calorie solution that can be infused into a patient via a peripheral vein without causing venous thrombosis, thrombophlebitis, or other undesired side effects is needed.

Process Conditions: If all the molecules of a 5% w/v glucose solution were individually replaced by glucose polymer molecules containing two glucose units (maltose), the osmotic pressure of the solution would remain the same, even though the weight of each molecule would have been doubled. Such a maltose solution would, therefore, constitute 100 g/1,000 ml, which is a 10% w/v solution.

Further, each molecule would provide two glucose units, so that if the human body can utilize intravenously administered maltose, such a maltose solution would provide twice as many calories as the 5% w/v glucose solution.

Low molecular weight glucose polymer mixtures which can be used to prepare substantially clear, nonpyrogenic, stable and sterile high calorie solutions suitable for administration to human patients, including diabetics, via a peripheral vein, are provided by this process.

The solutions can be isotonic to human blood at 20 to 50% w/v as the glucose polymer mixtures can have an average degree of polymerization ranging from 4 to 10. The high calorie solutions are suitable for admixture with amino acids, lipid emulsions, vitamins and electrolytes.

Maximum utilization of the glucose polymer mixtures has been found to occur at rates of infusion less than 5 mg/min/kg of patient body weight. Utilization of the glucose polymer mixtures is enhanced by concomitant subcutaneous injections of insulin for both diabetic and nondiabetic patients.

Preferred methods of preparing the glucose polymer are by either solvent extraction or molecular membrane filtration of certain commercially available dried cornstarch hydrolysates or their equivalents.

Nitrosugar

Highlights of the Technological Achievement: Disclosed are degradation products of the antibiotics rubradirin and rubradirin B and processes for their preparation.

Some of these products have antibacterial activity, and, thus, can be used in various environments to inhibit susceptible bacteria. Also, some of these products can be used as intermediates to make useful antibacterials.

Reference: H. Hoeksema; U.S. Patent 4,182,855; January 8, 1980; assigned to The Upjohn Company.

Background: The antibiotic rubradirin, and a microbiological process for preparing the same, are disclosed in U.S. Patent 3,335,057. The structure of rubradirin has been tentatively determined to be as follows:

(1)

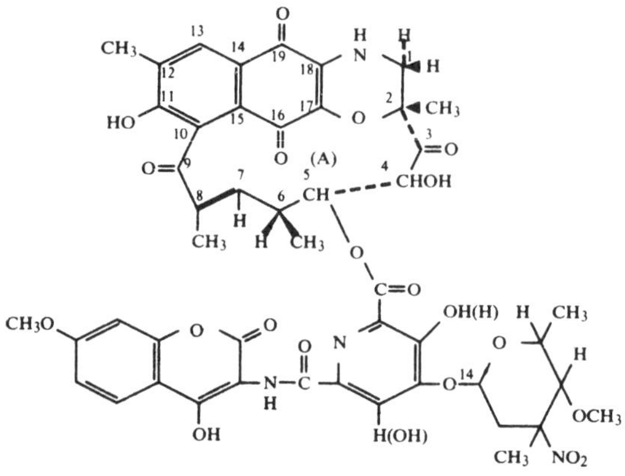

The structure of rubradirin B has been tentatively determined to be as follows:

(2)

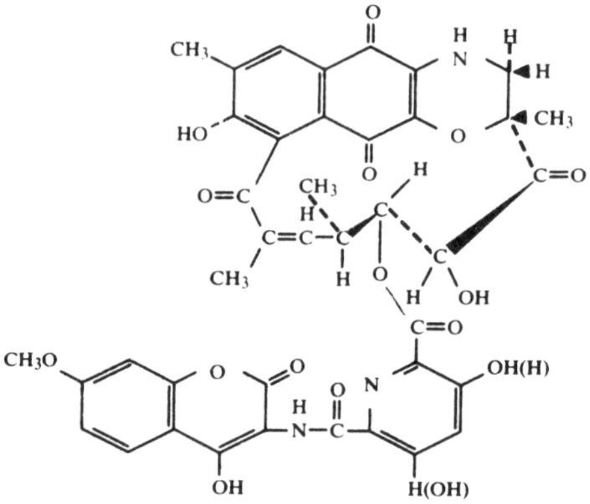

Process Conditions: Upon subjecting rubradirin to base hydrolysis there are obtained compounds which can be shown as follows:

(3)

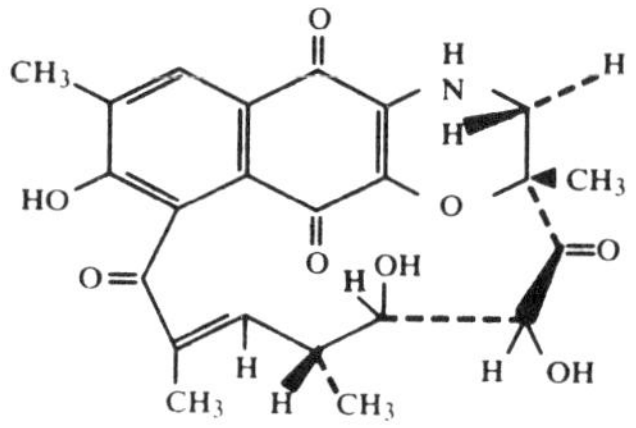

Rubransarol A

(4) 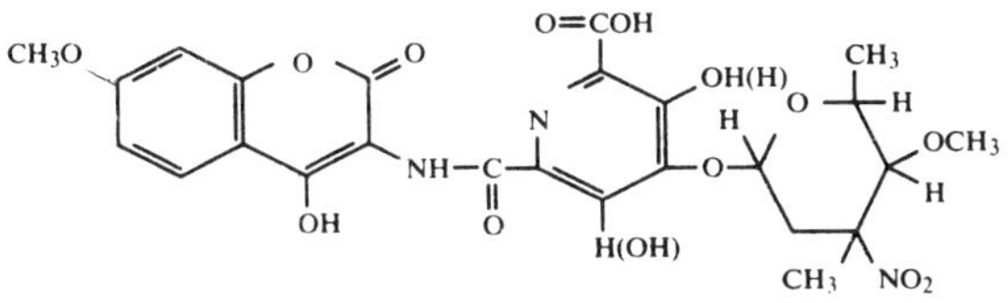

If rubradirin is reacted with an amine, e.g., aqueous methylamine, the amide of (4) is obtained.

Acid hydrolysis of rubradirin gives compounds which can be shown as follows:

(5) 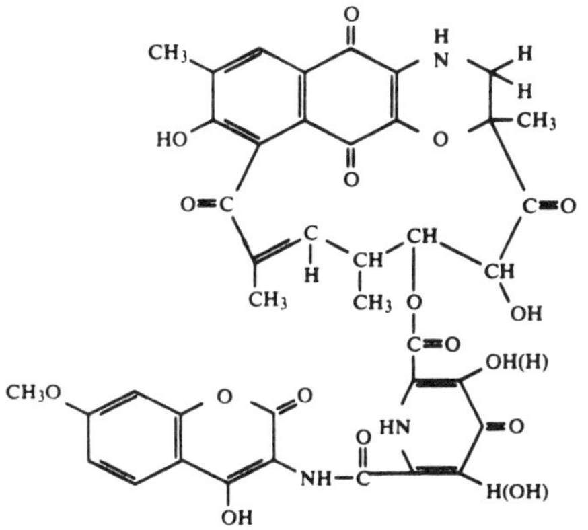

(6) 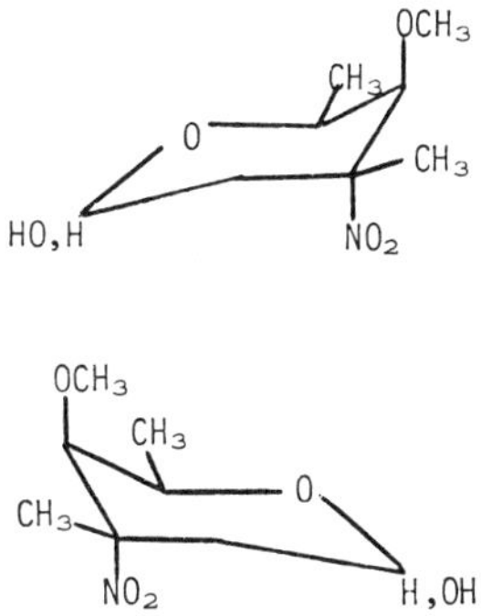

Upon reacting compound (5) with a saturated solution of ammonia and methanol, there are obtained compound (3) and also another compound which can be shown as follows:

(7) 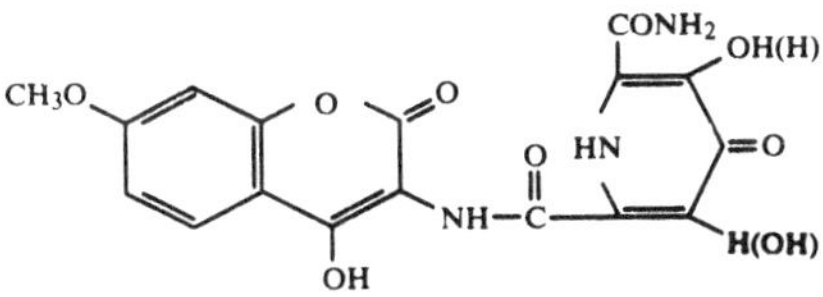

If compound (5) is reacted with primary amines such as ethylamine, the ethylamide corresponding to (7) is obtained.

Degradation of rubradirin in acetic anhydride and pyridine yields a compound which can be shown as follows:

(8) 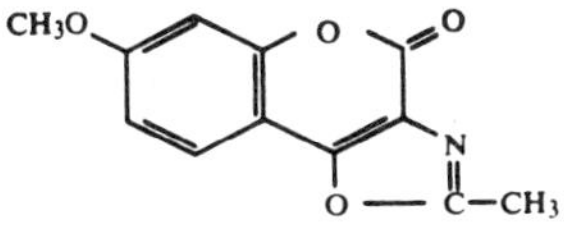

A similar compound having the structure:

(9) 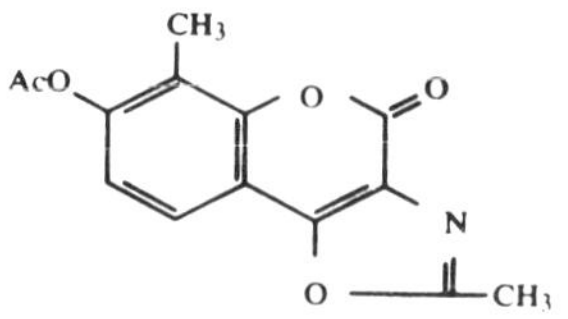

wherein Ac is acetyl is know in the prior art.

Base hydrolysis of rubradirin B gives compounds which can be shown as follows:

(10) 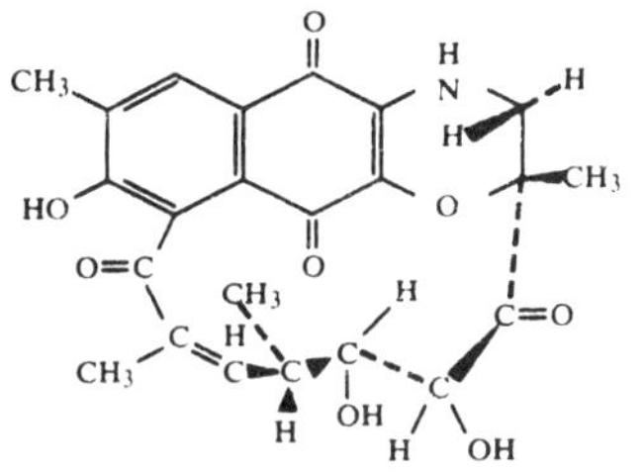

(11) 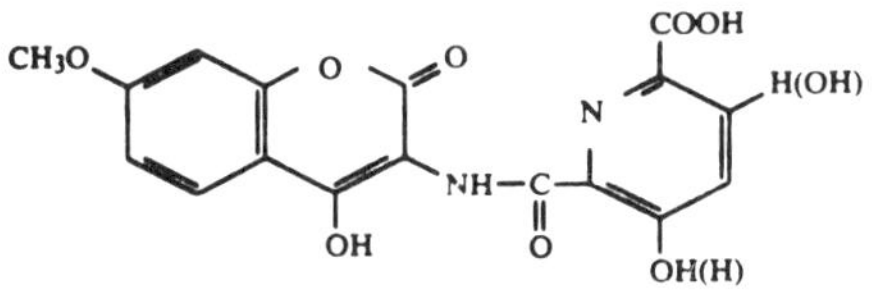

Example 1: *Degradation of Rubradirin in Acetic Anhydride and Pyridine: Preparation of Compound (8)* – A mixture of 10 g of rubradirin, 125 ml of pyridine, and 25 g of acetic anhydride is refluxed for 4 hours. It is then poured on ice and the resulting precipitate is removed by filtration and recrystallized from methanol twice to give 300 mg of white crystals of compound (8), mp 212° to 213°C. Analysis–Calculated for $C_{12}H_9NO_4$: C, 62.34%; H, 3.92%; N, 6.06%. Found: C, 62.49%; H, 3.78%; N, 6.05%.

Example 2: By substituting rubradirin B, or compound (4), or compound (5), or compound (7), or compound (11), for rubradirin in Example 1, there is obtained compound (8).

Starch or Starch Derivatives for Improved Paper Coating Compositions

Highlights of the Technological Achievement: The effectiveness of pigment coating compositions for printing paper stock is enhanced by the use of phosphate esters of aliphatic alcohols and/or their salts therein. The compositions consist of:

(A) 100 parts of a uniformly fine mesh kaolin clay pigment of paper coating grade (having less than 0.05% particles greater than 325 mesh);

(B) 5 to 30 parts of a binder consisting of:

  (1) 0 to 30 parts starch, oxidized starch, enzyme-converted starch, hydroxyethylated starch, hydroxypropylated starch and starch derivatives;

  (2) 0 to 14 parts of a synthetic latex binder;

  (3) 0 to 10 parts of water-soluble polymers of polyvinyl alcohol, protein, hydrolyzed protein, copolymers of maleic anhydride with butadiene, styrene, methyl vinyl ether and ethylene, polyacrylic acid, polymethacrylic acid; and water-soluble copolymers of vinyl carboxylic acids; and

(C) 0.3 to 20 parts of a gloss improver of phosphate esters of aliphatic alcohols.

Reference: F.E. Woodward; U.S. Patent 4,183,766; January 15, 1980.

Background: The phosphate esters of this process alter the surface and interfacial properties of liquids to an unusual degree even when these chemicals are present in small quantities, as manifested by energy changes in surface and interfacial tensions. As a result, they affect the dispersion of the coating ingredients, modify the rheological properties and afford a greater degree of control over the penetration and adhesiveness of the coating, thereby imparting properties to the coated sheet unobtainable without the addition of such phosphate esters.

Process Conditions: According to this process the effectiveness of existing paper coating compositions is greatly enhanced by the inclusion therein of certain aliphatic phosphate esters. These esters comprise the phosphate esters of aliphatic alcohols having a carbon atom content of 6 to 26. Aliphatic alcohols having 16 to 26 carbon atoms are particularly advantageous for preparing the esters for the purposes of this process. A mixture of alcohols of varying chain length may be used and, in fact, the usual, available commercial sources of raw materials will constitute such mixtures.

Alcohols that are highly effective for the purpose of this process include those produced by the Oxo process from monoolefinic straight chain hydrocarbons, e.g., Neodol (Shell Chemical Co.). These products have random chain lengths. The process may be optimized around any carbon length but in commercial practice a range from C_{12} to C_{15} alcohols are produced which have 20% branched (methyl) chain primary alcohols, the remainder being straight chain primary.

Alternatively, suitable alcohols may be produced by telomerization of ethylene with a Ziegler-type aluminum catalyst and with subsequent decomposition to fatty alcohols.

Reaction: The phosphate esters may be prepared by reacting an aliphatic alcohol, such as those referred to above, with an oxy acid of pentavalent phosphorus, such as orthophosphoric acid, pyrophosphoric acid, polyphosphoric acid, phosphorus pentoxide or mixtures of these. In general, the reaction is carried out by mixing the acid and alcohol while heating at 70° to 125°C. The reaction proceeds to completion within one-half to three hours.

Stabilization: Examples of products which are suitable particularly for stabilizing the alkaline earth salts of this process but which may also be beneficially used in other salts of this process are: sodium stearoyl-N-methyl taurate; sodium lauroyl isethionate; lauryl alcohol + 7 mols ethylene oxide; C_{11-15} straight chain alcohol ethoxylated with 60% by weight ethylene oxide; nonylphenol + 9 mols ethylene oxide; oleic acid + 15 mols of ethylene oxide; and polypropylene oxide (MW 600) + 40% ethylene oxide (by weight).

Example: Starch-clay paper coating formulations were prepared using the following formulation:

Parts (solids basis)	
100.6	Huber hydrasperse No. 2 70% dispersion
0.25	Tetrapotassium pyrophosphate
0.10	Sodium hydroxide
20.0	Dry-Flo 49-4951
1.0	Additive (solids basis) (or as indicated)
124.5	Water

The viscosities found were:

Additive	Concentration* (%)	Gel Strength (cp)	Plastic Viscosity (cp)
Calcium salt of PE** of stearyl alcohol	0.4	2	25
Calcium salt of PE of Alfol C_{20}	1.0	1.5	21
Calcium salt of PE of Epal C_{20}	1.0	1.5	18.5
Calcium salt of PE of C_{14}, C_{16}, C_{18} fatty alcohols	1.0	1.5	23.0
Calcium stearate	1.0	1.0	20.0

*100% active additive on weight of clay
**Phosphate ester

At higher concentrations of additive in the same clay-starch coating, the following viscosities were observed indicating that the gloss improver additives of this process have a pronounced tendency to improve the ratio of gel strength to plastic viscosity.

Additive	(%)	(Gel) Strength (cp)	Plastic Viscosity (cp)
PE of Epal C_{20}	5.0	18	58
PE of Alfol 18 (stearyl alcohol)	3.2	86	50
PE of Epal 1418 (C_{14}, C_{16}, C_{18})	5	38	18.5
PE of Alfol C_{20}	5	58	82
Calcium stearate	2.5	3	35

NEW TECHNOLOGIES

PROCESSING

Copolymerization in Water-Soluble Polyol with Controlled Reaction with the Polyol

Highlights of the Technological Achievement: Methods are disclosed for the production of a water dispersible solution of acidic copolymer in water soluble polyol with controlled limited reaction between the copolymer and the polyol.

Reference: K. Sekmakas and R. Shah; U.S. Patent 4,193,904; March 18, 1980; assigned to DeSoto, Inc.

Background: It is well-known to copolymerize monoethylenic monomers, including carboxyl-functional monomers, in volatile water-miscible organic solvent solution to produce acidic copolymers which can be dispersed in water with the aid of a base. The removal of the organic solvent is not practicable, so it carries over into the aqueous dispersion. When this dispersion is coated on a substrate and baked, the solvent evaporates. The result is unnecessary expense, the hazard of fire, and undesired pollution. The proportion of solvent is normally at least 50% of the weight of the copolymer and it is desired to reduce this to not more than about 20%, and preferably to eliminate volatile solvent entirely.

The difficulty with the prior technology is that the polymerization takes place in solution in the polyol which yields relatively low molecular weight copolymers. As a result, in the absence of extensive crosslinking with the aminoplast or phenoplast curing agent, the cured coatings are not as hard as desired. With more curing agent to obtain more extensive crosslinking and greater hardness, then brittleness becomes a factor. In either event, superior water resistance as is needed to maintain adhesion and resist blushing on elevated temperature steam treatment (as encountered in pasteurization) is hard to achieve.

These problems are illustrated in the coating of beer can exteriors which are being developed to provide a pigmented gloss coating which can be printed and used without overcoating.

Process Conditions: In this process monoethylenically unsaturated monomers including a proportion of carboxyl-functional monomer providing water dispersibility with the aid of a base and a monoethylenically unsaturated monomer carrying a reactive group selected from dicarboxylic acid anhydride functionality and N-methylol functionality in small proportion of from 0.5 to 5% of the total weight monomers, are copolymerized in solution in a water soluble polyol which comprises preferably an adduct of ethylene oxide with a bisphenol of the type taught in U.S. Patent 3,960,983. This copolymerization is conducted under conditions causing extensive reaction between the dicarboxylic acid anhydride functionality or the N-methylol functionality in the reactive monomer and the hydroxy groups in the polyol.

The monoethylenically unsaturated monomers other than the anhydride functional monomer, the N-methylol functional monomer, and the carboxyl-functional monomer, can be termed "nonreactive" which denotes that they do not react under the conditions of polymerization and subsequent cure.

Any water soluble polyhydric alcohol having a molecular weight in excess of 300 may be used such as polyethers and polyesters of polyhydric alcohol, such as ethylene glycol, propylene glycol, butylene glycol, glycerin, trimethylolpropane, or pentaerythritol. Ethylene oxide or propylene oxide adducts having a molecular weight appropriate for water solubility may be used alone or in combination to regulate the functionality in the final cured coating.

The copolymerization reaction including the N-methylol functional monomer, or the prereaction of this monomer with the diol is preferably carried out at a temperature of 125° to 140°C. This is higher than the customary copolymerization temperature which does not usually exceed 120°C. The product of reaction (water or alcohol) is removed to drive the reaction with the N-methylol group to substantial completion.

Example 1: Charge 1,410 pbw bisphenol A-ethylene oxide adduct (mol ratio 1:6)(XD 8025.01, American Cyanamid), set empty trap, and heat to 130°C. Premix 515 pbw styrene, 411 pbw isobutyl acrylate, 21 pbw isobutoxymethyl acrylamide, 110 pbw acrylic acid, 32 pbw benzoyl peroxide and 18.5 tert-butyl mercaptan and add over a 2½ hr period at 130°C. Collect butanol. Hold an additional hour at 130°C. Add 5 pbw tert-butyl perbenzoate and hold for 1 hr at 130°C. Add 5 pbw tert-butyl perbenzoate and hold for 1½ hr at 130°C, cool to 80°C. Add 155 pbw dimethylethanolamine and 300 pbw deionized water slowly over a 10 min period. Add 1,800 pbw deionized water slowly over a 15 min period to provide a product having the following characteristics: solids, 51.3%; viscosity, 7,000 cp; acid value (nonvolatiles), 36.6.

Example 2: Charge 1,410 pbw bisphenol A-ethylene oxide adduct (mol ratio 1:6) (see Example 1), heat to 110°C and set reflux condenser. Premix 515 pbw styrene, 432 pbw isobutyl acrylate, 110 pbw acrylic acid, 32 pbw benzoyl peroxide and 18.5 pbw tert-butyl mercaptan and add over a 2½ hr period at 110°C. After addition, hold for 1 hr at 110°C. Add 5 pbw tert-butyl perbenzoate and hold 1 hr at 110°C. Add 5 pbw tert-butyl perbenzoate and hold 1½ hr at 110°C, then cool to 80°C. Slowly add 155 pbw dimethylethanolamine and 300 pbw deionized water over a 10 min period. Slowly add 180 pbw deionized water over a 15 min period to provide a product having the following characteristics: solids, 51.5%; viscosity, 5,000 cp; and acid value (nonvolatiles), 35.9.

Comparing Example 1, which represents the process, with Example 2, which presents a control in which the N-methylol functional monomer is omitted, the viscosity in Example 1 is much higher than the viscosity in Example 2, indicating extensive coreaction between the N-methylol functionality and the hydroxyl group in the bisphenol-based polyol. When the aqueous dispersions of Examples 1 and 2 are coated and cured by baking, the Example 1 product was harder, possessed greater impact resistance, and greater resistance to blushing when subjected to beer pasteurization testing. More particularly, the pencil hardness was improved from 1H in Example 2 to 3H in Example 1.

Example 2 can be varied by replacing 20 parts of the acrylic acid with 20 parts of maleic anhydride, everything else remaining the same. The product will exhibit a higher viscosity and will cure to greater hardness.

Process and Apparatus for Mixing Viscous Liquids

Highlights of the Technological Achievement: A method and apparatus for fluidizing materials wherein a gas such as air is fed into conduits through which the materials to be fluidized flow under pressure, the air being fed into the material conducting conduits at angles in the range of 60° to 120° relative to the direction of material flow, the material conducting conduits being coupled with a third conduit for mixing the fluidized materials. The material conducting conduits are flushed by insertion in a container having means for effecting the flow of a solvent and a gas such as air therethrough.

Reference: H.F. Geise; U.S. Patent 4,186,256; January 29, 1980; assigned to Sealed Air Corporation.

Background: Prior known devices for fluidizing viscous chemicals which then expand to produce a foam-like structure have not been found to be satisfactory because of the failure to attain thorough intermixing of the chemicals to produce a uniform resultant foam and repeated clogging of the apparatus because of ineffective means for flushing with the result that the apparatus had to be frequently disassembled for cleaning.

This process has as one of its objects the provision of an improved method and apparatus which effectively fluidize the viscous materials to be mixed and thoroughly intermix the fluidized materials to produce a substantially homogenous and uniform resultant product.

In Figure 25.1 the two liquids to be fluidized and mixed are denoted by the letters **A** and **B** and fed in the direction of the arrows through conduits **10** and **11**. A fluidizing gas such as air is fed through conduit **12** which communicates with conduit **10** at a steep angle β which is preferably of the order of 60° to 120° relative to the flow of the liquid. A fluidizing gas is also fed through conduit **13** which communicates with conduit **11** at a steep angle β which is also preferably between 60° and 120°. The tubes or conduits **10** and **11** are angularly disposed at an angle α, generally in the range of 45° to 90°, and converge in and communicate with a mixing tube substantially larger in diameter than the tubes **10** and **11**.

In the illustrated embodiment of the process, the tubes **10** and **11** are approximately 3/16" i.d. while the tube **14** is approximately 3/4" i.d. The liquids **A** and

B, in the case of an isocyanate and polyol, may be fed into tubes **10** and **11** at the rate of about 0.03 to 0.04 cfm while air is introduced at the rate of about 2 cfm.

Figure 25.1: Cross-Sectional Diagrammatic View of the Mixing Apparatus

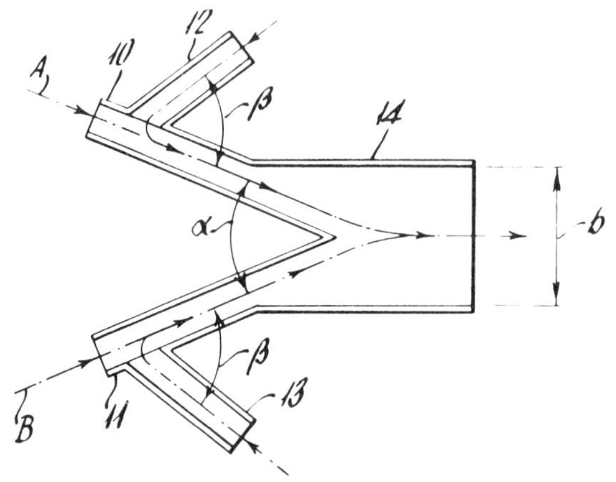

Source: U.S. Patent 4,186,256

It is evident that other ratios of liquid to air or gas may also be employed depending on the mode of operation and nature of the liquids.

With the structure thus far described, it has been found that substantially complete fluidization of each of the liquids is attained within the tubes **10** and **11** with the result that complete intermixture of the two liquids occurs within the tube **14**. This procedure when producing a polyurethane foam insures the production of a substantially uniform product and thus provides a more effective packaging material.

APPLICATIONS

Additives for Delaying the Release of Chemicals in Aqueous Fluids

Highlights of the Technological Achievement: Methods and additives are provided for bringing about the delayed release of a chemical such as a gel breaker or demulsifier in an aqueous fluid. The additives are pelletized solids comprised of the chemical to be released, a hydratable gelling agent and a breaker for the gel produced by the gelling agent when hydrated.

Reference: J.W. Burnham, J.E. Briscoe and E.A. Elphingstone; U.S. Patent 4,202,795; May 13, 1980; assigned to Halliburton Company.

Background: Heretofore, high viscosity gelled aqueous fluids and high viscosity water-hydrocarbon emulsions have been utilized as fracturing and fracture-acidizing fluids. Such high viscosity fluids are capable of suspending propping agent without excessive settling, bringing about the opening of one or more fractures in the formation to a greater width than is possible with low viscosity fluids and have good fluid loss properties as well as other desirable properties.

After a high viscosity aqueous gel or emulsion has been pumped into a subterranean formation and one or more fractures formed therein, it is desirable to convert the gel or emulsion into a low viscosity fluid, referred to herein as "breaking" the gel or emulsion, so that it can be recovered from the formation through the well bore. While a variety of chemicals which are added to high viscosity fracturing fluids have been utilized for breaking the fluids, hereinafter referred to as "breakers," problems caused by insufficient breaking or the breaking of the fluid too quickly are often experienced. Ideally, a high viscosity fracturing fluid or other well treating fluid has an initial high viscosity, stability during the well treatment and controlled breaking after the treatment.

Heretofore, the control of the breaking of gelled aqueous fluids, and particularly water-hydrocarbon emulsions, has been inadequate resulting in less than desirable treatment results and/or clean-up problems after the treatment has been carried out. For example, attempts to use demulsifiers for breaking water-hydrocarbon emulsions have generally been unsuccessful in that rapid or instantaneous breaking of the emulsion takes place when the demulsifier is added thereto decreasing the viscosity of the emulsion.

Process Conditions: By the process, methods and additives for delaying the release of chemicals such as demulsifiers, gel breakers and the like in aqueous fluids are provided whereby such release is accurately controlled and the desired results achieved. In accordance with the process, the chemical to be released in an aqueous fluid is combined with a solid hydratable gelling agent and a breaker for the gel formed by the gelling agent when hydrated. The mixture is then formed into prills or pellets, preferably having a size in the range of from about 20 to 40 mesh (U.S. Sieve Series).

Upon combining the pellets with an aqueous fluid into which the chemical is to be released, the gelling agent in the pellets hydrates and forms a protective gel around each of the pellets which prevents the release of the chemical into the aqueous fluid for the time period required for the protective gel to be broken by the gel breaker in the pellets. Once the gel breaker has broken the protective gel, the chemical in the pellets is released into the aqueous fluid. As will be understood, the time required for the protective gel to be broken can be varied by varying the quantities of hydratable gelling agent and gel breaker utilized in the pellets and by using different gelling agents and gel breakers.

Gelling Agents: A variety of solid hydratable gelling agents can be utilized in accordance with the process, such as hydratable polysaccharides, polyacrylamides, polyacrylamide copolymers and polyvinyl alcohol. Particularly preferred hydratable polysaccharides are galactomannan gums and derivatives thereof, glucomannan gums and derivatives thereof, and cellulose derivatives.

Examples of such compounds are guar gum, locust beam gum, karaya gum, sodium carboxymethylguar, hydroxyethylguar, sodium carboxymethylhydroxyethylguar, hydroxypropylguar, sodium carboxymethylhydroxymethylcellulose, sodium carboxymethylhydroxyethylcellulose, and hydroxyethylcellulose. The hydroxyethylcellulose derivatives used as gelling agents should be those having between 0.5 and 10 mols of ethylene oxide per anhydroglucose unit. The most preferred gelling agent for use in accordance with the process is hydroxypropylguar.

Breakers for the gels produced by the gelling agents mentioned above when hydrated which are preferred for use in accordance with the process are mild oxidizing agents, enzymes, acids and mixtures of such compounds. Examples of particularly suitable oxidizing agents are sodium persulfate and ammonium persulfate. Examples of suitable enzymes which can be utilized are alpha and beta amylases, amyloglucosidase, aligoglucosidase, invertase, maltase, cellulase and hemicellulase. An example of a suitable acid is fumaric acid. Of these, a mixture of sodium persulfate, hemicellulase and cellulase is most preferred.

Use: In carrying out the method of the process for breaking gelled aqueous fluids, a breaker for the gelled aqueous fluid is mixed with a solid hydratable gelling agent and a breaker for the gelling agent when hydrated and formed into pellets. As will be understood, depending on the particular gelled aqueous fluid to be broken, separate breakers for the gelled aqueous fluid and for the hydrated gelling agent can be included in the pellets or a single breaker can be included which breaks both the gelled aqueous fluid and the hydrated gelling agent.

In applications for hydraulic fracturing and/or fracture-acidizing fluids, the type of breaker used and the quantity of breaker included in the pellets is preferably such that complete breaking of the fluids takes place when pellets are combined with the fluids in an amount in the range of from 0.005% to 0.5% by weight of the fluid.

Example 1: In the laboratory, aqueous gel-hydrocarbon emulsions are prepared by first preparing gelled water by mixing tap water containing 1% by weight potassium chloride with guar gum in an amount of 50 lb/1,000 gal of water and sodium hydrogen phosphate in an amount of 10 lb/1,000 gal of water. The mixture is continuously stirred in a Waring blender and allowed to hydrate for 30 min. An emulsifying agent, i.e., alkyltrimethylammonium chloride (50 weight percent active in isopropyl alcohol/water/solvent) is added to the aqueous gel in an amount of 0.5% by volume while slowly stirring the aqueous gel in the Waring blender.

Kerosene is added to the aqueous gel in an amount of 2 times the volume of the aqueous gel and the mixture is pumped through a Jabsco self-priming pump Model 4270–½ inch standard for 1 min at a temperature of 80°F to form the emulsion.

An additive of the process is prepared by mixing hydroxypropylguar in an amount of 48.75% by weight of the additive, sodium lauryl sulfate in an amount of 48.75% by weight of the additive, hemicellulase enzyme in an amount of 0.5% by weight of the additive, and sodium persulfate in an amount of 2.0% by weight of the additive. The mixture is compacted and extruded through a 1/16" die. The extruded material is then cut into lengths of about 1/16" and screened to obtain pellets ranging in size from about 10 to 18 mesh (U.S. Sieve Series).

Portions of the aqueous gel-kerosene emulsion are placed in a modified heating cup with a pulley driven propeller for stirring, and the pelletized additive combined therewith in the amounts given in Table 1 below. The apparent viscosities of the emulsion at the temperatures indicated are measured with a Model 35 Fann viscometer at the time intervals indicated to determine when the emulsion is broken by the additive.

The Model 35 Fann viscometer is equipped with a No. 1 spring, standard bob and sleeve and is operated at 300 rpm. The results of these tests are given in Table 1 below.

Table 1

Quantity of Additive* (lb/1,000 gal)	Emulsion Temp. (°F)	Elapsed Time, min / Apparent Viscosity, cp: 0	5	10	15	30	60	90	120	24**
12	80	180	167	163	153	142	139	142	125	7
7	130	195	165	148	137	113	101	96	83	2
6.5	180	185	147	131.5	118	86.5	68	62.3	57	2

*Combined with emulsion.
**24 hours; viscosity of aqueous gel phase from broken emulsion.

Example 2: The procedure of Example 1 is repeated except that emulsions are formed using the quantities of kerosene given in Table 2 below at a temperature of 130°F.

Table 2

Quantity of Additive* (lb/1,000 gal)	Kerosene Used in Emulsion (vol %)	Elapsed Time, min / Apparent Viscosity, cp: 0	5	10	15	30	60	90	120	24**
9	60	140	121	114	110	98.5	86	79	74	2
5.5	75	233	190	169	154	144	135	125	120	2

*Combined with emulsion.
**24 hours; viscosity of aqueous gel phase from broken emulsion.

Example 3: The procedure described in Example 1 is repeated except that the emulsifying agent used in forming the emulsion is alkyltrimethylammonium chloride containing a lower active concentration in the solvent. The results of these tests are given in Table 3.

Table 3

Quantity of Additive* (lb/1,000 gal)	Emulsion Temp. (°F)	Elapsed Time, min / Apparent Viscosity, cp: 0	5	10	15	30	60	90	120	24**
6.5	80	150	141	140.5	139.5	137	133.5	132.5	127	10
4.0	130	150	127	123	118	106	92	43	43	–
3.5	180	150	122	110	109	74	60	51	10	5

*Combined with emulsion.
**24 hours; viscosity of aqueous gel phase from broken emulsion.

From the results of the tests shown in Tables 1, 2, and 3, it can be seen that the pelletized additives of the process effectively bring about the delayed breaking of aqueous gel-kerosene emulsions at various temperatures.

Etch-Resist Colloid

Highlights of the Technological Achievement: A radiation-sensitive material is described that comprises a hydrophilic colloid layer containing a dispersed phase of at least one radiation sensitive polymer the polymer chain of which comprises units with side substituents containing oxime ester groups.

Reference: H.V. Van Goethem, M. Stroobants and W.F. De Winter; U.S. Patent 4,202,697; May 13, 1980; assigned to Agfa-Gevaert NV, Belgium.

Background: Various kinds of etch resists are known. In the preparation of printing forms especially the preparation of gravure printing forms, the etch resist controls image-wise the action of the etchant so that the surface of the underlying printing member is modified image-wise so as to obtain more or less depressed areas.

The etch resist controlling the action of the etchant on the underlying printing member may consist, e.g., of a hardened relief pattern. The thinner the pattern, the more intense the etching, in other words the deeper the etchant attacks the element to be etched.

The overwhelming part of gravure printing is performed with the aid of carbon tissue or pigment paper. Carbon tissue or pigment paper comprises a gelatin layer, which is made sensitive to ultraviolet light by means of an aqueous solution of alkali- or ammonium dichromate. After careful drying, the so-called "dichromated gelatin" layer is exposed, e.g., through a continuous tone image and a gravure screen. The gelatin hardens in conformity with this exposure. The carbon tissue is then adhered to the printing surface, while pressure and moisture are applied.

By the action of water the gelatin layer swells and firmly adheres to the printing surface, which had been hydrophilized beforehand. Subsequently, the unexposed and thus unhardened gelatin is washed away with warm water, whereas the printing element remains covered with an exposed and hardened etch resist forming a gelatin relief pattern. At this stage an etchant can be applied. The action of such etchant on the underlying printing element is proportional to the thickness of the relief pattern.

Because of the dependency on many variables it is necessary to simplify the preparation of gravure printing forms. It has been proposed therefore to use more consistent products and methods.

Process Conditions: The process thus provides a method for image-wise modifying the surface of an element by making an etchant diffuse through a layer applied to the surface, this layer forming an etch resist with image-wise differentiations in permeability for an etchant. The etch resist is formed by the steps of image-wise exposing to actinic radiation a layer of a hydrophilic colloid binder containing, in the absence of any ethylenically unsaturated monomeric material, a dispersed phase of at least one radiation-sensitive polymer, the polymer chain of which comprises units with side-substituents containing oxime ester groups. By the exposure to actinic radiation the permeability of the layer for an etchant is reduced in conformity with the exposure.

The phenomena taking place in the conduct of the process are believed to be the following. The hydrophilic binder medium wherein a polymer, whose polymer chain comprises units with side-substituents containing oxime ester groups, is heterogeneously dispersed, constitutes the medium, through which an etchant, e.g., an Fe(III) chloride solution can diffuse in consequence of the capillary behavior of the medium. Upon exposure to actinic radiation the oxime ester units in the polymer form radicals which crosslink with the hydrophilic binder medium so that the capillarity of the layer is reduced and thus the permeability of the layer for a hydrophilic etchant is reduced.

By the term oxime ester group is meant an O-acyloxime group of the formula >C=N–O–CO–R wherein R is an organic group.

Particularly suitable polymeric compounds having side-substituents comprising an oxime ester group are polymers obtained by (1) the reaction of a monooxime of an α,β-diketone, e.g., 1-phenyl-1,2-propanedione-2-oxime, 1-(p-hydroxyphenyl)-1,2-propanedione-2-oxime, p-hydroxyphenylglyoxalaldoxime and 2,3-butanedione monooxime with an ethylenically unsaturated acid chloride, e.g., acryloyl or methacryloyl chloride and polymerization or copolymerization of the thus formed monomer carrying an oxime ester group, or (2) the reaction of an ethylenically unsaturated compound comprising an α,β-diketomonooxime, with a carboxylic acid chloride which includes an aliphatic, aromatic or heterocyclic acid chloride. These polymers are thus polymers or copolymers of monomers corresponding to the formula

$$M^1COON{=}\underset{\displaystyle R}{\underset{|}{C}}{-}COM^2$$

wherein one of M^1 and M^2 is or comprises an ethylenically unsaturated group and the other is an aliphatic, aromatic or heterocyclic group, e.g., alkyl including substituted alkyl or aryl including substituted aryl and R represents hydrogen, alkyl including substituted alkyl or aryl including substituted aryl.

Particularly suitable polymers are those comprising recurring units corresponding to one of the following general formulas (I) and (II):

(I)

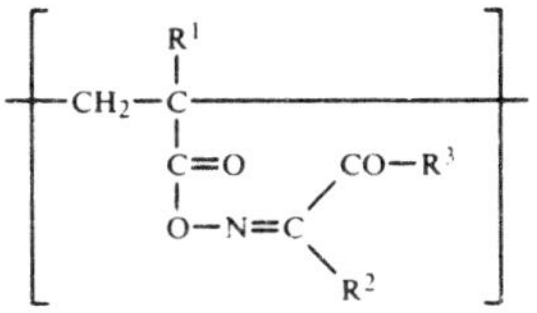

(II)

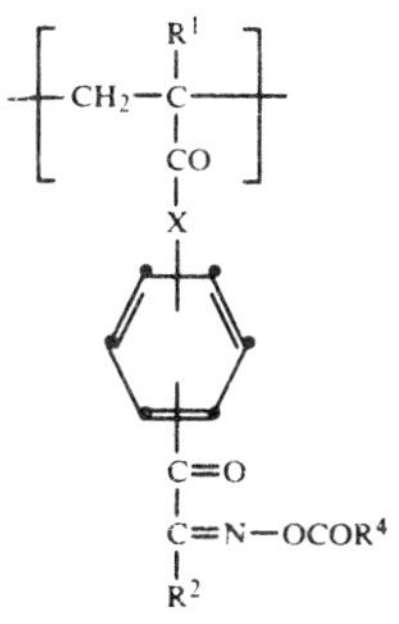

wherein:

R^1 is hydrogen or C_1–C_5 alkyl, e.g., methyl,

R^2 represents hydrogen, alkyl including substituted alkyl or aryl including substituted aryl,

R^3 represents alkyl including substituted alkyl, aryl including substituted alkyl, aryl including substituted aryl,

X represents –NH– or –O–, and

R^4 represents an aliphatic, aromatic, or heterocyclic group.

The polymers used according to the process can be homopolymers or copolymers. They are preferably copolymers comprising besides the units with oxime ester group in the side chain units derived from a variety of monomers, whereof preferably at least one is hydrophobic. Examples of comonomers are (meth)acrylamide, N-alkyl(meth)acrylamide, alkyl(meth)acrylate, styrene, acrylonitrile, N-vinylpyrrolidone, vinyl acetate and other vinyl esters. Preferred copolymers comprise from 5 to 50 mol percent recurring units with oxime ester groups in the side chain.

Representative examples of monomers from which the polymers used according to the process can be derived are:

(1)

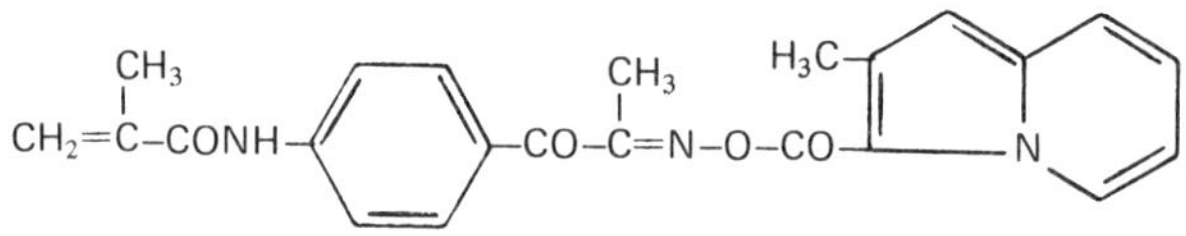

(2)

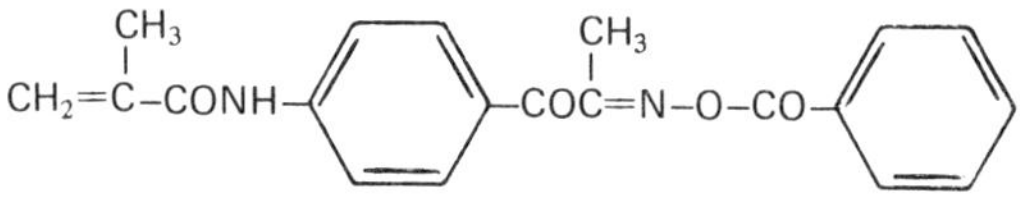

(3)

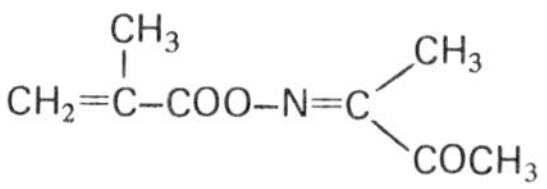

(4)

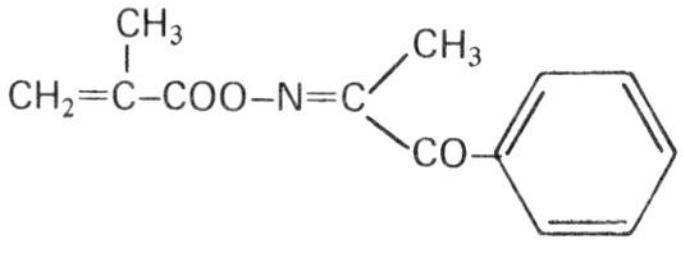

(5)

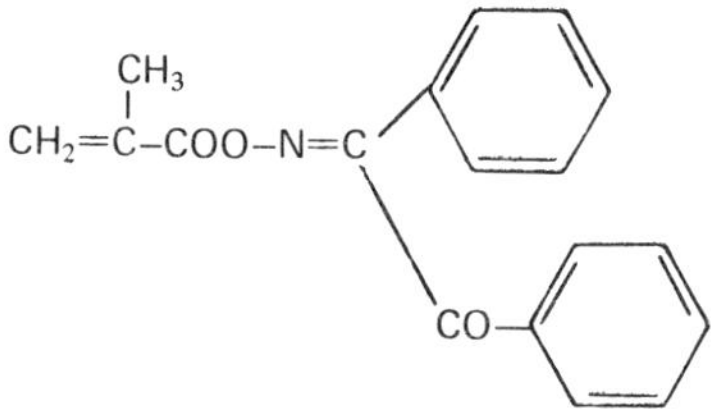

Aqueous Preparations for Treating Textile Substrates Comprising Regenerated Cellulose

Highlights of the Technological Achievement: Disclosed is a treatment process for textile substrates comprising or consisting of regenerated cellulose, which process comprises applying to the substrate an aqueous medium, subsequently drying the substrate and subjecting it to a temperature at which crosslinking takes place.

Reference: P. Komminoth, T. Robinson, M. Urosevic; U.S. Patent 4,200,564; April 29, 1980; assigned to Sandoz Ltd., Switzerland.

Background: The process relates to a treatment process for textile substrates comprising or consisting of regenerated cellulose, which process comprises applying to the substrate an aqueous medium comprising (A) a monomeric, hydrolysis stable, hydro-soluble, resin forming crosslinking agent containing at least two N-methylol or N-alkoxymethyl groups, (B) a hydro-soluble, prepolymerized, liner, filler resin forming, crosslinking agent, also containing at least two N-methylol or N-alkoxymethyl groups, (C) a reactive acetal of formula

$$\begin{array}{l} XO{-}(CH_2)_2{-}O{-}(CH_2)_2{-}O{-}CH_2{-}O{-}(CH_2)_2{-}O{-}(CH_2)_2{-}O \diagdown \\ \qquad\qquad\qquad\qquad\qquad\qquad\qquad\qquad\qquad\qquad\qquad\qquad\qquad CH_2 \\ YO{-}(CH_2)_2{-}O{-}(CH_2)_2{-}O{-}CH_2{-}O{-}(CH_2)_2{-}O{-}(CH_2)_2{-}O \diagup \end{array}$$

wherein X and Y, independently, are H or CH_2OH, (D) a first crosslinking catalyst, being an alkaline earth metal salt of a strong acid and (E) a second crosslinking catalyst, being an acidic aluminum salt, subsequently drying the substrate and subjecting it to a temperature at which crosslinking takes place.

Process Conditions: The hydrolysis stable N-methylol compounds employed as component (A) are the N,N'-dimethylol and N,N'-dialkoxymethyl derivatives of 4,5-dihydroxy or 4,5-dimethoxyethylene urea or of 4-methoxy-5,5-dimethylpropylene urea, and the N,N-dimethylol and N,N-dialkoxymethyl derivatives of carbamates, especially of ethyl or methyl carbamate.

The preferred compounds for use as component (B) are water soluble linear precondensates of N-methylol or N-alkoxymethyl derivatives of urea or melamine of a chain length sufficiently short to enable penetration into at least partially swollen regenerated cellulose fibers.

The preferred compounds for use as component (D) are magnesium and calcium chloride and sulfate, the magnesium compounds being especially preferred, particularly magnesium chloride. Suitable weight ratios of components A:B:C:D:E are 10–80:5–40:10–80:2.5–20:0.5–4.

Use: In the process of the method, the aqueous medium may be applied to the substrate in conventional manner, suitably the substrate being impregnated, e.g., using padding techniques. The pick-up is generally of the order of most preferably from 85 to 90%, depending on the method of application, the nature of the substrate, e.g., the amount of nonregenerated cellulose therein, and the concentration of the crosslinking agents in the medium. After application of medium, usually effected at room temperature, the substrate is dried and cured, i.e., crosslinking is caused to take place. Where drying and curing steps are carried out separately, the drying is at 70° to 120°C, the curing generally at 130° to 180°C, the latter step generally taking from 2 to 8 min, except where rapid curing techniques, at 180° to 200°C, are employed where from 30 to 60 sec only are required. Where the drying and curing steps are carried out simultaneously, as is often appropriate with lightweight fabrics, such is generally carried out at from 160° to 200°C, suitably on a tenter.

The substrate may consist solely of regenerated cellulose or comprise regenerated cellulose in blend form with other fibers, both natural and synthetic, e.g., cotton or polyester, and may be in any conventional, preferably textile, form, e.g., in fiber, thread, woven or other fabric form. The blend substrates preferably contain at least 20% by weight of regenerated cellulose.

By the method of the process, which is a dry crosslinking process, the crease recovery of the substrate, both wet and dry, is improved as is its resistance to swelling in neutral and alkaline aqueous media. The hydroelasticity thereof is also reduced. Of particular interest, however, is the wet crease resistance of the substrate and the relative permanence thereof even after several subsequent washes.

Anionic Paper Surface Sizing Agent

Highlights of the Technological Achievement: Paper surface sizing agents based on aqueous solutions of copolymers of maleic acid anhydride and dicyclopentadiene, which are further reacted after preparation to produce the half amides with ammonia, or primary or secondary aliphatic amines.

Reference: K. Peterlein, F. Schulde, K.M. Diedrich, and V.W. Kulisch; U.S. Patent 4,200,559; April 29, 1980; assigned to Chemische Werke Huls AG, Germany.

Background: It is generally known to size paper in the pulp form by use of rosin soaps and aluminum ions. Besides the known sizing agents based on rosin derivatives other solutions or dispersions of polymer compositions have been utilized for the sizing of paper. The advantage of such synthetic agents arises from the fact that they can also be brought to the already finished paper surface, while a normal resin glue with aluminum ions precipitates on the paper fibers.

The utilization of synthetic polymer sizing agents allows paper sizing without addition of metal salts.

It is known from DT-OS 1621693 that copolymers containing succinic acid imide groups can be used for the sizing of paper surfaces, wherein the copolymers contain vinylogous compounds as comonomers. As a possible vinylogous compound is mentioned, among others dicyclopentadiene. In order to make these copolymers soluble for the paper sizing, they have to be reacted with ammonia, bases or amines. This treatment with base, causes the formation of a variety of undesired derivatives because of the imide group content of the copolymer.

By the abovementioned method, it is necessary in order to form the imide groups, either to transform maleic acid anhydride into maleic acid imide or to form the imide groups from a copolymer composed of maleic acid anhydride and vinylogous compounds. Because of industrial interest it is desirable to simplify the preparation of the paper sizing agents containing copolymers and also based on a need for better reproducibility, to keep the number of the abovementioned derivatives obtained upon solution, under strict control.

Process Conditions: The statistical copolymers of the process can be obtained by a known method, in that the polymerization is carried out by suspension or solution polymerization in aliphatic or aromatic hydrocarbons, in alcohols or in ethers, at temperatures preferably between 70° and 90°C, in the presence of radical initiators. As initiators it is possible to use azo compounds, peroxides or hydroperoxides, such as, for example, azo bisisobutyronitrile, dibenzoyl peroxide or di-tert-butyl peroxide. The initiators are used preferentially in amounts of from 2 to 6% by weight of the comonomers.

Solubilizing: In order to make the copolymers of the process, which contain anhydride groups, soluble in water, it is necessary to transform them with bases into their salts. It is, according to this process, not necessary to neutralize all carboxyl groups, since in general a neutralization degree of about 80%–relative to the total carboxylic acid content–is sufficient. As cations, it is possible to use ammonium and/or mono-, di- or trialkylammonium with up to 6 carbon atoms as well as alkali metal ions. It is therefore possible to use the copolymers of the process in the form of their reaction products with ammonia or primary aliphatic or aromatic amines, as for example, ethylamine or aniline, or secondary amines as for example diethylamine, morpholine or pyrrolidine and/or aliphatic alcohols.

Paper Sizing: The quantity of copolymers of the process necessary for the paper sizing is preferably between 0.1 and 1% by weight, relative to the paper weight.

The solid contents of sizing agent to starch was 1.5:8.5. This liquor was applied by means of a labor sizing press onto a sample of 80 g of unsized paper, so that the amount of sizing agent corresponds to 0.6% of the paper weight. The drying of the paper in the sizing press followed at 140° to 160°C in 5 sec. The obtained sizing values are described in the table below.

	Sizing Agent			
	I	II	III	None
Water absorption (DIN 53 132),				
sec	30	10	<1	<1
g/m^2	20	12	10	89
Sizing degree (DIN 53 145), sec	$>$1,800	1,160	157	<1
Absorption height in 10 min				
(DIN 35 106), mm	3	<1	8	44
Surface strength (Dennison test)	18	16	14	11

Aqueous Fortified Rosin Dispersions

Highlights of the Technological Achievement: An inversion process for preparing paper size from fortified rosin wherein a water-soluble dispersing agent including an ethylene oxide sulfate group is used to provide a stable aqueous dispersion of fortified rosin; and an invert size produced by such a process.

Reference: R.M. Hughes and H. Warnecke; U.S. Patent 4,199,369; April 22, 1980; assigned to Hercules Incorporated.

Process Conditions: In accordance with this process there is provided a method for preparing aqueous dispersions of fortified rosin by the inversion process wherein the use of protective colloids, such as casein, heretofore used in the inversion process is not essential.

In one broad form the process provides in the inversion process of preparing paper size from fortified rosin that the molten fortified rosin and water containing a dispersing agent are first mixed to provide an emulsion wherein the molten fortified rosin is the continuous phase and the water is the disperse phase and the emulsion is subsequently inverted by the addition of water to provide an emulsion wherein the water is the continuous phase and the fortified rosin is the disperse phase. The improvement consists of employing as the dispersing agent one or more water-soluble compounds of the formulas

(1) 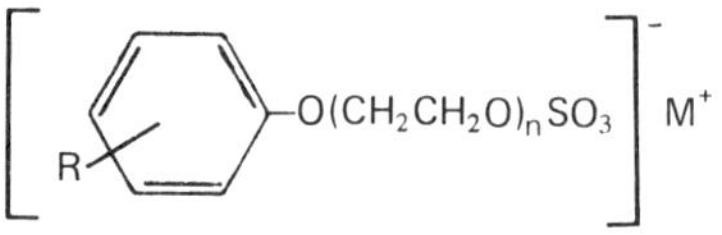

(2) $[R'{-}O(CH_2CH_2O)_nSO_3]^-\ M^+$

In formula (1), R is a normal or branched chain alkyl containing from 4 through 18 carbon atoms. In formula (2) R' is a normal or branched chain alkyl or alkenyl or fused ring cycloalkyl containing from 10 through 20 carbon atoms. In formulas (1) and (2), n is an integer such that of the molecular weight of each anion, from about 27 to 75% thereof is attributable to the (CH_2CH_2O) group and M^+ is a cation, as hereinafter defined.

Aqueous Dispersion: The aqueous dispersion prepared by the method of this process will comprise, by weight, from 5 to 70% fortified rosin or fortified rosin blend, preferably from 10 to 55%; from 1 to 10% dispersing agent, preferably from 2 to 8%, based on the fortified rosin or fortified rosin blend; and the balance water to 100%.

The fortified rosin employed is the adduct reaction product of rosin and an acidic compound containing the

$$>C{=}\underset{|}{C}{-}\underset{|}{C}{=}O$$

group and is derived by reacting rosin and the acidic compound at elevated temperatures, usually from 150° to 210°C.

The amount of acidic compound employed will be that amount which will provide fortified rosin containing preferably 4 to 9%, by weight, of adducted acidic compound based on the weight of the fortified rosin.

Examples of acidic compounds containing the

$$>C{=}\underset{|}{C}{-}\underset{|}{C}{=}O$$

group that can be used to prepare the fortified rosin include the alpha, beta-unsaturated organic acids and their available anhydrides.

In formulas (1) and (2), M^+ is any suitable monovalent cation, other than the hydrogen ion, that will allow substantial dissolution of the dispersing agent in water and will remain in solution under the conditions employed in preparation of the water-in-oil emulsion.

Preferably, in formulas (1) and (2), M^+ is an alkali metal ion, ammonium, dimethylammonium, morpholinium, etc.

It has been determined that dispersions of this process have good stability at a pH of 3.0 to 5.5. Adjustment of the pH can be made by adding an alkali, preferably sodium hydroxide, to the inversion water or to the aqueous solution of dispersing agent in an amount to bring the aqueous dispersion to the desired pH.

The initial temperature of the fused resin (fortified rosin or fortified rosin blend) will be in the range of 90° to 160°C. If the process is carried out at atmospheric pressure and the initial rosin temperature is above 100°C, the temperature will drop on addition of the aqueous solution of dispersing agent due to partial evaporation of the water.

The process can also be carried out in two stages by feeding the molten resin and the hot water solution of the dispersing agent to a mixer to provide the water-in-resin dispersion. Subsequently, the water-in-resin dispersion and hot inversion water are fed, in separate streams, to a second mixer maintained at a temperature of 70° to 150°C where the inversion takes place. The resulting aqueous dispersion is then recovered from the second mixer.

Example 1: This example illustrates the preparation of a fortified rosin-unfortified rosin blend. 92.3 parts of Chinese gum rosin and 7.7 parts of furmaric acid were heated in a reaction vessel at 200°C for 3½ hr. Some of the reaction mass, 20 parts, was removed from the reaction vessel and then 20 parts of unfortified Chinese gum rosin was added. Analysis of the resulting fortified rosin-unfortified rosin blend shows 6.2% combined fumaric acid, 0.175% free fumaric acid, and an acid value of 208.

Examples 2 through 18: In these examples, the following ingredients were employed: 360 g fortified rosin-unfortified rosin blend of Example 1; 115 g 13.2% aqueous solution of dispersing agent (15.2 g of dispersing agent and 100 parts deionized water); 539.6 g deionized water containing NaOH-inversion water (539.3 parts deionized water and 0.3 g NaOH).

The fortified rosin-unfortified rosin blend was heated in stirred autoclave to 170°C over a period of 80 minutes to provide a molten mass which was subsequently cooled to 135°C. At 150°C, stirring of the molten mass is commenced and continued throughout the process. The aqueous solution of dispersing agent was forced into the autoclave under 20 psi nitrogen. As a water-in-resin emulsion is formed, the temperature dropped to 125° to 127°C and the pressure in the autoclave rose to 15 psi. The autoclave pressure was then reduced to 5 to 8 psi by blowing out the nitrogen and some steam.

The inversion water, heated to a pressure of 25 psi (115°C to 118°C) was forced into the autoclave within 3 min of the aqueous solution of dispersing agent. The resin-in-water emulsion was cooled to 40°C. Formalin (0.2%) was added to the resulting dispersion to prevent bacterial or fungal growth. The dispersion was then strained through 200 mesh nylon. Table 1 shows 7 day stability results using formula (1) dispersing agent and Table 2 shows 7 day stability results using formula (2) dispersing agent.

Table 1: Formula (1) Dispersing Agent

Ex. No.	R	n	CH_2CH_2O Groups in Anion (%)	M	Total Solids (%)	Settled Solids After 7 Days (%)
2	nonyl	3	30.5	NH_4^+	37.4	0.5
3	nonyl	5	42.3	NH_4^+	38.0	6.3
4	nonyl	9	56.9	NH_4^+	36.9	0.0
5	nonyl	10	59.5	NH_4^+	37.7	0.5
6	nonyl	11	61.7	NH_4^+	36.7	1.0
7	nonyl	12	63.8	NH_4^+	36.9	0.5
8	nonyl	13	65.6	NH_4^+	36.8	1.3
9	nonyl	15	68.7	NH_4^+	36.9	1.9
10	octyl	5	43.5	NH_4^+	36.1	1.7
11	octyl	7	51.9	NH_4^+	37.3	1.3
12	dodecyl	9	53.6	NH_4^+	36.2	0.3
13	nonyl	5	42.3	Na^+	37.0	0.0
14	nonyl	20	74.6	Na^+	36.9	1.3
15	nonyl	20	74.6	NH_4^+	36.5	8.5
16	octyl	13	66.7	NH_4^+	37.8	27.2

Table 2: Formula (2) Dispersing Agent

Ex. No.	R'	n	CH_2CH_2O Groups in Anion (%)	M	Total Solids (%)	Settled Solids After 7 Days (%)
17	lauryl	16	72.6	NH_4^+	37.4	2.9
18	cetyl/oleyl	10	56.7	NH_4^+	37.2	1.6

Compositions Which Gel at Body Temperature for Drug Delivery to Mucous Membranes

Highlights of the Technological Achievement: Pharmaceutical vehicles for delivery of pharmacologically active chemical materials to mucous membranes, as well as pharmaceutically active compositions containing such vehicles, are provided. The pharmaceutical vehicles are clear, water-miscible, physiologically-acceptable, liquid compositions which gel to a thickened, nonflowing and adhering consistency at human body temperature.

Reference: J.Z. Krezanoski; U.S. Patent 4,188,373; February 12, 1980; assigned to Cooper Laboratories, Inc.

Background: Aqueous vehicles which are thickened by the addition of selected gums or cellulose-derived viscosity building agents are perhaps the most commonly used media for delivery of drugs or medicaments to mucous membranes. Generally, the viscosity of such preparation ranges from 25 cp to indeterminate values in stiff gels. Nearly uniform drug delivery is possible with such vehicles, and they frequently provide desirable protection to the mucous membranes.

In contrast, nonviscous aqueous suspensions have many disadvantages and are are not typically used. A major problem is rapid settling of the suspended drug. This gives rise to undesirable need for continuous stirring during administration in order to deliver a uniform dose.

While thick gels would seem to offer the best potential in terms of protection as well as holding and delivering medication, they in fact have some disadvantages. In some instances, they are difficult to apply from their respective commercial containers. Moreover, thick gels do not spread readily over the area being treated, and possibly painful spreading and rubbing may be necessary. Also, on evaporation of the water from the vehicle, a cosmetically unappealing hard granular or flaky residue often results at the site of the application.

Attempts to use oily vehicles to increase drug delivery and prolong ensuing pharmacologic action have not met with uniform success.

Another approach to the delivery of drugs or medicaments to mucous membranes is the recent development of silicone plastic devices which deliver drugs at predetermined, nearly uniform, zero order rates extending from a few days to several years. However, the usefulness of such devices depends upon a constant supply of tissue fluid or glandular secretion; in the absence of fluid, plastic devices are not operative. Such devices are not designed to offer any protection to an inflamed mucous membrane. Discomfort often associated with the devices, and inadvertent loss of the devices, are additional problems.

The existence of all these disparate approaches to drug delivery to mucous membranes evidences the need for new pharmaceutical vehicles. Against the background of this array of formulations and devices with all their attendant problems, the process fills that need.

Process Conditions: Pharmaceutical vehicles in accordance with the process are liquid at ambient temperatures preferably 25°C and below, have a sol-gel transition temperature of preferably 29° to 31°C and a pH of 6.2 to 7.8.

In accordance with the process, it has been discovered that aqueous solutions of certain polyoxyethylene-polyoxypropylene block copolymers are useful pharmaceutical vehicles having the properties set forth above. In particular, the process provides a pharmaceutical vehicle or base for carrying a pharmaceutically active material, i.e., a drug or medicament, which comprises:

(a) preferably from 17 to 26%, by weight of a polyoxyethylene-polyoxypropylene block copolymer in which the number of polyoxyethylene units is at least preferably 70%, of the total number of monomeric units in the total molecule, the block copolymer having an average molecular weight of preferably 11,500; a room temperature solubility in water of greater than 10 g/100 ml of water, and a cloud point in 1% aqueous solution of at least 100°C; and

(b) from 74 to 90% by weight water,

the vehicle having a sol-gel or gel transition temperature in the range of preferably from 29° to 31°C.

Water-Restricted, Water-Soluble Paint

Highlights of the Technological Achievement: A paint composition which can be applied from a spray can and can be cleaned with water. The composition comprises a water-soluble, neutralized resin film former contained in a water-soluble solvent, but has a maximum water content of 10 weight percent. Formulations can be provided to obtain coatings ranging from flat to high gloss.

Reference: J.D. Howard; U.S. Patent 4,187,204; February 5, 1980; assigned to Pactra Industries, Inc.

Background: The process provides a paint formulation which can be dispensed in aerosol form, but which is water-receptive and which can be readily cleaned up with water. Formulations can be provided in accordance with this process to obtain coatings ranging from flat to high gloss. Specifically, the formulations contain no water, or only minimal amounts of water, thereby eliminating can corrosion and stability problems, and permitting spraying from an aerosol can, yet the paint when applied can be cleaned up with water. After drying, the resultant coating is quite water resistant. Importantly, the foregoing environmental and health hazards are completely eliminated.

Process Conditions: A paint composition is provided comprising a water-soluble, neutralized resin film former in a water-soluble solvent. The film former is obtained by reaction of a resin containing carboxyl groups with an amine or ammonia as a basic neutralizer to form a salt. The resin contains sufficient carboxyl groups to be solubilized by the neutralizer as hereinafter described in more detail. Suitable resins for neutralizations are maleinized oils, oil modified alkyds, copolymer alkyds, oil-free alkyds, acrylic solutions, copolymer oils containing carboxyl groups, urethane polymers and epoxy esters.

Significantly, the paint composition is formulated with no water or with only a minimal amount of water, more specifically with a maximum content of 8 weight percent. Pigment can be added to provide any desired color and in an amount sufficient to provide hiding power. Wetting agents and antisettling agents can be used in conjunction with pigments to help disperse the pigments and maintain them in a dispersed state.

Dryers can be added to accelerate oxidation of the coating to facilitate drying and slip additives can be added so to impart mar resistance to the dried coating. Antirust inhibitors can be added as desired as can perfumes. The formulations are characterized by a high gloss, but a flattening agent, such as silica, can be incorporated to provide a flat coating.

The paint formulations can be applied with a brush, but find their greatest advantage in being dispensed in aerosol form. In this regard, the usual aerosol-type system comprises a pressure-tight spray can having a valved opening and containing a liquid paint composition and a volatile propellant for spraying the liquid composition through the valve opening.

Referring to Figure 25.2, the process finds its most advantageous use in being dispensed in aerosol form. To accomplish such dispensing, a pressure-tight spray can **10** of usual construction is provided. The can **10** incorporates a tubular shell **12** having an opened upper end mounted to the inside surface of the can top **14**. The bottom wall **16** of the shell **12** is provided with a central nipple **18** depending from the wall **16** and defining a central aperture **20**.

A syphon tube **22** has its upper end forced over the nipple **18** and extends to a position at the bottom of the container **10**. A coiled compression spring **24** is seated in the shell **12** and exerts a constant force against the bottom flange **26** of a floating valve **28** urging it upwardly toward sealing engagement with a gasket **30**. The floating valve **28** is slidable within an opening in the top can wall **14** and is formed with a channel **32** leading from the shell **12** to a nozzle opening **34**.

Figure 25.2: Pressure-Tight Spray Container for Dispensing Paint Compositions in Aerosol Form

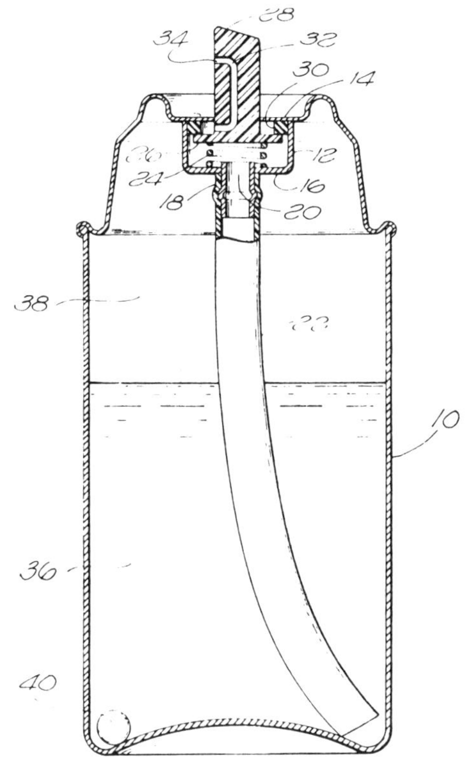

Source: U.S. Patent 4,187,204

The spray can **10** contains a liquid mixture **36** of paint composition and aerosol fluid, a region **38** above the liquid mixture being saturated with aerosol gas. As known, a steel ball **40** is placed in the mixture to facilitate mixing of paint and aerosol fluid. In accordance with usual practice, upon depressing of the valve **28**, the channel **32** is opened to the outside. The mixture **36** of paint and aerosol fluid is forced by pressure inside the can through the channel **32** and out of the nozzle **34**. As the aerosol fluid in the mixture reaches the atmosphere, it bursts into a gas turning the paint composition into a fine spray.

Examples 1 through 3: For purposes of comparison of the effect of water, clear paint compositions were prepared using the ingredients listed in the amounts by percentage weight indicated below:

	Example 1	Example 2	Example 3
		(percent)	
Aquamac 1200	23.29	23.29	23.29
Kelsol 3902	11.71	11.71	11.71
Ethylene glycol n-butyl ether	5.12	5.12	5.12
Propylene glycol methyl ether	5.12	5.12	5.12
Triethylamine	1.62	1.62	1.62
Acetone	47.38	42.53	37.70
Byk 301	0.40	0.40	0.40
Cobalt naphthenate, 6%	0.40	0.40	0.40
Activ 8	0.12	0.12	0.12
Water	4.84	9.69	14.52

The Aquamac 1200 is a urethane resin (International Minerals and Chemical Corporation). Kelsol 3902 is an alkyd resin (Spencer Kellogg Company). The ethylene glycol n-butyl ether, propylene glycol ether and acetone are water-soluble solvents, and the triethylamine is a neutralizer for the resin. Byk 301 is a slip additive (Byk Mallinckrodt Co.) The 6% solution of cobalt naphthenate is a drier and Activ 8 is a drier (R.T. Vanderbilt Co.). Water is present in the solutions to illustrate the effect of the various levels on sprayability of the composition and these effects will be described further below.

In preparing the foregoing formulations, the Kelsol 3902 can be dissolved in the ethylene glycol n-butyl ether and propylene glycol methyl ether and reacted with about one half of the triethylamine. In a separate vessel, the Aquamac 1200 can be dissolved in the acetone and reacted with the remaining portion of the triethylamine. The contents of the two vessels can then be combined and the remaining ingredients can then be added. The result in each case is a clear paint.

In order to demonstrate the effect of the water content on sprayability, each paint composition was placed in a loadable spray can along with half its volume of methylene chloride. A valve was inserted and each spray can was pressurized with a hydrocarbon propellant consisting of an equal volume mixture of propane and isobutane, until an initial pressure of 45 psig was obtained.

Under these conditions, amounts of hydrocarbon propellant approximately equal to the methylene chloride were charged to the spray cans. The methylene chloride, although liquid, also acts as a propellant in this mixture. By such means, spray cans containing paint compositions of Examples 1, 2 and 3 were obtained. Upon spraying of a metal panel, application of the composition of Example 1 was found to be excellent with no foam. Application of the composition of Example 2 was found to be from fair to good with some foam but an acceptable level for most applications. Application of the composition of Example 3 was found to be very poor with an unacceptable level of foam.

Mixed-Form Polyhalide Resins for Disinfecting Water

Highlights of the Technological Achievement: A strongly basic anion-exchange resin loaded with iodine and a less than stoichiometic amount of an iodide salt or bromine and a less than stoichiometric amount of bromide salt is an effective demand bactericide for disinfecting water. The bactericidal resin according to the process elutes only small amounts of halide ions and is especially suited for killing bacteria in feed water supplies having relatively high salt content.

Reference: G.L. Hatch; U.S. Patent 4,190,529; February 26, 1980; assigned to Aqua-Chem, Inc. (Other Reference: U.S. Patent 4,187,183).

Background: Several techniques are known for killing bacteria in water. One such method includes treating the water with halogens, but because this method leaves large residual concentrations of halogens in the treated water, its usefulness has been limited to the treatment of swimming pools and the like. Another method involves the use of polyhalide resin bactericides of one single form or another. For example, bromine can be eluted from a strong base anion-exchange resin at concentrations ranging from 10 to 10,000 ppm. Residual bromine is captured by a scavenger resin to reduce the final bromine level to 1.0 ppm or below. Another water disinfecting technique involves the use of an anion-exchange resin fully loaded with triiodide (I_3^-) ions.

The latter technique, while being useful for a wide variety of different feed waters, has one drawback when the feed solution contains relatively high salt concentrations, i.e., 250 ppm as sodium chloride or greater. While little free iodine is eluted from the resins, relatively large quantities of iodide ions are eluted. A resin which elutes reduced levels of iodide ions under these conditions would be a significant improvement in this water treatment technology.

Process Conditions: In accordance with one preferred form of the process a strong base anion-exchange resin, preferably in chloride form, is reacted with a mixture of iodine and a preselected amount of iodide salt to form a mixed-form polyhalide resin product. For purposes of the following description, iodine and iodide salt will be discussed, but bromine or an appropriate interhalogen may be substituted for iodine, and a bromide salt may be substituted for the iodide salt to produce other mixed-form resins within the scope of the process.

A number of commercially available strong base anion-exchange resins are useful as the starting material for preparation of the resin of the process including those containing strongly basic (cationic) groups such as quaternary ammonium groups, tertiary sulfonium groups, quaternary phosphonium groups or alkyl pyridinium groups.

The resin chosen for use in the process is measured on a wet volume basis to determine its ionic loading capacity and is transferred to a suitable container with a minimum volume of water. In a separate container, a preselected amount of crystalline iodine is mixed with a preselected amount of potassium or other water soluble iodide salt, and just enough water to dissolve the reactants. The amount of iodine and potassium iodide is selected according to either Table 1 or Table 2, which will be discussed below.

The iodine and iodide salt are mixed until the ingredients are completely dissolved. The resultant mixture consists of triiodide (I_3^-) ions and the superpolyiodide, I_5^-. The resulting mixture is added to the anion-exchange resin in chloride form and mixed therewith for a time sufficient to permit complete absorption of the iodine and iodide ions on the exchange sites of the resin, for example, 24 to 72 hr.

The anion forms shown below are believed to be predominant species when A, B, Y and Z = I for which n = 1 and m = 2. The data contained in Table 1 was derived from basic stability data for the iodide anions. A similar table could be prepared for the bromide or interhalogen anions using appropriate stability data.

Table 1

Resin:I_2:KI Ratio	A (I_3^-), %	B (I_5^-), %
1:1.2:1	80	20
1:1.1:1	90	10
1:1.0:0.9	80	10
1:1.0:0.8	60	20
1:1.0:0.7	40	30

The anions of the mixed-form polyhalide resin have the general formula:

$$a(A_{2n}Y)^- + b(B_{2m}Z)^-$$

wherein:

- $0 < (a \text{ and } b) < 1$;
- $a + b \leqslant$ total resin sites available for exchange;
- A and B = I or Br;
- Y and Z = Cl, Br or I;
- if A = B and Y = Z, then m cannot equal n;
- n and m are whole numbers in excess of 0.

As seen in Table 1, the stoichiometric ratio of I_2:KI is always greater than 1, so that the resultant bactericidal resin is of mixed form. The actual value of a and b depends on not only the resin:X_2:KX ratio but also on the relative stabilities of $\text{resin}^+\text{–}(A_{2n}Y)^-$ and $\text{resin}^+\text{–}(B_{2m}Z)^-$. For instance, in Table 1 for resin:I_2:KI = 1.0:1.0:0.8 the values for a and b are believed to be approximately 0.6 and 0.2, respectively. This belief is based on experimental evidence which indicates that the stability of

$$[\text{Resin}^+\text{–}(I_3^-)] > [\text{Resin}^+\text{–}I_5^-)] >>> [\text{Resin}^+\text{–}(I_2Cl^-)]$$

the latter resulting from reaction of I_2 with the original chloride form resin. As can be visualized, then, the resin can consist of three or more species of polyhalide anions. However, the satisfaction of stability requirements usually leads to the formation of two predominant species of which the resin is expected to be comprised and which results in the formation of only small equilibrium amounts of other polyhalide anions.

In Table 2 the release of I_2 and exchange of I^- may vary slightly depending on the base resin employed and may even vary for the same resin from batch to batch. For example, other measurements on different batches of one mixed-form resin having a resin:I_2:KI ratio of 1.0:1.0:0.8 yielded I_2 elution quantities of from 1.45 to 2.08 ppm and I^- exchange of from 0.55 to 0.66 ppm.

Table 2: Effect of Resin:I_2:KI Equivalent Ratio on Iodine Release*

Resin:I_2:KI Ratio	I_2 Released Initially	I_2 Released After 34 Bed Volumes of 2 M NaCl Wash	I^- Exchanged Initially	I^- Exchanged After 34 Bed Volumes of 2 M NaCl Wash
	(ppm)			
1.0:1.2:1.0	1.97	–	0.80	–
1.0:1.1:1.0	0.79	–	0.68	–

(continued)

Table 2: (Continued)

Resin:I_2:KI Ratio	I_2 Released Initially	I_2 Released After 34 Bed Volumes of 2 M NaCl Wash	I^- Exchanged Initially	I^- Exchanged After 34 Bed Volumes of 2 M NaCl Wash
	(ppm)			
1.0:1.0:0.9	0.82	0.99	3.78	0.5
1.0:1.0:0.8	1.23	1.34	1.90	0.27
1.0:1.0:0.7	1.42	1.56	0.21	0.0

*100 ml mixed-form resin bed; 160 ml/min flow rate.

It can be seen from Table 2 that the quantity of elemental iodine and iodide ions eluted from the resin of the process varies significantly depending on the Resin:I_2:KI stoichiometric ratio. Table 2 was prepared using a 412 ppm salt (NaCl) solution as the feed liquid to simulate extreme feed water conditions. One preferred Resin:I_2:KI ratio is 1.0:1.0:0.8 because the resin product elutes only approximately 1.34 ppm elemental iodine.

Table 2 shows that the properties of the resins prepared according to the method described above may be further modified by washing with 34 bed volumes (approximately one gallon) of 2 M sodium chloride solution. It is believed that the additional washings with concentrated salt solution remove exchangeable iodide ions from the resin, replacing them with chloride ions from the salt solution. This in effect decreases the I^-:Resin ratio which then allows an increase in the elution of elemental iodine.

Throughout the discussion of the process reference has been made to decreasing the halide elution, even though, as shown in Table 2, the halogen concentration is increased. The latter is permissible because the halogen released from the resins can be readily removed by a scavenger resin, the preferred ones of which are unloaded strong base anion-exchange resins as described above. These resins are much more efficient in removing halogens than the corresponding halide ions.

The bactericidal efficiency of the resin with a Resin:I_2:KI ratio of 1.0:1.0:0.8 is shown in Table 3. All effluents were negative for coliform bacteria as indicated by both the plate count and fermentation tube test methods.

Table 3

Test	Resin Volume Mix-Form Resin (ml)	Resin Volume Scavenger (ml)	Flow Rate (BV/min)	*E. Coli* Inoculum (ml)	*E. Coli* Effluent (ml)
1	100	–	1.0	3.0×10^5	1/100
2	150	75	1.3	3.0×10^5	1/100
3	100	–	1.6	3.5×10^5	1/100
4	100	–	1.6	1.5×10^6	1/100
5	50	–	3.7	1.8×10^5	1/100
6	50	–	3.7	3.3×10^5	1/100

Further research into the treatment of water according to the process has led to the discovery that when using the mixed-form resins having a KI:Resin ratio of less than one, especially with low salt feed solutions, a significant and possibly desirable amount of hypoiodous acid (HOI) is produced in the treatment system (Table 4).

Table 4: Hypoiodous Acid Production from Resin with Resin:I_2:KI Ratio of 1.0:1.0:0.8

Treatment System: Mixed-Form Resin (ml)	Scavenger Resin (ml)	Charcoal (ml)	Influent Water	Flow Rate (BV/min)	Effluent Analysis: HOI as I^+ (ppm)	I_2 (ppm)
150	–	–	deionized water	1.33	1.8	1.0
150	–	–	deionized water	4.67	2.2	1.4
150	75	–	deionized water	4.67	1.6	–
200	260	–	250 ppm NaCl	4.0	0.22	–
200	260	60	250 ppm NaCl	4.0	0.12	–

The proposed reaction scheme offers an explanation of the hypoiodous acid production:

(1) $Resin^+{-}I_5^- \rightarrow Resin^+{-}I_3^- + I_2$

(2) $I_2 + H_2O \rightleftharpoons H^+ + I^- + HOI$

(3) $I_2 + I^- \rightleftharpoons I_3^-$

(4) $Resin^+{-}Cl^- + I_3^- \rightarrow Resin^+{-}I_3^- + Cl^-$

the net result being the formation of $2\,Resin^+{-}I_3^- + H^+ + Cl^- + HOI$.

With $Resin^+{-}Cl^-$ having a high affinity for I_3^-, reaction (3) is shifted far to the right. With excess I_2 produced by reaction (1), and with reaction (3) consuming I^-, reaction (2) is shifted far to the right which results in excessive production of hypoiodous acid.

The hypoiodous acid is an effective virus killer, approximately four times more effective than molecular iodine itself, but if the final use of the treated water precludes even small concentrations of such acid, a substantial percentage of it may be effectively removed by employing activated charcoal as a third stage in the water treatment system.

Resins: The resins according to the process may be employed not only for such large scale jobs such as sterilizing water for swimming pools, but may also be used in either industrial or residential devices for purifying water for human consumption. The process solves a problem apparently not recognized in the prior art, i.e., the reduction of undesirable halide concentrations.

Photopolymerizable Latex Systems

Highlights of the Technological Achievement: A photopolymerizable system is provided as a latex system that is an oil-in-water polymer emulsion including particulates of water-insoluble polymers in combination with a water-soluble or water-dispersible, light-sensitive polymer having light-sensitive moieties.

Reference: T.M. Muzyczko and D.C. Thomas; U.S. Patent 4,186,069; Jan. 29, 1980; assigned to Richardson Graphics Company.

Background: It has been recognized that photopolymerizable systems having good durability when cured are composed of water-insoluble polymers. Heretofore, when using water-insoluble polymers in producing light-sensitive coatings for use on substrates such as planographic printing plates, these polymers are incorporated as part of a solvent system layer characterized by including organic materials for dissolving the polymer, which organic materials must be driven off of the plate onto which a light-sensitive coating layer is placed, this requirement being time and energy consuming, wasteful of the solvents, and hazardous to the operator and the environment.

Thus, there is need to provide photopolymerizable systems which do not require an organic solvent medium, preferably systems that are water-based latex systems.

Systems have been proposed in which water-soluble or water-dispersible polymers are used along with a water-soluble light-sensitive compound. Such systems avoid the disadvantages of the organic-based systems, but they are not particularly long-lasting because the polymers are water-soluble and thus subject to erosion of the image area when in use and/or through the action of atmospheric moisture. In such systems, since they are water-based, it was previously believed that the polymer must be water-soluble or at least water-dispersible in order to achieve the necessary degree of internal compatibility within the system.

In one such proposal, inherently water-insoluble polyvinyl acetate polymer is used in combination with water-soluble polyvinyl alcohol polymer so as to render the polyvinyl acetate polymer water-dispersible and thus compatible with the water-based system. Even after curing this type of system, the water-dispersible polymers are susceptible to inadvertent removal by ambient moisture.

Accordingly, it would be advantageous to provide photopolymerizable systems which are water-based but are not easily water-degraded because of the water-dispersible nature of the image-forming polymer.

Process Conditions: The process is a light-sensitive coating that is an emulsion mixture of a polymer emulsion component having water insoluble polymer particulates within an aqueous suspension medium together with a water-soluble or water-dispersible light-sensitive component including a polymer having at least two moieties per molecule that are sensitive to actinic radiation for generation of a free radical or other crosslinking species to form a matrix of crosslinked polymer having encapsulated or entrapped therein and dispersed therethrough the water-insoluble polymer. Prior to curing, the system is characterized as a three-phase precursor emulsion including the water-insoluble polymer particulates within the aqueous suspension medium with which is also mixed the light-sensitive polymer.

When the coated composition is applied, the aqueous suspension is substantially removed to realize a two-phase mixture of the polymer particulates and the light-sensitive polymer. When this coating is subjected to actinic radiation, the light-sensitive polymer cures or crosslinks to thereby insolubilize same and at the same time form a matrix in which the water-insoluble polymer particulates are entrapped and of which they can be a part. This provides a coating which is easily applied as a low-viscosity, water-based material but which, after cure, is especially resistent to deterioration upon exposure to water-containing materials.

In the system of this process, the water-insoluble polymer emulsion component and the light-sensitive component are present in quantities of the same order of magnitude. For example, they may be present in the emulsion mixture at a weight percent ratio on the order of 10:1 to 1:10, preferably 3:1 to 1:3.

The water-soluble, light-sensitive polymer present in the system prior to curing includes a multiplicity of functional groups per molecule which can react upon exposure to actinic radiation and thereby crosslink with itself and/or with other reactive polymers present, which may include the water-insoluble polymer emulsion particle itself that may or may not be also light-sensitive.

Water-Soluble Polymers: The water-soluble or water-dispersible, light-sensitive polymer included in the system of the process is one that will be in aqueous solution within the water phase of the system and also will be rendered insoluble or substantially insoluble upon exposure to actinic radiation. This polymer should contain at least two reactive moieties per molecule of the polymer, which reactive moieties are of the type capable of generating free radicals or other active species upon exposure to radiation.

The free radicals or other active species cause curing or crosslinking of the light-sensitive polymer with itself, with other light-sensitive polymers, or with a mixture of the same or other light-sensitive polymers and other polymers which can react upon being subjected to free radicals or other active species.

The light-sensitive polymers are in the form of dimers, trimers, tetramers, oligomers, etc., or they may alternatively be in the form of high molecular weight polymers having pendant reactive groups and in which the backbone contributes to producing a large-chain molecule. The molecular weights of such polymers can range up to one million, or even several million.

They may be, for example, chloral, bromal, water solubilized haloalkyl compounds such as ethylene chlorohydrin; dimethylaminoethylphenyl ketone or salt thereof; polyesters containing unsaturated groups and having a high acid number of salts thereof; alphaketo acids both aliphatic and aromatic and of compound or polymeric formation; xanthone carboxylic acid; water soluble acrylates such as hydroxyethyl acrylate, hydroxypropyl acrylate; diacetone acrylamides; the salt of diacrylated dimethylolpropionic acid; maleic anhydride copolymers reacted with hydroxy, thio or amino alkyl acrylates or with amino, thio or hydroxy allylic mono- or multifunctional compounds.

Inhibiting the Foaming of Aqueous Solutions Containing a Polymeric Dye

Highlights of the Technological Achievement: Foaming of aqueous solutions containing a polymeric dye can be inhibited by use of glycerol monooleate, glycerol dioleate and mixtures thereof.

Reference: T.E. Furia; U.S. Patent 4,185,122; January 22, 1980; assigned to Dynapol.

Background: Polymeric dyes have been recently proposed as satisfactory replacements for FD&C Red No. 2, which can no longer be used in foods, drugs and cosmetics in the United States. When polymeric colors are used in edibles, if the size of the molecules of polymeric color exceeds a certain limit—usually a molecular size of from about 1,000 to 2,000 daltons—and if the color compounds do not break down and thus maintain this size, the polymeric colors are not absorbed through the walls of the gastrointestinal tract. This means that when such materials are consumed, they essentially pass directly through the gastrointestinal tract. They are not taken into the body or its systemic circulation and thus any risk of possible systemic toxicity is eliminated. An excellent red polymeric colorant, which shows promise as an especially desirable replacement for existing red food colors such as FD&C Red No. 2, is an anthrapyridone known as Poly R-481. Poly R-481 is an acetylated version of a colorant having the structural formula:

(1)

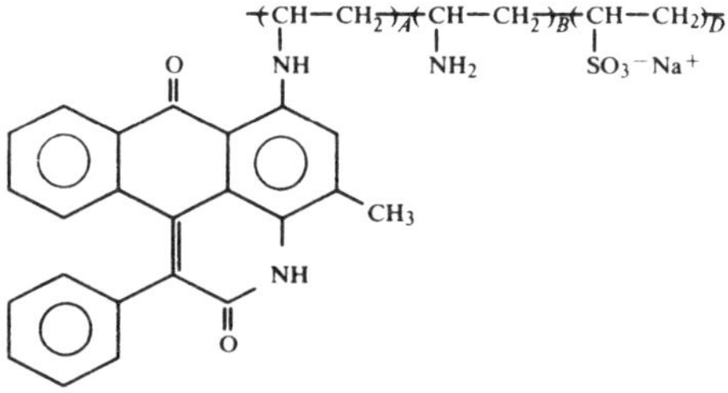

wherein about 95% of the free backbone amines have been converted to

$$-NH-CO-CH_3$$

groups and wherein A + B + D equals about 900 (A equals about 180, B equals about 360 and D equals about 360). Equivalent to Poly R-481 for the purposes of this process are the corresponding polymeric colorants wherein about 80 to 98% of the free backbone amines have been acetylated.

Poly R-481 can be prepared by a reaction sequence which begins with 1-amino-2-methyl-4-bromoanthraquinone. That material is reacted with phenylacetyl chloride, suitably in an aprotic organic solvent such as toluene, at an elevated temperature. The resultant addition product of the formula:

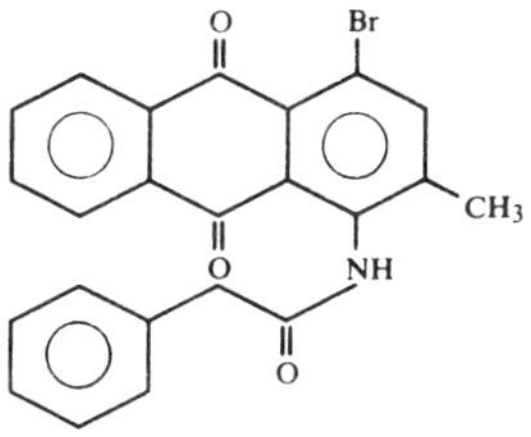

is then cyclized with base to afford a compound of the formula:

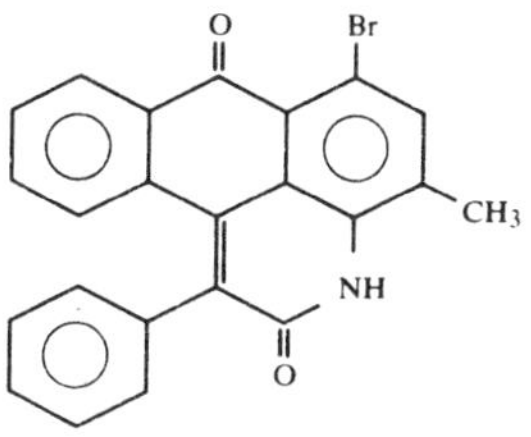

which is then reacted with vinylamine-vinylsulfonate copolymer. This amine displacement is an adaption of the classic Ullmann reaction and is carried out in water or mixed water/organic solvent and base and with a copper catalyst such as Cu_2Cl_2. The resultant polymer of formula (1) is then acetylated by reacting it with acetic anhydride in aqueous base. The red solution thus obtained yields, after lyophilization, a solid red product corresponding to formula (1) except that about 95% of the free backbone amines have been converted to

$$-NH-CO-CH_3.$$

Process Conditions: It has been found that the foaming of aqueous solutions, particularly carbonated beverages, containing the polymeric dye Poly R-481 can be prevented or eliminated by the addition of glycerol monooleate, glycerol dioleate or a mixture thereof to such solutions.

Glycerol monooleate (GMO) and glycerol mono- and dioleate (GMDO) are food-grade additives usually considered to be emulsifiers and solvents. They have been found to be surprisingly effective, at low concentrations, when used as defoamers for aqueous solutions of Poly R-481. The fact that other food grade additives, including those generally considered to be food grade defoamers, have been found to be ineffective or to actually increase foaming in aqueous solutions containing Poly R-481.

All water-soluble additives were dissolved directly in the makeup water used to dilute the 3,000 ppm Poly R-481 stock solution to 1,500 ppm. Water-insoluble or water-dispersible additives were incorporated via a 50/50 blend (w/w) in ethanol or, when applicable, into a portion of finished carbonated beverage containing essential flavoring oils.

In the tests described below, foam height was measured as a function of time by filling a 250 ml graduated glass cylinder, fitted with a ground glass stopper, with 20 ml of test solution. The test solution and a control were simultaneously agitated manually for 25 vertical strokes after which the initial foam height and its decay as a function of time were measured in ml.

In the examples, the GMDO used was a commercially available mixture of mono- and diglycerides of oleic acid containing 47 to 50% alpha monoglyceride, with a total monoglyceride content of 54 to 59%. It was a light amber clear liquid (< 1 Gardner unit) at 75°F, with an HLB value (±1) of 2.8, an iodine value of 74 to 78 and a viscosity at 75°F of 150 cp.

Example 1: The foam-inhibiting characteristics of various food grade additives in solutions containing 1,500 ppm Poly R-481 were compared. The results are tabulated below.

Additive, 100 ppm	.. Foam Height, ml .. Initial	After 30 Seconds
None	15	10
Glycerine	15	8
Propylene glycol	20	18
Polyethylene glycol 600	20	16
Polysorbate 20 (polyoxyethylene 20 sorbitan monolaurate)	25	20

(continued)

Additive, 100 ppm	. . Foam Height, ml . . Initial	After 30 Seconds
Polysorbate 80 (polyoxyethylene 20 sorbitan monooleate)	25	20
Dioctyl sodium sulfosuccinate	35	30
Ethanol*	15	10
GMO (glycerol monooleate) + ethanol	10	2
GMDO (glycerol mono- and dioleate) + ethanol	<0.5	<0.5
Octyl alcohol + ethanol	13	10
Polyoxyethylene 50 stearate + ethanol	17	15

*Solvent for water insoluble additives.

Thus, 100 ppm of GMDO was found to effectively inhibit the copious foam resulting from the agitation of a 1,500 ppm aqueous solution of Poly R-481. GMO was also found to be effective. Other additives, including those categorized as food grade defoamers (i.e., octyl alcohol, polyoxyethylene stearate, polyoxyethylene sorbitan monolaurate and polyoxyethylene sorbitan monooleate), were ineffective at the level tested or in some instances enhanced foaming. Dioctyl sodium sulfosuccinate and the polysorbates enhanced foaming of agitated solutions of Poly R-481 at 100 ppm.

Example 2: The prevention of foaming in a 1,500 ppm solution of Poly R-481 by various levels of GMDO was studied, with the results tabulated below:

	. . . . Foam Height After Agitation (ml)				
Time (min)	Control	10 ppm ETOH*	1 ppm ETOH*	10 ppm GMDO	1 ppm GMDO
0	15	15	15	<0.5	5
0.5	10	10	10	<0.5	2
1.0	10	10	8	<0.5	1
2.0	8	8	8	<0.5	<0.5
4.0	5	5	5	<0.5	<0.5
10	5	5	5	<0.5	<0.5
15	5	5	5	<0.5	<0.5

*Ethanol was the cosolvent for GMDO at equal parts concentration.

Although 1.0 ppm GMDO did reduce foaming, 10 ppm GMDO was found to be considerably more effective. The use of 1 and 10 ppm ethanol as a cosolvent for GMDO did not affect foam height.

This test also showed the persistence of the foam resulting from agitation of an aqueous Poly R-481; even 15 min after agitation, a 5 ml foam height was still evident.

Thixotropic Gel Fuels

Highlights of the Technological Achievement: A thixotropic gel fuel composition capable of being used in standard type oil burners comprising 5 to 75 volume percent of a solid carboniferous combustible material suspended in a liquid fuel and 1 to 10 weight percent of a substantially completely combustible gelling agent, the composition having a yield stress of 20 to 300 dyn/cm^2. The composition may also contain ash modifiers.

Reference: W.B. Tarpley, Jr.; U.S. Patent 4,156,594; May 29, 1979; assigned to Energy and Minerals Research Co.

Background: Recent events have underscored the need to conserve our natural resources, and more particularly, our dwindling supply of oil. One suggested manner of dealing with the energy problem has been to use more coal and solid fuels until more oil or alternative forms of energy can be discovered and commercially developed. The use of solid fuels could also reduce our dependence on foreign oil. However, most residential, commercial and industrial fuel users do not have equipment which can presently use solid fuels. Accordingly, the equipment, generally consisting of gas or oil furnaces, must either be replaced or converted to be able to burn the solid fuel.

The process comprises a stable thixotropic gel fuel composition which can be burned in standard types of oil furnaces, such as that described in J.J. Demeter et al, *Combustion of Coal-Oil Slurry in a 100 HP Firetube Boiler,* U.S. Energy Research and Development Administration publication PERC/RI-77/8, May 1977, with a minimal increase in cost. The Library of Congress Congressional Research Service has reported that the cost involved in converting a gas fired furnace or an oil fired furnace to one capable of burning known coal/oil mixtures would be about one-fourth of the cost of replacing a gas fired furnace or oil fired furnace. It is believed that the cost of conversion would be even less with the compositions of the process.

The thixotropic gel compositions of the process do not require complex suspending agents, such as coal tar or surface-active materials, which interfere with the atomization process in the furnace or burner. Rather, the present fuels are formulated of particular gelling agents to produce thixotropic gels which can immediately break apart upon injection in the furnace.

In addition, the thixotropic gel fuels made from the particulate gelling agents are far less sensitive to temperature thinning than prior art compositions. Thixotropic gelation provides gels having the advantage of solid-like immobility, similar to grease, during storage and transportation, but imparts to the compositions the ability to shear thin to apparent viscosities approximating those of the unmodified liquid fuel when pumped or atomized. Therefore, the thixotropic gel fuels of the process can be transported, delivered and used with the conventional equipment and methods used with liquid fuels, such as fuel oil.

Process Conditions: The main components of the compositions are a solid carboniferous combustible material, a liquid fuel and a substantially completely combustible gelling agent in one embodiment of the process. Another embodiment of the process includes the same components with the addition of an effective amount of an ash modifier to modify the melting point of any ash produced by the burning of the composition to prevent the ash from fouling the furnace or other burning apparatus.

The solid carboniferous combustible material which is incorporated in the thixotropic gel fuel may be any combustible material containing carbon which is insoluble in the liquid and capable of being suspended in the thixotropic gel. Suitable carboniferous materials include coal, coke, pitch, charcoal, tar sand, oil shale, wood fibers, wastepaper, waste carbon paper, bagasse, waste agricultural products, char, etc.

It is preferred for combustion efficiency that the solid carboniferous combustible material be finely divided, but it is not necessary that the particles be of colloidal size. In fact, in suitable furnaces, granules can be used. A suitable particle size would be wherein about 80 weight percent pass through a 200 mesh screen and are retained on a 325 mesh screen, although smaller particles could be used. The solid combustible material should comprise about 5 to 75 volume percent of the composition.

The liquid fuel in which the solid carboniferous combustible material is suspended need only be combustible and gelable. Preferably, the liquid fuel has a lower ignition point than the solid material to pilot the ignition of the solid material as the composition is atomized in the furnace.

Combustible Gelling Agents: Suitable substantially completely combustible gelling agents include hydroxypropylcellulose, polyvinyl sulfonate, methylcellulose, carrageenan and other natural and synthetic gums and resins, chemically modified castor oil Thixcin R (N.L. Industries, Inc.), butyl rubber, p-tert-butyl styrene polymers, amine reacted carboxypolymethylene resins, methyl vinyl ether-maleic anhydride copolymers, and the like.

With respect to the natural and synthetic cellulosic gums, hydration of these gelling agents with a small amount of water followed by crosslinking to form weak bonds permits the formation of gels in nonaqueous systems.

The function of the gelling agent is to keep the solid particles in the composition from contacting each other so that they do not adhere to each other, as by sintering or by van der Waals attraction.

The thixotropic gel fuel compositions according to the process demonstrate reversible pseudoplasticity. That is, under conditions of relatively short relaxation time, the yield stress value possessed by these compositions permit them to retain their gel or gel-like form when under little or no external force. However, when subjected to the type of external force encountered in pumping the compositions for delivery to a storage tank or forces caused by the flow through the small orifices typical of injection nozzles in standard type oil burners and other combustion equipment, the compositions will be in liquid form. The compositions of the process resume their gel or gel-like structure almost immediately after being subjected to flow forces.

Yield stress represents the force necessary to cause the gel to liquefy and start to flow. The yield stress at which the thixotropic gel fuels become fluid can be readily adjusted by varying the amount of the components of the compositions. It is necessary that this yield stress be sufficiently high that the gravitational force on a suspended particle will not cause that particle to liquefy the gel beneath it so that the particle will settle. On the other hand, to move the thixotropic gel fuel compositions conveniently, to pump them with standard equipment and to drain storage and transportation tanks as completely as possible, it is desirable that the yield stress be kept as low as possible.

The yield stress should be adjusted so that the self-supporting height of the gel will be very low, on the order of a few inches at most, whereby the tanks can be drained substantially completely of their contents without requiring special mechanical cleaners or drainers. Suitable yield stress values are 20 to 300 dyn/cm^2.

Another embodiment of the process includes an effective amount of an ash modifier in the thixotropic gel fuel compositions discussed above. The ash modifier is stably suspended throughout the thixotropic gel. The ash modifier modifies the melting point of the ash produced by the burning of the composition to minimize or prevent the ash from fouling the furnace or other burning apparatus.

The ash modifiers should be added in an amount equivalent in weight to the ash content of the solid carboniferous combustible material and the liquid fuel present in the composition. For example, if the net ash content of coal and fuel oil in a composition according to the process amount to, say, 5 to 10%, then 5 to 10% of the ash modifying additives should be incorporated in the composition. Preferably, the ash modifiers have a particle size of less than 100 mesh.

One type of ash modifier acts to increase the melting point of the fused ash in the combustion zone to prevent its persistent deposition as a viscous liquid or solid on the walls of the furnace and on the heat exchange surfaces. The modified ash will be in the form of dry dust which is exhausted along with the flue gases generated in the furnace. Conventional scrubbers and like equipment may be used to remove the ash from the flue gases. Suitable additives which raise the melting point of the composition include silica, alumina, titania, zirconia, magnesium oxide, calcium oxide, or chemical compounds which will give these oxides on combustion, dolomite, kaolin and attapulgite clays.

COMPANY INDEX

The company names listed below are given exactly as they appear in the patents, despite name changes, mergers and acquisitions which have, at times, resulted in the revision of a company name.

INVENTOR INDEX

U.S. PATENT NUMBER INDEX

Copies of U.S. patents are easily obtained
from the U.S. Patent Office at 50¢ a copy.

NOTICE